T0329308

# Tribology of Polymeric Nanocomposites

# Tribology of Polymeric Nanocomposites
## Friction and Wear of Bulk Materials and Coatings

**Klaus Friedrich**

**Alois K. Schlarb**

AMSTERDAM • BOSTON • HEIDELBERG • LONDON
NEW YORK • OXFORD • PARIS • SAN DIEGO
SAN FRANCISCO • SINGAPORE • SYDNEY • TOKYO

Butterworth-Heinemann is an Imprint of Elsevier

Butterworth-Heinemann is an imprint of Elsevier
The Boulevard, Langford Lane, Kidlington, Oxford OX5 1GB, UK
225 Wyman Street, Waltham, MA 02451, USA

First edition 2008
Second edition 2013

**Notice**
No responsibility is assumed by the publisher for any injury and/or damage to
persons or property as a matter of products liability, negligence or otherwise, or
from any use or operation of any methods, products, instructions or ideas contained
in the material herein. Because of rapid advances in the medical sciences, in
particular, independent verification of diagnoses and drug dosages should be made

**British Library Cataloguing in Publication Data**
A catalogue record for this book is available from the British Library

**Library of Congress Cataloging-in-Publication Data**
A catalog record for this book is availabe from the Library of Congress

ISBN–13: 978-0-444-59455-6

For information on all Butterworth-Heinemann publications
visit our web site at books.elsevier.com

Printed and bound in the UK

13 14 15 16 17   10 9 8 7 6 5 4 3 2 1

Working together
to grow libraries in
developing countries

www.elsevier.com • www.bookaid.org

# Contents

# Foreword to the 2nd Edition

The area of tribology deals with the design, friction, wear, and lubrication of interacting surfaces in relative motion. Polymer composite materials have been used increasingly for such tribological applications in recent years. Yet, by now, much of the knowledge on their tribological behavior is often empirical, and very limited predictive capability currently exists. Nevertheless, it has been attempted in several books and scientific papers from 1982 to 2012 to determine to what degree phenomena governing the friction and wear performance of polymer composites can be generalized (see, e.g. Refs [1–7] and references therein).

Within the last 10 years, many developments of new polymers composites have incorporated nanofillers as reinforcing agents, resulting in the term "polymeric nanocomposites". In fact, it has been demonstrated that these fillers of a very small dimension (as compared to the classical micrometer-sized fibers or particles) can also result in remarkable improvements in the friction and wear properties of both bulk materials and coatings. Therefore, it is the intention of this book to give a comprehensive description of polymeric nanocomposites, both as bulk materials and as thin surface coatings, and their behavior and potential use in tribological applications. The preparation techniques, friction and wear mechanisms, properties of polymeric nanocomposites, characterization, evaluation, and selection methodology in addition to application examples are described and discussed. One aim of the book is to bring together, systematically in a single volume, the state-of-the-art knowledge on the tribology of polymeric nanocomposites and coatings. This has previously been difficult to achieve and overview because the information has only been available in the form of numerous separate articles not linked together in a logical order.

More than 20 groups of authors worldwide, many of them well known in the tribology community since years, have agreed to write down their particular expertise on nanocomposite tribology in individual chapters. The latter cover not only different types of polymer matrices, that is, from thermosets to thermoplastics and elastomers but also a variety of microfillers and nanofillers, from ceramic nanoparticles to carbon nanotubes, in combination with traditional tribofillers, such as short carbon fibers, graphite flakes, and polytetrafluoroethylene (PTFE) particles. The coatings can be prepared on ceramic, metallic, or polymeric substrates and applied in many different applications, including automotive, aerospace, and mechanical engineering.

After a preface by Briscoe and Sinha, in which tribological trends for polymer composites, both traditional and nanocomposites, are presented, using data currently available in the literature, the book is structured into four main parts. The first one is dedicated to the tribology of bulk polymer composites against metallic counterparts, with particular emphasis on the use of spherical nanoparticles. S. Bahadur and C. J. Schwartz report on the influence of nanoparticle fillers in polymer matrices on the formation and stability of transfer films during wear. Synergistic effects of nanoparticles and traditional fillers on the sliding wear of polymeric hybrid composites are discussed by L. Chang et al. In more detail, Q. Wang and X. Pei illustrate that the volume content, the size, and the shape of the nanoparticles may exert a great influence on the friction and wear behavior of various polymer matrices. In the following chapter by Pesetski, Bogdanovich, and Myshkin, special emphasis is focused on the tribological behavior of thermoplastic nanocomposites, containing as fillers carbon nanomaterials, layered clays, metals, and metal-containing compounds. Sliding wear of thermosetting nanocomposites based on an epoxy resin matrix, is, on the other hand, the topic of the contribution by M. Q. Zhang and his group. In conclusion of this section, L. Kónya and K. Váradi present some wear simulation studies of a polymer–steel sliding pair by considering temperature- and time-dependent material properties.

The second part of this book concentrates on the use of carbon nanotubes and nanofibers as reinforcements in bulk nanocomposites against metallic counterparts. Ruckdäschel, Sandler, and Altstädt give a comprehensive overview on the friction and wear of carbon nanofiber-reinforced polyetheretherketone (PEEK)-based polymer composites. Based on their promising results, the performance of advanced nanocomposite hybrid materials for an intended industrial tribological application is discussed. The chapter of O. Jacobs and B. Schädel elucidates the effect of carbon nanotube reinforcement on the sliding wear of epoxy resin and of ultrahigh-molecular-weight polyethylene against two different steel counterparts. It also shows that the dispersion method of the carbon nanotubes has a remarkable influence on the tribological properties. Finally, the friction and wear characteristics of randomly dispersed vs. well-aligned carbon nanotubes in two different matrices, that is, epoxy vs. carbon, is outlined by Q. Gong et al.

The third main section of this book relates to the problem of scratch/wear resistance of nanocomposites and their coatings. In the spirit of the now-classical treatment of mechanical data in the form of fracture and failure maps, K. Kato and D. Diao describe, in general, wear and corresponding wear maps of hard coatings. In the following, E. Iwamura introduces two types of hybridized carbon films with different nanocomposite configurations with regard to their tribological properties. In particular, the structurally modified column/intercolumn films showed a high wear resistance in spite of a distinctively poor film hardness. Some new developments in the field of wear of "nanomodified" rubbers and their coatings, studied under three different test configurations, are presented in the chapter of J. Karger-Kocsis and D. Felhös. It is the intention to use these new developments in automotive seal applications. Another coating material, that is, sol–gel coatings on polymer substrates, has been widely used in optical lenses, safety windows, and flexible display panels. Z. Chen and L. Y. L. Wu review in their chapter the scratch failure modes of such coatings on polymeric substrates, the related failure mechanisms, and the parametric models related to these failure modes. A new scratch testing and evaluation method is presented by R. L. Browing, H. Jiang, and H.-J. Sue. The potential of this procedure is demonstrated with regard to the scratch behavior of acrylic coatings and epoxy nanocomposites. Finally, a wide survey of the existing literature on scratch and wear damage in polymer nanocomposites, followed by own results on the scratch behavior of various nanoclay- and other nanoparticle-filled polymer systems, is given by A. Dasari, Z.-Z. Yu, and Y.-W. Mai.

In the fourth part, chapters are found that especially focus on the tribological use of nanocomposites and their coating for special applications. The group of W. G. Sawyer gives a comprehensive review on the development of PTFE matrix nanocomposites for the use of these systems in moving mechanical devices. A. Gebhard, F. Haupert, and A. K. Schlarb report about the application of nanostructured slide coatings for automotive components, such as engine piston skirts and polymer/metal-slide bearings. A special topic, namely, the friction and wear of nanoparticle-filled PEEK and its composite coatings applied by flame spraying or painting on metallic parts is described by G. Zhang, H. Liao, and C. Coddet. In a more traditional com-posite-processing method, Bijwe, Sharma, Hufenbach, Kunze, and Langkamp report about the development of polymer composite bearings with engineered tribosurfaces. They can be used in extreme conditions of temperature (cryogenic to 300 °C), vac-uum to high-pressure environment, especially where liquid lubrications cannot be considered. On the contrary, well-lubricating coating and thin film applications are the focus of the chapter by N. R. Choudhury, A. G. Kannan, and N. Dutta. The mate-rials are organic–inorganic hybrids, their methods of preparation, their relevance to biomaterials' friction, their fundamental mechanics of tribology, and the lubrication mechanisms of such coatings and thin films. The last chapter of M. Busse and A. K. Schlarb provides new insights into the use of the artificial neural network approach for the description of the tribological behavior of nanocomposites. This approach requires a large amount of data. However, it can be shown that very com-plex relationships can be detected by using this method. Thus, this represents an excellent basis for a tribological materials development.

When considering the content of this book as a whole, it becomes clear that it is primarily intended for scientists in academia and industry, who are involved or want to become involved in tribology problems, and who look for new solutions in materi-als' development and for particular applications. The book will be, therefore, a refer-ence work and a guide to practice for those who are or want to become professional in the field of polymer composite tribology.

By preparing this book, we hope that we have managed to take a first step toward a systematic structure for this complex field of technology. At the same time, we believe that this is timely, but only a first attempt to cover a topic that is in the process of rapid development since the last couple of years. We are sure that many more interesting results on the tribology of nanocomposite materials will be published in the open literature within the near future.

Finally, we would like to thank all the contributors who managed to include their thoughts and results in this second edition of our book on the tribology of polymeric nanocomposites.

The Editors:
Klaus Friedrich
Alois K. Schlarb

# References

[1] K. Friedrich, R.B. Pipes (Eds.), Advances in Composite Tribology, Composite Materials Series, vol. 8, Elsevier Science Publications, Amsterdam, The Netherlands, 1993.

[2] K. Holmberg, A. Matthews, in: D. Dowson (Ed.), Coatings Tribology: Properties, Techniques and Applications in Surface Engineering, Tribology Series, vol. 28, Elsevier Science B. V., Amsterdam, The Netherlands, 1994.

[3] G.W. Stachowiak, A.W. Batchelor, Engineering Tribology, Elsevier Butterworth–Heinemann, Oxford, UK, 2005.

[4] A. Sethuramiah, in: D. Dowson (Ed.), Lubricated Wear: Science and Technology, Tribology Series, vol. 42, Elsevier Science B. V., Amsterdam, The Netherlands, 2003.

[5] G.W. Stachowiak, A.W. Batchelor, G.B. Stachowiak, in: D. Dowson (Ed.), Experimental Methods in Tribology, Tribology Series, vol. 44, Elsevier B. V., Amsterdam, The Netherlands, 2004.

[6] S.K. Sinha, B.J. Briscoe (Eds.), Polymer Tribology, Imperial College Press, London, UK, 2009.

[7] R.W. Bruce (Ed.), Handbook of Lubrication and Tribology, Theory and Design, vol. 2, CRC Press, Boca Raton, USA, 2012.

# Contributors

**Volker Altstädt**
Polymer Engineering, University of Bayreuth, Bayreuth, Germany

**Shyam Bahadur**
Department of Mechanical Engineering, Iowa State University, Ames, IA, USA

**Jayashree Bijwe**
Industrial Tribology Machine Dynamics & Maintenance Engineering Centre, Indian Institute of Technology, Delhi, Hauz Khas, New Delhi, India

**Qing Bing Guo**
College of Chemistry and Chemical Engineering, Zhongkai University of Agriculture and Engineering, Guangzhou, PR China

**Thierry A. Blanchet**
Rensselaer Polytechnic Institute, NY, USA

**Sergei P. Bogdanovich**
V.A. Belyi Metal–Polymer Research Institute of National Academy of Sciences of Belarus, Gomel, Belarus

**Brian J. Briscoe**
Department of Chemical Engineering & Chemical Technology, Imperial College, London, UK

**Robert L. Browning**
Polymer Technology Center, Texas A&M University, USA

**David L. Burris**
University of Florida, FL, USA

**Michael Busse**
Lehrstuhl für Verbundwerkstoffe, Composite Engineering cCe, University of Kaiserslautern, Kaiserslautern, Germany

**Li Chang**
Centre for Advanced Materials Technology, The University of Sydney, Sydney, NSW, Australia

**Zhong Chen**
School of Materials Science and Engineering, Nanyang Technological University, Singapore

**Namita Roy Choudhury**
Ian Wark Research Institute, ARC Special Research Centre for Particle and Material Interfaces, University of South Australia, Mawson Lakes, South Australia, Australia

**Christian Coddet**
University of Technology of Belfort-Montbeliard, Belfort, France

**Li Dan**
Department of Mechanical Engineering, Tsinghua University, Beijing,
PR China

**Aravind Dasari**
School of Materials Science & Engineering, Nanyang Technological University,
Singapore

**Dongfeng Diao**
School of Mechanical Engineering, Xi'an Jiaotong University, Xi'an, China

**Naba Dutta**
Ian Wark Research Institute, ARC Special Research Centre for Particle and
Material Interfaces, University of South Australia, Mawson Lakes, South
Australia, Australia

**Dávid Felhős**
Department of Polymer Engineering, Faculty of Materials Science and Engineering, University of Miskolc, Miskolc, Hungary

**Klaus Friedrich**
Institute for Composite Materials (IVW GmbH), Technical University of
Kaiserslautern, Kaiserslautern, Germany; and College of Engineering, King
Saud University, Riyadh, Saudi Arabia

**Andreas Gebhard**
NanoProfile GmbH, Kaiserslautern, Germany

**Frank Haupert**
Hochschule Hamm-Lippstadt, Hamm, Germany

**Werner Hufenbach**
Technische Universität Dresden, Institute für Leichtbau und Kunstofftechnik,
Dresden, Germany

**Eiji Iwamura**
R&D Center, Arakawa Chemical Industries, Ltd, Tsurumi-ku, Osaka, Japan

**Olaf Jacobs**
Fachhochschule Lübeck, Lübeck, Germany

**Liang Ji**
Department of Mechanical Engineering, Tsinghua University, Beijing, PR China

**Han Jiang**
Polymer Technology Center, Dept. Mechanical Engineering, Texas A&M
University, College Station, TX, USA

**Aravindaraj G. Kannan**
Ian Wark Research Institute, ARC Special Research Centre for Particle and Material Interfaces, University of South Australia, Mawson Lakes, South Australia, Australia

**József Karger-Kocsis**
MTA-BME Research Group for Composite Science and Technology and Department of Polymer Engineering, Faculty of Mechanical Engineering, Budapest University of Technology and Economics, Budapest, Hungary

**Koji Kato**
Department Mechanical Engineering, Nihon University, Koriyama City, Japan

**László Kónya**
Institute of Machine Design Budapest University of Technology and Economics, Budapest, Hungary

**Klaus Kunze**
Technische Universität Dresden, Institute für Leichtbau und Kunstofftechnik, Dresden, Germany

**Albert Langkamp**
Technische Universität Dresden, Institute für Leichtbau und Kunstofftechnik, Dresden, Germany

**Sarah L. Lewis**
Rensselaer Polytechnic Institute, NY, USA

**Hanlin Liao**
University of Technology of Belfort-Montbeliard, Belfort, France

**Xinxing Liu**
Rensselaer Polytechnic Institute, NY, USA

**Ying Luo**
College of Science, South China Agriculture University, Guangzhou, PR China

**Yiu-Wing Mai**
Centre for Advanced Materials Technology (CAMT), School of Aerospace, Mechanical and Mechatronic Engineering, The University of Sydney, Sydney, NSW, Australia

**Nikolai K. Myshkin**
V.A. Belyi Metal–Polymer Research Institute of National Academy of Sciences of Belarus, Gomel, Belarus

**Xianqiang Pei**
Laboratory of Solid Lubrication, Lanzhou Institute for Chemical Physics, Lanzhou, China

**Scott S. Perry**
University of Florida, FL, USA

**Stepan S. Pesetskii**
V.A. Belyi Metal–Polymer Research Institute of National Academy of Sciences of Belarus, Gomel, Belarus

**Gong Qianming**
Department of Mechanical Engineering, Tsinghua University, Beijing, PR China

**Min Zhi Rong**
Materials Science Institute, Sun Yat-sen (Zhongshan) University, Guangzhou, PR China

**Holger Ruckdäschel**
Polymer Engineering, University of Bayreuth, Bayreuth, Germany

**Jan K.W. Sandler**
Polymer Engineering, University of Bayreuth, Bayreuth, Germany

**Katherine Santos**
University of Florida, FL, USA

**W. Gregory Sawyer**
University of Florida, FL, USA

**Birgit Schädel**
Fachhochschule Lübeck, Lübeck, Germany

**Linda S. Schadler**
Rensselaer Polytechnic Institute, NY, USA

**Alois K. Schlarb**
Lehrstuhl für Verbundwerkstoffe, Composite Engineering cCe, University of Kaiserslautern, Kaiserslautern, Germany

**Cris J. Schwartz**
Department of Mechanical Engineering, Iowa State University, Ames, IA, USA

**Mohit Sharma**
Industrial Tribology Center, Indian Institute of Technology, New Delhi, India

**Sujeet K. Sinha**
Department of Mechanical Engineering, National University of Singapore, Singapore

**Hung-Jue Sue**
Polymer Technology Center, Dept. Mechanical Engineering, Texas A&M University, College Station, TX, USA

**Károly Váradi**
Institute of Machine Design Budapest University of Technology and Economics, Budapest, Hungary

**Qihua Wang**
Laboratory of Solid Lubrication, Lanzhou Institute for Chemical Physics, Lanzhou, China

**Linda Y.L. Wu**
Singapore Institute of Manufacturing Technology, Singapore

**Yi Xiaosu**
Beijing Institute of Aeronautical Materials, Beijing, PR China

**Lin Ye**
Centre for Advanced Materials Technology, The University of Sydney, Sydney, NSW, Australia

**Zhong-Zhen Yu**
State Key Laboratory of Organic-Inorganic Composites, Department of Polymer Engineering, College of Materials Science and Engineering, Beijing University of Chemical Technology, Beijing, China

**Zhong Zhang**
National Center for NanoScience and Technology, Beijing, China

**Ming Qiu Zhang**
Key Laboratory for Polymeric Composite and Functional Materials of Ministry of Education, School of Chemistry and Chemical Engineering, Sun Yat-sen (Zhongshan) University, Guangzhou, PR China; Materials Science Institute, Sun Yat-sen (Zhongshan) University, Guangzhou, PR China

**Ga Zhang**
Institute for Composite Materials (IVW GmbH), Technical University Kaiserslautern, Kaiserslautern, Germany

**Li Zhi**
Department of Mechanical Engineering, Tsinghua University, Beijing, PR China

# Tribological applications of polymers and their composites – past, present and future prospects

1

**Brian J. Briscoe\*, Sujeet K. Sinha†**

*\*Department of Chemical Engineering & Chemical Technology, Imperial College, London, UK,*
*†Department of Mechanical Engineering, National University of Singapore, Singapore*

## CHAPTER OUTLINE HEAD

## 1.1 INTRODUCTION

Polymers play an important part in materials and mechanical engineering, not just for their ease in manufacturing and low unit cost, but also for their potentially excellent tribological performance in engineered forms [1]. In the pristine or bulk form, only a few of the polymers would satisfy most of the tribological requirements; however, in the composite and hybrid forms, polymers often have an advantage over other materials such as metals and ceramics. Polymer tribology, as a research field, is now well mature given that roughly 50 plus years have seen publication of numerous research articles and reports dealing with a variety of tribological phenomena on a considerably large number of polymers, in bulk, composite and hybrid forms. Tribological applications of polymers include gears, a range of bearings, bearing cages, artificial

**Tribology of Polymeric Nanocomposites.** http://dx.doi.org/10.1016/B978-0-444-59455-6.00001-5

human joint bearing surfaces, bearing materials for space applications including coatings, tires, shoe soles, automobile brake pads, nonstick frying pans, floorings and various types of surfaces for optimum tactile properties such as fibers. The list is growing. For example, in the new area of microelectromechanical systems (MEMS), polymers (such as poly(methylemethacrylate) (PMMA) and poly(dimethylsiloxane) (PDMS)) are gaining popularity as structural materials over the widely used material, Si [2]. Often, Si is modified by a suitable polymeric film in order to enhance frictional, antiwear or antistiction properties [3].

Similar to the bulk mechanical responses, the tribological characteristics of polymers are greatly influenced by the effects of temperature, relative speed of the interacting surfaces, normal load and the environment. Therefore, to deal with these effects and for better control of the responses, polymers are modified by adding appropriate fillers to suit a particular application. Thus, they are invariably used in composite or, at best, blended form for an optimum combination of mainly friction and wear performances. In addition, pragmatically fillers may be less expensive than the polymer matrix. The composition of the filler materials, often a closely guarded secret of the manufacturer, is both science and art, for the final performance may depend upon the delicately balanced recipe of the matrix and filler materials. However, the last many years of research in the area of polymer tribology in various laboratories have shed much light into the mechanisms of friction and wear. This has somewhat eased the work of materials selection for any particular tribological application.

This chapter on the tribology of nanocomposite, an area which is still in its infancy, would endeavor to set the background of the research in polymer tribology. We will refer to the term "polymer" for synthetic organic solid in pristine form, with some additives but no fillers aimed at modifying mechanical properties. The word "composite" would be used when one or more than one fillers have been added to a base polymer with the aim of drastically changing mechanical and tribological properties. "Nanocomposite" will mean a composite in which at least one filler material has one of its dimensions in the range of a few to several nanometers.

The chapter would review some of the past, but now classical, works when much of the mechanisms of friction and wear for general polymers were studied and these explanations have stood the test of the time. The early works led to the area of polymer composites where polymers were reinforced with particles and/or short or long fibers. Often, the use of the filler materials has followed two trends that mainly reflect the actual function that the fillers are expected to perform. This type of work on the design of multiphase tribological materials continues mainly aimed at improving an existing formulation, or, using a new polymer matrix or novel filler. The present trend is expected to extend into the future but with much more refinement in materials and process selections. For example, the use of nanosized particles or fibers coupled with chemical enhancement of the interactions between the filler and the matrix seems to produce better tribological performance. Also, there have been some very recent attempts on utilizing some unique properties of polymers, often mimicking the biological systems in one way or the other, which have opened up new possibility of using polymers in tribological applications. One example of this is polymer brush that can be used as a boundary lubricant. This trend will definitely continue into the future with great promises for solving new tribological issues in micro, nano and bio systems.

## 1.2 CLASSICAL WORKS ON POLYMER TRIBOLOGY

### 1.2.1 Friction

The earliest works on polymer tribology probably started with the sliding friction studies on rubbers and elastomers [4,5]. Further work on other polymers (thermosets and thermoplastics) led to the development of the two-term model of friction [6]. The two-term model proposes that the frictional force is a consequence of the interfacial and the cohesive works done on the surface of the polymer material. This is assuming that the counterface is sufficiently hard in comparison to the polymer mating surface and undergoes only mild or no elastic deformation. Figure 1.1 shows a schematic diagram of the energy dissipation processes in the two-term model [7].

The interfacial frictional work is the result of adhesive interactions and the extent of this component obviously depends upon factors such as the hardness of the polymer, molecular structure, glass transition temperature and crystallinity of the polymer, surface roughness of the counterface and chemical–electrostatic interactions between the counterface and the polymer. For example, an elastomeric solid, which has its glass transition temperature below the room temperature and hence is very

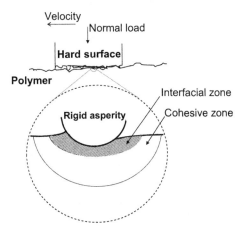

**FIGURE 1.1**

Two-term model of friction and wear processes. Total friction force is the sum of the forces required for interfacial and cohesive energy dissipations. Likewise, the distinction between interfacial and cohesive wear processes arises from the extent of deformation in the softer material (usually polymer) by rigid asperity of the counterface. For interfacial wear, the frictional energy is dissipated mainly by adhesive interaction while for cohesive wear the energy is dissipated by adhesive and abrasive (subsurface) interactions [7] (with publisher's permission).

soft, would have very high adhesive component leading to high friction. Beyond interfacial work is the contribution of the cohesive term, which is a result of the plowing actions of the asperities of the harder counterface into the polymer. The energy required for the plowing action will depend primarily upon the tensile strength and the elongation before fracture (or toughness) of the polymer and, the geometric parameters (height and the cutting angle) of the asperities on the counterface. The elastic hysteresis is another factor generally associated with the cohesive term for polymers that show large viscoelastic strains such as in the case of rubbers and elastomers [8]. Further, both the interfacial and the cohesive works would be dependent upon the prevailing interface and ambient temperatures, and the rate of relative velocity as these factors would in turn modify the polymer's other materials parameters. Pressure has some effect on the interfacial friction as normal contact pressure tends to modify the shear strength of the interface layer by a relation given as [9]

$$\tau = \tau_0 + \alpha_p \tag{1.1}$$

The implication of the above relation is that as the contact pressure increases, the shear stress would increase linearly leading to high friction. Equation (1.1), as simple as it may look in the form, hides the very complex nature of polymer. Also, it does not include the temperature and the shear rate effects on the shear stress.

In a normal sliding experiment, it is nontrivial to separate the two terms (interfacial and cohesive) and therefore most of the data available in the literature generally include a combined effect. Often, the practice among experimentalists is to fix all other parameters and vary one parameter to study its effect on the overall friction coefficient for a polymer. Looking at the published data one can easily deduce that depending upon other factors, the friction is greatly influenced by the class of polymers viz. elastomers, thermosets and thermoplastics (semicrystalline and amorphous). Semicrystalline linear thermoplastic would give the lowest coefficient of friction, whereas elastomers and rubbers show large values. This is because of the molecular architecture of the linear polymers that helps molecules stretch easily in the direction of shear giving least frictional resistance. Table 1.1 provides some typical values of the coefficient of friction for pristine or virgin polymers.

## 1.2.2 Wear

The inevitable consequence of friction in a sliding contact is wear. Wear of polymers is a complex process and the explanation of the wear mechanism can be most efficiently given if we follow one of the three systems of classification. Depending upon which classification we are following, wear of a polymer sliding against a hard counterface may be termed as interfacial, cohesive, abrasive, adhesive, chemical wear etc. Figure 1.2 describes the classification of polymer wear [7]. It is to be noted that, similar to the case of friction, polymer wear is also greatly influenced by the type (elastomer, amorphous, and semicrystalline) of the polymer. Of particular importance are the properties such as the elastic modulus, tensile strength and the percentage elongation at failure (toughness), which changes drastically as we

**Table 1.1** Friction Coefficient of Few Polymers When Slid against a Steel Disk Counterface (Surface Roughness, $R_a = 1.34$ μm) with the Corresponding Specific Wear Rates and the Pressure ($P$) × Velocity ($v$) Values

| Polymer | Coefficient of Friction | Specific Wear Rate ($\times 10^{-6}$ mm$^3$ Nm$^{-1}$) | PV Value (Pa m s$^{-1}$) |
|---|---|---|---|
| PMMA | 0.48 | 1315.9 | 145,560 |
| PEEK | 0.32 | 31.7 | 149,690 |
| UHMWPE | 0.19 | 15.5 | 187,138 |
| POM | 0.32 | 168.2 | 149,690 |
| Epoxy | 0.45 | 3506.6 | 153,997 |

*POM, poly(oxymethylene).*

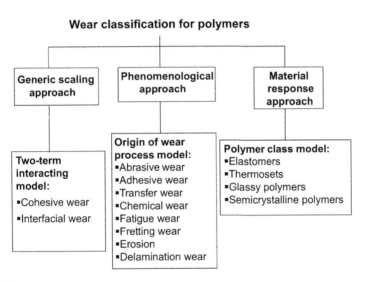

**FIGURE 1.2**

Simplified approach to classification of the wear of polymers [7] (with publisher's permission).

move from one type of polymer to another. Usually, high tensile strength coupled with high elongation at failure promotes wear resistance in a polymer. Therefore, given all other factors remain constant, some of the linear thermoplastic polymers with semicrystalline microstructure perform far better in wear resistance than thermosets or amorphous thermoplastics. These observations are in line with the idea that for polymers, surface hardness is not a controlling factor for wear resistance. In fact high hardness of a polymer may be harmful for wear resistance in dry sliding against hard counterface as hardness normally comes with low toughness for polymers. High extents of elongation at failure of a polymer means that the shear stress in a sliding event can be drastically reduced due to extensive plastic deformation of the polymer within a very thin layer close to the interface. This interfacial layer accommodates almost all the energy dissipation processes and thus the bulk of the polymer undergoes minimal deformation or wear. Frictional heat generated at the interface is the major impediment to high wear life of the polymer. Many classical and recent works suggest that the wear rate of polymers slid against metal counterface in abrasive wear condition may be given by a simple proportionality as [10,11]

$$W_{sp} \propto 1/Se \qquad (1.2)$$

The above equation, often described as the Ratner–Lancaster correlation, is supported by data obtained by several researchers (Ref [1] provides a summary of published data for the above relation) (Figure 1.3). Equation (1.2) is applicable across one type of polymer class such as semicrystalline thermoplastics. We do not have data to compare this rule for other types of polymers. Because of the low friction and high wear resistance, many of the thermoplastics can be used in tribological applications without any reinforcement and notable among them are the ultra-high molecular weight poly(ethylene) (UHMWPE) and poly(ether ether

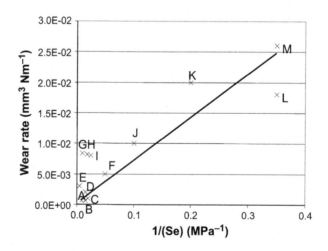

**FIGURE 1.3**

A plot of wear rate (cubic millimeters per millimeter per kilogram) as a function of the reciprocal of the product of ultimate tensile stress and elongation to fracture [1]. The data are taken from the literature. A – poly(ethylene); B – Nylon 66; C – PTFE; D – poly(propene); E – high-density poly(ethylene); F – acetal; G – poly(carbonate); H – poly(propylene); I – poly(ethyleneterephthalate glycol); J – poly(vinyl chloride); K – PMMA; L – poly(styrene); M – PMMA (refer to [1] for the sources of the data) (with publisher's permission).

ketone) (PEEK). UHMWPE has found extensive usage as bearing material for artificial human joints because of its excellent biocompatibility and wear resistance.
PEEK, which is a high-temperature polymer, tends to show low wear rate but the
coefficient of friction can be relatively high (~0.3). PEEK is now a popular polymer as matrix for some new composites with the aim of formulating wear-resistant
materials. Nylons are other tribological materials that show low friction and low
wear. Poly(tetrafluoroether) (PTFE), a linear fluorocarbon, normally shows very
low friction coefficient but, relative to many other thermoplastics, high wear rate
due to its unique characteristics of slippage in the crystalline formation of the
molecular bond structure. Due to low friction property, PTFE is a good solid lubricant if used in composite form. Amorphous thermoplastics such as PMMA and
poly(styrene) do not perform very well in a wear test. They show high coefficients
of friction and high wear rates.

Thermosetting polymers, although they posses high hardness and strength among
polymers, show very high wear rate and high coefficient of friction because of
very low elongation at failure values. Thermosets are normally used in the form
of composites as fiber strengthening can drastically reduce wear. Fiber strengthening sometimes improves the material's resistance to subsurface crack initiation and
propagation giving reduced plowing by the counterface asperities or fatigue cracking. Interface friction can also be optimized by adding a suitable percentage of a
solid lubricant. This trend has led to much research in recent times on producing
composite or hybrid materials for optimum wear and friction control using epoxy or
phenolic resins as the matrix [12].

## 1.3  TRIBOLOGY OF POLYMER COMPOSITES

Except for probably only UHMWPE and to some extent nylons, no other polymer
is currently being used in its pristine form for a tribological application. The reason
is that no polymer can provide a reasonable low working wear rate with optimum
coefficient of friction required. Hence, there is a need to modify most polymers by a
suitable filler that can reduce the wear rate and, depending upon the design requirement, either increase or decrease the coefficient of friction. Such a need was realized
quite early on [13] and this trend has continued.

The second component or the filler can perform a variety of roles depending upon the choice of the matrix and the filler materials. Some of these roles
are, strengthening of the matrix (high load carrying capacity), improvement in
the subsurface crack-arresting ability (better toughness), lubricating effect at the
interface by decreased shear stress and the enhancement of the thermal conductivity of the polymer. The entire aspects of the tribology of polymer composites
can be quite complex so as to defy any economic classification. Therefore, a
simple but efficient way to handle this topic is to classify the composites according to the role of the filler material in the composite, modifying the bulk or the
interface [14].

## 1.3.1 Bulk modification – "hard and strong" fillers in a "softer" matrix

A self-lubricating polymer, such as PTFE, can be made wear resistant by strengthening the bulk with hard or strong filler material such as particles of ceramics/metals or a suitable strong fiber (carbon, aramid or glass fibers). The function of the filler here is to strengthen the polymer matrix and thus increase the load bearing capacity of the composite. The coefficient of friction remains low or increases marginally but the wear resistance can be increased up to an order of magnitude. The disadvantage of using fillers (especially the particulate type) is that the composite material may become somewhat less tough in comparison to the pristine polymer and thus encourage wear by fatigue; however, this can be avoided by proper optimization of the mechanical and tribological properties. Strengthening by fibers, usually oriented normal to the sliding interface, has shown better result in terms of load-bearing capacity and toughness. A fiber that is nonabrading to the counterface, such as aramid and carbon fibers, is even more beneficial as it promotes the formation of a tenacious and thin transfer film on the counterface that can help in reducing the wear of the composite after a short running-in period.

Examples of a composite where a hard and strong filler has been added to a softer matrix are the PTFE/GF and Nylon 11/GF systems as shown in Figure 1.4 [1]. As we can see, the coefficient of friction for each case has increased slightly and there is considerable gain in the wear resistance as a result of fiber strengthening.

## 1.3.2 Interface modification – "soft" and "lubricating" fillers in a "hard and strong" matrix

This type of composite utilizes the low shear strength and self-lubricating properties of the filler to reduce the coefficient of friction and, as a result, wear and frictional heating is drastically reduced. The main requirement is the availability of the filler at the interface in sufficient amount such that a reduction in the coefficient of friction and an increase in the wear resistance can be realized. The disadvantage of this type of composite is obviously the reduction in the strength and load carrying capacity of the material in the composite form. Hence, adding this type of fillers beyond a certain percentage by volume or by weight would be counterproductive for tribological performance due to a drastic decrease in the bulk strength. Several researches have focused on finding an optimum ratio of the filler and the matrix to achieve maximum wear resistance [1]. PTFE and graphite in a variety of polymer matrices have been tried with good results; popular among matrices are epoxy, phenolic and PEEK.

For both types of composites discussed above, the properties of the transfer film formed on the counterface will define whether or not the composite can have low wear rate. A strongly adhering and tenacious, yet lubricating, transfer film would reduce wear after the formation of the film during the running-in period. Bulky and thick film has the tendency to detach itself from the counterface, which may increase the wear rate due to a continuous film formation and detachment mechanism (transfer wear) as well as promoting thermal effects.

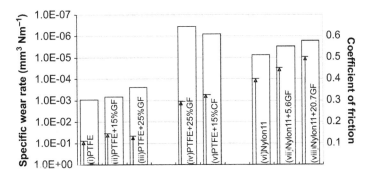

**FIGURE 1.4**

The effect of fiber addition on the specific wear rates of a few polymers [1]. The rectangular bar chart indicates specific wear rate (units on the left of the graph) and vertical arrows indicate the coefficient of friction (units on the right of the graph). Test conditions are given below: (i) 440C steel ball (diameter = 9 mm) sliding on polymer specimen, Normal load = 5 N, $v = 0.1$ m s$^{-1}$, roughness of polymer surface $R_a = 400$ nm, 30% humidity; (ii) test conditions same as for (i); (iii) test conditions same as for (i); (iv) reciprocating – pin – steel plate apparatus, counterface roughness $R_a = 0.051$ μm, $N_2$ environment; (v) test conditions same as for (iv); (vi) pin – on – steel (AISIO2 quench hardened) disk apparatus, counterface roughness $R_a = 0.11$ μm, $p = 0.66$ MPa, $v = 1$ m s$^{-1}$; (vii) test conditions same as for (vi); (viii) test conditions same as for (vi) (refer to [1] for the sources of the data; data for (iii) and (iv) are for the same formulation of the composite but the values are different as they have been taken from different research works) (with publisher's permission), GF - Glass Fiber.

# 1.4 **TRIBOLOGY OF POLYMER NANOCOMPOSITES**

The use of nanoparticles in polymers for tribology performance enhancement started around mid-1990s and this area has become quite promising for the future as newer nanomaterials are being economically and routinely fabricated. In most of the cases, a polymer nanocomposite relies for its better mechanical properties on the extremely high interface area between the filler (nanoparticles or nanofibers) and the matrix (a polymer). High interface leads to a better bonding between the two phases and hence better strength and toughness properties over unfilled polymer or traditional polymer composites. For all polymer/nanoparticle systems, there will be an optimum amount of the nanoparticles beyond which there will be a reduction in the toughness as the stiffness and strength increase. Table 1.2 summarizes friction and wear results of polymer nanocomposites with data taken from the published literature [15–49]. There are mainly two types of polymer nanocomposites that have been tested for tribological performance. One type is where ceramic nanoparticles, mainly metal and some nonmetal oxides, have been added with the aim to improve load-bearing capacity and wear resistance of the material against the counterface. Examples of polymer nanocomposite systems of this type include fillers such as $SiO_2$, $SiC$, $ZnO$, $TiO_2$, $Al_2O_3$, $Si_3N_4$, $CuO$ in polymer matrices such as epoxy, PEEK, PTFE and poly(phenylene

**Table 1.2** A Summary of the Tribological Results on Polymer Nanocomposites

| S. No | Matrix | Filler Material | Size of Filler Material (nm) | Optimum Filler Content | | Coefficient of Friction | | |
|---|---|---|---|---|---|---|---|---|
| | | | | With Lowest COF | With Lowest Wear | Without Filler | With Filler | Change (%) |
| 1 | PTFE | ZnO | 50 | 15 wt% | 15 wt% | 0.202 | 0.209 | +3.4 |
| 2 | PTFE | Al$_2$O$_3$ | 40 | 20 wt% | 20 wt% | 0.152 | 0.219 | +44.1 |
| 3 | PTFE | CNT | 20–30 | 30 vol.% | 20 vol.% | 0.2 | 0.17 | −15.0 |
| 4 | PTFE | Nanoattapulgite | 10–25 | 5 wt% | 5 wt% | 0.22 | 0.2 | −9.1 |
| | PTFE | 2 M acid-treated attapulgite | 10–25 | 5 wt% | 5 wt% | 0.22 | 0.2 | −9.1 |
| 5 | Epoxy | TiO$_2$ | 10 | 7 wt% | 3 wt% | 0.54 | 0.4 | −25.9 |
| 6 | Epoxy | TiO$_2$ | 300 | – | 4 vol.% | – | – | −25.9 |
| 7 | Epoxy | Al$_2$O$_3$ | 13 | – | 2 vol.% | – | – | −25.9 |
| 8 | Epoxy | Si$_3$N$_4$ | <20 nm | 0.8 vol.% | 0.8 vol.% | 0.57 | 0.38 | −33.3 |
| 9 | Epoxy | SiO$_2$ | 9 | 2.2 vol.% | 2.2 vol.% | 0.58 | 0.45 | −22.4 |
| | Epoxy | SiO$_2$-g-PAAM | 9 | 2.2 vol.% | 2.2 vol.% | 0.58 | 0.35 | −39.7 |
| 10 | Epoxy | MWCNT-untreated | 10–30 | – | | – | – | – |
| | Epoxy | MWCNT-acid treated | | | 1 wt% | – | – | – |
| | Epoxy | MWCNT-acid treated | | | 1 wt% | – | – | – |
| 11 | Epoxy | SiC | 61 | – | – | 0.6 | 0.44 | −26.7 |
| | Epoxy | SiC-g-PGMA | 61 | | | 0.6 | 0.41 | −31.7 |
| | Epoxy | SiC-g-PGMA | 61 | – | – | 0.6 | 0.39 | −35 |

| Specific Wear Rate ($\times 10^{-6}$ mm$^3$ Nm$^{-1}$) | | | | |
| Without Filler | With Filler | Change (%) | Equipment Used, Counterface Material, Load/Pressure Used, Sliding Velocity, Remarks (if any) | Reference |
| --- | --- | --- | --- | --- |
| 1125.3 | 13 | −98.8 | Block-on-ring tribometer, stainless steel, 200 N, 0.431 m s$^{-1}$ | [15] |
| 715 | 1.2 | −99.8 | Reciprocating tribometer, stainless steel, 260 N, 50 mm s$^{-1}$ | [16] |
| 800 | 2–3 | −99.6 | Block-on-ring tribometer, stainless steel, 200 N | [17] |
| 625.8 | 31.2 | −95 | Block-on-ring tribometer, steel, 200 N, 0.42 m s$^{-1}$ | [18] |
| 625.8 | 4.9 | −99.2 | Block-on-ring tribometer, steel, 200 N, 0.42 m s$^{-1}$, nanoattapulgite was treated with hydrochloric acid | [18] |
| 26×10$^3$ | 1.63×10$^3$ | −93.7 | Pin-on-ring tribometer, carbon steel, 1 MPa, 0.4 m s$^{-1}$ | [19] |
| 40 | 14 | −65 | Pin-on-ring tribometer, carbon steel, 1 MPa, 1 m s$^{-1}$ | [20] |
| 5.9 | 3.9 | −33.9 | Pin-on-ring tribometer, carbon steel, 1 MPa, 0.4 m s$^{-1}$ | [21] |
| 38 | 2 | −94.7 | Pin-on-ring tribometer, carbon steel, 3 MPa, 0.4 m s$^{-1}$ | [22] |
| 200 | 45 | −77.5 | Pin-on-ring tribometer, carbon steel, 3 MPa, 0.4 m s$^{-1}$ | [23,24] |
| 200 | 11 | −94.5 | Pin-on-ring tribometer, carbon steel, 3 MPa, 0.4 m s$^{-1}$, SiO$_2$ nanoparticles were modified with PAAM | [23,24] |
| 9 | 12.5 | +38.8 | Ball-on-prism tribometer, 30 N, 28.2 mm s$^{-1}$, mixed with four-blade stirrer | [25] |
| 9 | 4.5 | −50 | Ball-on-prism tribometer, 30 N, 28.2 mm s$^{-1}$, mixed with four-blade stirrer, CNTs were treated with nitric acid | [25] |
| 31 | 3 | −90.3 | Ball-on-prism tribometer, 30 N, 28.2 mm s$^{-1}$, mixed with four-blade stirrer, mixed using speed mixer, CNTs were treated with nitric acid | [25] |
| 290 | 6 | −98 | Pin-on-ring tribometer, 3 MPa, 0.4 m s$^{-1}$ | [26] |
| 290 | 2.9 | −99 | Pin-on-ring tribometer, 3 MPa, 0.4 m s$^{-1}$, dry mixing and SiC nanoparticles were modified with PGMA | [26] |
| 290 | 0.8 | −99.7 | Pin-on-ring tribometer, 3 MPa, 0.4 m s$^{-1}$, wet mixing and SiC nanoparticles were modified with PGMA | [26] |

*(Continued)*

**Table 1.2** A Summary of the Tribological Results on Polymer Nanocomposites—*Cont'd*

| S. No | Matrix | Filler Material | Size of Filler Material (nm) | Optimum Filler Content | | Coefficient of Friction | | |
|---|---|---|---|---|---|---|---|---|
| | | | | With Lowest COF | With Lowest Wear | Without Filler | With Filler | Change (%) |
| 12 | Epoxy | $Al_2O_3$ | 3.8 | 1.6 wt% | 1 wt% | 1.1 | 1 | −9.1 |
| | Epoxy | $Al_2O_3$-g-PAAM | 3.8 | 0.8 wt% | 0.3 wt% | 1.1 | 0.78 | −29.1 |
| | Epoxy | $Al_2O_3$-c-PAAM | 3.8 | 0.9 wt% | 0.3 wt% | 1.1 | 0.68 | −38.2 |
| 13 | Epoxy | SiC | 62.2 | 1.9 wt% | 0.3 wt% | 1.1 | 0.9 | −18.2 |
| | Epoxy | SiC-g-PAAM | 62.2 | 1.8 wt% | 0.3 wt% | 1.1 | 0.62 | −43.6 |
| | Epoxy | SiC-c-PAAM | 62.2 | 1.8 wt% | 0.3 wt% | 1.1 | 0.72 | −34.5 |
| 14 | Epoxy | $Si_3N_4$ | 16.8 | 1.7 wt% | 0.3 wt% | 1.1 | 0.95 | −13.6 |
| | Epoxy | KH550 treated $Si_3N_4$ | 16.8 | 1.0 wt% | 0.3 wt% | 1.1 | 0.8 | −27.3 |
| 15 | PEEK | $Si_3N_4$ | <50 | 7.5 wt% | 7.5 wt% | 0.38 | 0.25 | −34.2 |
| 16 | PEEK | $ZrO_2$ | 10 | 7.5 wt% | 7.5 wt% | 0.38 | 0.29 | −23.7 |
| 17 | PEEK | $SiO_2$ | <100 | 7.5 wt% | 7.5 wt% | 0.37 | 0.21 | −43.2 |
| 18 | PEEK | SiC | <100 | 20 wt% | 10 wt% | 0.38 | 0.2 | −47.4 |
| 19 | PEEK | CNF | 150 | – | 10 wt% | – | – | – |
| 20 | PEEK | $Al_2O_3$ | 15 | 5 wt% | 5 wt% | 0.32 | 0.35 | +9.4 |
| 21 | PEEK | $Al_2O_3$ | 15 | – | 5 wt% | – | – | – |

| Specific Wear Rate ($\times 10^{-6}$ mm$^3$ Nm$^{-1}$) | | | | |
|---|---|---|---|---|
| **Without Filler** | **With Filler** | **Change (%)** | **Equipment Used, Counterface Material, Load/Pressure Used, Sliding Velocity, Remarks (if any)** | **Reference** |
| 25 | 8 | −68 | Block-on-ring tribometer, carbon steel, 1 MPa, 1 m s$^{-1}$ | [27] |
| 25 | 0.18 | −99.3 | Block-on-ring tribometer, carbon steel, 1 MPa, 1 m s$^{-1}$, Al$_2$O$_3$ nanoparticles were modified with PAAM | [27] |
| 25 | 0.18 | −99.3 | Block-on-ring tribometer, carbon steel, 1 MPa, 1 m s$^{-1}$, Al$_2$O$_3$ nanoparticles were modified with PAAM | [27] |
| 25 | 8.5 | −66 | Block-on-ring tribometer, carbon steel, 1 MPa, 1 m s$^{-1}$, Al$_2$O$_3$ nanoparticles were modified with PAAM | [27] |
| 25 | 0.15 | −99.4 | Block-on-ring tribometer, carbon steel, 1 MPa, 1 m s$^{-1}$, SiC nanoparticles were modified with PAAM | [27] |
| 25 | 0.15 | −99.4 | Block-on-ring tribometer, carbon steel, 1 MPa, 1 m s$^{-1}$, SiC nanoparticles were modified with PAAM | [27] |
| 25 | 4.5 | −82 | Block-on-ring tribometer, carbon steel, 1 MPa, 1 m s$^{-1}$, SiC nanoparticles were modified with PAAM | [27] |
| 25 | 4.5 | −82 | Block-on-ring tribometer, carbon steel, 1 MPa, 1 m s$^{-1}$, Si3N4 nanoparticles were treated with aminopropyltrimethoxysilane | [27] |
| 7 | 1.3 | −81.4 | Block-on-ring tribometer, plain carbon steel, 196 N, 0.445 m s$^{-1}$ | [28] |
| 7.5 | 3.9 | −48 | Block-on-ring tribometer, plain carbon steel, 196 N, 0.445 m s$^{-1}$ | [29,30] |
| 7.5 | 1.4 | −81.3 | Block-on-ring tribometer, plain carbon steel, 196 N, 0.445 m s$^{-1}$ | [31] |
| 7.5 | 3.4 | −54.7 | Block-on-ring tribometer, plain carbon steel, 196 N, 0.445 m s$^{-1}$ | [32,33] |
| 2–3 | 0.1–0.2 | −94 | Ball-on-prism tribometer, steel, 21.2 N, 28.2 mm s$^{-1}$ | [34] |
| 13 | 3.5 | −73.1 | Block-on-ring tribometer, medium carbon steel, 196 N, 0.42 m s$^{-1}$ | [35] |
| 1.4 m$^2$ | 0.5 m$^2$ | −64 | Fretting test; wear scar area reduction; 10 mm diameter steel ball at 19.6 N normal load; fretting frequency of 27 Hz, amplitude 500 mm and $2 \times 10^5$ cycles | [36] |

*(Continued)*

**Table 1.2** A Summary of the Tribological Results on Polymer Nanocomposites—*Cont'd*

| | | | | Optimum Filler Content | | Coefficient of Friction | | |
| | | | | With Lowest COF | With Lowest Wear | Without Filler | With Filler | Change (%) |
| S. No | Matrix | Filler Material | Size of Filler Material (nm) | | | | | |
|---|---|---|---|---|---|---|---|---|
| 22 | PEEK | SiO$_2$ | 12 | – | 5 wt% | – | – | – |
| 23 | PEEK | SiO$_2$ | 13 | – | 1 wt% | 0.455 | 0.425 | –6.6 |
| 24 | PEEK | Microsized 10 wt% each of short carbon fiber, PTFE and graphite + nanosized SiO$_2$ | 13 | – | 1 wt% | 0.21 | 0.13 | –38 |
| 25 | Polyester | Nano clay platelets | | 3 wt% | 3 wt% | 0.52 | 0.35 | –32.6 |
| 26 | PET | Al$_2$O$_3$ | 38 | 2 wt% | 2 wt% | 0.32 | 0.3 | –6.3 |
| 27 | PS | MWCNT | 10–20 | 1.5 wt% | 1.5 wt% | 0.42 | 0.31 | –26.2 |
| 28 | PPESK | TiO$_2$ | 40 | 1 vol.% | 1.75 vol.% | 0.55 | 0.43 | –21.8 |
| 29 | BMI | SiC | <100 | 8 wt% | 6 wt% | 0.36 | 0.24 | –33.3 |
| 30 | UHM-WPE | CNT | 10–50 | 0.5 wt% | 0.5wt% | 0.05 | 0.11 | +120 |
| 31 | HDPE | Graphite nanoparticles | 400 | – | 8 wt% | – | – | – |
| 32 | UHM-WPE (80%) +HDPE (20%) | MWCNT | 60–100 nm diameter, 5–15 mm length, 95% purity | – | 2 wt% | 0.10–0.12 | 0.1–0.12 | 0 |
| 33 | UHM-WPE (80%) + HDPE (20%) | MWCNT | 60–100 nm diameter, 5–15 mm length, 95% purity | – | 2 wt% | 0.10–0.12 | 0.1–0.12 | 0 |
| 34 | PPS | CuO | 30–50 | 10 vol.% | 2 vol.% | 0.43 | 0.34 | –20.9 |

| Specific Wear Rate ($\times 10^{-6}$ mm$^3$ Nm$^{-1}$) | | | | |
|---|---|---|---|---|
| **Without Filler** | **With Filler** | **Change (%)** | **Equipment Used, Counterface Material, Load/Pressure Used, Sliding Velocity, Remarks (if any)** | **Reference** |
| 1.4 m$^2$ | 0.95 m$^2$ | −32 | Fretting test; wear scar area reduction; 10 mm diameter steel ball at 19.6 N normal load; fretting frequency of 27 Hz, amplitude 500 mm and $2 \times 10^5$ cycles | [36] |
| 29 | 9 | −69 | Block-on-ring test, 60 mm diameter 100Cr6 steel ring $R_a$ = 0.2 μm, apparent normal pressure 4 MPa and sliding speed 1 m s$^{-1}$, sliding time 20 h | [37] |
| 0.6 | 0.42 | −30 | Same as above with apparent normal pressure 7 MPa | [38] |
| $8 \times 10^2$ | $1 \times 10^2$ | −87.5 | Pin-on-disk tribometer, tool steel, 0.4 MPa, 0.3 m s$^{-1}$ | [39] |
| 17.4 | 9.5 | −45.4 | Reciprocating tribometer, steel, 340 N, 25 mm s$^{-1}$. The crystallinity of PET also effects the wear coefficient | [40] |
| 130 | 8 | −93.8 | Block-on-ring tribometer, plain carbon steel, 50 N, 0.431 m s$^{-1}$ | [41] |
| 80.12 | 4.86 | −93.9 | Block-on-ring tribometer, mild carbon steel, 200 N, 0.43 m s$^{-1}$ | [42] |
| 6.8 | 2.2 | −67.6 | Block-on-ring tribometer, carbon steel, 196 N, 0.42 m s$^{-1}$ | [43] |
| 0.35 mg | 0.02 mg | −94.3 | Ball-on-disk tribometer, Si$_3$N$_4$ ball, 5 N, 0.3 m s$^{-1}$ | [44] |
| 61 mg | 63 mg | +3.2 | Taber abrasion test: two rotating abrasive wheel CS-10, 50 mm diameter, 12.6 mm thickness, contact load 1 kg; rotational speed 72 rpm and 5000 cycles. Weight loss measured | [45] |
| 0.1 | 0.2 | +100 | Ball-on-prism test, ball diameter 12.7 mm, normal load 21.2 N, sliding speed 28.2 mm s$^{-1}$, wear test duration 60 h. Counterface 100Cr6 steel | [46] |
| 0.4 | 0.14 | −65 | Same as above with counterface stainless steel X5CrNi18-10 | [46] |
| 0.324 mm$^3$ km$^{-1}$ | 0.078 mm$^3$ km$^{-1}$ | −78.4 | Pin-on-disk tribometer, hardened tool steel, 0.65 MPa, 1 m s$^{-1}$ | [47] |

*(Continued)*

**Table 1.2** A Summary of the Tribological Results on Polymer Nanocomposites—*Cont'd*

| S. No | Matrix | Filler Material | Size of Filler Material (nm) | Optimum Filler Content | | Coefficient of Friction | | |
| | | | | With Lowest COF | With Lowest Wear | Without Filler | With Filler | Change (%) |
|---|---|---|---|---|---|---|---|---|
| 35 | PPS | TiO$_2$ | 30–50 | 1 vol.% | 2 vol.% | 0.43 | 0.35 | −18.6 |
| 36 | PPS | Al$_2$O$_3$ | 33 | 10 vol.% | 2 vol.% | 0.455 | 0.415 | −8.8 |
| 37 | PI | CNT | 10–50 | >5 wt% | >10 wt% | 0.38 | 0.28– 0.3 | −21.1 |

*Data are taken from the published literature [15–49].*
*PAAM, poly(acrylamide); PGMA, poly(glycidyl methacrylate); PI, poly(imide); PPESK, poly(phthalazine ethersulfoneketone); PS, poly(styrene); COF, coefficient of friction; MWCNT, multi-wall carbon nanotube; CNF, carbon nano fibre; BMI, bismaleimide; HDPE, high density poly(ethylene).*

sulfide) (PPS). The specific wear rates of these nanocomposites have been reported in the range of 10–100 times lower than the specific wear rates of the polymers without fillers when optimum weight percentage of the nanoparticles is introduced. The optimum composition depends upon the systems and is high for PTFE system where the nanoparticles are required for the mechanical strength while PTFE still promotes the reduction of the coefficient of friction. For this system, there is no change in the coefficient of friction after adding nanoparticles. For epoxy, PEEK and PPS polymers, the role of nanoparticles is to increase the load-bearing capacity of the material and thus the actual contact area is reduced leading to lower frictional stress for the nanocomposite. In addition, the presence of nanoparticles in the matrix improves the toughness to an extent that leads to lesser propensities for wear by subsurface fatigue or asperity plowing actions. It has also been reported that the gain in wear resistance for nanocomposite could also be attributed to an increase in the thermal conductivity of the composite compared to the pristine polymer. High thermal conductivity of the material facilitates a lower mean interfacial temperature as the frictional heat generated at the interface is more readily transferred (conducted) to other parts of the machine. It is not very clear if the nanoparticle can also change the properties of the transfer film, although it has been claimed by some researchers that the presence of nanoparticles render the transfer film thin and tenacious.

The second type of polymer nanocomposite system is a polymer filled with carbon nanotubes (CNTs). CNT is an excellent material for reinforcing polymers as CNT shows very high strength and stiffness. In addition, the nanometer size (length and diameter) ensures an extremely large interface area between the filler and the matrix providing a potentially excellent bonding between the two. Tribological tests on CNT-based polymer nanocomposites have started appearing in the literature only recently in the past nine years. CNT's role in a low-friction polymer such as PTFE is mainly of reinforcement type. Thus, the wear resistance is increased by at least three

| Specific Wear Rate ($\times 10^{-6}$ mm$^3$ Nm$^{-1}$) | | | | |
|---|---|---|---|---|
| **Without Filler** | **With Filler** | **Change (%)** | **Equipment Used, Counterface Material, Load/Pressure Used, Sliding Velocity, Remarks (if any)** | **Reference** |
| 0.324 mm$^3$ km$^{-1}$ | 0.162 mm$^3$ km$^{-1}$ | −50.6 | Pin-on-disk tribometer, hardened tool steel, 0.65 MPa, 1 m s$^{-1}$ | [47] |
| 0.46 mm$^3$ km$^{-1}$ | 0.3 mm$^3$ km$^{-1}$ | −34.8 | Pin-on-disk tribometer, hardened tool steel, 0.65 MPa, 1 m s$^{-1}$ | [48] |
| 4.4 mm$^3$ | 2.5 mm$^3$ | −43.2 | Ball-on-ring tribometer, plain carbon steel, 290 N, 0.431 m s$^{-1}$ | [49] |

orders of magnitude with slight decrease in the coefficient of friction. Application of CNT as filler for UHMWPE and epoxy has similar effects on the wear resistance; however, in quantitative terms the improvement is only in the range of a few times to about an order of magnitude. For coefficient of friction, either there is marginal decrease in the value due to low real contact area owing to high load-bearing capacity of the nanocomposite, or there is slight increase in the coefficient of friction. The increase in the coefficient of friction is often believed to be due to the changes in the topography and the increased resistance of the CNTs to shearing actions at the interface.

For both types of nanocomposites, the processing method has much significance in the improvement of mechanical and tribological performance of the material. A suitable surface pretreatment of the nanoparticles can lead to less filler agglomeration and thus better bonding between the nanoparticles and the matrix. Similarly, shear vigorous mixing in the case of CNT-filled polymer has given better results for a lower agglomeration and improved filler matrix bonding.

## 1.5 **FUTURE PROSPECTS**

The technological applications of polymers and their composites have steadily increased. Often such applications are motivated by the desire to replace an existing material such as a metal. The self-lubricating property of many linear thermoplastics, both in pristine and composite forms, has been well exploited for tribological applications such as gears, bearings, human hip/knee joints, nonstick cooking pans, etc. Thermosetting polymers such as epoxy and phenolic are used in composite forms as load-bearing materials and as matrices for composite brake materials for automobiles. These traditional applications will continue to require ever demand for high-performance materials. Materials that can sustain high pressure and velocity with low wear rate would be needed. In this aspect, the newly emerging area of polymer nanocomposites could be a great promise. The main advantage of nanocomposites over traditional composites is in their ability to improve both strength and toughness properties simultaneously and isotropically. The tribological research on polymer nanocomposites is still at a relatively early stage and we will see many innovative research work in this area in the coming years. It may be necessary to reconsider the concept of hybrid composites with nanoparticle or nanofibers as the strengthening agent while some other filler (PTFE or graphite) is adopted for the self-lubrication property. Various types of surface modification techniques (covalently grafting of an organic precursor on the nanoparticle or nitric acid treatment of CNTs) for the nanoparticles and nanofibers are other areas that need to be focused on as the bonding strength between the filler and the matrix can be greatly enhanced by the surface modification of the fillers. Agglomeration of nanoparticles or CNTs is a dominant practical problem that could be effectively minimized by surface chemical treatment or high mechanical shear mixing.

Polymers have begun to be used in micromachines such as MEMS as a structural material for which silicon has been traditionally used. A polymer that is being used frequently is PMMA, which shows very poor intrinsic tribological performance. Thus, there is some future prospect in trying to improve the wear resistance property of polymers such as PMMA.

Many tribological problems associated with metals or other materials, such as silicon, can be eliminated by having a mechanically robust polymer coating. High-performance (strength and thermal stability) polymers such as PEEK and poly(imide) when mixed with nanoparticle or nanofibers can be used as a thin film on a substrate for high wear resistance. Research papers currently available in the literature in this area are very few and hence this could be another major growth area for future polymer tribology research.

Summarizing, the tribological application of polymers is expected to grow as the usage of polymers expands to newer applications. Since polymers do not always follow the rule of better strength and hardness for better wear resistance, there will be always a need for fine-tuning tribological performances of the material without seriously compromising the bulk strength. Thus, we can expect a healthy rise in tribological research in the area of polymers in all forms (pristine, composite, nanocomposite and hybrid).

## 1.6 FINAL REMARKS

The past research in the area of tribology of polymers and their composites has established a good and secure understanding of the frictional energy dissipative processes and the associated damage and wear mechanisms. The two-term model can facilitate the separating of the contributions of the interface and the subsurface (cohesive) to total frictional force. This model can also help in the understanding of wear and in providing solutions to tribological problems. For example, the design of a polymer composite for a particular tribological application must consider, a priori, the main contributing factor (interface or bulk) to friction and wear. The selection of the scale of the filler is influenced by such considerations.

The present trend in tribological research in the area of polymers is moving toward polymer nanocomposites and polymer surface modifications and this trend will continue to grow. It may be important now that novel ways, chemical or physical, of improving polymers' tribology performance are researched.

Although we did not discuss the area of lubrication of polymer contacts, this area opens up new opportunities for those who are working on the use of polymers in biological systems. Surface energy modifications (hydrophilic as opposed to hydrophobic) can change the nature of lubrication and thus the wear life of the material. Some tests involving lubricated sliding on polymer brushes have shown that this concept can work well for any surface if we can grow a suitable polymer brush on the substrate [50–52]. In addition, lubrication of polymer surfaces by lubricants such as perfluoropolyether has opened up new possibilities of applying external lubricant to a polymer surface for low friction and extended wear life [3].

## Acknowledgment

The authors thank Dr N. Satyanarayana for help in producing a summary of the tribology research in the area of polymer nanocomposites as presented in Table 1.2.

## NOTATIONS AND ABBREVIATIONS

$\tau$  Shear stress

$\tau_o$  Shear strength of the interface

$\alpha$  Proportionality constant

$p$  Mean contact pressure

$S$  Ultimate tensile strength

$e$  % Elongation (plastic) at failure in tensile test

$W_{sp}$  Specific wear rate ($mm^3 Nm^{-1}$)

**$Al_2O_3$-c-PAAM**  Alumina nanoparticles with grafted polymer and some amount of homopolymer

**$Al_2O_3$-g-PAAM**  Alumina nanoparticles with grafted polymer

**Attapulgite**  Attapulgite (or palygorskite) is a clay mineral

**CNF**  Carbon nano fiber

**BMI**  Bismaleimide

**HDPE**  High density poly(ethylene)

**CNT**  Carbon nanotube

**COF**  Coefficient of friction

**KH550**  $\gamma$-Aminopropyl trimethoxysilane

**MWCNT**  Multi-wall carbon nanotube

**PMMA**  Poly(methylemethacrylate)

**PAAM** Poly(acrylamide)
**PDMS** Poly(dimethylsiloxane)
**PEEK** Poly(ether ether ketone)
**PET** Poly(ethylene terephthalate)
**PGMA** Poly(glycidyl methacrylate)
**PPESK** Poly(phthalazine ethersulfoneketone)
**PPS** Poly(phenylene sulfide)
**PI** Poly(imide)
**PS** Poly(styrene)
**POM** Poly(oxymethylene)
**PTFE** Poly(tetrafluoroether)
**UHMWPE** Ultra-high molecular weight poly(ethylene)

## References

[1] B.J. Briscoe, S.K. Sinha, Tribology of polymeric solids and their composites, in: G. Stachowiak (Ed.), Wear – Materials, Mechanism and Practice, John Wiley & Sons, Chichester, England, 2005, pp. 223–267.

[2] N.S. Tambe, B. Bhushan, Micro/nanotribological characterization of PDMS and PMMA used for BioMEMS/NEMS applications, Ultramicroscopy 105 (2005) 238–247.

[3] N. Satyanarayana, S.K. Sinha, B.H. Ong, Tribology of a novel UHMWPE/PFPE dual-film coated onto Si surface, Sensors and Actuators A: Physical 128 (1) (2006) 98–108.

[4] A. Schallamach, Abrasion of rubber by a needle, Journal of Polymer Science 9 (5) (1952) 385–404.

[5] J.A. Greenwood, D. Tabor, The friction of hard sliders on lubricated rubber: the importance of deformation losses, Proceedings of the Physical Society 71 (1958) 989–1001.

[6] B.J. Briscoe, Wear of polymers: an assay on fundamental aspects, Tribology International (August, 1981) 231–243.

[7] B.J. Briscoe, S.K. Sinha, Wear of polymers, Proceedings of the Institution of Mechanical Engineers Part J: Journal of Engineering Tribology 216 (2002) 401–413.

[8] D. Tabor, The physical meaning of indentation and scratch hardness, British Journal of Applied Physics 7 (1956) 159–166.

[9] B.J. Briscoe, D. Tabor, Rheology of thin organic films, ASLE Transactions 17 (1974) 158–165.

[10] S.N. Ratner, I.I. Farberoua, O.V. Radyukeuich, E.G. Lure, Correlation between wear resistance of plastics and other mechanical properties, Soviet Plastics 7 (1964) 37–45.

[11] J.K. Lancaster, Friction and wear, (Chapter 14) in: A.D. Jenkins (Ed.), Polymer Science, A Materials Science Handbook, North-Holland Publishing Company, Amsterdam, 1972.

[12] K. Friedrich, Z. Lu, A.M. Hager, Recent advances in polymer composites' tribology, Wear 190 (2) (1995) 139–144. (and references therein).

[13] B.J. Briscoe, A.K. Pogosian, D. Tabor, The friction and wear of high density polyethylene: the action of lead oxide and copper oxide fillers, Wear 27 (1974) 19–34.

[14] B.J. Briscoe, The tribology of composites materials: a preface, in: K. Friedrich (Ed.), Advances in Composite Tribology, Elsevier Science Publishers B. V., 1993, pp. 3–15.

[15] F. Li, K. Hu, J. Li, B. Zhao, The friction and wear characteristics of nanometer ZnO filled polytetrafluoroethylene, Wear 249 (2001) 877–882.

[16] W.G. Sawyer, K.D. Freudenberg, P. Bhimaraj, L.S. Schadler, A study on the friction and wear behavior of PTFE filled with alumina nanoparticles, Wear 254 (2003) 573–580.

[17] W.X. Chen, F. Li, G. Han, J.B. Xia, L.Y. Wang, J.P. Tu, Z.D. Xu, Tribological behavior of carbon-nanotube-filled PTFE composites, Tribology Letters 15 (3) (2003) 275–278.

[18] S.Q. Lai, T.S. Li, X.J. Liu, R.G. Lv, A study on the friction and wear behavior of PTFE filled with acid treated nano-attapulgite, Macromolecular Materials and Engineering 289 (2004) 916–922.

[19] M.Z. Rong, M.Q. Zhang, H. Liu, H. Zeng, B. Wetzel, K. Friedrich, Microstructure and tribological behavior of polymeric nanocomposites, Industrial Lubrication and Tribology 53 (2001) 72–77.

[20] B. Wetzel, F. Haupert, K. Friedrich, M.Q. Zhang, M.Z. Rong, Impact and wear resistance of polymer nanocomposites at low filler content, Polymer Engineering and Science 42 (2002) 1919–1927.

[21] B. Wetzel, F. Haupert, M.Q. Zhang, Epoxy nanocomposites with high mechanical and tribological performance, Composites Science and Technology 63 (2003) 2055–2067.

[22] G. Shi, M.Q. Zhang, M.Z. Rong, B. Wetzel, K. Friedrich, Friction and wear of low nanometer $Si_3N_4$ filled epoxy composites, Wear 254 (2003) 784–796.

[23] M.Q. Zhang, M.Z. Rong, S.L. Yu, B. Wetzel, K. Friedrich, Effect of particle surface treatment on the tribological performance of epoxy based nanocomposites, Wear 253 (2002) 1086–1093.

[24] M.Q. Zhang, M.Z. Rong, S.L. Yu, B. Wetzel, K. Friedrich, Improvement of tribological performance of epoxy by the addition of irradiation grafted nano-inorganic particles, Macromolecular Materials Engineering 287 (2002) 111–115.

[25] O. Jacobs, W. Xu, B. Schadel, W. Wu, Wear behaviour of carbon nanotube reinforced epoxy resin composites, Tribology Letters 23 (1) (2006) 65–75.

[26] Y. Luo, M.Z. Rong, M.Q. Zhang, Covalently connecting nanoparticles with epoxy matrix and its effect on the improvement of tribological performance of the composites, Polymers and Polymers Composites 13 (2005) 245–252.

[27] M.Z. Rong, M.Q. Zhang, G. Shi, Q.L. Ji, B. Wetzel, K. Friedrich, Graft polymerization onto inorganic nanoparticles and its effect on tribological performance improvement of polymer composites, Tribology International 136 (2003) 697–707.

[28] Q. Wang, J. Xu, W. Shen, W. Liu, An investigation of the friction and wear properties of nanometer Si3N4 filled PEEK, Wear 196 (1996) 82–86.

[29] Q. Wang, Q. Xue, H. Liu, W. Shen, J. Xu, The effect of particle size of nanometer $ZrO_2$ on the tribological behaviour of PEEK, Wear 198 (1996) 216–219.

[30] Q. Wang, Q. Xue, W. Shen, J. Zhang, The friction and wear properties of nanometer $ZrO_2$-filled polyetheretherketone, Journal of Applied Polymer Science 69 (1998) 135–141.

[31] Q. Wang, Q. Xue, W. Shen, The friction and wear properties of nanometer $SiO_2$ filled polyetheretherketone, Tribology International 130 (1997) 193.

[32] Q. Wang, J. Xu, W. Shen, Q. Xue, The effect of nanometer SiC filler on the tribological behavior of PEEK, Wear 209 (1997) 316–321.

[33] Q. Xue, Q. Wang, Wear mechanisms of polyetheretherketone composites filled with various kinds of SiC, Wear 213 (1997) 54–58.

[34] P. Werner, V. Altstadt, R. Jaskulka, O. Jacobs, J.K.W. Sandler, M.S.P. Shaffer, A.H. Windle, Tribological behaviour of carbon-nanofibre-reinforced poly (ether ether ketone), Wear 257 (2004) 1006–1014.

[35] H.B. Qiao, Q. Guo, A.G. Tian, G.L. Pan, L.B. Xu, A study on friction and wear characteristics of nanometer $Al_2O_3$/PEEK composites under the dry sliding condition, Tribology International 140 (2007) 105–110.

[36] G. Pan, Q. Guo, W. Zhang, A. Tian, Fretting wear behavior of nanometer $Al_2O_3$ and $SiO_2$ reinforced PEEK composites, Wear 266 (2009) 1208–1215.

[37] G. Zhang, A.K. Schlarb, S. Tria, O. Elkedim, Tensile and tribological behaviors of PEEK/nano-$SiO_2$ composites compounded using a ball milling technique, Composites Science and Technology 68 (2008) 3073–3080.

[38] G. Zhang, L. Chang, A.K. Schlarb, The roles of nano-$SiO_2$ particles on the tribological behavior of short carbon fiber reinforced PEEK, Composites Science and Technology 69 (2009) 1029–1035.

[39] P. Jawahar, R. Gnanamoorthy, M. Balasubramanian, Tribological behaviour of clay-thermoset polyester nanocomposites, Wear 261 (2006) 835–840.

[40] P. Bhimaraj, D.L. Burris, J. Action, W.G. Sawyer, C.G. Toney, R.W. Siegel, L.S. Schadler, Effect of matrix morphology on the wear and friction behavior of alumina nanoparticle/poly (ethylene) terephthalate composites, Wear 258 (2005) 1437–1443.

[41] Z. Yang, B. Dong, Y. Huang, L. Liu, F.-Y. Yan, H.-L. Li, Enhanced wear resistance and micro-hardness of polystyrene nanocomposites by carbon nanotubes, Materials Chemistry and Physics 94 (2003) 109–113.

[42] X. Shao, W. Liu, Q. Xue, The tribological behavior of micrometer and nanometer $TiO_2$ particle-filled poly (phthalazine ether sulfone ketone) composites, Journal of Applied Polymer Science 92 (2004) 906–914.

[43] H. Yan, R. Ning, G. Liang, X. Ma, The performance of BMI nanocomposites filled with nanometer SiC, Journal of Applied Polymer Science 95 (2005) 1246–1250.

[44] Y.S. Zoo, J.W. An, D.P. Lim, D.S. Lim, Effect of carbon nanotube addition on tribological behavior of UHMWPE, Tribology Letters 16 (4) (2004) 305.

[45] H. Fouad, R. Elleithy, High density polyethylene/graphite nano-composites for total hip joint replacements: processing and in vitro characterization, Journal of the Mechanical Behavior of Biomedical Materials 4 (2011) 1376–1383.

[46] Y. Xue, W. Wu, O. Jacobs, B. Schädel, Tribological behaviour of UHMWPE/HDPE blends reinforced with multi-wall carbon nanotubes, Polymer Testing 25 (2006) 221–229.

[47] S. Bahadur, C. Sunkara, Effect of transfer film structure, composition and bonding on the tribological behavior of polyphenylene sulfide filled with nano particles of $TiO_2$, ZnO, CuO and SiC, Wear 258 (2005) 1411–1421.

[48] C.J. Schwartz, S. Bahadur, Studies on the tribological behavior and transfer film-counterface bond strength for polyphenylene sulfide filled with nanoscale alumina particles, Wear 237 (2000) 261–273.

[49] H. Cai, F. Yan, Q. Xue, Investigation of tribological properties of polyimide/carbon nanotube nanocomposites, Materials Science and Engineering A 364 (2004) 94–100.

[50] J. Klein, E. Kumacheva, D. Mahalu, D. Perahia, L.J. Fetters, Reduction of frictional forces between solid surfaces bearing polymer brushes, Nature 370 (1994) 634–636.

[51] M. Müller, X. Yan, S. Lee, S. Perry, N.D. Spencer, Lubrication properties of a brushlike copolymer as a function of the amount of solvent absorbed within the brush, Macromolecules 38 (13) (2005) 5706–5713.

[52] R. Matsuno, H. Otsuka, A. Takahara, Polystyrene-grafted titanium oxide nanoparticles prepared through surface-initiated nitroxide-mediated radical polymerization and their application to polymer hybrid thin films, Soft Matter 2 (2006) 415–421.

# The effect of nanoparticle fillers on transfer film formation and the tribological behavior of polymers

2

**Shyam Bahadur, Cris Schwartz**

*Department of Mechanical Engineering, Iowa State University, Ames, IA, USA*

## CHAPTER OUTLINE HEAD

## 2.1 INTRODUCTION

The self-lubricating ability of polymers while sliding against metallic surfaces makes them desirable for many sliding situations. Thus, low coefficient of friction coupled with reasonable wear resistance may often be obtained in polymer–metal sliding systems without the use of an external lubricant. This is a great advantage particularly in situations where external lubrication is difficult. The challenge has been to increase the wear resistance without adversely affecting coefficient of friction. This has thus been the thrust of research in the area of polymer tribology over the past decades. To this end, the approaches tried have focused on the reinforcement of polymers with fibers and filling the polymers with inorganic fillers. This chapter deals with the latter aspect, in particular, the blending of polymers with inorganic fillers of nanosize. This research has built upon earlier studies performed with microsize inorganic fillers. The presentation in this chapter is basically an update of the chapter prepared by the authors that appeared in the earlier edition of the book.

**Tribology of Polymeric Nanocomposites. http://dx.doi.org/10.1016/B978-0-444-59455-6.01002-3**

The self-lubricating ability of many polymers arises from their ability to form a transfer film on the mating surface during sliding. Thus, later sliding occurs between the polymer component and its own transfer film. In this situation, low friction can be obtained because of the low shear strength of polymers as well as the low shear force needed to cause shearing of the adhesive bonds formed during sliding. In the absence of this transfer film, the polymer surface would be abraded by the asperities on the hard metal surface and this is detrimental for sliding because wear becomes excessive. The initial process of film deposition leading to a persistent transfer film contributes to the transient wear behavior observed in polymer–metal sliding [1]. In many systems, steady state is reached when the film is uniformly deposited on the metal counterface. The film offers protection to the bulk polymer from the counterface asperities. During this phase, small portions of the transfer film may be removed and promptly replaced by additional material transfer from the polymer bulk. The wear debris produced during sliding thus results either from the material removed from the transfer film or from the material being removed from the polymer. Therefore, the quality of the bond between the transfer film and counterface has a direct impact on the wear amount produced during sliding. The mechanisms of material transfer and transfer film development are difficult to investigate during the wear process. Therefore, descriptions of the transfer film have often been qualitative in nature. Investigators have described useful transfer films as thin, uniform, and well bonded to the counterface.

Polymer properties are typically changed by the addition of a number of materials including particles of inorganic compounds, clays, fibers, and even particles of other polymers. These materials in general are of the form of fibers or particulate fillers. Fibers are often more effective than the fillers in improving the mechanical properties but the processing cost can be higher. This makes the addition of fillers to polymers more feasible commercially where it results in improving the tribological properties. The prediction of mechanical properties such as stiffness modulus of filled polymers is often possible with the knowledge of the properties of the constituents, and volume fraction of the filler [2], whereas the effect of a particular filler material on wear resistance is usually not known a priori. To date, there have been no rigorous criteria developed to predict how a specific filler will affect the tribological behavior of the polymer. This is due to the complexity of the wear process, which is affected by the development and bonding of transfer film to the counterface. In many cases, the addition of filler particles to polymers leads to a more uniform and better bonded transfer film than that of unfilled polymer, while in other cases the filler hinders the film formation ability of the polymer. The result of this is improved wear resistance in the former case and decreased wear resistance in the latter.

One area of recent interest has been the addition of nanoparticle fillers to polymers for enhanced wear resistance. By definition, the size of nanoparticles is in the nanometer range and this results in a much higher surface area-to-volume ratio than for microsize particles. The physical and chemical properties of the nano materials are thus governed more by their surface behavior than by the bulk. This results in properties such as hardness and surface activity of nanoparticles being a strong function of their size [3]. The tribological studies on nanoparticle-filled polymers have shown some very interesting results. There are fillers that contribute to enhanced wear resistance while there are others that result in the deterioration of wear resistance. Interestingly, with the reduction in size to nanoscale, even hard abrasive fillers such as alumina and silicon nitride have demonstrated the improvement in the wear resistance of the polymer.

In order to understand better the role of nanoparticle fillers on tribological properties, this chapter will discuss the mechanics of transfer film formation and the results related to the transfer film development and stability as affected by the addition of nanoprticles. The goal of this pursuit is to utilize the understanding developed from research in this area to lead to more informed selection of nanoparticle compounds for use as fillers in polymers involved in tribological applications.

## 2.2 TRANSFER FILM DEVELOPMENT AND CHARACTERISTICS

The formation of the transfer film in polymer tribology is of crucial importance because of its direct effect on wear rate. This section discusses the mechanism of formation of transfer film and also the parameters that play a role in film deposition. It must be noted that much of the work presented here by various researchers uses higher loads than what is often encountered in practical sliding applications. The focus of much of this work has been on the steady state of wear, which takes hours to set in even under aggressive wear conditions. Therefore, to obtain steady state conditions within a reasonable amount of time, many researchers have decided to run the wear experiments at relatively high loads and high speeds. Even under these conditions, the wear experiments are typically run for 8–10 h to allow for meaningful analysis of the wear data. It is recognized that under mild sliding conditions, the wear rates would be expected to be lower than those reported here for aggressive sliding conditions. However, most of the conclusions drawn regarding the effects of sliding parameters and fillers are valid even at lower sliding speeds.

### 2.2.1 Transfer film deposition

In polymer-on-metal sliding, there is a natural propensity for polymeric material to be transferred to the metallic counterface. This was shown even in static conditions by the work of Brainard and Buckley [4] in their work on polytetrafluoroethylene (PTFE) against metal surfaces. In sliding wear, the adhesion of polymer to the metal counterface is the fundamental basis for transfer film development. The mechanisms proposed for adhesion of polymer to the metal surface are van der Waals forces of attraction, Coulomb electrostatic forces, and bonding resulting from chemical reactions. For instance, Yang et al. [5] determined by infrared spectrophotometry that fluoride ions were produced during the sliding of PTFE against stainless steel. These ions presumably reacted with the metal surface. Based on observations like these and others, it is thought that polymer particles adhere to the counterface due to the attraction of polymer chain ends or free radicals to the metallic surface. Although there is inherent attraction of polymer to the metal counterface, neat polymers generally produce poorly bonded and nonuniform transfer films [6].

Another mechanism that is responsible for transfer film formation is mechanical interlocking of the fragments of polymer generated during sliding into the crevices of the counterface surface asperities. This process initiates the formation of discrete patches of the polymeric material, which are augmented by cohesion between mutually compatible polymer fragments produced during sliding and the polymeric material initially transferred to the counterface. This can be seen in Fig. 2.1, which shows the gradual development of transfer film during sliding between polyetheretherketone (PEEK) and a tool steel counterface [6]. The figure shows the sequences in the development of transfer film as a function of the number of sliding traverses.

## 2.2.2 Parameters affecting the formation of transfer film

The presence of a uniform and thin transfer film has been shown to be beneficial in the sliding wear of polymers because the film protects the relatively soft polymer bulk from the abrasive action of the hard metallic counterface asperities. A condition of self-lubrication thus occurs when a stable and well-bonded film has been deposited, resulting in low coefficient of friction sliding of the polymer bulk against a polymeric transfer film. Therefore, the ultimate tribological behavior of a polymer depends greatly on the extent and texture of transfer film and the stability of this film in steady state sliding. Investigations of the transfer film in sliding wear suggest that the variables that are pertinent to the film characteristics are sliding parameters, polymer composition and counterface characteristics.

### 2.2.2.1 Sliding parameters and polymer composition

The sliding parameters pertinent to a tribological situation are the ambient temperature of the system, the sliding velocity, the normal load between polymer and counterface, and the atmosphere in which the system operates. It is well known that these parameters affect friction and wear behavior and also the transfer film.

Temperature plays a large role in the behavior of polymers, and thus strongly affects the amount of material transfer produced in polymer–metal sliding. However, a distinction must be made between the ambient temperature of the system, which can be directly measured, and that of the temperature distribution at the sliding interface, which does not easily lend itself to experimental measurement. The temperature at the interface is higher than the ambient temperature because of frictional heating induced during sliding. At higher temperatures, polymers become more compliant resulting in greater adhesion between the mating surfaces. Yang et al. [5] showed that at higher ambient temperatures the thickness of PTFE transfer film on a 316 stainless steel surface increased significantly (Fig. 2.2). An increase in the sliding velocity produces a temperature increase at the sliding interface, regardless of the ambient temperature. Due to softening and increased adhesion, the material transfer rate from the polymer to the counterface is thus increased.

The nature of the film formed during polymer-on-metal sliding also depends upon the polymer itself [7]. For example, low-density polyethylene, polypropylene, and polyamide 66 have been reported to produce relatively thick films of

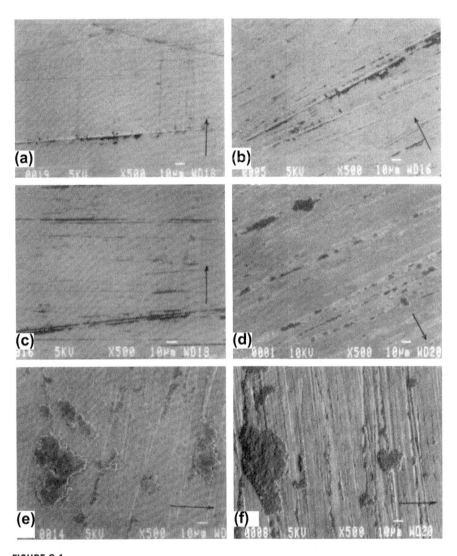

**FIGURE 2.1**

Scanning electron micrographs of PEEK transfer film after (a) 1 cycle, (b) 10 cycles, (c) 100 cycles, (d) 10,000 cycles, (e) 70,000 cycles, and (f) 141,000 cycles during sliding between PEEK and tool steel surface. Sliding conditions: pin cross-section 6 mm × 5 mm, load 19.6 N, speed 1.0 m s⁻¹, and counterface roughness 0.11 μm Ra [7].

0.10–1.00 μm, whereas high-density polyethylene, PTFE, and polyoxymethylene produce thinner films. This behavior was observed at low speeds, intermediate temperatures and for smooth surfaces. A change in sliding conditions would be expected to alter this behavior.

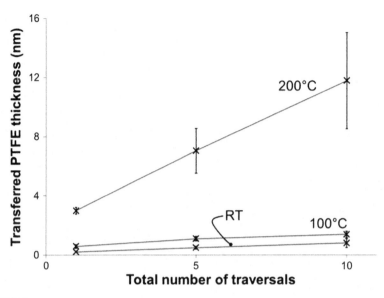

**FIGURE 2.2**

Variation of PTFE transfer film thickness with the number of traverses (1, 5, or 10) at three temperatures under a contact pressure of 89 kPa RT indicates room temperature.

*Adapted from [5].*

### 2.2.2.2 Counterface characteristics

The counterface characteristics related to transfer film quality are the surface texture, roughness, and material composition of the counterface. Most engineering surfaces are moderately rough with a surface roughness of 0.05–0.15 μm Ra. Surfaces with roughness in this range produce moderate abrasion in the initial stages of sliding that contributes to the generation of wear debris. The debris tends to interlock into the crevices of asperities, as indicated earlier. The development of initial transfer film is thus promoted by mechanical interlocking, as illustrated by Fig. 2.1. Once the initial film is formed, adhesion becomes the contributing factor to the development of a fully developed transfer film during later sliding. In the case of smoother surfaces, adhesion is the predominant mode of bonding as compared to mechanical interlocking. However, in the case of engineering surfaces, both moderate abrasion followed by adhesion govern the development and bonding of transfer film to the counterface.

The counterface material too plays a role in transfer film development and bonding. Due to their high thermal conductivities, most metallic counterfaces transfer heat from the sliding interface and thus reduce the detrimental effect of frictional heating indicated earlier. On the other hand, in case of the surfaces of materials with poor heat transfer characteristics, such as stainless steel and glass, the effect of temperature becomes dominant. This is shown in Fig. 2.3 for polyamide sliding on a glass surface [8]. It should be noted that the transfer film here has been affected by interface heating so that it is lumpy and nonuniform and is also discontinuous.

100 μm

**FIGURE 2.3**

Optical micrograph of the transfer film formed on a glass counterface of 0.04 μm Ra by Nylon 11. Sliding conditions: pin cross-section 6 mm × 5 mm, load 19.6 N, speed 1.0 m s$^{-1}$ [8].

## 2.3 EFFECT OF FILLERS ON WEAR AND TRANSFER FILMS

Filler materials are added to polymers to improve thermal, mechanical and tribological properties. In the context of tribological properties, the ability of fillers to influence transfer film development and adhesion to the counterface are important. The fillers used for modifying the tribological behavior are mostly inorganic compounds both in micro- and nanosizes. Initially the work was done with microsize fillers, but lately the focus has been on nanosize fillers. In most cases, investigators have focused on the effect of filler on wear resistance. In limited cases, the transfer film characteristics have been addressed as well with respect to their contribution to wear resistance. Some of the findings are reported below with respect to both wear and transfer film.

Inorganic Cu-based fillers in both micro- and nanosizes have been extensively studied in many polymer matrices because of their effectiveness in wear reduction. Microsize CuO has been shown to enhance the wear resistance of polyphenylene sulfide (PPS) [9], high-density polyethylene [10], polyamide [11], PEEK [12] and thermosetting polyester [13]. Another Cu-based filler, microsize CuS, has been shown to result in the reduction of wear in PTFE [14], polyamide [15], PEEK [6] and PPS [9,16]. These publications also describe the transfer films formed in these polymers with and without the microsize fillers.

Interestingly, some of the same benefits of microsize Cu-based fillers are also found with nanoparticulate fillers but at much smaller volume fractions. In fact, various investigators have shown that the major difference between microsize and nanosize particles used as fillers is that the maximum reduction in wear rate is obtained by 25–35 vol.% microsize filler proportions as opposed to 1–5 vol.% filler proportions for nanosize fillers. Figure 2.4 shows the wear and friction behavior of PPS filled with CuO nanoparticles in volume proportions of 1, 2 and 4% while sliding in a pin-on-disk configuration against a tool steel counterface of 0.10 μm Ra roughness, at 1 m s$^{-1}$

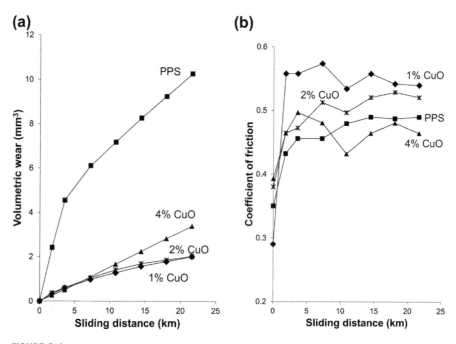

**FIGURE 2.4**

Variation of wear (a) and coefficient (b) of friction with sliding distance for PPS filled with nanoparticles of CuO.

*Adapted from [17].*

sliding speed, and under a nominal contact pressure of 0.65 MPa [17]. The curve representing the wear loss versus sliding distance for the unfilled PPS exhibits two states, the first being the transient state and the second the steady state wear. In the transient state, wear rate (defined as wear volume per unit sliding distance) is much higher than in the steady state because of the lack of transfer film, which is still being formed. Because of the smoothening of the counterface from the transfer film formed in the transient state, wear rate in the steady state is considerably reduced. Figure 2.4 shows that the transient state almost disappeared with the addition of 1–4 vol.% CuO filler. The steady state wear rate for unfilled PPS was 0.291 $mm^3$ $km^{-1}$, with the addition of 1 and 2 vol.% filler, whereas the wear rates were reduced to 0.061 and 0.047 $mm^3$ $km^{-1}$, respectively. With 4 vol.% filler proportion, the wear rate increased to 0.161 $mm^3$ $km^{-1}$, which is still lower than that of the unfilled PPS. The coefficient of friction for these compositions varied from 0.43 to 0.57.

The reduction in wear rate with the addition of nano-CuO filler in PPS, as described above, is attributed to the development of a uniform, well-bonded transfer film, which provides an effective coverage over the counterface with hard metal asperities [17]. Figure 2.5 provides the comparison of the transfer films corresponding to the steady state wear regimes for unfilled PPS and PPS + 2 vol.% CuO filler, as observed by atomic force microscopy. Furrows representing the grooving marks can be seen in Fig. 2.5(a) for unfilled PPS, whereas the surface in Fig. 2.5(b) does not exhibit the grooving marks and instead the polymer film deposited on the steel counterface surface is seen. The deposition of the polymer film and the smoothing effect of the metal counterface can also be seen in the transfer film micrographs in Fig. 2.6, obtained by optical and scanning electron microscopy for PPS + 2 vol.% CuO.

The use of hard and abrasive nanoparticle fillers in polymers has been found to have mixed results on wear. For instance, Shao et al. [18] found that both micro- and nano-sized $SiO_2$ fillers in poly(phthalazine ether sulfone ketone) led to significant abrasive wear loss. In contrast, $SiO_2$ nanoparticles were shown to increase the wear resistance of epoxy so long as the filler particles were modified in order to produce a strong bond to the polymer [19]. Wang et al. [20] reported that nanoparticles (<15 nm) of $ZrO_2$ as the filler were effective in reducing the wear rate of PEEK. Wang et al. [21] also reported that PEEK filled with $SiO_2$ nanoparticles provided lower wear rates and lower coefficients of friction than the neat polymer. Zhao et al. [22] found that 3 wt.% nano-$Al_2O_3$ in polyamide 6 produced wear rates lower than that of the unfilled polymer.

Schwartz and Bahadur [23] studied the tribological behavior of PPS filled with $Al_2O_3$ nanoparticles sliding against a tool steel counterface of 0.10 μm Ra. The variation of wear volume and the coefficient of friction with sliding distance for this case is shown in Fig. 2.7. The wear plots for all the compositions exhibit transient and steady states of wear. Since the change from transient state to steady state wear occurs due to the development of transfer film, it would indicate that the role of transfer film was significant in the wear process. There is reduction in the steady state wear with the addition of 1 and 2 vol.% $Al_2O_3$ filler, but wear rate increased with the higher filler proportion of 3 vol.%. The coefficient of friction for all the compositions, including the unfilled PPS, is about the same with a value of 0.5.

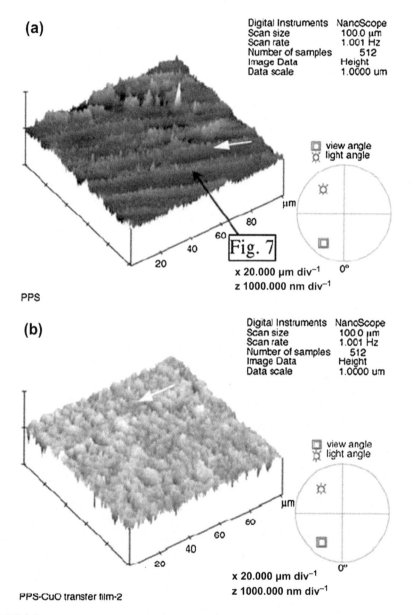

**FIGURE 2.5**

Atomic force micrographs of transfer films of (a) PPS and (b) PPS + 2 vol% CuO. Sliding conditions same as in Fig. 2.4 [17].

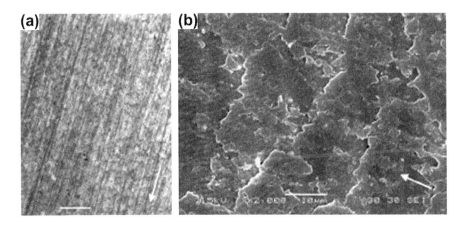

**FIGURE 2.6**

Optical micrograph of the transfer film of PPS + 2 vol.% CuO (a) and the scanning electron micrograph (b) [17].

In order to examine the dependence of wear on transfer film, Schwartz and Bahadur [23] also studied the wear and friction behavior of nanoparticle $Al_2O_3$-filled PPS composites sliding against counterfaces of three different roughnesses. All counterfaces were finished by abrasion against emery paper, with the smoothest surface being subsequently polished with a micron-scale diamond paste. The steady state wear values for these cases are given in Table 2.1. The trend in the variation of steady state wear rate with filler proportion for the counterface of 0.06 μm Ra is similar to that of 0.10 μm Ra, but no reduction in wear rate with the addition of filler is observed for the counterface of 0.027 μm Ra. The coefficient of friction values were in the 0.45–0.50 range. Figure 2.8 shows micrographs of the transfer films formed during steady state wear of PPS-2 vol.% $Al_2O_3$ sliding against the counterfaces of these three different roughnesses. As may be seen, the transfer film formed on the smoothest counterface of 0.027 μm Ra is nonuniform and leaves much of the metal surface exposed. On the other hand, the transfer films in the other two cases seem to cover the counterface completely. As the uncovered metal asperities cause aggressive damage to the softer pin material, the filler was not capable of reducing wear in the case of the smoothest counterface. The role of the counterface roughness on the capability of the development of transfer film and hence on the wear is thus obvious. Further work indicated that the spotty film development in the case of the smoothest counterface occurred because the transfer film did not bond well, likely because of poor mechanical interlocking. This was confirmed by measuring the bond (peel) strength of the transfer films, which are shown in Fig. 2.9. The results for the smoothest surface were not reported because the shear stress needed to strip the film was too small to be measured. This indicates that, due to the lack of bonding, wear in this case occurred by easy removal of the film together with the attrition of material from the pin surface due to the damage from exposed metal asperities. It should be

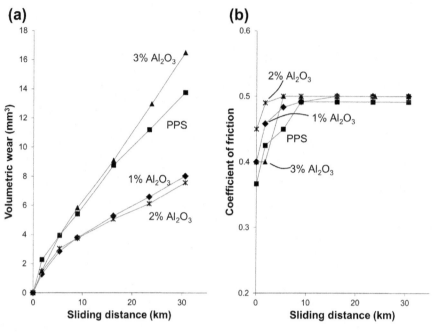

**FIGURE 2.7**

Variation of (a) wear and (b) coefficient of friction with sliding distance for nanoparticle-filled PPS composites sliding against a counterface of 0.10 μm roughness. Sliding conditions: sliding velocity 1 m s⁻¹, contact pressure 0.65 MPa.

*Adapted from [23].*

**Table 2.1** Steady State Wear Rates (Cubic Millimeters per Kilometer) of PPS and Nanoparticulate $Al_2O_3$-Filled Composites for Sliding against the Tool Steel Counterfaces of the Surface Roughnesses Given [23]

| Composition | Ra = 0.027 μm | Ra = 0.027 μm | Ra = 0.027 μm |
|---|---|---|---|
| PPS | 0.304 | 0.384 | 0.340 |
| PPS-1% $Al_2O_3$ | 0.328 | 0.384 | 0.220 |
| PPS-2% $Al_2O_3$ | 0.325 | 0.268 | 0.203 |
| PPS-3% $Al_2O_3$ | 0.425 | 0.496 | 0.394 |
| PPS-5% $Al_2O_3$ | | 0.749 | |
| PPS-10% $Al_2O_3$ | | 1.679 | |

further noted that the composite transfer film bond strength for the rougher counterface of 0.10 μm Ra was higher than that for the counterface of 0.060 μm Ra. The steady state wear rate for the former case was thus lower than that for the latter. It was also indicated in this study that the bond strength of the transfer film formed by

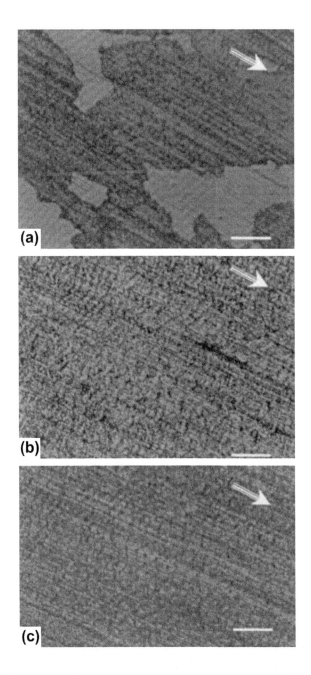

**FIGURE 2.8**

Micrographs showing transfer films formed in steady state sliding wear by 2 vol.% nanoparticle Al$_2$O$_3$-Filled PPS composite against tool steel counterfaces of (a) 0.027, (b) 0.060, and (c) 0.100 $\mu$m Ra roughnesses. Sliding directions are indicated by arrows and the bar indicates 50 $\mu$m [23].

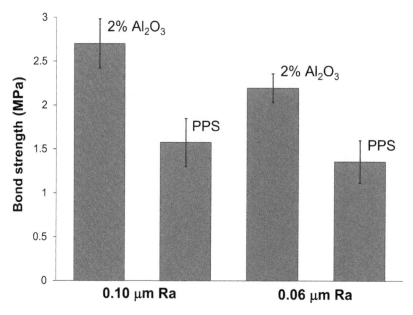

**FIGURE 2.9**

Comparison of transfer film bond strengths to the counterface for unfilled PPS and that filled with 2 vol.% $Al_2O_3$. Bars indicate mean standard error.

*Adapted from [23].*

PPS + 3 vol.% $Al_2O_3$ was too small to measure. As such, the transfer film formed for this composition did not stay on the counterface and contributed to high wear rate.

Bahadur and Sunkara [24] studied the wear of PPS filled with nanoparticles of $TiO_2$ (30–50 nm) and found that the filler in 2 vol.% proportion was effective in reducing the wear rate. A slight reduction in wear rate was also obtained with 1 vol.% $TiO_2$, but the wear rate increased when higher proportions of 3 and 5 vol.% were used. These results are shown in Fig. 2.10. The coefficient of friction for these compositions varied from 0.37 to 0.48. The steady state transfer films of PPS filled with 2 and 5 vol.% $TiO_2$, which formed on the steel counterfaces, are shown in Fig. 2.11. It may be seen that when the filler proportion is 2 vol.%, the transfer film is thin and uniform. The size of the wear debris produced in this case was also very small. With the increase in $TiO_2$ proportion to 5 vol.%, the transfer film became thick and lumpy, and it also did not cover the counterface completely. As a result of this, wear rate increased considerably for this composition and exceeded that of the unfilled PPS.

Increasing the bond between the filler and the polymer matrix has also been shown to have a beneficial effect on wear resistance. Rong et al. [25] grafted polyacrylamide onto the nanoparticles of $Al_2O_3$, $Si_3N_4$ and SiC to enhance bonding between the particles and an epoxy matrix and showed that the particles as the filler contributed to increased wear resistance. Similar benefits of grafting of $Al_2O_3$ nanoparticles were reported by Shi et al. [26] and Wetzel et al. [27] as well. It has been indicated that

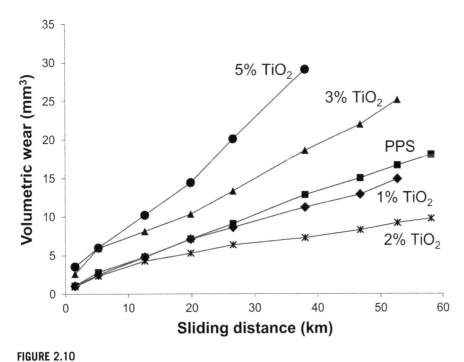

**FIGURE 2.10**

Wear vs. sliding distance for PPS filled with $TiO_2$. Sliding conditions: 1 m s$^{-1}$ sliding speed, 0.65 MPa pressure, 0.10 μm counterface roughness [24].

grafting the filler particles ensures that the nanoparticles are locked in position in the matrix, which prevents agglomeration of the particles at the sliding interface that would lead to abrasive wear.

Another variation used in the study of the nanoparticulate fillers on wear resistance is the incorporation of both nanofillers and fibers in the polymer matrix. Titania nanoparticle fillers have been shown to enhance the wear resistance of certain polymers, whereas the addition of 5 vol.% $TiO_2$ nanoparticles to poly(etherimide) containing short carbon fibers and graphite flakes produced composites with even higher wear resistance, as shown by Chang et al. [28]. These researchers also showed an increase in the wear resistance of polyamide 66 composites filled with short carbon fibers, graphite flakes and $TiO_2$ nanoparticles. These composites showed excellent wear resistance even at very high loads [29]. Zhang et al. [30] reported that an epoxy composite containing short carbon fibers, graphite flakes and 5 vol.% $TiO_2$ nanoparticles showed excellent wear resistance compared to that of neat epoxy. Breidt et al. [31] reported on the effects of nanoparticles of various materials as fillers on the mechanical properties and wear resistance of epoxy composites.

Chang et al. [32] demonstrated in their work with 5 vol.% nano $TiO_2$ + 15 vol.% short carbon fiber epoxy composites sliding against steel surfaces that the addition of lubricants such as 10 vol.% graphite or 10 vol.% PTFE affected the transfer film

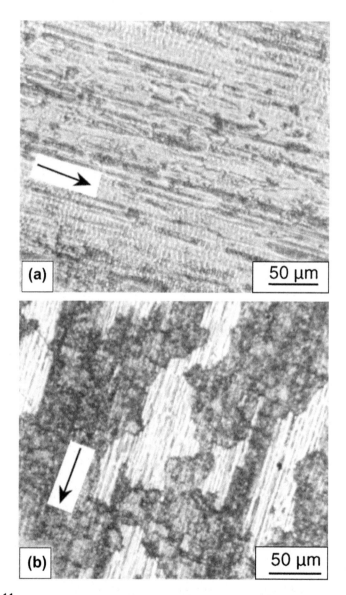

**FIGURE 2.11**

Optical micrographs of transfer films formed during sliding on steel counterface for PPS filled with (a) 2 vol.% $TiO_2$ and (b) 5 vol.% $TiO_2$. Arrows show sliding direction [24].

as well as the friction and wear behavior. Figure 2.12 shows the morphology of the transfer film sliding against metallic counterpart surfaces at 1 m s$^{-1}$ under a pressure of 1 MPa. It should be noted that the transfer film developed slightly discontinuously for the composite without lubricant. The transfer films for the composite with graphite

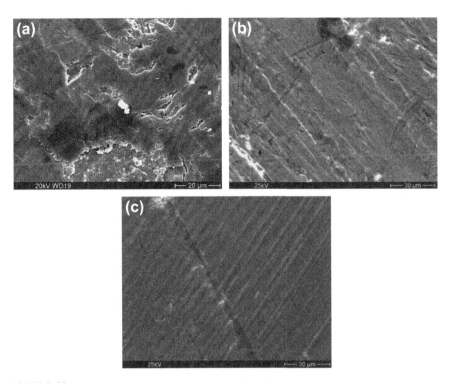

**FIGURE 2.12**

Scanning electron micrographs of transfer films of 5 vol.% nano $TiO_2$ + 15 vol.% short carbon fiber epoxy composites sliding against a steel surface at 1 MPa and 1 m s$^{-1}$, (a) without lubricant, (b) with graphite, and (c) with PTFE.

and PTFE seem to be more compact and smooth. The run-in time was reduced with the addition of the lubricants and the specific wear rate ($10^{-6}$ mm$^3$ Nm$^{-1}$) was reduced to 0.64 for the composite with 10 vol.% graphite and to 0.86 with 10 vol.% PTFE as against 3.95 for the composite without either lubricant. Bhimaraj et al. [33] found that small-volume fractions of nanoparticle $Al_2O_3$ as the filler in poly(ethylene terephthalate) enhanced the wear resistance of the polymer. Filler content up to 5 vol.% produced reductions in wear rate through the deposition of thin uniform transfer films; however, further increases led to abrasive wear of the filled polymer that was attributed to the agglomeration of $Al_2O_3$ nanoparticles in the deposited transfer films. Wetzel et al. [27] also cited the detrimental effects of the agglomeration of alumina nanoparticles in the transfer film of epoxy composites with nano-$Al_2O_3$ filler on wear resistance. With nanoparticle $Al_2O_3$ filler in phenolic composite coatings, Song et al. [34] reported optimal wear resistance with 3 wt.% filler and also the most uniform transfer film for this filler proportion. For higher filler proportions, the transfer film was reported to become discontinuous and poorly bonded to the counterface. These investigators theorized that once the filler proportion becomes too high, the filler particles are not able to be dispersed homogeneously in the transfer film and thus the agglomeration of the nanoparticles in the film generates crack-initiation sites, which lead to film removal.

Bhimaraj et al. [35] studied the tribological behavior of polyethylene terephthalate composites filled with alumina nanoparticles. The specimens were prepared by melt mixing the polymer pellets with the nanoparticles, breaking the blend into small pellets in cryogenic atmosphere, and using the pellets for compression molding of specimens. Friction and wear tests were conducted in a linear reciprocating tribometer using stainless steel counterface with an average surface roughness of 0.1–0.2 μm. They found that the wear rate and the coefficient of friction were the lowest at optimal loadings of 0.1–10 depending on the crystallinity and particle size. The wear rate decreased monotonically with the decreasing particle size.

Bahadur et al. [24] reported that ZnO nanoparticles as the filler in PPS led to degradation in wear resistance during sliding against a tool steel counterface. This was attributed to the discontinuous and poorly bonded transfer film produced during sliding. In contrast, 15 vol.% nano ZnO filler in PTFE was reported by Li et al. [36] to provide increased wear resistance during sliding on stainless steel counterface and it also produced a uniform transfer film. The difference between the transfer films in these two cases could be due to the differences in polymers, counterface materials, and/or sliding configurations (pin-on-disk vs. block-on-ring).

Wang and Pei reviewed the literature on the influence of nanoparticle fillers on the friction and wear behavior of polymer matrices [37]. Of other things, they attributed the modification of the friction and wear behavior by nanoparticles to the strengthening of the transfer film caused by anchoring of the particles, three-body rolling, and tribochemical reactions.

There is a recent review on nanocomposites by Díez-Pascual et al. [38] that deals with carbon-based nanofillers, such as nanotubes or nanofibers, and inorganic nanoparticles in polyetherketones and some other polymers. It covers strategies such as ultrasonication, ball milling, mechanochemical treatments in organic solvents, polymer functionalization, covalent grafting, and nanofiller wrapping in compatibilizing systems for uniform dispersion of the fillers in polymers. It is reported that the incorporation of inorganic nanoparticles such as $WS_2$, $SiO_2$ or $Al_2O_3$ in polyetherketones significantly enhances the tribological properties of the polymer. The effect of the nanofillers and processing techniques on the mechanical properties is also discussed.

## 2.4 FILLER AND TRANSFER FILM CHARACTERISTICS FOR REDUCTION IN WEAR

Friedrich et al. [39] reviewed the beneficial and detrimental effects of fillers on wear resistance for a broad array of filler–polymer combinations. They attributed the wear behavior mainly to the resulting transfer films formed during sliding, and noted that fillers could improve or degrade the quality of these films. This aspect has been discussed in an earlier section with experimental results on the specific filler–polymer combinations. With the realization that transfer film affects the wear behavior and that the filler material affects the transfer film, the fundamental issue is how the filler particles affect the transfer film.

There are a large number of particulate materials that can be considered as candidates for fillers in polymers. The limited amount of work that has been done with nanoparticulate materials as fillers has demonstrated that only a select few materials are effective in terms of the reduction in wear. Several nanofillers have been reported to reduce wear, whereas there are others that have been found to accelerate wear. It has not yet been possible to determine a priori whether a particular filler material will improve or deteriorate the wear resistance of a chosen polymer matrix. However, in all the cases studied involving both the micro- and the nanoparticulate fillers, it has been found that the necessary prerequisite for an effective filler for wear reduction is that it should contribute to the development of a thin and uniform transfer film that is well bonded to the counterface. The transfer film micrographs presented previously clearly demonstrated that wear was reduced if the transfer film formed on the counterface was thin and uniform. On the other hand, if the film was thick and discontinuous, wear increased. Uniform coverage of the counterface is needed to prevent the abrasion of the softer polymer pin surface from the harder metal asperities. As for the thickness of transfer film, thinner films tend to bond stronger to the counterface than thicker films, presumably due to mechanical interlocking. Experimental observations indicate that thicker films peel off the counterface surface easier than thinner films. It should be noted that the contribution to wear is provided by two sources: the loss of polymer material by attrition due to interaction with metal asperities, and the loss of transfer film due to debonding from the counterface. Hence, if the transfer film does not cover the counterface uniformly, wear would be caused by mechanical attrition and if the film is not bonded well to the counterface, it would be caused by the loss of transfer film itself. The wear particles in the first case are in the form of polymer fragments while in the second case they are in the form of rolled film.

As for the filler characteristics, much more understanding than what is known today is needed. Schwartz and Bahadur [16] proposed that the mechanical properties of the filler played an important role in affecting the wear behavior of the polymer. They introduced the term "deformability" to describe the ability of the filler particles themselves to flow and flatten under pressure as opposed to crumbling. They showed that PPS filled with deformable microsize fillers, AgS and CuS, had increased wear resistance over the neat polymer, while the addition of nondeformable microsize fillers, $ZnF_2$ and SnS, resulted in greatly reduced wear resistance. These results were confirmed on both tool steel and tantalum-tungsten alloy counterfaces.

It is known that hard fillers in microsize are detrimental to wear resistance because of their angularity and thus they contribute to abrasion during sliding. Since angularity decreases with particle size, many of the same hard fillers have been observed to contribute to increased wear resistance when the filler is in nanosize [21–23,25,32].

## 2.4.1 Transfer film bonding

In addition to the mechanical effects, the fillers also seem to interact with the counterface surface, as revealed by the X-ray photoelectron (XPSE) analyses of transfer films for a number of filled polymer systems using both the nano- and microparticle fillers. For example, Bahadur and Sunkara [24] analyzed the transfer films of PPS filled with 2 vol.% each of the nanoparticulates of CuO, $TiO_2$ and ZnO. The results are shown in Tables 2.2–2.4. Referring to Table 2.2, it is seen that the compounds $FeSO_4$ and $Fe_2O_3$ are detected in the transfer film of PPS + 2 vol.% CuO. $FeSO_4$ is the product of the reaction between the element Fe in the counterface and S in PPS. $Fe_2O_3$ is the product of reaction between the counterface element Fe and the filler CuO in the composite. Such a reaction would be expected to produce elemental Cu, which is the case here. However, because of the close binding energies, it is difficult to tell elemental Cu apart from $Cu_2O$ in the Cu(2p) spectrum, as the binding energies of both Cu and $Cu_2O$ listed in the handbook [40] are 929.8 eV. The identified species in the XPS spectra of the transfer film of PPS filled with 2 vol.% $TiO_2$ are listed in Table 2.3. The results here are similar to those obtained in the case of PPS filled with CuO. Thus it was concluded that Fe in the counterface reacted with $TiO_2$ to produce $Fe_2O_3$ and Ti. Fe in the counterface also reacted with PPS to give $FeSO_4$. As described in the earlier section, these two fillers in PPS were effective in reducing the wear of PPS. Contrary to this, nanoparticulate ZnO as the filler was not able to reduce the wear of PPS. The XPS results of the transfer film formed in this case are given in Table 2.4. Here, unlike the cases of CuO and $TiO_2$, no reduced Zn was detected. This showed that ZnO did not decompose during sliding. However, $Fe_2O_3$ was detected, which could have been formed by reaction between Fe and atmospheric oxygen. Furthermore, no $FeSO_4$ was detected indicating that PPS did not react with the counterface metal. The reactions as in the cases of CuO and $TiO_2$ have been indicated to

**Table 2.2** Compositions Identified in the XPS Spectra of the Transfer Film of 2 vol.% CuO-Filled PPS [24]

| Binding Energies of Peaks (eV) | | | | | |
|---|---|---|---|---|---|
| Compositions | C(1s) | O(1s) | S(2p) | Fe(2p) | Cu(2p) |
| Contaminated C | 284.8 | | | | |
| C in PPS | 284.8 | | | | |
| S in PPS | | | 163.7 | | |
| Fe | | | | 707.0 | |
| $FeSO_4$ | | 532.4 | 168.8 | 712.1 | |
| $Fe_2SO_3$ | | 530.2 | | 710.9 | |
| Cu, $Cu_2O$ | | 529.8 | | | 929.8 |
| CuO | | 526.8 | | | 932.3 |
| $Cu(OH)_2$ | | 528.6 | | | 933.3 |

promote bonding between the transfer film and the counterface. In support of this, the researchers provided the bond strength results for these systems, which are given in Fig. 2.13. It should be noted that increased bonding is achieved in the cases of CuO and $TiO_2$ but the bonding was adversely affected when ZnO was used as the filler.

In view of the XPS results on many microparticle fillers similar to those reported above for nanoparticulate CuO and $TiO_2$, Zhao and Bahadur [41] proposed the concept of Gibbs free energy change to determine if the interaction between the transfer film of the filled polymer composite and the metal counterface would occur or not. It stated that if the Gibbs free energy change for the chemical reaction between the filler and the counterface metal is negative, a reaction of the type indicated above is possible. If the free energy change for the contemplated reaction is positive, the chemical reaction is not feasible. These researchers demonstrated that this principle applied to all the fillers for which the tribological data existed, as shown in Table 2.5. In the absence of any other clue, this appears to be a good starting point in the selection of fillers for wear reduction. It should, however, be noted that it addresses only the chemical interaction between the filler in the polymer and the counterface elements. If the polymer by itself forms a good adherent transfer film, this may not be very important.

**Table 2.3** Identified Compositions in the XPS Spectra of the Transfer Film of 2 vol.% $TiO_2$-filled PPS [24]

| **Binding Energies of Peaks (eV)** | | | | | |
|---|---|---|---|---|---|
| Compositions | C(1s) | O(1s) | S(2p) | Fe(2p) | Ti(2p) |
| Contaminated C | 284.8 | | | | |
| C in PPS | 284.8 | | | | |
| S in PPS | | | 163.7 | | |
| $Fe_2O_3$ | | 530.2 | | 710.9 | |
| $FeSO_4$ | | 532.4 | 168.8 | 712.1 | |
| Ti | | | | | 454.1 |

**Table 2.4** Identified Compositions in the XPS Spectra of the Transfer Film of 2 vol.% ZO-Filled PPS [24]

| **Binding Energies of Peaks (eV)** | | | | | |
|---|---|---|---|---|---|
| Compositions | C(1s) | O(1s) | S(2p) | Fe(2p) | Cu(2p) |
| Contaminated C | 284.8 | | | | |
| C in PPS | 284.8 | | | | |
| S in PPS | | | 163.7 | | |
| Fe | | | | 707.0 | |
| $Fe_2O_3$ | | 530.2 | | 710.9 | |
| ZnO | | 530.4 | | | 1021.8 |

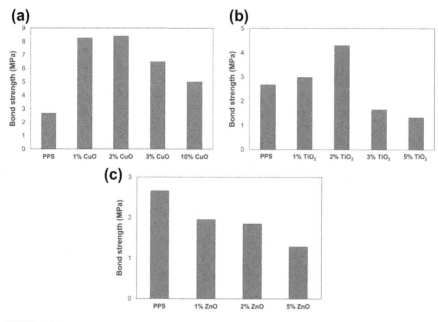

**FIGURE 2.13**

Transfer film bond strengths for PPS filled with varying proportions of CuO (a), TiO$_2$ (b) and ZnO (c). Siding conditions: 1 m s$^{-1}$ sliding speed, 0.65 MPa pressure, 0.10 μm counterface roughness [24].

The above results indicate that both mechanical and chemical aspects must be considered in determining whether a particular nanoparticle filler material will enhance the wear resistance of a polymer. Because sliding wear is a very complex process, there are many other factors hitherto unknown that will govern the final outcome.

## 2.5  CONCLUDING REMARKS

Polymer nanocomposites are a relatively new variety of materials that have shown great promise in many applications because of their excellent balance between strength and toughness as well tribological behavior. It is recognized that the key to the enhancement of these properties is the distribution of the fillers as well as interfacial bonding between the particles and the polymer. Various techniques in processing the composites have been tried to accomplish this. It has been found that filler materials can drastically affect the tribological performance of polymers sliding against metals. In the case of beneficial effect, the filler material is able to increase the uniformity of the transfer film and its adhesion to the counterface. These benefits are due both to the mechanical and chemical action of the filler particles during the sliding process. This involves the ability of the filler particles to aid in the interlocking of polymer in the

**Table 2.5** Gibbs Free Energy Changes ($\Delta G$) in Chemical Reactions between Various Fillers and the Counterface Metal Fe[a]

| Polymer | Filler | Wear | $\Delta G$ (kJ mol$^{-1}$) | Chemical Reaction Equation | XPS Result | References |
|---|---|---|---|---|---|---|
| PPS | Ag$_2$S | Down | -59.7 | Ag$_2$S + Fe = FeS + 2Ag | Ag, FeS, Fe$_2$O$_3$ | [42] |
| | | | -780.1 | Ag$_2$S + Fe + 2O$_2$ = FeSO$_4$ + 2Ag | FeSO$_4$ | |
| PPS | NiS | Down | -20.9 | NiS + Fe = FeS + Ni | Ni, FeS, Fe$_2$O$_3$ | [41] |
| | | | -741.3 | NiS + Fe + 2O$_2$ = FeSO$_4$ + Ni | FeSO$_4$ | |
| PPS | PbSe | Up | +26.4 | PbSe + Fe = FeSe + Pb | Fe$_2$O$_3$ | [41] |
| PPS | PbTe | Up | +6.7 | PbTe + Fe = FeTe + Pb | Fe$_2$O$_3$ | [42] |
| PEEK | CuO | Down | -353.1 | 3CuO + 2Fe = Fe$_2$O$_3$ + 3Cu | Cu, Fe$_2$O$_3$ | [43] |
| | | | -212.7 | Cu + H$_2$O = Cu(OH)$_2$ | Cu(OH)$_2$ | |
| Nylon11 | CuO | Down | -402.0 | 6CuO + 2Fe = 3Cu$_2$O + Fe$_2$O$_3$ | Cu, Cu$_2$O | [11] |
| | | | -212.7 | CuO + H$_2$O = Cu(OH)$_2$ | Cu(OH)$_2$, Fe$_2$O$_3$ | |
| Nylon11 | CuS | Down | -767.2 | CuS + Fe + 2O$_2$ = FeSO$_4$ + Cu | Cu, Fe$_2$O$_3$ | [8,44] |
| PEEK | CuS | Down | -46.8 | CuS + Fe = FeS + Cu | FeSO$_4$ | [45] |
| | | | -1141.6 | 2FeS + 3.5O$_2$ = 2SO$_2$ + Fe$_2$O$_3$ | | |
| Nylon11 | CuF$_2$ | Down | -168.6 | CuF$_2$ + Fe = FeF$_2$ + Cu | Cu, FeF$_2$, Fe$_2$O$_3$ | [44] |
| Nylon11 | ZnS | Up | +100.9 | ZnS + Fe = FeS + Zn | FeO, Fe$_2$O$_3$ | [46] |
| | | | -272.0 | Fe + 0.5O$_2$ = FeO | | |
| Polyester | ZnO | Up | +219.3 | 3ZnO + 2Fe = Fe$_2$O$_3$ + 3Zn | Fe$_2$O$_3$ | [47] |
| Polyester | CaF$_2$ | Up | +507.0 | CaF$_2$ + Fe = FeF$_2$ + Ca | Fe$_2$O$_3$ | [47] |

[a]Gibbs free energy data are from [48] in standard conditions (298.15 K, 0.1 MPa).
Fe$_2$O$_3$ is detected for all cases and the related reaction equation (with $\Delta G$ = −742.2 kJ mol$^{-1}$) is not listed in this table. Species in polymers or fillers and Fe in the counterface metal are not included in the XPS results.
Adapted from [41].

metallic surface asperities, the chemical reaction of the filler material with the counter-face, or the chemical degradation of the filler material as it reacts with the atmosphere. However, because of their large surface area-to-volume ratio, nanoparticles tend to offer benefits in much smaller weight fractions than do microparticles. The benefit of nanoparticle fillers in terms of the improvement in tribological behavior cannot yet be predicted a priori, but lessons have been learned that may aid in the filler selection process for the most optimal size and proportion of nanoparticle filler in a polymer matrix.

# References

[1]   Q. Zhao, S. Bahadur, Investigation of the transition state in the wear of polyphenylene sulfide sliding against steel, Tribology Letters 12 (1) (2002) 23–33.

[2]   S. Ahmed, F.R. Jones, A review of particulate reinforcement theories for polymer composites, Journal of Materials Science 25 (12) (1990) 4933–4942.

[3]   G. Cao, Nanostructures & Nanomaterials: Synthesis, Properties & Applications. vol. xiv, Imperial College Press, London, 2004, 433.

[4]   W.A. Brainard, D.H. Buckley, Adhesion and friction of PTFE in contact with metals as studied by Auger spectroscopy, field-ion and scanning electron-microscopy, Wear 26 (1) (1973) 75–93.

[5]   E.L. Yang, J.P. Hirvonen, R.O. Toivanen, Effect of temperature on the transfer film formation in sliding contact of PTFE with stainless-steel, Wear 146 (2) (1991) 367–376.

[6]   J. Vande Voort, S. Bahadur, The growth and bonding of transfer film and the role of CuS and PTFE in the tribological behavior of PEEK, Wear 181 (1995) 212–221.

[7]   S. Bahadur, The development of transfer layers and their role in polymer tribology, Wear 245 (1–2) (2000) 92–99.

[8]   S. Bahadur, D. Gong, The transfer and wear of nylon and CuS-nylon composites: filler proportion and counterface characteristics, Wear 162–164 (Part 1) (1993) 397–406.

[9]   L. Yu, S. Bahadur, An investigation of the transfer film characteristics and the tribological behaviors of polyphenylene sulfide composites in sliding against tool steel, Wear 214 (2) (1998) 245–251.

[10]  S. Bahadur, D. Tabor, Role of fillers in the friction and wear of high-density polyethylene, in: L.H. Lee (Ed.), Polymer Wear and Its Control, ACS Symposium, Ser. 287, American Chemical Society, Washington, D.C., 1984, pp. 253–268.

[11]  S. Bahadur, D. Gong, J.W. Anderegg, The role of copper compounds as fillers in transfer film formation and wear of nylon, Wear 154 (2) (1992) 207–223.

[12]  J. Vande Voort, S. Bahadur, Effect of PTFE addition on the transfer film, wear, and friction of PEEK–CuO composite, in: Tribology Symposium, Houston, Texas, 1996.

[13]  S. Bahadur, L. Zhang, J.W. Anderegg, The effect of zinc and copper oxides and other zinc compounds as fillers on the tribological behavior of thermosetting polyester, Wear 203 (1997) 464–473.

[14]  S. Bahadur, D. Tabor, The wear of filled polytetrafluoroethylene, Wear 98 (1–3) (1984) 1–13.

[15]  A. Kapoor, S. Bahadur, Transfer film bonding and wear studies on CuS-nylon composite sliding against steel, Tribology International 27 (5) (1994) 323–329.

[16]  C.J. Schwartz, S. Bahadur, The role of filler deformability, filler-polymer bonding, and counterface material on the tribological behavior of polyphenylene sulfide (PPS), Wear 251 (1–12) (2001) 1532–1540.

[17] M.H. Cho, S. Bahadur, Study of the tribological synergistic effects in nano CuO-filled and fiber-reinforced polyphenylene sulfide composites, Wear 258 (5–6) (2005) 835–845.

[18] X. Shao, Q.-J. Xue, Effect of nanometer and micrometer SiO(2) particles on the tribology properties of Poly (phthalazine ether sulfone ketone) composites, Materials and Mechanical Engineering (China) 28 (6) (2004) 39–42. 45.

[19] M.Q. Zhang, M.Z. Rong, S.L. Yu, B. Wetzel, K. Friedrich, Effect of particle surface treatment on the tribological performance of epoxy based nanocomposites, Wear 253 (9–10) (2002) 1086–1093.

[20] Q. Wang, Q. Xue, H. Liu, W. Shen, J. Xu, The effect of particle size of nanometer $ZrO_2$ on the tribological behaviour of PEEK, Wear 198 (1–2) (1996) 216–219.

[21] Q. Wang, Q. Xue, W. Shen, The friction and wear properties of nanometre $SiO_2$ filled polyetheretherketone, Tribology International 30 (3) (1997) 193–197.

[22] L.X. Zhao, L.Y. Zheng, S.G. Zhao, Tribological performance of nano-$Al_2O_3$ reinforced polyamide 6 composites, Materials Letters 60 (21–22) (2006) 2590–2593.

[23] C.J. Schwartz, S. Bahadur, Studies on the tribological behavior and transfer film-counterface bond strength for polyphenylene sulfide filled with nanoscale alumina particles, Wear 237 (2) (2000) 261–273.

[24] S. Bahadur, C. Sunkara, Effect of transfer film structure, composition and bonding on the tribological behavior of polyphenylene sulfide filled with nano particles of $TiO_2$, ZnO, CuO and SiC, Wear 258 (9) (2005) 1411–1421.

[25] M.Z. Rong, M.Q. Zhang, G. Shi, Q.L. Ji, B. Wetzel, K. Friedrich, Graft polymerization onto inorganic nanoparticles and its effect on tribological performance improvement of polymer composites, Tribology International 36 (9) (2003) 697–707.

[26] G. Shi, M.Q. Zhang, M.Z. Rong, B. Wetzel, K. Friedrich, Sliding wear behavior of epoxy containing nano-$Al_2O_3$ particles with different pretreatments, Wear 256 (11–12) (2004) 1072–1081.

[27] B. Wetzel, F. Haupert, M.Q. Zhang, Epoxy nanocomposites with high mechanical and tribological performance, Composites Science and Technology 63 (14) (2003) 2055–2067.

[28] L. Chang, Z. Zhang, H. Zhang, K. Friedrich, Effect of nanoparticles on the tribological behaviour of short carbon fibre reinforced poly(etherimide) composites, Tribology International 38 (11–12) (2005) 966–973.

[29] L. Chang, Z. Zhang, H. Zhang, A.K. Schlarb, On the sliding wear of nanoparticle filled polyamide 66 composites, Composites Science and Technology 66 (16) (2006) 3188–3198.

[30] Z. Zhang, C. Breidt, L. Chang, F. Haupert, K. Friedrich, Enhancement of the wear resistance of epoxy: short carbon fibre, graphite, PTFE and nano-$TiO_2$, Composites Part A Applied Science and Manufacturing 35 (12) (2004) 1385–1392.

[31] C. Breidt, L. Chang, Z. Zhang, Tribological and mechanical characteristics of nano-particle-strengthened composite materials. Influence of the different surface treatment of nano-particles on the matrix particle adhesion, Tribologie und Schmierungstechnik 52 (2) (2005) 14–17.

[32] L. Chang, Z. Zhang, L. Ye, K. Friedrich, Tribological properties of epoxy nanocomposites – III. Characteristics of transfer films, Wear 262 (5–6) (2007) 699–706.

[33] P. Bhimaraj, D.L. Burris, J. Action, W.G. Sawyer, C.G. Toney, R.W. Siegel, L.S. Schadler, Effect of matrix morphology on the wear and friction behavior of alumina nanoparticle/poly(ethylene) terephthalate composites, Wear 258 (9) (2005) 1437–1443.

[34] H.-J. Song, Z.-Z. Zhang, X.-h. Men, Effect of nano-$Al_2O_3$ surface treatment on the tribological performance of phenolic composite coating, Surface and Coatings Technology 201 (6) (2006) 3767–3774.

[35] P. Bhimaraj, D. Burris, W.G. Sawyer, C.G. Toney, R.W. Siegel, L.S. Schadler, Tribological investigation of the effects of particle size, loading and crystallinity on poly(ethylene) terephthalate nanocomposites, Wear 264 (7–8) (2008) 632–637.

[36] F. Li, K.A. Hu, J.L. Li, B.Y. Zhao, The friction and wear characteristics of nanometer ZnO filled polytetrafluoroethylene, Wear 249 (10–11) (2001) 877–882.

[37] Q. Wang, X. Pei, Chapter 4. The influence of nanoparticle fillers on the friction and wear behavior of polymer matrices, in: F. Klaus, K.S. Alois (Eds.), Tribology and Interface Engineering Series, Elsevier, 2008, pp. 62–81.

[38] A.M. Díez-Pascual, M. Naffakh, C. Marco, G. Ellis, M.A. Gómez-Fatou, High-performance nanocomposites based on polyetherketones. Progress in Materials Science 57 (7) (2012) 1106–1190.

[39] K. Friedrich, Z. Zhang, A.K. Schlarb, Effects of various fillers on the sliding wear of polymer composites, Composites Science and Technology 65 (15–16) (2005) 2329–2343.

[40] J.F. Moulder, J. Chastain, Handbook of X-ray Photoelectron Spectroscopy: A Reference Book of Standard Spectra for Identification and Interpretation of XPS Data, Physical Electronics Division, Perkin-Elmer Corp., Eden Prairie, Minn., 1992 p. 261.

[41] Q. Zhao, S. Bahadur, The mechanism of filler action and the criterion of filler selection for reducing wear, Wear 225–229 (Part 1) (1999) 660–668.

[42] Q. Zhao, S. Bahadur, A study of the modification of the friction and wear behavior of polyphenylene sulfide by particulate $Ag_2S$ and PbTe fillers, Wear 217 (1) (1998) 62–72.

[43] S. Bahadur, D. Gong, The role of copper compounds as fillers in the transfer and wear behavior of polyetheretherketone, Wear 154 (1) (1992) 151–165.

[44] S. Bahadur, D. Gong, J.W. Anderegg, Tribochemical studies by XPS analysis of transfer films of Nylon 11 and its composites containing copper compounds, Wear 165 (2) (1993) 205–212.

[45] S. Bahadur, D. Gong, J.W. Anderegg, The investigation of the action of fillers by XPS studies of the transfer films of PEEK and its composites containing CuS and $CuF_2$, Wear 160 (1) (1993) 131–138.

[46] S. Bahadur, A. Kapoor, The effect of $ZnF_2$, ZnS and PbS fillers on the tribological behavior of Nylon-11, Wear 155 (1) (1992) 49–61.

[47] L. Zhang, The Role of Particulate Inorganic Fillers on the Tribological Behavior of Polyester, Iowa State University, Ames, Iowa, 1995, p. 85 leaves.

[48] R. Lied (Ed.), CRC Handbook of Chemistry and Physics, seventyfirst ed., CRC Press, 1991.

# Synergistic effects of nanoparticles and traditional tribofillers on sliding wear of polymeric hybrid composites

3

Li Chang*, Zhong Zhang‡, Lin Ye*, Klaus Friedrich†

*Centre for Advanced Materials Technology, The University of Sydney, Sydney, NSW, Australia, †Institute for Composite Materials (IVW GmbH), Technical University of Kaiserslautern, Kaiserslautern, Germany; and College of Engineering, King Saud University, Riyadh, Saudi Arabia, ‡National Center for NanoScience and Technology, Beijing, China

## CHAPTER OUTLINE HEAD

Tribology of Polymeric Nanocomposites. http://dx.doi.org/10.1016/B978-0-444-59455-6.00003-9

## 3.1 INTRODUCTION

Over the past decades, polymer composites have been increasingly applied as structural materials in the aerospace, automotive, and chemical industries, providing lower weight alternatives to metallic materials. A number of these applications are concentrated on tribological components, such as gears, cams, bearings, and seals, where the self-lubrication of polymers is of special advantage. To overcome the inhibited weakness of polymers, various fillers were used to develop polymer composites for high wear resistance (cf. Fig. 3.1). In particular, short fiber reinforcements, such as carbon, glass, and aramid fibers, have been successfully used to improve the strength and therefore the load-carrying capacity of polymer composites subjected to various wear modes [1–4]. Solid lubricants, such as polytetrafluorethylene (PTFE), graphite, and $MoS_2$, have proved to be generally helpful in reducing the frictional coefficient and consequently the wear rate [5–7]. More recently, with the booming of nanophased materials, nanosized fillers such as nanoparticles and carbon nanotubes (CNTs) have also come under consideration, and results have shown that such fillers are promising for improving the wear resistance of polymers even at very low filler content (~1–4 vol.%) [8–12]. In this chapter, developments in the design of such tribomaterials are briefly reviewed. Special emphasis is given to recent attempts in the development of wear-resistant polymeric hybrid composites blended with both nanoparticles and conventional fillers. All cases discussed here relate mainly to the sliding wear behavior of polymer composite specimens against polished steel counterparts.

To characterize the wear behavior of polymers in the laboratory, standard tests are used, corresponding to real service conditions. The pin-on-disc (P-o-D) and block-on-ring are two commonly applied configurations for sliding wear tests according to American Society for Testing and Materials (ASTM) standards [13]. These tests allow the ranking of the most important tribological property of materials, the specific wear rate, which is determined by the equation,

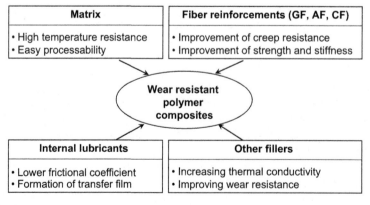

**FIGURE 3.1**

Systematic of the structural components of composite materials.

$$W_s = \frac{\Delta m}{\rho F_N L} [\text{mm}^3 \text{ Nm}^{-1}] \tag{3.1}$$

where $F_N$ is the normal load applied on the specimen during sliding, $\Delta m$ is the specimen's mass loss, $\rho$ is the density of the specimen, and $L$ is the total sliding distance. The specific wear rate is often supposed to be a material constant as long as changes in contact pressure and sliding velocity do not influence this value to a great extent. However, tribological properties are actually not real material parameters, but depend on the system in which these materials have to function [8,14,15]. When the pressure and/or velocity increase under certain conditions, the specific wear rate of the material may also increase, due to an elevation of the temperature in the contact area and consequently to changes in the dominant wear mechanisms [2]. To evaluate the wear behavior of materials under different sliding conditions, the time-related depth wear rate, $W_t$, is frequently used:

$$W_t = k^* \cdot pv = \frac{\Delta h}{t} [\text{nm s}^{-1}] \tag{3.2}$$

and

$$k^* = W_s [\text{mm}^3 \text{ Nm}^{-1}] \tag{3.3}$$

in which $k^*$ is called the wear factor, which is the same as the specific wear rate determined by Eqn (3.1) $p$ is the nominal pressure (normal load divided by the apparent contact area), $v$ is the sliding velocity, $t$ is the test time, and $\Delta h$ is the height loss of the specimen. On the basis of Eqn (3.2), the pv factor could be considered as a tribological criterion of the load-carrying capacity for bearing materials, and results in two evaluation parameters [2]:

1. the basic wear factor, $k^*$, which remains constant in a certain range of pv factor, and
2. the "limiting pv," above which the increase of the wear rate of materials is too rapid, so that the material cannot be applied any more.

The general objectives in the design of wear-resistant polymer composites are to reduce the basic wear factor $k^*$ and to enhance the "limiting pv" value.

## 3.2 THE TRIBOLOGICAL ROLES OF VARIOUS FILLERS IN SLIDING WEAR OF POLYMER COMPOSITES

### 3.2.1 Short fiber reinforcements

Various short fibers are known to considerably enhance not only the mechanical properties but also the tribological performance of polymer composites. Moreover, in comparison with continuous fiber-reinforced polymers, short fiber-reinforced polymer (SFRP) composites have the advantage of rapid, lower-cost processability by injection/compression molding or by extrusion [16]. The beneficial effect on the

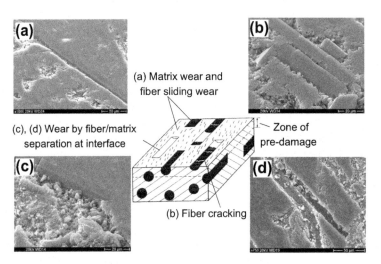

(a) Matrix wear and fiber sliding wear

(c), (d) Wear by fiber/matrix separation at interface

Zone of pre-damage

(b) Fiber cracking

**FIGURE 3.2**

A schematic of the wear process of fiber-reinforced epoxy composite: (a) matrix wear and fiber thinning, (b) fiber fracture, (c) matrix/fiber interfacial debonding, and (d) fiber removal. The P-o-D configuration for which the wear tests and the temperature calculations are performed.

wear behavior of polymer composites by short fibers can be attributed to a reduced ability of plowing, tearing, and other nonadhesive components of wear [17,18]. A schematic of the wear process of the SFRP is represented in Fig. 3.2 [2], depicted by the characteristic micrographs of worn surfaces of epoxy matrix composite filled with short carbon fiber (SCF), PTFE particles, and graphite flakes. The following stages of wear mechanisms can be recognized: (1) matrix wear and fiber thinning, (2) fiber breakage, (3) interfacial debonding, and (4) removal of fiber fragments. The last two stages occur sequentially, that is, interfacial debonding is followed by the removal of fiber fragments. The latter step is induced by the asperities of the counterface, having a large mass and leaving an open cavity on the worn surface.

For randomly oriented short fibers, the tribological properties of SFRPs are greatly influenced by the type of fibers. For instance, SCFs have been reported to be more effective than short glass fibers (SGFs) in improvement of the wear resistance of various polymers [18,19]. This can be explained by the different wear mechanisms introduced by the two types of fibers. For SCF-reinforced polymer fiber pulverization and consequently abrasive processes induced by fiber fragments are much reduced in comparison with SGF-reinforced materials. Besides, the ability of SCF to transfer a smooth carbon film onto the counterface and its contribution to the improved thermal conductivity of the composites may also lead to superior wear performance of materials [20,21]. To further optimize wear performance of SFRP, the influence of SCF volume fraction has been studied in various polymeric systems. Figure 3.3 shows the dependence of the specific wear rate on the volume contents of the SCF of

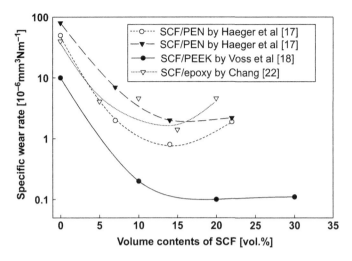

**FIGURE 3.3**

Influence of the volume contents of the SCF on the sliding wear behavior of various polymeric matrices.

three polymers, that is, polyetheretherketone (PEEK) [17], polyethernitrile [18], and epoxy [22]. It is obvious that a content of approximately 15 vol.% of SCF exhibits an optimum effect in all these cases. With a further increase in the amount of CFs, however, the wear rate of the composites does not change dramatically any more. It may even slightly increase, due to possible changes in wear mechanisms [18]. Recently, the influence of interfacial bonding between fibers and matrix on the wear behavior of SCF-reinforced epoxy composites was investigated by Zhang et al. [23]. It was found that the wear resistance of SFRPs can be enhanced under lower pv conditions using CFs treated by either air oxidation or a cryogenic treatment, which results in improved fiber–matrix interfacial bonding. However, at high pv conditions, oxidation treatment could induce negative effects due to fiber damage, whereas fibers with previous cryogenic treatment show better performance.

Voss and Friedrich [18] recognized the significance of fiber orientation with respect to the sliding plane. They found that if the short fibers in the core region of moldings were aligned orthogonally to the sliding plane (normal direction), the wear resistance was much better than when the skin fibers were at the worn surface (parallel or antiparallel direction). This outstanding tribological behavior of normally oriented fibers was attributed to the reduction of fiber breakage and pulverization. However, it has been indicated in other works of Friedrich et al. that normally oriented fibers may suffer the risk of severe wear or sudden seizure, which is related to an abrupt increase of friction as well as contact temperature. In a comparative study on sliding wear of unidirectional continuous aramid, carbon, and glass fiber composites, CFs proved to be more favorable under parallel, in plane orientation, whereas aramid fibers showed better results under normal orientation [24]. When designing high wear-resistant materials, it is therefore necessary to take all these factors into consideration. Combining favorable factors, for example, orientation and type of fibers, may lead to further improvements or to the so-called "synergistic" effects [25].

### 3.2.2 Effects of solid lubricants: development of a transfer film layer

The importance of a "transfer film layer" (TFL) for the tribological performances of polymers has long been realized and widely studied [5]. When polymers slide against metal surfaces or other harder counterfaces, material transfer generally occurs, and sometimes continuous films can develop in certain wear processes. These films thereafter act as soft shields for the polymers from the harder asperities of the counterfaces, and therefore, the wear loss of polymeric specimen can be sometime dramatically reduced [5,26]. It is agreed that the tribological characteristics of the TFL are mostly determined by the bonding strength between the TFL and the counterface, as well as the morphology of the film, depending on the sliding conditions.

In order to achieve a uniform and continuous TFL, solid lubricants, for example, graphite and PTFE are commonly used [27,28]. The lubricating effects of these materials result from their special band or layer molecular structure, which makes them relatively free to slip over each other and transfer to the metallic counterface [14,28]. In addition to these two solid lubricants, some inorganic particles, for example, $TiO_2$, SiC, $MoS_2$, and copper compounds (CuO, CuS, $CuF_2$), have also been used as lubricating fillers in different polymer matrices, such as PTFE [28–31], polyamide (PA) [32,33], PEEK [34], polyphenylene sulfide (PPS) [35–38], and polyoxymethylene [39]. They improve the wear performance of the system by mechanical and/or chemical actions. For instance, Bahadur et al. [29] observed a remarkable reduction in the wear rate of PTFE by using graphite, CuS, etc. They explained this reduction by changes in the shape and size of the worn aggregates as well as their bonding to the counterface, which were helpful for the development of a uniform and coherent TFL. Using X-ray photoelectron spectroscopy analysis, Gao [31] highlighted the effect of the tribochemical reaction of PTFE and its particle fillers on the formation of a TFL. Zhao and Bahadur [36,37] also pointed out that the tribochemical reaction between the fillers and the metallic counterface was needed for wear reduction in PPS. Schwartz and Bahadur [38] filled various particles into PPS and concluded that the formation of a good TFL, associated with a higher wear resistance of the composite, was determined by the ability of the particles to plastically deform, and by the quality of the bond strength between the filler and the matrix.

Despite numerous studies, understanding of the growth of TFLs and their tribological behavior in the steady state is still limited owing to the lack of quantitative techniques [5]. Therefore, new methods and devices have recently been applied in this field to obtain more reliable material data of the TFL. Schwartz and Bahadur et al. [40,41] developed an interesting novel device to measure the tangential shear stress needed to peel the TFL, which was taken as a measurement of the bonding strength between TFL and counterface. Friedrich et al. [42,43] studied the mechanical properties of films by microhardness/nanohardness measurement, and accordingly evaluated the thickness of TFLs. In particular, nanotechniques such as atomic force microscopy (AFM)/friction force microscopy and scanning probe microscopy have also been introduced to study highly localized damage and/or topographical changes of very thin films [43–45]. These studies entailing high accuracy are expected to provide an insight into failure mechanisms of thin TFLs and to be of help in better understanding the contribution of lubricants.

### 3.2.3 Nanosized fillers: polymer nanocomposites

Polymer nanocomposites, that is, polymeric matrices reinforced with nanosized fillers (at least one dimension of the fillers being in the range of ~10–100 nm [46]), have been the focus of current research in materials science [47–49]. Research has shown that such small reinforcing dimensions can lead to special effects, which cannot be reached so easily with traditional microsized fillers. It is believed that the structure of polymer nanocomposites is composed of a much larger interface between the matrix and the nanofillers, resulting in better mechanical properties (including tribological properties) of the material in comparison to a traditional composite with the same filler volume fraction [50–52]. Up to now, various kinds of nanofillers have been employed for improving the tribological properties of polymer composites, such as nanosized particulate fillers [9,50–52], CNTs/fibers [11,12], and single-layered graphene [53]. In this chapter, we will particularly discuss the tribological role of nanoparticles in sliding wear performance of polymer nanocomposites. More discussions on CNTs/fibers can be found in other chapters of this book. Here, the term "nanocomposites" is also used when the size of fillers are in the submicron range, for example, 300 nm.

Inorganic particles are well known to enhance the mechanical properties of polymers, which have been widely investigated in the past decades [54]. It has been found that the size of the particles plays an important role in improving the mechanical properties, as well as the tribological properties of polymer composites [50]. In general, smaller particles can contribute better to the improvement of tribological properties under sliding wear conditions than larger particles. One example of this relationship was detected by Xue et al. [55]. They found that various kinds of SiC particles, that is, nano, micron, and whisker-type particles, could reduce friction and wear when incorporating them into a PEEK matrix at a constant filler content, for example, 10 wt.% ($\approx$ 4vol.%). The use of nanoparticles resulted in the most effective reduction, which was attributed to their beneficial effect on the formation of a thin, uniform, and continuous TFL. Wang et al. [56] investigated the influence of the size of $ZrO_2$ nanoparticles, varying from 10 to 100 nm. Effective reduction of the wear of filled PEEK by nanoparticles was observed only when the size of the particles was <15 nm. Xing et al. [57] compared the wear properties of spherical particle-filled epoxy, of which the particle size varied from 120 to 510 nm. They also confirmed a similar trend, that is, the smaller the particles used as fillers were, the better was the wear resistance of the composites. The relationship between the size of the nanoparticles and asperity size was also studied by Schwartz and Bahadur et al. [40,41]. Regarding the wear performance of PPS filled with various kinds and amounts of nanoparticles, it was found that 2 vol.% of $Al_2O_3$ resulted in an optimum reduction of the wear rate of the composites when the surface roughness of the steel counterpart was in the range between $R_a = 60$ and 100 nm. However, with a surface roughness of 27 nm, that is, being smaller than the particle size (33 nm on average), any amount of nanoparticles increased the wear rate.

Although in most of the studies cited above the morphologies of the nanoparticle dispersions were not provided in detail, it should be clear that high nanofiller contents lead to a reduction in wear properties, which may be due to a tendency of particle agglomeration. In general, it is necessary that nanoparticles are uniformly dispersed rather than agglomerated, in order to yield a good property profile. Agglomeration is considered to be a common problem of polymer nanocomposites, especially at higher nanofiller contents. To disintegrate the agglomerated nanoparticles, different processing routes have been developed, depending on the type of the polymer matrices. For instance, the sol–gel process has been successfully used to disperse near spherical nanoparticles (typically metal–oxide particles) in epoxy matrices [58,59]. In this case, a coupling agent was usually added to provide bonding between the organic and inorganic phases, thereby preventing macroscopic phase separations. Owing to the excellent dispersion of nanoparticles, the nanocomposites prepared in this way could achieve desirable optical properties. Meanwhile, inorganic nanoparticles could improve the hardness of polymer matrix and therefore enhance its wear resistance under abrasive/scratch conditions [60–62]. It was proposed that the materials could be particularly useful for transparent and scratch-resistant coatings [60]. Nevertheless, as nanoparticles were prepared in situ, the applicability of the sol–gel process is restricted to a limited number of polymers as the base materials. Zhang et al. [63–65] successfully dispersed nanoparticles in the epoxy matrix by using an irradiation grafting method. In their work, nanoparticles were grafted with a copolymer in terms of emulsion graft polymerization, in order to generate network structures by connecting the nanoparticles with the matrix covalently. The results showed that the frictional coefficient and the wear rate of the material could be effectively reduced by the treated nanoparticles at very low filler content (<1 vol.%) [64] (Fig. 3.4). More recently, Knör et al. [66,67] achieved dispersions of nanoparticles in PEEK by using an optimized multipass extrusion technique. A masterbatch of the PEEK composite was prepared with a relatively high content of nanoparticles, for example, 9 vol.% during the extrusion process. Then, the material was extruded again with an intensive distribution screw setup. Finally, the dispersed masterbatch was diluted in a third extrusion step by a low shearing setup to produce the necessary volume content variations. It was found that the wear resistance of PEEK could be improved nearly four times with only 3 vol.% nanoparticles. However, the enhancement was less pronounced with a further increase in nanoparticle content [66].

Up to the present, various inorganic nanoparticles, for example, $Al_2O_3$, $TiO_2$, ZnO, CuO, SiC, $ZrO_2$, $Si_3N_4$, $SiO_2$, and $CaCO_3$, have been incorporated into

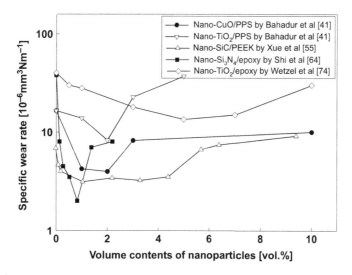

**FIGURE 3.4**

Influence of the volume contents of different nanoparticles on the sliding wear behavior of various polymeric matrices.

PPS [40,41], PEEK [55,56,66], epoxy [57,63–65,68], poly(methyl methacrylate) [69], PTFE [70–72], and polyimide (PI) [73] matrices, to improve their wear performance. In most cases, optimum nanoparticle filler contents could be identified at which highest wear resistances of these polymers were found. It was noticed that the optimum filler content of small particles was mostly in a range between 1 and 6 vol.% (Fig. 3.4). However, in comparison to microfillers such as short fibers, the improvement in wear resistance by using nanoparticles is modest (cf. Figs 3.3 and 3.4). On the other hand, the critical volume contents of nanofillers are significantly lower than those of microfillers. Therefore, nanocomposites can obtain improved wear performance while retaining or even improving other mechanical properties of the polymer matrix.

## 3.3 DEVELOPMENT OF POLYMERIC HYBRID NANOCOMPOSITES FOR HIGH WEAR RESISTANCE

### 3.3.1 State of the art

To achieve high wear-resistant polymer composites, it is a traditional route to integrate various functional fillers (cf. Fig. 3.1) [8]. In fact, more and more evidence has shown that the concept of hybrid composites, integrating inorganic nanoparticles with various traditional fillers, could be a new promising way to develop high performance wear-resistant polymeric materials. For example, Wetzel et al. [74] reported that the combination of nano-$Al_2O_3$ (13 nm) and micro-$CaSiO_3$ (4–15 μm) could induce some kind of synergistic effect and improved both the wear resistance and the stiffness of epoxy. Wang et al. [75] also found that the tribological properties of PI composites could be more effectively enhanced by using both nano-CuO and microsized graphite, compared with the composite filled only with one kind of these fillers. The synergistic effect between nanoparticles and short fibers on the sliding wear behavior of polymer composites was observed by Cho et al. [76] and Lin et al. [77], respectively. In particular, it was proposed that nanoparticles could effectively inhibit the failure of the CFs by reducing both the stress concentration on the CF interfaces and the shear stresses between the two sliding surfaces [77]. Zhang and Su et al. [78–80] and Wang et al. [81] indicated that the incorporation of nanoparticles with optimized contents could further increase the wear resistance of carbon fabric composites. The beneficial effect of additional nanoparticles on wear performance of the composites could be caused by an increased mechanical strength of the fabric composites and a better bonding strength of the TFL developed on the metallic counterfaces.

In summary, the research works reported have clearly shown that a simultaneous incorporation of nanosized inorganic particles and traditional microsized tribofillers into a polymer matrix can provide a synergism in terms of an improved wear resistance. It is therefore worthwhile to thoroughly study the improvement mechanisms by using nanoparticles for the development of high wear-resistant polymeric hybrid nanocomposites.

### 3.3.2 Integration of nanoparticles with short fibers and solid lubricants in selected polymer matrices

#### 3.3.2.1 Materials and experiments

To fully promote the effect of nanoparticles, systematic studies on the combinative effect of nanoparticles with traditional fillers have been recently carried out by Zhang et al. [82–85]. Three different polymers, that is, epoxy, polyamide 66 (PA 66) and polyetherimide (PEI) were chosen as matrices. The SCF and two solid lubricants, graphite and PTFE, were used as traditional tribofillers. The average diameter of the SCF was approximately 14.5 μm, with an average fiber length of approximately 90 μm. The sizes of the graphite flakes and the PTFE powders amounted to approximately 20 and approximately 4 μm, respectively. Nanosized $TiO_2$-inorganic particles were applied as additional fillers. The average diameter of the particles was 300 nm.

Wear tests were performed on a Wazau P-o-D apparatus according to ASTM D3702. To evaluate the load-carrying capacity of these materials, selected compositions were tested in a wide range of pv factors, for example, the nominal pressure in a range from 0.5 to 12 MPa, and the sliding velocity from 0.5 to 3 m/s. During the tests, the frictional coefficient was recorded and calculated by the ratio between the tangential and the normal force. The test temperature was monitored by an iron–constantan thermocouple positioned at the edge of the disc. After the tests, measurements of the mass loss of the specimens took place in order to calculate the specific wear rate, using Eqn (3.1).

### 3.3.2.2 Synergistic effect of nanoparticles and traditional fillers on the sliding wear of different polymer composites

The tribological performance of a series of epoxy-based composites was preliminarily investigated at a pressure of 1 MPa and a sliding speed of 1 m/s [83,84]. The results in Fig. 3.5 exhibit a synergistic effect of nano-TiO$_2$ particles and traditional tribofiller on the wear resistance of epoxy-based composites. The wear rate of the hybrid composites filled with both nanoparticles and traditional fillers (point ③ in Fig. 3.5) is significantly lower than the values linearly interpolated between those of the composites filled only with nano-TiO$_2$ (point ①) or traditional fillers (point ②). Moreover, the addition of nanoparticles could further reduce the coefficient of friction and the wear rate of the epoxy composites filled with traditional fillers, especially under extreme sliding conditions. The interrelationship between time-related depth wear rate (calculated with Eqn (3.2)), wear factor (specific wear rate), and load-carrying capacity (pv) for two epoxy composites (one filled only with traditional fillers and another one filled with both nanoparticles and traditional fillers) is given in Fig. 3.6 (cf. curves ① and ②). Baseline ③ (corresponding to curve ①) illustrates that the wear factor of the composite filled only with traditional fillers was clearly increased with an increase of pv, due to changes in the dominant wear mechanisms. For the composite with both nanoparticles and traditional fillers, however, the wear factor of the nanocomposite was relatively stable at ~1×10$^{-6}$mm$^3$ Nm$^{-1}$ even under high pv conditions (cf. baseline ④, with corresponding to curve ②). The slope ⑤ indicates that the time-related depth wear rate of the nanocomposite is much lower than that of the composite filled only with traditional fillers under the same sliding conditions. Therefore, the "limiting pv"

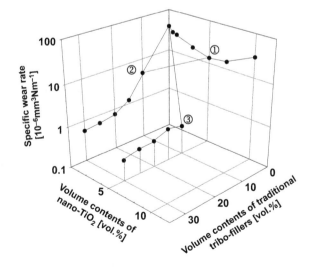

**FIGURE 3.5**

Synergistic effect of nano-TiO$_2$ (5 vol.%) and traditional tribofillers (graphite + PTFE + SCF) on the sliding wear of epoxy-based composites. Test condition: 1 MPa, 1 ms$^{-1}$.

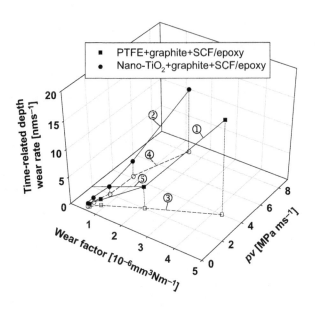

**FIGURE 3.6**

Enhancement effect of the wear performance of epoxy-based composite (filled with 5 vol.% graphite + 15 vol.% SCF) by 5 vol.% nano-TiO$_2$.

of the composite is clearly improved, which will promote the material for the triboapplications under more severe wear conditions.

The wear synergism detected in epoxy composites was also found in further investigations of thermoplastic composite, that is, with PA6.6 [85]. The composition of the PA 66 filled with only conventional fillers, that is, 5 vol.% graphite and 15 vol.% SCF, was used as a benchmark. Nano-TiO$_2$ particles served as additional fillers, at a content of 5 vol.%. Figure 3.7 summarizes the wear test results of polymeric composites filled with only traditional fillers and with additional nanoparticles. It was clear that the coefficient of friction of the fiber-reinforced thermoplastic composite was remarkably reduced by the addition of nano-TiO$_2$. Correspondingly, the wear resistance of the composites was improved through the incorporation of nanoparticles, especially under high contact pressure and high sliding speed conditions.

Therefore, the experimental data have well demonstrated that there is a synergistic effect of nanoparticles and traditional fillers on the sliding wear of different polymer composites. The desirable high load-carrying capacity can be achieved by the use of the combination of both nanoparticles and microsized traditional microsized fillers alone, but cannot be obtained by using one kind of fillers alone (Fig. 3.5). In particular, the addition of nanoparticles can effectively reduce the friction coefficient of the polymeric composites filled with short fibers and solid lubricants, under all the testing conditions (Figs 3.6 and 3.7). Correspondingly, the wear resistance of polymer composites was generally improved, especially under extreme sliding conditions.

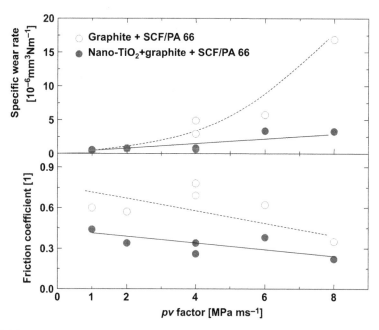

**FIGURE 3.7**

Wear and friction results of PA 66-based composites filled without and with additional nanoparticles and tested under different sliding conditions. For color version of this figure, the reader is referred to the online version of this book.

Although there is no general relationship between the trends in friction coefficient and specific wear rate [15], a high friction force/coefficient is normally undesirable for polymeric materials. This is not only because a high friction force may accelerate the wear loss of the materials but it will also lead to a high contact temperature due to the friction heating, and thus a thermal–mechanical failure of the material.

In the following sections, the wear mechanisms of hybrid polymer nanocomposites will be further discussed based on microscopic observations. In particular, the mechanisms for the favorable effects of nanoparticles will be discussed in more detail.

### 3.3.2.3 The tribological role of the nanoparticles in wear performance of polymeric hybrid composites

#### 3.3.2.3.1 On the removal process of short fibers

It is known that the wear performance of SFRPs is to a great extent determined by the properties of the fibers [16,25]. This is also true for hybrid SFRPs filled with additional particulate fillers, such as the materials considered in this study. Figure 3.8 shows the microscopic view of fibers exposed on the worn surfaces of epoxy-based SFRP composites filled without and with nanoparticles. It can be seen that fibers clearly stand out from the polymeric matrix and are fully exposed to the counterparts. By using AFM, the height from the matrix surface to the exposed fiber can be accurately

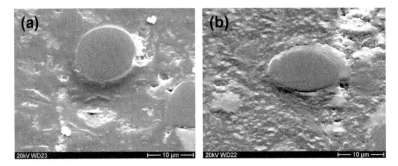

**FIGURE 3.8**

A tilted SEM image with an angle of 45° for the short fibers in the worn surfaces of (a) graphite + SCF +PTFE/epoxy and (b) nano-TiO$_2$ + graphite + SCF + PTFE/epoxy. Loading conditions: normal pressure = 1 MPa; sliding velocity = 1 ms$^{-1}$.

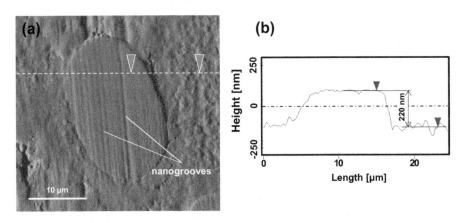

**FIGURE 3.9**

An AFM picture of the exposed fiber in the worn surface of nano-TiO$_2$ + graphite + SCF + PTFE/epoxy (a), and a cross-sectional measurement (b). Loading conditions: normal pressure = 1 MPa; sliding velocity = 1 ms$^{-1}$.

measured (Fig. 3.9). After examination of more samples, it was found that the height of the exposed fibers mostly agreed with the original surface roughness of the steel counterpart (which is ~220 nm), regardless of the type of the matrix [85,86]. Hence, during the wear process, the short fibers had to carry most of the load. To fully explore the strengthening effect of short fibers, it is critical to ensure that the fibers are only gradually removed from the polymer matrix, that is, without serious breakage. In other words, the wear behavior of SFRPs is governed by the removal process of the fibers.

Figure 3.10 compares the patterns of fiber removal for the epoxy composite filled without and with additional nanoparticles tested under different loading conditions at a sliding speed of 1 m/s. For the composite without nanoparticles, the worn surface

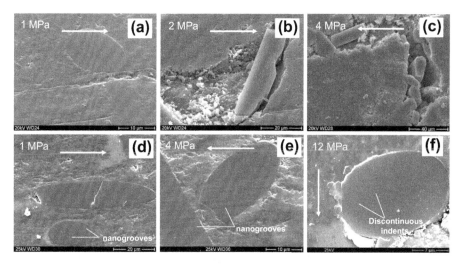

**FIGURE 3.10**

Comparisons of the damage characteristics of the fibers in the worn surfaces of epoxy-based composites without and with additional nanoparticles: (a)–(c) are the representative SEM images for graphite + SCF + PTFE/epoxy tested under 1, 2, and 4 MPa, respectively; (e)–(f) are the representative SEM images for nano-TiO$_2$ + graphite + SCF + PTFE/epoxy tested under 1, 4, and 12 MPa, respectively. The sliding velocity was kept constant at 1 ms$^{-1}$. The white arrows indicate the sliding direction of the counterpart.

is relatively smooth under the low loading condition (Fig. 3.10(a)), which indicates a gradual removal process of the fibers [86]. However, with an increase in the applied contact pressure, breakage of the epoxy matrix occurs, especially at the interfacial region around the fibers (Fig. 3.10(b)). As a result, the fibers are removed more easily, because the local support of the matrix is missing. The large fiber debris can further reduce the wear resistance of the composite because of a third body abrasive wear effect. Consequently, the wear rate of the material progressively increases (cf. Fig. 3.6). When the pressure increases to 4 MPa, the specimen failed by the formation of macrocracks (Fig. 3.10(c)), which contributed to a further increase in the specific wear rate. With the addition of nanoparticles, on the other hand, the situation was rather different. As shown in Figs 3.10(d)–(f), the worn surfaces appear to be much smoother, even at a severe wear condition of 12 MPa and 1 ms$^{-1}$. The fibers were always removed gradually and fully contributed to the wear resistance of the composites [86]. As a result, the specific wear rate of the material was much more stable (cf. Fig. 3.6), and the load-carrying capacity of the material was significantly improved.

A similar behavior was observed for the thermoplastic composites. As shown in Fig. 3.11, the fiber removal in the PA 66-based composite filled only traditional fillers was also very much aggravated with an increase of applied pressure. Nevertheless, in comparison with the brittle epoxy system, the breakage of the matrix at the interfacial regions was much more limited because of the higher ductility of

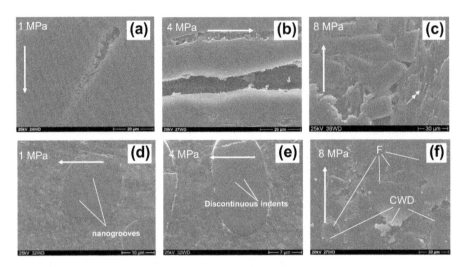

**FIGURE 3.11**

Comparisons of the damage characteristics of the fibers in the worn surfaces of PA 66-based composites without and with the additional nanoparticles: (a)–(c) are the representative SEM images for graphite + SCF/PA 66 tested under 1, 4, and 8 MPa, respectively; (e)–(f) are the representative SEM images for nano-TiO$_2$ + graphite + SCF/PA 66 tested under 1, 4, and 8 MPa, respectively. The sliding velocity was always the same (1 ms$^{-1}$). The letters on (f) have the following meaning: F = fibers; CWD = compacted wear debris, piled up in front of upstanding fiber edges.

the PA matrix. The latter deformed through elongation rather than breakage [85]. But, due to the thermal softening of the polymer matrix caused by frictional heating, serious fiber removal also happened at higher loading conditions, leaving large grooves on the worn surfaces (Fig. 3.11(b)). At 8 MPa, even some melting features (double arrow) of the PA 66 matrix could be observed (Fig. 3.11(c)). As a result, the wear rate increased because the reinforced fibers could not contribute to wear resistance any more. Again, the addition of nanoparticles resulted in smoother wear surfaces under all test conditions (cf. Fig. 3.11(d)–(f)). Accordingly, the specific wear rates of the nanocomposites were much lower than those of the composites without nanoparticles, especially under extreme loading conditions (cf. Fig. 3.7).

Thus, with the addition of nanoparticles, the fibers could be maintained in the polymeric matrix (with a gradual wear process even at high pv conditions), which finally led to an enhanced load-carrying capacity of the material (Figs 3.6 and 3.7).

### 3.3.2.3.2 On the formation of TFLs

As mentioned before, the sliding wear of polymeric specimens is greatly affected by the TFLs developed on the steel counterparts [5]. Therefore, to fully understand the tribological role of nanoparticles, it is crucial to examine their effect on the formation of TFLs.

The beneficial effect of the TFL on the wear performance of a polymer-on-steel tribosystem is normally attributed to its protection of the polymeric partner from severe abrasive wear caused by the hard roughness tips of the metallic counterface [5,87]. In order to describe the effectiveness of such protection, a new term, namely, the "TFL efficiency factor," $\lambda$, was introduced [43]:

$$\lambda = \frac{R_a}{t} \tag{3.4}$$

where $t$ is the thickness of the TFL, and $R_a$ is the surface roughness of the steel counterface. The factor considers mainly the relative contributions of the TFL and the metallic counterpart to the wear process of the sliding system. The thickness of the localized TFL can be further determined by using nanoindentations according to the following equation [43]:

$$h_f = h_t - h_s = h_t - h_t \times \sqrt{\frac{(H_c - H_f)}{(H_s - H_f)}} = h_t[1 - \sqrt{\frac{(H_c - H_f)}{(H_s - H_f)}}] \tag{3.5}$$

where $h_f$ is the film thickness, $h_t$ is the total indentation depth, and $h_s$ is the indentation depth into the substrate. $H_f$ and $H_s$ are the intrinsic hardnesses of the film and the substrate, respectively.

Table 3.1 summarized the results for the TFLs formed by both epoxy and PA 66-based composites, compared with their wear properties tested under different sliding conditions. Here, the thickness of TFLs was determined by the mean value of the thickness of localized TFLs calculated from 80 indentations. It should be noted that the distribution of the TFLs on the steel counterpart was not always uniform [43]. However, the average value of the scanned indentations at different positions on the same steel counterface showed a good repeatability. As shown in Table 3.1, the value of $\lambda$ typically ranged from 0 to 1. At a given value of $R_a$, a larger value of $\lambda$ suggests a thicker TFL, so that the surface properties are more governed by the TFL. With the factor, $\lambda$, the effect of the TFL on the tribological properties of the polymer composites tested can be quantitatively studied.

Figure 3.12 shows the friction coefficient and the wear rate of polymer composites as a function of $\lambda$. It is interesting to note that the $w_s$– and $\mu$–$\lambda$ curves showed apparent resemblance to the well-known Stribeck curves, which describe the effect of a liquid film (in terms of the lubricant thickness related to the average roughness between the contacting partners) on the friction and wear of a lubricated sliding system [88]. In general, the higher values of $\lambda$ were associated with a desirable tribological performance of the polymer composites, that is, a low friction coefficient and a low wear rate. However, if the amount of TFL becomes too large, this may again result in a relatively high wear rate, for example, for the PA 66-based nanocomposites tested at 8 MPa, 1 m/s (which belongs to the category of "rich wear debris" shown in Fig. 3.12). In this case, the thick TFL can act as a thermal barrier, since the thermal conductivity of the steel (in the range of 58 Wm$^{-1}$ K$^{-1}$ [89]) is much higher than that of polymers (e.g. PA 66 has a value of about 0.25 Wm$^{-1}$ K$^{-1}$ [90]). As a

**Table 3.1** The Thickness of TFLs Formed by Various Polymer Composites Under Different Sliding Conditions, Compared with Their Tribological Properties

| | | TFLs | | Tribological Properties | |
| | | Thickness of the TFL [nm] | Transfer Film Efficiency Factor, $\lambda$ | Friction Coefficient, $\mu$ | Specific Wear Rate [$10^{-6}$ mm$^3$ Nm$^{-1}$] |
| Composition | pv Factors | | | | |
|---|---|---|---|---|---|
| SCF/Gr/PA 66 | 1 MPa, 1 m s$^{-1}$ | 48.2 | 0.210 | 0.60 | 0.53 |
| | 2 MPa, 1 m s$^{-1}$ | 21.9 | 0.095 | 0.57 | 0.80 |
| | 4 MPa, 1 m s$^{-1}$ | 18.2 | 0.079 | 0.69 | 2.93 |
| | 8 MPa, 1 m s$^{-1}$ | 8.1 | 0.035 | 0.35 | 16.88 |
| | 2 MPa, 2 m s$^{-1}$ | 9.0 | 0.039 | 0.78 | 4.88 |
| | 2 MPa, 3 m s$^{-1}$ | 14.6 | 0.064 | 0.62 | 5.73 |
| Nano-TiO$_2$/ SCF/Gr/PA 66 | 1 MPa, 1 m s$^{-1}$ | 37.4 | 0.163 | 0.44 | 0.50 |
| | 2 MPa, 1 m s$^{-1}$ | 56.4 | 0.245 | 0.34 | 0.72 |
| | 4 MPa, 1 m s$^{-1}$ | 81.4 | 0.354 | 0.26 | 0.80 |
| | 8 MPa, 1 m s$^{-1}$ | 181.9 | 0.791 | 0.22 | 3.26 |
| | 2 MPa, 2 m s$^{-1}$ | 101.6 | 0.442 | 0.34 | 0.63 |
| | 2 MPa, 3 m s$^{-1}$ | 40.4 | 0.176 | 0.38 | 3.35 |
| PTFE/SCF/ Gr/EP | 1 MPa, 1 m s$^{-1}$ | 23.6 | 0.103 | 0.59 | 0.81 |
| | 2 MPa, 1 m s$^{-1}$ | 20.7 | 0.090 | 0.78 | 2.08 |
| | 4 MPa, 1 m s$^{-1}$ | 25.8 | 0.112 | 0.63 | 4.28 |
| | 1 MPa, 0.5 m s$^{-1}$ | 34.4 | 0.150 | 0.47 | 0.54 |
| | 1 MPa, 2 m s$^{-1}$ | 16.0 | 0.070 | 1.08 | 3.14 |

**Table 3.1** The Thickness of TFLs Formed by Various Polymer Composites Under Different Sliding Conditions, Compared with Their Tribological Properties—*Cont'd*

| Composition | pv Factors | TFLs | | Tribological Properties | |
|---|---|---|---|---|---|
| | | Thickness of the TFL [nm] | Transfer Film Efficiency Factor, λ | Friction Coefficient, μ | Specific Wear Rate [$10^{-6}$ mm$^3$ Nm$^{-1}$] |
| Nano-TiO$_2$/ PTFE/SCF/ Gr/EP | 1 MPa, 1 m s$^{-1}$ | 28.3 | 0.123 | 0.49 | 0.89 |
| | 4 MPa, 1 m s$^{-1}$ | 34.5 | 0.150 | 0.33 | 0.98 |
| | 8 MPa, 1 m s$^{-1}$ | 71.0 | 0.308 | 0.21 | 1.22 |
| | 12 MPa, 1 m s$^{-1}$ | 120.4 | 0.524 | 0.14 | 0.95 |
| | 4 MPa, 2 m s$^{-1}$ | 64.8 | 0.282 | 0.21 | 1.09 |
| | 4 MPa, 3 m s$^{-1}$ | 135.5 | 0.589 | 0.20 | 1.40 |

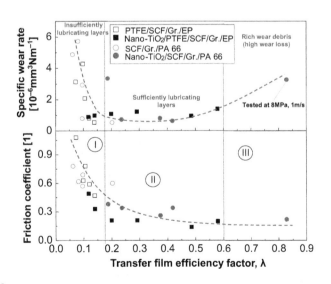

**FIGURE 3.12**

Specific wear rate and friction coefficient as a function of the TFL efficiency factor, λ, as defined by Eqn (3.4).

result, the local contact temperature in such a thick TFL region would be relatively high, resulting in strong adhesive wear between the polymer specimen and the TFL. Consequently, a back transfer of compacted wear debris (CWD) could be observed on the worn surface of the polymer nanocomposite (cf. Fig. 3.11(f)). Nevertheless, the presence of sufficient TFL could protect short fibers from sever fiber breakage or pulverization, and thus contributes to a relative stable wear performance of the polymeric specimen under different sliding conditions.

As shown in Fig. 3.12, effective TFLs were more likely formed with the addition of nanoparticles under all the tested conditions. Accordingly, polymer nanocomposites showed improved triboperformance, especially under extreme sliding conditions (Figs 3.6 and 3.7).

### 3.3.2.3.3 Considerations of the contact mechanisms

On the basis of the experimental observations shown above, Fig. 3.13 gives a schematic of the failure mechanisms during sliding wear of SFRP composites without nanoparticles. A thin, discontinuous TFL formed during the running-in stage can somewhat reduce the "direct contact" of the composite with asperities of the hard metallic counterface. Meanwhile, the worn fibers are exposed to most of the normal load and the resulting shear forces during the wear process. They often slide

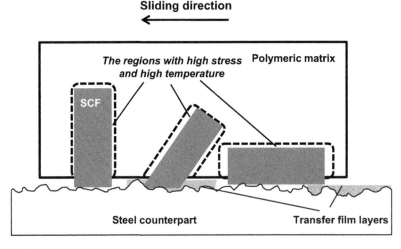

**FIGURE 3.13**

A schematic of the failure mechanism for the sliding wear of SFRP composites without nanoparticles (for better illustration purposes: fiber diameter too small relative to roughness of steel counterpart). During the wear process, the short fibers are subjected to the major part of the load when sliding against the counterpart. In this case, the polymeric matrix in the interfacial region around the fibers suffers from higher stresses and temperatures. With an increase of pv, thermal–mechanical failure of the material in this region may occur, due to the high friction heating. As a result, the fibers are removed more easily, associated with a progressive increase of the wear rate of the composites (cf. Figs 3.4(c) and 3.5(c)).

directly against the counterpart, which results in special stress concentrations at the interfacial regions between fibers and matrix. When the pv product is relatively low, fibers can carry the load with a gradual removal process, resulting in relatively smooth wear surfaces (cf. Figs 3.10(a) and 3.11(a)). However, under high pv conditions, the polymer composites may risk severe wear loss, as the short fibers are more likely pulverized and quickly removed. In particular, due to the frictional heating induced by the fibers, the temperature of the matrix around the fibers is relatively high. As a result, with an increase in the pv factor, failure of the material occurs first in the interfacial region. The specific failure mechanism is dependent on the thermal–mechanical properties of the polymeric matrix, for example, brittle fractures occurred in the case of the epoxy-based composites (cf. Fig. 3.10(c)), whereas thermal softening/melting was evident for the PA 66-based composites (cf. Fig. 3.11(c)). When the matrix failed to support the short fibers, fast fiber removal could occur, and thus a rapid increase in the composite's wear rate. Finally, the material can no longer be employed.

For the nanocomposites, a three-body contact condition was induced by the additional nanoparticles between the contact surfaces. The presence of nanoparticles in the contact region was confirmed by wavelength dispersive X-ray (WDX) spectrometry analysis. As shown in Fig. 3.14, the peaks of Titanium elements (representing nano-TiO$_2$) can be clearly observed in the wear region. Accordingly, a three-body contact mode was proposed for the sliding wear of hybrid nanocomposites, as illustrated in Fig. 3.15. It is known that the hard particles tend to be embedded in the softer surface, thus scratching the harder one [91]. This was also proved by our scanning electron microcopy (SEM; Figs 3.10(d), (e), and Fig. 3.11(d)) and AFM (Fig. 3.9) observations, which showed nanogrooves caused by additional nanoparticles. The results, in turn, support the three-body contact mode given in Fig. 3.15. Based on this contact mode, three beneficial effects of nanoparticles were further proposed [92]:

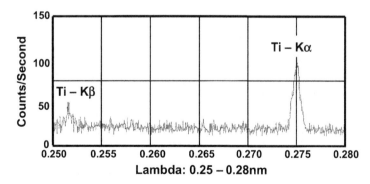

**FIGURE 3.14**

WDX analysis of Ti-K$_\alpha$ and Ti-K$_\beta$ on the steel counterpart covered with the TFL. Scanning area: 100 μm × 100 μm. Polymeric pin: nano-TiO$_2$ + graphite + SCF + PTFE/Epoxy. Loading conditions: normal pressure = 1 MPa; sliding velocity = 1 ms$^{-1}$.

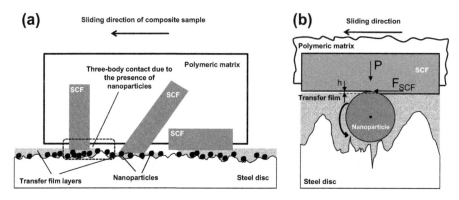

**FIGURE 3.15**

The contact mode for the SFRP composites reinforced with additional nanoparticles (for better illustration purposes: fiber diameter too small relative to roughness of steel counterpart and size of nanoparticles) (a), and the movement patterns for a single nanoparticle embedded in TFLs (b). The nanoparticle can scratch the fiber under relatively low pv conditions, but tends to roll/tumble under high pv conditions [43].

1. With the presence of nanoparticles, the distance between the steel and the composite material was also enhanced, that is, the particle acted as "spacers". This could cause a reduction in the adhesion between the contacting surfaces, which explains the fact that the friction coefficient of the nanocomposites was always lower than that of the composites without nanoparticles (Figs 3.6 and 3.7).
2. As the nanoparticles were free to move, they tended to be dispersed uniformly over the TFL during the wear process. As a result, the contact stress was more uniform between the contacting surfaces, which minimized the stress concentration on the individual fibers. As a result, the thermal failure of polymer matrix in the interfacial region between SCF/matrix was avoided, and a more gradual removal process of short fibers occurred. This ensured that the specific wear rate of the nanocomposites was stable, even under extreme loading conditions.
3. The rolling ability of nanoparticles could restrict the further increase of the friction force of the system, especially under extreme loading conditions. Accordingly, the temperature rise for the nanocomposites was much less than that for the composites without nanoparticles [85,86]. The rolling or tumbling behavior is evidenced by the discontinuous indents on the short fibers (cf. Figs 3.10(f) and 3.11(f)). Very recently, such friction reduction caused by rolling behavior of nanoparticles was also experimentally confirmed by Anantheshwara et al. [93].

Therefore, the additional nanoparticles can contribute to a low friction coefficient, owing to their "spacer" effect and their rolling ability. These factors can also reasonably explain the beneficial effect of nanoparticles on the formation of TFLs under high pv conditions. Although earlier results had shown that the wear performance of TFLs was little affected by nanoparticles under a relatively low pv condition, for

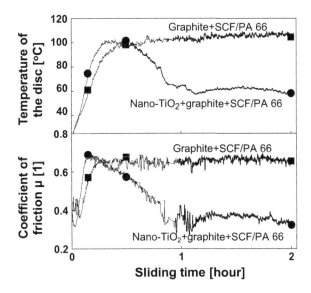

**FIGURE 3.16**

Comparisons of the sliding process of the two PA 66 composites with and without nanoparticles tested at 2 MPa and 1 ms$^{-1}$, with different durations, that is, 10, 30, and 120 min.

example, 2 MPa, 1 ms$^{-1}$ [92], a low friction status is generally helpful for the formation of TFLs on the steel counterparts, since wear debris could be more likely captured in the worn regions [43]. On the other hand, soft TFLs could hold the nanoparticles, as shown in Fig. 3.15. In this way, TFLs can partly cover the nanoparticles and minimize their abrasiveness. Such a synergistic action between the nanoparticles and TFLs is critical for an enhanced wear performance of the polymer hybrid nanocomposites, as presented in Fig. 3.15 [43,92].

Finally, to further illustrate the beneficial interaction between the nanoparticles and TFLs on the sliding wear behavior, a study was conducted on the running-in stage of the wear process, in which TFLs were gradually developed.

Figure 3.16 compares the typical curves of the coefficient of friction and the contact temperature as a function of time for two PA6.6 composites at 2 MPa and 3 ms$^{-1}$. Accordingly, the worn surfaces of the two composites after various sliding durations, that is, 10, 30, and 120 min, were given in Fig. 3.17. It is clear that at the beginning of the running-in stage, the sliding performance is similar for the two composites. Correspondingly, the worn surfaces of the two compositions were also very similar after the initial sliding duration of 30 min (Figs 3.17(a–d)). However, after about an hour, the coefficient of friction of the nanocomposite was significantly reduced and then became steady, suggesting the formation of stable TFLs. Accordingly, the hybrid nanocomposite achieved much smoother worn surface after 2 h (Fig. 3.17(f)), compared with that of the composite without nanoparticles (Fig. 3.17(e)). The experiment well demonstrated that the low friction was gradually achieved in the

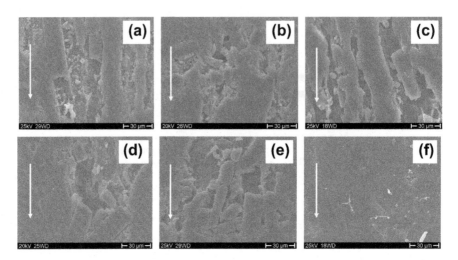

**FIGURE 3.17**

Worn surfaces of graphite + SCF/PA 66 after (a) 10, (b) 30, and (c) 120 min, and that of nano-TiO$_2$+graphite+ SCF/PA 66 after (e) 10, (f) 30, and (f) 60 min.

running-in wear process due to the beneficial interaction between nanoparticles and TFLs, which finally led to a desirable contact status for SRFPs. In this case, short fibers were protected from a severe removal process and could fully contribute to the wear resistance, even under extreme sliding conditions. As a result, the load-carrying capacity of the hybrid nanocomposite was effectively enhanced.

### 3.3.3 Recent advances in high temperature-resistant polymer composites

It is a current trend in the development of polymers to seek materials retaining reliable properties at high temperatures. One example for such a requirement is a new generation of control arm mountings or ball joints in the car chassis technology. In these cases, polymer composites have to operate as triboelements at relatively high environmental temperature, for example, 120 °C, and the demand for high wear resistance becomes increasingly important. High temperature polymers such as PEEK or PEI are particularly interesting candidates for these tribological applications. In preliminary investigations, it was found that nanoparticle additives can further improve the tribological performance of the above-mentioned polymers, especially at elevated temperatures [94,95]. For example, the influence of nano-TiO$_2$ (300 nm) and two micro-CaSiO$_3$ (2.5 and 3.5 μm) particles on the wear behavior of PEI composites was investigated by Xian et al. [94]. The wear tests were performed at both room temperature and 150 °C. At room temperature, all additional particles improved the wear resistance of PEI composites filled with SCF and graphite flakes by a factor of about two. Once the sliding temperature was increased up to 150 °C, the special potential of the nanoparticles became more obvious. The nanoparticle-filled composites possessed the lowest wear rate and the coefficient of friction. The wear rate was only about one-ninth of that of the PEI composites filled only with traditional fillers. The beneficial effects of the nanoparticles on the tribological behavior of PEI composite were attributed to their assistance in the formation of a more stable TFL.

Another comparative study on PEEK-based composites was performed by Chang et al. [95]. SCF and graphite were used as traditional fillers, without or in combination with two different types of nanoparticles, that is, $TiO_2$ (300 nm) and ZnS (300 nm). Wear tests were performed on a P-o-D apparatus under different sliding conditions. To evaluate the durability of materials at elevated temperatures, the steel disc was heated, and the temperature was controlled by a heating device.

Figure 3.18 compares friction and wear results of two PEEK-based composites tested under different pressures at 150 °C. It can be seen that the coefficients of friction of the two composites remained almost constant under different contact pressures at this high temperature level, indicating that the physical properties of the PEEK itself were not dramatically changed. Nevertheless, the coefficient of friction of the composite filled with additional nanoparticles was significantly lower than that of the composite filled only with traditional fillers, similarly as was observed in previous cases. The same is true for the depth wear rate of the two different composite systems.

To conclude, some successes have been achieved in designing polymeric hybrid nanocomposites for high wear resistance in extreme sliding conditions. It is thus worthwhile to further explore the tribopotential of such compositions blended with both nanoparticles and conventional fillers, especially for some emerging high temperature-resistant polymers such as polybenzimidazole (which can offer very high heat resistance with the heat deflection temperature of 427 °C) [96,97]. However, before this potential can be fully realized, a good understanding of the role of fillers, especially the additional fine particles, in modifying the wear behavior of polymer composites is essential. Such an understanding will facilitate the formulation of optimal criteria for the design and selection of materials subject to specific tribological applications. More research is still ongoing in the field. For instance, Zhang et al. [98] recently studied the size effect of fine particles in their interaction with the surface roughness of the steel counterparts. It was found that both nanoparticles (20 nm) and submicron particles (300 nm) could enhance the sliding wear performance of SFRPs against a steel disc with the surface roughness of 0.3 μm. With a polished counterface (with the surface roughness 0.01 μm), however, only nanoparticles could reduce the friction coefficient and the wear rate of SFRPs. The work advanced the knowledge about the tribological role of additional fine particles and also highlighted the complexity of the wear behavior of polymer hybrid nanocomposites. The wear problem is normally system behavior and depends on many variables, such as particle characteristics, the counterpart properties, lubricating and loading conditions, as well the interactions among them.

It is a vital and continuing purpose in tribological studies to model the wear behavior of materials and to establish the mathematical equations to predict the wear rate of materials. However, owing to the complexity of the wear process, wear predictions for hybrid polymer nanocomposites remains challenging. In order to overcome these problems, a new mathematical approach, the use of artificial neural networks (ANN), has recently been introduced to provide a calculable solution for wear problems.

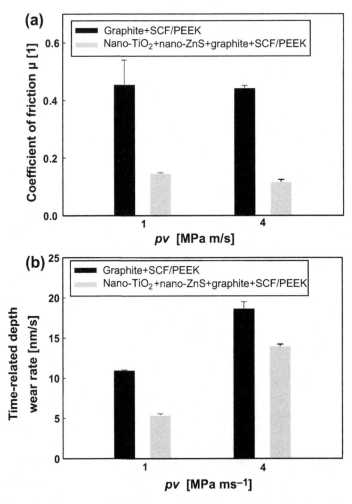

**FIGURE 3.18**

Comparison of (a) frictional coefficient and (b) specific wear rate of PEEK-based composites with and without additional nanoparticles under different pressures at an elevated temperature of 150 °C. Wear conditions: sliding velocity = 1 ms⁻¹; duration = 20 h.

## 3.4 ARTIFICIAL NEURAL NETWORKS APPLIED IN THE TRIBOLOGY OF POLYMERIC HYBRID COMPOSITES

### 3.4.1 General remarks

Artificial neural networks (ANNs) are computational systems that simulate the microstructure (neurons) of a biological nerve system [99]. Inspired by biological neurons, ANNs are composed of interconnected units, which are called artificial neurons. As shown in Fig. 3.19, some of the neurons interface with the real world to receive its input (input layer), and other neurons provide the real world with the network's output (output layer), and all the rest of the neurons are hidden from view (hidden layers). As in nature, the network function is determined largely by the interconnections between neurons, which are not simple connections, but certain nonlinear functions. Each input to a neuron has a weight factor of the function that determines the strength of the interconnection and thus the contribution of that interconnection to the following neurons. ANNs can be trained to perform a particular function by adjusting the values of these weight factors between the neurons, either by using information from outside the network or by the neurons themselves in response to the input. For materials research, a certain amount of experimental results is always needed first to develop a well-performing neural network, including its architecture, training functions, training algorithms, and other parameters, followed by the training process and the evaluation method. After the network has learned to solve the problems based on these datasets, new data from the same knowledge domain can then be put into the trained neural network to output realistic solutions.

The greatest advantage of ANNs is their ability to describe complex nonlinear, multidimensional functional relationships without any prior assumptions about the nature of the relationships, in that the network is built directly from experimental data by the ANNs' self-organizing capabilities. As a potential mathematical tool,

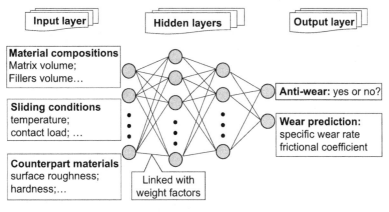

**FIGURE 3.19**

Input data, output data, and schematic construction of an artificial neural network for correlating tribological properties with various parameters.

ANNs have recently been applied to polymer composites (as reviewed by Zhang and Friedrich [100]). The usage of ANN in modeling the mechanical behavior of fiber-reinforced polymer composites was discussed in a review by Kadi [101]. Recently, ANN has been increasingly used for wear prediction to develop high wear performance polymeric composites for different tribological applications such as erosion wear [102] and the brake system [103]. The current work, however, is a concentrated effort only addressing the applications of ANN to the sliding wear of polymeric composites.

### 3.4.2 Wear prediction

Velten et al. [104] were among the pioneers in exploring this approach in polymer composites, using an ANN to predict the wear volume of short-fiber/particle reinforced thermoplastics. A total dataset of 72 independent wear measurements from fretting tests was used to train and test the neural network. Some successes have been achieved with this first attempt of property analysis using an ANN to deal with wear problems of polymer composites, although the predictive quality still needs to be improved. Zhang et al. [105] developed further improvements based on an enlarged dataset of 103 independent measurements. The database contained (1) the material composition (volume fractions of the matrix, SGFs, pitch-based CFs, polyacrylonitrile (PAN) CFs, and PTFE and Graphite fillers), (2) mechanical properties of the composites studied (compression modulus and strength, impact strength, etc., all tested at related testing temperatures), and (3) testing conditions (temperature, normal force, and sliding speed) as the input parameters. Wear characteristics such as specific wear rate or frictional coefficient were chosen as the output data. It was found that the predictive quality was clearly improved when compared to reference [104], due to an enlarged dataset.

ANNs have also been used to deal with sliding wear data coming from block-on-ring measurements [22]. These data were obtained for epoxy composites that were modified with different amounts of SCFs, PTFE, graphite, and $TiO_2$. To obtain an optimized neural network construction, the dataset given was divided into a training dataset and a test dataset. The training dataset was used to adjust the weights of all the connecting nodes until the desired error level was reached. In order to evaluate the quality of ANNs, the coefficient of determination $B$ (also called the $R^2$ coefficient in some publications) was introduced:

$$B = 1 - \frac{\sum_{i=1}^{M} \left(O(p^{(i)}) - O^{(i)}\right)^2}{\sum_{i=1}^{M} \left(O^{(i)} - O\right)^2} \tag{3.6}$$

where $O(p^{(i)})$ is the $i$th predicted property characteristic, $O^{(i)}$ is the $i$th measured value, $O$ is the mean value of $O^{(i)}$, and $M$ is the number of test data. The coefficient $B$ describes the fit of the ANNs output variable approximation curve to the actual test data output variable curve. A higher $B$ coefficient indicates an ANN with better output approximation capabilities. As an example, Fig. 3.20 exhibits the predictive results of

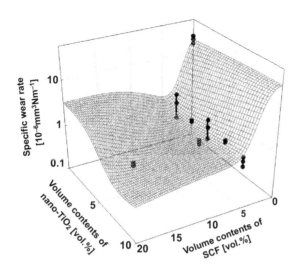

**FIGURE 3.20**

Specific wear rate of epoxy-based composites as a function of SCF and nano-TiO$_2$ volume content. The experimental data points with error scatters, whereas the rest of the 3D-plane was calculated by an artificial neural network approach. Wear test configuration: block-on-ring; Wear conditions: sliding velocity = 1 ms$^{-1}$; contact pressure = 1 MPa; duration = 20 h.

the specific wear rate as a function of SCF and nano-TiO$_2$ vol.%, whereas the contents of graphite and PTFE remained fixed at 5 vol.% and 0 vol.% respectively. Compared to the real test results (dots with error bars in the figure), the predictive results are very well acceptable. Therefore, once a well-trained ANN has been obtained, new data can be predicted without performing too many, long experiments, so that an ANN can significantly reduce the time for designing new polymer composites for special purposes. From the foregoing, the predictive quality could be further improved by enlarging the datasets for input data and optimizing the ANN configuration.

Gyurova and Friedrich [106] recently applied ANNs for the prediction of sliding friction and wear properties of PPS matrix composites by using a new measured dataset of 124 independent P-o-D sliding wear tests. It was also confirmed that the ANN prediction profiles for the characteristic tribological properties exhibited very good agreement with the experimental results demonstrating that a well-trained network had been created.

### 3.4.3 Ranking the importance of characteristic properties of polymer composites to wear rate

Although ANNs are a purely phenomenological method and do not directly produce a mechanistic understanding of the process being modeled, importance analysis of the ANN inputs is possible [100,105]. Ranking the importance of material properties to wear characteristics could provide some useful information about which properties have a stronger relationship to wear in each case. For example, Zhang et al. [107] have applied this approach for dealing with the erosive wear data of three polymeric materials. An epoxy modified by hygrothermally decomposed polyure-thane (EP-PUR) was studied. The impact angle for solid particle erosion and some characteristic properties (material composition for the case of EP-PUR as well) were selected as ANN input variables for predicting the erosive wear rate. In order to investigate the correlations between erosive wear rate and characteristic properties of these polymers, each characteristic property was used only with the necessary ero-sive conditions together as input variables for training the ANN. The qualities were analyzed by a percentage of $B \geq 0.9$, which were used to rank the importance of these characteristic properties to erosion for various polymer samples. It was found that the ranking of the importance of characteristic properties to the erosive wear rate could provide information about which property has a stronger relationship to the wear of polymers, which is of considerable assistance to a deep understanding of the wear mechanisms involved.

To continue these efforts, ANNs were recently applied to a total dataset of 56 wear measurements obtained from a block-on-ring machine [22]. To investigate the correlation between sliding wear rate and characteristic properties of these materials, each of these properties was used separately together with the volume contents of the composites. The prediction results are shown in Fig. 3.21. The configuration of the ANN applied was 9-[25]-1, which means nine inputs, 25 neurons in the hidden layer, and one output. According to this analysis, it seems that the fracture behavior of the materials (represented here by the Charpy impact energy) plays a dominant role in

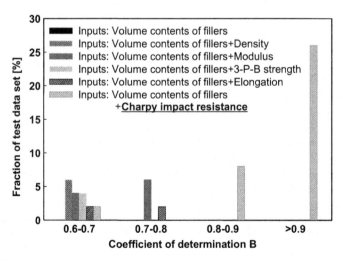

**FIGURE 3.21**

Ranking the importance of the input variables to wear loss of epoxy-based composites by ANN. Wear test configuration: block-on-ring; Wear conditions: sliding velocity = 1 ms$^{-1}$; contact pressure = 1 MPa; duration = 20 h.

sliding wear of epoxy composites. This is consistent with the proposition that crack generation and propagation are important damage mechanisms during wear of brittle epoxy matrix composites (cf. Figs 3.2 and 3.7). Therefore, importance analysis by ANN for investigating the possible correlations between simply measurable parameters (e.g. Charpy) and more complex properties (e.g. wear) can be of additional help to materials researchers for a better understanding of the mechanisms involved in the wear process [105].

### 3.4.4 Online wear monitoring

An ANN approach has been also used for online tool wear monitoring [108]. However, little work has been conducted for sliding wear problem. Hence, it is worth mentioning a recent work, using ANN for wear prediction as the function of the temperature of the metallic counterpart. In practice, it is relatively easy to measure the surface temperature of metallic counterparts, in comparison with the friction/wear of triboparts. Therefore, the work is expected to be helpful in online wear monitoring. A total dataset of 42 wear measurements was used. The wear results came from a P-o-D study using various material compositions (epoxy-based composites reinforced with SCF, graphite, PTFE, and nano-$TiO_2$) under different sliding conditions [22,84]. The temperature of the disc and known parameters, such as filler contents and pv factors, were used as inputs. The output was the time-related depth wear rate under steady state conditions. All the experimental data are given in Table 3.2. The ANN contains only one hidden layer, and the neurons in this layer were optimized based on the datasets involved. A Bayesian regularization was selected as a training algorithm, which is appropriate for determining the optimal regularization parameters in an automated fashion. The experimental data and the predicted values are compared in Fig. 3.22. Each predicted valued was independently calculated by an ANN that was trained with the remaining experimental data (i.e. excluding the one to be predicted). It is noticeable that a satisfied ANN prediction quality was reached with a relatively small dataset, due to a relatively simple and reliable relationship between inputs and outputs. Moreover, the relative simple construction required for training ANN offers a possibility with quite fast computing speed, which will be certainly useful for online wear monitoring in practical applications.

It is evident that the relatively new ANN approach, based on a limited amount of datasets, can be used for different approaches:

1. predicting tribological properties of new compositions that have not been previously tested experimentally within a given polymeric composite system,
2. a well-trained ANN is expected to be very helpful to predict the material properties before manufacturing/testing the real composites,
3. analyzing relationships between some simple properties and other complex properties (e.g. the ranking of the importance), which will be of help in understanding the importance of various parameters on wear properties,
4. online wear monitoring with measurable parameters (e.g. the temperature of the metallic counterparts) to reduce/avoid the frequent inspection and replacement of triboparts.

**Table 3.2** Wear Results of a Series of Epoxy-Based Composites Used for ANN Prediction

| No. | Inputs | | | | | | | | Time-Related Depth Wear Rate $w_t$* [nm s$^{-1}$] | Output | |
|---|---|---|---|---|---|---|---|---|---|---|---|
| | Contact Pressure [MPa] | Sliding Velocity [m s$^{-1}$] | Epoxy | PTFE | Graphite | SCF | TiO$_2$ | Temperature of Disc $T_d$ [°C] | | Wear Rate in Steady Stage $w_{s\text{-}t}$** [nm s$^{-1}$] | Predicted Value of Wear Rate in Steady Stage $w_{s\text{-}t}$** [nm s$^{-1}$] |
| 1 | 1 | 1 | 75 | 5 | 5 | 15 | 0 | 34.07 | 0.81 | 0.71 | 0.92 |
| 2 | 2 | 1 | 75 | 5 | 5 | 15 | 0 | 67.98 | 4.16 | 4.57 | 3.17 |
| 3 | 4 | 1 | 75 | 5 | 5 | 15 | 0 | 82.78 | 17.12 | 16.21 | 16.76 |
| 4 | 1 | 2 | 75 | 5 | 5 | 15 | 0 | 62.27 | 6.28 | 7.08 | 9.52 |
| 5 | 4 | 1 | 70 | 5 | 5 | 15 | 5 | 50.99 | 3.92 | 3.83 | 2.00 |
| 6 | 8 | 1 | 70 | 5 | 5 | 15 | 5 | 57.24 | 9.79 | 9.05 | 9.34 |
| 7 | 12 | 1 | 70 | 5 | 5 | 15 | 5 | 60.65 | 11.36 | 10.03 | 10.99 |
| 8 | 4 | 2 | 70 | 5 | 5 | 15 | 5 | 54.9 | 8.68 | 8.04 | 8.26 |
| 9 | 4 | 3 | 70 | 5 | 5 | 15 | 5 | 67.2 | 16.8 | 16.26 | 19.69 |
| 10 | 4 | 1 | 70 | 10 | 0 | 15 | 5 | 52.33 | 4.49 | 3.85 | 2.90 |
| 11 | 8 | 1 | 70 | 10 | 0 | 15 | 5 | 69.12 | 13.65 | 11.58 | 12.93 |
| 12 | 12 | 1 | 70 | 10 | 0 | 15 | 5 | 74.92 | 15.13 | 13.15 | 18.11 |
| 13 | 4 | 2 | 70 | 10 | 0 | 15 | 5 | 71.45 | 21.16 | 22.2 | 15.76 |
| 14 | 4 | 3 | 70 | 10 | 0 | 15 | 5 | 78.57 | 27.71 | 24.86 | 23.75 |
| 15 | 4 | 1 | 70 | 0 | 10 | 15 | 5 | 38.84 | 3.27 | 2.97 | 6.06 |

| | | | | | | | | | | | |
|---|---|---|---|---|---|---|---|---|---|---|---|
| 16 | 8 | 1 | 70 | 0 | 10 | 15 | 5 | 44.12 | 12.32 | 11.77 | 9.79 |
| 17 | 12 | 1 | 70 | 0 | 10 | 15 | 5 | 43.14 | 11.47 | 10.09 | 10.49 |
| 18 | 4 | 2 | 70 | 0 | 10 | 15 | 5 | 40.91 | 12.28 | 11.29 | 10.83 |
| 19 | 4 | 3 | 70 | 0 | 10 | 15 | 5 | 55.77 | 31.73 | 26.23 | 21.05 |
| 20 | 1 | 1 | 90 | 10 | 0 | 0 | 0 | 28.28 | 1.02 | 0.95 | 0.56 |
| 21 | 1 | 1 | 90 | 0 | 10 | 0 | 0 | 30.29 | 7.07 | 1.89 | 1.59 |
| 22 | 1 | 1 | 90 | 0 | 0 | 10 | 0 | 32.01 | 4.6 | 1.16 | 2.38 |
| 23 | 1 | 1 | 70 | 10 | 10 | 10 | 0 | 38.58 | 0.75 | 0.69 | 1.55 |
| 24 | 1 | 1 | 85 | 5 | 5 | 5 | 0 | 38.17 | 0.94 | 1.58 | 1.14 |
| 25 | 1 | 1 | 70 | 5 | 10 | 15 | 0 | 46.67 | 0.88 | 0.62 | 0.97 |
| 26 | 1 | 1 | 85 | 0 | 5 | 5 | 5 | 26.97 | 0.55 | 0.63 | 0.56 |
| 27 | 1 | 1 | 65 | 0 | 15 | 15 | 5 | 26.51 | 0.47 | 0.79 | 0.80 |
| 28 | 1 | 1 | 68 | 0 | 15 | 15 | 2 | 24.3 | 0.53 | 0.62 | 0.71 |
| 29 | 1 | 1 | 62 | 0 | 13 | 15 | 10 | 23.06 | 0.52 | 0.8 | 1.59 |
| 30 | 1 | 1 | 80 | 0 | 5 | 5 | 10 | 27.93 | 0.67 | 0.85 | 2.68 |
| 31 | 1 | 1 | 64 | 0 | 15 | 15 | 6 | 24.56 | 0.57 | 0.91 | 0.72 |
| 32 | 2 | 1 | 75 | 0 | 5 | 15 | 5 | 32.09 | 1.44 | 1.18 | 0.57 |
| 33 | 4 | 1 | 75 | 0 | 5 | 15 | 5 | 41.67 | 3.32 | 3.2 | 4.81 |
| 34 | 8 | 1 | 75 | 0 | 5 | 15 | 5 | 50.36 | 11.92 | 11.3 | 10.77 |
| 35 | 4 | 2 | 75 | 0 | 5 | 15 | 5 | 46.41 | 13.76 | 12.78 | 11.65 |
| 36 | 1 | 2 | 75 | 0 | 5 | 15 | 5 | 30.35 | 1.38 | 0.88 | 1.57 |

(Continued)

**Table 3.2** Wear Results of a Series of Epoxy-Based Composites Used for ANN Prediction—*Cont'd*

| | | Inputs | | | | | | | | Output | Predicted |
|---|---|---|---|---|---|---|---|---|---|---|---|
| No. | Contact Pressure [MPa] | Sliding Velocity [m s⁻¹] | Epoxy | PTFE | Graphite | SCF | TiO₂ | Temperature of Disc $T_d$ [°C] | Time-Related Depth Wear Rate $w_t$* [nm s⁻¹] | Wear Rate in Steady Stage $w_{s-t}$** [nm s⁻¹] | Predicted Value of Wear Rate in Steady Stage $w_{s-t}$** [nm s⁻¹] |
|---|---|---|---|---|---|---|---|---|---|---|---|
| 37 | 1 | 1 | 75 | 0 | 5 | 15 | 5 | 26.06 | 0.45 | 0.41 | 0.30 |
| 38 | 1 | 0.5 | 75 | 5 | 5 | 15 | 0 | 25.95 | 0.27 | 0.43 | 2.29 |
| 39 | 1 | 0.5 | 75 | 0 | 5 | 15 | 5 | 24.56 | 0.26 | 0.3 | 0.66 |
| 40 | 1 | 1 | 70 | 5 | 5 | 15 | 5 | 29.74 | 0.89 | 0.69 | 0.23 |
| 41 | 1 | 1 | 70 | 10 | 0 | 15 | 5 | 29.41 | 0.86 | 1.01 | 0.15 |
| 42 | 1 | 1 | 70 | 0 | 10 | 15 | 5 | 25.97 | 0.64 | 1.02 | 0.23 |

*$w_t$: the time-related depth wear rate calculated by mass loss after the wear test according to Eqn (3.2).
**$w_{s-t}$: the time-related depth wear rate of the material in the steady stage determined by the slope of height loss during the steady stage. It is noticed that the value of $w_{s-t}$ agrees well with the $w_t$ calculated by the total mass loss, since the duration of the running-in stage is much shorter than the whole testing duration.

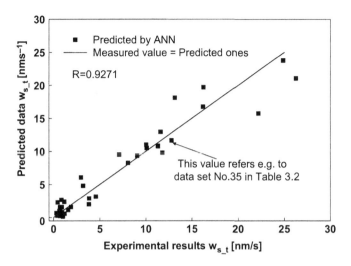

**FIGURE 3.22**

Wear rate prediction using artificial neural networks, with the inputs of the temperature of steel disc, material compositions, and pv factors. Wear test configuration: P-o-D.

However, more work in this field has to be done in order to obtain more robust and faster prediction qualities with ANNs.

## 3.5 SUMMARY

The main aim of this chapter is to strengthen the importance of integrating various functional fillers in developing wear-resistant polymer composites. The role of different fillers in modifying the wear behavior of the materials is discussed. In particular, wear synergisms between nanoparticles and traditional tribofillers are reported. It was found that the combination of nanoparticles and traditional fillers, that is, SCF and graphite could significantly improve the tribological performance of both thermosetting and thermoplastic composites. Further, this concept allowed the use of these materials under more extreme wear conditions, that is, higher normal pressures, faster sliding velocities, and elevated temperatures.

To obtain a calculable solution for the wear problem of polymer composites, ANN approaches have been recently introduced into this field. They have become a practical method for wear prediction of polymers and composites (using existing experimental results) and for a better understanding of the dominating wear mechanisms (with importance analysis).

## Acknowledgments

L. Chang gratefully acknowledges the financial support of the Australian Research Council (DP0877750). Further thanks are due to the group of eight Australia–Germany Joint Research Cooperation Scheme (DAAD NO. 507 533 07) for the exchange of young researchers.

## References

[1]   H. Czichos, Introduction on friction and wear, in: K. Friedrich (Ed.), Friction and Wear of Polymer Composites, Elsevier Science Publishers, B.V, 1986, pp. 1–23.

[2]   K. Friedrich, Wear of reinforced polymers by different abrasive counterparts, in: K. Friedrich (Ed.), Friction and Wear of Polymer Composites, Elsevier Science Publishers, B.V, 1986, pp. 233–287.

[3]   S.N. Kukureka, C.J. Hooke, M. Rao, P. Liao, Y.K. Chen, The effect of fiber reinforcement on the friction and wear of polyamide 6,6 under dry rolling-sliding contact, Tribology International 32 (1999) 107–116.

[4]   J. Bijwe, J.J. Rajesh, A. Jeyakumar, A. Ghosh, U.S. Tewari, Influence of solid lubricants and fiber reinforcement on wear behavior of polyethersulphone, Tribology International 33 (2000) 697–706.

[5]   S. Bahadur, The development of transfer layers and their role in polymer tribology, Wear 245 (2000) 92–99.

[6]   C. Donnet, A. Erdemir, Historical developments and new trends in tribological and solid lubricant coatings, Surface and Coatings Technology 180–181 (2001) 76–84.

[7]   D.G. Teer, New solid lubricant coatings, Wear 251 (2001) 1068–1074.

[8]   K. Friedrich, Z. Zhang, P. Klein, Wear of polymer composites, Edited by P. Sydenham and R. Thorn in: G.W. Stachowiak (Ed.), Wear-Materials, Mechanisms and Practice, as Part of Handbook of Measuring System Design, John Wiley & Sons, 2005, pp. 269–290.

[9]   Z. Zhang, K. Friedrich, Tribological characteristics of micro- and nanoparticle filled polymer composites, in: K. Friedrich, S. Fakirov, Z. Zhang (Eds.), Polymer Composite— from Nano- to Macro-scale, Springer, 2005, pp. 169–185.

[10]  J. Karger-Kocsis, Z. Zhang, Structure–property relationships in nanoparticles/semicrystalline thermoplastic composites, in: J.F. Balta Calleja, G. Michler (Eds.), Mechanical Properties of Polymers Based on Nanostructure and Morphology, CRC Press, New York, 2005, pp. 547–596.

[11]  Z. Yang, B. Dong, Y. Huang, L. Liu, F.Y. Yan, H.L. Li, Enhanced wear resistance and micro-hardness of polystyrene nanocomposites by carbon nanotubes, Materials Chemistry and Physics 94 (2005) 109–113.

[12]  Q. Jacobs, W. Xu, B. Schädel, W. Wu, Wear behaviour of carbon nanotube reinforced epoxy resin composites, Tribology Letters 23 (2006) 65–75.

[13]  P.J. Blau, K.G. Budinski, Development and use of ASTM standards for wear testing, Wear 225–229 (1999) 1159–1170.

[14]  G.W. Stachowiak, A.W. Batchelor, Engineering Tribology, second ed., Butterworth-Heinemann, Woburn, 2001 pp. 619–667.

[15]  K. Kato, Wear in relation to friction—a review, Wear 241 (2000) 151–157.

[16] U.S. Tewari, J. Bijwe, Recent development in tribology of fiber reinforced composites with thermoplastic and thermosetting matrices, in: K. Friedrich (Ed.), Advances in Composites Tribology, Elsevier Science Publishers, B.V, 1993, pp. 159–207.

[17] A.M. Häger, M. Davies, Short-fiber reinforced, high-temperature resistant polymers for a wide field of tribological applicants, in: K. Friedrich (Ed.), Advances in Composites Tribology, Elsevier Science Publishers, B.V, 1993, pp. 104–157.

[18] H. Voss, K. Friedrich, On the wear behavior of short-fiber-reinforced PEEK composites, Wear 116 (1987) 1–18.

[19] K. Friedrich, Z. Lu, A.M. Hager, Recent advances in polymer composites' tribology, Wear 190 (1995) 139–144.

[20] W. Bonfield, B.C. Edwards, A.J. Markham, J.R. White, Wear transfer films formed by carbon fiber reinforced epoxy resin sliding on stainless steel, Wear 37 (1976) 113–121.

[21] V.K. Jain, Investigation of the wear mechanism of carbon-fiber-reinforced acetal, Wear 92 (1983) 279–292.

[22] L. Chang, Friction and Wear of Nanoparticle Filled POLYMER Composites, PhD thesis, TU Kaiserslautern, ISBN 3-934930-56-5, 2005.

[23] H. Zhang, Z. Zhang, Comparison of short carbon fiber surface treatments on epoxy composites. II. Enhancement of the wear resistance, Composites Science and Technology 64 (2004) 2031–2038.

[24] T. Tsukizoe, N. Ohmae, Friction and wear performance of unidirectionally oriented glass, carbon, aramid and stainless steel fiber-reinforced plastics, in: K. Friedrich (Ed.), Friction and Wear of Polymer Composites, Elsevier Science Publishers B.V, 1986, pp. 205–231.

[25] K. Friedrich, Wear models for multiphase materials and synergistic effects in polymeric hybrid composites, in: K. Friedrich (Ed.), Advances in Composite Tribology, Elsevier Science Publishers, B.V, 1993, pp. 209–273.

[26] M. Vaziri, R.T. Spurr, F.H. Stott, An investigation of the wear of polymeric materials, Wear 122 (1988) 329–342.

[27] A.A. Cenna, P. Dastoor, A. Beehag, N.W. Page, Effects of graphite particle addition upon the abrasive wear of polymer surfaces, Journal of Materials Science 36 (2001) 891–900.

[28] K. Tanaka, Effects of various fillers on the friction and wear of PTFE-based composites, in: K. Friedrich (Ed.), Friction and Wear of Polymer Composites, Elsevier Science Publishers, B.V, 1986, pp. 137–174.

[29] S. Bahadur, D. Tabor, The wear of filled polytetrafluoroethylene, Wear 98 (1984) 1–13.

[30] J. Khedkar, I. Negulescu, E.I. Meletis, Sliding wear behavior of PTFE composites, Wear 252 (2002) 361–369.

[31] J. Gao, Tribochemical effects in formation of polymer transfer film, Wear 245 (2000) 100–106.

[32] S. Bahadur, D. Gong, The role of copper compounds as fillers in the transfer film formation and wear of nylon, Wear 154 (1992) 207–223.

[33] S. Bahadur, A. Kapoor, The effect of $ZnF_2$, ZnS and PbS fillers on the tribological behavior of nylon 11, Wear 155 (1992) 49–61.

[34] J.V. Voort, S. Bahadur, The growth and bonding of transfer film and the role of CuS and PTFE in the tribological behavior of PEEK, Wear 181–183 (1995) 212–221.

[35] L. Yu, S. Bahadur, An investigation of the transfer film characteristics and the tribological behaviors of polyphenylene sulfide composites in sliding against tool steel, Wear 214 (1998) 245–251.

[36] Q. Zhao, S. Bahadur, A study of the modification of the friction and wear behavior of polyphenylene sulfide by particulate $Ag_2S$ and PbTe fillers, Wear 217 (1998) 62–72.

[37] Q. Zhao, S. Bahadur, The mechanism of filler action and the criterion of filler selection for reducing wear, Wear 225–229 (1999) 660–668.

[38] C.J. Schwartz, S. Bahadur, The role of filler deformability, filler-polymer bonding, and counterface material on the tribological behavior of polyphenylene sulfide (PPS), Wear 251 (2001) 1532–1540.

[39] L. Yu, S. Yang, H. Wang, Q. Xue, An investigation of the friction and wear behaviors of micrometer copper particle- and nanometer copper particle-filled polyoxymethylene composites, Journal of Applied Polymer Science 77 (2000) 2404–2410.

[40] C.J. Schwartz, S. Bahadur, Studies on the tribological behavior and transfer film-counterface bond strength for polyphenylene sulfide filled with nanoscale alumina particles, Wear 237 (2000) 261–273.

[41] S. Bahadur, C. Sunkara, Effect of transfer film structure, composition and bonding on the tribological behavior of polyphenylene sulfide filled with nano particles of $TiO_2$, ZnO, CuO and SiC, Wear 258 (2005) 1411–1421.

[42] K. Friedrich, J. Flöck, K. Váradi, Z. Néder, Experimental and numerical evaluation of the mechanical properties of compacted wear debris layers formed between composite and steel surfaces in sliding contact, Wear 251 (2001) 1202–1212.

[43] L. Chang, K. Friedrich, L. Ye, Study of transfer film layer in sliding contact between polymer composites and steel disks using nanoindentation, Journal of Tribology – Transactions of the ASME submitted for publication.

[44] B. Bhushan, Nano- to microscale wear and mechanical characterization using scanning probe microscopy, Wear 251 (2001) 1105–1123.

[45] N.X. Randall, J.L. Bozet, Nanoindentation and scanning force microscopy as a novel method for the characterization of tribological transfer films, Wear 212 (1997) 18–24.

[46] H.S. Nalwa, Handbook of Organic-inorganic Hybrid Materials and Nanocomposites. vol. 2, American Scientific Publ., Stevenson Ranch, California, USA, 2003.

[47] E. Thostenson, C. Li, T. Chou, Nanocomposites in context, Composites Science and Technology 65 (2005) 491–516.

[48] J. Coleman, U. Khan, W. Blau, Y. Gun'ko, Small but strong: a review of the mechanical properties of carbon nanotube polymer composites, Carbon 44 (2006) 1624–1652.

[49] D. Paul, L. Robeson, Polymer nanotechnology: nanocomposites, Polymer 49 (2008) 3187–3204.

[50] M. Rong, M. Zhang, H. Liu, H. Zeng, B. Wetzel, K. Friedrich, Microstructure and tribological behavior of polymeric nanocomposites, Industrial Lubrication & Tribology 53 (2001) 72–77.

[51] B. Wetzel, F. Haupert, M.Q. Zhang, Epoxy nanocomposites with high mechanical and tribological performance, Composites Science and Technology 63 (2003) 2055–2067.

[52] M.Q. Zhang, M.Z. Rong, K. Friedrich, Processing and properties of non-layered nanoparticle reinforced thermoplastic composites, in: H.S. Nalwa (Ed.), Handbook of Organic–Inorganic Hybrid Materials and Nanocomposites, Nanocomposites, vol. 2, American Scientific Publ., Los Angeles, USA, 2003, pp. 113–150.

[53] S.S. Kandanur, M.A. Rafiee, F. Yavari, M. Schrameyer, Z.Z. Yu, T.A. Blanchet, N. Koratkar, Suppression of wear in graphene polymer composites, Carbon 50 (9) (2012) 3178–3183.

[54] K. Friedrich, Z. Zhang, A.K. Schlarb, Effects of various fillers on the sliding wear of polymer composites, Composites Science and Technology 65 (2005) 2329–2343.

[55] Q. Xue, Q. Wang, Wear mechanisms of polyetheretherketone composites filled with various kinds of SiC, Wear 213 (1997) 54–58.

[56] Q. Wang, Q. Xue, H. Liu, W. Shen, J. Xu, The effect of particle size of nanometer $ZrO_2$ on the tribological behavior of PEEK, Wear 198 (1996) 216–219.

[57] X.S. Xing, R.K.Y. Li, Wear behavior of epoxy matrix composites filled with uniform sized sub-micron spherical silica particles, Wear 256 (2004) 21–26.

[58] F. Bondioli, V. Cannillo, E. Fabbri, M. Messori, Epoxy-silica nanocomposites: preparation, experimental characterization, and modeling, Journal of Applied Polymer Science 97 (2005) 2382–2386.

[59] M. Sangermano, G. Malucelli, F. Amerio, R. Bongiovanni, A. Priola, A.D. Gianni, B. Voit, G. Rizza, Preparation and characterization of Nanostructured $TiO_2$/Epoxy polymeric films, Macromolecular Materials and Engineering 291 (2006) 517–523.

[60] H. Zhang, H. Zhang, L.C. Tang, Z. Zhang, L. Gu, Y.Z. Xu, C. Eger, Wear-resistant and transparent acrylate-based coating with highly filled nanosilica particles, Tribology International 43 (2010) 83–91.

[61] S. Turri, L. Torlaj, F. Piccinini, M. Levi, Abrasion and Nanoscratch in Nanostructured epoxy coatings, Journal of Applied Polymer Science 118 (2010) 1720–1727.

[62] D. Morselli, F. Bondioli, M. Sangermano, M. Messori, Photo-cured epoxy networks reinforced with $TiO_2$ in-situ generated by means of non-hydrolytic sol gel process, Polymer 53 (2012) 283–290.

[63] M. Zhang, M. Rong, S. Yu, B. Wetzel, K. Friedrich, Improvement of tribological performance of epoxy by the addition of irradiation grafted nano-inorganic particles, Macromolecular Materials and Engineering 287 (2002) 111–115.

[64] G. Shi, M. Zhang, M. Rong, B. Wetzel, K. Friedrich, Friction and wear of low nanometer $Si_3N_4$ filled epoxy composites, Wear 254 (2003) 784–796.

[65] G. Shi, M. Zhang, M. Rong, B. Wetzel, K. Friedrich, Sliding wear behavior of epoxy containing nano-$Al_2O_3$ particles with different pretreatments, Wear 256 (2003) 1072–1081.

[66] N. Knör, A. Gebhard, F. Haupert, A.K. Schlarb, Polyetheretherketone (PEEK) nanocomposites for extreme mechanical and tribological loads, Mechanics of Composite Materials 45 (2009) 199–206.

[67] N. Knör, R. Walter, F. Haupert, Mechanical and thermal properties of nano-titanium dioxide-reinforced polyetheretherketone produced by optimized twin screw extrusion, Journal of Thermoplastic Composite Materials 24 (2011) 185–205.

[68] M.S. Sreekala, C. Eger, Property improvements of an epoxy resin by nanosilica particle reinforcement, in: K. Friedrich, S. Fakirov, Z. Zhang (Eds.), Polymer Composites—From Nano- to Macro-Scale, Springer, 2005, pp. 91–105.

[69] L.Y. Lin, D.E. Kim, Tribological properties of polymer/silica composite coatings for microsystems applications, Tribology International 44 (2011) 1926–1931.

[70] F. Li, K. Hu, J. Li, B. Zhao, The friction and wear characteristics of nanometer ZnO filled polytetrafluoroethylene, Wear 249 (2002) 877–882.

[71] W.G. Sawyer, K.D. Freudenberg, P. Bhimaraj, L.S. Schadler, A study on the friction and wear behavior of PTFE filled with alumina nanoparticles, Wear 254 (2003) 573–580.

[72] S. Beckford, Y.A. Wang, M. Zou, Wear-resistant PTFE/$SiO_2$ nanoparticle composite films, Tribology Transactions 54 (2011) 849–858.

[73] B.X. Liu, X.Q. Pei, Q.H. Wang, X.J. Sun, T.M. Wang, Effects of atomic oxygen irradiation on structural and tribological properties of polyimide/$Al_2O_3$ composites, Surface and Interface Analysis 44 (2012) 372–376.

[74] B. Wetzel, F. Haupert, K. Friedrich, M. Zhang, M. Rong, Impact and wear resistance of polymer nanocomposites at low filler content, Polymer Engineering and Science 42 (2002) 1919–1927.

[75] Q.H. Wang, X.R. Zhang, X.Q. Pei, A synergistic effect of graphite and nano-CuO on the tribological behavior of polyimide composites, Journal of Macromolecular Science, Part B: Physics 50 (2011) 213–224.

[76] M.H. Cho, S. Bahadur, Study of the tribological synergistic effects in CuO-filled and fiber-reinforced polyphenylene sulfide composites, Wear 258 (2005) 835–845.

[77] G.M. Lin, G.Y. Xie, G.X. Sui, R. Yang, Hybrid effect of nanoparticles with carbon fibers on the mechanical and wear properties of polymer composites, Composites: Part B 43 (2012) 44–49.

[78] Z.Z. Zhang, F.H. Su, K. Wang, W. Jiang, X.H. Men, W.M. Liu, Study on the friction and wear properties of carbon fabric composites reinforced with micro- and nano-particles, Materials Science and Engineering A 404 (2005) 251–258.

[79] F.H. Su, Z.Z. Zhang, K. Wang, W. Jiang, X.H. Men, W.M. Liu, Friction and wear properties of carbon fabric composites filled with nano-$Al_2O_3$ and nano-$Si_3N_4$, Composites: Part A 37 (2006) 1351–1357.

[80] F.H. Su, Z.Z. Zhang, W.M. Liu, Mechanical and tribological properties of carbon fabric composites filled with several nano-particulates, Wear 260 (2006) 861–868.

[81] Q.H. Wang, X.R. Zhang, X.Q. Pei, T.M. Wang, Investigations of the tribological properties of carbon fabric reinforced phenolic polymer composites filled with several nanoparticles, Journal of Applied Polymer Science 123 (2012) 3081–3089.

[82] Z. Zhang, F. Haupert, K. Friedrich, Enhancement of the Wear Resistance of Polymer Composites by Nano-Fillers, German Patent 2005 103 29 228.4-43.

[83] Z. Zhang, C. Breidt, L. Chang, F. Haupert, K. Friedrich, Enhancement of the wear resistance of epoxy: short carbon fiber, graphite, PTFE and nano-$TiO_2$, Composites: Part A 35 (2004) 1385–1392.

[84] L. Chang, Z. Zhang, C. Breidt, K. Friedrich, Tribological properties of epoxy nanocomposites: I. Enhancement of the wear resistance by nano-$TiO_2$ particles, Wear 258 (2005) 141–148.

[85] L. Chang, Z. Zhang, H. Zhang, A.K. Schlarb, On the sliding wear of nanoparticles filled polyamide 6,6, Composites Science and Technology 66 (2006) 3188–3198.

[86] L. Chang, Z. Zhang, Tribological properties of epoxy nanocomposites: II. A combinative effect of short carbon fiber and nano-$TiO_2$, Wear 260 (2006) 869–878.

[87] K. Tanaka, Transfer of semicrystalline polymers sliding against a smooth steel surface, Wear 75 (1) (1982) 183–199.

[88] H. Czichos, Tribology and its many facets: from macroscopic to microscopic to nanoscale phenomena, Meccanica 36 (2001) 605–615.

[89] M. Kalin, J. Vizintin, Comparison of different theoretical models for flash temperature calculation under fretting conditions, Tribology International 34 (2001) 831–839.

[90] M.I. Kohan, Nylon Plastic Handbook, Hanser/Gardner Publications, Inc., Cincinnati, 1995 p. 344.

[91] R.S. Dwyer-Joyce, R.S. Sayles, E. Ioannides, An investigation into the mechanisms of closed three-body abrasive wear, Wear 175 (1994) 133–142.

[92] L. Chang, K. Friedrich, Enhancement effect of nanoparticles on the sliding wear of short fiber-reinforced polymer composites: a critical discussion of wear mechanisms, Tribology International 43 (2010) 2355–2364.

[93] K. Anantheshwara, A.J. Lockwood, R.J. Mishra, B.J. Inkson, M.S. Bobji, Dynamical evolution of wear particles in nanocontacts, Tribology Letters 45 (2012) 229–235.

[94] G.J. Xian, Z. Zhang, K. Friedrich, Tribological properties of micro- and nanoparticles-filled Poly(etherimide) composites, Journal of Applied Polymer Science 101 (2006) 1678–1686.

[95] L. Chang, Z. Zhang, L. Ye, K. Friedrich, Tribological properties of high temperature resistant polymer composites with fine particles, Tribology International 40 (2007) 1170–1178.

[96] K. Friedrich, H.J. Sue, P. Liu, A.A. Almajid, Scratch resistance of high performance polymers, Tribology International 44 (2011) 1032–1046.

[97] X. Pei, K. Friedrich, Sliding wear properties of PEEK, PBI and PPP, Wear 274–275 (2012) 452–455.

[98] G. Zhang, R. Sebastian, T. Burkhart, K. Friedrich, Role of mono-dispersed nanoparticles on the tribological behavior of conventional epoxy composites filled with carbon fibers and graphite lubricants, Wear 292–293 (2012) 176–187.

[99] K. Swingler, Applying Neural Networks: A Practical Guide, Academic Press, Burlington, USA, 1996.

[100] Z. Zhang, K. Friedrich, Artificial neural networks applied to polymer composites: a review, Composites Science and Technology 63 (2003) 2029–2044.

[101] H.E. Kadi, Modeling the mechanical behavior of fiber-reinforced polymeric composite materials using artificial neural networks—A review, Composite Structures 73 (2006) 1–23.

[102] A. Suresh, Erosion studies of short glass fiber-reinforced thermoplastic composites and prediction of erosion rate using ANNs, Journal of Reinforced Plastics and Composites 29 (2010) 1641–1652.

[103] D. Aleksendric, Neural network prediction of brake friction materials wear, Wear 268 (2010) 117–125.

[104] K. Velten, R. Reinicke, K. Friedrich, Wear volume prediction with artificial neural networks, Tribology International 33 (2000) 731–736.

[105] Z. Zhang, K. Friedrich, K. Velten, Prediction of tribological properties of short fiber composites using artificial neural networks, Wear 252 (2002) 668–675.

[106] L.A. Gyurova, K. Friedrich, Artificial neural networks for predicting sliding friction and wear properties of polyphenylene sulfide composites, Tribology International 44 (2011) 603–609.

[107] Z. Zhang, N.M. Barkoula, J. Karger-Kocsis, K. Friedrich, Artificial neural network predictions on erosive wear of polymers, Wear 255 (2003) 708–713.

[108] B. Sick, On-line and indirect tool wear monitoring in turning with artificial neural networks: a review of more than a decade of research, Mechanical Systems and Signal Processing 16 (4) (2002) 487–546.

# The influence of nanoparticle fillers on the friction and wear behavior of polymer matrices

# 4

**Qihua Wang, Xianqiang Pei**

*Laboratory of Solid Lubrication, Lanzhou Institute for Chemical Physics, Lanzhou, China*

## CHAPTER OUTLINE HEAD

## 4.1 INTRODUCTION

Over the past decades, polymer composites have been increasingly used as structural materials in aerospace, automotive, and chemical industries, because they provide potential lower weight alternatives to traditional metallic materials. Among these applications, numerous are related to tribological components, such as gears, cams, bearings, and seals, where the self-lubrication properties of polymers and polymer-based composites are of special advantage [1]. The feature that makes polymer composites promising in industrial applications is the possibility to tailor their properties by adding special fillers with different volume fractions, shapes, and sizes. To tailor the mechanical and/or tribological performance of polymers, short-cut fibers and/or fabrics have been incorporated into some polymer matrices [2–14]. Besides, it has been found that nanosized or microsized inorganic particles are quite effective in improving the performance of polymer composites, including their friction and wear properties [15–24].

In comparison with the widely used conventional microscale particles, nanoparticles have some unique features [25–28]. First, the much higher specific surface area can promote stress transfer from matrix to nanoparticles. Second, the required amounts of nanoparticles in polymer matrices are usually much lower than those of the corresponding microparticles. Therefore, many intrinsic merits of pure polymers, such as low weight, ductility, and good processability, will be retained after additivation with nanoparticles. Third, the mechanical behavior of the bulk materials can be improved while the often-disturbing abrasiveness of the hard microparticles decreases remarkably by reduction of their angularity.

It has been well established that the dispersion state of nanoparticles is a crucial factor for the final properties of nanocomposites. Unfortunately, due to their high surface energy, nanoparticles tend to form agglomerates or clusters in a polymer matrix, which can result in property degradations. This is especially the case when the fabrication of composites with high nanoparticle content is attempted. In order to obtain perfect dispersion of nanoparticles in polymer matrices, several methods have been applied to break down the clusters or agglomerates, including ultrasonic vibration [29–35], special sol–gel techniques [36], high shear energy dispersion process [24,28], and melt mixing [37–39]. Among these methods, ultrasonic vibration is most widely used.

Recently, many attempts have been made to develop nanoparticle-filled polymeric composites for tribological applications. The tribology of nanoparticle-filled polymers is of significant interest because of the ability of nanoparticles in altering the properties of the matrix and the surfaces involved while keeping many intrinsic merits of pure polymers. Despite these efforts, the influence of nanoparticle fillers on friction and wear behavior of polymers is not yet thoroughly understood. In this context, this chapter gives some information on the influences of nanoparticles on the friction and wear behavior of polymer matrices, based on the authors' as well as external research results.

## 4.2 INFLUENCE OF VOLUME CONTENT OF NANOPARTICLES ON THE FRICTION AND WEAR BEHAVIOR OF POLYMERS

Generally, tribological characteristics, that is, friction and wear properties, are not intrinsic material properties, but depend on the whole tribological system. Consequently, the design of any kind of tribological material must depend on the requirements of the particular application. As mentioned above, it is possible to tailor the properties of polymer composite with appropriate fillers. It is reasonable to expect that different volume content of nanoparticles added into polymers may exert a different influence on the final material's friction and wear properties.

Up to now, much attention has been paid to the influence of the volume content of different kinds of nanoparticles on the friction and wear behavior of polymer nanocomposites. Various inorganic nanoparticles have been used to fill different polymeric matrices. For example, polyetheretherketone (PEEK) has been filled with nanoparticles of $Si_3N_4$, $SiO_2$, SiC, $ZrO_2$ [40–43], polyphenylene sulfide (PPS) has been filled with nanometer $Al_2O_3$, $TiO_2$, CuO, ZnO, SiC [44,45], polytetrafluoroethylene (PTFE) was filled with nanometer ZnO and $Al_2O_3$ [46,50], (poly(phthalazine ether sulfone ketone)) (PPESK) was filled with $SiO_2$, $TiO_2$, and $Al_2O_3$ [47–49]. Table 4.1 summarizes the friction and wear of some of these nanoparticle-filled polymer composites. It can be seen that an optimal filler content exists from the point of view of wear resistance. The coefficient of friction as a result can then either decrease or increase compared to that of the original pure polymer. Additionally, it can be seen that the optimal content of these nanoparticles is usually in a low range of 1–4 vol.%, except for PTFE and PPESK matrix filled with nano-ZnO and nano-$SiO_2$, respectively. This lower optimal nanoparticle content can be explained considering the fact that at high content nanoparticles may agglomerate and rather deteriorate friction and wear behavior of the resultant nanocomposite.

As mentioned above, a maximum wear resistance of polymer nanocomposites can be obtained at the optimal nanoparticle content. For example, the wear rate of pure PEEK decreases gradually from $7.4 \times 10^{-6}$ to $1.3 \times 10^{-6}$ $mm^3$ $Nm^{-1}$ at an optimal nano-$Si_3N_4$ content of 3.3 vol.% [40]. A similar tendency can also be found for the other composites listed in Table 4.1. On the other hand, the changes of the coefficients of friction of nanoparticle-filled polymers are not the same for all polymers, even not at an optimal filler content. For example, in $SiO_2$/ PEEK nanocomposites [41], the coefficient of friction decreases compared to that of the pure PEEK at the optimal filler content, while for $Al_2O_3$/PPS nanocomposite [44] it increases, and for ZnO/PTFE nanocomposite it remains more or less constant [46]. Clearly, changes in wear rate are not linearly correlated to the changes in the coefficient of friction. The influence of nanoparticles on the friction behavior compared to that on wear behavior can thus be different. For example, in SiC/PEEK nanocomposites [42], the coefficient of friction decreases sharply with nanometer SiC content up to 3.3 vol.%. Then, it decreases smoothly with increasing SiC content up to 9.4 vol.%. The change in the wear rate does not exactly follow the change in the coefficient of friction. As can be seen in Fig. 4.1, with SiC content <1.06 vol.%, the wear rate decreases almost linearly. In the range between 1.06 and 4.4 vol.%, the wear rate reaches a minimum and remains constant. Above 4.4 vol.%, the wear rate almost linearly increases with increasing SiC content and is higher in comparison to that of the unfilled PEEK. Based on these observations, an optimal SiC content is recommended, that is, 1.06–4.4 vol.%.

The different influence of nanoparticles on the friction and wear behavior can also be found for PEEK filled with nanometer $Si_3N_4$, $SiO_2$, $ZrO_2$ etc. [40,41,43]. It can generally be concluded that nanoparticles alter the friction and wear behavior in different ways and that a lower coefficient of friction does not necessarily correspond to a decreased wear rate.

**Table 4.1** Summary of the Friction and Wear of Some Nanoparticle-Filled Polymer Composites

| Matrix/ Nanoparticle | Nanoparticle Size (nm) | Content for Optimal Wear Resistance (vol.%) | Change of Coefficient of Friction | Minimum Wear Rate ($10^{-6}$ mm$^3$ Nm$^{-1}$) | Wear Rate of Pure Polymer ($10^{-6}$ mm$^3$ Nm$^{-1}$) | Reference |
|---|---|---|---|---|---|---|
| PEEK/Si$_3$N$_4$ | <50 | 3.3 | ↓ | 1.3 | 7.4 | [40] |
| PEEK/SiO$_2$ | <100 | 4.4 | ↓ | 2.4 | 7.4 | [41] |
| PEEK/SiC | <100 | 1.06–4.4 | ↓ | 3.4 | 7.4 | [42] |
| PEEK/ZrO$_2$ | 10 | 1.8 | ↓ | 3.9 | 7.4 | [43] |
| PPS/Al$_2$O$_3$ | 33 | 2 | ↑ | 12 | 23 | [44] |
| PPS/TiO$_2$ | 30–50 | 2 | ↓ | 8 | 16.6 | [45] |
| PPS/CuO | 30–50 | 2 | ↓ | 4.6 | 16.6 | [45] |
| PTFE/ZnO | 50 | 15 | Less change | 13 | 1125.3 | [46] |
| PPESK/SiO$_2$ | 20 | 14.5 | ↓ | 2.95 | 89.06 | [47] |
| PPESK/TiO$_2$ | 40 | 1.75 | ↑ | 4.86 | 80.12 | [48] |
| PPESK/Al$_2$O$_3$ | 40 | 1 | ↑ | 7.31 | 80.12 | [49] |

Notes: PEEK—Polyetheretherketone, PPS—Polyphenylene sulfide, PTFE—Polytetrafluoroethylene, PPESK—Polyphthalazine ether sulfone ketone).

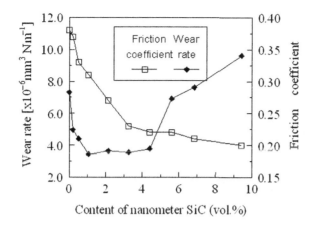

**FIGURE 4.1**

Effect of the content of nanometer SiC on the friction coefficient and wear rate of the filled PEEK (load: 196 N, sliding speed: 0.445 m s⁻¹).

*Reprinted from Ref. [42] with permission from Elsevier.*

## 4.3 INFLUENCE OF THE SIZE AND SHAPE OF NANOPARTICLES ON THE FRICTION AND WEAR BEHAVIOR OF POLYMERS

It is well known that inorganic filler particles can enhance mechanical and tribological properties of polymers. The effectivity of the reinforcement is determined by several factors such as the basic properties of the composite matrix, microstructure of the composite, filler size and shape, homogeneity of particle distribution, and the quality of the filler–matrix interface [23].

Among these factors, the interface between the filler and matrix plays a critical role for the ultimate nanocomposite's properties. The quality of the filler–matrix interface may to some extent be affected by the filler's size and shape, because of different specific surface areas and surface reactivity. Unfortunately, these aspects have not yet been extensively investigated.

In the case of PEEK, the effect of particle size of nanometer $ZrO_2$ on its tribological behavior has been investigated (Fig. 4.2) [51]. It is found that nanometer $ZrO_2$ at any size reduces the friction of PEEK. But the reduction of the coefficient of friction is slightly more important with smaller sized $ZrO_2$ particles. The wear rate of nanometer $ZrO_2$ filled PEEK decreases gradually with decreasing size of nanometer $ZrO_2$ particles. The wear rate of the filled PEEK significantly decreases with particle size <15 nm. Above 50 nm, however, the wear rate gradually increases with increasing $ZrO_2$ particle size and becomes even higher than that of pure PEEK. This indicates that the nanometer $ZrO_2$ is very effective in reducing the wear of filled PEEK, but only with the particle size of nanometer $ZrO_2$ <15 nm. Xing et al.

[52] investigated the wear behavior of an epoxy matrix filled with 120- and 510-nm silica particles. They found that the wear resistance of the composite dramatically increases by adding a small amount of either type of silica filler. But, anyway, the smaller sized particles seemed to be more effective in improving the wear resistance, which is consistent with the results of PEEK filled with different sized nanometer $ZrO_2$ particles [51].

As for the influence of the nanoparticle shape, Sawyer et al. [53] recently carried out some research on PTFE filled with irregularly shaped nanometer alumina. They found that irregularly shaped filler particles are effective in reducing the wear of PTFE, at the expense of an increase of the coefficient of friction. Spherical alumina nanoparticles of "40 nm" at a concentration of 20 wt.% improve the wear resistance by a factor 600 [50], while "80-nm" irregularly shaped alumina nanoparticles at a concentration of 1 wt.% improve the wear resistance by a factor of 3000 [53]. Jia et al. [54] reported the tribological behavior of epoxy resin (EP) modified by differently shaped nanofillers, such as spherical silica ($SiO_2$), layered organomodified montmorillonite (oMMT) and oMMT-$SiO_2$. They showed that adding these nanofillers significantly improves the tribological properties of pure EP. The average coefficient of friction of EP/$SiO_2$ obviously decreases, but friction loss only decreases a little. Adding oMMT into the EP matrix clearly reduces friction loss, and the average coefficient of friction decreases a little. However, when the $SiO_2$-oMMT is added to the EP matrix, the coefficient of friction and friction loss decrease at the same time.

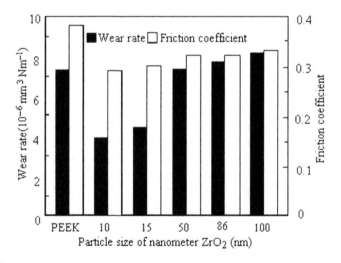

**FIGURE 4.2**

Effect of the particle size of nanometer $ZrO_2$ on the friction coefficient and wear rate of the filled PEEK (content of nanometer $ZrO_2$: 1.8 vol.%, load: 196 N, sliding velocity: 0.445 m s$^{-1}$).

*Reprinted from Ref. [51] with permission from Elsevier.*

Conclusively, it is clear that both the size and shape of nanoparticles may have a significant influence on the friction and wear behavior of polymers. Further research, however, is needed to clarify these aspects.

## 4.4 INFLUENCE OF NANOPARTICLES IN COMBINATION WITH TRADITIONAL TRIBOFILLERS ON THE FRICTION AND WEAR BEHAVIOR OF POLYMERS

As known up to now, nanoparticles can be effective in reducing the wear rate of polymers, and in some cases, the coefficient of friction can also be reduced. However, in practice, polymers filled with only nanofillers still do not meet practical requirements, due to their relatively high wear rate. In view of this, a combination of various functional fillers is necessary. For example, internal lubricants, such as PTFE powders and graphite flakes, are commonly incorporated to reduce adhesion of polymers sliding against metallic counterparts. Unfortunately, few efforts have been made in the study of such combinations.

For PEEK, PTFE powders have been applied as microfillers in combination with nanometer SiC [55]. It was found that the incorporation of PTFE into 3.3 vol.% nanometer SiC-filled PEEK had a detrimental effect on the tribological properties of the final PTFE/SiC/PEEK composite (Fig. 4.3). Namely, the coefficient of friction of the PTFE/SiC/PEEK composite increased sharply when a small amount of PTFE (1–5 vol.%) was added. As the PTFE content went >5 vol.%, the coefficient of friction of PTFE/SiC/PEEK slightly decreased with an increase in the PTFE content (Fig. 4.3(a)). The wear rate of the PTFE/SiC/PEEK composite increased significantly with increasing PTFE <5 vol.%. With PTFE content >10 vol.%, the wear rate of PTFE/SiC/PEEK decreased slightly with an increase in the PTFE content (Fig. 4.3(b)). However, the wear rate of PTFE/SiC/PEEK composites is still higher

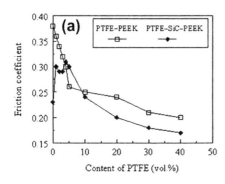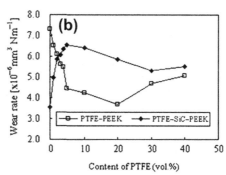

**FIGURE 4.3**

Effect of the content of PTFE on the (a) friction coefficient and (b) wear rate of PTFE–PEEK and PTFE–SiC (3.3 vol.%)–PEEK (load: 196 N; sliding velocity: 0.445 m s$^{-1}$).

*Reprinted from Ref. [55] with permission from Elsevier.*

than that of PTFE/PEEK. This indicates that the incorporation of PTFE into 3.3 vol.% nanometer SiC-filled PEEK weakens the friction reduction and wear resistance abilities of the filled PEEK composite. Also for the PEEK matrix, it was found that the addition of nano-$SiO_2$ remarkably reduces the friction coefficient of the short carbon fiber (SCF)/PTFE/graphite/PEEK composite, while the modification in the wear resistance was associated with the *pv* factor [56,57].

As stated above, the incorporation of nanoparticles may positively or negatively influence the tribological properties of traditional tribofiller-filled polymers. It is worth noting that, in some cases, the negative effects of single nanoparticle addition can be shifted to positive influences when nanoparticles and traditional fillers are incorporated simultaneously. As is revealed by Wang et al. [58], the best tribological properties were registered when nano-$Si_3N_4$, SCF and Gr were incorporated in combination into PI (polyimide) matrices, although nano-$Si_3N_4$ deteriorated the wear resistance of the PI composite drastically as a single filler. Similarly, nano-CuO was found to impair the wear resistance of the PI matrix when incorporated alone, while the synergistic effect was found for the combination of nano-CuO and graphite, which led to the best tribological properties [59].

In the case of an epoxy matrix, nano-$TiO_2$/PTFE/graphite/SCF with different volume contents have been incorporated by Chang et al. [60,61]. The results of this research indicate that the addition of nano-$TiO_2$ apparently can reduce the coefficient of friction and that the wear resistance is certainly improved, especially at high contact pressure and high sliding speed. Under standard sliding conditions (1 MPa and 1 m $s^{-1}$), the lowest wear rate was achieved for epoxy composites with composition of 5/0/5/15 (volume content of nano-$TiO_2$/PTFE/graphite/SCF). However, a composition of 5/5/5/15 with combined lubricants exhibited the highest wear resistance at an extreme high *pv* factor for polymers of 12 MPa m $s^{-1}$. But in the case of a very smooth countersurface ($R_a$ = 0.03 μm), the addition of nanoparticles was not able to reduce the coefficient of friction. Besides, a series of additional fillers such as SCF, Aramid, and PTFE particles have been added into nano-$TiO_2$/graphite/epoxy nanocomposites by Xian et al. [62]. It was found that the sliding wear resistance of the composite can be remarkably improved by the incorporation of SCF and Aramid particles. However, the low amplitude oscillating wear resistance was impaired by SCF. Besides, the friction and wear under both adhesive and low-amplitude oscillating wear conditions can further be reduced by the addition of PTFE to the SCF-filled nanocomposites. However, for the Aramid particle-filled nanocomposites, an adverse effect of PTFE was obtained.

Apart from the above, the effect of nanoparticles on the tribological behavior of polyamide 66 (PA66) and poly(etherimide) (PEI) composite have also been studied by Chang et al. [63,64]. For the PA66 matrix [63], it was found that conventional fillers, such as SCF and graphite flakes, could effectively reduce the coefficient of friction and wear rate of PA66 at lower *pv* conditions. With the addition of nano-$TiO_2$, the coefficient of friction of the composite was further decreased, especially at very high *pv* products. In the case of PEI composites [64], a further reduction in the coefficient of friction was also found, which was also more pronounced at higher *pv* conditions.

## 4.5 **THE ROLE OF NANOPARTICLES IN MODIFYING THE FRICTION AND WEAR BEHAVIOR OF POLYMERS**

As discussed above, nanoparticles have a great influence on the friction and wear behavior of polymers, which may depend on the type, volume content, size and shape of nanoparticles, matrix, test conditions, etc. But what exact role do nanoparticles play in modifying the friction and wear behavior of polymers? The authors proposed different answers to this question. In what follows, some widely accepted mechanisms are summarized.

Studies on PEEK nanocomposites have shown that nanometer $Si_3N_4$, $SiO_2$, and SiC are very effective in reducing the friction and wear of PEEK (Figs 4.4 and 4.5) [40–42]. In order to understand the action of nanometer fillers, the wear tracks of pure PEEK and its composites were studied. As an example, Fig. 4.6 shows the scanning electron microscope (SEM) micrographs of the worn surfaces of pure PEEK and nano-PEEK/$SiO_2$ (7.5 wt.%) [41]. It can be seen that plucked and plowed marks appeared on the wear scar of pure PEEK, while the scuffing in the wear traces on the nanofilled PEEK was obviously abated. The transfer films formed on the steel countersurfaces are shown in Fig. 4.7. A thick, lumpy, and incoherent transfer film was formed in the case of pure PEEK, while a thin, uniform, and coherent transfer film was deposited with nanofilled composite. This can be brought into relation with differences observed in the wear traces' morphology of Fig. 4.6. The obvious scuffing on the pure PEEK indicates the severe plowing of the polymer surface by the countersurface during the friction process. In other words, the transfer film in that case was of poor quality. In contrast, no scuffing or only slight scuffing was observed on the surface of the nanofilled composite. This can be ascribed to the tenacious character of the transfer film formed on the countersurface. This is in agreement

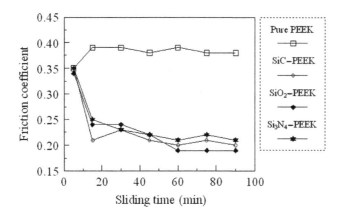

**FIGURE 4.4**

Variation in friction coefficient of PEEK and its composites rubbing against a plain carbon steel ring with sliding time (load: 196 N, sliding velocity: 0.445 m s$^{-1}$, the proportion of nanometer fillers was 7.5 wt.%).

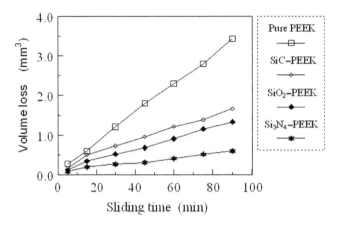

**FIGURE 4.5**

Variation in wear volume of PEEK and its composites rubbing against a plain carbon steel ring with sliding time (load: 196 N, sliding velocity: 0.445 m s$^{-1}$, the proportion of nanometer fillers was 7.5 wt.%).

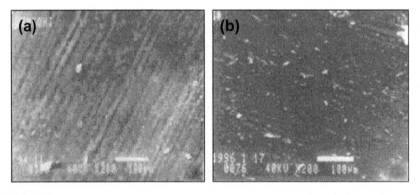

**FIGURE 4.6**

SEM micrographs of the worn surfaces of the pure PEEK and its composites filled with 7.5 wt.% nanometer SiO$_2$ (load: 196 N; sliding velocity: 0.445 m s$^{-1}$; test duration: 90 min) (a) pure PEEK; (b) SiO$_2$/PEEK.

*Reprinted from Ref. [41] with permission from Elsevier.*

with the observations of Fig. 4.7. The transfer film is responsible for the improved tribological properties of the nanofilled PEEK. With the formation of a uniform and tenacious transfer film, the subsequent sliding occurs between the PEEK surface and the transfer film covering the countersurface. Consequently, a lowered wear rate and coefficient of friction were obtained. For PEEK composite-filled with nanometer Si$_3$N$_4$, SiC, and ZrO$_2$, the same mechanism has been found [40,42,43]. So, it can be concluded that the transfer films play a dominant role in improving the friction and

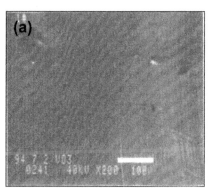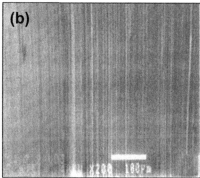

**FIGURE 4.7**

SEM micrographs of the transfer films formed by running the steel ring against the pure PEEK block and the filled PEEK block with 7.5 wt.% nanometer $SiO_2$ (load: 196 N; sliding velocity: 0.445 m s$^{-1}$; test duration: 90 min) (a) pure PEEK; (b) $SiO_2$/PEEK.

*Reprinted from Ref. [41] with permission from Elsevier.*

wear behavior of nanoparticle-filled polymers, by altering basic friction and wear mechanisms. The role of the transfer film in modifying the tribological properties of filled polymers has also been described in other studies [44–51,65–72].

By comparing the transfer films formed by pure and nanofilled PEEK, it is concluded that with the addition of nanoparticle fillers, the transfer films are strengthened. There may be several factors governing this strengthening action.

First, a certain roughness of the counterface is necessary, which can supply interlocking with the surface asperities. The dependence of the transfer film bonding strength on the counterface surface roughness has been studied by Schwartz et al. [44], who believed that nanoparticles in the transfer film provide stronger anchoring and hence increase bonding strength by virtue of their preferential location in the surface asperities of the counterface. Transfer films are caused by both adhesion and interlocking of fragments of polymer with countersurface (metal) asperities. In nanofilled polymers, the tiny nanometer particles can easily transfer together with the polymer matrix to the countersurface. Nanoparticles remain embedded in the transfer film due to the strong bonding between the nanoparticles and polymer matrix. Once transferred, the uniformly distributed nanoparticles in the transfer films will provide strong anchoring in the surface asperities of the counterface, and hence increase the bonding strength between the transfer film and steel surface. But, when the counterpart surface is very smooth, it cannot provide enough asperities for the nanoparticles to anchor and consequently the bonding strength of the transfer film is rather poor.

Second, tribochemical reaction during friction may exert some influence on the strengthening of transfer films. The wear traces of PEEK composites filled with nanometer SiC and $Si_3N_4$ were analyzed by X-ray photoelectron spectroscopy (XPS) to detect the chemical changes occurring during the complex wear process. The Si2p spectra of the wear traces and original surfaces of the nanometer SiC- and

Si$_3$N$_4$-filled PEEK are shown in Figs 4.8 and 4.9. For comparison, the XPS spectra of Si2p in pure SiC, SiO$_2$, and Si$_3$N$_4$ are given in Fig. 4.10. From Figs 4.8 and 4.9, it can be seen that a large portion of SiO$_2$ was contained in the transfer films, while

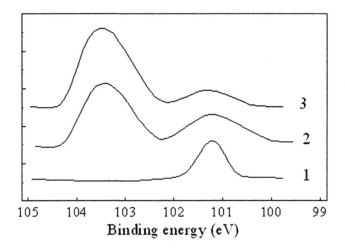

**FIGURE 4.8**

The XPS spectra of Si$_2$p on (1) original nanometer SiC filled PEEK surface; (2) wear scar surface of SiC/PEEK; (3) surface of the counterpart ring.

*Reprinted from Ref. [42] with permission from Elsevier.*

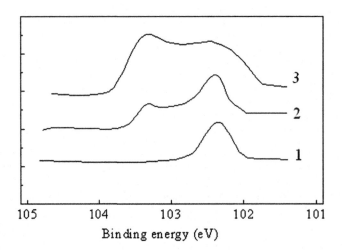

**FIGURE 4.9**

The XPS spectra of Si$_2$p on (1) original nanometer Si$_3$N$_4$ filled PEEK surface; (2) wear scar surface of Si$_3$N$_4$/PEEK; (3) surface of the counterpart ring.

*Reprinted from Ref. [40] with permission from Elsevier.*

only a small portion of $SiO_2$ existed in the wear traces of the composite specimens. This indicates that nanometer SiC and $Si_3N_4$ were partly oxidized to $SiO_2$ during the rubbing process. This hypothesis is in good agreement with the work reported in Refs [73–75]. Thus, it can be concluded that the oxidation of nanometer SiC and $Si_3N_4$ and the formation of $SiO_2$ determine the friction and wear behavior of the pure PEEK and its composites filled with nanometer SiC, $SiO_2$, and $Si_3N_4$.

In contrast to the oxidation of nanoparticle fillers as explained above, the reduction of nanometer CuO and $TiO_2$ has also been found [45]. It was indicated that the reduction of these fillers into elemental Cu and Ti provides attraction between the nascent Cu or Ti elements and the steel surface, due to the high reactivity of these elements. As a result, a strong adhesion between the transfer films containing pure Cu and Ti and steel surface occurs, which improves the tribological properties of the filled PPS as studied by Bahradur et al. Similarly, red metallic Cu instead of copper oxide was found in the optical micrographs of nano-CuO/Gr/PI composites' worn surface [59], indicating the reaction of nano-CuO and the counterpart steel. However, the above-mentioned effect only took place when Gr was incorporated simultaneously.

Once a uniform and tenacious transfer film is formed, the friction goes on between the filled polymer surface and the transfer film. Other wear mechanisms are then introduced, altering the friction and wear behavior.

As explained before, the change in the coefficient of friction and wear rate does not necessarily follow the same tendency. For example, it has been found that nanometer $Si_3N_4$, $SiO_2$, SiC, and $ZrO_2$ were very effective in reducing both friction and wear of PEEK [40–43]. In this case, the transfer film prevents direct contact between

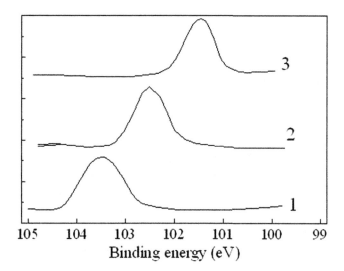

**FIGURE 4.10**

The XPS spectra of $Si_{2p}$: (1) $SiO_2$; (2) $Si_3N_4$; (3)SiC.

*Reprinted from Refs [40,42] with permission from Elsevier.*

the filled PEEK and steel surface, which reduces the plowing of the counterface asperities and results in a decrease of coefficient of friction and wear rate. But in other cases, such as 5 vol.% nanometer SiC-filled PPS [45], the transfer film is not able to cover the counterface completely, which results in an increased coefficient of friction and wear rate. For some polymers, the variation of tribological properties does not conform to the results of the above two cases. An example is the PPESK matrix. When filled with nanometer $Al_2O_3$, the nanocomposite showed an increased coefficient of friction compared to that of the PPESK matrix, while greatly increased wear resistance was achieved with only small additions of nanometer $Al_2O_3$ [49]. In this case, there is hardly any transfer film observed. The absence of a transfer film can be explained by the crosslinking of the PPESK. After crosslinking, the composite is difficult to transfer from the composite to the surface of the steel surface, which can account for the increased coefficient of friction and improved wear resistance of the nanofilled PPESK.

For polymers filled with nanoparticles in combination with traditional tribofillers, some authors also investigated the role of nanoparticles. The effect of the fillers can be either detrimental or positive.

For PEEK filled with 3.3 vol.% nanometer SiC and PTFE [55], it was found that the incorporation of PTFE had a detrimental effect on the tribological properties of the resultant PTFE/SiC/PEEK composite. The reason for this was the formation of $SiF_x$ on the polymer surface and worn surface during both production (compression molding) and sliding, as a result of the chemical reaction between nanometer SiC and PTFE. This chemical reaction and accompanying formation of $SiF_x$ dominate the tribological behavior of the PTFE/SiC/PEEK composite. With low PTFE content, the $SiF_x$ causes the friction and wear of the PTFE/SiC/PEEK composite to rise. However, the intrinsic low friction of PTFE dominates the global friction and wear behavior of the composites at high PTFE volume percentages. Friction further decreases with an increase of the PTFE content. The chemical reaction and the formation of $SiF_x$ also lead to changes in the worn surface morphologies and have a detrimental effect on the characteristics of the transfer films. Conclusively, nanoparticles may chemically react with traditional tribofillers and result in the deterioration of global tribological behavior.

Besides the detrimental effects of nanoparticles in combination with conventional tribofillers, positive effects resulting from the combination of nanofillers and traditional fillers have also been found. Studies by Chang et al. [60,61] indicate that a rolling effect of the nanometer $TiO_2$ between the material pairs helped to reduce the coefficient of friction during sliding. This rolling effect also protects the SCFs from more severe wear mechanisms, which can be evidenced by comparison of the atomic force microscopy (AFM) images of the fibers in the worn surfaces of composites with composition of 0/5/5/15 and 5/0/5/15 (volume content of nano-$TiO_2$/PTFE/graphite/SCF) [61]. For the composite without nanoparticles, the fiber surface is quite smooth but obviously tilted toward the worn surface. Impaction by the counterpart asperities has caused tilting of the fiber surface and interfacial damage between SCF and matrix (Fig. 4.11(a)). For the composite with nanoparticles, the AFM images of the fibers in the worn surface show a different appearance. Nanoscale grooves are found on the

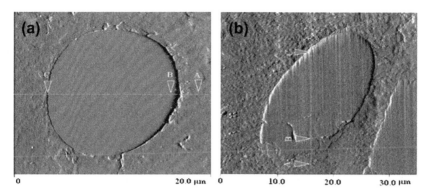

**FIGURE 4.11**

AFM images of the fiber in the worn surface of epoxy composites with different volume contents of nano-$TiO_2$/PTFE/graphite/SCF: (a) 0/5/5/15 and (b) 5/0/5/15 at 1 MPa and 1 m s$^{-1}$.

*Reprinted from Ref. [61] with permission from Elsevier.*

fiber surfaces of the nanocomposite and the width of the grooves is comparable to the size of the nanoparticles (Fig. 4.11(b)). However, the depth of the grooves is smaller than the particle size, which indicates that some of the nanoparticles might have been embedded in the counterpart surface. Here once more, it must be emphasized that a certain roughness of the steel counterpart is necessary in order to obtain the positive rolling effect of nanoparticles (cf. formation of transfer films). The rolling effect of nanometer $TiO_2$ has also been found in PA66 and PEI composite [63,64]. In addition, a combinative effect of nanoparticles with SCFs was suggested by Friedrich [76]. In this combinative effect, a topographic smoothening and a possible rolling effect of the nanoparticles are supposed, together with the formation of a transfer film. In a study of the tribological behavior of clay–thermoset polyester nanocomposites, the coefficient of friction and wear loss are found to decrease significantly with the addition of organoclay. The improvement in wear resistance is ascribed to the nanosized clay platelets dispersed in the polymer matrix, which acts as a barrier and prevents the large-scale fragmentation of polyester matrix. The decrease in the coefficient of friction is due to the combined effect of the three-body roller bearing action of both the nanoclay and nanoclay-reinforced wear debris and the formation of a thin transfer film on the counterpart surface [77].

Apart from the role of nanoparticles as described above, nanofillers are known to be able to improve the mechanical properties of polymers in various ways [78–82]. In the following paragraph, this will be dealt with the question whether there exists some relation between mechanical properties of nanoparticle-filled polymers and their friction and wear behavior.

For PEEK filled with nanometer $Si_3N_4$ [40], it has been found that the hardness of the composite almost linearly increases with increasing $Si_3N_4$ content, which indicates that nanometer $Si_3N_4$ improves the load-carrying capacity of PEEK.

However, a maximum bending and compressive strength were obtained with 0.52 vol.% $Si_3N_4$ (Fig. 4.12). The changes in mechanical properties of the filled PEEK and the variation of their friction and wear behavior do not follow the same trend. Figure 4.13 shows the coefficient of friction and wear rate of the filled PEEK as a function of $Si_3N_4$ content. By comparing the relation between mechanical properties and tribological behavior of the filled PEEK, it was inferred that optimal mechanical strength does not coincide with optimal tribological properties. Mechanical strength is thus not the only factor determining the tribological behavior of polymer nanocomposites, which has been confirmed by different authors [44,45]. In fact, there are several other factors that affect transfer film behavior, friction, and wear of polymeric nanocomposites [56,57,83,84].

## 4.6 INFLUENCE OF NANOPARTICLES ON THE FRICTION AND WEAR BEHAVIOR OF POLYMERS UNDER DIFFERENT TESTING CONDITIONS

Depending on the application, polymer nanocomposites may be subjected to different operating conditions, such as different loads, ambient conditions, temperatures, sliding velocities. It is thus necessary to study the influence of nanoparticles on the friction and wear behavior of polymers under different relevant testing conditions in order to meet the different requirements of any particular application.

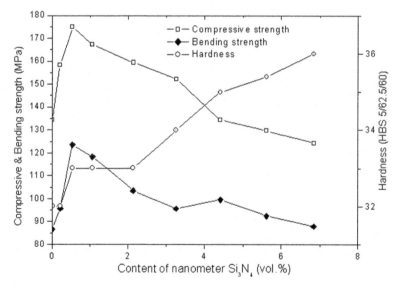

**FIGURE 4.12**

Three mechanical properties of the filled PEEK composite as a function of composition.

*Reprinted from Ref. [40] with permission from Elsevier.*

In the case of PEEK nanocomposites, the influence of testing conditions, such as load and lubricant, on friction and wear behavior is summarized below. Figures 4.14 and 4.15 show the effect of load on the coefficient of friction and wear rate of pure PEEK and its composites filled with 7.5 wt.% nanometer SiC, $SiO_2$, and $Si_3N_4$.

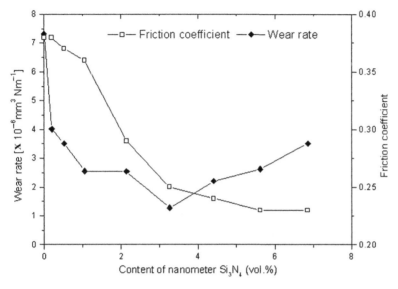

**FIGURE 4.13**

Effect of the content of nanometer $Si_3N_4$ on the friction coefficient and wear rate of the filled PEEK (load: 196 N, sliding velocity: 0.445 m s$^{-1}$).

*Reprinted from Ref. [40] with permission from Elsevier.*

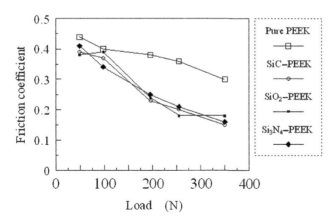

**FIGURE 4.14**

The friction coefficient of pure PEEK and its composites filled with 7.5 wt.% nanometer silicon compounds as a function of load (sliding velocity: 0.445 m s$^{-1}$).

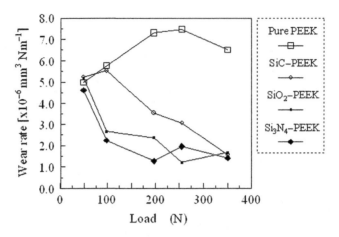

**FIGURE 4.15**

The wear rate of pure PEEK and its composites filled with 7.5 wt.% nanometer silicon compounds as a function of load (sliding velocity: 0.445 m s$^{-1}$).

It can be seen that the coefficient of friction of pure PEEK decreases with increasing load. Especially, the lowest coefficient of friction is obtained at a load of 350 N. The coefficient of friction as a function of load for nanofilled PEEK is fairly similarly independent of nanometer fillers. According to Fig. 4.14, nanometer silicon compounds are more effective for friction reduction at a higher load. On the other hand, the wear rate of the nanometer filler-filled composites also decreases considerably with increasing load, in opposition to pure PEEK. This shows that nanometer silicon compounds are also more effective for wear reduction at a higher load. It can thus be concluded that nanometer SiC-, SiO$_2$-, and Si$_3$N$_4$-filled PEEK composites bear better tribological properties at a higher load. The improved tribological properties at higher loads have also been found for SiO$_2$/PPESK nanocomposites by Shao [47], and is explained as follows: First, at higher load, the polishing action of nanoparticles can be strengthened, which reduces the scuffing of the composites. Second, the transfer films on the counterpart surface may be of a higher quality at higher loads compared to that formed at lower loads [40]. With the formation of higher quality transfer films, the plowing and scuffing will be abated, and the tribological behavior is improved. Also, the formation rate of transfer films may be enhanced at higher loads, which can shorten the running-in period and is favorable for improving the tribological properties of polymer composites.

However, the changes of coefficient of friction and wear rate do not always follow the same tendency. For example, for epoxy composites filled with different volume contents of nano-TiO$_2$/PTFE/graphite/SCF [60], the wear rate with nanoparticles (volume content: 5/0/5/15) and without nanoparticles (volume content: 0/5/5/15) increase with increasing loads (Fig. 4.16(a)), while the coefficient of friction changes differently. For the composite without nanoparticles, the coefficient of friction goes

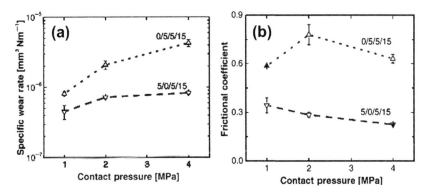

**FIGURE 4.16**

Influence of load on the (a) wear rate and (b) friction coefficient of epoxy composite with different volume content of nano-TiO$_2$/PTFE/graphite/SCF (sliding velocity: 1 m s$^{-1}$).

*Reprinted from Ref. [60] with permission from Elsevier.*

through a maximum. For composites with nanoparticles, the coefficient of friction monotonously decreases with increasing load (Fig. 4.16(b)).

It is assumed that both the increased load-carrying capacity and hence reduced real contact area, and the enhanced polishing effect at higher loads contribute to the decrease in the coefficient of friction.

In a study on Al$_2$O$_3$/PI nanocomposites [68], it was found that the coefficient of friction of pure PI increased with increasing load, while that of nanocomposites varied in a more complex manner (Fig. 4.17(a)). Nanometer Al$_2$O$_3$ can also be used to reduce wear at a relatively high load. For example, with proper concentration of nanometer Al$_2$O$_3$, the nanocomposite shows a lower wear volume loss than PI does at loads of 200 and 290 N (Fig. 4.17(b)). Simultaneously, at the loads of 200 and 290 N, the coefficient of friction of the nanocomposites decreases sharply with the mass fraction of nanometer Al$_2$O$_3$ around 3%. So at a higher load, the Al$_2$O$_3$/PI nanocomposite shows better tribological behavior compared to that shown by pure PI. This can be ascribed to the improved characteristics of the transfer film. At higher loads, the nanofillers can improve the quality of the transfer film, which then plays a dominant role in improving the tribological properties of polymer nanocomposites. Besides, the enhanced polishing action of nanoparticles at higher loads can also contribute to better tribological behavior.

In some cases, components of polymer composites are supposed to function under water. Therefore studies on the tribological behavior of Si$_3$N$_4$/PEEK and SiC/PEEK nanocomposites under water lubrication were performed [70,71]. Different tribological behavior was found for these nanocomposites. For Si$_3$N$_4$/PEEK composite [70], the coefficient of friction increased continuously with increasing Si$_3$N$_4$ content, in opposition to what happens in dry conditions. The wear rates of both the unfilled and filled PEEKs were much higher under water lubrication than under dry sliding. So, nanometer Si$_3$N$_4$ fillers could not really improve the friction

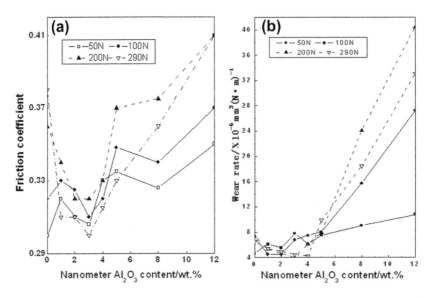

**FIGURE 4.17**

The (a) friction coefficient and (b) wear volume loss of $Al_2O_3$/PI nanocomposites under various loads as a function of nanometer $Al_2O_3$ content (sliding speed: 0.431 m s$^{-1}$).

*Reprinted from Ref. [68] with permission from Elsevier.*

and wear behavior of PEEK under water lubrication. Severely plowed and plucked marks were found on the PEEK surface (Fig. 4.18(a)), and a rough, discontinuous and thick transfer film was observed on the steel countersurface (Fig. 4.18(b)). This indicates that the filled PEEK under water lubrication cannot form uniform and tenacious films on the steel surface, which is necessary to protect the PEEK against the asperities of the steel counterface. For SiC/PEEK nanocomposites another behavior appears [71]. Nanometer SiC as a filler is effective in improving the friction-reduction ability and in increasing the wear resistance of PEEK under water lubrication. It was found that the adhesion, scuffing, and water erosion of both the composite block and the steel counterpart were effectively abated after the addition of nanometer SiC in PEEK (Figs 4.19(c) and (d)), as a result of the formation of a thin, uniform, and protective transfer film on the counterpart surface (Figs 4.19(a) and (b)). Again, it must be concluded that transfer films play a dominant role in the friction and wear behavior of polymer nanocomposites, even under water lubrication.

In addition to load and lubricating conditions, temperature may be another important factor influencing friction and wear behavior of polymer nanocomposites. Xian et al. [72] investigated the tribological properties of PEI composites at room temperature and at 150 °C and found different friction and wear behavior. At room temperature, there was no obvious difference between nanoparticles and microparticles in modifying the coefficient of friction. The wear resistance of

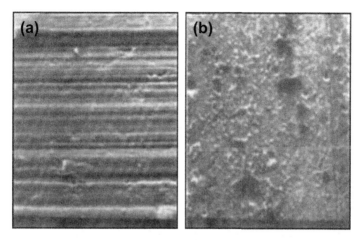

**FIGURE 4.18**

SEM pictures of the worn surfaces for the frictional couple of the carbon steel ring and the 4.4 vol.% nanometer $Si_3N_4$-filled PEEK under water lubrication (196-N load, 0.445 m s$^{-1}$ sliding velocity, 90-min test duration): (a) the worn surface of the filled PEEK, (b) the worn surface of the steel ring.

*Reprinted from Ref. [70] with permission from Wiley.*

nanofilled PEI was worse than that of microparticle-filled PEI. But at an elevated temperature of 150 °C, the advantage of nanocomposites is clear. The nanocomposite shows the lowest coefficient of friction and the best wear resistance at 150 °C compared to that of the microparticle-filled composite. The better tribological properties of nanocomposites at elevated temperatures are ascribed to the easier formation of transfer films with nanoparticles. At an elevated temperature, the mobility of molecular chain of PEI is improved and plastic deformation tends to happen, both with positive effect on the formation of transfer films. In a study on the tribological properties of $Al_2O_3$/PPESK nanocomposites [49], the coefficient of friction of a 1.0 vol.% nanometer $Al_2O_3$-filled PPESK was found to be stable, and its wear rate was little affected by increasing the temperature from room temperature to 270 °C. The reason for this phenomenon may be due to the increased thermal properties of the composite after the addition of nanometer $Al_2O_3$, which may have resulted in the possible crosslinking of the matrix to some extent.

As to the effect of sliding speed on the tribological properties of polymer nanocomposites, work has been done by several authors. For $Al_2O_3$/PI nanocomposites [68], it was found that the coefficient of friction of the nanocomposite decreased at a higher sliding velocity of 0.862 m s$^{-1}$ compared to that at 0.431 m s$^{-1}$, while the wear volume loss increases. At a higher sliding velocity, a higher surface temperature occurs, which results in micromelting of the composite surface and accompanying degradation of its load-carrying capacity.

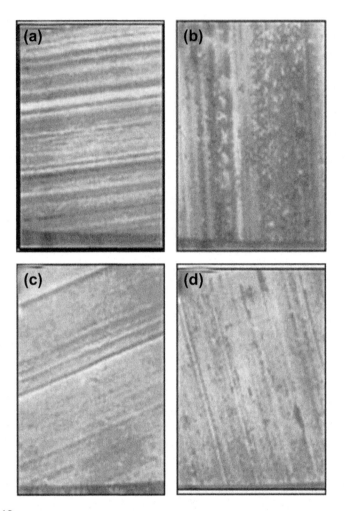

**FIGURE 4.19**

SEM pictures of the worn surfaces of unfilled PEEK and the 4.4 vol.% nanometer SiC filled PEEK as well as their counterpart rings under water lubrication (196-N load, 0.445 m s$^{-1}$ sliding velocity, 90-min test duration): (a) the worn surface of the unfilled PEEK, (b) the worn surface of the steel ring friction against unfilled PEEK, (c) worn surface of SiC/PEEK composite, and (d) the worn surface of the steel ring friction against filled PEEK.

*Reprinted from Ref. [71] with permission from Wiley.*

Both mechanisms lead to a decrease of coefficient of friction due to decreased shear strength of the surface. Although the formation of transfer films may be enhanced at a higher sliding velocity, the prolonged sliding distance at a higher sliding velocity and the micromelting of the composite surface induced by higher friction heat result in a higher wear volume loss. Moreover, the $Al_2O_3$/PI composites with different nanofiller contents showed different tribological behavior compared with pure PI.

For the case of the combination of nanoparticles with traditional fillers, Chang et al. [60] investigated the tribological properties of epoxy composites at different sliding velocities. They found that the addition of nano-$TiO_2$ effectively reduced the coefficient of friction and increased the wear resistance of nanocomposites, especially at a high sliding velocity. But with increasing sliding velocity, the coefficient of friction and wear rate of both the epoxy composites with and without nanoparticles tended to increase. However, the increase for composite with nano-$TiO_2$ was smaller compared to that without nanoparticles, which may be related to the positive rolling effect of nanometer $TiO_2$.

For a full understanding of the different parameters on the tribological behavior of nanocomposites, however, more systematic studies are needed.

## 4.7 SUMMARY

Polymers are important matrices for forming advanced composites. Inorganic particles are popular forms of reinforcing materials in advanced composites. It has been found that particle size has a great influence on the performance of polymer composites. When the particle sizes reduce to nanoscale levels, the friction and wear behavior of filled polymers will be greatly affected at very low nanoparticle contents compared to those of the corresponding microfillers. But, the influence of nanoparticles in the polymer matrix on friction and wear behavior depends on many factors, such as the type of polymer, type of nanoparticle, filler content, size and shape, as well as operating conditions. The real fundamental understanding of the influence of nanoparticles on tribological properties of filled polymers is still an open field for further research. Especially, this is the case when nanoparticles in combination with conventional fillers are considered.

Up to now, it is generally accepted that nanoparticles in polymer matrices function in several ways. First, by adding nanoparticles in polymer matrices, the transfer film formed on the steel countersurface can be strengthened due to anchoring and/or tribochemical reaction. The formation of a uniform and tenacious transfer film is important for improved tribological properties. Second, nanoparticles may act as a third-body rolling element, which can decrease friction and increase wear resistance of polymer composites. Third, there may be combined but a still not understood effect between nanofillers and other traditional fillers. In this case, a detrimental or positive effect may occur depending on the characteristics of the nanofillers and traditional tribofillers.

The further understanding of the influence of nanoparticle fillers on the friction and wear behavior of polymers will help in developing advanced composites that can meet the increasingly higher needs of practical applications.

# References

[1]  K. Friedrich, L. Chang, F. Haupert, in: L. Nicolais, M. Meo, E. Milella (Eds.), Current and Future Applications of Polymer Composites in the Field of Tribology Composite Materials, Springer, London, 2011, pp. 129–167.

[2]  X. Zhang, X. Pei, Q. Jia, Q. Wang, Effects of carbon fiber surface treatment on the tribological properties of 2D woven carbon fabric/polyimide composites, Applied Physics A Materials Science and Processing 95 (2009) 793–799.

[3]  X. Zhang, X. Pei, B. Mu, Q. Wang, Effect of carbon fiber surface treatments on the flexural strength and tribological properties of short carbon fiber/polyimide composites, Surface and Interface Analysis 40 (2008) 961–965.

[4]  X. Zhang, X. Pei, Q. Wang, Friction and wear properties of combined surface modified carbon fabric reinforced phenolic composites, European Polymer Journal 44 (2008) 2551–2557.

[5]  X. Zhang, X. Pei, Q. Wang, Friction and wear properties of basalt fiber reinforced/solid lubricants filled polyimide composites under different sliding conditions, Journal of Applied Polymer Science 114 (2009) 1746–1752.

[6]  X. Zhang, X. Pei, Q. Wang, The tribological properties of acid- and diamine-modified carbon fiber reinforced polyimide composites, Materials Chemistry and Physics 115 (2009) 825–830.

[7]  X. Zhang, X. Pei, Q. Wang, Friction and wear properties of polyimide matrix composites reinforced with short basalt fibers, Journal of Applied Polymer Science 111 (2009) 2980–2985.

[8]  X. Zhang, X. Pei, Q. Wang, Friction and wear behavior of basalt-fabric-reinforced/solid-lubricant-filled phenolic composites, Journal of Applied Polymer Science 117 (2010) 3428–3433.

[9]  X. Zhang, X. Pei, J. Zhang, Q. Wang, Effects of carbon fiber surface treatment on the friction and wear behavior of 2D woven carbon fabric/phenolic composites, Colloids and Surfaces A Physicochemical and Engineering Aspects 339 (2009) 7–12.

[10] X.R. Zhang, X.Q. Pei, Q.H. Wang, The effect of fiber oxidation on the friction and wear behaviors of short-cut carbon fiber/polyimide composites, Express Polymer Letters 1 (2007) 318–325.

[11] X.R. Zhang, X.Q. Pei, Q.H. Wang, Effect of solid lubricant on the tribological properties of polyimide composites reinforced with carbon fibers, Journal of Reinforced Plastics and Composites 27 (2008) 2005–2012.

[12] X.R. Zhang, X.Q. Pei, Q.H. Wang, Tribological properties of $MoS_2$ and carbon fiber reinforced polyimide composites, Journal of Materials Science 43 (2008) 4567–4572.

[13] X.R. Zhang, P. Zhao, X.Q. Pei, Q.H. Wang, Q. Jia, Flexural strength and tribological properties of rare earth treated short carbon fiber/polyimide composites, Express Polymer Letters 1 (2007) 667–672.

[14] X.-R. Zhang, X.-Q. Pei, Q.-H. Wang, Friction and wear studies of polyimide composites filled with short carbon fibers and graphite and micro $SiO_2$, Materials and Design 30 (2009) 4414–4420.

[15] Q. Wang, Q. Xue, W. Liu, W. Shen, Tribological properties of micron silicon carbide filled poly(ether ether ketone), Journal of Applied Polymer Science 74 (1999) 2611–2615.

[16] J.M. Durand, M. Vardavoulias, M. Jeandin, Role of reinforcing ceramic particles in the wear behaviour of polymer-based model composites, Wear 181–183 (Part 2) (1995) 833–839.

[17] M.Q. Zhang, M.Z. Rong, S.L. Yu, B. Wetzel, K. Friedrich, Effect of particle surface treatment on the tribological performance of epoxy based nanocomposites, Wear 253 (2002) 1086–1093.

[18] L. Jin-Chein, Compression and wear behavior of composites filled with various nanoparticles, Composites Part B: Engineering 38 (2007) 79–85.

[19] S. Bahadur, D. Gong, The role of copper compounds as fillers in the transfer and wear behavior of polyetheretherketone, Wear 154 (1992) 151–165.

[20] S. Bahadur, D. Gong, J.W. Anderegg, The investigation of the action of fillers by XPS studies of the transfer films of PEEK and its composites containing CuS and $CuF_2$, Wear 160 (1993) 131–138.

[21] B.J. Briscoe, A.K. Pogosian, D. Tabor, The friction and wear of high density polythene: the action of lead oxide and copper oxide fillers, Wear 27 (1974) 19–34.

[22] N.V. Klaas, K. Marcus, C. Kellock, The tribological behaviour of glass filled polytetrafluoroethylene, Tribology International 38 (2005) 824–833.

[23] B. Wetzel, F. Haupert, K. Friedrich, M.Q. Zhang, M.Z. Rong, Impact and wear resistance of polymer nanocomposites at low filler content, Polymer Engineering and Science 42 (2002) 1919–1927.

[24] B. Wetzel, F. Haupert, M. Qiu Zhang, Epoxy nanocomposites with high mechanical and tribological performance, Composites Science and Technology 63 (2003) 2055–2067.

[25] M.Z. Rong, M.Q. Zhang, H. Liu, H. Zeng, B. Wetzel, K. Friedrich, Microstructure and tribological behavior of polymeric nanocomposites, Industrial Lubrication and Tribology 53 (2001) 72–77.

[26] M.Z. Rong, M.Q. Zhang, Y.X. Zheng, H.M. Zeng, K. Friedrich, Improvement of tensile properties of nano-$SiO_2$/PP composites in relation to percolation mechanism, Polymer 42 (2001) 3301–3304.

[27] H. Zhang, Z. Zhang, K. Friedrich, C. Eger, Property improvements of in situ epoxy nanocomposites with reduced interparticle distance at high nanosilica content, Acta Materialia 54 (2006) 1833–1842.

[28] B. Wetzel, P. Rosso, F. Haupert, K. Friedrich, Epoxy nanocomposites – fracture and toughening mechanisms, Engineering Fracture Mechanics 73 (2006) 2375–2398.

[29] M.C. Kuo, C.M. Tsai, J.C. Huang, M. Chen, PEEK composites reinforced by nano-sized $SiO_2$ and $Al_2O_3$ particulates, Materials Chemistry and Physics 90 (2005) 185–195.

[30] N. Chisholm, H. Mahfuz, V.K. Rangari, A. Ashfaq, S. Jeelani, Fabrication and mechanical characterization of carbon/SiC-epoxy nanocomposites, Composite Structures 67 (2005) 115–124.

[31] V.M.F. Evora, A. Shukla, Fabrication, characterization, and dynamic behavior of polyester/$TiO_2$ nanocomposites, Materials Science and Engineering: A 361 (2003) 358–366.

[32] C.K. Lam, K.T. Lau, Localized elastic modulus distribution of nanoclay/epoxy composites by using nanoindentation, Composite Structures 75 (2006) 553–558.

[33] J.-C. Lin, L.C. Chang, M.H. Nien, H.L. Ho, Mechanical behavior of various nanoparticle filled composites at low-velocity impact, Composite Structures 74 (2006) 30–36.

[34] V. Balaji, A.N. Tiwari, R.K. Goyal, Fabrication and properties of high performance PEEK/$Si_3N_4$ nanocomposites, Journal of Applied Polymer Science 119 (2011) 311–318.

[35] A. Chatterjee, M.S. Islam, Fabrication and characterization of $TiO_2$–epoxy nanocomposite, Materials Science and Engineering: A 487 (2008) 574–585.

[36] L. Matĕjka, O. Dukh, J. Kolařík, Reinforcement of crosslinked rubbery epoxies by in-situ formed silica, Polymer 41 (2000) 1449–1459.

[37] A. Ashter, S.-J. Tsai, J.S. Lee, M.J. Ellenbecker, J.L. Mead, C.F. Barry, Effects of nanoparticle feed location during nanocomposite compounding, Polymer Engineering and Science 50 (2010) 154–164.

[38] S. Barus, M. Zanetti, M. Lazzari, L. Costa, Preparation of polymeric hybrid nanocomposites based on PE and nanosilica, Polymer 50 (2009) 2595–2600.

[39] P. Bhimaraj, D. Burris, W.G. Sawyer, C.G. Toney, R.W. Siegel, L.S. Schadler, Tribological investigation of the effects of particle size, loading and crystallinity on poly(ethylene) terephthalate nanocomposites, Wear 264 (2008) 632–637.

[40] Q. Wang, J. Xu, W. Shen, W. Liu, An investigation of the friction and wear properties of nanometer $Si_3N_4$ filled PEEK, Wear 196 (1996) 82–86.

[41] Q. Wang, Q. Xue, W. Shen, The friction and wear properties of nanometre $SiO_2$ filled polyetheretherketone, Tribology International 30 (1997) 193–197.

[42] Q.-H. Wang, J. Xu, W. Shen, Q. Xue, The effect of nanometer SiC filler on the tribological behavior of PEEK, Wear 209 (1997) 316–321.

[43] Q. Wang, Q. Xue, W. Shen, J. Zhang, The friction and wear properties of nanometer $ZrO_2$-filled polyetheretherketone, Journal of Applied Polymer Science 69 (1998) 135–141.

[44] C.J. Schwartz, S. Bahadur, Studies on the tribological behavior and transfer film–counterface bond strength for polyphenylene sulfide filled with nanoscale alumina particles, Wear 237 (2000) 261–273.

[45] S. Bahadur, C. Sunkara, Effect of transfer film structure, composition and bonding on the tribological behavior of polyphenylene sulfide filled with nano particles of $TiO_2$, ZnO, CuO and SiC, Wear 258 (2005) 1411–1421.

[46] F. Li, K.-a. Hu, J.-l. Li, B.-y. Zhao, The friction and wear characteristics of nanometer ZnO filled polytetrafluoroethylene, Wear 249 (2001) 877–882.

[47] X. Shao, J. Tian, W. Liu, Q. Xue, C. Ma, Tribological properties of $SiO_2$ nanoparticle filled–phthalazine ether sulfone/phthalazine ether ketone (50/50 mol %) copolymer composites, Journal of Applied Polymer Science 85 (2002) 2136–2144.

[48] X. Shao, W. Liu, Q. Xue, The tribological behavior of micrometer and nanometer $TiO_2$ particle-filled poly(phthalazine ether sulfone ketone) composites, Journal of Applied Polymer Science 92 (2004) 906–914.

[49] X. Shao, Q. Xue, W. Liu, M. Teng, H. Liu, X. Tao, Tribological behavior of micrometer- and nanometer-$Al_2O_3$-particle-filled poly(phthalazine ether sulfone ketone) copolymer composites used as frictional materials, Journal of Applied Polymer Science 95 (2005) 993–1001.

[50] W.G. Sawyer, K.D. Freudenberg, P. Bhimaraj, L.S. Schadler, A study on the friction and wear behavior of PTFE filled with alumina nanoparticles, Wear 254 (2003) 573–580.

[51] Q. Wang, Q. Xue, H. Liu, W. Shen, J. Xu, The effect of particle size of nanometer $ZrO_2$ on the tribological behaviour of PEEK, Wear 198 (1996) 216–219.

[52] X.S. Xing, R.K.Y. Li, Wear behavior of epoxy matrix composites filled with uniform sized sub-micron spherical silica particles, Wear 256 (2004) 21–26.

[53] D.L. Burris, W.G. Sawyer, Improved wear resistance in alumina-PTFE nanocomposites with irregular shaped nanoparticles, Wear 260 (2006) 915–918.

[54] Q.M. Jia, M. Zheng, C.Z. Xu, H.X. Chen, The mechanical properties and tribological behavior of epoxy resin composites modified by different shape nanofillers, Polymers for Advanced Technologies 17 (2006) 168–173.

[55] Q.-H. Wang, Q.-J. Xue, W.-M. Liu, J.-M. Chen, The friction and wear characteristics of nanometer SiC and polytetrafluoroethylene filled polyetheretherketone, Wear 243 (2000) 140–146.

[56] G. Zhang, Structure–tribological property relationship of nanoparticles and short carbon fibers reinforced PEEK hybrid composites, Journal of Polymer Science Part B: Polymer Physics 48 (2010) 801–811.

[57] G. Zhang, L. Chang, A.K. Schlarb, The roles of nano-$SiO_2$ particles on the tribological behavior of short carbon fiber reinforced PEEK, Composites Science and Technology 69 (2009) 1029–1035.

[58] Q. Wang, X. Zhang, X. Pei, Study on the synergistic effect of carbon fiber and graphite and nanoparticle on the friction and wear behavior of polyimide composites, Materials and Design 31 (2010) 3761–3768.

[59] Q. Wang, X. Zhang, X. Pei, A synergistic effect of graphite and nano-CuO on the tribological behavior of polyimide composites, Journal of Macromolecular Science Part B 50 (2010) 213–224.

[60] L. Chang, Z. Zhang, C. Breidt, K. Friedrich, Tribological properties of epoxy nanocomposites: I. Enhancement of the wear resistance by nano-$TiO_2$ particles, Wear 258 (2005) 141–148.

[61] L. Chang, Z. Zhang, Tribological properties of epoxy nanocomposites: part II. A combinative effect of short carbon fibre with nano-$TiO_2$, Wear 260 (2006) 869–878.

[62] G. Xian, R. Walter, F. Haupert, Friction and wear of epoxy/$TiO_2$ nanocomposites: influence of additional short carbon fibers, Aramid and PTFE particles, Composites Science and Technology 66 (2006) 3199–3209.

[63] L. Chang, Z. Zhang, H. Zhang, A.K. Schlarb, On the sliding wear of nanoparticle filled polyamide 66 composites, Composites Science and Technology 66 (2006) 3188–3198.

[64] L. Chang, Z. Zhang, H. Zhang, K. Friedrich, Effect of nanoparticles on the tribological behaviour of short carbon fibre reinforced poly(etherimide) composites, Tribology International 38 (2005) 966–973.

[65] Q.-J. Xue, Q.-H. Wang, Wear mechanisms of polyetheretherketone composites filled with various kinds of SiC, Wear 213 (1997) 54–58.

[66] P. Bhimaraj, D.L. Burris, J. Action, W.G. Sawyer, C.G. Toney, R.W. Siegel, L.S. Schadler, Effect of matrix morphology on the wear and friction behavior of alumina nanoparticle/poly(ethylene) terephthalate composites, Wear 258 (2005) 1437–1443.

[67] H. Yan, R. Ning, G. Liang, X. Ma, The performances of BMI nanocomposites filled with nanometer SiC, Journal of Applied Polymer Science 95 (2005) 1246–1250.

[68] H. Cai, F. Yan, Q. Xue, W. Liu, Investigation of tribological properties of $Al_2O_3$-polyimide nanocomposites, Polymer Testing 22 (2003) 875–882.

[69] M. García, M. De Rooij, L. Winnubst, W.E. van Zyl, H. Verweij, Friction and wear studies on nylon-6/$SiO_2$ nanocomposites, Journal of Applied Polymer Science 92 (2004) 1855–1862.

[70] Q.-H. Wang, Q.-J. Xue, W.-M. Liu, J.-M. Chen, Tribological characteristics of nanometer $Si_3N_4$ filled poly(ether ether ketone) under distilled water lubrication, Journal of Applied Polymer Science 79 (2001) 1394–1400.

[71] Q.-H. Wang, Q.-J. Xue, W.-M. Liu, J.-M. Chen, Effect of nanometer SiC filler on the tribological behavior of PEEK under distilled water lubrication, Journal of Applied Polymer Science 78 (2000) 609–614.

[72] G. Xian, Z. Zhang, K. Friedrich, Tribological properties of micro- and nanoparticles-filled poly(etherimide) composites, Journal of Applied Polymer Science 101 (2006) 1678–1686.

[73] T.E. Fischer, H. Tomizawa, Interaction of tribochemistry and microfracture in the friction and wear of silicon nitride, Wear 105 (1985) 29–45.

[74] Y. Yamamoto, A. Ura, Influence of interposed wear particles on the wear and friction of silicon carbide in different dry atmospheres, Wear 154 (1992) 141–150.

[75] J. Wei, Q. Xue, Tribochemical mechanisms of $Si_3N_4$ with additives, Wear 162–164 (Part B) (1993) 1068–1072.

[76] K. Friedrich, Z. Zhang, A.K. Schlarb, Effects of various fillers on the sliding wear of polymer composites, Composites Science and Technology 65 (2005) 2329–2343.

[77] P. Jawahar, R. Gnanamoorthy, M. Balasubramanian, Tribological behaviour of clay–thermoset polyester nanocomposites, Wear 261 (2006) 835–840.

[78] M. Zhang, R.P. Singh, Mechanical reinforcement of unsaturated polyester by $Al_2O_3$ nanoparticles, Materials Letters 58 (2004) 408–412.

[79] J. Che, B. Luan, X. Yang, L. Lu, X. Wang, Graft polymerization onto nano-sized $SiO_2$ surface and its application to the modification of PBT, Materials Letters 59 (2005) 1603–1609.

[80] S.C. Tjong, Structural and mechanical properties of polymer nanocomposites, Materials Science and Engineering: R: Reports 53 (2006) 73–197.

[81] F.R. Costa, B.K. Satapathy, U. Wagenknecht, R. Weidisch, G. Heinrich, Morphology and fracture behaviour of polyethylene/Mg–Al layered double hydroxide (LDH) nanocomposites, European Polymer Journal 42 (2006) 2140–2152.

[82] G. Ragosta, M. Abbate, P. Musto, G. Scarinzi, L. Mascia, Epoxy-silica particulate nanocomposites: chemical interactions, reinforcement and fracture toughness, Polymer 46 (2005) 10506–10516.

[83] G. Zhang, A.K. Schlarb, Correlation of the tribological behaviors with the mechanical properties of poly-ether-ether-ketones (PEEKs) with different molecular weights and their fiber filled composites, Wear 266 (2009) 337–344.

[84] G. Zhang, A.K. Schlarb, S. Tria, O. Elkedim, Tensile and tribological behaviors of PEEK/nano-$SiO_2$ composites compounded using a ball milling technique, Composites Science and Technology 68 (2008) 3073–3080.

# Tribological behavior of polymer nanocomposites produced by dispersion of nanofillers in molten thermoplastics

**Stepan S. Pesetskii, Sergei P. Bogdanovich, Nikolai K. Myshkin**

*V.A. Belyi Metal–Polymer Research Institute of National Academy of Sciences of Belarus,*
*Gomel, Belarus*

## CHAPTER OUTLINE HEAD

## 5.1 INTRODUCTION

The advances in technology are connected with a widening application of polymer engineering materials (engineering plastics) in such industries as automobile and tractor building, aviation, petroleum production, and manufacture of consumer goods. Polymer materials (PMs), unlike metals or ceramics, afford to lower the weight and increase output of parts along with their resistance to chemically hostile conditions. PMs (other symbols and abbreviations see nomenclature list) have been advantageous for making triboengineering components such as gears, bearings, cams, and parts covered with thin polymer films [1,2]. In terms of application volume and prospects for further production increase, the leading role belongs to thermoplastic PMs, which can be processed into final products by considerably more efficient and ecologically friendly processes [3,4].

An improved set of triboengineering properties of thermoplastic materials has been, as a rule, ensured by filling the polymer matrix with special substances. Reinforcement with short fibers (glass or carbon), for example, is most often used for increasing the mechanical strength and, as a consequence, the load-bearing capacity of composites [5–7]. Solid lubricants such as poly(tetrafluoroethylene) (PTFE), graphite, and molybdenum disulfide ($MoS_2$) added to a polymer significantly affect the formation of transfer layers on the counterface at the conditions of frictional interaction and decrease the coefficient of friction [8–10].

Often, engineering plastics have been blended with other polymers besides PTFE in order to control the impact strength and processability, as well as to decrease the cost of the materials [11–13]; in a number of situations, the triboengineering characteristics of plastics can be upgraded [14].

In recent years, owing to rapid advances in nanotechnologies [15–17], polymer nanocomposites become more and more attractive [18]. Nanocomposites have been studied extensively, and the number of published works constantly increases. The nanofillers most often used in plastics are carbon nanomaterials (e.g. fullerene or its derivatives) [19–22], layered clayey minerals [23], and nanoparticles of metals or their organic and inorganic compounds [24]. Information is available on polymer–polymer nanoblends in which the dispersed phase is broken up into nanometer dimensions within the matrix polymer [25,26].

The results of several research works indicate that nanofilling of polymers is an advantageous method of making novel wear-resistant plastics at relatively low levels of filling (~1–5 vol.%) [27–29]. The mechanism by which nanoparticles affect the triboengineering characteristics of polymers has been understood inadequately.

In view of the above, the present review is an attempt to survey and analyze the information available on the tribology of polymer nanocomposites. Special attention is focused on the consideration of composites prepared by dispersion of nanoparticles in melts of thermoplastic matrices. Some consideration is made of nanostructurized thermoplastics and polymer blends on their base.

## 5.2 **TECHNOLOGY OF NANOCOMPOSITES**

When the dimension of filler particles decreases in the polymer matrix, the surface area of the filler enlarges along with the share of the boundary layer (its thickness can reach 0.02–0.5 μm) [30]. With particle dimensions of the dispersed phase between 10 and 100 nm, the surface area of the nanoparticles is so large that the whole polymer becomes a boundary layer with reduced molecular mobility. The particle concentration may come to some decimal fractions of a percent.

The major difficulty of making nanocomposites is that ultradispersed particles tend to aggregate; besides, metal particles tend to oxidize [23]. Therefore, special processes are used to prepare composites that ensure the required distribution of the particles in the polymer matrix and prevent them from coagulation.

### 5.2.1 **Polymer–clay nanocomposites**

Today, polymer–clay nanocomposites (PCNs) have been studied most thoroughly [18,23,31,32]. Their methods of preparation can be divided into three groups: (1) introduction of a polymer from solution, (2) polymerization (in situ polycondensation), and (3) blending in polymer melt [32]. The latter method is more advantageous than the former two. It is more ecologically friendly because no organic solvents are present. Also, it can be successfully used with polymers unsuitable for in situ synthesis in the presence of clayey minerals. Finally, it can be realized with commercial equipment used to compound and process thermoplastics, which makes this method profitable (this method was first used by Vaia et al.) [33–35].

Figure 5.1 is a schematic representation of nanocomposites formed by the direct intercalation of a polymer melt into the galleries of a layered clay mineral, the particles of which were treated with a specialty surfactant. The important factors affecting the extent of penetration of macromolecules into galleries of mineral particles and the level of particle exfoliation in the polymer relate to the nature of surfactants and polar interactions of the organoclay and polymer matrix.

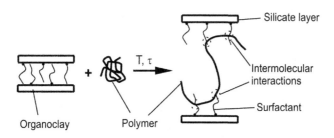

**FIGURE 5.1**

Interaction of polymer macromolecules – in melt – with layered clayey mineral modified with an organic surfactant.

Layered silicates (montmorillonite (MMT), hectorite, and saponite) that are usually used to make nanocomposites belong, to the 2:1 type of clay minerals or phillosilicates. They have two important characteristics: (1) ability to disperse into separate layers (layer thickness about 1 nm; lateral dimensions between 30 nm and several micrometers) and (2) ability to adjust the reactivity by ion exchange with organic or inorganic cations. The gap between layers (called interlayer space or gallery) can alter owing to ion exchange.

Parent layered silicates usually contain hydrated $Na^+$- and $K^+$-ions [36]. In their original (unmodified) conditions, layered silicates can only be combined with hydrophilic polymers. To be well moistened by other polymer matrices (among them weak-polar ones), the surface of a layered mineral must be made organophilic by means of exchange reactions with cationic surfactants that usually contain primary, secondary, ternary, or quaternary cations of ammonium or alkylphosphonium [37]. Chemisorption of ionogenic surfactants reduces the surface energy of the silicate and raises its wettability by the polymer, leading to larger galleries (the gap between layers increases).

The surfactants, used to treat layered silicates, should provide for macromolecule intercalation into galleries, as well as ensure adequate interphase adhesion in a polymer–mineral surface arrangement. Depending on the degree of interphase interaction, three types of PCNs can be produced [38] (Fig. 5.2):

1. Intercalated nanocomposites (a polymer is intercalated in a layered structure giving the latter a crystallographic regularity).
2. Intercalated and flocculated nanocomposites (with distorted crystal arrangement).
3. Exfoliated nanocomposites (separate layers are chaotically distributed in the polymer).

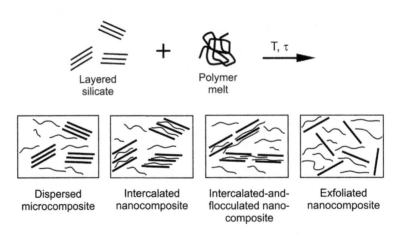

**FIGURE 5.2**

Structure of nanocomposites formed by the dispersion of a layered clayey mineral in the polymer melt.

The type of nanocomposite depends on the nature and concentration of the clayey mineral, the nature of the polymer and surfactant, and also the compounding conditions in the melt.

Most often, nanocomposites are produced in twin-screw mixers reactors [32], although single-screw extruders can be employed if equipped with special static and dynamic mixers, which can provide a required level of shearing in the polymer melt [39,40]. It should be noted, however, that single-screw extruders are less advantageous than are twin-screw ones [41].

Polyamides and polyesters (PET, poly(ethylene terephthalate); PBT poly(butylene terephthalate)) have been most frequently investigated as matrices for nanocomposites. Some authors [42] first made nanocomposites based on polyamide 6 (PA6) and MMT (concentrations between 1 and 18 wt%) in a twin-screw extruder. It was found that <10 wt% of MMT gives nanocomposites with an exfoliated structure; with higher amounts of MMT, intercalated structures are formed. Further studies with PA6 [41,43–48] showed that organoclay minerals could be thoroughly exfoliated in polyamide (PA) matrix with the help of various surfactants, by molecular weight variations of the polymer, and by using extruders with corotating and counterrotating screws. Much consideration has been given to surfactant selection because this factor determines the structure of composites. It was shown that alkylammonium surfactants [48], bis(hydroxyethyl) (methyl) rapeseed alkyl ammonium chloride [48], and trialkylamidazole [46–48] are most suitable for aliphatic PA (PA6, PA66, PA11, and PA12).

A more important role is played by surfactants in PCNs. This is explained by the high processing temperature of polyester melts (PET melts at ≈225 °C), together with the sensitivity of ester bonds in macromolecules toward thermal, mechanical, and hydrolytic degradation [49]. These features complicate the right choice of surfactants.

In an earlier work [50] on nanocomposites based on PET, 1,2-dimethyl-3-$N$-alkylimidazole was used as a surfactant to treat MMT particles. Being compounded in melt on the twin-screw laboratory-type extruder at 285 °C, the nanocomposites were wide-angle X-ray diffraction (WAXD) and transmission electron microscopy (TEM) observed and showed mixed intercalated and exfoliated nanostructures. In subsequent studies on the production of nanocomposites with molten PET and PBT, much consideration was made of surfactants with the purpose to optimize both the structure and properties of the materials [51–56]. Hexadecylamine [50], pentaerythrytol and maleic anhydride (MAH) [52], and cetyl pyridium chloride [55] have been tested for treating clayey mineral particles with the aim of controlling their interaction with the polymer. Unlike the situation with aliphatic polyamides, the use of alkylammonium surfactants with PBT, and particularly with PET, was less effective probably because of aminolysis of the ester bonds and resulted in excessive macromolecular degradation.

There are published works considering the technological details, structure, and properties of blends of aliphatic polyamides and polyesters with thermoplastic polymers [57–59] and liquid crystalline [60] polymers, and also with organosilicon oligomers [61]. Some authors [62] reported the advantages, for example, increased mechanical properties of combining carbon fiber and a layered clayey mineral in PA composites.

The compounding process in the melt has already been quite widely exercised for making nanocomposites based on PP, a nonpolar polymer, which has also found numerous applications, for example, tribosystems [63]. To ensure good dispersion of clays in molten PP during compounding, the particles of layered minerals are usually treated with ammonium surfactants that contain a long hydrocarbon radical, for example, stearylammonium [64] or dioctadecylammoniumbromide [65]. Exfoliation of the mineral becomes easier, while the bond between the polymer matrix and nanoparticles becomes stronger, when PP has been modified by grafting products of MAH [66–68].

The process of intercalation in the melt also gives polyethylene (PE)-g-MAH/clay nanocomposites [69]. Polar fragments incorporated in polyolefin (PO) macromolecules lead to their stronger adhesion with nanoparticles, and increase strength characteristics.

Gerasin et al. [65] pay attention to the concentration of the surfactant being introduced into the gallery when making organic days. It is determined in proportion to the cation-exchange capacity (CEC). For good intercalation of nonpolar polymers, it is necessary that the surfactant aliphatic chains be oriented in the galleries at an angle close to 90°, which ensures maximum separation of plates in layered silicates. The surfactant concentration must be less than one CEC, which allows one to obtain adsorption layers of the surfactant with large free space [65].

When making PCNs by compounding in melt, it is important to remember that on heating an organic clay in a molten polymer, the distance between planes reduces owing to the removal of the adsorbed water from the clay at $T = 160–180\ °C$ the water leaves the space between layers for a few minutes [70,71]. These factors should be taken into consideration when real technologies are being developed for making PCNs.

This process had been successful in preparing nanocomposites using polycarbonate (PC) [66] (dimethyl ditallow ammonium was the surfactant), poly(ether imide) (PEI) [72–74] (hexadecylamine as surfactant), liquid crystalline polymer in a nematic condition [75] (dioctadecyldimethylammonium as surfactant), and thermotropic liquid crystalline polyester [76] (organic clay Cloisite 25A, Southern Clay Product, USA, was used as filler).

There are reports describing the preparation of nanocomposites by combining organoclays with a molten polymer in a static condition at hydrostatic pressure (pressing process). In another work [77], polystyrene (PS)–clay nanocomposites were prepared by mixing PS powder with MMT treated with alkylaluminum, followed by melt molding in vacuum at 165 °C, where the PS turned to a molten state. However, the molding time was long (≈25 h) before the clay particles became exfoliated. Therefore, the method of direct melt molding is almost unsuitable for practical applications. Besides, further research [78] showed that nanocomposite formation, in this situation, depends on the type of surfactant to a greater degree compared to that by blending in an extruder.

### 5.2.2 **Polymer–carbon nanocomposites**

The major types of carbon nanomaterials characterized by a regular structure cover ultradispersed diamonds (UDDs) ($sp^3$ hybridization of atom orbitals, three-dimensional spatial arrangement of carbon atoms), fullerenes $C_{60}$ (an individual molecule – cluster – is in the shape of a truncated icosahedron of $I_h$-symmetry), and $C_{70}$ (cluster is in the shape of ellipsoid of $D_{5h}$-symmetry), and also tubules (carbon tubes of different diameters and structures) [19].

One of the most important and promising carbon nanofillers is *graphite* possessing unique mechanical and electrophysical properties of $sp^2$-hybridized carbon layers, called "graphene plates," are separated from each other [79–82]. Being chemically identical to carbon nanotubes (CNTs) and similar in structure to layered silicates, graphite has the potential of a unique nanofiller in the form of separate graphite layers or nanodimensional-layered blocks [83].

UDDs can be produced from explosives by detonation synthesis [84]. Several levels of structural self-arrangement in UDDs have been detected [85]. The first level comprises compact clusters (primary particles) of an average diameter of about 4 nm. The primary particles are combined into secondary clusters of 20–60 nm, which, in turn, get arranged into still larger aggregates. Nondiamond versions of carbon are removed from UDDs by thermal treatment at $T > 400$ °C.

It was reported elsewhere [86] that UDDs break up to nanoparticles in molten PC when being mixed in a Brabender device for 5–15 min at 260–280 °C.

Of particular interest is the application – for controlling polymer properties, among them triboengineering ones – of novel allotropic forms of carbon detected in the course of discovering fullerenes [87], and the possibility of their physicochemical modification [88], as well as of CNTs, the formation of which was first observed by Iijima [89].

Other surveys [90,91] have presented information on polymer nanocomposites containing CNTs and nanostructures. The process of melt compounding, particularly on the twin-screw extruder at higher shear rates, provides for satisfactory distribution of carbon nanomaterials in the bulk of both nonpolar (polypropylene, PP [92]), and polar thermoplastics (PA 6 [22,93,94], PC [95], and polyimide (PI) [96]). To improve the distribution of carbon nanomaterials in polymer matrices, it is advantageous to apply combined processes, for example, treatments of the polymer and nanofiller by a solvent (with its subsequent evaporation) and compounding in melts [97], collective dispersion of nanofiller with a powdery polymer by sonicating in an easily evaporated liquid [22], the use of special compatibilizers for improving nanoparticle distribution in a polymer melt [98]; the use of special chain extenders being simultaneously adhesionally active toward the CNTs and chemically modifying their surface [99].

Reports were published on the introduction of carbon nanomaterials into high-viscous rubber melts [100,101]. Jurkowska et al. [100] introduced fullerene into a natural rubber by stirring in Banbury-type device at 195 °C for 15 min. They achieved a satisfactory distribution of the filler.

Natural graphite exists in three states: (1) in the form of graphite flakes, (2) amorphous graphite, and (3) crystalline graphite. Graphite flakes (Fig. 5.3) are naturally

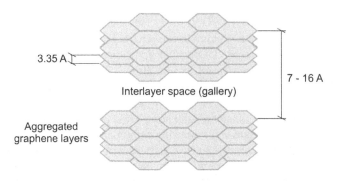

**FIGURE 5.3**

Layered structures of graphite.

abundant, and they show high electrical conductivity (~$10^4$ Ohm m$^{-1}$) at room temperature [102]. Owing to weak Van der Waals bonds between graphene layers (layer thickness equals the carbon atom diameter), suitably intercalated graphite flakes can be produced by the introduction of different low-molecular weight compounds of atoms or ions into the interlayer space (gallery) to make it larger; in this way, the antifriction characteristics of the graphite can be controlled. The calculated surface area of a single graphene layer is 2630–2965 m$^2$ g$^{-1}$ [103]. Graphene is a material with the highest ratio of strength to weight. In theory, Young's modulus for an individual graphite layer is 1600 GPa, which makes graphite nanolayers promising nanofillers in the production of high-strength polymer nanocomposites [104].

As a rule, the intercalated (or expanded) graphite can only be used to produce polymer–graphite nanocomposites, because high-molecular-weight polymers cannot enter the extremely small interlayer spaces in the source graphite [4,105]. There are available a great number of ways to intercalate graphite: by vapors of alkaline metals (the galleries are filled with Na or K atoms; e.g. $KC_8$ intercalated mix can be formed by heating graphite powder with K in vacuum at 200 °C) [105–107]; in solutions or melts of substances (soaking of graphite in molten metal chlorides or in halogens) [107–110]; electrochemical introduction from solutions of $H_2SO_4$, $HNO_3$, organic acids, alkaline metals, or metal chlorides [111–114]; cointercalation and methods of step-be-step intercalation (simultaneous introduction of $H_2$ and K with formation, in the gallery, of compounds like $C_{16}K_2N_{4/3}$) [115,116].

The simplest method of preparing expanded graphite is fast heating of its intercalated mix; a most often used method is the intercalation of $H_2SO_4$. Graphene layers become easily exfoliated by the introduction into the gallery $H_2SO_4$, HNO, $FeCl_3$, or K/tetrahydrofuran blends, or Na/tetrahydrofuran blends [105]. The $FeCl_3$ intercalation is preferable for making triboengineering materials [105].

The hydrophilic–hydrophobic characteristics of graphite allow to disperse it in molten polar and nonpolar polymer matrices. When preparing a nanocomposite, however, it is necessary to employ special mixing equipment and intercalated (expanded) graphite [117,118]. The thermodynamic processes influencing the penetration of a molten polymer into the interlayer space in an expanded graphite had been reported elsewhere [34]. The intercalation degree of the polymer depends on the combination of enthalpy and entropy factors. Although bonding of chain macromolecules within the graphite interlayer space causes a reduction in their total entropy, this reduction can be compensated by a greater interlayer distance in the expanded graphite [34].

Unlike the layered silicates, graphite does not participate in the processes of ion exchange. However, its oxidation causes the physicochemical characteristics to vary. Graphite oxide $C_7O_4N_2$ is hydrophilic, easily adsorbs water and other polar liquids, contains chemisorbed hydroxyl, ketone, carboxyl, and ether groups, as well as double bonds in the structure of chemisorbed compounds [119,120]. After a graphite oxide has been formed, its layered structure remains the same. Each layer is formed by a dense two-dimensional lattice that consists of a greater number of $sp^3$-carbon atoms and a fewer number of $sp^3$-carbon atoms. Like the case with layered silicates, graphite oxide layers become expanded through one-dimensional swelling, which favors intercalation, into interlayer spaces, of different substances (alkylammonium salts or macromolecules) [119,121–127]. The intercalated oxidized graphite has been used to prepare various nanocomposites [105].

### 5.2.3 Metal-containing polymer nanocomposites

This group of nanocomposites includes PM-containing metals, their salts, oxides, organometallic, or other compounds. Such materials are produced by introducing (immobilizing) particles, the size of which is comparable with those of macromolecules; such nanocomposites can be prepared in numerous ways [24,128], for example, by impregnation, deposition and codeposition, sorption from solutions, spraying, microencapsulation, dispersion in solutions or melts of polymers, sorption of metalocomplex compounds with their subsequent decomposition. In addition to differences in nanoparticle size and their level of dispersion in a polymer matrix, metal–polymer nanocomposites differ in the interaction of macromolecules with the nanoparticle surface. The interaction can be of physical character, when chemically inert polymers (POs, PTFE) are used; be based on chemical bonding of nanoparticles, if the macromolecules contain reactive functional groups; or it can be of a combined physicochemical character. Immobilization of metal nanoparticles on linear polymers leads to partial intermolecular cross-linking at the expense of intermolecular complex formation [128]; however, it would not alter the thermoplastic nature of the materials prepared. The formation of crosslinks between molecules is reversible. Mutual influencing was observed in metals and functional groups of macromolecules on their reactivity, as well as their activity in various macromolecular transformations and processes. Metals and their compounds can act catalytically on contact reactions, the role of which in metal–polymer friction contact is rather significant.

Most technological, easily realized and cost efficient are the methods of making nanocomposites based on the introduction of nanoparticles directly into a molten thermoplastic [129]; mixing of nanoparticles with nanodispersed polymers followed by melting and molding [130,131]; swelling of a polymer and nanofiller in the same solvent, followed by processing the composition from melt [129,132]; and decomposition of thermally unstable salts and other metal compounds, in a polymer melt [128,129]. The advantages of extrusion compounding (using a single-screw extruder equipped with a static mixer [39]), for making nanocomposites of PA with metal-containing compounds containing an active complex-forming ion of the metal, were described elsewhere [133]. The extrusion technologies, as has been mentioned above, are most preferable for producing thermoplastic nanocomposites.

## 5.2.4 Solid-phase extrusion in nanocomposite technology

Although nanofillers are subjected to special pretreatment before being introduced into polymer, their blending with a molten polymer, in many instances, does not ensure the required level of nanoparticle distribution. This is the case with almost all types of the nanocomposites discussed here.

It had been reported elsewhere [134–137] that the dispersion of nanoparticles in a polymer matrix can be enhanced by operating the screw plasticizing device in the regime of solid-phase extrusion (SPE), where the material cylinder temperature in certain mixing zones is much lower than that of the matrix polymer melting (solid-state shear pulverization (SSSP)). Usually the SPE is followed by mechanochemical effects leading to nanoscale blending without thermodynamic restrictions [137].

Other authors [33,134,137–141] have shown that shear and compression, acting upon the components being blended at the SSSP, lead to nanoscale dispersion in PCNs, as well as to better compatibilization and to fine-dispersion distribution of the disperse phase within blends of incompatible polymers.

Composites of polypropylene/CNT (tube diameters of 8–50 nm, lengths of 10–30 μm) have been studied [142] to show that the two-step technology (compounding by the SPE regime with subsequent processing from the melt) affords a high degree of dispersion, in the polymer matrix, of originally intertwined nanotubes. With a CNT concentration of 1 wt%, the Young's modulus can be increased by 50–57%. It is also assumed that shear compounding of PP/CNT nanocomposites in the solid state is followed by chemical grafting (the radical mechanism) of macromolecular fragments onto the nanoparticle surface.

Wakabayashi et al. [143]. reported that the SPE method gave a nanodispersed unmodified (taken as supplied) graphite in PP, which resulted in a twice as high Young's modulus and approximately 60% increase in the yield point in comparison with the pure PP. The graphite concentration in the nanocomposites was 3 wt%; it was dispersed in the PP matrix in the form of nanoplates consisting of approximately 30 graphene layers. These results are indicative of promise for the SPE method in making various polymer composited using a simplified pretreatment technology for nanoparticles.

## 5.3 FRICTION OF NANOCOMPOSITES

The specificity of tribology for polymer nanocomposites results from chain arrangements of macromolecules, as well as effects of nanofillers on the physical structure and physicochemical properties of macromolecules. It is believed that the determinative role belongs to phenomena taking place in thin surface layers of the contacting elements.

### 5.3.1 Friction of a polymer over a hard counterface

Several authors [2,8,144–153] have studied the state of the art in polymer tribology. Similar to other materials, the frictional interaction of polymers and hard counterparts depends on two independent components of friction force: deformation and adhesion.

The mechanical deformation component results from resistance to "plowing" of the softer material by asperities of the harder one [2,144]. The asperities experience elastic, plastic, and viscoelastic deformation, depending on the material properties. At initial loading on the polymer, plastic deformation mainly occurs, if the polymer is in a glassy state; viscoelastic deformation, or even viscoplastic one occurs, if the polymer is in a highly elastic state.

The adhesion component comes from adhesive bonds formed between the surfaces in friction contact – on spots of the real contact. The work of friction force equals that of the breaking energy of bonds, which prevent the parts from moving relatively to each other.

It is believed that for polymers, the adhesion (molecular) component much exceeds the mechanical one [154]. This is explained by transfer films generated on the metal counterface. As a result, the role of the mechanical component of friction force reduces, but the contribution of the molecular component rises. Special consideration is given to transfer films as the key factor that determines triboengineering properties of PMs [8,144]. The following factors considerably affect the friction force: contact load, sliding velocity, and temperature (Table 5.1). Besides, the data reported by different authors may be contradictory, which is explained by nonidentical test methods and different levels of influence of one or other parameters [144]. For example, depending on the contact load and velocity, the temperature in a contact zone may vary and affect the friction mechanism.

Theoretically, friction force should not depend on the sliding velocity. For polymers, however, this statement is only true if the contact temperature grows negligibly; the relaxation behavior of the surface material, as a result, does not change. Usually, however, there is a complex dependence of the friction coefficient on velocity, explained by variations in the relaxation properties and physicochemical activity of macromolecules [162]. When, in the course of testing, the temperature of the polymer sample approaches that of glass transition, at which segmental mobility unfreezes, a strong dependence of friction coefficient on the velocity is observed. At lower temperatures (frozen segmental mobility), $f$ does not practically depend on velocity.

**Table 5.1** Effect of Load, Sliding Velocity, and Temperature on Friction Coefficient [97]

| Authors | Materials | Values of parameters; friction pair | Illustrations |
|---|---|---|---|
| Shooter, Thomas [155] | PTFE, PMMA, PC | 10–40 N Steel–polymer | |
| Shooter, Tabor [156] | PTFE, PE, PMMA, PVC, nylon | 10–100 N Steel–polymer | |
| Rees [157] | PTFE, PMMA, nylon | Steel–polymer | |
| Shooter, Thomas [155] | PTFE, PE, PMMA, PC | 0,01–1,00 cm/s Steel–polymer Limited loading | |
| Milz, Sargent [159] | 1 – Nylon; 2 – PC | 4–183 cm/s Polymer–polymer | |
| White [158] | 1 – PTFE; 2 – nylon | 0,1–10,0 cm/s Steel–polymer | |
| Ludema, Tabor [160] | 1, 2 – PTFCE; 3 – PP | −50 to +150 °C Steel–polymer $1 - v = 3,5$ $10^{-5}$ cm/s; $2 - v = 3,5$ $10^{-2}$ cm/s | |
| King, Tabor [161] | 1 – PE; 2 – PTFE | −40 to +20 °C Steel–polymer | |

In polymer friction, the main mechanisms of wear are adhesion, abrasion, and fatigue [8,144]. The abrasive wear is caused by hard asperities acting upon the counterface and/or by hard particles that move over the polymer surface. Hard asperities (and/or particles) penetrate into the PM and remove its surface layer by microcutting or by repeated deformation. This type of wear occurs when roughness is the determinative parameter in frictional interaction.

The adhesive mechanism of wear [162] is observed in sliding of a polymer over a foreign surface, where the strength of adhesive bonds formed between the contacting materials can exceed that of the cohesive strength of the PM. As a result, the polymer fails within the bulk of the surface layer. Some part of the separated material gets transferred onto the counterface and forms a transfer layer; another part of the plucked off material is removed from the friction zone as wear debris. Depending on the polymer characteristics and friction conditions, the transferred fragments may have different shapes. In the presence of a liquid lubricant, which prevents adhesional interaction, this mechanism contributes but negligibly to total wear.

The fatigue mechanism in polymers is caused by crack propagation at repeated deformation of the material during friction. Fatigue results in pitting, crack generation, and delamination. Wear debris are formed as a result of propagation and intersection of small surface cracks that are oriented perpendicularly to the sliding direction. The fatigue wear occurs after prolonged friction; its contribution may be insignificant at short times of testing. It may be of importance in the absence of adhesive wear, when the counterface is smooth.

It should be noted that none of the above friction mechanisms of failure can be achieved without the participation of the others.

In a general case, three stages can be distinguished in frictional interaction [163]: (1) running-in, (2) steady regime, and (3) catastrophic wear. As applied to metal–polymer friction systems, the first stage implies the generation of a transfer film on the metal counterface. For most metal–polymer friction systems, the triboengineering characteristics strongly depend on the properties of the transfer film, that is, its nature, strength of bonding to the metal surface; the part of counterface area covered with the transfer film; and the intensity of adhesive interaction on the polymer–transfer layer interface. This has been supported experimentally and reported by several authors [8,145,146].

When considering polymer friction, a number of circumstances should be mentioned that are important in subsequent treatments of the data on friction of nanocomposites. First, as polymers adhere to hard surfaces through adsorption [164], it can be assumed that the adhesion component of the friction force depends on the adsorptivity of macromolecules. The adsorption of polymers is determined by their functional groups and molecular mobility that depend on the temperature. Therefore, immobilization of the functional groups, ensured by introducing into the polymer – or surface layers of the sample – a substance capable of adsorptive (chemisorptive) interaction with the macromolecules, and constraining the molecular (segmental) mobility, must strongly affect the triboengineering properties of the polymer.

Second, in friction of polymers over metals, the tribochemistry in the interaction zone greatly influences the triboengineering properties [165,166]. Tribochemical transformations in macromolecules follow the free-radical mechanism consisting of three main stages: (1) initiation (formation of macroradicals), (2) development of chain processes of radical breakup, and (3) termination of the chain processes. These reactions occur – at friction – in thin surface layers of the material, even if it is in a solid state.

The direction and kinetics of chemical processes in solid polymers depend not only on the chemical structure of macromolecules but also on the physical structure of the material (arrangement of molecules), perfection and size of crystallites (for partially crystalline polymers), the degree of molecular orientation, and molecular mobility. A structural heterogeneity in polymers leads to a nonuniform distribution of additives and reagents in their bulk. In partially crystalline polymers, low-molecular-weight substances (oxygen, products of oxidation, inhibitors, plasticizers, dyes, fillers, etc.) tend to concentrate in amorphous regions of the polymer; most reactive fragments of macromolecules (oxidized groups, branches, unsaturated bonds, etc.) are also localized there. The local concentrations of reagents may much differ from an average concentration; consequently, the local rates of tribochemical reactions must be different from average rates [165]. Thus, the introduction – into a polymer – of additives that alter its physical structure, unavoidably affects the triboengineering properties.

The tribochemical transformations in macromolecules and wear of the polymer can be affected by the metal composition of the counterface. The role of this factor increases with the load and velocity becoming more severe [165].

Effective catalysts of chemical processes in polymers, which follow the radical mechanisms (destruction, crosslinking, and others), are Pt, Pd, Rh, Mo, Ta, Cr, and Ti. No effect is caused by Au and Ag. Quite typical tribochemical processes are hydrogenation (catalyzed by alloys containing Ni, Co, Fe, Cu, Pt, and Re), dehydrogenation (catalyzed by oxides of alkali-earth elements, transition metals, and rare-earth elements; sulfides, tellurides, stibides, arsenides, selenides of Mg, Ca, Zn, Cd, Cr, Ni, Mo, etc.; borides, nitrides, carbides, silicides, phosphides of V, Ti, Cr, Mo, W, etc.; by metals such as Ni, Cu, Ru, Os, Pd, Pt); oxidation (partial oxidation is catalyzed by simple or complex systems based on metal oxides (V–VIII groups of metals; most reactive are oxides with high energy of the metal–oxygen bond, e.g. Mo, Bi, Co, Fe; deep oxidation up to $CO_2$ and $H_2O$ is catalyzed by metals of the platinum group, viz. Pt, Pd, Ph, and also by Ni and Co; simple and complex oxides of metals of VI–VIII groups). The compounds resulting from the interaction of macromolecules and products of tribochemical reactions have various chemical structures. Often, especially at friction in air, these may be compounds of the coordination type. Coordination compounds generated on a surface seem typical enough for metal–polymer systems, as this process is observed in static contact between a polymer and metal [165].

## 5.3.2 Effect of nanofillers on frictional behavior of PMs

The reported data evidence to a strong influence of nanoadditives on the structure and properties of polymers. The variations taking place in them are essential from the tribological point of view. Let us consider the most important features related to the influence of nanoadditives on PM frictional behavior.

One of the most useful results achieved with nanodimensional fillers is increased mechanical properties of PMs. This is especially pronounced with nanoclays and carbon nanomaterials [22,32,100,101]. It is clearly proved at relatively low contents of the nanofiller (usually from a few tenths of a percent to a few percent by weight). One of the main reasons for this is strong adsorptive interaction of macromolecules with quite developed filler surface, and also polymer transition to some particular conditions of "an interphase layer" with constricted molecular mobility. The increased cohesive strength of the polymer binder can result in improved triboengineering properties of the nanocomposites, such as abrasion resistance and resistance to fatigue wear.

Nanoparticle fillers can markedly raise the heat distortion temperature (HDT) of PMs. For example, in the case of Nylon 6, HDT under loading increases from 65 °C – for the pure polymer – up to 152 °C for the nanocomposite containing MMT – 14.7 wt% [167]. An increase in heat resistance is also typical of nonpolar polymers. For PP, for example, HDT rises from 109 up to 152 °C at a clay content of 6 wt% [168]. As a high temperature may develop in the friction contact zone, a rise in HDT must be favorable for triboengineering parameters.

It was learned that clay, added to PMs, enhances their heat resistance, as well as thermal resistance, including the resistance to thermooxidative degradation [32]. Improved barrier features of nanoparticles prevent oxygen from diffusion into the polymer, and escape of the volatile products of thermolysis. It was also noticed that prolonged heating of a nanocomposite can strengthen the barrier quality – owing to baking of the silicate in the surface layer of the sample – and result in a networked structure [169].

The inhibitive action of fullerenes and carbon nanomaterials, with respect to thermal and thermooxidative degradation of PMs, has been studied and is still under intensive study [19,21,170]. Fullerenes are electophilic, being active acceptors of electrons. They tend to accept nucleophilic reagents and also hydrogen, free-radical, and carbenoidal particles [19]. Therefore, fullerenes – present in the surface layer of the rubbed PM – must influence the course of tribochemical transformations and, consequently, the friction parameters.

Similar to fullerenes, nanoparticles of metals and their compounds must greatly influence the mechanism of contact tribochemical transformations [165], the development of transfer layers on the counterface, and wear resistance. The nature of nanoparticles and the dimensional effects are of importance as well.

At strong adsorptive interaction of nanofillers and macromolecules, the molecular mobility becomes abruptly frozen in the PM amorphous phase. The functional groups in macromolecules become *blocked* by interaction with the filler surface [32]. As a result, obstacles are created and prevent from adhesional interaction with the counterface, so that the adhesion component of friction force decreases.

The presence of nanoparticles within a friction contact must unavoidably influence the processes of mass transfer, accumulation of static electricity, heat transfer on individual microsites (nanosites) of contacting surfaces, etc.

Thus, effects of nanofillers on triboengineering properties of PMs can alter the set of both bulk and surface qualities of the polymer.

### 5.3.3 Frictional interaction of metals with different polymer nanocomposites

The objects, dealt with in this review, are characterized by that the polymer binder is thermoplastic, and the nanoparticles – incorporated in a molten polymer – get distributed in a chaotic manner within the polymer matrix, and become encapsulated in the polymer shell. A review of information on friction of certain types of such materials will be discussed in the following sections.

### 5.3.3.1 Polymer–clay nanomaterials

Despite a great volume of research in the field of PCNs, only a few of the studies have considered their tribological behavior [14,171,172]. However, these PMs find wider and wider application in the automobile industry, as well as some other branches of engineering, where serviceability of mechanisms depends on wear resistance and other friction parameters [171,173].

The research indicated that triboengineering properties can be increased by filling thermoplastics with organoclays. The friction behavior of nanocomposites is influenced by a number of factors, which should be considered at designing mix formulations. It is shown [171] that interaction of clay nanoparticles and polymer matrix is determinative for wear resistance of nanocomposites. In that work, Na-MMT was mixed with molten PA6 in the twin-screw extruder. Three ways were used for introducing 5 wt% of Na-MMT: (1) original powder was added to polymer melt, (2) before mixing the polymer, MMT powder had been treated with a surfactant (dioctadecylammonium), and (3) MMT powder together with PA6 was treated with water in the course of extrusion processing. Water, as a reagent, causes plasticization of PA6 and swelling of MMT, easily penetrates into the polymer melt and intercalates clay particles.

When the melt in the material cylinder of the extruder is under shear stresses, the intercalated layers get dispersed within the polymer bulk. After the water has been removed from the nanocomposite, the exfoliated morphology is retained. The dispersion degree of MMT – treated either with the surfactant or water – in the polymer is approximately the same as it had been detected by TEM technique [171]. Untreated MMT added to PA6 does not lead to exfoliation of layered particles; therefore, tactoids and large aggregates of the clay are present in the bulk of the polymer. The mechanical strength of the composites prepared using water-treated MMT did not deteriorate, unlike the case where MMT had been treated with the surfactant [171].

Sliding friction tests were conducted using a sphere of silicon nitride (~6.3 mm in diameter) and a polymer block. It was found that the wear resistance of PCN strongly depended on the way by which MMT had been incorporated. Increased wear resistance (in comparison with PA6) was only observed for nanocomposites containing surfactant-treated MMT; this effect was only achieved at an elevated sliding velocity, 0.2 m s$^{-1}$. The wear rates were as follows: for PA6 = 107 ± 5.3; PA6/MMT = 195.8 ± 14; PA6/MMT + surfactant = 76.3 ± 0.65; and PA6/MMT + H$_2$O = 119 ± 7.3 (10$^{-6}$ mm$^3$ N$^{-1}$ m$^{-1}$) [171]. The lowest wear resistance was measured for PA6/MMT composite in which the clay was in the form of large particles. The latter migrated to the friction zone and caused abrasive wear of the PCN [171].

The higher wear resistance of PA6/MMT + surfactant composite in comparison with PA6/MMT + H$_2$O results from adhesion of the polymer to the surface of clay nanoparticles exfoliated within the bulk of PA6, and not from their morphology [171]. The treatment of MMT with the surfactant, unlike that with water, ensured a higher level of adhesion between the phases, hence, increased wear resistance.

Thus, much improved wear resistance of PCNs can only be achieved by combining two factors, such as fine dispersion of layered mineral platelets in the polymer matrix and a high level of adhesion between the phases. The latter feature depends on the right choice of surfactants for treating the clay.

The above arguments have been supported by the data of testing such nanocomposites as PA6/Nanofil and PA6/Grade F-160 [14], in which Nanofil is bentonite Nanofil SE 3010 treated with a surfactant (supplied by Sud-Chemie, Germany); Grade F-160 is acid-activated powdery bentonite (supplied by Engelhard, USA), and treated with surfactants being products of refinement of vegetable oils [59]. The friction tests have been run using the block-on-ring arrangement. The ring was made of Steel 40Cr (Cr, 0.4%; hardness, HRC, 45–50; surface roughness, $R_a \approx 0.2$ µm; diameter, 40 mm). The polymer block was in the form of segments (width, 10 mm; thickness, 10 mm; length, 20 mm; work surface area, 2 cm$^2$). Nanocomposites had been prepared by the extrusion technique using the static mixer [133].

Figure 5.4 shows that the addition of organoclays to PA6 causes a strong effect on the friction parameters. Of importance are both the concentration and type of the organoclay used. The initial PA6 and nanocomposites containing 0.5 wt% of clay follow the wear mechanism of heat failure owing to a high level of adhesion component of friction, as well as surface warming up of the polymer specimen [14]. It should be mentioned that $f$ of a PA6–clay composite containing 3–7 wt% of clay is 1.5 times lower than that of the initial polymer. This can be explained by a decreased adhesion component of friction [14]. The marked drop in the wear rate of PA6/Grade F-160

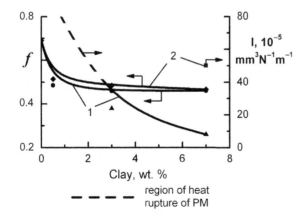

**FIGURE 5.4**

Dependence of the friction coefficient and wear rate on the concentration of organoclay Grade F-160 (1) and Nanofil (2) in PA6/clay nanocomposites: $\upsilon = 0.63$ m s$^{-1}$; $p = 1.0$ MPa.

nanocomposite, in comparison with original PA6/Nanofil, results from the enhanced adhesion of the polymer to clay particles. This fact has been indirectly proved by a higher level of the mechanical properties of PA6/Grade F-160 composite [14].

Addition of layered clayey materials also affects tribochemical processes taking place in the friction contact zone, which, in turn, affects the friction parameters of thermoplastics. It was reported elsewhere [172] that the presence of nanodispersed clay particles (palygorskite) in PI can inhibit the degradation of macromolecules. As a result, wear resistance of the nanocomposite, sliding over metal, is enhanced. At increased concentrations of palygorskite, the triboengineering properties ($f$ and $I$ grow) deteriorated. This was associated with the formation of large clay aggregates and their abrasive action upon the contacting materials [172].

The recent research prompted prospects for the application of PCNs in tribosystems. The available test data, however, are not enough for a detailed understanding, generalization, or working-out practical recommendations for a wide range of PMs.

### 5.3.3.2 Thermoplastic-containing carbon nanofillers

Among the novel allotropic forms of carbon nanomaterials, UDDs are most easily available, and their tribological aspects have been investigated. The nanomaterials investigated contained fullerenes, or their derivatives; the binders used were expensive special polymers (PTFE, PI, ultra-high molecular weight polyethylene (UHMWPE)), or superengineering thermoplastics (poly(etherether ketone), PEEK). As increased life service of such materials leads to technical and financial advantages, we shall discuss the main experimental results, which are of tribological importance.

For partially crystalline thermoplastics, the effectiveness of carbon nanoparticles – in their influence on property values – much depends on the physicochemical changes in the amorphous phase of the polymer. With properly distributed nanoparticles, they can reinforce the amorphous phase, even in such an inert polymer as PTFE. Therefore, the role of nanocarbon materials in triboengineering characteristics of this polymer is significant.

The authors [174] report 100–150 times decreased wear rates for PTFE containing 2 wt% of UDD, which had been prepared by the detonation synthesis. It was emphasized that UDD nanoparticles undergo coordination bonding among themselves – in the PTFE-amorphous phase – and reinforce the latter. As a result, the compressive strength increased by approximately 42% and hardness by 14%.

Besides, reactive oligomers like R–COOH, where R is the fluorine-containing radical, added into PTFE/UDD nanocomposites increased the effective energy of thermooxidative degradation from 59.0 up to 68.4 kJmol$^{-1}$ and decreased the wear rate by approximately 25% [175].

Still greater decrease (>300 times) in wear rates of PTFE sliding over steel can be achieved by the addition of CNTs into the polymer [176]. An optimal CNT concentration is between 15 and 25 wt% (Fig. 5.5). Increased wear resistance and load-bearing capacity can also be reached by introducing CNTs into UHMWPE or PI [177,178]. This is explained by reinforcement of the polymer matrix with CNT. CNTs are 6–7 times as light as steel; their specific strength approximately 100 times as high as that of steel [179], which makes them extremely promising reinforcing fillers for PMs.

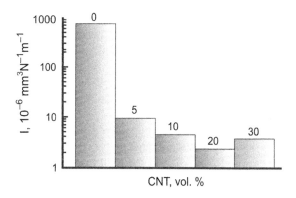

**FIGURE 5.5**

Dependence of wear rate of PTFE on CNT concentration.

The favorable effect of CNT on the friction parameters is explained by enhanced PM resistance to shear stresses at friction, raised hardness of surface layers on the contacting test piece, accumulation of CNT in the film transferred onto the metal counterface, and, as a result, shallower penetration of metal asperities into the polymer specimen [176–178].

It should be noted that – depending on the matrix polymer – addition of CNT to a PM can favor either the decrease (PTFE or PI) [176,178] or increase in the friction coefficient $f$ [121]. The decrease in $f$ is caused by improved strength properties of the polymer along with the lubricating action of CNT [176]. The increased $f$ results from higher losses by wear owing to the stronger resistance of the transfer films to shearing and abrasive behavior of CNT, which break up by friction [177].

Of academic and practical interest is research on triboengineering properties of PTFE modified by fullerene black (FB). It was learned earlier [180] that fullerene $C_{60}$ added to liquid lubricants improves their antifrictional and antiwear qualities at sliding of metals. Besides, FB is much cheaper than original fullerenes are [181,182]. FB can be produced in the electric arc plasma followed by extraction of $C_{60}$- and $C_{70}$-fullerenes in toluene. The share of fullerenes in the remaining FB is usually <1 wt%.

It was reported that 1 wt% of FB was first mechanically mixed with PTFE and then the molten stock was pressed [182]. The test piece was in the form of a rectangular plate of thickness 10 mm. A roller of chromium–nickel–molybdenum steel, 46 mm in diameter, was the counterface. The friction tests were run at a speed $v = 1$ m s$^{-1}$ without lubrication. It was found that FB added to PTFE markedly decreased $f$ over the whole range of contact pressures. Simultaneously, the PM wear rate was reduced several times [182]. The authors' explanation was that macroradicals, generated at PTFE tribological degradation, become chemically grafted to fullerene molecules and give a fullerene–polymer network of higher wear resistance [182]. It is also believed that microcracks – originated at friction in the specimen surface – propagate but do not envelope the fullerene-containing particles, like they do with ordinary inclusions, and become "healed" on the fullerenes owing to high electronegative ability and inhibition of the free radicals.

Substantially lower $f$, as well as $I$ (~25 times), values, were observed after fullerenes (78:22 wt% mix of $C_{60}$ and $C_{70}$) had been added to partially crystalline PI (based on dianhydride 3,3′,4,4′-benzophenontetracarboxylic acid and 3,3′-diaminobenzophenone) [183]. It is presumed that largely enhanced wear resistance results from the raised values of mechanical properties for fullerene-filled PI, as well as retarded thermal degradation of macromolecules, which may act as "traps" for free radicals [183].

It was reported that the addition of carbon nanofibers (CNFs) to such superengineering thermoplastics as PEEK was a promising way of their modification [184]. It was found that 10 wt% of CNF added to PEEK reduced unlubricated wear over alloy steel as much as three times. It was surprising that the addition of 10 wt% of ordinary CF, or PTFE, reduced wear resistance of the composition. Simultaneous adding of all three fillers led to a PM that showed the highest wear resistance. The reported information evidences to the practical advantages of using hybrid filling for improving triboengineering properties of PMs.

In similar composites, CNF and their debris anchor in counterface microirregularities and reduce the roughness, as well as insulate the part of surface occupied by the metal being adhesionally active toward the polymer. It is quite probable that CNF can reduce to nanoparticles of graphite structure that can act as a solid lubricant. Unlike CNF, ordinary carbon fiber can initiate tribochemical transformations in macromolecules and result in higher wear rates.

Thus, the information available prompts a wide range for application of CNF for enhancing the triboengineering characteristics of PMs. Their commercial use will increase with mastering the production technology for CNF, along with their price reduction.

### 5.3.3.3 Metal-containing nanocomposites

Recently, a number of interesting reviews have considered tribology of polymer composites containing, as nanofillers, metals or their derivatives like chalcogenides, oxides, and some other compounds of metals [27,185–187]. Therefore, we shall only attempt to systematize the information available with reference to polymer matrices and technologies being dealt within this chapter. The main specificities of our objects are the thermoplastic character of the matrix polymer and the fact that individual properties of nanoparticles within a composite remain unchanged owing to macromolecular screens present on each of them. These screens are created by preferable covalent bonding of polymer chains and nanoparticles [24,128,130]. Obviously, many of the physicochemical properties of such systems may depend on the polymer–particle bond nature, as well as on the nature of the contacting materials. It is, therefore, useful to discuss triboengineering characteristics of metal-containing nanocomposites based on nonpolar thermoplastics like PTFE and POs, conventional engineering thermoplastics like PA, poly (oxymethylene), PBT, PET, polar thermoplastics with comparatively low $T_g$, as well as thermoplastics with high $T_g$ (>150 °C), among them superengineering plastics like PEEK, PEI, poly(phenylene sulfide) (PPS), and PI. We shall mostly deal with the latest data covered in the reviews mentioned above, along with the results of original studies.

PTFE has a low coefficient of friction and is insufficiently wear resistant since its hardness is relatively low. Introduction of metal oxide ($ZrO_2$, $TiO_2$, $Al_2O_3$, or ZnO) nanoparticles into PTFE leads to enhanced wear resistance with simultaneously rising $f$ [185]. The major explanation is a greater toughness and creep resistance of PTFE provided by incorporated hard particles. It should be noted that a maximum rise in PTFE-wear resistance can be achieved at relatively high oxide concentrations (up to 20 wt%). Obviously, at these concentrations, it is difficult to prevent nanoparticles from agglomeration, which would negatively affect the PM-strength values.

Somewhat different effects had been observed when nanocomposites had been prepared using PTFE particles of a nanodimensional size [130]. An increased reactivity of the chemically inert PTFE was observed; the effect of PTFE on properties of the other components in the nanocomposite became more pronounced. A significant drop (by more than an order of magnitude) in the wear rate of the nanocomposite over steel occurred at 2 wt% of nano-$Al_2O_3$, $Cr_2O_3$, $ZrO_2$ in PTFE. Still more marked drop in wear rates occurred after 2–5 wt% of nanodimensional particles of layered oxides – magnesium or cobalt spinel ($MgAl_2O_4$ or $CoAl_2O_4$) produced by mild mechanochemical synthesis – had been added to PTFE [187]. The values of $f$ became 3.5–4.8 times as high as that of unfilled PTFE. The effects observed are explained by the reinforcement and increased hardness of PTFE containing nanofillers. The use of nanodispersed particles, instead of coarse-dispersed powdery PTFE for making nanocomposites, seems to ensure stronger bonding between the filler and matrix, and largely reduces the filler concentration that is required to ensure high wear resistance of the composite.

The distribution of nanoparticles in the bulk of the polymer matrix can be improved, and the interphase adhesion strengthened, by special modification, for example, grafting to nanoparticle surface of functional monomers, oligomers, silanes, etc. Such treatment is particularly important when polymers of low adhesiveness are used (PTFE or PO). The reports dealing with these problems had been reviewed earlier [186].

Other authors [185,186] have summarized information that corroborates the favorable role of metal-containing compounds, such as $TiO_2$, ZrO, CuO, CuS, and $CuF_2$, in enhancing wear resistance of aliphatic polyamides. Nano-CuO, CuS, $CaF_2$, $SnS_2$, and PbTe appeared to effectively reduce the PA-11 wear rate, while nano-$ZnF_2$, ZnS, PbS, on the contrary, increase it. This is because nanoparticles of metal compounds differently influence the creation of transfer films, as well as the interaction of the latter with the metal counterface and polymer specimen surfaces [185].

Several studies have corroborated the advantages of nanofillers used in combination with conventional fillers. For example, Chang et al. [29] had studied the triboengineering properties of high-polar PA66 filled with 5 vol. % of nano-$TiO_2$, 5 vol. % of graphite, and 15 vol. % of CF. The wear tests were conducted using the arrangement: PA66 pin over a disc of polished steel (initial roughness ~220 nm). The normal pressure was varied between 1 and 8 MPa; the sliding velocity, between 1 and 3 m s$^{-1}$. It was found that $TiO_2$ nanofillers added to PA66 could reduce the wear rate by approximately 10 times in comparison with the original PA66. Addition of

$TiO_2$ reduced $f$, contact temperature and wear rate especially at high $pv$, at which the contact temperature exceeded $T_g$ of PA66 ($\approx 70$ °C).

Chang et al. [29] claimed that the favorable effect of $TiO_2$-spherical nanoparticles on friction parameters came from their rolling within the transferred film. Obviously, for this polar polymer (PA66), the said effect can only be realized at a contact temperature that exceeds $T_g$ at which the amorphous phase of the polymer changes to a high elastic state, and the adhesional interaction with nanoparticles weakens.

The positive role of hybrid filling of PA with nanoparticles of metal-containing compounds, and that of conventional fillers (PTFE, CF, glass fiber (GF) or graphite), has been emphasized repeatedly [136,137].

The effect of Cu nanoparticles (Cu was obtained by the decomposition of cupric formate in molten polymer) on $f$ and $I$ of PA6 had been evaluated elsewhere [14]. The addition of 1.5 wt% of Cu-nano to PA6 eliminated heat failure of the test pieces. Figure 5.6 shows that 1.5–5 wt% of Cu did not in fact change $f$, while the wear rate dropped more than three times. This effect is believed to result from increased heat conduction of the material and heat removal from the microsites where the local temperatures are high. A certain contribution is probably made by inhibition of thermooxidative degradation of PA macromolecules in the friction contact zone by Cu nanoparticles.

Numerous studies reported advantages of using nanoparticles of metal-containing compounds for enhancing triboengineering properties of thermostable polymers [185,186,188–192].

The wear rate of PEEK was observed to reduce by introducing nano-CuO, CuS, $TiO_2$, $ZrO_2$, $Al_2O_3$, ZnO, or $CaCO_3$. The optimal concentration of the nanofiller was within 1–4 vol. %. It is worth mentioning that $f$ dropped from 0.4 – for the original PEEK – up to approximately 0.2 for that containing an optimal concentration of the nanofiller, while the wear resistance increased. The wear mechanism, at that,

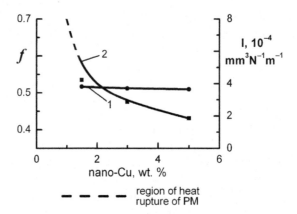

FIGURE 5.6

Coefficient of friction (1) and wear rate (2) for PA6/steel pair depending on nano-Cu concentration: $p = 1.0$ MPa; $v = 0.63$ m/s.

changes from adhesion to fatigue (unfilled PEEK) to mild abrasion (nanocomposite). In some cases, only wear rate dropped, while friction coefficient did not vary with nano-$Al_2O_3$, 5 wt% [189]. For a specimen of PEEK filled with $Al_2O_3$ (particle size ~15 nm) – when it moved over a steel counterface – a thin uniform film was formed. It was strongly adhered to the metal, preventing the polymer specimen from wearing. Larger $Al_2O_3$ particles (up to 90 or up to 500 nm) led to lower wear resistance. Addition of 10 wt% of PTFE to PEEK/nano-$Al_2O_3$ composite results in a lower $f$ and higher $I$ indicating a lack of synergism from joint introduction of a conventional filler and a nanofiller into PEEK.

Like the case with PA, filling PEEK with nanospheres (e.g. ZrO) improves the triboengineering properties owing to nanosphere rolling in the friction zone [186].

When added to PPS, nanoparticles can either improve (CuO, $TiO_2$) or lower (ZnO) the triboengineering properties of the PPS [190]. An optimal concentration of nano-CuO or $TiO_2$ is 2 vol. %. Both $f$ and $I$ were reduced. This observation can be explained by the effect of nanofillers on the transfer film formation. In the presence of CuO or $TiO_2$, the transferred film is thin, uniform, and adhered strongly to the metal. A further improvement in wear resistance (~ 4 times as much) of PPS/CuO (2 vol. %) could be achieved by introducing 15 vol. % of aramide fiber Kevlar [141]. The lowest $f$ (0.2–0.23) was shown by PPS/CuO–2 vol. %/CF–15 vol. %. For the other materials, $f \approx 0.45$ [191]. Thus, by testing PPS, the favorable effect of hybrid fillers on the triboengineering parameters of PMs was confirmed. Similar effects have been observed at the joint addition of nano-$TiO_2$ (5 vol.%), CF (15 vol.%) and graphite (5 vol.%) to PEI [188].

This brief review of works on the tribology of metal-containing nanocomposites shows numerous possibilities of nanofillers in affecting the friction parameters. The triboengineering properties appear rather sensitive to the nature and size of nanoparticles, their concentration, the type of matrix polymer, as well as the operation regimes for friction pairs. In several instances, it seems advantageous to combine conventional fillers and nanofillers in a PM. It was found that nanoparticles of metals, or their compounds, strongly affect the formation of transfer films, their morphology, stability, interaction with the counterface and PM. It is quite clear, however, that more experimental data are required to optimize the formulas for metal-containing PMs, as well as for predicting their serviceability in friction systems.

### 5.3.3.4 Polymer blends

In modern engineering, polymer blends find more and more applications. By blending unlike polymers, it seems possible to design a wide range of competitive and promising PMs, without synthesizing novel macromolecules. In future, the industry of polymer blends will be developing faster in comparison with other types of PMs [193]. It is not incidental that some of the blended PMs become the subject matter of tribological investigation [14,194–201]. Like the case with homopolymers, nanomodifiers introduced into blended PMs have an effect on PMs frictional properties. The published works confirm this expectation.

It was reported that the PA6/high-density polyethylene (HDPE) blend was filled with nano-CuO, PTFE, or short GF [202]. It was found that PM containing any of these components showed an optimal wear resistance.

It was found earlier [14] that nano-Cu added to PA6/HDPE – 15 wt%/C–5 wt% led to increased wear resistance at sliding over steel (Fig. 5.7). The compatibilizer was HDPE-g-IA (~1 wt%); the grafting process had been described elsewhere [133]. Nano-Cu had been prepared by the thermolysis of copper formate in PA melt using the single-screw reactor-extruder equipped with the static mixer [39].

At Cu 1.5 wt% concentration, the wear rate of the blended material dropped as much as three times; the value of $f$ did not actually vary (Fig. 5.7). These facts have been explained by the active interaction of Cu nanoparticles with macromolecules of PA6/HDPE blended systems, which led to the generation of a dense and strong transfer film. Of advantage may be the stabilizing behavior of nano-Cu toward the thermo-oxidative degradation of macromolecules [14], and also somewhat greater values of the mechanical properties of the blend after the nanofiller had been added (Table 5.2).

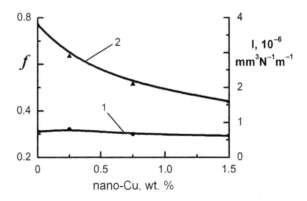

**FIGURE 5.7**

Dependence of the friction coefficient (1) and wear rate (2) of PA6/HDPE – 15 wt. %/C – 5 wt.% on nano-Cu-concentration: $p = 1.0$ MPa; $\upsilon = 0.63$ m s$^{-1}$. Counterface was steel (~Cr 1.0%); HRC = 48.

**Table 5.2** Mechanical Properties of PA6/HDPE—15 wt%/C—5 wt% Blends Modified with Nano-Cu

| Cu-concentration, wt% | Property | | | | |
|---|---|---|---|---|---|
| | $\sigma_y$, MPa | $\varepsilon_n$, % | $\sigma_b$, МПа | $E_b$, GPa | $a$, кJ m$^{-2}$ |
| 0 | 54 | 200 | 78 | 1,64 | 19 |
| 0,25 | 55 | 100 | 89 | 1,76 | 23 |
| 0,75 | 56 | 55 | 90 | 1,78 | 27 |
| 1,5 | 55 | 15 | 88 | 1,72 | 26 |

The effect of layered clayey nanofillers on the triboengineering properties of compatibilized blends – PA6/HDPE/C – had been estimated earlier [14]. It appeared that nanoclay, Grade F-160, was detrimental for the triboengineering parameters of the blends. Addition of 1–3 wt% of the clay caused 1.3–1.4 times increase in friction coefficient and approximately 1.5 times increase in the wear rate. However, after chopped GF had been added to the blended composition (PA6/HDPE/C + GF), the friction parameters were much improved [14]. The composition containing, additionally, 20 wt% GF and 3 wt% clay Grade F-160, in comparison with the base blend material (PA6/HDPE – 15 wt%/C – 5 wt%), gave $f = 0.21$; $I = 0.74 \times 10^{-6}$ g m$^{-1}$ against 0.31 and $0.8 \times 10^{-6}$ g m$^{-1}$ for the original blend [14].

Thus, like the situation with homopolymers, maximum improvements in the triboengineering properties of thermoplastic blends can be achieved, in our opinion, by combining conventional fillers and nanodimensional fillers.

As there are few reports available on the friction of polymer nanoblends, the necessity is obvious to intensify research in the field of tribology of this advantageous and promising type of PMs.

Of interest are blended materials in which both the nanofiller and the polymer, which form the dispersed phase, have nanodimensional parameters [25,26,203].

### 5.3.3.5 Elastomeric materials

Preferably organoclays and carbon nanomaterials, among them UDDs and fullerenes, have been used to modify elastomers covering rubbers, thermoplastic elastomers (TEEs), and dynamic vulcanizates that possess the qualities of TEEs [31,32,100,101,204–206].

For rubbers, fullerene materials seem promising as modifiers, because small quantifies of them cause considerable variations in the triboengineering parameters, as well as in values of the mechanical properties. It should be noted that depending on the purpose of elastomer modification, it is sometimes required to reduce or increase friction coefficient, while high wear resistance must be ensured. Tire rubbers, for example – also natural rubbers – must have high friction coefficient to provide for adequate grip of a tire with the road. The elastomers intended for operation in movable seals, on the contrary, must suffer minimum loss at friction.

It had been found [100] that small quantities (0.065–0.75 wt%) of C$_{60}$ fullerene added to the natural rubber mixture (100 wt%) containing commercial carbon, 5 wt%, caused great variations in the friction coefficient and dynamic–mechanical properties of the rubber (Figs 5.8 and 5.9). The coefficient of friction over steel was measured using the pendulum tribometer at 0.2 and 0.5 N; the temperature dependences of dynamic–mechanical properties (tg $\delta$ and G$'$) were measured using the reverse torsion pendulum [100].

Figure 5.8 shows that the $f$ of the rubbers depends on both contact load, $p$, and fullerene concentration. Independently of $\tau$ and $p$, the $f$-values are the highest at the maximum concentration of fullerene (0.75 wt%) in the rubber mixture.

A decrease in tg $\delta$ with higher fullerene concentrations over a wide temperature interval (from −20 to 230 °C) was observed (Fig. 5.9(a)). The fullerene-containing

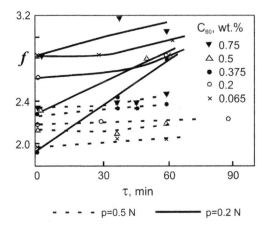

**FIGURE 5.8**

Kinetic dependences of friction coefficient for natural rubber-based materials on $C_{60}$-fullerene concentration in the rubber mixture.

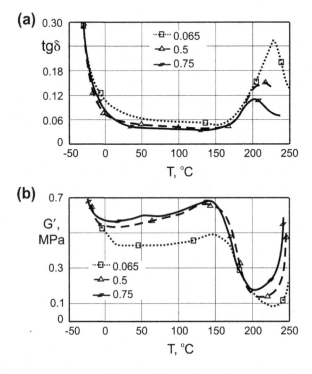

**FIGURE 5.9**

Temperature dependences of the mechanical loss tangent (a) and dynamic shear modulus (b) for NR-based materials containing different quantities of $C_{60}$-fullerene.

rubbers, therefore, appeared less susceptible to warming-up at dynamic loading in the course of service [100]. The patterns of temperature dependence of dynamic shear modulus ($G'$) are identical for rubbers containing different quantities of fullerene (Fig. 5.9(b)): advance to the plateau of high elasticity at $-15$ °C; raised $G'$-values owing to rubber vulcanization at 100–150 °C, and reduced modulus values, because physical nodes of the primary vulcanization network fail at 150–220 °C; raised $G'$ at further heating owing to thermooxidative crosslinking of rubber macromolecules [100]. The comparison of $G'$-values for the rubbers, containing different amounts of fullerene indicated a markedly enhanced effect of the nanofiller in the region of high elastic state of the rubber. In the region of glassy state of the rubber ($T_g$(NR)≈ $-54$ °C), however, the dynamic–mechanical properties did, not in fact, alter in the presence of fullerene [100]. Consequently, fullerene intensifies the interaction of macromolecules with the nanofiller, which imposes some constraints on the segmental mobility in the high elastic state. Besides, fullerene can affect the vulcanization process of the rubber, as well as the density of nodes in the vulcanization network [100]. It is believed that fullerene can catalyze vulcanization, favor creation of more physical nodes for anchoring in the vulcanization network, and reduce the degradation extent of the networked structure, when the rubber undergoes heating >150 °C [100].

Some studies [101,204] considered the effect of UDD on the fluoroplastic friction parameters. The tests were conducted using the disk-on-disk arrangement; the counterface of chilled cast iron – surface roughness $R_a = 0.16$–0.2 – was fixed. It was found [101] that approximately 20 wt% of UDD added to fluoro-co-polymer SKF-32 resulted in a PM with $f = 0.02$, which is lower than $f$ of PTFE (0.04) and also of standardized materials, based on SKF-32, that contained 20 wt% of graphite ($f = 0.58$), or the mixture of graphite and PTFE ($f = 0.53$). It was observed for different types of fluoro-co-polymers that 10–15 wt% of UDD – introduced in the vulcanizates on their base – increased the resistance to abrasive surfaces approximately twice as much. An extremal pattern (with a minimum at ~10 wt% UDD) of the concentration dependence of abrasive wear was established [101].

In view of the above, further research in the tribological field of elastomeric materials seems worth doing. Of particular interest are polyester TEEs [205,206] that have a domain with the nanodimensional hard blocks. The range of applications for such materials, among them dynamic vulcanizates [205] and TEEs of other types, in triboengineering is quite wide.

The information on how the SPE influences the formation and properties of different nanocomposites, presented in Section 5.2.4, was taken into consideration when the estimation was made of variations in the tribological parameters, and also of important characteristics, depending on the compounding conditions. The experiments on making the materials – the results of which are presented in Table 5.3 – were run using the twin-screw extruder TSSK 35/40 (China; screw diameter 35; the ratio $L/D = 40$; 10 independent zones for heating the material cylinder). For compounding in the molten conditions, the material cylinder temperature in the zones (IV–VIII zones of the material cylinder) exceeded by approximately 20 °C the temperature of polymer melting, whereas under SPE conditions (zones IV–VIII), it was

**Table 5.3** Effect of Composition and Compounding Conditions of Nano-PCMs on their Rheological, Mechanical and Trioengineering Properties

| Composition, components in wt% | Mode of compounding | MFI (g 10min⁻¹) | $\eta$ (Pa s) | $\sigma_T$ (MPa) | $\varepsilon$ (%) | $I_{abr.} \times 10^3$ (g m⁻¹) | $\Delta I \times 10^6$ (g m⁻¹) |
|---|---|---|---|---|---|---|---|
| (I) PA 6/MMT—3%/ Rh-1098—0.2% | In melt | 26.1 | 384.2 | 75 | 76 | 2.8 | 1.2 |
| | SPE | 12.8 | 781.4 | 76 | 107 | 2.3 | 0.9 |
| (II) PA 6/MMT(ODA)—3%/ Rh-1098—0.2% | In melt | 30.4 | 352.3 | 86 | 53 | 4.0 | 1.7 |
| | SPE | 18.0 | 596.6 | 82 | 68 | 3.7 | 1.9 |
| (III) PA 6/CNT—0.5%/ Rh-1098—0.2% | In melt | 15.0 | 678.6 | 74 | 90 | 2.5 | 0.5 |
| | SPE | 14.8 | 688.2 | 80 | 40 | 2.6 | 0.6 |
| (IV) TEE/Graphite—13%/ Rh-1010—0.3% | In melt | 22.0 | 515.2 | 15 | 490 | 2.2 | 48.7 |
| | SPE | 18.0 | 629.8 | 16 | 361 | 3.5 | 51.5 |
| (V) TEE/Graphite— 13%/L-24—2%/ Rh-1010—0.3% | In melt | 25.7 | 440.4 | 15 | 463 | 2.7 | 3.4 |
| | SPE | 21.8 | 519.3 | 15 | 474 | 2.2 | 4.0 |

*Notes: MFI was determined at P = 21.6 N, T = 250 °C (Compositions I–IV), and T = 230 °C (Composition V); test conditions for finding I: wear pattern—shaft/sector; counterface—steel 40 Cr (HRC = 48); p = 2 MPa, and u = 0.1 m s⁻¹; I_{abr}: wear pattern—shaft/pin; counterface—abrasive paper; u = 0.35 m s⁻¹.*

lower by 70–80 °C. The base polymers were PA 6 (melting temperature 220 °C) and a polyester TEE – being a polyblock copolymer of PBT and poly(tetramethylene oxide) (PTMO) [205,206] (the molecular weight of PTMO blocks being 1000; their concentration, 25 wt%, melting temperature, 210 °C).

PA6 was used to prepare PA6–clay and PA6–CNT composites. The clay used was $N^+$-MMT with a CEC of 95 mg-eq/100 g; it was used in two versions: (1) original substance and (2) that modified by octadecylamine following the procedure described elsewhere [23,40]. The CNTs were multiwall nanotubes of a diameter of approximately 30 nm and a length of 1–3 μm [99].

The TEE was used to compound polymer–graphite composites; one version of the material contained only graphite (Grade GLS-3, GOST17022), 13 wt%; another version contained graphite and 2 wt% of lubricant grease (Litol-24, GOST 21,150).

The polyamide composites were stabilized by $N,N'$-hexane-1,6-diyl-bis[3-(3,5-di-tert-butyl)-4-hydroxyphenylpropyonamide] (I-1098). The TEE-based materials were stabilized by tetra-cis(3-(3,5-di-tert-butyl-4-hydroxyphenyl)propyonate) (I-1010); the two stabilizers were supplied by Ciba Co., Switzerland.

It can be seen in Table 5.3 that the thermal conditions of compounding significantly influence the set of material characteristics. It is worth mentioning that lower values of melt flow index (MFI; higher melt viscosity) are typical of the composites produced under the SPE conditions, irrespective of their formulas. This fact is a consequence of less severe macromolecular degradation at the compounding stage. Especially noticeable variations in the melt viscosity versus the compounding conditions have been observed for PA6–clay composites (Table 5.3, compositions I and II): the change from the classical theory of compounding in melt for the SPE regime lead to, approximately, a twofold increase in viscosity (a decrease in MFI). The PA6–clay composites produced by the SPE method show a much higher deformability (a high value of $\varepsilon_r$), high abrasive resistance and wear resistance when sliding over steel.

The PA6–CNT material (Table 5.3, composition III), prepared under SPE conditions, is characterized by enhanced mechanical strength, but its wear resistance – under the test conditions – would not change with variations in the compounding conditions.

The results (Table 5.3) are indicative of prospects for using the SPE technology to make high-performance triboengineering nanocomposites. But the mechanism of influence of SPE on tribological properties of the polymer composites needs more detailed investigation.

## 5.4 **APPLICATION OF NANOCOMPOSITES IN FRICTION UNITS**

The number of fields for the application of polymer nanocomposites constantly increases. This has been much favored by the progress made in the technologies of nanodispersed materials. First of all, this concerns UDD synthesis; UDDs are usually obtained by detonation decomposition of powerful explosives [207]. A number of industrial companies produce commercial organoclays [32]. Mechanochemistry, plasma synthesis, and other methods have been employed for making nitrides, as well as oxides, of Al, Cr, Zr , Ti, etc.; solid solutions based on metal nitrides and oxides; complex oxides (spinels) and other metal-containing compounds [174,208]. Wide prospects are being opened for working out industrial technologies for making nanomaterials on modern twin-screw extrusion-granulating equipment. Such equipment provides for controlled regimes for heating and shearing polymer melts, which is important for synthesis of polymer nanomaterials. No capital expenses are required for industrial implementation of these technologies, because in many instances conventional equipment may be suitable.

Some authors [130,174] reported results on the application of nanocomposites based on PTFE and metal-containing compounds. Such materials are used as thin coatings on friction surfaces, or as readymade components for tribological systems. They have also been used successfully as seals in hydraulic and pneumatic systems of quarry excavators and autodumpers. The service life of seals made from the nanomaterials, based on PTFE, is 1.5–3 times as long as that of conventional composites.

Materials based on PTFE and hybrid nanofillers, which had been subjected to mechanical activation, or treated with functional fluorine oligomers have served successfully in sliding bearings in raw diamonds sizers (Russian Diamonds Co., Saha, Russia). They appeared helpful in twofold raising the equipment service life [174], ensured high wear resistance at elevated (up to 340 °C) temperatures.

The introduction of metal oxides, nitrides, and oxynitrides into PTFE can result in tough and hard materials suitable for application as matrices for abrasive tools intended for grinding and polishing precious and semiprecious stones, steels and alloys [174,208]. This enhances the tool's service life threefold to fourfold, as well as ensures a high quality of treatment without lubricating-cutting liquids.

The results of an earlier work [14] were used to design nanomaterials based on PA6 and containing a hybrid nanofiller (organoclay and GF), and also some quantities of HDPE, and a compatibilizer [59]. These materials have been used to make numerous parts for friction pairs in modern automobiles and tractors manufactured in Commonwealth of Independent States countries (Fig. 5.10). The service tests of these materials showed the wear resistance to be like that of polyacetal materials (e.g. Delrin 500P, DuPont, USA) commonly used for the same purposes, but the novel materials are much cheaper.

## 5.5 CONCLUSIONS

The preparation of polymer nanocomposites by nanofiller dispersion in molten thermoplastics is one of the most advanced and fast-developing methods. Compounding of nanodispersed particles with high-viscous melts of rubbers at vigorous shearing can give vulcanizates with a three-dimensional network structure. Compounding in melt, however, seems more promising for extrusion technologies of making nanocomposites. Twin-screw extrusion mixer reactors are quite suitable for these purposes.

1, 5 - Shells for ball joints;

2 - Bush for steering stub of front axle;

3 - Washer for differential self-blocking;

4 - Rod cap for steering wheel

**FIGURE 5.10**

Machine parts for Belarus tractors (1–3) and VAZ cars (4, 5) made from PA 66-based nanocomposites.

The theoretical substantiation of the extrusion technology of producing various thermoplastic nanocomposites prompts an easy change to fast production of triboengineering nanocomposites, as it does not require great expense.

The scientific substantiation of the solid-phase extrusion technology appears promising because it allows one to simplify considerably the preparation of nanoscale fillers such as layered silicates, graphite, or CNTs.

The nanocomposites produced by the above process contain nanoparticles stabilized within the polymer matrix by a screen of macromolecules. The properties of macromolecules, which surround nanoparticles, much differ from their properties in the bulk of the polymer, because macromolecules are affected by the vectorial field of the nanoparticle surface, along with adsorptive interaction. The variations in triboengineering properties of PMs studied are explained by the special structure of the polymer within interphase layers, as well as by physicochemical reactivity of nanoparticles enveloped in a polymer shell. Most of the studies have confirmed the advantages of combining conventional fillers, among them being nucleators of polymer crystallization and nanofillers in a PM with the aim of obtaining optimal results in reducing losses on friction and wear.

# NOMENCLATURE

**CNF** carbon nanofiber
**CNT** carbon nanotube
**EP** engineering plastic
**FB** fullerene black
**GF** glass fiber

**HDPE**  high-density polyethylene
**MMT**  montmorillonite
**MFI**  melt flow index
$\eta$  melt viscosity
**NR**  natural rubber
**PA**  polyamide
**PA6**  polyamide 6
**PBT**  poly(butylene terephthalate)
**PC**  polycarbonate
**PE**  polyethylene
**PEEK**  poly(etherether ketone)
**PEI**  poly(ether imide)
**PET**  poly(ethylene terephthalate)
**PI**  polyimide
**PM**  polymer material
**PMMA**  poly(methyl methacrylate)
**PO**  polyolefin
**POM**  poly(oxymethylene)
**PP**  polypropylene
**PPS**  poly(phenylene sulfide)
**PS**  polystyrene
**PTFCE**  poly(trifluorochloroethylene)
**PTMO**  poly(tetramethyl oxide)
**PTFE**  poly(tetrafluoroethylene)
**TEE**  polyester thermoplastic elastomer
**PVC**  poly(vinyl chloride)
**SPE**  solid-phase extrusion
**UDD**  ultradispersed diamond
**UHMWPE**  ultra-high molecular weight polyethylene
**WAXD**  wide-angle X-ray diffraction
**XRD**  small-angle X-ray diffraction
**MAH**  maleic anhydride
**TEM**  transmission electron microscopy
$a$  Charpy notched impact strength
$f$  coefficient of friction
$G'$  dynamic shear modulus
$I$  wear rate
$I_{abr}$  abrasive wear against abrasive paper
$I_m$  mass wear rate
$p$  contact load
$\upsilon$  sliding velocity
$T$  temperature
$T_g$  glass-transition temperature
**tg$\delta$**  mechanical loss tangent

**THD** temperature of heat deformation
$\varepsilon_r$ relative elongation at rupture
$\sigma_y$ yield point in elongation
$\sigma_b$, **and** $E_b$ bending strength and elastic modulus
$\tau$ shear stress

# References

[1] G.W. Stachowiak, A.W. Batchelor, Engineering Tribology, second ed., Butterworth-Heinemann, Oxford, 2001.

[2] N.K. Myshkin, M.I. Petrokovets, A.V. Kovalev, Tribology of polymers: adhesion, friction, wear and mass-transfer, Tribology International 38 (2005) 910–921.

[3] V.E. Starzhynsky, A.M. Farberov, S.S. Pesetskii, S.A. Osipenko, V.A. Braginsky, Precision Plastics Parts and Their Production Technology, Nauka i Tekhnika, Minsk, 1992 (in Russian).

[4] T.A. Osswald, L.S. Turng, P.J. Gramman, Injection Molding Handbook, Hanser Publishers, Munich, 2002.

[5] K. Friedrich, Wear of reinforced polymers by different abrasive counterparts, in: K. Friedrich (Ed.), Friction and Wear Polymer Composites, Elsevier, Amsterdam, 1986, pp. 233–287.

[6] J. Bijwe, J.J. Rajesh, A. Jeyakumar, A. Ghosh, V.S. Tewari, Influence of solid lubricants and fibre reinforcement on wear behavior of polyethersulphone, Tribology International 33 (2000) 697–706.

[7] S.N. Kukureka, C.J. Hooke, M. Rao, P. Liao, Y.K. Chen, The effect of fibre reinforcement on the friction and wear of polyamide 66 under own rolling-sliding contact, Tribology International 32 (1999) 107–116.

[8] S. Bahadur, The development of transfer layers and their role in polymer tribology, Wear 245 (2000) 92–99.

[9] S. Bahadur, V.K. Polineni, Tribological studies of glass fiber reinforced polyamide composites filled with CuO and PTFE, Wear 200 (1996) 95–104.

[10] A.A. Cenna, P. Dastoor, A. Beehag, N.V. Page, Effect of graphite particle addition upon the abrasive wear of polymer surfaces, Journal of Material Science 36 (2001) 891–900.

[11] D.R. Paul, S. Newman, Polymer Blends, Academic Press, New York, 1978.

[12] L.A. Utracki, Commercial Polymer Blends, Hapman and Hall, London, 1998.

[13] O. Olabisi, L.M. Roberson, M.T. Shaw, Polymer–polymer Miscibility, Academic Press, New York, 1979.

[14] S. P. Bogdanovich, Triboengineering Composite Materials Based on Compatibilized Blends of Aliphatic Polyamides and Polyolefins. Candidate of Scientific (Technical) Dissertation, Gomel, 2006 (in Russian).

[15] H.S. Nalwa, Encyclopedia of Nanoscience and Nanotechnology, American Scientific Publishers, California, 2004.

[16] G.B. Sergeev, Nanochemistry, Moscow State University Press, Moscow, 2003.

[17] P. Alivisatos, M.C. Roco, R.S. Williams, Nanotechnology Research Directions: Vision for Nanotechnology in the Next Decade, Springer, Berlin, 2000.

[18] Y.-W. Mai, Zh.-Zh. Yu, Polymer Nanocomposites, Woodhead Publishing Limited, Cambridge, England, 2006.

[19] V.I. Sokolov, I.V. Stankevich, The fullerenes – new allotropic forms of carbon: molecular and electronic structure, and chemical properties, Russian Chemical Reviews 62 (5) (1993) 419–435.

[20] B.B. Troitskii, L.S. Troitskaya, A.S. Yakhnov, A.A. Dmitriev, L.I. Anikina, M.A. Novikova, V.N. Denisova, International Journal of Polymeric Materials 46 (1–2) (2000) 301.

[21] B.M. Ginzburg, L.A. Shibaev, O.F. Kireenko, A.A. Shepelevskii, E. Yu. Melenevskaya, V.L. Ugolkov, Thermal degradation of fullerene-containing polymer systems and formation of tribopolymer films, Polymer Science A47 (2) (2005) 160–174.

[22] S.S. Pesetskii, S.A. Zhdanok, I.F. Buyakov, S.P. Bogdanovich, A.P. Solntsev, A.V. Krauklis, Structure and properties of polyamide 6 modified in melt with carbon nanomaterials, Doklady Natsionalnoi Academii Nauk Belarusi (Reports Natl. Acad. Sci. Belarus) 48 (6) (2004) 102–107. (in Russian).

[23] S.S. Ray, M. Okamoto, Polymer/layered silicate nanocomposites: a review from preparation to processing, Progress in Polymer Science 28 (11) (2003) 1539–1641.

[24] A.D. Pomogailo, A.S. Rozenberg, I.E. Ufliand, Metal Nanoparticles, in Polymers, Khimia, Moscow, 2000 (Russian translation).

[25] G.-H. Hu, H. Cartier, C. Plummer, Reactive extrusion: toward nanoblends, Macromolecules 32 (1999) 4713–4718.

[26] G.-H. Hu, L.-F. Feng, Extruder processing for nanoblends and nanocomposites, Macromolecular Symposium 195 (2003) 303–308.

[27] Z. Zhang, K. Friedrich, Tribological characteristics of micro-and nanoparticle filled polymer composites, in: K. Friedrich, S. Fakirov, Z. Zhang (Eds.), Polymer Composites – from Nano-to Macroscale, Springer, Berlin, 2005, pp. 169–185. (Chapter 10).

[28] J. Karger-Kocsis, Z. Zhang, Structure–property relationships in nanoparticles/ semicrystalline thermoplast composites, in: J.F. Balta Calleja, G. Michler (Eds.), Mechanical Properties of Polymers Based on Nanostructure and Morphology, CRC Press, New York, 2005, pp. 547–596.

[29] L. Chang, Z. Zhang, H. Zhang, A.K. Schlarb, On the sliding wear of nanoparticle filled polyamide 66 composites, Composites Science and Technology 66 (2006) 3188–3198.

[30] Yu. S. Lipatov, Interphase Phenomena in Polymers, Navukova dumka, Kiev, 1980.

[31] P.C. Le Baron, Z. Wang, T.J. Pinnavaia, Polymer-layered silicate nanocomposites: an overview, Applied Clay Science 15 (1999) 11–29.

[32] S. Pavlidou, C. Papaspyrides, A review on polymer–layered silicate nanocomposites, Progress in Polymer Science 33 (12) (2008) 1119–1198.

[33] R.A. Vaia, K.D. Jandt, E.J. Kramer, E.P. Giannelis, Kinetics of polymer melt intercalation, Macromolecules 28 (24) (1995) 8080–8085.

[34] R.A. Vaia, E.P. Giannelis, Lattice model of polymer melt intercalation in organically-modified layered silicates, Macromolecules 30 (25) (1997) 7990–7999.

[35] R.A. Vaia, E.P. Giannelis, Polymer melts intercalation in organically-modified layered silicates: model preparations and experiment, Macromolecules 30 (1997) 8000–8009.

[36] S.W. Brindly, G. Brown, Crystal Structure of Clay Minerals and Their X-ray Diffraction, Mineralogical Society, London, 1980.

[37] F. Bergaya, G. Lagaly, Surface modification of clay minerals, Applied Clay Science 19 (2001) 1–3.

[38] R.S. Sinha, K. Okamoto, M. Okamoto, Structure–property relationship in biodegradable poly(butylene succinate)/layered silicate nanocomposites, Macromolecules 36 (7) (2003) 2355–2367.

[39] S.S. Pesetskii, B. Jurkowski, Yu. M. Krivoguz, R. Urbanowicz, Itaconic acid grafting on LDPE blended in molten state, Journal of Applied Polymer Science 65 (1997) 1493–1502.

[40] S.S. Pesetskii, B. Jurkowski, Yu. M. Krivoguz, S.S. Ivanchev, S.P. Bogdanovich, On the Structure and Properties of Polymer–clay Nanocomposites Obtained by Compounding in Polymer Melt, European Polymer Congress-2005, Moscow, 2005 p. 146.

[41] J.W. Cho, D.R. Paul, Nylon 6 nanocomposites by melt compounding, Polymer 42 (2001) 1083–1094.

[42] D.L. VanderHart, A. Asano, J.W. Gilman, Solid-state NMR investigation of paramagnetic nylon-6 clay nanocomposites. 1. Crystallinity, morphology, and the direct influence of Fe3+ on nuclear spins, Chemistry of Materials 13 (2001) 3781–3795.

[43] T.D. Fornes, P.J. Yoon, H. Keskkula, D.R. Paul, Nylon 6 nanocomposites: the effect of matrix molecular weight, Polymer 42 (2001) 9929–9940.

[44] T.D. Fornes, P.J. Yoon, D.L. Hunter, H. Keskkula, D.R. Paul, Effect of organoclay structure on nylon-6 nanocomposite morphology and properties, Polymer 43 (2002) 5915–5933.

[45] N. Hasegawa, H. Okamoto, M. Kato, A. Usuki, N. Sato, Nylon 6-montmorillonite nanocomposites prepared by compounding nylon 6 with Na-montmorillonite slurry, Polymer 44 (2003) 2933–2937.

[46] J.W. Gilman, W.H. Awad, R.D. Davis, J. Shields, R.H. Harris Jr., C. Davis, A.B. Morgan, T.E. Sutto, J. Callahan, P.C. Trulove, H.C. DeLong, Polymer/layered silicate nanocomposites from thermally stable trialkylimidazolium-treated montmorillonite, Chemistry of Materials 14 (2002) 3776–3785.

[47] T.D. Fornes, D.R. Paul, Crystallization behavior of nylon 6 nanocomposites, Polymer 44 (2003) 3945–3961.

[48] M. Kato, A. Usuki, Polyamide/clay nanocomposites, in: Y.-W. Mai, Zh.-Zh. Yu (Eds.), Polymer Nanocomposites, Woodhead Publishing Limited, Cambridge, England, 2006, pp. 3–28.

[49] V.V. Korshak, S.V. Vinogradova, Heterochain Polyesters, AN SSSR Press (USSR AS Press), Moscow, 1958.

[50] C.H. Davis, L.J. Mathias, J.W. Gilman, D.A. Schiraldi, J.R. Shields, P. Trulove, T.E. Sutto, H.C. Delon, Effect of melt-processing conditions on the quality of poly (ethylene terephthalate) montmorillonite clay nanocomposite, Journal of Polymer Science Part B: Polymer Physics 40 (2002) 2661–2666.

[51] B.J. Chisholm, R.B. Moore, Q. Barber, F. Khouri, A. Hempstead, M. Larsen, E. Olson, J. Kelley, G. Balch, J. Caraher, Nanocomposites derived from sulfonated poly(butylene terephthalate), Macromolecules 35 (14) (2002) 5508–5516.

[52] A. Sanchez-Solis, A. Garcia-Rejon, O. Manero, Production of nanocomposites of PET-montmorillonite clay by an extrusion process, Macromolecular Symposium 192 (1) (2005) 281–292.

[53] D. Acierno, P. Scarfato, E. Amendola, G. Nocerino, G. Costa, Preparation and characterization of PBT nanocomposites compounded with different montmorillonites, Polymer Engineering and Science 44 (2004) 1012–1018.

[54] Y.-W. Chang, S. Kim, Y. Kyung, Poly(butylene terephthalate)-clay nanocomposites prepared by melt interaction: morphology and thermomechanical properties, Polymer International 54 (2) (2004) 348–363.

[55] J. Xiao, Y. Hu, Z. Wang, Y. Tang, Z. Chen, W. Fan, Preparation and characterization of poly(butylene terephthalate) nanocomposites from thermally stable organic-modified montmorillonite, European Polymer Journal 41 (5) (2005) 1030–1035.

[56] C.-S. Ha, Poly(butylene terephthlate) (PBT) based nanocomposites, in: Y.-W. Mai, Zh.-Zh. Yu (Eds.), Polymer Nanocomposites, Woodhead Publishing Limited, Cambridge, England, 2006, pp. 234–255.

[57] X. Li, H.-M. Park, J.-O. Lee, C.S. Ha, Effect of blending sequence on the microstructure and properties of PBT/EVA-g-MAH/organoclay ternary nanocomposites, Polymer Engineering and Science 42 (11) (2004) 2156–2164.

[58] J.I. Velasko, M. Ardanuy, L. Miralles, S. Ortiz, M.S. Soto, Poly(propylene)/PET/undecyl ammonium montmorillonite nanocomposites synthesis and characterization, Macromolecular Symposium 221 (1) (2005) 63–74.

[59] S. S. Pesetskii, S. P. Bogdanovich, A. N. Paduchin, I. M. Krymsky, N. A. Malkova, Polyamide nanomaterial, Eurasian Patent 007560 (2006).

[60] J.H. Chang, B.-S. Seo, S.H. Kim, Blends of thermotropic liquid-crystalline polymer and a poly(butylene terephthalate) organoclay nanocomposite, Journal of Polymer Science Part B: Polymer Physics 42 (20) (2004) 3667–3676.

[61] K.H. Yoon, M.B. Polk, J.H. Park, B.G. Min, D.A. Schiraldi, Properties of poly(ethylene terephthalate) containing epoxy-functionalized polyhedral oligomeric silsesquioxane, Polymer International 54 (1) (2005) 47–53.

[62] S.-H. Wu, F.-Y. Wang, C.-C.M. Ma, W.-C. Chang, C.-T. Kuo, H.-C. Kuan, W.-J. Chen, Mechanical, thermal and morphological properties of glass fiber and carbon fiber reinforced polyamide-6/clay nanocomposites, Materials Letters 49 (2001) 327–333.

[63] K. Jayaraman, S. Kumar, Polypropylene layered silicate nanocomposites, in: Y.-W. Mai, Zh.-Zh. Yu (Eds.), Polymer Nanocomposites, Woodhead Publishing Limited, Cambridge, England, 2006, pp. 130–150.

[64] M. Kawasumi, L. Hasegawa, M. Kato, A. Usuki, A. Okada, Preparation and mechanical properties of polypropylene-clay hybrids, Macromolecules 30 (20) (1997) 6333–6338.

[65] ВА Герасин, ТА Зубова, ФН Бахов, АА Баранников, НД Меркалова, ЮМ Королев, ЕМ Антипов, Структура нанокомпозитов полимер/Na⁺-монтмориллонит, полученных смешением в расплаве, Российские Нанотехнологии 2 (1–2) (2007) 90–105.

[66] N. Hasegawa, M. Kawasumi, M. Kato, A. Usuki, A. Okada, Preparation and mechanical properties of polypropylene-clay hybrids using a maleic anhydride-modified polypropylene oligomer, Applied Polymer Science 67 (1998) 87–92.

[67] P.H. Nam, P. Maiti, M. Okamote, T. Kataka, N. Hagesawa, A. Usuki, A hierarchical structure and properties of intercalated polypropylene/clay nanocomposites, Polymer 42 (2001) 9633–9640.

[68] D. Kaempfer, R. Thomman, R. Mulhaupt, Melt compounding of syndiotactic polypropylene nanocomposites containing organophilic layered silicates and in situ formed core/shell nanoparticles, Polymer 43 (2002) 2909–2916.

[69] K.H. Wang, M.H. Chai, C.M. Koo, Y.S. Choi, I.J. Chung, Synthesis and characterization of maleated polyethylene/clay nanocomposites, Polymer 42 (2001) 9819–9826.

[70] D. Dharaiya, S.C. Jana, Thermal decomposition of alkyl ammonium ions and its effects on surface polarity of organically treated nanoclay, Polymer 46 (2005) 10139–10147.

[71] W. Xie, Z. Gao, K. Liu, W.-P. Pan, R. Vaia, D. Hunter, A. Singh, Thermal characterization of organically modified montmorillonite, Thermochimica Acta 367-368 (2001) 339–350.

[72] X. Huang, S. Lewis, W.J. Brittain, R.A. Vaia, Synthesis of polycarbonate-layered silicate nanocomposites via cyclic oligomers, Macromolecules 33 (6) (2000) 2000–2004.

[73]  J. Le, T. Takekkoski, E.P. Giannelis, Fire retardant polyether imide nanocomposites, Materials Research Society Symposium Proceedings 457 (1997) 513–518.

[74]  J.C. Huang, Z.K. Zhu, J. Yin, X.F. Qian, Y.Y. Sun, Poly(etherimide)/montmorillonite nanocomposites prepared by melt intercalation: morphology, solvent resistance properties and thermal properties, Polymer 42 (2001) 873–877.

[75]  M. Kawasumi, N. Hasegawa, A. Vsuki, A. Okada, Nematic liquid crystal/clay mineral composites, Materials Science and Engineering C 6 (1998) 135–143.

[76]  J.H. Chang, B.S. Seo, D.H. Hwang, An exfoliation of organoclay in thermotropic liquid crystalline polyester nanocomposites, Polymer 43 (2002) 2969–2974.

[77]  R.A. Vaia, H. Ishii, E.P. Giannelis, Synthesis and properties of two-dimensional nanostructures by direct intercalation of polymer melt in layered silicates, Chemistry of Materials 5 (1993) 1694–1706.

[78]  R.A. Vaia, E.P. Giannelis, Polymer melt intercalation in organically-modified layered silicates: model predictions and experiment, Macromolecules 30 (25) (1997) 8000–8009.

[79]  A.K. Geim, K.S. Novoselov, The rise of graphene, Nature Materials 6 (2007) 183–191.

[80]  J. van den Brink, Graphene - from strength to strength, Nature Nanotechnology 2 (2007) 199–201.

[81]  D.A. Dikin, S. Stankovich, E.J. Zimney, R.D. Piner, G.H.B. Dommett, G. Evmenenko, S.T. Nguyen, R.S. Ruoff, Preparation and characterization of graphene oxide paper. London Nature 448 (2007) 457–460.

[82]  S. Stankovich, D.A. Dikin, G.H.B. Dommett, K.M. Kohlhaas, E.J. Zimney, E.A. Stach, R.D. Piner, S.T. Nguyen, R.S. Ruoff, Graphene-based composite materials. London Nature 442 (7100) (2006) 282–286.

[83]  R. Setton, P. Bernier, S. Lefrant, Carbon Molecules and Materials, Taylor and Francis, London, 2002.

[84]  A.I. Lyamkin, E.A. Petrov, A.P. Ershov, G.V. Sakovich, A.M. Staver, V.M. Titov, Making of diamonds from explosives, Doklady SSSR (USSR AS Reports) 302 (3) (1988) 611–613. (in Russian).

[85]  G.V. Sakovich, V.D. Gubarevich, F.Z. Badaev, P.M. Brylyakov, O.A. Besedina, Aggregation of diamonds made from explosives, Doklady SSSR (USSR AS Reports) 310 (2) (1990) 402–404. (in Russian).

[86]  A.P. Korobko, S.V. Krasheninnikov, I.V. Levakova, L.A. Ozerina, S.N. Chvalun, Nanocomposites based on polycarbonate and ultrafine diamonds, Polymer Science A43 (11) (2001) 1163–1170.

[87]  H.W. Kroto, J.R. Heath, S.C. O'Brien, R.F. Curl, R.E. Smalley, C60: buckminsterfullerene, Nature 318 (1985) 162–163.

[88]  J.R. Heath, S.C. O'Brien, Q. Zhang, Y. Liu, R.F. Curl, F.K. Tittel, R.E. Smalley, Lanthanum complexes of spheroidal carbon shells, Journal of the American Chemical Society 107 (25) (1985) 7779–7780.

[89]  S. Iijima, Helical microtubules of graphitic carbon, Nature 354 (1991) 56–58.

[90]  E.T. Thostenson, Z. Ren, T.-W. Chou, Advances in the science and technology of carbon nanotubes and their composites: a review, Composites Science and Technology 61 (2001) 1899–1912.

[91]  M. Moniruzzaman, K.I. Winey, Review: polymer nanocomposites containing carbon nanotubes, Macromolecules 39 (16) (2006) 5194–5205.

[92]  S. Kumar, H. Doshi, M. Srinivasarao, J.O. Park, D.A. Schiraldi, Fibers from polypropylene/nano carbon fiber composites, Polymer 43 (2002) 1701–1703.

[93] W.-D. Zhang, L. Shen, I.Y. Phang, T. Liu, Carbon nanotubes reinforced nylon-6 composite prepared by simple melt-compounding, Macromolecules 37 (2) (2004) 256–259.

[94] T. Liu, I.Y. Phang, L. Shen, S.Y. Chow, W.-D. Zhang, Morphology and mechanical properties of multiwalled carbon nanotubes reinforced nylon-6 composites, Macromolecules 37 (16) (2004) 7214–7222.

[95] P. Poetschke, A.R. Bhattacharyya, A. Janke, H. Goering, Melt mixing of polycarbonate/multi-wall carbon nanotube composites, Composite Interfaces 10 (2003) 389–404.

[96] E.I. Siochi, D.C. Working, Ch. Park, P.T. Lillehei, J.H. Rouse, C. Topping, A.R. Bhattacharyya, S. Kumar, Melt processing of SWCNT-polyimide nanocomposite fibers, Composites B35 (2004) 439–446.

[97] R. Haggenmueller, H.H. Gommans, A.G. Rinzler, J.E. Fischer, K.I. Winey, Aligned single-wall carbon nanotubes in composites by melt processing methods, Chemical Physics Letters 330 (2000) 219–225.

[98] Zh. Jin, K.P. Pramoda, S.H. Goh, G. Xu, Poly(vinylidene fluoride)-assisted melt-blending of multi-walled carbon nanotube/poly(methyl methacrylate) composites, Materials Research Bulletin 37 (2002) 271–278.

[99] V.E. Agabekov, V.V. Golubovich, S.S. Pesetskii, Effect of nanodisperse carbon fillers and isocyanate chain extender on structure and properties of poly(ethylene terephthalate), Journal of Nanomaterials (2012) in press.

[100] B. Jurkowska, B. Jurkowski, P. Kamrowski, S.S. Pesetskii, V.N. Kowal, L.S. Pinchuk, Y.A. Olkhov, Properties of fullerene-containing natural rubber, Journal of Applied Polymer Science 100 (2006) 390–398.

[101] A.P. Voznyakovsky, V. Yu. Dolmatov, E.A. Levintova, T.M. Gubarevich, Composite materials based on polyfluorinated copolymers and industrial diamond carbon from explosion synthesis. Moscow Proceedings of International Rubbers Conference 2 (1994) 80–87. (in Russian).

[102] D.D.L. Chung, Exfoliation of graphite, Journal of Materials Science 22 (1987) 4190–4198.

[103] B. Li, W.-H. Zhong, Review on polymer/graphite nanoplatelet nanocomposites, Journal of Materials Science 46 (17) (2011) 5595–5614.

[104] L.S. Schadler, S.C. Giannaris, P.M. Ajayan, Load transfer in carbon nanotube epoxy composites, Applied Physics Letters 73 (26) (1998) 3842–3844.

[105] Y. Meng, Polymer/graphite nanocomposites, in: Y.-W. Mai, Zh.-Zh. Yu (Eds.), Polymer Nanocomposites, Woodhead Publishing Limited, Cambridge England, 2006, pp. 510–539.

[106] M.S. Dresselhaus, G. Dresselhaus, Intercalation compounds of graphite, Advances in Physics 30 (2) (1981) 139–326.

[107] C. Underhill, T. Krapchev, M.S. Dresselhaus, Synthesis and characterization of high stage alkali metal donor compounds, Synthetic Metals 2 (1–2) (1980) 47–53.

[108] M. Inagaki, Formation and stability of new $FeCl_3$-graphite intercalation compounds, Solid State Ionics 63-65 (1993) 523–527.

[109] J. Walter, H. Shioyama, Boron trichloride graphite intercalation compound studied by selected area electron diffraction and scanning tunneling microscopy, Journal of Physics and Chemistry of Solids 60 (6) (1999) 737–741.

[110] M. Inaba, Z. Ogumi, Y. Mizutani, E. Ihara, T. Abe, M. Asano, T. Harada, Preparation of alkali metal graphite intercalation compounds in organic solvents, Journal of Physics and Chemistry of Solids 57 (6–8) (1996) 799–803.

[111] F. Kang, Y. Leng, T.-Y. Zhang, B. Li, Electrochemical synthesis and characterization of ferric chloride-graphite intercalation compounds in aqueous solution, Carbon 36 (4) (1998) 383–390.

[112] H. Shioyama, M. Crespin, A. Seron, R. Setton, D. Bonnin, F. Beguin, Electrochemical oxidation of graphite in an aqueous medium: intercalation of $FeCl_4^-$, Carbon 31 (1) (1993) 223–226.

[113] F. Kang, Y. Leng, T.Y. Zhang, Electrochemical synthesis of graphite intercalation compounds in $ZnCl_2$ aqueous solutions, Carbon 34 (7) (1996) 889–894.

[114] I. Izumi, J. Sato, N. Iwashita, M. Inagaki, Electrochemical intercalation of bromide into graphite in an aqueous electrolyte solutions, Synthetic Metals 75 (1) (1995) 75–77.

[115] D. Billaud, A. Herold, F.L. Vogel, The synthesis and resistivity of the ternary graphite-potassium-sodium compounds, Materials Science and Engineering 45 (1) (1980) 55–59.

[116] P. Lagrange, A. Herold, Chimisorption de l'hydrogene par les composes d'insertion graphite-potassium, Carbon 16 (4) (1978) 235–240.

[117] F.M. Uhl, Q. Yao, H. Nakajima, E. Manias, C.A. Wilkie, Expandable graphite/polyamide-6 nanocomposites, Polymer Degradation and Stability 89 (2005) 70–84.

[118] T.G. Gopakumar, D.J.Y.S. Pagé, Polypropylene/graphite nanocomposites by thermo-kinetic mixing, Polymer Engineering and Science 44 (6) (2004) 1162–1169.

[119] T. Nakajima, Y. Matsuo, Formation process and structure of graphite oxide, Carbon 32 (3) (1994) 469–475.

[120] I. Dékány, R. Krüger-Grasser, A. Weiss, Selective liquid sorption properties of hydrophobized graphite oxide nanostructures, Colloid Polymer Science 276 (7) (1998) 570–576.

[121] A. Lerf, H. He, M. Forster, J. Klinowski, Structure of graphite oxide revisited, Journal of Physical Chemistry B 102 (23) (1998) 4477–4482.

[122] Y. Matsuo, T. Niwa, Y. Sugie, Preparation and characterization of cationic surfactant-intercalated graphite oxide, Carbon 37 (6) (1999) 897–901.

[123] Y. Matsuo, K. Tahara, Y. Sugie, Structure and thermal properties of poly(ethylene oxide)-intercalated graphite oxide, Carbon 35 (1) (1997) 113–120.

[124] Y. Matsuo, K. Tahara, Y. Sugie, Synthesis of poly(ethylene oxide)-intercalated graphite oxide, Carbon 34 (5) (1996) 672–674.

[125] Y. Matsuo, K. Hatase, Y. Sugie, Preparation and characterization of poly(vinyl alcolol)- and $Cu(OH)_2$-poly(vinyl alcohol)-intercalated graphite oxides, Chemistry of Materials 10 (8) (1998) 2266–2269.

[126] J. Xu, Y. Hu, L. Song, Q. Wang, W. Fan, Preparation and characterization of poly-acrylamide-intercalated graphite oxide, Materials Research Bulletin 36 (10) (2001) 1833–1836.

[127] P.G. Liu, P. Xiao, M. Xiao, K.C. Gong, Synthesis of poly(vinyl acetate)-intercalated graphite oxide by an in situ intercalative polymerization, Chinese Journal of Polymer Science 18 (5) (2000) 413–418.

[128] A.D. Pomogailo, Polymer Immobilized Metalocomplex Catalysts, Nauka, Moscow, 1988.

[129] E.M. Natanson, Z.R. Ulberg, Colloid Metals and Metallopolymers, Naukova Dumka, Kiev, 1971 (in Russian).

[130] V.M. Buznik, V.M. Fomin, A.P. Alkhimov, L.I. Ignateva, A.K. Tsvetkov, V.G. Kudry-avyi, V.F. Kosarev, S.P. Gubin, O.I. Lomovskii, A.A. Okhlopkova, N.F. Uvarov, S.V. Klinkov, I.I. Shabalin, Metal–polymer Nanocomposites (Production, Properties, Application), Siberian Branch of Russian Academy of Sciences Publication, Novosibirsk, 2005 (in Russian).

[131] S.P. Gubin, M.S. Korobov, G. Yu. Yurkov, A.K. Tsvetnikov, V.M. Buznik, Nanomet-alization of ultradispersed poly(tetrafluoroethylene), Doklady Akademii Nauk Rossii (Rep. Russ. Acad. Sci.) 388 (4) (2003) 493–496. (in Russian).

[132] S.P. Gubin, Yu. A. Koksharov, G.B. Khomutov, G. Yu. Yurkov, Magnetic nanoparti-cles: preparation, structure and properties, Russian Chemical Reviews 74 (6) (2005) 489–520.

[133] S.S. Pesetskii, B. Jurkowski, A.A. Davydov, Y.M. Krivoguz, S.P. Bogdanovich, Metal–polymer nanocomposites produced by melt-compounding: interaction of aliphatic poly-amide with metal particles, Journal of Applied Polymer Science 105 (2007) 1366–1376.

[134] A.H. Lebovitz, K. Khait, J.M. Torkelson, Stabilization of dispersed phase to static coarsening: polymer blend compatibilization via solid-state shear pulverization, Macromolecules 35 (23) (2002) 8672–8675.

[135] A.H. Lebovitz, K. Khait, J.M. Torkelson, In situ block copolymer formation dur-ing solid-state shear pulverization: an explanation for blend compatibilization via Interpolymer radical reactions, Macromolecules 35 (26) (2002) 9716–9722.

[136] K.L. Brinker, A.H. Lebovitz, J.M. Torkelson, W.R. Burghardt, Porod Scattering study of Coarsening in Immiscible polymer blends, Journal of Polymer Science Part B: Polymer Physics 43 (2005) 3413–3420.

[137] Y. Tao, J. Kim, J.M. Torkelson, Achievement of quasi-nanostructured polymer blends by solid-state shear pulverization and compatibilization by gradient copolymer addition, Polymer 47 (19) (2006) 6773–6781.

[138] N. Furgiuele, A.H. Lebovitz, K. Khait, J.M. Torkelson, Novel Strategy for polymer blend Compatibilization: Solid-state shear pulverization, Macromolecules 33 (2) (2000) 225–228.

[139] N. Furgiuele, A.H. Lebovitz, K. Khait, J.M. Torkelson, Efficient mixing of polymer blends of extreme viscosity ratio: elimination of phase inversion via solid-state shear pulverization, Polymer Engineering and Science 40 (6) (2000) 1447–1457.

[140] A.M. Walker, Y. Tao, J.M. Torkelson, Polyethylene/starch blends with enhanced oxygen barrier and mechanical properties: effect of granule morphology damage by solid-state shear pulverization, Polymer 48 (4) (2007) 1066–1074.

[141] A.H. Lebovitz, K. Khait, J.M. Torkelson, Sub-micron dispersed-phase particle size in polymer blends: overcoming the Taylor limit via solid-state shear pulverization, Polymer 44 (1) (2003) 199–206.

[142] J. Masuda, J.M. Torkelson, Dispersion and major property enhancements in polymer/multiwall carbon nanotube nanocomposites via solid-state shear pulverization followed by melt mixing, Macromolecules 41 (16) (2008) 5974–5977.

[143] K. Wakabayashi, C. Pierre, D.A. Dikin, R.S. Ruoff, T. Ramanathan, L.C. Brinson, J.M. Torkelson, Polymer–Graphite nanocomposites: effective dispersion and major prop-erty Enhancement via solid-state shear pulverization, Macromolecules 41 (6) (2008) 1905–1908.

[144] I.V. Kragelskii, Friction and Wear, Pergamon Press, Elmsford, 1982.

[145] L. Yu, S. Bahadur, An investigation of the transfer film characteristics and tribological behaviors of polyphenylene sulfide composites in sliding against tool steel, Wear 214 (1998) 245–251.

[146] G. Jintang, Tribochemical effects in formation of polymer transfer film, Wear 245 (2000) 100–106.

[147] H. Unal, U. Sen, A. Mimaroglu, Dry sliding wear characteristics of some industrial polymers against steel counterface, Tribology International 37 (2004) 727–732.

[148] H. Unal, A. Mimaroglu, Friction and wear behavior of unfilled engineering thermoplastics, Materials Design 24 (2003) 183–187.

[149] Y.K. Chen, O.P. Modi, A.S. Mhay, A. Chrysanthou, J.M. O'Sullivan, The effect of different metallic counterface materials and different surface treatments on the wear and friction of polyamide 66 and its composites in rolling-sliding contact, Wear 255 (2003) 714–721.

[150] A.J. Schwartz, S. Bahadur, The role of filler deformability, filler-polymer bonding, and counterface material on the tribological behavior of polyphenylene sulfide (PPS), Wear 251 (2001) 1532–1540.

[151] Y.M. Xu, B.G. Mellor, The effect of fillers on the wear resistance of thermoplastic polymer coatings, Wear 251 (2001) 1522–1531.

[152] P.N. Bogdanovich, V. Ya. Prushak, Friction and Wear in Machines: A Textbook for Institutions of Higher Education, Vysshaya Shkola Press, Minsk, 1999 (in Russian).

[153] P.J. Blau, The significance and use of the friction coefficient, Tribology International 34 (2001) 585–591.

[154] V.A. Bely, A.I. Sviridenok, M.I. Petrokovets, V.G. Savkin, Friction and Wear in Polymer-based Materials, Pergamon Press, Oxford, 1982.

[155] K. Shooter, R.H. Thomas, Frictional properties of some plastics, Research 2 (1952) 533–539.

[156] K. Shooter, D. Tabor, The frictional properties of plastics, Proceedings of the Royal Society B65 (1952) 661–673.

[157] A.L. Rees, Static friction of bulk polymers, Research 10 (1957) 331–338.

[158] N.S. White, Small oil-free bearings, Journal of Research of National Bureau of Standards 57 (1956) 185–189.

[159] W.C. Milz, L.E. Sargent, Frictional characteristics of plastics, Lubrication Engineering 11 (1955) 313–317.

[160] K.C. Ludema, D. Tabor, The friction and visco-elastic properties of polymeric solids, Wear 9 (1966) 329–348.

[161] R.T. King, D. Tabor, The effect of temperature on the mechanical properties and the friction of plastics, Proceedings of the Physical Society B66 (1953) 728–737.

[162] G.V. Vinogradov, G.M. Bartenev, A.I. El'kin, V.K. Mikhaylov, Effect of temperature on friction and adhesion of crystalline polymers, Wear 16 (1970) 213–219.

[163] I.V. Kragelskii, M.N. Dobychin, V.S. Kombalov, Friction and Wear Calculation Methods, Pergamon Press, Oxford, 1982.

[164] V.L. Vakula, L.M. Pritykin, Physical Chemistry of Polymer Adhesion, Khimia, Moscow, 1984 (in Russian).

[165] V.A. Goldade, V.A. Struk, S.S. Pesetskii, Wear Inhibitors for Metalopolymeric Systems, Khimia, Moscow, 1993 (in Russian).

[166] G. Heinicke, Tribochemistry, Akademie-Verlag, Berlin, 1984.

[167] Y. Kojima, A. Usuki, M. Kawasumi, A. Okada, T. Kuraucki, O. Kamigaito, Synthesis of nylon-6 hybrid by montmorillonite intercalated with ε-caprolactam, Journal of Polymer Science Part A: Polymer Chemistry 31 (1993) 983–986.

[168] P.H. Nam, P. Maiti, M. Okamoto, T. Kotaka, Foam Processing and Cellular Structure of Polypropylene/Clay Nanocomposites. Processing Nanocomposites, ESM Publication, Chicago, 2001.

[169] I.W. Gilman, T. Kashiwagi, E.P. Giannelis, E. Manias, S. Lomakin, J.D. Lichtenhan, P. Jones, Flammability properties of polymer-layered silicate nanocomposites, (Chapter

14) in: S. Al-Malaika, A. Golovoy, C.A. Wilkie (Eds.), Chemistry and Technology of Polymer Additives, Blackwell Science, Oxford, 1999.

[170] B.B. Troitskii, G.A. Domrachev, L.V. Khokhlova, L.I. Anikina, Thermooxidative degradation of poly(methyl methacrylate) in the presence of C60 fullerene, Polymer Science A43 (2001) 964–969.

[171] A. Dasari, Zh.-Zh. Yu, Y.-W. Mai, G.-H. Hu, J. Varlet, Clay exfoliation and organic modification on wear of nylon 6 nanocomposites processed by different routes, Composites Science and Technology 65 (2005) 2314–2328.

[172] Sh.-Q. Lai, L. Yue, T.-Sh. Li, X.-J. Liu, R.-G. Lv, An investigation of friction and wear behaviors of polyimide/attapulgite hybrid materials, Macromolecular Materials and Engineering 290 (2005) 195–201.

[173] I.M. Garces, D.J. Moll, J. Bicerano, R. Fibiger, D.G. McLeod, Polymeric nanocomposites for automotive applications, Advanced Materials 12 (2000) 1835–1839.

[174] A.A. Okhlopkova, Physico-chemical principles of developing triboengineering materials based on polytetrafluoroethylene and ultradispersed ceramics, Doctor's Thesis, Gomel, 2000 (in Russian).

[175] A.M. Malevich, E.V. Ovchinnikov, Yu. S. Boiko, V.A. Struk, Tribological properties of PTFE modified by ultra-dispersed clusters of synthetic carbon, The Journal of Friction and Wear 19 (3) (1998) 71–74.

[176] W.X. Chen, F. Li, G. Han, J.B. Xia, L.Y. Wang, J.P. Tu, Z.D. Xu, Tribological behavior of carbon-nanotube-filled PTFE composites, Tribology Letters 15 (2003) 275–278.

[177] Y.-S. Zoo, J.-W. An, D.-Ph. Lim, D.-S. Lim, Effect of carbon nanotube addition on tribological behavior of HMWPE, Tribology Letters 16 (2004) 305–309.

[178] H. Cai, F. Yan, Q. Xue, Investigation of tribological properties of polyimide/carbon nanotube nanocomposites, Materials Science and Engineering A364 (2004) 94–100.

[179] K.-T. Lau, D. Hui, Effectiveness of using carbon nanotubes as nano-reinforcements for advanced composite structure, Carbon 40 (2002) 1605–1606.

[180] A.M. Ginzburg, O.F. Kireenko, D.G. Tochilnikov, V.P. Bulatov, Development of wear resistant structure at steel-over-copper sliding in presence of fullerene black, Journal: Applied Physics Letters 21 (23) (1995) 35–38. pp.3 8–42 (in Russian).

[181] A.M. Ginzburg, D.G. Tochilnikov, Effect of fullerene-containing admixes in fluoroplastics on their load-bearing capacity at friction, Zhurnal Tekhnicheskoi Fiziki (J. Appl. Phys.) 71 (2) (2001) 120–124. (in Russian).

[182] B.M. Ginzburg, D.G. Tochil'nikov, A.A. Shepelevskii, A.M. Leksovskii, Sh. Tuichiev, Tribological properties of polytetrafluoroethylene modified with fullerene black in dry sliding friction, Russian Journal of Applied Chemistry 79 (9) (2006) 1518–1521.

[183] G.N. Gubanova, T.K. Meleshko, V.E. Yudin, Y.A. Fadin, Y.P. Kozyrev, V.I. Gofman, A.A. Mikhailov, N.N. Bogorad, A.G. Kalbin, Y.N. Panov, G.N. Fedorova, V.V. Kudryavtsev, Fullerene-modified polyimide derived from 3,3,4,4-benzophenonetetracarboxylic acid and 3,3diaminobenzophenone for casted items and its use in tribology, Russian Journal of Applied Chemistry 76 (2003) 1156–1163.

[184] Ph. Werner, V. Altstadt, R. Jaskulka, O. Jacobs, J.K.W. Sandler, M.S.P. Shaffer, Al. H. Windle, Tribological behaviour of carbon-nanofibre-reinforced poly(ether ether ketone), Wear 257 (2004) 1006–1014.

[185] I. Friedrich, Zh. Zhong, A.K. Schlarb, Effects of various fillers on the sliding wear of polymer composites, Composites Science and Technology 65 (2005) 2329–2343.

[186] I.Q. Zhang, M.Z. Rong, K. Friedrich, Wear resisting polymer nanocomposites: preparation and properties, in: Y. Mai, Z. Yu (Eds.), Polymer Nanocomposites, Woodhead Publishing, Cambridge, 2006.

[187] A.G. Awakumov, M. Senna, N. Kosova, Soft Mechanochemical Synthesis: A Basis for New Chemical Technologies, Kluwer Academic Publishers, London, 2001.

[188] L. Chang, Z. Zhang, H. Zhang, K. Friedrich, Effect of nanoparticles on the tribological behaviour of short carbon fibre reinforced poly(etherimide) composites, Tribology International 38 (2005) 966–973.

[189] H.-B. Qiao, Q. Guo, A.-G. Tian, G.-L. Pan, L.-B. Xu, A study on friction and wear characteristics of nanometer Al2O3/PEEK composites under the dry sliding condition, Tribology International 40 (2007) 105–110.

[190] S. Bahadur, C. Sunkara, Effect of transfer film structure, composition and bonding on the tribological behavior of polyphenylene sulfide filled with nanoparticles of TiO2, ZnO, CuO and SiC, Wear 258 (2005) 1411–1421.

[191] I.H. Cho, S. Bahadur, Study of the tribological synergistic effects on nano CuO-filled and fiber-reinforced polyphenylene sulfide composites, Wear 258 (2005) 835–845.

[192] M.H. Cho, S. Bahadur, A.K. Pogosian, Observations on the effectiveness of some surface treatments of mineral particles in polyphenylene sulfide for tribological performance, Tribology International 39 (2006) 249–260.

[193] A. Ciardelli, S. Penczek, Modification and Blending of Synthetic and Natural Macromolecules, Kluwer Academic Publishers, London, 2003.

[194] M. Palabiyik, S. Bahadur, Mechanical and tribological properties of polyamide 6 and high density polyethylene polyblends with and without compatibilizers, Wear 246 (2000) 149–158.

[195] A.Z. Liu, L.Q. Ren, J. Tong, T.J. Joyce, S.M. Green, R.D. Arnell, Statistical wear analysis of PA-6/UHMWPE alloy, UHMWPE and PA-6, Wear 249 (2000) 31–36.

[196] C.Z. Liu, L.Q. Ren, J. Tong, S.M. Green, R.D. Arnell, Effect of operating parameters on the lubricated wear behavior of a PA-6/UHMWPE blend: a statistical analysis, Wear 253 (2002) 878–884.

[197] A.F. Zyuzina, A.A. Gribova, A.P. Krasnov, G.L. Slonimskii, A.A. Askadskii, L.V. Dubrovina, L.I. Komarova, O.G. Nikolskii, Study of the structure and friction behavior of materials produced from blends of polyarylates and polycarbonates with limited compatibility, The Journal of Friction and Wear 21 (2) (2000) 62–70.

[198] A.P. Krasnov, Yu. M. Pleskachevsky, V.N. Aderikha, V.A. Mit', O.V. Afonicheva, Compatibility and triboengineering properties of UHMWPE/PA-6 blends, Plasticheskie Massy (Plastics) 12 (2001) 12–13. (in Russian).

[199] A.Z. Liu, J.Q. Wu, J.Q. Li, L.Q. Ren, J. Tong, A.D. Arnell, Tribological behaviours of PA/UHMWPE blend under dry and lubricating condition, Wear 260 (2006) 109–115.

[200] S.P. Bogdanovich, S.S. Pesetskii, Friction of polyamide 6 – HDPE blends against steel, The Journal of Friction and Wear 22 (5) (2001) 95–102.

[201] S.P. Bogdanovich, S.S. Pesetskii, Effect of metallic counterbody on tribological behavior of compatibilized polyamide 6 – polyethylene blend: analysis of mass transfer, The Journal of Friction and Wear 25 (5) (2005) 70–76.

[202] M. Palabiyik, S. Bahadur, Tribological studies of polyamide 6 and high-density polyethylene blends filled with PTFE and copper oxide and reinforced with short glass fibres, Wear 253 (2002) 369–376.

[203] L.A. Utracki, Clay-containing Nanocomposites. vol. 1, Rapra Technology Limited, UK, 2004 2.

[204] A.P. Voznyakovsky, Polymer composite materials based on nanocarbons, in Polymer Science for 21st Century, Thesis of V A Kargin Conference, 2007, Moscow.

[205] Yu.M. Mozheiko, S.S. Pesetskii, I.P. Storozhuk, Preparation and study of polyester thermoplastic elastomers and their blended composition, in Polymer Science for 21st Century, Thesis of V A Kargin Conference-2007, 2007, Moscow.

[206] S.S. Pesetskii, B. Jurkowski, Y.A. Olkhov, O.M. Olkhova, I.P. Storozhuk, Yu. M. Mozheiko, Molecular and topological structures in polyester block copolymers, European Polymer Journal 37 (2001) 2187–2199.

[207] A.A. Okhlopkova, A.V. Vinogradov, L.S. Pinchuk, Plastics Filled with Ultradispersed Inorganic Compounds, MPRI NAS B Press, Gomel, 1999.

[208] O.A. Andrianova, A.A. Okhlopkova, S.N. Popov, I.N. Chersky, A polymer composite material for abrasive tools, Russian Patent 2064942, (1996).

# Sliding wear performance of epoxy-based nanocomposites

6

**Ming Qiu Zhang\*[†], Min Zhi Rong[†], Qing Bing Guo[‡], Ying Luo[§]**

*\*Key Laboratory for Polymeric Composite and Functional Materials of Ministry of Education, School of Chemistry and Chemical Engineering, Sun Yat-sen (Zhongshan) University, Guangzhou, PR China, [†]Materials Science Institute, Sun Yat-sen (Zhongshan) University, Guangzhou, PR China, [‡]College of Chemistry and Chemical Engineering, Zhongkai University of Agriculture and Engineering, Guangzhou, PR China, [§]College of Science, South China Agriculture University, Guangzhou, PR China*

## CHAPTER OUTLINE HEAD

## 6.1 INTRODUCTION AND THE STATE OF THE ART

In recent years, many projects have been launched on polymer composites containing particles at a nanometer scale with variable extent of success. Owing to the ultrafine dimension of the nanoparticles, a large fraction of atoms can reside at the interface, leading to strong interfacial interaction if a homogeneous dispersion of the nanofillers in the polymer matrix is guaranteed. As a result, the nanocomposites coupled with a great number of interfaces can be expected to provide remarkable performance [1,2].

Tribology of Polymeric Nanocomposites. http://dx.doi.org/10.1016/B978-0-444-59455-6.00006-4

The value of polymer nanocomposite technology is not solely based on the mechanical enhancement of the neat resin nor the direct replacement of current filler or blend technology. Rather, its importance comes from providing value-added properties not present in the neat resin, without sacrificing the resin's inherent processability and mechanical properties, or by adding excessive weight [3]. Traditionally, production of "multifunctional" materials in terms of blend or composite imposes a tradeoff between desired performance, mechanical properties, cost, and processability. However, over and over again, improved properties as compared to those of the pure polymers or conventional particulate composites are reported for polymer nanocomposites containing substantially less filler (typically 1–5 vol.%) and thus enabling greater retention of the inherent processability of the neat resin [4].

On the other hand, polymer and polymer composites are used increasingly often as engineering materials for technical applications in which tribological properties are of considerable importance. Fillers (e.g. glass, carbon, asbestos, oxides, and textile fibers) are incorporated within many polymers to improve their tribological performance [5–7]. The reduction in wear is mainly due to preferential load support offered by the reinforcement components. The contribution of abrasive mechanisms to the wear of the materials is highly suppressed. Besides, the solid lubricant particles transfer fairly readily to a metal counterface, leading to the formation of a strengthened transfer film. Finally, some fillers are used to improve the poor thermal conductivity possessed by polymers, thereby facilitating dissipation of frictional heat. However, the composites filled with the micron particles usually need a quite large amount of filler to attain an evident improvement of wear resistance. Yamaguchi found that the wear rates of unsaturated polyesters and epoxy (EP, further called Ep) resins filled with different proportions of $SiO_2$ decreased significantly only at a high loading of 40 wt.% [8]. In this case, some inherent defects are inevitable. For example, disintegration of the fillers and detached particles were frequently observed to be torn out of both the phenolic and Ep resin composites when an indentor moved over the systems, resulting in sharp fluctuations in the measured coefficient of friction [9]. Furthermore, the high loading of fillers must be detrimental to the processability of polymer, especially for the thermosetting polymer used as coating. Considering the disadvantages imparted by microfillers, utilization of nanoparticles or nanofibers would be an optimum alternative to make the most of the technique based on the filler incorporation. The wear mechanisms involved in the nanocomposites might be different from those of conventional composites because the fillers have the same size as the segments of the surrounding polymer chains. Severe wear caused by abrasion and particle pullout associated with the accumulation of detached harder bulk particulate fillers adherent to the frictional surface would be replaced by rather mild wear resulting from the fine individual debris that act as a lubricant and contribute to material removal by polishing.

Thermosetting resins like Ep are finding increasing use in a wide range of engineering applications because of their on-the-spot processing characteristics, good affinity to heterogeneous materials, considerable creep and solvent resistance, and higher operating temperature. For example, the polymers have been frequently mixed with various inorganic particles to formulate wear resistant composite materials. The tribological properties of these composites, however, are factually far from the specifications demanded by consumers for various working conditions, especially in the case of protective coatings that should shield the substrates against mechanical wear. This is due to both the poor interfacial adhesion around the particle boundaries and the heterogeneous dispersion of the particles. Yet, there are some exceptions. Symonds and Mellor [10] stated that the silica particles (10 μm) in an Ep matrix, exposed to wear loading, support a large fraction of the load. Since the particles fracture before they plastically deform, the true contact area and hence the frictional coefficient remain constant. Besides, the silica particles also reduced the wear of the coating by blocking the penetration of the steel asperity tips.

Relatively, only a few works concerning Ep-based nanocomposites have been reported. Rong et al. compared the effects of micro-$TiO_2$ (44 μm) and nano-$TiO_2$ (10 nm) particles on the wear resistance of Ep [11]. Their results revealed that the $TiO_2$ nanoparticles can remarkably reduce the wear rate of Ep, while the micron-$TiO_2$ particles cannot. A similar conclusion was made by Ng et al. in an earlier report [12]. Shi and coworkers suggested that unlike the severe wear observed in unfilled Ep dominated by fatigue-delamination mechanism, the wear mode of nano-$Si_3N_4$/Ep composites was characterized by mild polishing. It was believed that strong interfacial adhesion between $Si_3N_4$ nanoparticles and the matrix, reduced the damping ability and enhanced resistance to thermal distortion of the composites, and tribochemical reactions involving $Si_3N_4$ nanoparticles accounted for the reduced frictional coefficient and wear rate of the composites [13]. Lin studied sliding wear performance of Ep nanocomposites with organically treated Na-montmorillonite and inorganically treated titanium dioxide nanoparticles [14]. The wear rate of the nanocomposites with montmorillonite may improve nearly 30% due to its large agglomerated particle that will prevent the entrapping of nanoparticle and Ep debris between two sliding materials. The properties of the friction coefficient depicted that appropriate proportion of titanium dioxide nanofiller may provide certain lubricating effects, but the higher filler content will lead to an abrasively interactive wear result. Since the counterbody of a steel ring will first be abraded by wear debris and then part of these wear particles will be removed to the wear surface of nanocomposites, the wear loss of the pure Ep matrix had to be increased. With the formation of uniform and tenacious transfer particles, the wear resistance and friction coefficient of the composite were improved. Zhu and coworkers pretreated silica nanoparticles with a silane coupling agent, which were then mixed with Ep using a Mannich amine as the hardener [15]. They found that the composite with 3% nano-$SiO_2$ loading presented the best mechanical properties. The tribological performance and thermal stability of the materials were also improved with the addition of the nanoparticles.

In addition to the works that demonstrated the effectiveness of nanoparticles in improving tribological performance of Ep, the influence of geometry of the nanofillers was also investigated. Spirkova et al. prepared transparent colorless Ep nanocomposite coatings and studied the influence of mechanical properties and the size, shape, and concentration of additives (colloidal silica nanoparticles and montmorillonite sheets) on the measured surface characteristics [16]. They found that surface morphology is influenced by both types of admixtures: at the nanometer scale by colloidal nanosilica and at the micrometer scale by montmorillonite platelets. Only 1 wt.% montmorillonite can increase the friction coefficient and wear resistance without distinctive changes of tensile properties. However, the addition of 20 wt.% of silica nanoparticles was necessary for the increase of wear and scratch resistances. Jia et al. also investigated the mechanical properties and tribological behavior of Ep resin and its nanocomposites containing different shape nanofillers, that is, spherical nanosilica, layered organomodified montmorillonite (oMMT) and spherical-layered oMMT-$SiO_2$ [17]. The experiments proved that $SiO_2$/Ep had the lowest average friction coefficient and oMMT/Ep the lowest wear loss, while oMMT-$SiO_2$/Ep nanocomposite exhibited excellent wear resistance. Worn morphology observations indicated that both Ep and its nanocomposites underwent similar wear mechanisms, including the formation of cracks, development from cracks to waves, fatigue wear, and production of debris.

For purposes of maximizing the use of nanoparticles, hybrid particulate fillers were employed in Ep nanocomposites. In the work by Wetzel et al. [18], various amounts of microscale and nanoscale particles (calcium silicate $CaSiO_3$, 4–15 µm, alumina $Al_2O_3$, 13 nm) were introduced into the Ep. The influence of these particles on the impact energy, flexural strength, dynamic mechanical thermal properties, and block-on-ring wear behavior was investigated. If only the nanoparticles were incorporated, an improvement of the wear resistance of the Ep resin can be yielded at a nanoparticle content of 1–2 vol.%. Choosing the nanocomposite with the highest performance as a matrix, conventional $CaSiO_3$ microparticles were further added in order to achieve additional enhancements in the properties. As a result, synergistic effects were found in the form of a further increase in wear resistance and stiffness. McCook et al. compounded nanoparticles of zinc oxide and polytetrafluoroethylene (PTFE) with Ep in small volume percents [19]. Tribological testing indicated that the sample with the optimum wear rate (i.e. $1.79 \times 10^{-7}$ mm$^3$ Nm$^{-1}$, over 400 times lower than that of neat Ep) consisted of 1 vol.% of zinc oxide nanoparticles and 14.5 vol.% of PTFE nanoparticles. The sample with the optimum friction coefficient consisted of 3.5 vol.% of zinc oxide nanoparticles and 14.5 vol.% of PTFE nanoparticles and had a friction coefficient of 0.113, which is almost a 7× decrease in the friction coefficient from that of neat Ep. Xian et al. examined the influence of incorporated 300 nm $TiO_2$ (4 vol.%), graphite (7 vol.%), or combination of both fillers on the tribological performance of an Ep resin under various sliding loads (10–40 N) and velocity conditions (0.2–3.0 m s$^{-1}$) [20]. Under mild sliding conditions (i.e. low sliding velocity and load), the incorporation of nanometer-sized $TiO_2$ enhanced the wear resistance significantly, but led to a higher coefficient of friction than that of the neat Ep. Under severe sliding conditions, the wear rate-reducing effect of nano-$TiO_2$ markedly diminished, while the coefficient of friction was lower than that of the neat Ep. The reinforcing effect and the formation of transfer films due to the addition of the nanoparticles were assigned to the variation of the tribological behavior. A synergistic effect was found for the combination of both $TiO_2$ and graphite on the wear performance of the neat Ep. The corresponding composite exhibited the lowest wear rate and coefficient of friction under the investigated sliding conditions. The effective transfer films formed on sliding pair surfaces and the reinforcing effect of the nanoparticles were interpreted as the main mechanisms.

Fiber reinforcement is a traditional way to produce high-performance composite materials. When nanoparticles are incorporated, both the microstructure and the properties of the composites can be optimized. Zhang and coworkers systematically studied wear resistance of Ep filled with short carbon fiber (SCF), graphite, PTFE, and nano-$TiO_2$ under different sliding conditions [21–23]. The spherical $TiO_2$ nanoparticles (300 nm in diameter) were able to apparently reduce the frictional coefficient during sliding, and consequently to reduce the shear stress, contact temperature and wear rate of fiber-reinforced Ep composites. Based on microscopic observations of the worn surfaces, a positive rolling effect of the nanoparticles between the material pairs was proposed, which led to the remarkable reduction of the frictional coefficient. The rolling effect protects the SCFs from more severe wear mechanisms, especially at high sliding pressure and speed situations. Moreover, the addition of nanoparticles could not reduce the frictional coefficient when the nanocomposites are sliding against a very smooth disk.

As nanosized version of fiber reinforcer, carbon nanotubes (CNTs) are increasingly attracting scientific and industrial interest by virtue of their outstanding characteristics. Development of CNTs/polymer nanocomposites opens new perspectives for multifunctional materials. Dong et al. investigated the influence of multiwalled carbon nanotubes (MWNTs) on friction and wear behaviors of Ep [24]. They found that wear rate of the nanocomposites sharply decreased from $2.7 \times 10^{-5}$ to $6.0 \times 10^{-6}$ mm$^3$ Nm$^{-1}$ at MWNTs concentration of 1.5 wt.%. Moreover, Jacobs et al. indicated that hybrid filling in CNTs/Ep composites helped to make full use of CNTs [25]. When 10 wt.% graphite was added into 1 wt.% CNTs/Ep composites, for example, specific wear rate became as low as $3.5 \times 10^{-8}$ mm$^3$ Nm$^{-1}$, which is about 100 times lower than the composite without graphite and >300 times lower than that of the unfilled Ep.

With respect to the preparation of nanocomposites, it is worth noting that nanoparticles are very difficult to be uniformly dispersed in polymers because of the strong attraction between the particles and the limited shear force during compounding. Consequently, nanoparticle-filled polymers used to contain a number of loosened clusters of nanoparticles, which would lead to extensive material loss in terms of disintegration and crumbling of the particle agglomerates under wearing. To avoid these drawbacks, a series of methods have been attempted, which fall into two categories: development of new compounding (dispersion) techniques and surface pretreatment approaches for nanoparticles. Comparatively, surface modification of nanofillers is more effective and easier to be applied [26].

In our previous works, graft polymerization onto nanoparticles proved to be feasible in improving the tribological behavior of polymer nanocomposites [27–32]. The grafted polymers changed the hydrophilic surface feature of the particles to hydrophobic and broke apart the nanoparticle agglomerates during polymerization process. Besides, filler–matrix interfacial interaction in the composites was enhanced mainly due to chain entanglement of grafted polymer with matrix polymer.

Table 6.1 summarizes the representative results of Ep-based nanocomposites published by various groups in the world. It is seen that the wear rate of Ep can be

**Table 6.1** Friction and Wear Properties of Epoxy Filled with Nanoparticles and Other Additives

| Fillers | Surface Treatment | Lowest Wear Rate[a] ($10^{-6}$ mm$^3$ Nm$^{-1}$) | Filler Loading Corresponding to the Lowest Wear Rate | Lowest Frictional Coefficient[a] | Filler Loading Corresponding to the Lowest Frictional Coefficient | Reference |
|---|---|---|---|---|---|---|
| ZnO (20 nm) | Silane treated | 71.4 (0.29) | 10 wt.% | 0.48 (0.81) | 4 wt.% | [33] |
| Si$_3$N$_4$ (<20 nm) | – | 1.9 ($5.1 \times 10^{-2}$) | 0.83 vol.% | 0.35 (0.61) | 1.38 vol.% | [13] |
| γ-Al$_2$O$_3$ (3.8 nm) | – | 6.7 (0.18) | 1.2 vol.% | 0.50 (0.86) | 0.48 vol.% | [28] |
| γ-Al$_2$O$_3$ (3.8 nm) | PAAM grafted | 1.6 ($4.3 \times 10^{-2}$) | 0.24 vol.% | 0.35 (0.61) | 0.24 vol.% | [28] |
| Al$_2$O$_3$ (10.4 nm) | – | 7.8 ($3.3 \times 10^{-2}$) | 1.23 vol.% | 0.49 (0.90) | 0.74 vol.% | [32] |
| Al$_2$O$_3$ (10.4 nm) | PS grafted | 1.6 ($6.9 \times 10^{-3}$) | 0.48 vol.% | 0.46 (0.85) | 1.47 vol.% | [32] |
| Al$_2$O$_3$ (13 nm) | – | 3.8 (0.66) | 2.0 vol.% | – | – | [18] |
| SiC (61 nm) | – | 2.3 (0.55) | 0.40 vol.% | 0.90 (0.80) | 0.20 vol.% | [34] |
| SiC (61 nm) | PAAM grafted | 1.1 (0.27) | 0.20 vol.% | 0.74 (0.66) | 0.20 vol.% | [34] |
| SiO$_2$ (9 nm) | – | 23.5 (0.11) | 6.25 vol.% | 0.41 (0.71) | 6.25 vol.% | [30] |
| SiO$_2$ (9 nm) | PAAM grafted | 11.0 ($5.3 \times 10^{-2}$) | 2.17 vol.% | 0.37 (0.64) | 2.17 vol.% | [30] |
| TiO$_2$ (unknown size) | Inorganically treated | 25000 (0.49) | 7.0 vol.% | 0.21 (0.45) | 6.0 vol.% | [14] |
| TiO$_2$ (200–400 nm) | – | 14.0 (0.35) | 5.0 vol.% | – | – | [35] |
| Fe$_3$O$_4$ (14–20 nm) | – | – | – | 0.2–0.3 | – | [36] |
| Montmorillonite (unknown size) | Organomodified | 1.75 (0.17) | 5 wt.% | – | – | [37] |
| Clay (unknown size) | 3-Aminopropyl-triethoxysilane | 0.26 (0.12) | 6 wt.% | 0.17 (0.31) | 6 wt.% | [38] |
| Nanorubber (5 nm) | – | 149 (0.58) | 5 wt.% | 0.56 (0.66) | 5 wt.% | [39] |
| CNTs (10–30 nm in diameter) | – | 0.70 | 1 wt.% | – | – | [40] |
| MWNTs (10–20 nm in diameter) | – | 0.60 (0.22) | 1.5 wt.% | 0.21 (0.66) | 1.5 wt.% | [24] |

| | | | | | | |
|---|---|---|---|---|---|---|
| MWNTs (10–30 nm in diameter) | HNO$_3$ treated | — | 2.65 (0.09) | 1 wt.% | — | [25] |
| ZnO (53 nm) + PTFE (200 nm) | — | 0.18 (2.5 × 10$^{-3}$) | 1 vol.% ZnO + 14.5 vol.% PTFE | 0.11 (0.14) | 3.5 vol.% ZnO + 14.5 vol.% PTFE | [19] |
| TiO$_2$ (300 nm) + Graphite | — | 20 (4.2x10$^{-3}$) | 4 vol.% TiO$_2$ + 7 vol.% Graphite | 0.27 (0.42) | 4 vol.% TiO$_2$ + 7 vol.% Graphite | [20] |
| Al$_2$O$_3$ (13 nm) + CaSiO$_3$ (4–15 µm) | — | 170 (0.29) | 2 vol.% Al$_2$O$_3$ + 6.7 vol.% CaSiO$_3$ | — | — | [18] |
| CuO (30 nm) + PTFE | — | 12.0 (1.26) | 0.4 vol.% CuO + 7.5 vol.% PTFE | 0.36 (0.65) | 0.4 vol.% CuO + 7.5 vol.% PTFE | [41] |
| SiO$_2$ (12 nm) + short carbon fiber | SMA grafted | 1.65 (5.3 × 10$^{-3}$) | 4 wt.% SiO$_2$ + 6 wt.% SCF (1 mm) | 0.19 (0.34) | 4 wt.% SiO$_2$ + 6 wt.% SCF (1 mm) | [42] |
| MWNTs (10–30 nm in diameter) + graphite (4 µm) | — | 3.5×10$^{-2}$ (1.2 × 10$^{-3}$) | 1 wt.% MWNTs + 10 wt.% Graphite (4 µm) | — | — | [25] |
| MWCNTs (unknown size) + carbon fiber | Acid (H$_2$SO$_4$ + HNO$_3$) treated | — | — | 0.20–0.45 | 2 wt.% MWCNTs + 77 vol.% woven carbon fiber | [43] |
| MWCNTs (unknown size) + carbon fiber | Silane treated | — | — | 0.15–0.4 | 2 wt.% MWCNTs + 77 vol.% woven carbon fiber | [43] |
| TiO$_2$ (300 nm) + Graphite (20 µm) + carbon fiber (90 µm long) | — | 0.32 (1.3 × 10$^{-2}$) | 5 vol.% TiO$_2$ + 15 vol.% Graphite + 15 vol.% carbon fiber | — | — | [21] |

$^a$The numerals in the parentheses stand for the normalized properties of the composites relative to the unfilled matrices.

significantly reduced by the addition of nanoparticles. Surface grafting treatment of nanoparticles or incorporation of hybrid fillers is an effective measure for enhancing the effect of nanoparticles. The greatest decrement in the wear rate lies in three orders of magnitude. Comparatively, the reduction in frictional coefficient is not attractive enough. In some cases, the values of the nanocomposites are even higher than those of the neat matrices. Therefore, much more efforts have to be made to understand the mechanisms and to further take advantage of nanoparticles. For the moment, it is hard to make an explicit conclusion.

In fact, the interfacial adhesion can be further increased if a chemical reaction between the filler and matrix could take place. This concept is similar to the reactive compatibilization employed in making polymer blends. So far, however, there are few reports describing this approach for manufacturing wear resistant nanocomposites. The current chapter reviews our attempts in this direction. Polyglycidyl methacrylate (PGMA) was adhered to SiC nanoparticles through emulsifier-free emulsion graft polymerization [44], and then the treated nano-SiC particles were mixed with Ep. The grafted polymer (i.e. PGMA) was selected because the epoxide groups on PGMA would take part in the curing reaction of Ep resin forming a structure of nano-SiC-PGMA-Ep, so that the nano-SiC particles could be covalently connected to the matrix (Fig. 6.1), which is believed to be beneficial to the tribological performance of the composites. In addition, styrene (St) was used as a copolymerized monomer to adjust the amount of the reactive groups of the grafted macromolecular chains, so that the compatibility of the grafted copolymers with the composites matrix can be tuned. Its effect on the composite's properties was also studied.

On other hand, the incorporation of lubricant oil-loaded microcapsules into Ep composites [45] proved to integrate the merit of fluid lubrication but exclude the drawbacks of external lubrication. During sliding wear, the capsules were damaged by the asperities of the counterface, releasing the oil to the contact area. Owing to the lubrication

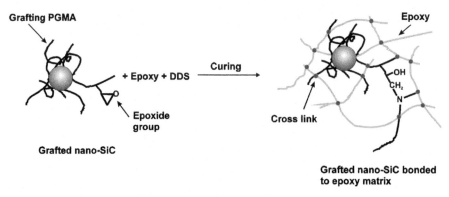

**FIGURE 6.1**

The possible structure of PGMA-grafted nano-SiC chemically bonded with epoxy resin through the reaction between the epoxide groups and the curing agent DDS.

effect of the librated oil, and the entrapment of wear particles in the cavities left by the ruptured capsules, significant reduction in frictional coefficient and specific wear rate was observed. Owing to the fact that the embedded soft capsules resulted in significantly deteriorated mechanical performance of Ep, however, load-bearing capability of the composites had to be lowered as a result. To solve the problem, we employed hybrid fillers including oil-loaded microcapsules, SCF, and silica nanoparticles. Similarly, graft treatment of $SiO_2$ nanoparticles with poly(styrene-co-maleic anhydride) (SMA) was conducted for the enhancing interaction between Ep matrix and the nanofillers. The interfacial reaction between anhydride and epoxide groups occurred in the course of curing. Wear performance of the hybrid composites is also described hereinafter.

## 6.2 EP FILLED WITH GRAFTED SiC NANOPARTICLES

### 6.2.1 Friction and wear performance of Ep filled with untreated SiC nanoparticles

Friction and wear properties of nano-SiC/Ep composites are plotted as a function of filler content in Fig. 6.2. Incorporation of the nanoparticles can drastically decrease the wear rate and frictional coefficient of Ep. Under a pressure 3 MPa, for example, the wear rate of Ep is decreased from ~200 × $10^{-6}$ to ~6 × $10^{-6}$ mm$^3$ Nm$^{-1}$ by the addition of 0.6 vol.% of untreated nano-SiC. This means that the nanoparticles are very effective in improving the tribological performance of Ep. These effects can be related with the enhancement of Ep surface stiffness and wear resistance by addition of nanoparticles. With a rise in the nanoparticle concentration, the declining trend of both wear rate and frictional coefficient is gradually subsiding or replaced by a slight increase. This might be due to the fact that the increased amount of the nanoparticles is unfavorable to filler dispersion [46]. Agglomeration of the nanoparticles reduces their lubricating effect. On the other hand, when the contact pressure increases from 2 to 3 MPa, the wear rate of the nanocomposites decreases and remains nearly unchanged as the pressure is further raised from 3 to 4 MPa, while the friction coefficient keeps a continuously decreasing tendency.

It is well known that the load-carrying capacity of unfilled Ep is poor. When the wear pressure increases from 1 to 3 MPa, the wear rate of unfilled Ep increases by two orders of magnitude. At a pressure of 5 MPa, the sample failed within a short time, and the test had to be stopped [32]. However, in this work, it seems that the higher pressure facilitates the nanoparticles to take effects. To have more information about the variation in wear behaviors of bulk Ep under different pressures due to the addition of the nano-SiC, morphologies of the worn pins' surfaces were examined by a scanning electron microscope (SEM; Figs 6.3 and 6.4). Severe damage characterized by the disintegration of the top layer was observed in the unfilled Ep (Fig. 6.3(a,b)). The higher contact pressure leads to a more substantial material loss (Fig. 6.3(c,d)), revealing that the adhesive wear mechanism plays the main role. Usually, higher pressure would cause higher frictional temperature at the surface, and hence more serious adhesive wear occurs.

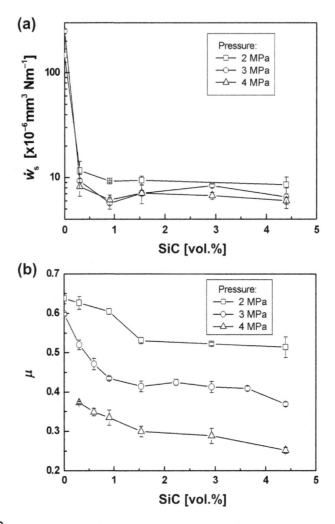

**FIGURE 6.2**

(a) Specific wear rate, $\dot{w}_s$, and (b) frictional coefficient, $\mu$, of untreated nano-SiC (61 nm) filled epoxy as a function of the SiC volume fraction.

    In the case of nano-SiC-filled composites, the worn appearances are completely different and become rather smooth (Fig. 6.4). The plowing grooves and cracks across the wear tracks are perceivable on the composites' worn surfaces. The former corresponds to the abrasive wear, while the latter is related to the fatigue-delamination process. The cracks might be nucleated at the subsurface layer as a result of shear deformation induced by the traction of the harder asperities, and then coalesce and cause scalelike damage patterns. It can thus be concluded that the adhesive wear of the unfilled Ep is replaced by a mixed mode of abrasive wear and fatigue wear when

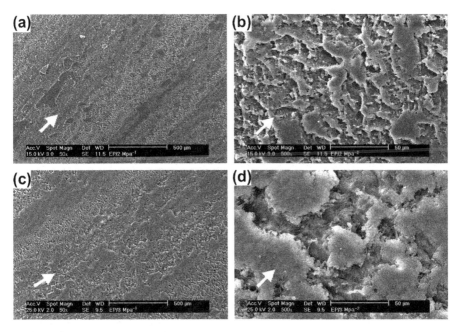

**FIGURE 6.3**

SEM micrographs of the worn surfaces of the unfilled epoxy after wear tests under pressures of (a, b) 2 MPa and (c, d) 3 MPa. The arrows indicate the sliding direction.

the untreated nano-SiC particles are incorporated. Under higher pressure, the higher frictional temperature certainly arouses large-scale plastic deformation, which consequently merges the cracks or debris and even helps to form a transferred film on the counterpart surface. All these factors bring about a smooth surface and smaller cracks on the worn surfaces (Fig. 6.4(e,f)).

## 6.2.2 Friction and wear performance of Ep filled with SiC-g-PGMA and SiC-g-P(GMA-co-St)

The purpose of grafting PGMA onto nano-SiC is to introduce chemical bonding between particles and matrix. Therefore, the reactivity of PGMA-grafted nano-SiC (SiC-g-PGMA) with curing agent 4,4-diaminodiphenysulfone (DDS) should be firstly explored even though the possible reaction has been proved in our previous work [47]. Accordingly, a model mixture consisting of stoichiometrical SiC-g-PGMA and DDS was heated and monitored by a differential scanning calorimeter (DSC) at different heating rates (Fig. 6.5). It is seen that the melting peaks of DDS appear at about 179 °C, while the exothermic peaks at 213 °C should be ascribed to the reaction between PGMA and DDS. By using Kissinger equation [48], the activation energy was estimated at 58.5 kJ mol$^{-1}$, which is much lower than the value

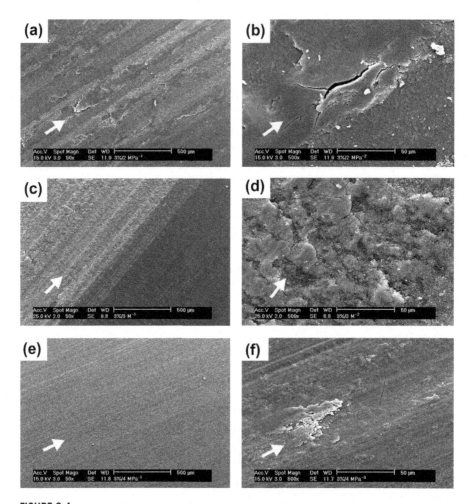

**FIGURE 6.4**

SEM micrographs of the worn surfaces of untreated nano-SiC-filled epoxy composites (SiC = 0.91 vol.%) after wear tests under pressures of (a, b) 2 MPa, (c, d) 3 MPa, and (e, f) 4 MPa. The arrows indicate the sliding direction.

of Ep/DDS system (62.5 kJ mol$^{-1}$). It means that the epoxide groups on the grafted PGMA should easily take part in the curing reaction.

The advantages of the grafted nanoparticles on tribological performance of Ep composites are revealed in Fig. 6.6. It is seen that within the entire range of filler content of interests, the wear resistance of SiC-g-PGMA/Ep composites is higher than that of SiC/Ep. Meanwhile, the grafted nanoparticles also lead to more significant reduction in friction coefficient of Ep as compared with the untreated version. The difference in improving the wear-resisting and friction-reducing ability

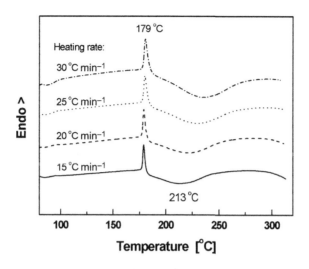

**FIGURE 6.5**

DSC traces of SiC-g-PGMA/epoxy system (2/1 by mole) recorded at various heating rates.

of Ep between the composites with untreated and treated SiC nanoparticles should mainly result from the filler/matrix adhesion strength. As stated above, the grafted PGMA employed in the present work can build up chemical bonding between the nanoparticles and Ep resin, which should be responsible for the higher resistance to periodic frictional stress. As a result, the strong interfacial adhesion is able to prevent the friction-induced crack initiation, coalescence, and propagation. Accordingly, the composites with grafted nano-SiC show higher wear resistance and lower frictional coefficient than do those with untreated nano-SiC.

Figure 6.7 compares the worn surfaces of the composites tested. Clearly, SiC/ Ep shows a much rougher appearance accompanied by many cracks, while SiC-g-PGMA/Ep possesses a rather smooth surface without obvious cracking. This suggests that the polishing effect predominates the wear process of the latter composites, evidencing the change in the wear mode described above.

The influence of contact pressure on the tribological performance of SiC-g-PGMA/Ep composites presents similar trend as untreated nano-SiC/Ep (Fig. 6.8), further demonstrating that fatigue wear mechanism takes place in this system. Due to the stronger bonding at the interface, the treated nanoparticles perform better in the composites (cf. Figs 6.2 and 6.8).

In our preliminary study, it was found that SiC nanoparticles with a moderate percent grafting ($\gamma_g$ = 8 wt.%) gave the best tribological performance for SiC-g-PGMA/Ep composites [47]. This means only a suitable amount of GMA groups is required for establishing the interfacial adhesion, while excessive GMA groups might interfere with the dispersion of particles by the possible reaction inside the particle agglomerates. To confirm this estimation, St was used as a copolymerized monomer to adjust the amount of the reactive groups of the grafted macromolecular

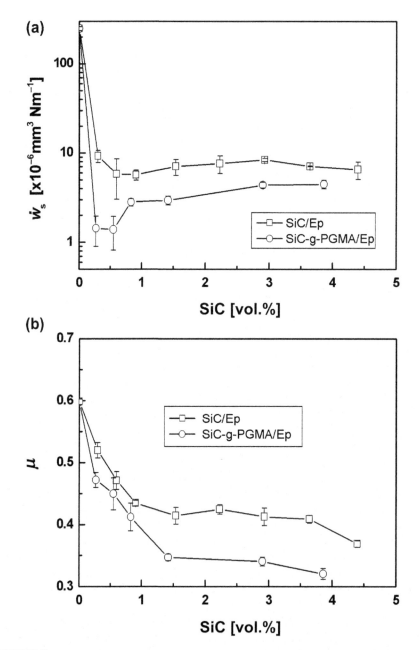

**FIGURE 6.6**

(a) Specific wear rate, $\dot{w}_s$, and (b) friction coefficient, $\mu$, of untreated nano-SiC-filled epoxy and SiC-g-PGMA-filled epoxy composites (percent grafting $\gamma_g = 8$ wt.%) as a function of the SiC volume fraction.

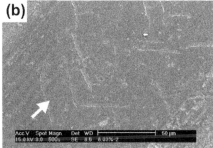

**FIGURE 6.7**

SEM micrographs of the worn surfaces of (a) untreated nano-SiC-filled epoxy (SiC = 0.91 vol.%) and (b) SiC-g-PGMA-filled epoxy (SiC = 0.83 vol.%, $\gamma_g$ = 8 wt.%) composites. The arrows indicate the sliding direction.

chains, and hence the compatibility of the grafted copolymers with the matrix Ep. Its effect on the composite's properties is shown in Fig. 6.9. When the percent grafting of PGMA is 36.72 wt.%, the specific wear rate and frictional coefficient of SiC-g-PGMA/Ep are even higher than those of untreated SiC/Ep. However, with the introduction of polystyrene into the grafted polymer, both specific wear rate and frictional coefficient of the composites are evidently lowered in the case of similar percent grafting. The results strongly suggest that excessive GMA groups are detrimental to the enhancement of the composites' wear resistance. Figure 6.10 further compares the tribological behaviors of SiC/Ep, SiC-g-PGMA/Ep, and copolymer of GMA and St grafted nano-SiC (SiC-g-P(GMA-co-St))/Ep composites as a function of nano-SiC content. Within the filler loading range of interests, the composites containing copolymer grafted nanoparticles are superior to the others despite the fact that they possess higher percent grafting.

The morphologies of the worn pin surfaces are shown in Fig. 6.11. In comparison with the flakelike wear failures on the surface of SiC-g-PGMA/Ep, the wear scars on the surface of SiC-g-P(GMA-co-St)/Ep composites are characterized by tiny scratches without any detached flakes. It can be attributed to the fact that the treatment via graft copolymerization onto the nanoparticles further improves the miscibility between the grafted polymer and matrix Ep. It should be beneficial to the dispersion of SiC nanoparticles, and improve the composites' integrity. As a result, the composites' wear resistance was raised accordingly.

### 6.2.3 Effect of the reaction between grafted polymer and Ep matrix on the nanocomposites' surface feature

The above results have demonstrated the efficiency of surface grafting treatment of SiC nanoparticles in improving sliding wear properties of Ep composites. Nevertheless, some details of the role of the chemical bonding on the composites' surface characteristics should be further investigated. Figure 6.12 shows the microhardness

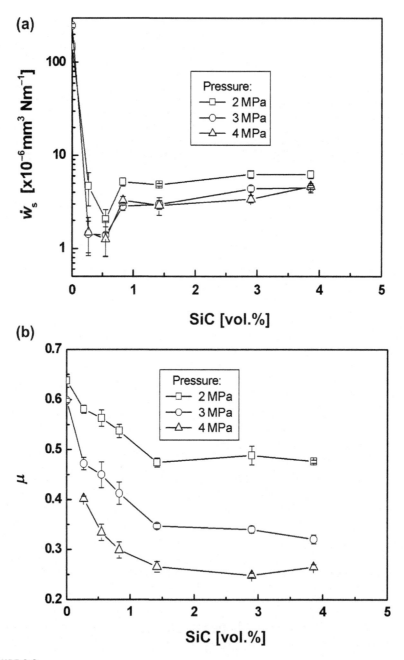

**FIGURE 6.8**

(a) Specific wear rate, $\dot{w}_s$, and (b) frictional coefficient, $\mu$, of SiC-g-PGMA-filled epoxy composites as a function of the SiC volume fraction.

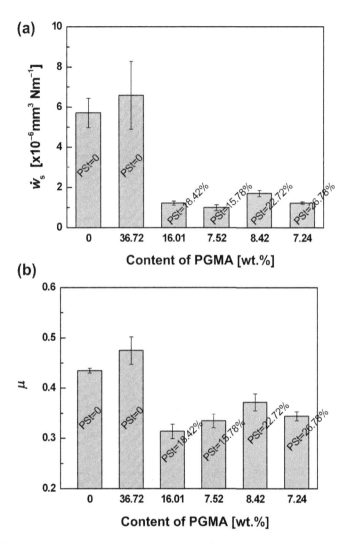

**FIGURE 6.9**

Effect of the PGMA/PS ratio of P(GMA-co-St) attached to SiC nanoparticles on (a) specific wear rate, $\dot{w}_s$, and (b) frictional coefficient, $\mu$, of SiC-g-P(GMA-co-St)-filled epoxy composites (SiC = 0.68 vol.%), PS - polystyrene.

of the materials' surfaces before and after the sliding wear tests. For unfilled Ep, the microhardness of the worn pin surface is remarkably lower than the value of the unworn one. This means that the repeated frictional force and high frictional temperature have deteriorated the microstructure of Ep. When nano-SiC is added, the microhardness of the worn pin surface of the composites is almost the same as that of the unworn one. Besides, it is noted that SiC-g-P(GMA-co-St)/Ep composite

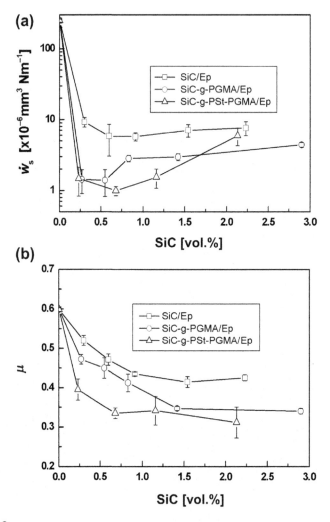

**FIGURE 6.10**

(a) Specific wear rate, $\dot{w}_s$, and (b) frictional coefficient, $\mu$, of untreated nano-SiC-filled epoxy, SiC-g-PGMA-filled epoxy ($\gamma_g = 8$ wt.%), and SiC-g-P(GMA-co-St)-filled epoxy ($\gamma_g = 23.3$ wt.%, PS/PGMA = 15.78/7.52) composites as a function of SiC volume fraction.

possesses a higher hardness, which is indicative of the strong interfacial interaction (chemical bonding) and coincides with its tribological behavior.

On the other hand, the surface feature of Ep and its composites was also characterized by load–unload curves recorded during the microhardness measurements (Fig. 6.13). In the case of unfilled Ep, the worn surface exhibits a greater proportion of plastic deformation as compared with the unworn one. This reflects destruction of the Ep network due to the high frictional temperature. However, the addition of SiC

 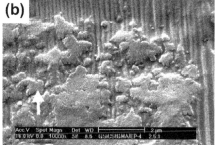

**FIGURE 6.11**

SEM micrographs of the worn surfaces of (a) SiC-g-PGMA-filled epoxy (SiC = 0.83 vol.%, $\gamma_g$ = 8 wt.%) and (b) SiC-g-P(GMA-co-St)-filled epoxy (SiC = 0.68 vol.%, $\gamma_g$ = 23.3 wt.%, PS/PGMA = 15.78/7.52) composites. The arrows indicate the sliding direction.

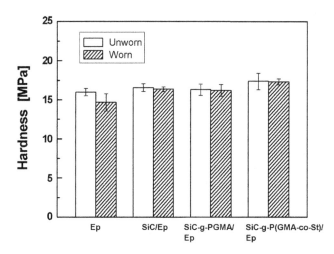

**FIGURE 6.12**

Microhardness of epoxy, untreated nano-SiC-filled epoxy (SiC = 0.3 vol.%), SiC-g-PGMA-filled epoxy (SiC = 0.27 vol.%, $\gamma_g$ = 8 wt.%) and SiC-g-PS-PGMA-filled epoxy (SiC = 0.68 vol.%, $\gamma_g$ = 23.3 wt.%, PS/PGMA = 15.78/7.52) composites before and after the wear tests.

nanoparticles exerts a restraint effect on the plastic deformation, and even results in a rise in the elastic deformation after the wear test, proving that the interaction between SiC nanoparticles and Ep is rather strong. When the SiC had been grafted with PGMA or P(GMA-co-St), the elastic deformation of the composites is further improved due to the chemical bonding at the interface. In this sense, the nanoparticles might act as crosslinking sites.

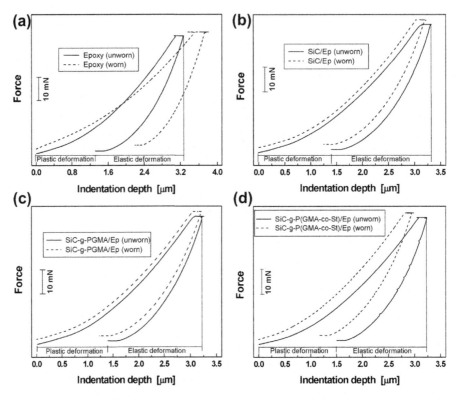

**FIGURE 6.13**

Load–unload curves measured during the microhardness tests of (a) epoxy, (b) untreated nano-SiC-filled epoxy (SiC = 0.91 vol.%), (c) SiC-g-PGMA-filled epoxy (SiC = 0.83 vol.%, $\gamma_g$ = 8 wt.%), and (d) SiC-g-P(GMA-co-St)-filled epoxy (SiC = 0.68 vol.%, $\gamma_g$ = 23.3 wt.%, PS/PGMA = 15.78/7.52) composites before and after the wear tests.

Quantitative description of the worn pin surface profile by atomic force microscope (AFM) presents information about the improvement of the composites' resistance to surface shearing due to the incorporation of SiC nanoparticles from another angle (Fig. 6.14). Clearly, the wear processes in the SiC-filled nanocomposites are dominated by polishing as the worn pin surfaces appear rather smooth. When the particles were grafted with PGMA or P(GMA-co-St), the average roughness of the composites is further reduced by two to three times. The results agree with the aforesaid enhanced integrity and higher wear resistance of the composites.

To illustrate the influence of the chemical bonding on the materials transfer from the specimens to the steel counterpart during wear tests, elements on the steel ring surface were examined (Table 6.2). It is seen that when the steel counterface had rubbed against either the unfilled Ep or the Ep-based composites, sulfur was detected. Besides, the amount of Si on the steel ring surface rubbing against the

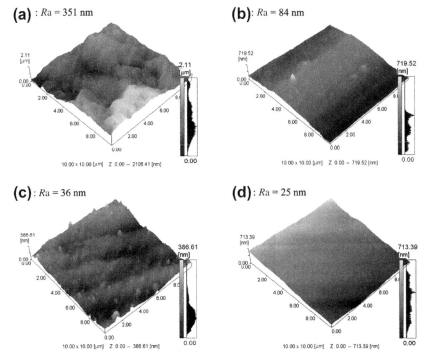

**FIGURE 6.14**

AFM 3D height trace images and average roughness, $R_a$, of the worn surfaces of (a) epoxy, (b) untreated nano-SiC-filled epoxy (SiC = 0.91 vol.%), (c) SiC-g-PGMA-filled epoxy (SiC = 0.83 vol.%, $\gamma_g$ = 8 wt.%), and (d) SiC-g-P(GMA-co-St)-filled epoxy (SiC = 0.68 vol.%, $\gamma_g$ = 23.3 wt.%, PS/PGMA = 15.78/7.52) composites.

**Table 6.2** X-Ray Energy Dispersive Spectroscopy (EDS) Analysis of the Steel Counterpart Surface ($p$ = 3 MPa, $v$ = 0.42 m s$^{-1}$, $t$ = 3 h)

|  | Elements and Contents (wt.%) | | |
|---|---|---|---|
|  | Fe | Si | S |
| Steel ring before wear | 98.76 | 0.48 | 0 |
| Steel ring rubbed against unfilled epoxy | 98.55 | 0.42 | 0.21 |
| Steel ring rubbed against untreated nano-SiC-filled epoxy composites (SiC = 0.6 vol.%) | 98.02 | 0.87 | 0.42 |
| Steel ring rubbed against SiC-g-PGMA-filled epoxy composites (SiC = 0.55 vol.%, $\gamma_g$ = 8.03 wt.%) | 98.37 | 0.51 | 0.28 |
| Steel ring rubbed against SiC-g-PS-PGMA-filled epoxy composites (SiC = 0.68 vol.%, $\gamma_g$ = 23.3 wt.%, PS/PGMA = 15.78/7.52) | 97.91 | 1.08 | 0.33 |

composites is higher than that on the steel ring rubbing against unfilled Ep. These data prove that the grafted polymer facilitates the adhesion of SiC particles on the steel ring, and strengthens the transferred film. Besides, the low-resolution X-ray photoelectron spectra (XPS) spectra of the pin surfaces of Ep and nano-SiC/Ep composites before and after the wear tests reveal that the atomic ratios of C/O have changed (Fig. 6.15). The decrease in the C/O ratios is indicative of oxidation of the Ep surface during wearing process. Incorporation of nano-SiC increases the resistance of Ep against oxidation, while the grafted nanoparticles are able to bring about a more prominent improvement. For SiC-g-P(GMA-co-St)/Ep system, the C/O ratios remain unchanged after wearing.

Dynamic mechanical analysis has long been used as a sensitive method to identify the interfacial interaction in composites. Generally, the strong interaction between the nanoparticle and matrix might restrict the movement of polymer

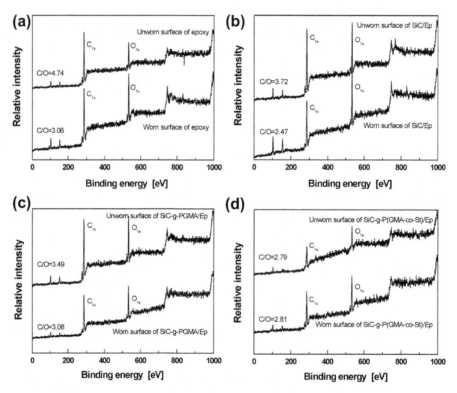

**FIGURE 6.15**

Low-resolution XPS spectra of pin surfaces of (a) epoxy, (b) untreated nano-SiC-filled epoxy (SiC = 0.91 vol.%), (c) SiC-g-PGMA-filled epoxy (SiC = 0.83 vol.%, $\gamma_g$ = 8 wt.%), and (d) SiC-g-P(GMA-co-St)-filled epoxy (SiC = 0.68 vol.%, $\gamma_g$ = 23.3 wt.%, PS/PGMA = 15.78/7.52) composites before and after the wear tests.

segments nearby the particulate fillers, resulting in a rise in the α-peak temperature and intensity [49,50]. Figure 6.16 shows the temperature dependence of the storage modulus and mechanical loss of Ep and its composites. As expected, the addition of either untreated nano-SiC or grafted nano-SiC enhances the storage modulus of Ep, especially at low-temperature ranges (Fig. 6.16(a)). The composite

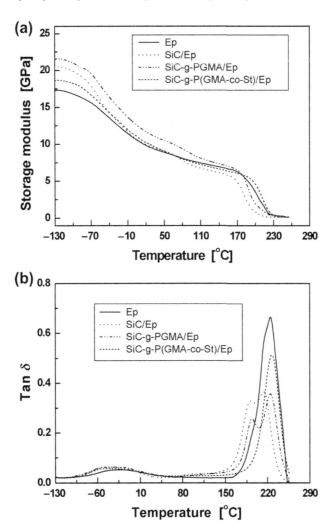

**FIGURE 6.16**

(a) Storage modulus and (b) mechanical loss spectra of epoxy, untreated nano-SiC-filled epoxy (SiC = 0.91 vol.%), SiC-g-PGMA-filled epoxy (SiC = 0.83 vol.%, $\gamma_g$ = 8 wt.%), and SiC-g-P(GMA-co-St)-filled epoxy (SiC = 0.68 vol.%, $\gamma_g$ = 23.3 wt.%, PS/PGMA = 15.78/7.52) composites.

with SiC-g-PGMA shows the highest storage modulus because the particles contain a higher content of GMA groups. In contrast to conventional composites, however, the incorporation of either the untreated or treated nanoparticles results in the reduction of both peak temperatures and intensities of glass transition of the matrix Ep (Fig. 6.16(b)). For the composites with untreated SiC and SiC-g-PGMA, the α-peaks were split into two parts, while the SiC-g-P(GMA-co-St)/Ep composite still exhibits a single α-peak. These phenomena should be related to the change of the crosslinking density at the interface rather than to the effect of the nanoparticles. The extra epoxide groups due to the grafted PGMA might lead to a lower crosslinking density at the interface and hence a decreased peak temperature and height of α-transition. This means that the looseness of Ep at the interface is so serious that the reinforcing effect of the fillers has been counteracted. When St was copolymerized into the grafted polymer, compatibility between the grafted polymer and matrix Ep would be increased, while the amount of GMA groups was decreased. As a result, only a single α-peak with a lower intensity is observed.

## 6.3 EP FILLED WITH LUBRICANT OIL-LOADED MICROCAPSULES, GRAFTED SiO₂ NANOPARTICLES, AND SCFS

Due to the fact that multicomponents are added to the Ep composites, the influence of filler contents on the tribological properties has to be studied by means of an orthographic factorial design, a technique to effectively reduce the number of tests and to determine the optimal conditions. Accordingly, an orthogonal $L_{16}(4)^5$ test design was used to find out the optimal filler contents of the composites (Tables 6.3 and 6.4). The term $L_{16}(4)^5$ represents an orthogonal array that handles up to five factors (i.e. five columns in the array) at four levels each, and requires 16 trial runs. When the number of factors is <5, the remaining columns (blank columns) can be used to calculate experimental errors. Here, the numbers 1–4 (Table 6.4, second to sixth columns) stand for the values of a designed factor at levels 1–4, respectively. In other words, these numbers designate each special trial run of the experiment. For example, in the

**Table 6.3** Creation of an Orthographic Factorial Design of Three Factors and Four Levels

| Factor | Level | | | |
|---|---|---|---|---|
| | 1 | 2 | 3 | 4 |
| Content of oil-loaded microcapsules (diameter = 110 µm; core content = 82.9 wt.%), $C_m$ (phr) | 3 | 6 | 8 | 10 |
| Content of nano-SiO₂ (12 nm), $C_p$ (phr) | 1 | 3 | 5 | 10 |
| Content of SCF (SCF, length = 1 mm), $C_f$ (phr) | 0.5 | 1 | 2 | 5 |

first row of Table 6.4 (test No.1), the levels of factor $C_m$ (which is assigned to the first column of the array) and those of factors $C_p$ and $C_f$ are 1. As a result, the first trial run of the experiment is assigned as a level set of $\{1, 1, 1\}$ for factors $C_m$, $C_p$, and $C_f$. The other trial runs perform the experiments in the same way in order to follow the expected property of orthogonality. In this study, the sliding wear experiments were carried out with three factors and four levels (Table 6.3), namely, within the filler

**Table 6.4** Results Analysis of the Orthographic Factorial Design

| Test No. | Factor | | | | | $\mu$ | $\dot{w}_s$ ($\times 10^{-6}$ mm³ Nm⁻¹) |
|---|---|---|---|---|---|---|---|
| | $C_m$ | $C_p$ | $C_f$ | _[a] | _[a] | | |
| 1 | 1 | 1 | 1 | 1 | 1 | 0.51 | 18.92 |
| 2 | 1 | 2 | 2 | 2 | 2 | 0.42 | 10.74 |
| 3 | 1 | 3 | 3 | 3 | 3 | 0.41 | 9.65 |
| 4 | 1 | 4 | 4 | 4 | 4 | 0.48 | 8.23 |
| 5 | 2 | 1 | 2 | 3 | 4 | 0.23 | 4.78 |
| 6 | 2 | 2 | 1 | 4 | 3 | 0.20 | 5.66 |
| 7 | 2 | 3 | 4 | 1 | 2 | 0.18 | 3.22 |
| 8 | 2 | 4 | 3 | 2 | 1 | 0.25 | 4.13 |
| 9 | 3 | 1 | 3 | 4 | 2 | 0.14 | 1.01 |
| 10 | 3 | 2 | 4 | 3 | 1 | 0.13 | 0.96 |
| 11 | 3 | 3 | 1 | 2 | 4 | 0.12 | 1.46 |
| 12 | 3 | 4 | 2 | 1 | 3 | 0.16 | 1.25 |
| 13 | 4 | 1 | 4 | 2 | 3 | 0.13 | 0.87 |
| 14 | 4 | 2 | 3 | 1 | 4 | 0.12 | 0.98 |
| 15 | 4 | 3 | 2 | 4 | 1 | 0.11 | 0.75 |
| 16 | 4 | 4 | 1 | 3 | 2 | 0.15 | 1.08 |
| $\mu$ | | | | | | | |
| $K1$[b] | 0.455 | 0.253 | 0.245 | 0.242 | 0.250 | | |
| $K2$ | 0.215 | 0.217 | 0.230 | 0.230 | 0.223 | | |
| $K3$ | 0.138 | 0.205 | 0.230 | 0.230 | 0.225 | | |
| $K4$ | 0.128 | 0.260 | 0.230 | 0.232 | 0.237 | | |
| $R$[c] | 0.327 | 0.055 | 0.015 | 0.012 | 0.027 | | |
| $\dot{w}_s$ ($\times 10^{-6}$ mm³ Nm⁻¹) | | | | | | | |
| $K1$ | 11.885 | 6.395 | 6.780 | 6.093 | 6.190 | | |
| $K2$ | 4.448 | 4.585 | 4.380 | 4.300 | 4.013 | | |
| $K3$ | 1.170 | 3.770 | 3.943 | 4.117 | 4.358 | | |
| $K4$ | 0.920 | 3.672 | 3.320 | 3.913 | 3.863 | | |
| $R$ | 10.965 | 2.723 | 3.460 | 2.180 | 2.327 | | |

[a]Dummy factors.
[b]Ki denotes the average of friction coefficient or specific wear rate with level i.
[c]R denotes the result of extreme analysis.

content ranges from 3 to 10 phr for the oil-loaded microcapsules, from 1 to 10 phr for SMA-grafted nano-SiO2 (SiO$_2$-g-SMA), and from 0.5 to 5 phr for SCF. The selection of each factor level was based on the results of our previous works [42,45]. The friction coefficient, μ, and specific wear rate, $\dot{w}_s$, act as the dependent variables.

The experimental data and related analysis are given in Fig. 6.17 and Table 6.4. The plots in Fig. 6.17 indicate that the composite with the highest microcapsule content offers the lowest wear rate and friction coefficient. Considering that the

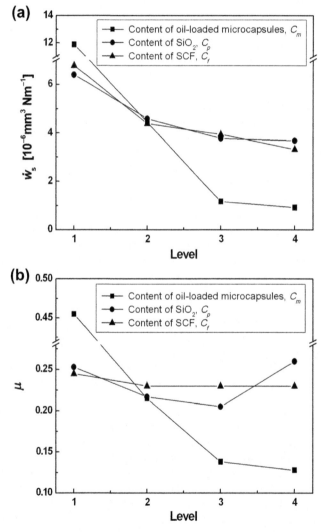

**FIGURE 6.17**

Influence of the different levels of the three factors on (a) specific wear rate, $\dot{w}_s$, and (b) friction coefficient, μ of the composites. The data were collected from $K_i$ in Table 6.4.

mechanical properties of Ep significantly decrease with the content of the oil-loaded microcapsules [42], however, level 3 (i.e. 8 phr) is chosen as the optimal content of the capsules. For $SiO_2$-g-SMA, the lowest specific wear rate appears at level 4 (10 phr), while the lowest friction coefficient at level 3 (5 phr). As a result, level 3 (5 phr) is selected in view of cost effectiveness. Similarly, the content of SCF is fixed at level 2 (1 phr). In summary, the optimal filler contents for formulating the wear-resisting and friction-reducing composite are 8 phr for the oil-loaded microcapsule, 5 phr for the grafted nano-$SiO_2$ and 1 phr for SCF.

As exhibited in Fig. 6.17, the content of the oil-loaded microcapsules has a very evident influence on both specific wear rate and friction coefficient. The oil released from the broken microcapsules due to repeated friction during sliding wear provides a remarkable lubricating effect. Comparatively, effects of $SiO_2$ nanoparticles and carbon fibers that are known as solid lubricants [42,51] are less prominent. Results of extreme analysis (Table 6.4) indicate that, for specific wear rate, the influence of the SCF content is more obvious than that of nano-$SiO_2$, while for friction coefficient, the situation is reversed. These might be related to the characteristics of sliding wear and the nature of the fillers as well. Since the wear rate is closely related to surface integrity, the strengthening effect resulting from fiber reinforcement must make a greater contribution than that of $SiO_2$ nanoparticles. On the other hand, nanoparticles proved to be able to protect the fibers by the rolling effect [23], suggesting that the former would be more favorable for reducing friction between the composite pin and the steel counterpart. As for the increase in the friction coefficient when the content of nano-$SiO_2$ exceeds 5 phr (Fig. 6.17(b)), it must be attributed to the worse dispersion of the particles in the composite, which causes particle pullout associated with the accumulation of detached harder bulk particulate fillers adherent to the frictional surface [29].

To check whether the formulation concluded in the above works, control experiments were done using recipes that are different from the optimal one (Table 6.5, Fig. 6.18). It can be seen from Fig. 6.18 that the addition of the oil-loaded microcapsules alone (recipe 1#) can increase wear resistance of the Ep by >60 times and reduce friction coefficient by about 74%. Further incorporation of $SiO_2$-g-SMA (recipe 2#) or SCF (recipe 3#) to the oil-loaded microcapsule/Ep composite leads to continuous reduction in the specific wear rate and friction coefficient. Compared to

**Table 6.5** Recipes for Comparing the Tribological Performance of the Epoxy Composites[a]

| Recipe ID | Content of Oil-Loaded Microcapsules, $C_m$ (phr) | Content of Nano-$SiO_2$, $C_p$ (phr) | Content of SCF, $C_f$ (phr) |
|---|---|---|---|
| 1# | 8 | 0 | 0 |
| 2# | 8 | 5 | 0 |
| 3# | 8 | 0 | 1 |
| 4# | 8 | 5 | 1 |

[a]Recipe 4# represents the optimal one concluded from the orthographic factorial design.

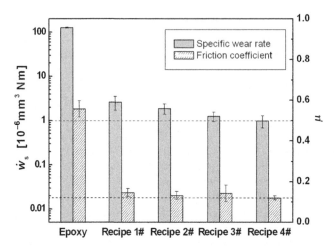

**FIGURE 6.18**

Specific wear rate, $\dot{w}_s$, and friction coefficient, μ, of epoxy and its composites.

SiO$_2$-g-SMA, SCF is more effective in reducing the wear rate although its content is lower. Likewise, SiO$_2$-g-SMA is able to decrease friction coefficient more remarkably than SCF can. When all the three types of additives are added to the composite (recipe 4#), a positive synergetic effect is observed as revealed by the lowest wear rate ($9.8 \times 10^{-7}$ mm$^3$ Nm$^{-1}$) and friction coefficient (0.12).

The worn surfaces of the unfilled Ep and its composites were observed by SEM (Fig. 6.19). In the case of neat Ep (Fig. 6.19(a)), fatigue wear should be the main mechanism responsible for material loss in the form of thin sheets, so that the worn surface is characterized by scalelike damage with cracks. For the composite filled with the oil-loaded microcapsules, a two-body abrasive should be the main mechanism, cavities from ruptured capsules are perceived (Fig. 6.19(b–e)), but the worn surfaces are much smoother than that of the unfilled Ep. When SiO$_2$-g-SMA or SCF is incorporated, the worn surface becomes less coarse (Fig. 6.19(c,d)) as compared to the case of oil-loaded microcapsules/Ep composite (Fig. 6.19(b)). The smaller wear debris further prevents bulk material from possible three-body abrasive wear. Furthermore, with the incorporation of oil-loaded microcapsules, SiO$_2$-g-SMA and SCF, the worn surface of the composite is rather flat (Fig. 6.19(e)), which is completely different from the others (Fig. 6.19(a–d)). Almost all the wear particles are captured by the cavities left by the broken microcapsules. Accordingly, the wear mode is changed to mild polishing [52,53]. Carbon fibers are able to reduce the wear rate and friction coefficient of polymers mainly due to their remarkable load-carrying capability. Wang and coworkers indicated that in nano-SiO$_2$ filled thermoplastic composites, the nanoparticles facilitated the formation of a thin, uniform, and tenacious transfer film, which was largely contributed to the decreased friction coefficient and wear rate of the composites [54].

In the present work, however, no obvious transfer of polymer to the steel counterpart occurred. Instead, the liberated oil from the broken microcapsules helped to

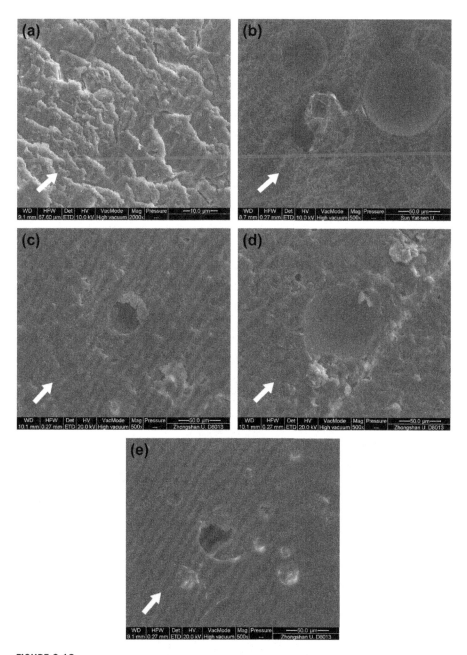

**FIGURE 6.19**

SEM micrographs of the worn surfaces of (a) unfilled epoxy, (b) oil-loaded microcapsules/epoxy composite (recipe 1#), (c) $SiO_2$-g-SMA/oil-loaded microcapsules/epoxy composite (recipe 2#), (d) SCF/oil-loaded microcapsules/epoxy composite (recipe 3#), and (e) $SiO_2$-g-SMA/SCF/oil-loaded microcapsules/epoxy composite (recipe 4#). The arrows show the sliding direction.

establish an oil film that greatly reduces the frictional force. This is evidenced by the changes in the surface energy of the counterface in terms of contact angle. As shown in Fig. 6.20, contact angle of the steel ring rubbed against neat Ep is 85°, slightly higher than the value of clean one (i.e. 82°). Having rubbed against the composite containing oil-loaded capsules (recipe 1#), however, the counterface has a significantly increased contact angle of 96° as a result of greatly raised hydrophobicity. The result explains why the orthogonal tests suggest 8 phr of the microcapsules is indispensable for yielding a low friction coefficient. On the other hand, the role of SiO$_2$-g-SMA and SCF should lie in the enhancement of structural integrity of the composites, and hence they can improve the tribological properties [42,55]. Moreover, the increased stiffness of Ep by incorporating nanoparticles can reduce stress concentration on SCF. During sliding wear, stress transfer from matrix to nanoparticles takes place in the frictional layer, and hence, the stress concentration on fibers is lowered [56]. This can be proved by observing morphologies of the worn surface

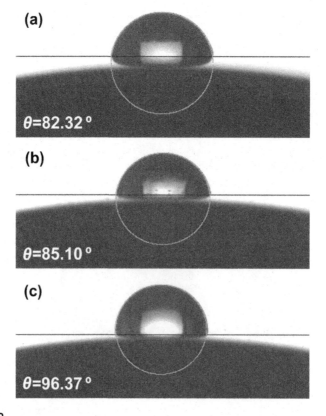

**(a)**

$\theta=82.32°$

**(b)**

$\theta=85.10°$

**(c)**

$\theta=96.37°$

**FIGURE 6.20**

Water contact angle measurements of (a) unworn counterface, (b) counterface rubbed against unfilled epoxy, and (c) counterpart rubbed against oil-loaded microcapsules/epoxy composite (recipe 1#).

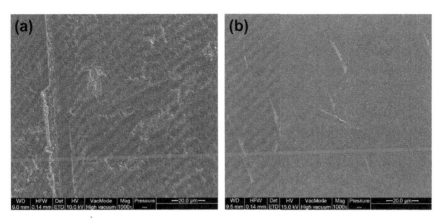

**FIGURE 6.21**

Morphologies of worn surfaces of (a) SCF/epoxy (content of SCF: 10 phr) and (b) $SiO_2$-g-SMA/SCF/epoxy composites (contents of $SiO_2$ and SCF: 4 and 6 phr, respectively).

of SCF/Ep and $SiO_2$-g-SMA/SCF/Ep composites (Fig. 6.21). Fiber breakage and pulloff are clearly seen in the worn SCF/Ep (Fig. 6.21(a)), while these are remarkably alleviated when $SiO_2$-g-SMA is added (Fig. 6.21(b)).

Our earlier works demonstrated that thermal decomposition characteristics of frictional surface layer helps to understand the underlying mechanism of the tribological performance [57,58]. Accordingly, pyrolytic habits of worn pin surfaces of Ep and its composites were measured in comparison to those of bulk materials (i.e. unworn versions). As shown in Table 6.6, the peak pyrolysis temperature of the worn surface of unfilled Ep is obviously lower than that of the unworn version. This means that structural degradation must have occurred in terms of damage of the molecular structure due to high frictional temperature and contact pressure [13,58]. With respect to the composites (recipes 1–4#), however, the differences between the peak pyrolysis temperature of Ep in the unworn specimens (i.e. the bulk materials)

**Table 6.6** Peak Pyrolytic Temperatures of Epoxy and its Composites before and after Sliding Wear Tests[a]

| Recipe ID | Peak Pyrolytic Temperature (°C) | |
| --- | --- | --- |
| | **Bulk** | **Worn Surface** |
| Unfilled epoxy | 430.4 | 425.8 |
| 1# | 429.6 | 428.5 |
| 2# | 431.1 | 429.1 |
| 3# | 430.1 | 429.5 |
| 4# | 430.0 | 430.1 |

[a]The listed peak pyrolytic temperatures of the composites only refer to those of the epoxy in the composites.

and the worn surface layers are rather small or even negligible. Clearly, the frictional heat accumulated in the rubbing zones must be greatly decreased in these cases, and hence, the possibility of surface failure is lowered. This in turn reduces wear rates and friction coefficients of the composites to a much greater extent.

To have further information about the mechanochemically induced changes in the surface layer during wear, XPS studies of the pin surfaces of Ep and its composites were conducted. It is seen that the ratios of C/O of the worn surfaces are higher than those of the unworn ones (Tables 6.7 and 6.8). The phenomenon is indicative of oxidation and carbonization, reflecting partial chain scission and reorganization of Ep macromolecules at the surface of the specimen pins in the course of wear tests [13]. With the appearance of oil-loaded microcapsules, the difference in the C/O ratio between the worn and unworn surfaces decreases as a result of reduced degree of oxidation and carbonization. When $SiO_2$-g-SMA and/or SCF are added, the differences in C/O ratio between the worn and unworn surfaces are significantly reduced. It manifests that the tribochemically induced structure variation on top layers of the composites pins had become rather mild. The deduction receives support from the above thermal stability study of the worn surfaces, and explains the improved tribological properties from another angle.

**Table 6.7** Survey XPS Spectra of Pin Surfaces of Epoxy and its Composites before and after Sliding Wear

| Recipe ID | | $C_{1s}$ (%) | $O_{1s}$ (%) | $N_{1s}$ (%) | $S_{2p}$ (%) | $Si_{2p}{}^a$ (%) | Fe2p (%) |
|---|---|---|---|---|---|---|---|
| Unfilled epoxy | Unworn | 72.37 | 23.57 | 2.77 | 1.18 | 0.11 | N/A |
| | Worn | 77.24 | 17.76 | 3.57 | 1.24 | 0.19 | N/A |
| 1# | Unworn | 77.01 | 18.87 | 2.97 | 0.98 | 0.16 | N/A |
| | Worn | 78.72 | 17.00 | 3.47 | 0.54 | 0.27 | N/A |
| 2# | Unworn | 71.88 | 18.20 | 2.49 | 0.59 | 6.04 | N/A |
| | Worn | 73.88 | 17.97 | 3.35 | 0.78 | 4.01 | N/A |
| 3# | Unworn | 77.55 | 17.51 | 3.64 | 0.95 | 0.35 | N/A |
| | Worn | 76.97 | 16.66 | 4.11 | 0.87 | 0.16 | 1.23 |
| 4# | Unworn | 75.56 | 17.29 | 2.70 | 0.93 | 3.51 | N/A |
| | Worn | 74.61 | 16.62 | 2.08 | 1.01 | 4.91 | 0.77 |

$^a$The Si element in the epoxy and its composites (recipe 1# and 3#) is derived from SiC abrasive paper during the preworn process (see Experimental).

**Table 6.8** C/O Ratios of Pin Surfaces of the Epoxy and its Composites before and after Sliding Wear

| Recipe ID | Unfilled Epoxy | 1# | 2# | 3# | 4# |
|---|---|---|---|---|---|
| Unworn | 3.07 | 4.08 | 3.95 | 4.43 | 4.37 |
| Worn | 4.35 | 4.63 | 4.11 | 4.62 | 4.49 |

On the other hand, our XPS experiments indicate that Fe2p spectra are only collected from the worn surfaces of the composites of recipes 3# and 4# (Table 6.7 and Fig. 6.22). It is believed that the Fe element had been generated by some abrasive action of the SCF on the steel counterpart asperities, transferred from the counterface and then oxidized, forming FeO and $Fe_3O_4$ that favor reduction in friction and wear [59]. In this

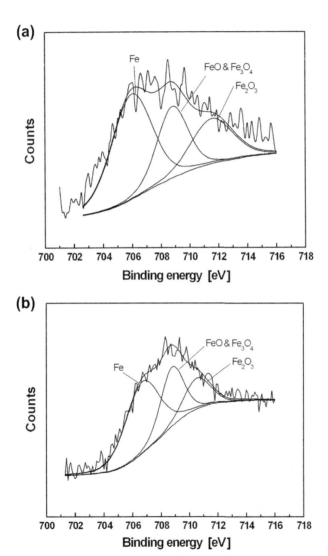

**FIGURE 6.22**

High-resolution Fe2p spectra of worn surfaces of (a) SCF/oil-loaded microcapsules/epoxy composite (recipe 3#) and (b) $SiO_2$-g-SMA/SCF/oil-loaded microcapsules/epoxy composite (recipe 4#).

context, the decreased wear rate and friction coefficient of the composites of recipes 3# and 4# can be partly attributed to the appearance of Fe oxide on the counterfaces [60].

Since mechanical properties are critical for the practical applications of tribological materials, flexural strength and modulus of the composites were measured (Fig. 6.23). As reported in our previous work [45], the oil-loaded microcapsules greatly deteriorate mechanical properties of Ep (recipe 1#). When $SiO_2$-g-SMA is added, both flexural strength and modulus are increased (recipe 2#). This improvement is attributed to the high interfacial stress transfer efficiency of the nanocomposites [61], resulting from the covalent bonding at the grafted $SiO_2$/Ep interface. With respect to the SCF/oil-loaded microcapsules/Ep composite (recipe 3#), its flexural strength and modulus are also higher than those of the oil-loaded microcapsules/Ep composite (recipe 1#). Although the strengthening effect of SCF is not evident because of the low concentration (i.e. only 1 phr), the increase in stiffness is superior to the case with nano-$SiO_2$, probably owing to the high modulus nature of the fibers. It is interesting to see that flexural strength of the quaternary composite (recipe 4#) is only 13% lower than that of unfilled Ep, while flexural modulus of the former is about 13% higher than that of the latter. Clearly, the loss of mechanical properties due to the addition of the oil-loaded microcapsules can be compensated by introducing nano-$SiO_2$ and SCF, as desired in the Introduction part.

By comparing Fig. 6.18 with Fig. 6.23, it can be seen that sliding wear performance is closely related to mechanical properties of the material. Incorporation of $SiO_2$-g-SMA into the oil-loaded microcapsules/Ep composite leads to a greater enhancement of flexural strength than SCF does. Accordingly, the composite made from recipe 2# exhibits a lower friction coefficient than the one from recipe 3#. In addition, the composite made from recipe 3# is evidently stiffer than that from recipe 2#, and hence, the former offers a lower specific wear rate. The findings coincide

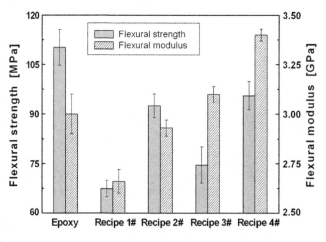

**FIGURE 6.23**

Mechanical properties of epoxy and its composites.

with the results reported by Zhang et al. [55], and can partly explain the very low wear rate and friction coefficient of the quaternary composite (recipe 4#). Therefore, the optimal formulation (i.e. recipe 4#) yielded from the orthographic factorial design using tribological performance as the response of the designed experiments also meet the requirements of mechanical properties.

Figure 6.24 shows the maximal loading ability of the materials (the highest normal load at which the composites could endure at the fixed sliding velocity of 0.42 m s$^{-1}$). Unfilled Ep specimens failed after sliding for 10 min under the load of 4 MPa. For the oil-loaded microcapsules/Ep composite (recipe 1#), it can only survive the load of 4 MPa in spite of its low friction coefficient. Relatively speaking, the inclusion of $SiO_2$-g-SMA or SCF can significantly enhance the loading ability of the composites (recipe 2# and 3#). This is particularly true when both the nanoparticles and carbon fibers are added (recipe 4#). The quaternary composite has a maximal loading ability of as high as 7 MPa, which is more than double that of unfilled Ep. Once again, the results show the importance of hybrid filling with $SiO_2$ and carbon fibers.

## 6.4 SUMMARY

Frictional coefficient and wear rate of Ep can be reduced by the addition of inorganic nanoparticles at low filler content. Grafting of polymeric chains with reactive groups onto the nanoparticles, which helps to establish chemical bonding with matrix Ep during composite manufacturing, proves to be an effective way to produce nanocomposites with further improved tribological performance.

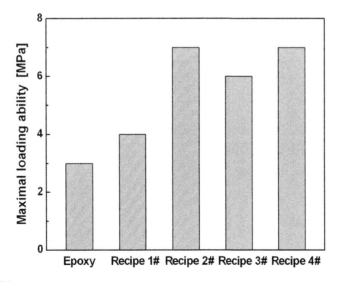

**FIGURE 6.24**

Maximal loading abilities of epoxy and its composites.

In fact, the surface treatment method mentioned above is also feasible for other inorganic nanoparticle–polymer composites aiming at improving particle dispersion and interface interaction. Continuous exploration of interplay between nanoparticles and the matrix and a comprehensive understanding of the underlying parameters determining the role of nanoparticles would facilitate to design a new generation of nanocomposites with balanced performance.

For purposes of preparing Ep material with extremely low frictional coefficient and wear rate, hybrid fillers including oil-loaded microcapsules, grafted nanoparticles and SCFs can be added. Oil-loaded microcapsules are very effective in reducing wear rate and friction coefficient of cured Ep by releasing trace quantity of lubricant to the rubbing surface, while grafted nanoparticles and SCFs play dual roles of both solid lubricant and reinforcement. Consequently, the weakness brought by the addition of oil-loaded microcapsules (i.e. mechanical properties deterioration of the composites) can be overcome by introducing the grafted nanoparticles and SCFs. During sliding wear, molecular structural degradation of frictional surface is greatly inhibited, and wear rate and friction coefficient of the composite are decreased to a very low level accordingly.

By surveying the Ep-based nanocomposites prepared so far, it can be concluded that inorganic nanoparticles are capable of providing the matrix resin with significantly enhanced tribological properties. Similar achievements due to the addition of a small amount of fillers are impossible to be perceived in microcomposites. The developments in this aspect have broadened the application possibility of particulate composites and solved the dilemma arising from the contradiction between tribological performance improvement and processability deterioration, as often observed in microparticle-filled composites. Nevertheless, Ep nanocomposites are still in their infancy. The reported results do not always coincide with each other. Even in the case of the same nanoparticle species, the observed "nanoeffects" are sometimes different. A general understanding of the role of nanoparticles and selection criterion for nanocomposites' components in terms of wear and friction performance has not yet been established. The situation in turn leaves room for continuous researches in the related areas.

## LIST OF ABBREVIATIONS

**AFM** atomic force microscope
**Al$_2$O$_3$** aluminum oxide
**C** carbon
**CaSiO$_3$** calcium silicate
**CNTs** carbon nanotubes
**CuO** copper oxide
**DDS** 4,4-diaminodiphenysulfone

**DMA**  dynamic mechanical analyzer (or analysis)
**DSC**  differential scanning calorimeter (or calorimetry)
**EDS**  X-ray energy dispersive spectroscopy
**Ep**  epoxy
**Fe$_3$O$_4$**  triiron tetraoxide
**FTIR**  Fourier transform infrared
**GMA**  glycidyl methacrylate
**GPC**  gel permeation chromatograph
**MWNTs**  multiwalled carbon nanotubes
**O**  oxide
**oMMT**  organomodified montmorillonite
**PGMA**  polyglycidyl methacrylate
**PSt**  polystyrene
**PTFE**  polytetrafluoroethylene
**SCF**  short carbon fiber
**SDS**  sodium dodecyl sulfonate
**SEM**  scanning electron microscope
**Si**  silicon
**SiC**  silicon carbide
**SiC-g-PGMA**  polyglycidyl methacrylate grafted silicon carbide nanoparticles
**SiC-g-P(GMA-co-St)**  copolymer of glycidyl methacrylate and styrene grafted silicon carbide nanoparticles
**Si$_3$N$_4$**  silicon nitride
**SiO$_2$**  silicon dioxide
**SiO$_2$-g-SMA**  poly(styrene-co-maleic anhydride) grafted silicon dioxide nanoparticles
**SMA**  poly(styrene-co-maleic anhydride)
**St**  styrene
**TGA**  thermogravimetric analyzer (or analysis)
**TiO$_2$**  titanium dioxide
**XPS**  X-ray photoelectron spectroscopy
**ZnO**  zinc oxide

# Acknowledgments

The financial support by the National Natural Science Foundation of China (Grant: 50273047) is gratefully acknowledged. Further thanks are due to the Team Project of the Natural Science Foundation of Guangdong, China (Grant: 20003038), the Key Program of the Science and Technology Department of Guangdong, China (Grant: 2004A10702001), and Sino-Hungarian Scientific and Technological Cooperation Project (Grant: 2009DFA52660).

# References

[1] M.Q. Zhang, M.Z. Rong, K. Friedrich, in: H.S. Nalwa (Ed.), Processing and Properties of Nonlayered Nanoparticle Reinforced Thermoplastic Composites, Handbook of Organic–Inorganic Hybrid Materials and Nanocomposites, vol. 2, American Science Publishers, California, 2003, pp. 113–150.

[2] H. Fischer, Polymer nanocomposites: from fundamental research to specific applications, Material Science and Engineering C 23 (2003) 763–772.

[3] Z.M. Huang, Y.Z. Zhang, M. Kotaki, S. Ramakrishna, A review on polymer nanofibers by electrospinning and their applications in nanocomposites, Composites Science and Technology 63 (2003) 2223–2253.

[4] M.Z. Rong, M.Q. Zhang, S.L. Pan, K. Friedrich, Interfacial effects in polypropylene–silica nanocomposites, Journal of Applied Polymer Science 92 (2004) 1771–1781.

[5] B.J. Briscoe, The tribology of composite materials: a preface, in: K. Friedrich (Ed.), Advance in Composite Tribology, Elsevier, Amsterdam, 1993, pp. 3–15.

[6] S. Bahadur, D. Gong, The action of fillers in the modification of the tribological behaviour of polymers, Wear 158 (1992) 41–59.

[7] J.M. Duand, M. Vardavoulias, M. Jeandin, Role of reinforcing ceramic particles in the wear behaviour of polymer-based model composites, Wear 181–183 (1995) 826–832.

[8] Y. Yamaguchi, in: Y. Yamaguchi (Ed.), Improvement of Lubricity, Tribology of Plastic Materials, Tribology Series, vol. 16, Elsevier, Amsterdam, 1990, pp. 143–202.

[9] V.A. Bely, A.I. Sviridenok, M.I. Petrokovets, V.G. Savkin, Friction and Wear in Polymer-Based Materials, Pergamon Press Ltd., Oxford, 1982.

[10] N. Symonds, B.G. Mellor, Polymeric coatings for impact and wear resistance I: Wear 225–229 (1999) 111–118.

[11] M.Z. Rong, M.Q. Zhang, H. Liu, H.M. Zeng, B. Wetzel, K. Friedrich, Microstructure and tribological behavior of polymeric nanocomposites, Industrial Lubrication and Tribology 53 (2001) 72–77.

[12] C.B. Ng, L.S. Schadler, R.W. Siegel, Synthesis and mechanical properties of $TiO_2$-epoxy nanocomposites, Nanostructured Materials 12 (1999) 507–510.

[13] G. Shi, M.Q. Zhang, M.Z. Rong, B. Wetzel, K. Friedrich, Friction and wear of low nanometer $Si_3N_4$ filled epoxy composites, Wear 254 (2003) 784–796.

[14] J.C. Lin, Compression and wear behavior of composites filled with various nanoparticles, Composites Part B-Engineering 38 (2007) 79–85.

[15] J.B. Zhu, X.J. Yang, Z.D. Cui, S.L. Zhu, Q. Wei, Preparation and properties of nano-$SiO_2$/epoxy composites cured by Mannich amine, Journal of Macromolecular Science, Physics 45 (2006) 811–820.

[16] M. Spirkova, M. Slouf, O. Blahova, T. Farkacova, J. Benesova, Submicrometer characterization of surfaces of epoxy-based organic–inorganic nanocomposite coatings. A comparison of AFM study with currently used testing techniques, Journal of Applied Polymer Science 102 (2006) 5763–5774.

[17] Q.M. Jia, M. Zheng, C.Z. Xu, H.X. Chen, The mechanical properties and tribological behavior of epoxy resin composites modified by different shape nanofillers, Polymers for Advanced Technologies 17 (2006) 168–173.

[18] B. Wetzel, F. Haupert, M.Q. Zhang, Epoxy nanocomposites with high mechanical and tribological performance, Composites Science and Technology 63 (2003) 2055–2067.

[19] L. McCook, B. Boesl, D.L. Burris, W.G. Sawyer, Epoxy, ZnO, and PTFE nanocomposite: friction and wear optimization, Tribology Letters 22 (2006) 253–257.

[20] G.J. Xian, R. Walter, F. Haupert, A synergistic effect of nano-TiO$_2$ and graphite on the tribological performance of epoxy matrix composites, Journal of Applied Polymer Science 102 (2006) 2391–2400.

[21] Z. Zhang, C. Breidt, L. Chang, F. Haupert, K. Friedrich, Enhancement of the wear resistance of epoxy: short carbon fibre, graphite, PTFE and nano-TiO$_2$, Composites Part A: Applied Science and Manufacturing 35 (2004) 1385–1392.

[22] L. Chang, Z. Zhang, C. Breidt, K. Friedrich, Tribological properties of epoxy nanocomposites. I. Enhancement of the wear resistance by nano-TiO$_2$ particles, Wear 258 (2005) 141–148.

[23] L. Chang, Z. Zhang, Tribological properties of epoxy nanocomposites. Part II. A combinative effect of short carbon fibre with nano-TiO$_2$, Wear 260 (2006) 869–878.

[24] B. Dong, Z. Yang, Y. Huang, H.L. Li, Study on tribological properties of multi-walled carbon nanotubes/epoxy resin nanocomposites, Tribology Letters 20 (2005) 251–254.

[25] O. Jacobs, W. Xu, B. Schädel, W. Wu, Wear behaviour of carbon nanotube reinforced epoxy resin composites, Tribology Letter 23 (2006) 65–75.

[26] M.Z. Rong, M.Q. Zhang, W.H. Ruan, Surface modification of nanoscale fillers for improving properties of polymer nanocomposites: a review, Materials Science and Technology 22 (2006) 787–796.

[27] M.Z. Rong, M.Q. Zhang, G. Shi, Q.L. Ji, B. Wetzel, K. Friedrich, Graft polymerization onto inorganic nanoparticles and its effect on tribological performance improvement of polymer composites, Tribology International 36 (2003) 697–707.

[28] G. Shi, M.Q. Zhang, M.Z. Rong, B. Wetzel, K. Friedrich, Sliding wear behavior of epoxy containing nano-Al$_2$O$_3$ particles with different pretreatments, Wear 256 (2004) 1072–1081.

[29] M.Q. Zhang, M.Z. Rong, S.L. Yu, B. Wetzel, K. Friedrich, Effect of particle surface treatment on the tribological performance of epoxy based nanocomposites, Wear 253 (2002) 1086–1093.

[30] M.Q. Zhang, M.Z. Rong, S.L. Yu, B. Wetzel, K. Friedrich, Improvement of the tribological performance of epoxy by the addition of irradiation grafted nano-inorganic particles, Macromolecular Materials and Engineering 287 (2002) 111–115.

[31] Q.L. Ji, M.Z. Rong, M.Q. Zhang, K. Friedrich, Graft polymerization of vinyl monomers onto nanosized silicon carbide particles, Polymers and Polymer Composites 10 (2002) 531–539.

[32] Q.L. Ji, M.Q. Zhang, M.Z. Rong, B. Wetzel, K. Friedrich, Tribological properties of surface modified nano-alumina/epoxy composites, Journal of Materials Science 39 (2004) 6487–6493.

[33] Y.H. Hu, H. Gao, F.Y. Yan, W.M. Liu, C.Z. Qi, Tribological and mechanical properties of nano ZnO-filled epoxy resin composites, Tribology 23 (3) (2003) 216–220. (in Chinese).

[34] Q.L. Ji, M.Q. Zhang, M.Z. Rong, B. Wetzel, K. Friedrich, Tribological properties of nanosized silicon carbide filled epoxy composites, Acta Materiea Compositea Sinica 21 (6) (2004) 14–20. (in Chinese).

[35] B. Wetzel, F. Haupert, K. Friedrich, M.Q. Zhang, M.Z. Rong, Impact and wear resistance of polymer nanocomposites at low filler content, Polymer Engineering and Science 42 (2002) 1919–1927.

[36] J.O. Park, K.Y. Rhee, S.J. Park, Silane treatment of Fe$_3$O$_4$ and its effect on the magnetic and wear properties of Fe$_3$O$_4$/epoxy nanocomposites, Applications of Surface Science 256 (2010) 6945–6950.

[37] N.M. Rashmi, B. Renukappa, R.M. Suresha, K.N. Devarajaiah, Shivakumar, dry sliding wear behaviour of organo-modified montmorillonite filled epoxy nanocomposites using Taguchi's techniques, Materials and Design 32 (2011) 4528–4536.

[38] S.R. Ha, K.Y. Rhee, Effect of surface-modification of clay using 3-aminopropyl-triethoxysilane on the wear behavior of clay/epoxy nanocomposites, Colloid Surface A 322 (1–3) (2008) 1–5.

[39] S. Yu, H. Hu, J. Ma, J. Yin, Tribological properties of epoxy/rubber nanocomposites, Tribology International 41 (2008) 1205–1211.

[40] H. Chen, O. Jacobs, W. Wu, G. Rüdiger, B. Schädel, Effect of dispersion method on tribological properties of carbon nanotube reinforced epoxy resin composites, Polymer Testing 26 (2007) 351–360.

[41] T.Ø Larsen, T.L. Andersen, B. Thorning, A. Horsewell, M.E. Vigild, Changes in the tribological behavior of an epoxy resin by incorporating CuO nanoparticles and PTFE microparticles, Wear 265 (1–2) (2008) 203–213.

[42] Q.B. Guo, M.Z. Rong, G.L. Jia, K.T. Lau, M.Q. Zhang, Sliding wear performance of nano-SiO$_2$/short carbon fiber/epoxy hybrid composites, Wear 266 (7–8) (2009) 658–665.

[43] M.T. Kim, K.Y. Rhee, J.H. Lee, D. Hui, A.K.T. Lau, Property enhancement of a carbon fiber/epoxy composite by using carbon nanotubes, Composites Part B Engineering 42 (5) (2011) 1257–1261.

[44] Y. Luo, M.Z. Rong, M.Q. Zhang, K. Friedrich, Surface grafting onto SiC nanoparticles with glycidyl methacrylate in emulsion, Journal of Polymer Science Part A Polymer Chemistry 42 (2004) 3842–3852.

[45] Q.B. Guo, K.T. Lau, B.F. Zheng, M.Z. Rong, M.Q. Zhang, Imparting ultra-low friction and wear rate to epoxy by the incorporation of microencapsulated lubricant? Macromolecular Materials and Engineering 294 (2009) 20–24.

[46] C.L. Wu, M.Q. Zhang, M.Z. Rong, K. Friedrich, Tensile performance improvement of low nanoparticles filled-polypropylene composites, Composites Science and Technology 62 (2002) 1327–1340.

[47] Y. Luo, M.Z. Rong, M.Q. Zhang, Covalently connecting nanoparticles with epoxy matrix and its effect on the improvement of tribological performance of the composites, Polymers and Polymer Composites 13 (2005) 245–252.

[48] H.E. Kissinger, Reaction kinetics in differential thermal analysis, Analytical Chemistry 29 (1957) 1702–1706.

[49] H.H. Huang, G.L. Wilkes, J.G. Carlson, Structure-property behaviour of hybrid materials incorporating tetraethoxysilane with multifunctional poly(tetramethylene oxide), Polymer 30 (1989) 2001–2012.

[50] L. Nicolais, G. Carotenuto, X. Kuang, Synthesis and characterization of new polymer–ceramic nanophase composite materials, Applied Composites Materials 3 (1996) 103–116.

[51] G. Xian, R. Walter, F. Haupert, Friction and wear of epoxy/TiO$_2$ nanocomposites: influence of additional short carbon fibers, Aramid and PTFE particles, Composites Science and Technology 66 (2006) 3199–3209.

[52] H.G. Lee, H.Y. Hwang, D.G. Lee, Effect of wear debris on the tribological characteristics of carbon fiber epoxy composites, Wear 261 (2006) 453–459.

[53] H.G. Lee, S.S. Kim, D.G. Lee, Effect of compacted wear debris on the tribological behavior of carbon/epoxy composites, Composite Structures 74 (2006) 136–144.

[54] Q. Wang, Q. Xue, W. Shen, The friction and wear properties of nanometre SiO$_2$ filled polyetheretherketone, Tribology International 30 (1997) 193–197.

[55] Z. Zhang, C. Breidt, L. Chang, K. Friedrich, Wear of PEEK composites related to their mechanical performances, Tribology International 37 (2004) 271–277.

[56] G. Zhang, L. Chang, A.K. Schlarb, The roles of nano-$SiO_2$ particles on the tribological behavior of short carbon fiber reinforced PEEK, Composites Science and Technology 69 (2009) 1029–1035.

[57] M.Q. Zhang, Z.P. Lu, K. Friedrich, Thermal analysis of the wear debris of polyetheretherketone, Tribology International 30 (1997) 103–111.

[58] J. Lu, M.Q. Zhang, M.Z. Rong, S.L. Yu, B. Wetzel, K. Friedrich, Thermal stability of frictional surface layer and wear debris of epoxy nanocomposites in relation to the mechanism of tribological performance improvement, Journal of Materials Science 39 (2004) 3817–3820.

[59] H. Kong, E.-S. Yoon, O.K. Kwon, Self-formation of protective oxide films at dry sliding mild steel surfaces under a medium vacuum, Wear 181–183 (1995) 325–333.

[60] M. Nakamura, K. Hirao, Y. Yamauchi, S. Kanzaki, Tribological behaviour of uni-directionally aligned silicon nitride against steel, Wear 252 (2002) 484–490.

[61] B. Wetzel, P. Rosso, F. Haupert, K. Friedrich, Epoxy nanocomposites – fracture and toughening mechanisms, Engineering Fracture Mechanics 73 (2006) 2375–2398.

# Wear simulation of a polymer–steel sliding pair considering temperature- and time-dependent material properties

7

**László Kónya, Károly Váradi**

*Institute of Machine Design Budapest University of Technology and Economics.*
*Budapest, Hungary*

## CHAPTER OUTLINE HEAD

## 7.1 INTRODUCTION

Friction behavior and wear are one of the most characteristic features of polymer–metal components transferring load under sliding motion. The wear behavior is mostly studied by experimental techniques using, for example, a Pin-on-Disc Configuration (PoDC). In a PoDC (Fig. 7.1), the length reduction of the pin is measured to represent the volume of the material loss and to calculate the specific wear rate,

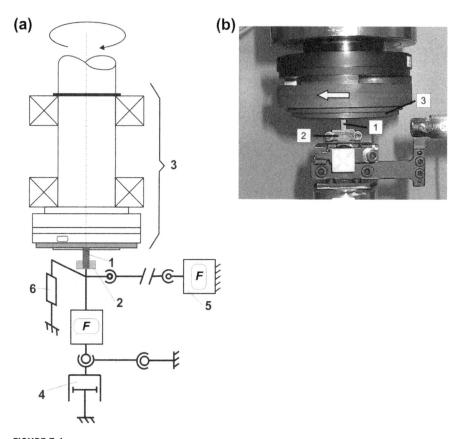

**FIGURE 7.1**

A schematic of PoDC studies (the notation: (1) pin, (2) pin holder, (3) disc side, (4) pneumatic cylinder, (5) force sensor, and (6) LVDT).

which is a widely used parameter of the wear behavior of different sliding pairs under given conditions.

To simulate the wear process, not so many attempts have been undertaken so far. Pödra [1] applied an incremental wear model to characterize the wear rate at room temperature. Results were presented for sphere-on-plate and cone-on-cone sliding contact applications.

For a PoD configuration, Yan et al. [2] prepared a wear simulation by "scaling the plane strain wear rate with a conversion factor related to the predicted shape of the wear track".

In order to have more information on friction and wear behavior, the heat generation should also be considered. The temperature distribution in a PoDC was studied by various analytical and numerical techniques mostly for metal sliding pairs, for example, in Refs [3,4]. In the case of polymers, the low thermal conductivity produces very different heat transportations. For a polyetheretherketone (PEEK)–steel sliding pair, in a PoDC, Kónya et al. [5] developed an FE moving heat source model and a substituting distributed one for the disc side and also a steady-state one for the pin side (see details in Section 7.3.3).

There are discussions over the role of the heat partition at the contact interface, especially if it is correct to assume the same contact temperature values at the opposing contacting nodes in the range of high sliding speeds [6,7]. In the present study, this assumption was applied, due to the moderate sliding speed. The other approach is to introduce a contact thermal resistance between the components of the engineering system, so as to consider the effect of any clearances [8]. In some cases, this was important to be considered.

Recently, also different wear simulation techniques are frequently used. The latest ones couple the thermal analysis, the time-dependent behavior of the materials, and the damage analysis. In Ref. [9], a simulation technique of deactivating elements was used for modeling wear, if the "extent" of wear is larger than the size of the elements in the FE mesh.

The present analysis aims to study the wear process of a PEEK–steel sliding pair at a 20 °C and 150 °C temperature level in a PoDC, based on the linear wear theory. The temperature-dependent material properties and the creep behavior of the pin are also taken into consideration. In the study at 150 °C, the disc side of the system had been preheated providing the elevated temperature condition. This temperature level represents the real operating conditions for PEEK material applications, since PEEK possesses excellent wear and reduced creep properties, which allow its maximum continuous use temperature to amount to 250 °C [10].

In the wear simulation algorithm, the contact calculations repeated, while the initial clearance is changed according to the wear depth increment and the thermal expansion of the system. The incremental algorithm has a cyclic operation using FE contact and thermal analysis of the COSMOS/M FE system [11].

Experimental data (coefficient of friction, temperature at different locations and wear depth change) are obtained for the given conditions in order to verify the simulation technique developed.

## 7.2 CHARACTERIZING THE WEAR PROCESS IN A PODC

The main elements of the PoDC, operating at the Institute for Composite Materials (IVW GmbH, Kaiserslautern, Germany), are illustrated in Fig. 7.1. The pin (1) and pin holder (2) are stationary components, while the complete disc side (3) is rotating. The pin is subjected to a normal load $F_n$ produced by a pneumatic cylinder (4), and the frictional force $F_s$ (measured by the horizontal force sensor (5)) is generated by the rotating motion (Fig. 7.2).

In the PoD arrangement, the wear reduces the length of the pin, which is proportional to the material loss. This length reduction is measured between the pin holder and the frame structure of the test configuration using a Linear Variable Differential

Transformer (LVDT) (6) shown in Fig. 7.1. Therefore, the deformation of the pin due to the compression and bending behavior as well as the thermal expansion of both sides are also involved, that is, not only the "neat wear" of the pin. Especially at the beginning of the test, the thermal expansion produces an additional effect, because at the momentary wear depth change and the thermal expansion, acting in the opposite direction, are in the same range. This means, the length reduction measurement yields to false wear rate results at the beginning of the wear process. Later, the wear depth change becomes the dominating term, while the thermal expansion reaches an equilibrium stage.

Additionally to the thermal expansion, the frictional heat generation along the contact area affects the actual value of the material properties. This feature is more dominant at an elevated temperature.

Figure 7.3 illustrates the wear process in a PoDC. At the beginning, the edge-like contact is produced by the compression and bending of the pin, followed by a

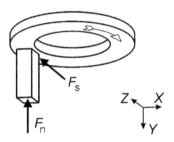

**FIGURE 7.2**

Forces acting on a PoDC ($F_n$ is the normal force, $F_s$ is the frictional force).

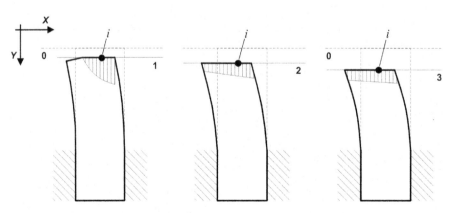

**FIGURE 7.3**

Contact behavior of the pin in the wear process: (1) edgelike contact, (2) full contact phase, and (3) steady wear phase.

full contact phase. After the running-in phase, the wear depth is increased, while the creep will further modify the deformed shape (not shown in Fig. 7.3).

## 7.3 **THE ALGORITHM OF THE WEAR SIMULATION**

The wear process is studied by a linear wear model, considering the thermal expansion of the system and the creep behavior of the pin. These features of the algorithm are illustrated in Fig. 7.4. The coupled solution of wear prediction follows an incremental technique by evaluating the actual wear depth and contact pressure distribution, from time to time, according to the temperature- and time-dependent material properties and the actual thermal expansion of the PoDC. These steps are explained in Sections 7.3.1–7.3.4.

### 7.3.1 **Temperature-dependent creep parameters**

In order to consider the effect of the creep behavior of the PEEK material, the creep properties were measured on standard specimens [12], using a universal testing machine, assuming constant load level at different temperatures. The tensile stress was 10 MPa, applied for a 10-h period, while the total strain was measured. The strain–time behavior, as well as the corresponding creep modulus curves, is presented in Fig. 7.5, at 20, 90, 120 and 150 °C. One can conclude that a limited creep behavior exists at room temperature and a characteristic creep exists at 150 °C.

To present an elastic creep analysis by the COSMOS/M system, the Classical Power Law (CPL)—also known as the Bailey–Norton law [13]—approach was followed at each temperature level. The approximation for the uniaxial creep strain is

$$\varepsilon^{creep} = C_0 \cdot \sigma^{C_1} \cdot t^{C_2}, \tag{7.1}$$

where

- $\sigma$ is the uniaxial stress,
- $t$ is the time,
- $C_0$, $C_1$, and $C_2$ are the creep constants.

The elastic strain is

$$\varepsilon^{elastic} = \frac{\sigma}{E_0}, \tag{7.2}$$

where $E_0$ is the instantaneous (unrelaxed) modulus.

In order to approximate the data measured for CPL, constants $C_0$, $C_1$, and $C_2$ of Eqn (7.1) were fitted (Table 7.1).

### 7.3.2 **Contact modeling**

To study the contact behavior in the wear process, 2D FE contact models (assuming plain-strain conditions in the $X–Y$ plane in Fig. 7.3) had been developed using

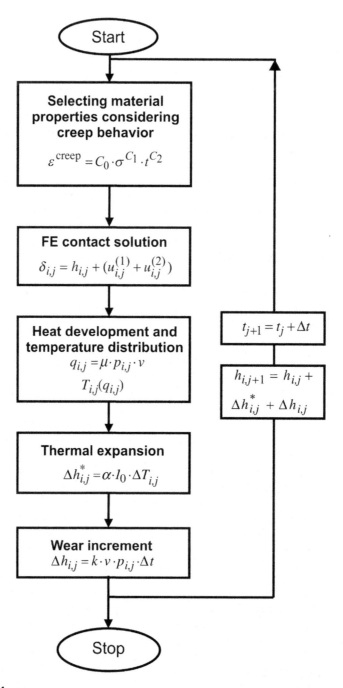

**FIGURE 7.4**

Main elements of the wear simulation.

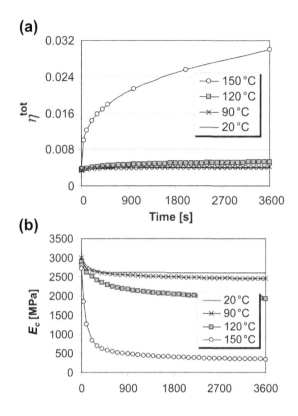

**FIGURE 7.5**

Measured uniaxial strain–time curves (a) and the creep modulus (b) at different temperatures.

**Table 7.1** Creep Constants of the CPL Approach at Different Temperatures

|  | $C_0$ | $C_1$ | $C_2$ |
|---|---|---|---|
| 20 °C | 3.5e$^{-7}$ | 1 | 0.12 |
| 90 °C | 1.9e$^{-7}$ | 1 | 0.185 |
| 120 °C | 1.6e$^{-7}$ | 1 | 0.3 |
| 150 °C | 2e$^{-4}$ | 1 | 0.315 |

node-to-node gap elements as well as considering the effect of friction. In Fig. 7.6, the bottom of the pin is fixed. The gap elements between the pin and disc are oriented in the direction of the resultant force of $F_n$ and $F_t$ in order to consider the effect of sliding friction; therefore, in this model, the angle of the gap elements is proportional to the magnitude of the coefficient of friction. The normal and tangential forces are applied on the top of the model, in the form of distributed loads. The disc is fixed in one direction perpendicular to the direction of the gap elements.

This frictional contact model will produce compression and bending of the pin, according to Fig. 7.3. The initial clearance is zero at the beginning of the wear process. The geometric condition of contact (Fig. 7.7) is defined in the gap direction, that is, the normal approach, the representing displacement and the gap distance (i.e. the worn depth and the thermal expansion) are specified in this direction. The contact condition within the contact area is

$$\delta_{i,j} = (u_{i,j})_g + (h_{i,j})_g,$$ (7.3)

where

- $i$ represents the location (Fig. 7.3),
- $j$ is the time step.

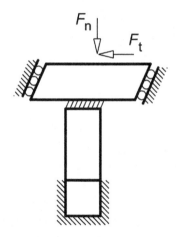

**FIGURE 7.6**

Schematic of the contact model.

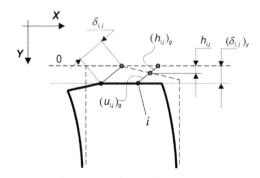

**FIGURE 7.7**

The geometric contact condition.

The vertical component of Eqn (7.3) is

$$(\delta_{i,j})_y = (u_{i,j})_y + h_{i,j},\qquad(7.4)$$

where

- $\delta_{i,j}$ is the normal approach,
- $(\delta_{i,j})_y$ is the vertical component of $\delta_{i,j}$,
- $(u_{i,j})_g$ is the elastic displacement in the gap direction,
- $(u_{i,j})_y$ is the vertical component of $(u_{i,j})_g$,
- $(h_{i,j})_g$ is the wear depth including thermal expansion,
- $h_{i,j}$ is the vertical component of $(h_{i,j})_g$.

The normal approach $\delta_{i,j}$ is constant at any time step within the contact area (Fig. 7.7).

### 7.3.3 Heat generation and thermal expansion of the PoDC

The frictional heat generation produces a higher temperature in the vicinity of the contact area and a nominal temperature rise in the body, yielding to thermal expansion. Studying the thermal expansion requires a complete model of the disc side as well as a model of the pin and pin holder (Fig. 7.8). At first, a thermal calculation is prepared (according to the heat partition between the

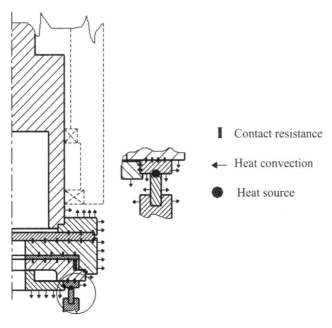

**∎** Contact resistance

◀— Heat convection

● Heat source

**FIGURE 7.8**

Schematic of the thermal model of the testing rig.

pin and disc sides), followed by an elastic calculation in order to find the thermal expansion of the components of the model. In the wear process, the thermal expansion is changing ($\Delta h^*_{i,j}$ in Fig. 7.7) according to the actual temperature distribution.

Frictional heat generated over the contact area produces a contact temperature rise in the vicinity of the pin and disc. The intensity of the heat source at location $i$ and time step $j$ is

$$q_{i,j} = \mu v p_{i,j}, \tag{7.5}$$

where

- $\mu$ is the coefficient of friction,
- $v$ is the sliding speed, and
- $p_{i,j}$ is the contact pressure.

The heat generated is partitioned between the pin and the disc according to the condition of the same contact temperature, both from the pin and disc sides, at the same locations of the contact area.

### 7.3.4 The wear modeling

The linear wear model [1] is based on the following approach. The increment of wear depth is

$$\Delta h_{i,j} = k p_{i,j} v \, \Delta t, \tag{7.6}$$

where

- $k$ is the wear coefficient,
- $p_{i,j}$ is the contact pressure,
- $v$ is the sliding speed, and
- $t$ is the time increment.

The wear coefficient can be obtained by experiments for a certain material pair under given operating conditions.

The linear wear model produces a wear depth increment in each increment of the sliding distance. This wear depth increment modifies the initial clearance in the actual step of wear simulation. The wear simulation is a repeated solution of the contact problem, for different initial clearances. In Eqn (7.6) and Fig. 7.4, subscript $i$ refers to the discretized point of the contact area (Fig. 7.3), and $j$ represents the time step. In the next time step, the initial clearance is modified by the effect of thermal expansion and the new wear increment at the representing point.

Using the linear wear model, at different time steps, starting from the beginning of the wear experiment, allows one to describe the changes in the initial clearance between the polymer sample and the steel disc (Fig. 7.3).

## 7.4  WEAR SIMULATION RESULTS

### 7.4.1  Running-in phase

The present study aims to simulate the wear process of a PEEK pin sliding against a steel disc under the following conditions:

- Nominal pressure $p = 4$ MPa,
- Pin cross-section $A = 4 \times 4 = 16$ mm$^2$,
- Sliding speed $v = 1$ m s$^{-1}$,
- Time to reach the total load $t^* = 36$ s,
- Coefficient of friction $\mu = 0.3$.

The specific wear rate (equivalent to the wear coefficient) for a PEEK pin sliding on a steel disc was measured as $k = 7 \times 10^{-6}$ mm$^3$ Nm$^{-1}$ for the operating conditions of $v = 1$ m s$^{-1}$ and $p = 1$ MPa [14,15]. It is assumed that this value is also valid under the testing conditions chosen here.

To get an idea of the thermal expansion of the PoD configuration, at first, a transient FE thermal calculation was prepared for the period of 72 s, considering the loading process. The thermal expansion of the disc in the 0 to 72-s period was in the range of 1 μm, since the disc had a much higher heat capacity; therefore, the pin side was considered only. The contact temperature result is presented in Fig. 7.9.

The thermal behavior of the pin in this time period is shown in Fig. 7.10, based on both experimental and FE thermal data. Temperature measurements of the pin were carried out at various positions away from the contact surface (2, 4, and 6 mm).

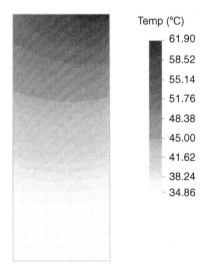

Temp (°C)

- 61.90
- 58.52
- 55.14
- 51.76
- 48.38
- 45.00
- 41.62
- 38.24
- 34.86

**FIGURE 7.9**

Calculated contact temperature distribution incide the polymer pin.

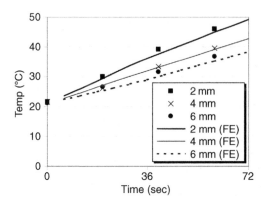

**FIGURE 7.10**

Experimental and FE thermal data in the pin.

In the FE thermal model, 0.1% of the total heat generated was applied to the pin [5]. This resulted in the final temperature distribution, which was similar to the experimental data.

The contact and wear conditions, at two time steps, are presented in Fig. 7.11 comparing results at 72 s to the ones at 36 s. The contact area is slightly wider, and the contact pressure maximum is reduced due to the wear process (Fig. 7.11(a)).

The initial wear is shown in Fig. 7.11b at 36 s, at the full load. After 36 s, the further wear process, until 72 s, produced a maximum of near 5-μm wear at the right corner of the pin.

Figure 7.11c shows the deformed shape of the worn profile of the pin in the $Y$ direction (Fig. 7.3). The horizontal parts of these curves illustrate the location of the contact area. The elastic deformation of the contact area is practically the same as the normal approach of the disc in the vertical direction, because the disc has a modulus of elasticity, being two orders of magnitude greater than that of the pin, as well as the disc does not show any wear.

To find experimental results for the starting phase of the wear process, wear tests for a period of 72 s were prepared. Figure 7.12 illustrates two worn pin surfaces, preworn (before the actual wear test) parallel and perpendicular to the final sliding direction, respectively. The worn surface segments in both cases are about the same in size, near to 3 mm, according to the dashed lines. These results are in good agreement with the calculated length of the contact area by FE in Fig. 7.11.

The actual worn depth at the corner of the pin could be verified by a surface roughness measurement (Fig. 7.13). The central part of the worn surface is about flat apart from its right edge, where a slightly deeper region can be identified. This feature is in good agreement with the wear depth change in Fig. 7.11b, where the dashed line shows the nearly flat part of the worn surface as well as the increased wear region at the right side.

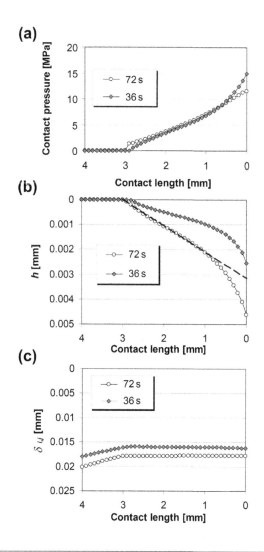

**FIGURE 7.11**

Contact and wear results: (a) contact pressure distribution, (b) worn profile of the pin, and (c) normal approach of the pin.

The results of the wear process in the running-in phase are summarized in Fig. 7.14. Three curves show the wear development at the right corner point (in 2D) of the pin, the thermal expansion at the same point and their resultant behavior. The results can be compared to the measured values. The measured length reduction between the pin holder and the frame of the test rig (Fig. 7.1) contains the wear and thermal expansion, as well as the compression of the pin, pin holder and other components (like bearing clearance) of the PoD configuration. To filter out these latter effects, the two curves represent the experimental data (Test 1 and Test 2 in Fig. 7.14) were shifted to the position of

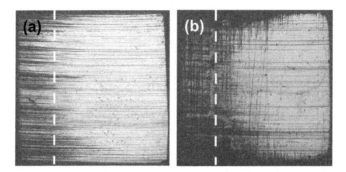

**FIGURE 7.12**

Worn pin surfaces after 72 s (preworn in parallel (a) and perpendicular (b) to the sliding direction).

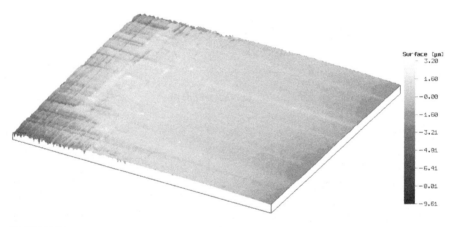

**FIGURE 7.13**

Surface roughness measurement on the worn surface of the pin.

"Wear + Th.exp." curve at the time to reach the total load (36 s). Taking the experimental data and considering 36 s as a reference state, we can compare the experimental and simulation data after 36 s. In this phase, the thermal expansion itself is more dominant than the wear process yielding to a slightly "increasing length" behavior in time. These two tendencies represent the running-in phase. Later, the wear depth change dominates.

## 7.4.2 Wear simulation at 150 °C

The present study aims to simulate the wear process of a PEEK pin sliding against a steel disc under the following conditions:

- Nominal pressure $p = 4$ MPa,
- Pin cross-section $A = 4 \times 4 = 16$ mm$^2$,

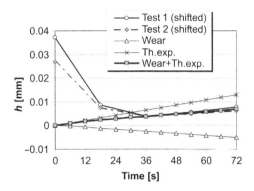

**FIGURE 7.14**

The length reduction of the pin due to the wear and the thermal expansion.

- Sliding speed $v = 1$ m s$^{-1}$,
- Coefficient of friction $\mu = 0.5$ (measured data),
- Preheated temperature $T = 150$ °C,
- Specific wear rate applied $k = 30 \times 10^{-6}$ mm$^3$ Nm$^{-1}$ [15].

The thermal material properties for the PEEK 450G materials are as follows [16]:

- Thermal conductivity $K = 0.30$ W m$^{-1}$ K$^{-1}$
- Specific heat $c = 1850$ J kg$^{-1}$ K$^{-1}$
- Density $\varrho = 1220$ kg/m$^3$

Further properties for the PoDC were listed in Ref. [5].

### 7.4.2.1 Temperature development and thermal expansion

At first, the PoDC was preheated by increasing the temperature of the disc up to 150 °C. In this phase, the pin was slightly pressed to the disc (in a stationary position) to provide heat conduction between them. The temperature distribution in the pin, at 2, 4, and 6 mm beneath the contact surface are presented in Fig. 7.15, obtained by thermocouples and FE evaluations. Period (−1000–0 s) represents the preheating conditions, while period (0–700 s) is the frictional heating phase produced by the applied load, speed, and coefficient of friction.

The FE thermal results are illustrated in Figs 7.16 and 7.17 at the end of the frictional heating. According to the experimental and numerical results, the temperature of the disc remained at 150 °C due to the setting of the preheated temperature, while at the pin side, the change of the temperature is much bigger. In Fig. 7.16, the temperature difference along the length of the pin is >100 °C due to the weak thermal conductivity of the pin. The maximum temperature of the pin is >150 °C representing a higher local temperature over the sliding contact area due to the presence of a transfer film layer. According to a previous study [17], the contact temperature is significantly higher in the transfer film layer, than between the original PEEK–steel surfaces.

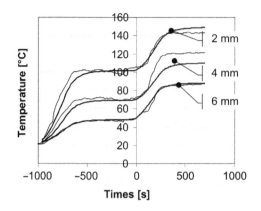

**FIGURE 7.15**

The temperature distribution in the pin solid lines: FE results thin lines: measured values.

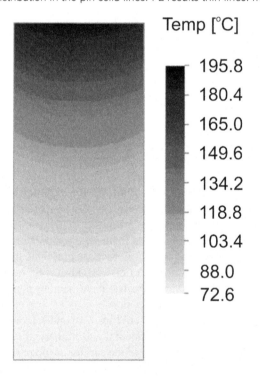

**FIGURE 7.16**

Temperature distribution in the pin at a steady state.

It should be noted here, that the incorporation of nanoparticles into the polymer matrix usually improves the thermal conductivity, thus resulting in a reduction of the temperature development. This finally leads also to a longer wear life of the polymer

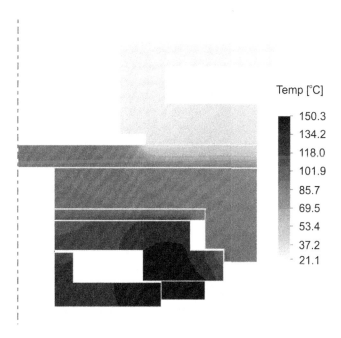

Temp [°C]

150.3
134.2
118.0
101.9
85.7
69.5
53.4
37.2
21.1

**FIGURE 7.17**

Temperature distribution in the disc at a steady state.

composites tested. This has been shown by recent papers of Chang et al. [18,19], who mainly used Titanium dioxide nanoparticles as additional fillers in different polymer matrices, containing also classical tribofillers, such as short carbon fibers, graphite. A promising alternative seems to be also the use of carbon nanotubes, to enhance the thermal conductivity of polycarbonate (having the same thermal conductivity as the PEEK used here) just by a relatively small volume fraction. Similar results have been found also by other authors for powderlike fillers [20].

### 7.4.2.2 Results of wear modeling

During the wear simulation, temperature-dependent creep properties were used for the pin according to the temperature ranges shown in Fig. 7.18 and Table 7.1. Wear simulation results are plotted in Fig. 7.19. The wear depth distribution (Fig. 7.19(a)) at 214 s illustrates a state where the full contact just occurred. The contact pressure distribution (Fig. 7.19(b)) at 214 s shows how the pressure increases at the left corner of the contact area. Finally, Fig. 7.19c illustrates the displacement of the worn pin profile in the gap direction, containing the effects of compressing and bending too. Figure 7.19 shows the results at 140 and 300 s for wear states before and after the full contact state occurred.

The wear simulation results are collected in Fig. 7.20. Based on the FE thermal results, the thermal expansion of the system was evaluated next. Considering the preheated thermal expansion as the initial state, we observed that the additional thermal expansion is in the range of 50 μm (Fig. 7.20), which should be compared later to the length reduction

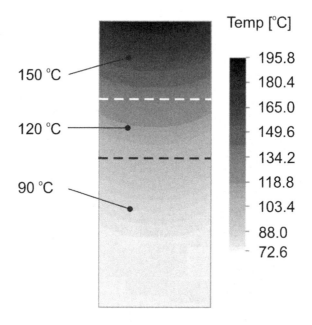

**FIGURE 7.18**

The assumed temperature ranges for selecting creep properties.

due to the wear process. The thermal expansion of the components reaches an almost steady-state condition after 600 s. The numerical wear result shows a slightly larger wear depth than the measured ones do. After superposition of the calculated wear and the thermal expansion effect, the results are comparable with the measured ones. These latter curves represent two independent experimental wear tests with the conditions above.

## 7.5 CONCLUSIONS

The wear simulation technique presented is based on a linear wear model, considering the frictional heat generation as well as the thermal expansion of the PoDC. These effects are involved in a series of FE contact solutions by changing the initial clearance time to time.

Due to the bending of the pin, the wear starts at the edge environment, yielding to an asymmetric contact pressure distribution and wear profile. The location of the worn profile evaluated and observed is in good agreement.

At the beginning of the wear process, the thermal expansion of the PoD configuration and the wear itself produce opposite effects on the length change of the pin specimen; as a fact, the thermal expansion is slightly greater than the length reduction due to the wear. Later, the wear is becoming more dominant, and the thermal expansion remains constant.

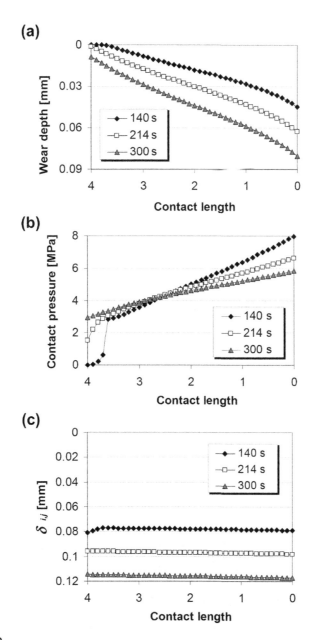

**FIGURE 7.19**

Wear simulation results, (a) wear depth, (b) contact pressure distribution, and (c) displacement of the pin profile.

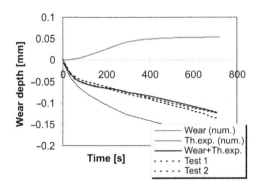

**FIGURE 7.20**

The length reduction of the pin due to the wear and the thermal expansion.

The algorithm developed for wear simulation at elevated temperatures is applicable, the numerical results and the experimental data show good agreement. The role of creep is characteristic during the complete wear process at a temperature level of 150 °C.

The addition of nanofillers to the polymers tested should improve their thermal conductivity and therefore reduce the temperature development and wear rate.

## Acknowledgment

The presented research was supported by the Deutsche Forschungsgemein schaft (DFG FR675/19-2). Additional help by the Hungarian National Scientific Research Foundation(F 046,229) and by the BMBF-TéT as part of the German–Hungarian research cooperation on contact mechanics of different materials is also gratefully esteemed.

## References

[1] P. Pödra, S. Andersson, Simulating sliding wear with finite element method, Tribology International 32 (1999) 71–81.

[2] W. Yan, N.P. O'Dowd, E.P. Busso, Numerical study of sliding wear caused by a loaded pin on a rotating disc, Journal of the Mechanics and Physics of Solid 50 (2002) 449–470.

[3] Y. Wang, C.M. Rodkiewicz, Temperature maps for Pin-on-Disk configuration in dry sliding, Tribology International 27 (No. 4) (1994) 259–266.

[4] H. So, Characteristics of wear results tested by Pin-on-Disk at moderate to high speeds, Tribology International 29 (No. 5) (1996) 415–423.

[5] L. Kónya, K. Váradi, K. Friedrich, J. Flöck, Finite element heat transfer analysis of a PEEK–steel sliding pair in a Pin-on-Disk configuration, Tribotest Journal 8-1 (2001) 3–26.

[6] R. Komanduri, Z.B. Hou, Analysis of heat partition and temperature distribution in sliding systems, Wear 251 (2001) 925–938.

[7] T.C. Kennedy, C. Plengsaard, R.F. Harder, Transient heat partition factor for a sliding rail car wheel, Wear 261 (2006) 932–936.

[8] L.S. Fletcher, Recent developments in contact conductance heat transfer, Journal of Heat Transfer 110 (1988) 1059–1070.

[9] N. Békési, K. Váradi, D. Felhős, Wear simulation of a reciprocating seal, Journal of Tribology 133 (3), 2011. 031601-1–031601-6.

[10] Material Properties Guide—PEEK™, Victrex Plc. UK.

[11] COSMOS/M User's Guide V2.85. Structural Research and Analysis Corporation, Santa Monica, USA, 2003.

[12] Prüfung von Kunststoffen: Zugversuch, DIN 53455, August 1981.

[13] J.A. Hult, Creep in Engineering Structures, Blaisdell Pub., Waltham, MA, 1966.

[14] J. Flöck, K. Friedrich, Q. Yuan, On the friction and wear behaviour of PAN- and pitch-carbon fiber reinforced PEEK composites, Wear 225–229 (1999) 304–311.

[15] Z.P. Lu, K. Friedrich, On sliding friction and wear of PEEK and its composites, Wear 181–183 (1995) 624–631.

[16] F.N. Cogswell, Thermoplastic Aromatic Polymer Composites, Butterworth–Heinemann Ltd., Oxford, 1992.

[17] K. Friedrich, J. Flöck, K. Váradi, Z. Néder, Real contact area, contact temperature rise and transfer film formation between original and worn surfaces of cf/peek composites sliding against steel, in: D. Dowson (Ed.), Lubrication at the Frontier, Tribology Series, 36, Elsevier Scientific Publishers, Amsterdam, 1999, pp. 241–252.

[18] L. Chang, Z. Zhang, L. Ye, K. Friedrich, Tribological properties of epoxy nanocomposites: III. Characteristics of transfer films, Wear 262 (2007) 699–706.

[19] L. Chang, Z. Zhang, H. Zhang, K. Friedrich, Effect of nanoparticles on the tribological behavior of short carbon fiber reinforced poly(etherimide) composites, Tribology International 38 (2007) 966–973.

[20] K.W. Garrett, H.M. Rosenberg, The thermal conductivity of epoxy-resin/powder composite materials, Journal of Physics D: Applied Physics 7 (1974) 1247–1258.

# On the friction and wear of carbon nanofiber-reinforced PEEK-based polymer composites

8

**Holger Ruckdäschel, Jan K.W. Sandler[1], Volker Altstädt**

*Polymer Engineering, University of Bayreuth, Bayreuth, Germany*

## CHAPTER OUTLINE HEAD

[1] Now at: BASF Aktiengesellschaft, Polymer Physics, GKP/R-B1, D-67056 Ludwigshafen, Germany.

Tribology of Polymeric Nanocomposites. http://dx.doi.org/10.1016/B978-0-444-59455-6.00008-8

## 8.1  INTRODUCTION AND MOTIVATION

Although the terms nanomaterial and nanocomposite represent new and exciting fields in materials science, such materials have actually been used for centuries and have always existed in nature. However, it is only recently that the means to characterize and control structure at the nanoscale have stimulated rational investigation and exploitation. A nanocomposite is defined as a composite material with at least one of the dimensions of one of the constituents on the nanometer-size scale [1]. The term usually also implies the combination of two (or more) distinct materials, such as a ceramic and a polymer, rather than spontaneously phase-segregated structures. The challenge and interest in developing nanocomposites is to find ways to create macroscopic components that benefit from the unique physical and mechanical properties of the very small objects within them.

Carbon nanotubes (CNTs) and carbon nanofibers (CNFs) especially have attracted particular interest because they are predicted, and indeed observed, to have remarkable mechanical and other physical properties. The combination of these properties with very low densities suggests that CNTs, for example, are ideal candidates for high-performance polymer composites, in a sense the next generation of carbon fibers (CFs). Although CNTs are currently produced in kilogram quantities per day, the development of high-strength and high-stiffness polymer composites based on these carbon nanostructures has been hampered so far by the lack of availability of high-quality (high-crystallinity) nanotubes in large quantities. In addition, a number of fundamental challenges arise from the small size of these fillers that yet need to be overcome. Some quality reviews of the current state of the art of nanotube/nanofibers and their composites can be found in Refs. [2–5].

Although significant advances have been made in recent years to overcome difficulties with the manufacture of polymer nanocomposites, processing remains a key challenge to fully utilize the properties of the nanoscale reinforcement. A primary difficulty is the achievement of a good dispersion of the nanoscale filler in a composite, independent of filler shape and aspect ratio. Without proper dispersion, filler aggregates tend to act as defect sites that limit the mechanical performance; such agglomerates also adversely influence the physical composite properties such as optical transmissivity.

Commercial nanotube–polymer composites exist today, yet they almost exclusively employ relatively low loadings (3–5 wt%) within thermoplastic matrices for the purposes of antistatic dissipation, particularly in the automotive and electronics industries [6]. One major uncertainty still is the type and quality of nanotubes that should be used. A wide variety of synthesis methods have been employed, yielding nanotubes and nanofibers of different size, aspect ratio, crystallinity, crystalline orientation, purity, entanglement, and straightness. All these factors affect the processing and properties of the resulting composites, but it has not yet emerged as to what the "ideal" carbon nanotube would be; the answer may vary with the matrix and application.

CNTs and CNFs may not produce practical replacements for existing structural high-performance materials in the near future. However, there is a continuing market for electrically conducting polymer compounds, and immediate potential to develop the reinforcement of delicate composite structures such as thin films, fibers [7,8], foams [9], and the matrices of conventional fiber composites. For example, such nanofillers provide the means to improve the wear properties of microstructured parts in which more conventional fillers physically cannot be accommodated, and in which problems of erosion are particularly problematic. As an example, Endo and colleagues helped Seiko develop a CNF-filled nylon watch gear, <200 μm in diameter.

Taking into account the emerging potential of nanocomposites for tribological applications especially, this chapter is aimed at providing the reader with a comprehensive overview of a model nanocomposite system based on semicrystalline poly(ether ether ketone) (PEEK) and commercial CNFs intended for industrial applications. Following an initial introduction to nanocomposites for tribological purposes in general (Section 8.2), the structure–property relationships of carbon nanostructures are briefly introduced (Section 8.3). Section 8.4 provides a detailed overview of the PEEK–CNF nanocomposite microstructures and properties, prior to demonstrating the potential of such systems under dry sliding and wear conditions. However, at early stages, the observed experimental findings are correlated with the unique features of the nanocomposites. Following a short introduction to hybrid composites based on a combination of microscale and nanoscale reinforcements, the significant potential of nanofiber-reinforced PEEK hybrids for tribological applications is demonstrated (Sections 8.5 and 8.6).

## 8.2 COMMON STRATEGIES TO ENHANCE THE TRIBOLOGICAL BEHAVIOR OF POLYMERS

Today, more and more advanced technical applications of polymeric materials involve friction and wear, often at elevated operation temperatures. In order to further exploit the economical advantages of polymers in general and to tailor the performance of devices or components to these ever-increasing demands regarding the overall tribological performance, a fundamental assessment not only of the intrinsic materials properties but also of the complete tribosystem is required. On the one hand, material properties such as the degree of crystallinity, glass transition temperature, mechanical properties, molecular weight, orientation, hardness, and surface energy are factors that have been shown to influence both the friction and wear behavior of polymers under various experimental conditions. On the other hand, the tribological system itself, more precisely the loading characteristics, the counterpart material, as well as the external conditions including the temperature or the presence of lubricants, for example, play a major role for the active wear mechanism and, subsequently, for the overall wear performance. An overview of the various factors influencing the wear behavior of polymers is shown in Fig. 8.1 [10].

Modern day applications often require a more specific adjustment of the tribological properties to meet all relevant demands. As such, significant efforts have been aimed at using different fillers and/or reinforcements in order to tailor the overall performance of polymer sliding elements previously composed of metallic components to the specific application and tribological conditions. Nevertheless, further developments are

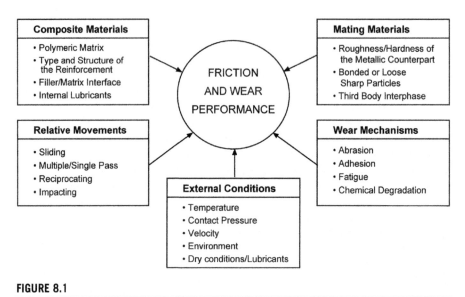

**FIGURE 8.1**

Factors determining the wear behavior of polymers.

*According to Ref. [10]).*

still underway to explore other fields of application and to tailor the performance of polymeric materials and composites for extreme loading and surrounding temperature conditions. One example for automotive applications is a new generation of control arm mountings or ball joints in the car chassis technology in which high loads and temperatures are acting on a tribocouple. A fundamental and concise understanding of the wear and friction properties of such materials at severe operating conditions is directly related to the safety and service life of the technical components.

Besides the traditional polymeric materials used for friction and wear loaded components, such as polyoxymethylene, polyamide (PA), and polytetrafluorethylene (PTFE) [11–15], newer high-performance polymers have found entrance into special tribological applications in which service temperatures >150 °C are a critical issue (polyetherimide (PEI), polybenzimidazole (PBI), liquid crystal polymers, polyimide (PI)) [16–19]. Special candidates in this respect are the polyaryletherketones that allow tailoring of the glass transition range and melting temperature by variation of the ratio and position of the ether and ketone groups. This special feature of the thermal properties as a function of the composition of polyaryletherketones is summarized in Fig. 8.2 [20].

Among this group, the semicrystalline PEEK has found a special interest as it is characterized by a comparatively good processability as well as outstanding mechanical properties such as toughness and strength. Consequently, PEEK is commercially produced today in a small number of modifications including different molecular weights.

Although PEEK has been identified as a good tribological polymer in general, past efforts [10,21,22] have clearly highlighted the influence of morphological

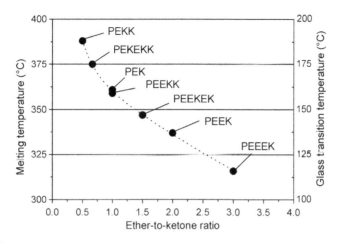

**FIGURE 8.2**

Melting point and glass transition temperature of PEEKs as a function of the ether-to-ketone ratio.

*Reproduced from Ref. [20].*

parameters, such as crystallite size and degree of crystallinity as well as orientations, on the resulting friction and wear performance. In addition, the matrix molecular weight especially has a pronounced influence on the tribological performance of PEEK [11,23,24], a feature commonly observed for tribological polymers [25].

While the frictional coefficient under adhesive sliding generally remains rather constant as a result of the surface-free energy being nearly independent of the molecular weight [23], the wear resistance deteriorates with decreasing molecular weight; in particular, at elevated velocities and contact pressures. This dependence of the wear rate on the molecular weight is highlighted in Fig. 8.3 [11,23], where the melt viscosity is taken as a measure for the molecular weight. Such a behavior can be attributed to the microstructure of the semicrystalline PEEK, where the lower molecular weight leads to a reduced number of interspherulitic links and a decreased entanglement density of the amorphous regions. Subsequently, plastic deformation is promoted by the enhanced sliding of molecules and a reduced shear strength with a lower molecular weight. Similarly, a high molecular weight increases the resistance to local cracking.

Reviewing the expanding literature, three distinct strategic approaches can be extracted:

1. Incorporation of solid lubricants such as PTFE and graphite (reduced friction);
2. Addition of fiber reinforcements such as carbon and glass fibers (GFs, increased stiffness and strength);
3. Incorporation of hard microscale and nanoscale particles (increased mechanical properties and/or reduced friction).

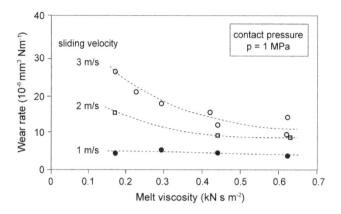

**FIGURE 8.3**

Influence of molecular weight on the wear behavior of PEEK.

*According to Ref. [23].*

[26] systems often do not allow a straightforward correlation between the induced pure material property variations and the performance enhancement. The selection of an appropriate material strategy must therefore always include a detailed consideration of the intended application.

## 8.2.1 Enhancing the wear resistance by addition of microscale additives

A common additive for reducing the coefficient of friction against various metallic counterparts is the incorporation of PTFE acting as a solid lubricant. During compounding, a disperse phase morphology must be obtained. Optimum increases in wear resistance and minimum frictional coefficients in PEEK have been found at around 10–25 wt% of PTFE [11,23,26,27]. In many cases, the low abrasion resistance due to the soft nature of the PTFE induces a pronounced wear of the disperse domains, which subsequently leads to the formation of a uniform and continuous transfer film on the counterpart [28]. Such materials often exhibit an enhanced running-in behavior, an important feature for many tribological applications. Yet, the addition of PTFE also induces a pronounced reduction of the mechanical properties of the matrix, a problem for mechanically demanding applications.

Another solid lubricant acting in a similar overall manner is graphite. Given a good dispersion of individual graphite flakes, there is some reduction in the overall mechanical performance accompanying the enhanced wear behavior. Under sliding conditions, stacks of graphene layers are easily sheared off particles exposed at the surface. However, this debris can lead to the formation of a stiffer transfer film as compared to the PTFE, effectively providing a stable reduction of the coefficient of friction over longer periods of time [29].

In contrast to this addition of solid lubricants, such as PTFE, graphite, molybdenum disulfide ($MoS_2$), the incorporation of reinforcing fibers is aimed at providing a significant enhancement of stiffness and strength, impact resistance, and creep behavior [21]. In a similar manner, the wear resistance of the polymeric matrix can be considerably improved, while the wear mechanism is completely different. In such cases, the fibers lead to significantly enhanced shear moduli, strengths, and hardness values as compared to the neat matrix, thereby limiting the contact temperature and abrasion [10,11]. However, interfacial properties between fiber and matrix are important, as the size of the wear debris and fiber debonding should be minimized. For such fiber-reinforced composites, the main wear mechanisms also include fiber cracking, fiber sliding wear and matrix wear [30]. Again, there is an optimum in fiber content between 20 and 30 wt%; effectively providing a balance between matrix- and fiber-dominated wear resistance.

Common fibers used in PEEK are carbon and GFs. Research has already indicated that both the selected fiber type as well as processing-related issues such as fiber length are further factors influencing the overall performance. For example, CFs are generally observed to be less abrasive reinforcements due to the essentially graphitic structure of the wear debris. Glass fibers are nevertheless commonly used in systems where the chemical inertness of the reinforcement is important. In the case of martensitic steels especially, carbon-based systems often show a tribochemical reaction [31] combined with rust formation; an effect that severely influences the long-term stability of the wear performance.

In addition to reinforcing fibers and solid lubricants, various kinds of microscale additives such as titanium dioxide ($TiO_2$), zirconium dioxide ($ZrO_2$), silicon carbide (SiC) were incorporated in polymeric materials in order to enhance their wear resistance [32–38]. Depending on the material system used, an improved tribological behavior has often been reported. The observed enhancement was mainly attributed to the positive influence of such particles on the mechanical properties such as strength and hardness, or to tribochemical reactions leading to an improved adhesion between the transfer film and the counterpart material.

Based on various studies, the size of the particles appears to play a major role on the wear performance of particulate-reinforced polymers. Although other trends were also reported [39,40], fine particles seem to lead to a better property profile under dry sliding wear conditions, as observed, for example, by Friedrich [41] and Xing et al. [42]. However, a significant reduction of the particle size down to the nanoscale level (<100 nm) leads to a completely distinct wear behavior. Considerably enhanced wear properties of polymeric materials reinforced with such nanoscale particles demonstrated the excellent potential of the nanoapproach [43–45].

### 8.2.2 Enhancing the wear resistance by the addition of nanoscale additives

When comparing the effects of nanoparticles and microscale additives on the wear behavior of polymers, some main advantages of adding nanoadditives can be pointed out:

1. Generally, lower abrasiveness due to a reduced angularity;
2. Reinforcing effects possible, that is, enhanced strength, modulus, and toughness;
3. Higher specific surface areas and, thus, improved adhesion;
4. In general, high effectiveness at very low contents.

The significantly enhanced wear behavior at very low volume contents of the nanoscale additive is highlighted in Table 8.1, which summarizes the tribological properties of nanocomposites reported in the literature [46–59]. Depending on the polymeric matrix, the nanoparticle and its size in particular, the optimum content of the additive is in the range of 1–4 vol% for most systems. Only in the case of PTFE, a maximum reduction of the wear rate is achieved at elevated volume contents of 12–15 vol%, probably related to the peculiar structure of the fluoropolymer.

For all nanocomposites, the interfacial area between the matrix polymer and the nanoparticles is tremendously increased as compared to that of microcomposites and can be identified as one of the primary reasons for the altered materials behavior. As can be seen in Fig. 8.4, for spherical particles, the specific surface area as well as the particle density significantly increases when reducing the particle size. For example, comparing identical volume fractions of the particulate additives with a diameter of 1 μm and 1 nm, respectively, the total internal interfacial area increases by three orders of magnitude and the particle density by nine orders of magnitude in the nanocomposite.

**Table 8.1** Wear of Nanoparticle-Filled Polymeric Systems

| Matrix | Nanoadditive (Particle Size) | Lowest Wear Rate [$10^{-6}$ mm$^3$ Nm$^{-1}$] | Optimum Particle Content [vol%] | Reference |
|---|---|---|---|---|
| PEEK | $Si_3N_4$ (<50 nm) | 1.3 | 2.8 | [46] |
| PEEK | $SiO_2$ (<100 nm) | 1.4 | 3.4 | [47] |
| PEEK | SiC (80 nm) | 3.4 | 1–3 | [48] |
| PEEK | $ZrO_2$ (10 nm) | 3.9 | 1.5 | [49] |
| PPS | $Al_2O_3$ (<35 nm) | 12 | 2 | [50] |
| PPS | $TiO_2$ (30–50 nm) | 8 | 2 | [51] |
| PPS | CuO (30–50 nm) | 4.6 | 2 | [51] |
| Epoxy | $SiO_2$ (9 nm) | 45 | 2.2 | [52,53] |
| Epoxy | $SiO_2$ -g-PAAM (9 nm) | 11 | 2.2 | [52,53] |
| Epoxy | $Si_3N_4$ (<20 nm) | 2.0 | 0.8 | [54] |
| Epoxy | $TiO_2$ (300 nm) | 14 | 4 | [55] |
| Epoxy | $Al_2O_3$ (13 nm) | 3.9 | 2 | [56] |
| Epoxy | $SiO_2$ (13 nm) | 22 | 3 | [57] |
| PTFE | ZnO (50 nm) | 13 | 15 | [58] |
| PTFE | $Al_2O_3$ (40 nm) | 1.2 | 12 | [59] |

*According to Ref. [17].*

Although these numbers by themselves are impressive already, the filler size needs to be regarded relative to the size of the polymer molecules to capture the full potential impact of nanoscale fillers on composite properties. Many properties are related to the size of the polymer chain, which is often expressed as the radius of gyration, and is typically on the order of 3–50 nm depending on the molecular weight and the composition of the macromolecule. Depending on the interfacial interaction, the polymer chains will be perturbed with respect to those in the bulk (i.e. away from the interface). Taking the interfacial region that surrounds the particle as independent of the particle size in a first approximation, the ratio between the interfacial volume and the particle volume for typical interfacial thicknesses of 1 and 10 nm as a function of particle size and content can be appreciated in Fig. 8.5.

These considerations highlight the pronounced impact of nanoparticles on the surrounding polymer. For example, by dispersing 2 vol% of spherical nanoparticles (diameter 10 nm) in a polymer matrix (interfacial thickness 10 nm), the volume fraction occupied by the interfacial region is 52%, which implies that more than half of the composite is affected by the presence of the nanoparticles. Slight changes of the particle diameter (e.g. 50 nm), however, would result in a significant reduction of the interfacial region. Besides the unique properties of nanocomposites provided by the aforementioned structural features, a good dispersion as well as distribution of the nanoparticles is a first priority.

For PEEK, as an example, it was found that even very small amounts of inorganic nanoparticles reduce the frictional coefficient from 0.4 to 0.2 [46–49]. In addition, the wear mechanism changed from adhesive and fatigue wear, as typically observed for neat PEEK, to mild abrasive wear of the nanocomposites. In the case of particle agglomeration, however, strong abrasive wear of the nanocomposites was detected by investigating the worn surfaces, accompanied with a significantly increased wear rate.

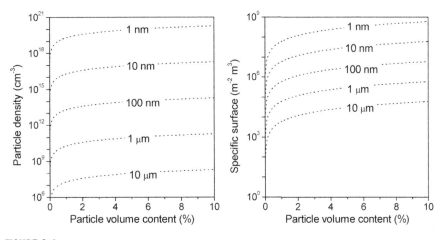

**FIGURE 8.4**

Interfacial area and particle density of polymeric composites with spherical additives.

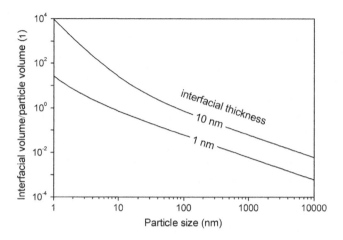

**FIGURE 8.5**

Ratio between the interfacial volume and the interfacial area as a function of particle size for two typical interfacial thicknesses of 1 and 10 nm, respectively.

Similar trends were observed by Bahadur et al. [50,51] for polyphenylenesulfide (PPS) nanocomposites containing various nanoparticles. As an example, for inorganic aluminum oxide ($Al_2O_3$) particles with an average size of 33 nm, an optimum in the wear reduction was demonstrated for 2 vol% of nanoparticles only when the surface roughness of the steel counterpart was 60 and 100 nm, respectively. In the case of a significantly reduced surface roughness of 27 nm, smaller than the particle diameter, any content of nanoparticles led to an increase in the wear rate. Here, the formation of the transfer film was strongly deteriorated. Thus, a strong dependence of the wear behavior on the ratio between the particle size and surface roughness can be anticipated. In addition, the composition of the nanoparticles was identified as a crucial factor for enhancing the tribological behavior. While both $TiO_2$ and copper oxide (CuO) could potentially reduce the wear rate, contrary results were reported for zinc oxide (ZnO) and SiC.

Finally, it is worth noting that also the interfacial adhesion between the nanoparticles and the polymer matrix plays an important role. Zhang et al. [52,53] recently reported epoxies filled with either unmodified or polyacrylamide (PAAM)-grafted silicon dioxide ($SiO_2$). In the latter case, the chemical bonding between the matrix and the nanoparticles considerably increased the interfacial region and, in turn, led to lower wear rates.

Most of the previously discussed nanocomposites have spherical nanoparticles, and therefore are relatively common or show a low degree of anisotropy. In order to effectively reinforce polymer materials, however, a fiberlike structure can be regarded as highly advantageous, in particular to enhance not only the composite strength but also the toughness as well as the modulus. Here, nanofibers and nanotubes are a promising approach to meet these requirements.

Table 8.2 Tribological Studies of CNT- and CNF-Reinforced Polymer Systems

| Matrix | Nanoadditive (Diameter) | Lowest Wear Rate [$10^{-6}$ mm$^3$ Nm$^{-1}$] | Optimum Particle Content [wt%] | Reference |
|---|---|---|---|---|
| Epoxy | MWCNT (10–25 nm) | n.a. | n.a. | [60] |
| PI | MWCNT (10–50 nm) | n.a. | > 15 | [61] |
| PMMA | MWCNT (10–20 nm) | 80 | 1.0 | [62] |
| PS | MWCNT (10–20 nm) | 15 | 1.5 | [63] |
| UHMWPE | MWCNT (10–50 nm) | n.a. | n.a. | [64] |
| UHMWPE | CNF (100–200 nm) | 2.0 | 5.0 | [65] |
| UHM-WPE/ HDPE | MWCNT (60–100 nm) | 13 | 2.0 | [66] |

Due to their peculiar structural features as well as their intrinsic properties derived from the essentially graphitic structure, CNFs and CNTs possess a significant potential in order to enhance the wear behavior of polymers in general; not only by improving the mechanical properties as well as the microhardness, but also by acting as a solid lubricant. However, the literature systematically discussing the tribological properties of polymer materials reinforced with either CNT or CNF is still limited, as highlighted in Table 8.2 [60–66].

As can be seen, the optimum particle content revealing the lowest wear rate (as far as available by these studies) strongly depends on the structural features of the nanofibers and nanotubes, respectively, and on the matrix system. In most cases, the addition of relatively low contents of CNFs or CNT already leads to a significant reduction of both the wear rate and the frictional coefficient of the polymeric matrix. This considerably enhanced tribological behavior of such materials can generally be attributed to several reasons [60–66]. On the one hand, the microhardness as well as the mechanical properties, for example, the strength and the modulus, are strongly improved by the highly effective reinforcement of such fiberlike nanoadditives. On the other hand, the structural similarity to graphite allows the nanofibers and nanotubes to act as a solid lubricant. Last but not least, beneficial effects on the formation and the stability of the transfer film have been reported, which were indicated by its continuity and smoothness. In summary, all these phenomena contribute to the enhanced wear behavior under dry sliding conditions.

These studies indicate an excellent potential of CNFs and CNTs to enhance the wear performance of polymers. Due to the large variety of the carbon nano-materials and their special features, the following section is aimed at briefly introducing the synthesis procedure as well as the resulting structures of common nanofibers and nanotubes. This information is relevant to understand the impact of such reinforcements on the behavior of the PEEK-nanofiber composites in general, and to evaluate the wear behavior of these nanocomposites described in subsequent sections.

## 8.3 CNTs AND CNFs

CNTs and CNFs can be seen as a bridge between traditional CFs and the fullerene family [67]; this intermediate position between the molecular and continuum domains is the classic signature of a nanomaterial. Research on these structures blossomed only recently, after the electric-arc synthesis of multiwalled nanotubes by Iijima, in 1991 [68]; since then, in excess of 10,000 papers have appeared discussing the science of CNTs, including a large fraction on polymer composites. This interest was initially stimulated by his recognition of the relationship with the closed, curved, carbon shells of the fullerene family that had been discovered a few years previously, in 1985 [69]. Although Iijima is often credited with the discovery of CNTs, there are earlier reports in the literature, notably by Endo in 1976, of the synthesis of tubular carbon structures using hydrocarbon decomposition [70], as well as earlier in the catalysis literature of the 1950s, and possibly even the late nineteenth century [71]. In fact, nanotubes are now known to occur naturally, having been observed in 10,000-year-old ice cores [72] and swords [73]. It was the advance in modern electron microscopy especially that allowed detailed evaluations of the structure of these particular graphitic nanostructures. The investigation of nanotubes today still is driven by their elegant and diverse structures, and their remarkable intrinsic properties.

However, CNTs were only produced in sufficient quantities for composite studies in the mid-1990s. There exists a range of different production techniques today leading to distinctly different nanotubes with unique material properties such as yield, degree of entanglement as well as structural quality. As in the case of established CF-reinforced composites, it is therefore important to select the appropriate starting material in the framework of developing nanotube-based polymer composites when aiming at exploiting their unique mechanical and physical properties.

### 8.3.1 Structures and appearances

Many crystalline graphitic materials are available today, including naturally occurring graphite flakes and several synthetic forms of graphite that can have an even higher structural perfection [67]. One important class of such graphite-related materials are CFs. Despite the many precursors such as hydrocarbons, polyacrylonitrile (PAN), pitch, rayon, and nylon that can be used to manufacture CFs with a variety of cross-sectional morphologies, the preferred orientation of the fiber axis for all kinds of fibers is parallel to the in-plane direction of the graphene layers. In fact, there are several kinds of CFs, each with its own kind of morphology and resulting properties. The final structure–property relationships of CFs are largely determined by the precursor material and the processing and subsequent treatment conditions. The two most common processing techniques are extrusion and chemical vapor deposition (CVD) as well as modifications of the two.

Commercial mesophase pitch-based CFs usually exhibit a high bulk modulus but, due to very high-temperature processing, they are rather expensive. PAN-based CFs on the other hand are widely used for their high strength. The differences in mechanical performance are related to different microstructures: the high modulus is a result of a high degree of in-plane orientation of adjacent graphene layers to the fiber axis, whereas the high strength is related to defects in the structure that inhibit slippage of adjacent planes relative to each other.

Figure 8.1 summarizes the superior mechanical properties of CFs compared to steel, which typically has a modulus of ~210 GPa and a strength of ~1.4 GPa. As can be seen, the mechanical properties of CFs can be tailored to suit the specifications of any given application.

Another class of fibers are vapor-grown carbon fibers (VGCFs) which, in the as-prepared state, have an onion-skin or tree-ring morphology [67]. After a subsequent heat treatment, such VGCFs show some degree of faceting. Of interest are also graphite whiskers that essentially consist of a rolled-up graphene layer to form a scroll. Since their discovery [74], such graphite whiskers have provided the standard by which the performance of CFs is measured.

It was only recently understood that solids of elemental carbon with $sp^2$ hybridization can form a wide variety of graphitic structures. The first suggestion of the existence of fullerenes as closed structures [69] and the subsequent discovery of their synthesis in a carbon arc [75], opened the door to a whole new understanding of carbon materials. A graphitic needlelike structure, discovered on the cathode surface in a carbon arc discharge apparatus that was used to produce C60 and other fullerenes, was given the name carbon nanotube because of its nanometer size and tubular morphology [68]. Typical dimensions of these tubes are approximately 1–50 nm in diameter and many microns in length (although centimeters have also been observed [76]). They can consist of one or more concentric layers. Commercial PAN and mesophase pitch-based CFs on the other hand typically are in the 7- to 20-μm diameter range, whereas VGCFs can have intermediate diameters ranging from a few 100 nm up to 1 mm. This variation in diameter of fibrous graphitic materials is summarized in Fig. 8.2.

A single-wall carbon nanotube (SWCNT) can be viewed as a conformal mapping of the two-dimensional hexagonal lattice of a single graphene sheet onto the surface of a cylinder. As-grown, each nanotube is closed at both ends by a hemispherical cap formed by the replacement of hexagons with pentagons in the graphite sheet that induces curvature. A multiwall carbon nanotube (MWCNT) consists of more than two coaxial cylinders, each rolled out of single sheets, separated by approximately the interlayer spacing in graphite. Each shell is similar to a single-wall nanotube. The outer diameter of such MWCNTs can vary between two and a somewhat arbitrary upper limit of about 50 nm, while the inner hollow core is often (though not necessarily) quite large with a diameter about half of that of the whole tube.

Structurally, carbon nanotubes are very diverse depending on their origins. Compared to CFs, the best nanotubes can have almost atomistically perfect structures; indeed, there is a general question as to whether the smallest CNTs should be regarded as a very small fiber or heavy molecule, especially as the diameters of the smallest nanotubes are similar to those of common polymer molecules. In this

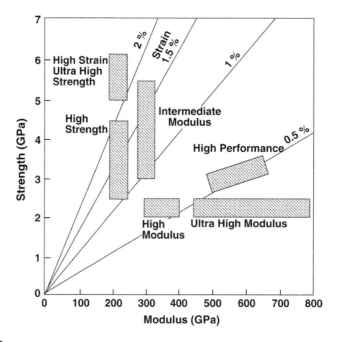

**FIGURE 8.6**

Schematic representation of the range of CFs available today.

*According to Ref. [67].*

context, it is not yet clear to what extent conventional fiber composite understanding can be extended to CNT composites. These intrinsic possibilities are further magnified by the range of defects that can exist: vacancies, extraneous nonhexagonal rings, edge dislocations, local sp³ hybridization, and noncarbon functional groups have all been observed. These defects can give rise to longer range morphological changes in the structure, such as kinks/bends, and changes in diameters. The formation of kinks during synthesis is particularly significant as it encourages the formation of an entangled network of nanotubes that is difficult to disperse [77]. Straight nanotubes, on the other hand, are less likely to be entangled, can be aligned more easily, and are likely to have a better performance in composites [78]. Figures 8.6–8.8 show comparative representative scanning electron micrographs of aligned and entangled MWCNTs following synthesis. In addition to variations in the structure of the nanotubes, other contaminating materials can be present; for example, amorphous carbon, graphitic nanoparticles, and catalyst metals, which can be difficult to remove.

CNFs are mainly differentiated from nanotubes by the orientation of the graphene planes: whereas the graphitic layers are parallel to the axis in nanotubes, nanofibers can show a wide range of orientations of the graphitic layers with respect to the fiber axis. They can be visualized as stacked graphitic discs or (truncated) cones, and are intrinsically less perfect as they have graphitic edge terminations on their surface.

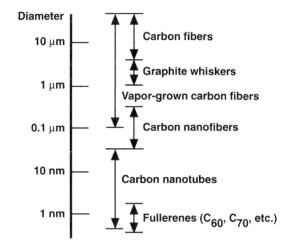

**FIGURE 8.7**

Comparison of diameters of various forms of fibrous carbon-based materials.

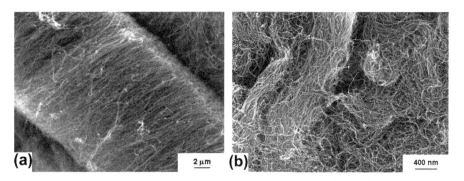

**FIGURE 8.8**

Scanning electron micrographs of (a) aligned and (b) commercial entangled MWCNTs produced by CVD methods.

Nevertheless, these nanostructures can be in the form of hollow tubes with an outer diameter as small as approximately 5 nm, although 50–100 nm is more typical. The stacked cone geometry is often called a "herringbone fiber" due to the appearance of the longitudinal cross-section. Slightly larger (100–200 nm) fibers are also often called CNFs, even if the graphitic orientation is approximately parallel to the axis. An example of the complicated structure of a commercial CNF is shown in Fig. 8.9.

These carbon nanostructures show remarkably different properties as compared to CFs. Notably different are the electronic structures and transport properties of nanotubes that are dominated by a quantum size effect. The mechanical properties

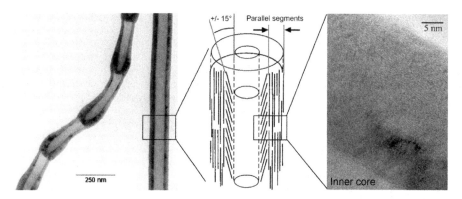

**FIGURE 8.9**

Representative transmission electron micrographs of commercial CNFs, highlighting structural variations both in overall morphology and in the orientation of the graphitic planes. The leftmost image shows a "bamboo" and a "cylindrical" CNF, while the rightmost image shows a high magnification image of one wall of the cylindrical fiber, which reveals the graphitic arrangement sketched in the central panel.

of nanotubes are even more different from those of CFs, mainly with regard to their flexibility. Nanotubes show a remarkable flexibility on bending, whereas CFs are more brittle and tend to crack when subjected to bending forces. Prior to considering the structure–property relationships established so far, a closer look at available and commonly used synthesis approaches is required.

## 8.3.2 Synthesis

A variety of synthesis methods now exist to produce CNTs and CNFs. However, these carbon nanostructures differ greatly in terms of diameter, aspect ratio, crystallinity, crystalline orientation, purity, entanglement, surface chemistry, and straightness. These structural variations dramatically affect intrinsic properties, processing, and behavior in composite systems. However, it is not yet clear as to which type of nanotube material is most suitable for composite applications, nor is there much theoretical basis for rational design. Ultimately, the selection will depend on the matrix material, processing technology, and the property enhancement required.

Both MWCNTs and SWCNTs can be produced by a variety of different processes that can broadly be divided into two categories: high-temperature evaporation using arc discharge [79–83] or laser ablation [77,84–94], and various CVD or catalytic growth processes [77,86–88]. In the high-temperature methods, MWCNTs can be produced from the evaporation of pure carbon, but the synthesis of SWCNTs requires the presence of a metallic catalyst. The CVD approach requires a catalyst for both types of CNTs but also allows the production of CNFs. The products of the high-temperature routes tend to be highly crystalline, with low defect concentrations, but are relatively impure, containing other, unwanted carbonaceous impurities. These methods usually work on the gram scale and are, therefore, relatively expensive. For the use of nanotubes in composites, large quantities of nanotubes are required at low cost, ideally without the requirement for complicated purification. At present, only CVD-grown nanotubes satisfy these requirements and, as such, tend to be the materials of choice for composite work, both in academia and in industry [89]. A number of companies have scaled up such processes to 100 tonnes per year or more. The main contaminants in CVD materials are residual catalyst particles that are mostly incorporated into the nanotubes. On the other hand, these gas-phase processes operate at lower temperatures and lead to structurally more imperfect nanotubes.

### 8.3.3 Intrinsic mechanical properties

A number of experimental studies have focused on the direct determination of the mechanical properties of carbon nanotubes. Experimental fittings to measurements of Young's modulus and elastic constants of nanotubes have mostly been made by assuming nanotubes to be elastic beams. It should be noted that concepts such as Young's modulus and elastic constants belong to the framework of continuum elasticity, an estimate of these material parameters for nanotubes therefore implies the continuum assumption. Since each individual SWCNT only involves a single layer of a rolled-up graphene sheet, the thickness $t$ will not make any sense until it is given based on the continuum assumption [90]. In the following, it is assumed that the effective thickness of an SWCNT is close to the interlayer spacing of graphite.

For example, thermal vibrations of nanotubes in transmission electron microscopy (TEM) led to average stiffness values of MWCNTs and SWCNTs of around 1.8 TPa [91] and 1.25 TPa [92], respectively. For MWCNTs, the estimated nanotube stiffness appeared to depend on the diameter [93], an effect that was explained by the occurrence of wavelike distortions for nanotubes with diameters of >12 nm, as predicted by a combination of finite element analysis and nonlinear vibration analysis [94]. Falvo et al. [95] showed that MWCNTs could be repeatedly bent to large angles with an AFM (Atomic Force Microscopy) tip without undergoing catastrophic failure, as verified by high-resolution TEM observations [96–98]. The effects of interlayer forces on the buckling and bending of nanotubes have been addressed by a shell model [99–104]. For MWCNTs, the critical axial strain is decreased from that of an SWCNT of the same outside diameter [99,101,102], in essence because the van der Waals forces between the layers cause an inward force. Note, that although the critical axial strain is reduced, the critical axial force may be increased due to the larger cross-section. This phenomenon is also predicted for nanotubes embedded in an elastic matrix [103,104].

Static models of beam bending have also been used to measure mechanical properties of nanotubes. AFM measurements lead to an average bending stiffness of arc-grown MWCNTs of about 1 TPa [105,106]; however, catalytic nanofibers with a higher defect concentration were found to have a substantially lower stiffness of only 10–50 GPa [106]. Whereas point defects do not affect the nanotube stiffness, a herringbone arrangement of the graphitic layers as in nanofibers leads to rather low overall properties.

More recently, a mechanical loading stage operating inside an SEM was used to perform the first in situ tensile tests on individual MWCNTs and ropes of SWCNTs. Individual arc-grown MWCNTs were found to fracture by a so-called sword-in-sheath mechanism in the outermost shell. Strength values ranging from 11 to 63 GPa at fracture strains of up to 12% and modulus values from 270 to 950 GPa were obtained [107]. Assuming that the load is carried by the SWCNTs on the perimeter of the rope, fracture strength values ranging from 13 to 52 GPa and a modulus between 320 and 1470 GPa were obtained [108]. It is interesting to note that the maximum fracture strain was found to be 5.3%, which is close to the theoretical value of ~5% for defect nucleation in individual SWCNTs [109].

The experimental results for structurally nearly perfect nanotubes (produced by high-temperature methods) show that such nanotubes can indeed have a Young's modulus approaching the theoretical limit of 1.06 TPa [97], the in-plane modulus of graphite, in agreement with theoretical studies [44,46] (Young's modulus values of around 5.5 TPa [91] relate to an assumed effective SWCNT wall thickness of 0.066 nm). It should also be kept in mind that a single value of Young's modulus should not be uniquely used to describe both the tension/compression and bending behavior of carbon nanotubes. Tension and compression are mostly governed by the in-plane σ-bonds, while pure bending is affected by the out-of-plane σ-bonds. Yet continuum elasticity is applicable to describe the elastic properties of such nano-structures up to the point were local instabilities occur, given that the geometry of the nanotubes is properly taken into account.

A number of theoretical studies have addressed the structural stability of nano-tubes in tension, compression, bending, and torsion. Under axial loads, abrupt changes in nanotube morphology were observed, which depended on the nanotube length [110,111]. Nanotube buckling due to bending has also been demonstrated [110,112], which is characterized by a collapse of the cross-section in the middle of the tube, confirming the experimental observations [96–98]. Lourie et al. captured the buckling of SWCNTs in compression and bending by embedding the nanotubes in a polymer film [113].

Partial nanotube flattening due to van der Waals forces in the contact region between adjacent and parallel nanotubes was observed in TEM images, indicating that nanotubes are not necessarily perfectly cylindrical [114]. The deformation pattern of SWCNTs in a closest-packed crystal showed that rigid nanotubes with a diameter of <1 nm are less affected by the van der Waals attraction and deform the least [115]. In contrast, nanotubes with a diameter >2.5-nm facet against each other and form a honeycomblike structure. Fully collapsed MWCNTs have also been seen in transmission electron micrographs, with the collapse depending on the overall diameter and wall thickness [116]. Furthermore, nanotubes deform against substrates, an effect observed experimentally and addressed theoretically for MWCNTs [117–121]. An effect that cannot be described by conventional mechanics is the interlayer interaction between opposing walls in fully collapsed nanotubes that might effectively stabilize the collapsed structure, although no external load is present.

The strength of nanotubes likely depends on the distribution of defects and geometric factors such as the interlayer interaction in bundles of SWCNTs and MWCNTs. Unlike in bulk materials, the defect density is presumably low in these nanostructures and distributed over large distances due to the high aspect ratio. However, these factors will depend on the growth process, and the strength should only approach the theoretical limit for high-temperature nanotubes. The few experimental results mentioned above indeed indicate that the strength of nanotubes can be one order of magnitude higher than that of current high-strength CFs. Further evidence for the high strength of high-temperature nanotubes has been found in other tensile tests [122], although the strength decreased significantly for 2-mm long bundles of MWCNTs grown in a CVD process [123]. In this case, the average strength of about 1.7 GPa might be related to the higher defect concentration as a result of the lower growth temperature, but could also be attributed to gauge length effects with individual nanotubes being shorter than the overall length of the bundle. Initial fragmentation tests of nanotubes embedded in thin film polymer composite films also led to an estimated high tensile strength of nanotubes [124,125], although an accurate determination of the fragment length of an embedded nanotube in a TEM is challenging.

In addition to the nanotube strength being dominated by bond breakage, another contribution stems from the interlayer sliding forces. MWCNTs appear to fall victim to their own in-plane structural perfection, which prohibits covalent bonds between the different shells. As a result, no significant load is transferred to the inner shells when only the outermost shell is strained in tension. Similarly, individual SWCNTs in a bundle would have to be very long in order to allow sufficient load transfer to the innermost nanotubes via shear so that, eventually, the strength of the whole bundle is reached.

## 8.4 CNF-REINFORCED PEEK NANOCOMPOSITES

The literature on processing and evaluating macroscopic nanotube/nanofiber–polymer composites is still in its infancy but developing rapidly. This situation is not surprising, given that initial attempts to produce such nanocomposites were hindered by the small quantities of nanotubes available. The focus on CVD synthesis techniques has opened the door to the manufacture of large-scale polymer nanocomposites. However, as yet, no standard analysis techniques have been defined to characterize the resulting nanocomposite properties and to correlate the observed findings with the intrinsic nanotube properties. For example, a large number of studies have focused on the effect of nanotubes and nanofibers on the composite stiffness, failing to report other, more relevant properties such as strength and strain to failure. Such mechanical properties are more dependent on the state of filler dispersion, alignment, and interfacial issues than the stiffness. Nevertheless, a number of interesting observations have been reported, which help to develop a fundamental understanding of such novel materials and to assess their true potential.

Taking into account the still high raw material prices for CNTs and CNFs, surprisingly little effort has been spent on the evaluation of their performance enhancement in high-performance thermoplastic polymers. Some of the commercial fillers are now available in large quantities coupled with a good quality and purity at a price level comparable to that of expensive high-temperature thermoplastics such as PEEK. The work described in the subsequent sections was therefore aimed at the evaluation of CNF-reinforced PEEK composites for a range of industrial applications. In order to establish the true potential of polymer nanocomposites in general for a given application, any assessment of the mechanical and physical properties must be accompanied by a fundamental investigation of the composite microstructure. This procedure also accounts for nanocomposite materials intended for the use in tribological applications. The following sections provide the required overview of the PEEK-nanofiber composite microstructures and mechanical properties prior to the assessment of the tribological performance.

## 8.4.1 Materials and specimen preparation

The starting vapor-grown CNF material (Pyrograph III, Pyrograf Products Inc., USA) consists of loosely aggregated hollow nanofibers with a mean diameter of 155 ± 30 nm. Two distinctive structures are present in a given as-received sample: relatively straight cylindrical tubes and so-called bamboolike structures, shown in Fig. 8.10. In both types of nanofibers, the hollow core region is surrounded by graphitic planes that are arranged in a herringbone style at ±15° with respect to the tube axis in the inner part of the nanofiber wall. The outer part consists of short irregular graphitic planes parallel to the tube axis. The as-produced CNF have an aspect ratio of $10^2$–$10^3$.

Various nanofiber composites based on Victrex PEEK powder grade 450p, obtained from Lehmann & Voss, Hamburg, Germany, containing up to 15 wt% of nanofibers were produced using a Berstorff corotating twin-screw extruder with a length-to-diameter ($L/D$) ratio of 33, operating at 380 °C. Prior to the melt compounding, the as-received nanofiber material and the polymer were dry mixed and added to the

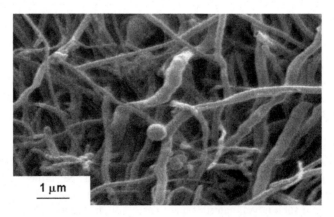

**FIGURE 8.10**

Representative SEM micrograph of the as-received CNF material.

hopper feeding the extruder. More precisely, three different experimental composites containing nanofiber weight fractions ranging from 5 to 15 wt% were prepared. For comparison, the unfilled polymer was processed under identical conditions.

In general, the addition of up to 15 wt% of nanofibers did not significantly alter the processing behavior of the thermoplastic matrix. The extruded strands of the pure PEEK were opaque, indicating that the cooling conditions did not suppress crystallization of the matrix. In contrast, all nanofiber composites were black in color and revealed an equally good surface finish.

Prior to the subsequent injection molding, the PEEK-based composite granulates were dried at 150 °C for 4 h. Tensile bars according to the ISO 179A standard were manufactured using an Arburg Allrounder injection-molding machine at processing temperatures of 390 °C, with the mold temperature set to 150 °C. All PEEK composites showed an excellent surface finish and a homogeneous color.

Following a standard industrial procedure to ensure a maximized and, more importantly, a uniform degree of crystallinity of the polymer matrix, all PEEK-based specimens were heat treated at 200 °C for 30 min, followed by 4 h at 220 °C, prior to testing.

In order to keep processing times of PEEK composites short, the mold temperature is usually set rather low compared to the melt temperature. The fast cooling at the cavity walls effectively suppresses crystallization, which only takes place in the core of injection-molded specimens [126]. It is standard industrial practice to remove this skin-core effect by subsequent annealing at temperatures above the glass transition temperature. However, this crystallization process reduces molecular orientations imposed by the shear flow and can lead to thermal shrinkage of components. Such thermal shrinkage during the heat treatment was observed for all PEEK-based specimens, but was significantly reduced for the nanofiber composites. Figure 8.11 summarizes the measured shrinkage of PEEK-based composites during annealing in

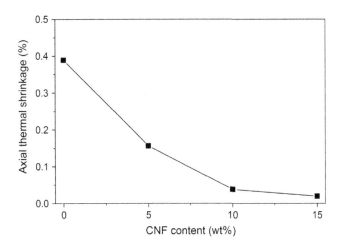

**FIGURE 8.11**

Axial thermal shrinkage of PEEK nanocomposites as a function of nanofiber content.

the axial direction. The stabilization effect levels off at higher nanofiber concentrations, but the overall dimensional stability of these composites is rather high.

### 8.4.2 Rheological behavior

To date, most scientific work regarding the rheological properties of nanotube and nanofiber composites has focused on the shear behavior [127–132], although more detailed elongational measurements important for foaming, film-blowing or fiber-spinning processes are slowly emerging [133]. In general, fundamental rheological investigations are expected to aid the understanding of the, often complex, microstructural evolution seen during processing of such nanocomposites, relating both to the filler and the polymer matrix, especially when dealing with a semicrystalline matrix.

Therefore, dynamic oscillatory shear rheological measurements of the PEEK–CNF nanocomposites were performed using an Advanced Rheometric Expansion System (ARES, Rheometric Scientific) with a 25-mm diameter parallel-plate fixture. In order to prepare specimens for the shear rheological investigations, the extruded pellets were compression molded at 360 °C for about 5 min into disc-shaped specimens of 1-mm thickness and 25-mm diameter. Prior to testing at 360 °C, all compression-molded specimens were dried at 120 °C for 24 h under vacuum. All measurements were conducted under a nitrogen atmosphere in order to prevent oxidative degradation of the specimens. In order to ensure the thermal stability of the melt during the measurements, a dynamic time sweep measurement was performed and verified constant rheological properties for at least 45 min. The temperature control was accurate to within ±1 °C.

From dynamic frequency sweep experiments of the pure PEEK and the various nanofiber-reinforced PEEK composites, the complex viscosity $|\eta^*|$, the dynamic storage modulus, $G'$, and the dynamic loss modulus, $G''$, of the systems were obtained as a function of the angular frequency, $\omega$. The frequency ranged from 0.1 to 500 rad s$^{-1}$. A strain range of 1–5% was used to ensure that the measurements were performed within the linear viscoelastic regime of all materials. In addition, dynamic temperature sweeps were performed at a fixed frequency (1 rad s$^{-1}$), and the temperature was varied starting from the molten phase at 380 °C down to the solidification of the sample (at ~300 °C), at a cooling rate of 5 °C/min.

The observed $G''$ data, normalized by the values of the linear region at low strain amplitude and plotted as a function of the strain, clearly showed a linear region at low strain amplitudes and a nonlinear region at high strain amplitudes. As expected, the linear viscoelastic behavior of the PEEK was altered significantly by the presence of the nanofibers. An increasing nanofibre content led to a continuous reduction of the critical strain amplitude, gc; indicating a shift of the onset of shear thinning to lower strain values with increasing nanofibre weight fraction and an associated reduction in the linear viscoelastic range. Related work [134] on clay-loaded systems suggested that the anisotropic clay platelets could be aligned to the flow direction at the critical strain amplitude. Similarly, in the current CNF system, the nanofibers are expected to gradually orient themselves and to align, thus reducing the fiber–fiber

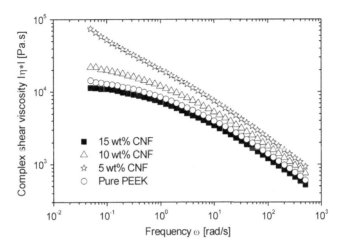

**FIGURE 8.12**

Complex shear viscosity of nanofiber-reinforced PEEK composites as function of frequency, at a temperature of 360 °C in the linear viscoelastic regime.

interactions [135]. As the formation of a network structure gradually develops with increasing nanofiber content, the dependence on the strain amplitude becomes more pronounced at elevated CNF contents. In order to explore network formation and its evolution, angular frequency sweeps were performed.

The frequency dependence of the complex shear viscosity, $|\eta^*|$, at 360 °C, of all PEEK composites is summarized in Fig. 8.12. The complex viscosity was calculated according to

$$|\eta^*| = \sqrt{\left(G'/_\omega\right)^2 + \left(G''/_\omega\right)^2},$$

(8.1)

where $G'$, $G''$, and $\omega$ denote the storage and loss modulus and the angular frequency, respectively. As can be seen, the viscosity of the PEEK nanocomposites generally increases with increasing nanofiber content; the modest magnitude of the increase at a higher frequency explains the good processability of the nanocomposite systems. Looking in more detail, the addition of 5 wt% CNF to PEEK already has an effect on the viscosity over the shear rate regime studied. However, up to 10 wt% CNF, the nanocomposites show a Newtonian behavior at low frequencies. At 15 wt% of CNF, no Newtonian regime can be detected, and a shear thinning behavior is observed; compared to the pure matrix, the viscosity increase is significantly more pronounced at low shear rates than at higher shear rates.

The pronounced increase in viscosity for the 15 wt% composite, at low frequencies, can be attributed to both polymer–fiber and fiber–fiber interactions. Such a behavior is typical for highly filled nanoparticulate systems, independent of filler type and geometry [128,130,132], whereas at high frequencies, the rheological response of the PEEK–CNF composites is dominated by the polymer matrix. More detailed investigations [129,131,136] of the low-frequency regime reveal that, below the threshold nanofiber concentration, the observed enhancement in the storage modulus can be attributed to a straightforward reinforcement of the melt by the nanofibers due to favorable matrix–filler interactions in the molten state. Besides the interaction between the polymer and the filler, above the critical concentration threshold filler–filler interactions begin to dominate as a filler network becomes established.

Pötschke et al. [130] reported that the rheological percolation threshold observed for melt-blended polycarbonate (PC)–MWCNT nanocomposites varies between 0.5 and 5 wt%, with a strong temperature dependency; the behavior was related to the existence of a combined nanotube–polymer network. Kharchenko et al. [137] reported similar rheological and electrical percolation thresholds of as low as 0.25–1 vol% for polypropylene (PP)–MWCNT nanocomposites. In this context, the higher percolation threshold (between 10 and 15 wt%) observed here for the PEEK/CNF system seems surprising, given that high aspect ratio particles are generally expected to percolate relatively easily.

The significance of this high threshold can be explored by comparison to theory. In particular, Smallwood [138] provided a model for the storage modulus of filled systems by extending Einstein's equation for spherical particles. For nonspherical filler particles such as fibers and platelets, this expression was modified by Guth et al. [139]

$$G'_{comp} = G'_m \left(1 + 0.67 f_s \phi + 1.62 f_s^2 \phi^2\right), \tag{8.2}$$

where $G'_{comp}$ and $G'_m$ are the shear modulus of composite and matrix, respectively $\phi$ is the volume fraction of the filler, and is the aspect ratio of the filler particles. This model holds only for particles that are well wetted by the polymer and have negligible particle–particle interactions. A deviation from the Guth relationship is an indicator of filler network formation. The volume fraction of nanofibers was calculated from the following equation

$$\phi_{CNF} = \left[(W_f/\rho_f) / (W_f/\rho_f + W_m/\rho_m)\right], \tag{8.3}$$

where $W_m$ and $W_f$ are the weight proportion of the PEEK matrix and the nanofibers, respectively, and $\rho_m$ and $\rho_f$ are matrix and filler density, respectively. The density of PEEK at 360 °C was estimated to be 1.096 g cm$^{-3}$ from the Sanchez–Lacombe equation of state [140] using the dimensionless parameter given by Brandrup et al. [141]. Assuming that the thermal expansion of the nanofibers is negligible, the nanofiber density at 360 °C was taken as the room temperature value of approximately 2 g cm$^{-3}$ [142].

Figure 8.13 shows the relative shear modulus versus the volume fraction of the filler. The solid lines represent the relative increase in modulus for various filler aspect ratios calculated based on Eqn (8.3). Below a filler volume fraction of 0.06 (corresponding to around 10 wt%), the modified Guth equation fits well with =15. This aspect ratio is much lower than that expected from microscopy of the raw CNFs, and indicates a significant length reduction during processing. A similar degradation of aspect ratio during processing has been reported by Kuriger et al. [143] and

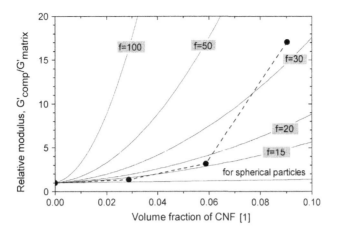

**FIGURE 8.13**

Plot of ratio of shear modulus of PEEK–CNF composites at low frequency ($w = 0.1$ rad/s), $G'_{complex}/G'_{Matrix}$, versus the filler volume fraction.

Kharchenko et al. [137], investigating PP filled with nanofibers and multiwall nanotubes, respectively. Assuming a constant fiber length for all compositions, the rheological response of the PEEK nanocomposite clearly deviates form the Guth equation (Eqn (8.2)) above a CNF volume fraction of 0.09; an effect that relates to the formation of a percolated filler network structure, in agreement with the previous discussion.

In summary, under shear conditions, the PEEK matrix viscosity generally increases with increasing nanofiber loading fraction, with a clear rheological percolation threshold appearing between 10 and 15 wt% of nanofibers. Increasing the CNF content both increases the shear thinning behavior and pushes the onset of shear thinning to lower shear rates, due to the direct shear response of the CNFs and the interaction with the polymer. Comparison with theory indicates that the length of the nanofibers is severely degraded during processing, resulting in a rather low aspect ratio of about 15. This degradation explains the relatively high rheological percolation threshold observed, but may also enable a good dispersion.

## 8.4.3 Microstructural assessment

As discussed before, any evaluation of the mechanical and/or physical properties of nanocomposites must involve a fundamental analysis of the composite microstructure, especially when dealing with a semicrystalline polymer. In addition to the previously introduced break down of the filler aspect ratio during processing, the presence of the significant particle surface area can have a pronounced effect on the crystallization kinetics of the matrix, effectively inducing altered properties of the polymer that need to be considered when attempting to describe the observed composite performance by established theories. The following subsections therefore provide a short overview of the composite microstructures.

### 8.4.3.1 Electron microscopy

Field emission gun scanning electron microscopy (SEM; JEOL 6340 FEGSEM) was carried out on fracture surfaces of all PEEK-based composites after tensile testing. All fracture surfaces were coated with a thin layer of gold for imaging. Similarly, TEM analysis was performed on PEEK-based samples microtomed at room temperature, using either a Philips 400T or JEOL 2000FX electron microscope operating at 180 kV.

Fracture surfaces of the injection-molded PEEK nanocomposites were examined in order to assess the degree of nanofiber dispersion and alignment. In all samples analyzed, the nanofibers appeared to be well dispersed, and no voids were observed. In addition, the nanofibers appeared partly aligned with the direction of flow during processing, verifying the rheological observations discussed above.

Scanning electron micrographs of PEEK–CNF composite fracture surfaces of samples containing (a) 10 and (b) 15 wt% are shown in Fig. 8.14. For most nanofibers, the pullout length was significantly <1 μm, reflecting the severe reduction of the intrinsic nanofiber aspect ratio as indicated by the shear rheological data. Occasional pullout lengths exceeding 2 μm were seen; however, these cases appear to correspond to thicker nanofibers lying within the fracture plane. In general, the interfacial bonding between the PEEK and the nanofibers appeared to be good, although debonded nanofibers were also observed, most likely reflecting interfacial failure and plastic deformation of the polymer during the tensile tests. Figure 8.14 reveals another interesting feature: some of the CNFs appeared to be coated by a cylindrical shell of polymer (solid squares). Such coated nanofibers were observed independent of loading fraction. In addition, Fig. 8.14 (b) shows broken nanofiber ends protruding from such a coating (dotted squares).

The SEM micrographs at higher magnifications shown in Fig. 8.15 of a sample containing 15 wt% of CNFs further verify the observations mentioned above. These images correspond to the core region of the tensile bar and highlight the typical radial features of the polymer around individual coated nanofibers (solid square).

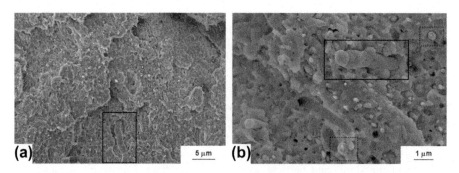

**FIGURE 8.14**

Representative SEM micrographs of fracture surfaces of PEEK nanocomposites containing (a) 10 wt% and (b) 15 wt% CNFs after tensile testing.

*With permission from Ref. [142].*

In addition, there are obvious variations in nanofiber orientation. The nanofiber in (b), lying in the image plane perpendicular to the direction of the tensile stress, exhibits a uniform polymer shell, features of the nanofiber bamboo structure are fully replicated. There also exists clear evidence for the presence of taut tie molecules (TTMs) [144] between this coating and the surrounding polymer matrix. Such TTMs usually form between crystalline regions upon drawing of a semicrystalline polymer and might indicate a crystalline nature of the coating.

The alignment of individual CNFs in the PEEK composites processed under shear flow conditions was further assessed qualitatively by TEM. Figure 8.16 shows two representative TEM micrographs of a PEEK composite containing 5 wt% of nanofibers. The direction of injection molding is normal to the image plane. Specimens (a) and (b) represent the skin and the core of the injection-molded sample, respectively. As can be seen, there are more circular cross-sections in the skin regime, whereas nanofiber segments within the plane are evident in the core. This reduced degree of alignment in the core of the injection-molded samples, where the shear flow is lower, is a well-known effect for injection-molded short fiber composites [145], which also accounts for injection-molded short fiber-reinforced PEEK composites [146].

Thin slices microtomed parallel to the flow direction were used to characterize the interfacial bonding between the filler and the matrix. Figure 8.17 shows a high-resolution transmission electron micrograph of a nanofiber wall (left) and the PEEK matrix. The fact that the interface was not destroyed during sample preparation can be taken as another indication of good interfacial bonding between the PEEK and this particular nanofiber. In contrast to the SEM observations, the TEM micrographs verify the generally good bonding between the PEEK matrix and the CNFs.

The general observations with regard to nanofiber dispersion and partial alignment are in agreement with published data for such nanocomposites. Furthermore, the electron micrographs do not indicate any variations in matrix morphology as a result of the nanofiber addition other than the crystalline coating of some individual nanofibers.

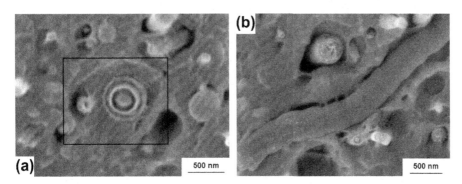

**FIGURE 8.15**

Magnified SEM micrographs of the 15 wt% PEEK–CNF composite after tensile testing.

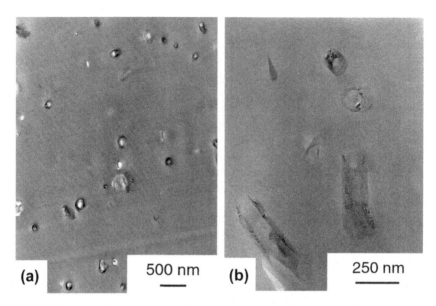

(a)  500 nm

(b)  250 nm

**FIGURE 8.16**

Representative TEM micrographs of a nanocomposite tensile specimen containing 5 wt% of CNFs perpendicular to the direction of injection molding at (a) the skin and (b) the core.

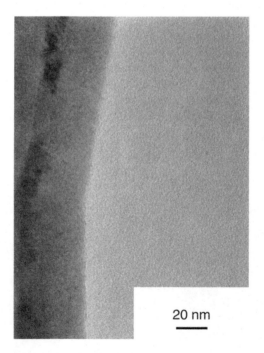

20 nm

**FIGURE 8.17**

High-resolution TEM micrograph of the interfacial bonding between a nanofiber wall (left) and the PEEK matrix.

*With permission from Ref. [142].*

### 8.4.3.2 *Thermal analysis*

The polymer matrix morphology of the injection-molded and subsequently annealed specimens was further assessed by means of differential scanning calorimetry (DSC), using a Perkin Elmer DSC 7 and a Thermal Advantage DSC 2920. All crystallization and melting thermograms were recorded at 10 °C min$^{-1}$ between 70 and 400 °C, with a 5-min hold in the molten state prior to cooling to erase the thermal history. In addition, isothermal thermograms were recorded at 303 °C, after the samples were quenched from the molten state at 200 °C min$^{-1}$.

This detailed DSC study was performed on all fully processed composite samples in order to further assess the matrix morphology and the overall degree of crystallinity of the matrices. Figure 8.18 shows the specific heat flows during (a) heating and (b)

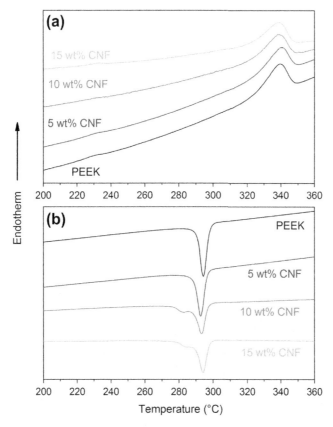

**FIGURE 8.18**

DSC thermograms at 10 °C min$^{-1}$ for injection-molded and annealed PEEK–CNF composites as a function of nanofiber content. (a) Heating and (b) cooling runs [142]. Traces are shifted vertically for clarity. (For color version of this figure, the reader is referred to the online version of this book.)

**Table 8.3** Thermal Parameters of PEEK–CNF Composites Determined from Nonisothermal DSC Scans at 10 °C min⁻¹

| Material | Nanofiber Content (wt%) | $T_m$ (°C) | $X_{c(m)}$ (%) | $T_c$ (°C) | $X_{c(c)}$ (%) |
|---|---|---|---|---|---|
| PEEK | 0 | 339.1 | 30 | 294.3 | 36 |
| PEEK | 5 | 339.7 | 29 | 294.8 | 36 |
| PEEK | 10 | 339.2 | 30 | 294.7 | 36 |
| PEEK | 15 | 338.9 | 30 | 295.1 | 36 |

cooling runs for all PEEK nanocomposites as a function of the temperature. A small melting peak >220 °C during the heating ramp was observed for all samples. This melting feature is related to the used annealing temperature and reflects melting of small crystals that have formed during the heat treatment. The enthalpies were similarly small and were neglected for the evaluation of the overall degree of crystallization. From these recorded melting and crystallization patterns, the thermal parameters such as crystallization temperature $T_c$, melting temperature $T_m$, crystallization enthalpy $H_c$, melting enthalpy $H_m$, and degree of crystallization $X_c$ were obtained and are summarized for the PEEK composites in Table 8.3. The degree of crystallinity was calculated from the peak enthalpies normalized to the actual weight fraction of polymer according to

$$X_c = \frac{\Delta H_c}{\Delta H_c^0} * \phi_{Mm} * 100,$$

(8.4)

where $\Delta H_c^0$ is 130 J g⁻¹, the theoretical value of enthalpy for 100% crystalline PEEK [147], and φMm is the weight fraction of polymer matrix.

The onset temperature to melting at 320 °C as well as the melting peak temperature varied by <1 °C, which is within experimental error. As can be seen, all the samples revealed a normalized degree of crystallinity of about 30% after the heat treatment. In addition, the glass transition temperature at 155 °C (not shown in Fig. 8.18) was not affected by the presence of the nanofibers.

An interesting feature was observed during the cooling experiments, Fig. 8.18(b). The onset temperature to crystallization at 300 °C and the crystallization peak temperature at 295 °C were not affected by the presence of the CNFs. Nucleation appeared to start homogeneously throughout the bulk matrix, without the nanofibers acting as heteronucleation sites. However, there was a small broadening of the crystallization peak on the lower temperature side for the PEEK composite containing 5 wt% of nanofibers, and a clear secondary peak appeared for the samples containing higher loading fractions.

CFs are known to nucleate crystal growth in PEEK. The resulting matrix morphology depends on the nature of the CF, the fiber volume fraction, as well as the processing conditions. Two types of on-fiber nucleation in CF-reinforced PEEK have been described: nucleation from discrete, well-separated points on the fiber surface and from thin layers of polymer sandwiched between touching or very close fibers [148]. The former nucleation event is influenced by the morphology of the CF, with high modulus CFs providing more nucleation sites than high-strength fibers. This high nucleation density can lead to a transcrystalline layer around the CF [148]. The formation of transcrystallinity strongly depends on the processing conditions, with slow cooling from the melt [149] and high isothermal crystallization temperatures favoring the development of transcrystallinity [150]. In contrast, bulk nucleation limits or even prevents the formation of transcrystalline layers around reinforcing fibers.

The second type of on-fiber nucleation that originates between fibers is believed to be caused by stresses induced in the PEEK at the fiber surface as a result of the different thermal contractions during cooling [151]. An additional contribution to this effect arises from short-range interactions between CF surfaces and aromatic polymers such as PEEK [152], which restrict chain movement during the cooling process from the melt [153].

Such heterogeneous nucleation effects would be indicated by a shift of the crystallization exotherm to higher temperatures, in contrast to the behavior observed for the nanocomposites. In general, nonisothermal DSC traces of PEEK samples cooled from the melt reveal a single crystallization peak. However, crystallization of this polymer is a two-stage process, clearly evidenced by thermal mechanical analysis experiments [154]. Primary crystallization is normally attributed to the formation and growth of spherulites, whereas the secondary crystallization process might be related to crystallization taking place within the spherulites (intraspherulitic crystallization) as well as to a thickening of individual lamellae.

The more pronounced separation of the two crystallization processes with increasing nanofiber content might be attributed to a delayed secondary crystallization of those PEEK molecules that are constrained by the finely dispersed CNFs and cannot contribute to the initial crystallization starting within the bulk of the matrix. This process might be assisted by local stresses as a result of the mismatched thermal contractions. Eventually, the constraint molecules will join the growing spherulites for small loading fractions of nanofibers. With increasing nanofiber content more molecules will be excluded from the initial crystallization and will crystallize around the nanofibers and in small gaps between them, given a slow enough cooling rate. Hence, it is not surprising that the nanocomposites did not show a secondary melting peak during the first heating ramp.

In order to confirm this explanation, nonisothermal measurements at a lower rate of 5 °C min$^{-1}$ were performed on the nanocomposite containing 10 wt% of nanofibers. During the first heating ramp, no changes in the melting peak were observed. Upon cooling, a pronounced appearance of the secondary crystallization peak was found. Reheating the fully crystallized sample then revealed a broader melting peak compared to the as-processed sample, a clear indication of a broader distribution in spherulite size and perfection as a result of the slow cooling. This observation indicates an interaction between the nanofibers and the matrix occurring during slow crystallization. Note that the cooling rate in these experiments is more than an order of magnitude lower than that experienced during injection molding.

In contrast to the experimental findings shown here, significant nucleation effects have been reported for CNFs [155], MWCNTs [156] as well as elastomeric nanoparticles [157] in isotactic PP (i-PP). In this semicrystalline matrix, these fillers nucleated the α-crystal structure. Furthermore, a significant reduction in average spherulite size was reported in case of the MWCNTs, even at a low filler concentration of 1 wt%. In contrast, the presence of SWCNTs [158] as well as of carbon black particles [159] in i-PP led to the nucleated formation of the α-crystal structure at the expense of the α-form [158]. With increasing content of SWCNTs, the average spherulite size was found to decrease [160], an effect that could be easily identified by a pronounced broadening of the melting endotherm at an SWCNT concentration as low as 0.8 wt% [161].

### 8.4.4 Mechanical behavior

Macroscopic tensile tests were performed at room temperature with a Zwick Universal tensile testing machine. The crosshead speed was set to 0.5 mm min$^{-1}$ in the 0.05–0.25% strain range and was then increased to 10 mm min$^{-1}$ until specimen fracture occurred.

Representative engineering stress–strain diagrams of the nanofiber–PEEK composites as a function of nanofiber loading are shown in Fig. 8.19. At least five specimens were tested for each composite composition and CNF loading fraction, and very small deviations in static mechanical properties such as Young's modulus $E$, yield stress $\sigma_{max}$, ultimate strength $\sigma_f$, and $\varepsilon_f$ were observed. These experimental observations indicate a uniform dispersion of nanofibers in all PEEK-based composites, independent of CNF loading fraction up to 15 wt%. The resulting average values and standard deviations for all composites are listed in Table 8.4.

Under the chosen testing conditions, the static mechanical properties such as tensile stiffness, maximum stress at the yield point, and strength of the PEEK composites increased linearly up to a nanofiber content of 15 wt%. Furthermore, the strain

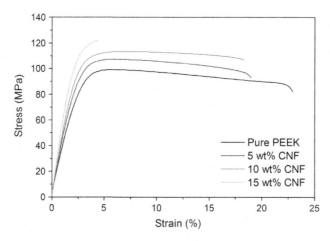

**FIGURE 8.19**

Representative engineering stress–strain diagrams of injection-molded PEEK–CNF composites [142]. (For color version of this figure, the reader is referred to the online version of this book.)

to failure was found not to decrease significantly for nanofiber loading fractions up to 10 wt%. Optical microscopy of fractured PEEK composite samples confirmed that the ability of the nanocomposites to contract and flow at the maximum stress decreased with increasing nanofiber content. This effect was most pronounced for the 15 wt% sample, which showed a brittle fracture behavior. Similarly, the toughness, a measure of the energy a sample can absorb before it breaks (as given by the area under the stress–strain curve), was only significantly reduced for the highest nanofiber loading fraction, further highlighting the ductile–brittle transition.

The linear increase in composite stiffness over the nanofiber concentration range investigated in this study agrees well with other published data, as summarized in Fig. 8.20. Furthermore, the significant increases in PEEK nanocomposite yield stress and strength with increasing nanofiber content reported here, represent an

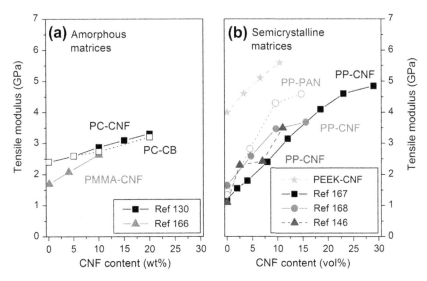

**FIGURE 8.20**

Comparative assessment of improvement in tensile modulus of nanofiber-reinforced thermoplastics. (For color version of this figure, the reader is referred to the online version of this book.)

**Table 8.4** Mechanical Properties of PEEK-Based Nanocomposites from Static Tensile Tests as a Function of Nanofiber Content

| Nanofiber Content (wt%) | $E$ (GPa) | $\sigma_{max}$ (MPa) | $\sigma_f$ (MPa) | $\varepsilon_f$ (%) |
|---|---|---|---|---|
| 0 | 4.0 ± 0.1 | 99.2 ± 0.4 | 80.4 ± 2.2 | 21.9 ± 1.2 |
| 5 | 4.6 ± 0.1 | 107.5 ± 0.5 | 93.1 ± 3.2 | 19.6 ± 1.3 |
| 10 | 5.1 ± 0.1 | 113.3 ± 0.5 | 103.9 ± 2.5 | 17.2 ± 2.5 |
| 15 | 5.6 ± 0.2 | 124.1 ± 2.8 | $= \sigma_{max}$ | 5.2 ± 1.1 |

**Table 8.5** Mechanical Properties of PEEK-Based Nanocomposites from Three-Point Bending Tests at Room Temperature as a Function of Nanofiber Content

| Nanofiber Content (wt%) | $E$ (GPa) | $\sigma_{max}$ (MPa) | $R_{p\,3.0}$ (MPa) | $\varepsilon_f$ (%) |
|---|---|---|---|---|
| 0 | 3.7 ± 0.1 | 154.7 ± 0.9 | 118.2 ± 1.1 | > 15 |
| 5 | 4.1 ± 0.1 | 165.3 ± 1.7 | 129.1 ± 1.7 | > 15 |
| 10 | 4.4 ± 0.1 | 172.8 ± 1.8 | 137.5 ± 1.4 | > 15 |
| 15 | 5.1 ± 0.1 | 187.3 ± 2.9 | 152.5 ± 2.5 | 9.7 ± 1.6 |

improvement in the overall nanocomposite performance as compared to the current state of the art. This overall enhancement in mechanical behavior is most likely due to an improved filler dispersion as a result of the large-scale compounding.

In addition to the tensile behavior, the bending properties of the nanocomposites were tested at room temperature according to ISO 178. As can be seen in Table 8.5, the mechanical properties such as bending stiffness, the bending stress at 3.0% strain, and strength of the PEEK composites increased in a similar manner as the tensile properties up to a nanofiber content of 15 wt%. For CNF contents up to 10 wt%, no fracture of the specimens could be detected up to the maximum bending strain of 15% (limited by the experimental setup); the sample containing 15 wt% of CNF again revealed a brittle behavior and showed premature failure accompanied with a significantly lower maximum bending strain.

The bending elastic modulus and damping properties of the PEEK–CNF composites were additionally characterized by dynamic mechanical thermal analysis (DMTA) as a function of nanofiber loading and temperature. The DMTA measurements in a bending configuration were carried out using Perkin Elmer equipment on all heat-treated injection-molded PEEK-based composites. A peak displacement of 64 mm at a frequency of 10 Hz was applied over the temperature range from −100 to 300 °C, at a rate of 2 °C/min. The nanofibers were oriented parallel to the long axis of the injection-molded rectangular bars and perpendicular to the direction of deformation in all cases.

Figure 8.21(a) shows an increase in PEEK composite stiffness with increasing nanofiber content, both below and above the softening point of the matrix. The stiffening effect was more pronounced above the glass transition temperature $T_g$ at around 164 °C, as determined by the maximum in tan delta, the tangent of the ratio of the loss to storage modulus (Fig. 8.21(b)). Tan delta is a measure of the damping within the system. In general, the temperature of this relaxation process depends on the experimental probe and is very sensitive to the semicrystalline morphology of PEEK [162–164], which, in turn, is governed by the processing history. The observed $T_g$ of 164 °C is in excellent agreement with literature data for injection-molded PEEK annealed above $T_g$ [162].

As verified by the peak position of tan delta, the $T_g$ itself was not affected by the addition of nanofibers. This observation indicates that the semicrystalline matrix morphology was not significantly altered by the nanofibers under the chosen processing conditions. The decreasing height of the tan delta peak with increasing nanofiber

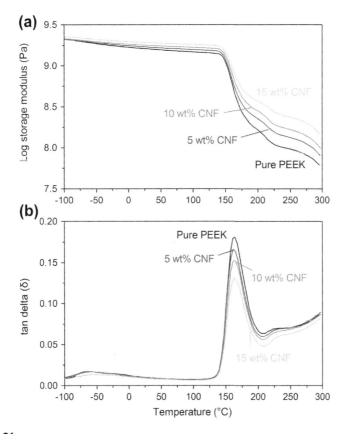

**FIGURE 8.21**

Dynamic mechanical analysis of PEEK nanocomposites as a function of temperature.
(a) Log storage modulus and (b) tan delta (δ) [142]. (For color version of this figure, the reader is referred to the online version of this book.)

loading relates to the reduced fraction of polymer matrix. The onset of the glass transition was not affected by the presence of the CNFs; however, there was a small broadening on the higher temperature side. A likely explanation is that unconstrained segments of the polymer retain the $T_g$ of the bulk PEEK but that those segments in the vicinity of the nanofibers are less mobile, and, therefore, lead to an increase in $T_g$. Similar effects have been previously reported for other nanotube [165] and nanofiber-thermoplastic composites [155] and are typical for polymer systems filled with finely dispersed materials [166]. In contrast, such a broadening of $T_g$ on the high-temperature side is not observed for short GF-reinforced PEEK [167]. The feature occurring at around 230 °C in all curves relates to the heat-treatment temperature.

There exists a sub-$T_g$ relaxation in amorphous and semicrystalline PEEK whose occurrence strongly depends on the experimental probe used [163,168]. The so-called β-relaxation at around −65 °C represents molecular motions, related to phenyl ring flipping [169], and is also sensitive to the morphology [163,168]. It is interesting to note that only the pure PEEK nanocomposite containing 15 wt% CNFs showed an effect on this transition (Fig. 8.21(b)). The peak was broadened and the peak temperature shifted to about −50 °C, indicating constraints on individual molecules from the presence of the nanofibers at higher loadings. This association also led to a more pronounced increase in composite stiffness at these low temperatures for the 15 wt% nanocomposite, and may relate to the brittle fracture behavior of these samples.

The data presented in this section indicate that the stiffening effect of nanofibers is more pronounced above the softening point of the matrix. This effect is not surprising, given that the matrix stiffness decreases significantly above $T_g$, and agrees with the general observations for nanotube and nanofiber-reinforced composites.

In the absence of significant morphological variations in the final composites, the mechanical performance can be further analyzed by conventional composite theories. Despite the fact that most practical short fiber composites are mixed according to weight fractions, the analysis of composite properties is normally performed considering fiber volume fractions, since many underlying structure–property relationships are linear as a function of volume fraction.

Assuming zero void content in the solid parts and identical degrees of crystallinity in all respective matrices, the composite density $\rho_c$ can be expressed as

$$\rho_c = \rho_m \phi_{Vm} + \rho_f \phi_{Vf} = \rho_m (1 - \phi_{Vf}) + \rho_f \phi_{Vf}, \tag{8.5}$$

where $\phi_{Vf}$ and $\phi_{Vm}$ are the volume fractions and $\rho_f$ and $\rho_m$ the density of filler and matrix, respectively. Second, the filler volume fraction can be expressed as

$$\phi_{Vf} = \frac{V_f}{V_f + V_m} = \frac{\dfrac{\phi_{Mf}}{\rho_f}}{\left(\dfrac{\phi_{Mf}}{\rho_f} + \dfrac{1 - \phi_{Mf}}{\rho_m}\right)} = \frac{1}{1 + \dfrac{\rho_f}{\rho_m}\left(\dfrac{1}{\phi_{Mf}} - 1\right)}, \tag{8.6}$$

where $V$ and $M$ correspond to the volume and mass of filler and matrix, respectively. With the known mass fractions $\phi_M$ and the experimentally determined composite densities as a function of filler content, this analysis allows the determination of the two unknowns, the density of the CNFs $\rho_f$ and the nanofiber volume fraction $\phi_{Vf}$ in the solid composites. The resulting nanofiber density was about 2 g cm$^{-3}$ [142]. The slightly lower value compared to graphite (2.2 g cm$^{-3}$) is expected, given the inner hollow core. The linear increases in composite tensile modulus $E_c$ as a function of nanofiber volume fraction up to 10 vol% can then be further evaluated by Krenchel's expression for short fiber composites [170]

$$E_c = \eta_O \eta_L \phi_{Vf} E_f + (1 - \phi_{Vf}) E_m, \tag{8.7}$$

where $E_f$ and $E_m$ are the fiber and matrix moduli, and $\eta_O$ and $\eta_L$ are efficiency factors relating to the orientation and length of the fibers, respectively. Simple rearrangement of Eqn (8.7) allows the determination of the effective nanofiber modulus $\eta_O * \eta_L * E_f$, the product of the intrinsic nanofiber modulus and the two orientation factors. The resulting average value of 21.5 ± 1.9 GPa can be used for comparative purposes. The value characterizes

**Table 8.6** Comparison of the Effective Nanofiber Moduli Determined for Semicrystalline Matrices Under Shear-Flow Conditions. In Addition, Data for Short CF and GF Composites Processed Under Similar Conditions are Included

| Material | Filler Content Range (vol%) | $\eta_O\eta_L E_f$ (GPa) | Reference |
|---|---|---|---|
| PEEK + CNF | 0–10.4 | 21.3 ± 1.7 | [143] |
| PP + CNF | 0–11.0 | 30.7 ± 16.0 | [143] |
| PP + CNF | 0–9.7 | 21.0 ± 0.8 | [171] |
| PP + CNF | 0–23.0 | 16.0 ± 1.0 | [172] |
| PP + CF | 0–9.6 | 33.1 ± 1.5 | [171] |
| PEEK + CF | 0–23.1 | 44.8 ± 2.2 | [173] |
| PEEK + GF | 0–17.9 | 29.6 ± 2.8 | [173] |

the effective nanofiber modulus and is in good agreement with the PP data shown in Fig. 8.20. In a simplified view, the value is describing the average slope of the curves in Fig. 8.20, neglecting the contribution of the matrix. As an example, if the modulus increases from 1.5 to 3.5 GPa via adding 10% of nanofibers, this relates to a slope of 20 GPa. It should be remembered that these composites were processed under similar conditions involving shear forces. However, PP is more susceptible to nanofiber nucleation effects that might have influenced the matrix morphology in the reference studies and might reflect the deviation from a linear behavior in Ref. [143]. All calculated effective moduli for the different studies are given in Table 8.6, as well as the loading fraction range used for the linear regression. Furthermore, some effective moduli for short fiber-reinforced composites processed under similar conditions are presented as further references.

These data highlight that the effective reinforcement capability of nanofibers is lower than that of standard short carbon or GFs. However, it should be kept in mind that the ductility of the nanocomposites was maintained up to a filler volume fraction of at least 6.6 vol% in the PEEK case. In contrast, the addition of a similar volume fraction of CFs leads to a brittle behavior already [173].

Upper and lower bounds for the true nanofiber modulus can be obtained by further analysis of the two efficiency factors. For a perfectly aligned short fiber composite $\eta_O$ equals 1, while a random two-dimensional arrangement of fibers yields a value of 0.375 [174]. The length efficiency factor for short carbon fiber (SCF) composites is given as $\eta_L = 0.9$ for an aspect ratio of >100 and $\eta_L = 0.2$ for an effective aspect ratio of 10 [174]. Taking these values as limits, the true nanofiber modulus lies in the range of $20 < E_f < 300$ GPa. The average length of nanofibers in thermoplastic matrix composites after similar shear-intensive processing has been determined as around 10 μm [143]. Nevertheless, structural characteristics of the as-supplied CNF material, such as intrinsic curvature and the presence of the bamboolike structures, as well as the rheology data introduced before suggest that the effective load bearing capacity could be as low as 10. Based on the experimental observations, the best estimate of the true modulus might be around 100 GPa. Both stiffness bounds appear to be far below values for the stiffness of nanotubes grown at high temperatures [106,107] and are closer to those experimentally determined for catalytically grown nanotubes with a higher defect concentration [106,175].

An assessment of the strength of discontinuous fiber composites is more complex and is omitted due to problems of obtaining statistically valid data for the various model parameters necessary. There is a significant spread in nanofiber pullout length visible in the SEM micrographs, which partly relates to the variation in nanofiber diameter and alignment. An intrinsic strength distribution along the nanofiber length is likely, given their low structural quality. The evidence for variations in the interphase around individual nanofibers prohibits the use of a constant interfacial shear strength since it is sensitive to the matrix morphology in CF-reinforced PEEK [176].

### 8.4.5 Tribological performance

For the specific wear tests described below, the injection-molded PEEK–CNF nanocomposite tensile bars were cut into $10 \times 10 \times 3$ mm samples parallel to the nanofiber direction. As for the mechanical testing discussed before, all samples were annealed at 200 °C for 30 min, followed by a 4 h cycle at 220 °C prior to testing to ensure maximum matrix crystallinity.

For comparative purposes, a range of reference PEEK composites were prepared under identical processing conditions as those used for the CNF composites as outlined in Section 8.4.1. Commercially available PEEK composites (LUVOCOM®, Lehmann & Voss, Hamburg) containing macroscopic glass (diameter 14 μm) and PAN-based carbon (diameter 9 μm) fibers with a length of approximately 3 mm prior to processing as well as a compound containing 10 wt% PTFE, 10 wt% graphite, and 10 wt% of the CFs were manufactured; the compositions and labels are listed in Table 8.7. It should be noted that the pure macroscopic fiber composites are optimized for excellent mechanical properties and thus have a higher fiber content than the compound optimized for the tribological applications.

Two different commercially available steels were used as counterparts for the wear tests: 100Cr6 (chromium content 1.35–1.6 wt%) steel balls were used for a

**Table 8.7** Overview of Sample Designation and Compositions

| Sample Designation | CNF Content (wt%) | CF Content (wt%) | GF Content (wt%) | PTFE Content (wt%) | Graphite Content (wt%) |
|---|---|---|---|---|---|
| PEEK | - | - | - | - | - |
| 5CNF | 5 | - | - | - | - |
| 10CNF | 10 | - | - | - | - |
| 15CNF | 15 | - | - | - | - |
| 30CF | - | 30 | - | - | - |
| 40GF | - | - | 40 | - | - |
| CF/PTFE/Gr | - | 10 | - | 10 | 10 |

comparison to data obtained from the literature, whilst X5CrNi18-10 (CrNi) stainless steel (chromium content 17.0–19.5 wt%) was investigated as it is frequently used as a relatively inexpensive counterpart for composite materials under corrosive conditions. The surface topography of the steel balls was measured with a profile tester from Perthen using a diamond probe. The Vickers hardness method was used to characterize the wide range of hardness observed in these materials. Table 8.8 lists the ball materials and their relevant properties. Prior to testing, the balls were cleaned in acetone.

The wear tests were performed according to ISO 7148-2/Section 8.5.2 with a ball-on-prism test system "Tribodata", from Dr Tillwich GmbH, Horb, Germany (Fig. 8.22), described in more detail elsewhere [177]. A metallic prism with an opening angle of 90° containing the specimens was pressed against a steel-bearing ball. A normal load component of FN = 21.2 N at 45° with respect to the load axis was applied to the specimen surface by dead weights via a lever. Due to the spherical contact geometry, the pressure*velocity ($p*v$) values decreased during the tests and cannot be quoted exactly. However, the wear marks had a diameter of about 1–3 mm leading to a contact pressure of 3–27 MPa and a $p*v$ value of <0.76 MPa m s$^{-1}$. The bearing balls had a diameter $d$ of 12.7 mm and rotated uniformly at a frequency of $f = 1$s$-1$, which results in a continuous sliding speed of $v = \pi\, d \cos 45° f = 28.2$ mm s$^{-1}$.

Each material combination was tested three times. The tests lasted 60 h. Before starting the wear tests, some specimens were loaded statically in the test rig for a

**Table 8.8** Counterpart Materials used in this Study and Their Relevant Properties ($R_a$; $R_z$ according to DIN 4768)

| Material | Vickers Hardness | $R_a$ (mm) | $R_z$ (mm) |
|----------|-----------------|-----------|-----------|
| 100Cr6 | 850 | 0.02 | 0.17 |
| X5CrNi18-10 | 375 | 0.04 | 0.31 |

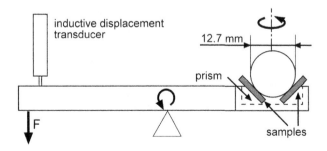

**FIGURE 8.22**

Schematics of a commonly used ball-on-prism tribometer.

*With permission from Ref. [178].*

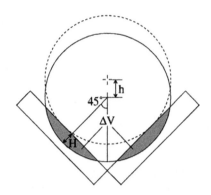

**FIGURE 8.23**

Geometric fundamentals for the calculation of the volume loss.

*With permission from Ref. [178].*

duration of about 30 h. Under this static load, the balls penetrated into the specimen surface due to plastic deformation and creep of the polymeric samples. During this test phase, the contact area increased continuously, reducing the contact pressure asymptotically until the rate of creep deformation became insignificant [177]. However, in case of the PEEK composites, the total creep deformation was on the order of a few microns and thus negligible in comparison to the magnitude of penetration caused by wear, which was on the order of some 100 microns. The creep effects were smaller than the scatter of the data; for this reason, the creep period was neglected for the tests performed in this study.

The vertical displacement signals, $h$, received every 10 min from inductive displacement transducers were converted into the penetration depth of the ball into each specimen surface, $H$, which was calculated according to (Fig. 8.23):

$$H = \frac{1}{\sqrt{2}} h.$$

(8.8)

From the penetration depth, the wear volume for each wear mark, $V$, was calculated:

$$V = \frac{\pi}{6} H^2 (3d - 2H).$$

(8.9)

The sliding distance, $s$, was calculated from the time, $t$, the frequency, $f$, and the ball diameter, $d$, by the following equation:

$$s = \pi d \cos 45^\circ \text{ ft.}$$

(8.10)

Finally, the wear volume was plotted versus the sliding distance. In all cases, the wear curves reached a steady state with a linear wear course. This part of the curve was fitted by a linear function of the type:

$$V = \dot{w}_{v/s} s + V_{in},$$

(8.11)

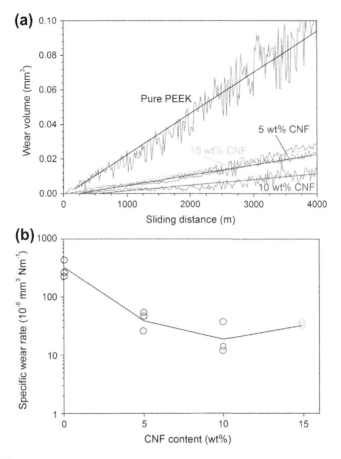

**FIGURE 8.24**

(a) Wear volume as a function of sliding distance and (b) semilog plot of specific wear rate of PEEK–CNF composites versus 100Cr6 steel, as a function of CNF content [178]. (For color version of this figure, the reader is referred to the online version of this book.)

where $\dot{w}_{v/s} = dV/ds$ represents the steady-state wear rate and $V_{in}$ is a measure for the running-in wear volume. Normalizing $\dot{w}_{v/s}$ with respect to the normal load acting on a single specimen surface, $F_N$, results in the specific wear rate, $\dot{w}_{v/s,n}$, which was taken as a quantitative measure for comparing the different material combinations:

$$\dot{w}_{v/s,n} = \frac{1}{F_N} \frac{dV}{ds}. \tag{8.12}$$

Figure 8.24(a) shows the results of the wear tests for the pure PEEK and the PEEK–CNF nanocomposites with CNF contents of 5–15 wt%, tested against the 100Cr6 steel. Both the raw data, shown as light gray scatter, and the fitting to the steady-state part of the wear test (solid line) are included. There is a clear reduction in wear volume for the PEEK–CNF nanocomposites as compared to the pure matrix, although it appears only weakly depended on filler weight content.

The specific wear rates for these materials show similar trends to the total wear volume, as summarized in Fig. 8.24(b). The CNF significantly reduce the specific wear rate of PEEK, however, given the error bars, the effect is not strongly dependent on the CNF content within the range of 5–15 wt%. Nevertheless, there appears a small optimum in the specific wear rate for a loading of 10 wt% of nanofibers. Such an optimum, at about 10 wt%, is also described in the literature for glass and CF-reinforced PEEK [179–182], while lower values were reported for CNT-reinforced polymers [62,63,66].

Returning to the PEEK–CNF nanocomposites, postulating a rigorous mechanism for the observed wear reduction by the addition of the nanofibers is not yet possible, but a number of alternatives may be considered. A change in the crystallinity of the polymer cannot be responsible for this improved wear behavior since the CNFs do not act as nucleation sites under the processing conditions used as verified above. Thus, likely reasons for the observed wear reduction are an increased shear strength of the CNF composites compared to the pure PEEK matrix, or a lubricating effect of the CNF wear debris. Unfortunately, neither the frictional coefficients nor the shear strength could be measured directly with the equipment available. Nevertheless, the mechanisms reported in other studies of CNFs and CNTs can also be considered for the present system [60–66]. For the given PEEK–CNF nanocomposites, nanofibers that are pulled out of the matrix must break up, since no CNFs were observed on the wear surfaces, using high-resolution SEM (not shown). Presumably, the CNFs degrade to small graphitic debris particles, which essentially act as a lubricant similar to graphite flakes [60,62,63,182]. Moreover, fragmentation of the CNFs steadily should readily take place due to the high contact stresses between the two wear partners. Given the small size and the intrinsic graphitic structure of individual nanofibers, CNF pullout is not expected to lead to any macroscopic surface roughness on either bearing surface. In addition, the CNFs may act to reinforce the transfer film, as observed with other nanofillers [46,47].

Light microscopy images of the polymer nanocomposite wear surfaces after testing against 100Cr6 steel are shown in Fig. 8.25. The image of the pure PEEK reveals depositions of polymer debris forming large plaques on the wear surface. These depositions were described by Voss and Friedrich [179] as backtransferred matrix particles and can be related to pronounced adhesive wear of the neat polymer. Moreover, the scratches in parallel to the sliding direction also indicate the presence of additional abrasive wear [66]. Similar observations were also made for short fiber-reinforced PEEK. The CNFs, however, tend to reduce the accumulation of the PEEK wear debris, such that smaller chips of matrix were found on the nanocomposite wear surfaces. Furthermore, the sliding surfaces become smoother with increasing CNF content due to the lubricating effect of the nanofibers. In addition, the enhanced stability and uniformity of the transfer film by the presence of the nanofibers helps to restrain the scuffing and adhesion of the PEEK matrix in sliding against the steel. It should be noted that also the creep-induced wear can be significantly lowered by the presence of the CNFs [66]. As a result of the aforementioned effects, the wear rate of PEEK can be considerably reduced by CNFs.

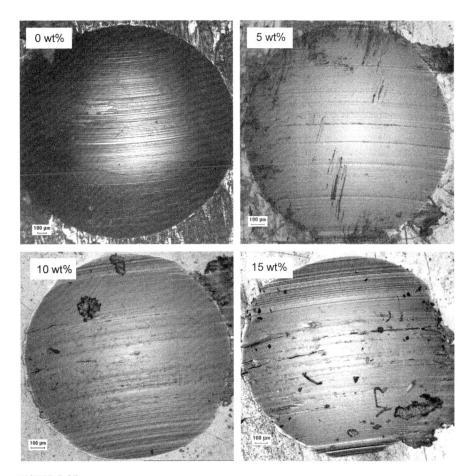

**FIGURE 8.25**

Surfaces of PEEK samples with 0, 5, 10, and 15 wt% CNFs after 60 h of sliding test against 100Cr6 steel.

*With permission from Ref. [178].*

The sharpness of the sliding tracks seems to show a minimum for the 10 wt% sample mirroring the trend for the specific wear rate. The 15 wt% composite reveals a greater number of larger dimples on the worn surface which, in turn, corresponds to a slightly increased specific wear rate as compared to the 10 wt% sample. Similar trends were also reported for polymer nanocomposites containing MWCNTs [62–64], where a minimum or leveling-off of the wear rate was observed at a certain CNT content.

In Fig. 8.26, the wear rate of the 10 wt% CNF composite is compared to that of pure PEEK and the three different commercially available PEEK composites that have been investigated by Jaskulka and Jacobs [183] for the two different steel counterparts. The difference in the wear behavior of these commercial composites,

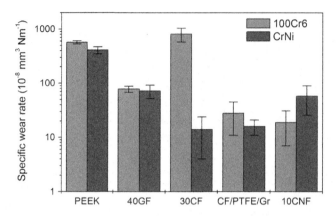

**FIGURE 8.26**

Specific wear rate of PEEK and its composites sliding against 100Cr6- and X5CrNi18-10-steels [178]. (For color version of this figure, the reader is referred to the online version of this book.)

with either metal counterpart, is influenced by several factors. Reducing the hardness of the metal counterpart often results in a higher wear rate of polymer composites [181,184]. Harder metal counterparts are less easily roughened by the fiber debris and, in turn, show a lower abrasive effect on the polymer surface. A reduction of the surface topography results in a reduced wear rate until a minimum is reached; a further reduction increases the wear rate again [185]. Certain fillers, such as CFs, can introduce a tribochemical reaction into the wear process [186,187] that leads to a sudden increase in the wear rate, some time after the wear process has reached its initial stable state. This reaction is easily identified, in the plots of wear volume against sliding distance, as a dramatic change in slope after a critical running-in time.

Independently of the nature of the steel counterpart, pure PEEK shows a rather high specific wear rate as compared to its composites. The addition of 40 wt% of GFs results in a specific wear rate reduction by a factor of 7 due to the reinforcing effect of the fibers. The main wear mechanism for PEEK–GF composites is known to be fiber pulverization [179], independent of the steel counterpart.

In this study, the wear behavior of CF-reinforced PEEK was found to be clearly controlled by the counterpart material. Tests against the 100Cr6 steel showed a strong time dependence of the specific wear rate under the chosen wear conditions. After 600 m sliding distance, a change in the slope of the wear curve can be observed in Fig. 8.27(a), accompanied by the onset of rust formation. A tribochemical reaction seems to take place, which leads to an increase in the specific wear rate by 140% compared to the pure PEEK (Fig. 8.26). A similar trend has been reported for epoxy-based composites, though in an even more pronounced manner [186]. The wear transition always occurred when CF-containing composites were tested against a low alloyed martensitic steel, and was associated with rust formation. Obviously, the CFs or their debris are able to induce tribochemical corrosion of unalloyed steels.

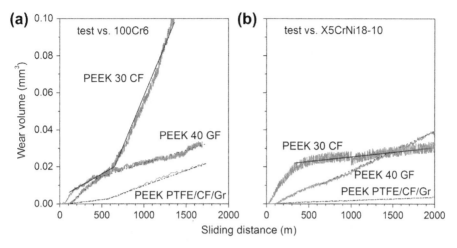

**FIGURE 8.27**

Comparative wear volume versus sliding distance curves for glass and CF-modified PEEK compared to an optimized PEEK compound containing PTFE, CFs, and graphite when tested against (a) 100Cr6 and (b) X5CrNi18-10 steels [178].

Conversely, tests performed against the more corrosion resistant CrNi steel did not show this unfavorable time dependence, as shown in Fig. 8.27(b). Following a pronounced running-in phase, the wear curve flattened gradually and reached a notably lower wear rate during steady state. In this case, the lower abrasiveness of the CFs and their debris in comparison to the previously described GFs obviously results in a substantial improvement against the less reactive CrNi steel.

In the case of the PEEK–GF composites, there was neither a transition in the wear curves (Fig. 8.27) nor any rust formation that could be evidence of a tribochemical reaction with either counterpart steel. This result is not surprising and can be attributed to the chemical inertness of the GFs.

Often, PEEK is optimized for tribological applications by the addition of a mixture of fillers, namely, 10 wt% PTFE, 10 wt% graphite, and 10 wt% CF. Although a change in the slope of the wear curve to higher wear rates (Fig. 8.27(a)) still provides evidence for the tribochemical reaction discussed previously, the effect is not as pronounced as for the composite containing solely 30 wt% CF. Interestingly, for the 30CF and the CF/PTFE/Gr samples, the steady-state wear rate against the CrNi steel is about the same (Fig. 8.26), although the total wear volume is drastically reduced for the optimized CF/PTFE/Gr compound. The reduced wear volume is due to the absence of a pronounced running-in phase for the optimized compound. The lubricating effects of the PTFE and the graphite seem to compensate for the higher mechanical strength of the composite containing 30 wt% CF, resulting in the similar specific wear rate.

Returning to Fig. 8.26, it is apparent that the specific wear rate of the PEEK–CNF composite tested against the standard bearing steel 100Cr6 is superior to all other tested commercial PEEK-based materials. Unlike the other CF containing composites,

**FIGURE 8.28**

Surface of a PEEK composite containing 10 wt% CNF run against X5CrNi18-10-steel for 60 h.

*With permission from Ref. [178].*

the PEEK–CNF nanocomposite show no wear transition during the entire test cycle and thus no evidence for a tribochemical reaction can be found. In this situation, the CNFs seem to be as chemically inert as the GFs, while being less abrasive due to their smaller size. It should be noted that the absence of tribochemical reactions was also reported for nanocomposite systems containing MWCNTs [62,63], while contrary results were reported for UHMWPE/HDPE (Ultra High Molecular Weight Polyethylene / High Density Polyethylene) blends reinforced CNTs [66]. Here, further investigations are needed to clarify the dependence of tribochemical reactions on the structural features of the CNTs and CNFs, respectively. Figure 8.28 finally shows that, on changing the steel counterpart to the softer CrNi steel, the sliding traces are more pronounced than for the 100Cr6 steel (Fig. 8.25) and thus a higher wear rate of the PEEK nanocomposite containing 10 wt% of CNF is observed.

## 8.5 HYBRID MATERIALS—CF-REINFORCED PEEK–CNF COMPOSITES

### 8.5.1 Hybrid materials—motivation and state of the art

During the last decades, the ever-increasing demands for polymers with enhanced wear characteristics have been met by various strategies, in particular the addition of solid lubricants as well as the reinforcement using microscale fibers and hard

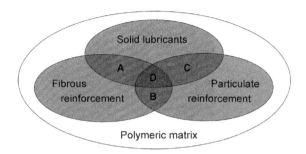

**FIGURE 8.29**

Strategies to develop hybrid materials for tribological applications.

particles. As discussed in the previous section for the CNF-reinforced PEEK, polymer nanocomposites especially can be regarded as a promising alternative for providing an improved tribological behavior. However, the application of microscale or nanoscale particles in polymeric matrices only is often not sufficient, regarding neither the wear rate nor other properties such as mechanical strength and toughness. Novel approaches for the customized materials development are still needed in order to keep pace with the increasing requirements of today's applications.

Hybrid materials composed of more than one additive in a polymer matrix provide an interesting approach to satisfy these demands by combining the properties of different fillers and even introduce synergisms. Nevertheless, the number of systematically developed hybrids is as yet still limited. This fact may be attributed to the complexity of the behavior of such materials as well as to the countless combinations of fillers available.

In general, hybrid materials can be classified by the purpose of the individual additives, as schematically shown in Fig. 8.29 (compositions A–D). It should be noted that individual additives might also act both as solid lubricant as well as reinforcement. Graphite, for example, mainly reduces the frictional coefficient and increases the modulus and the strength of the polymer matrix to some extent.

It is worth mentioning that hybrid materials attract increasing attention for tribological applications. The combination of different additives in a polymeric matrix can also be exploited to tailor other materials properties, for example, the electrical conductivity. A recent example for PP composites containing both CFs and carbon black is given by Drubetski et al. [188].

Returning to tribological compounds, morphologies corresponding to the hybrid systems introduced above are schematically shown in Fig. 8.30. While the overall behavior of such multiphase systems can often be considered as a function of the respective contribution of each phase, a synergism in wear resistance has often been reported. For hybrids prepared of microscale additives (Table 8.9), the combination of a fibrous reinforcement and solid lubricants, in particular graphite, belongs to the most investigated and also commercially successful systems (hybrid A). On the one hand, the addition of fibrous, microscale reinforcements such as GFs or CFs enhances the compressive and shear strength as well the creep resistance, and thus

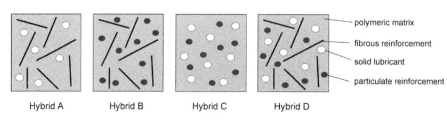

**FIGURE 8.30**

Schematic illustration of morphologies of hybrid materials for tribological applications.

**Table 8.9** Hybrid Composites for Tribological Applications—Combination of Microscale Additives

| Matrix | Microscale Additives | | Wear Conditions (Setup/counterpart/ lubrication/ temperature) | Reference (+Type of hybrid) |
|---|---|---|---|---|
| | Fibrous (type/ content) | Particulate (type/ content) | | |
| Epoxy | GF | Graphite/ 2.5–7.5 wt.% | n.a./hardened steel En 32/dry/RT | [192] (A) |
| Epoxy | CF/10–30 vol.% | SiC/4–12 vol.% graphite/7 vol.% | Block-on-ring/100Cr6/ dry/RT | [201] (D) |
| PEEK | CF/10 | PTFE/10 graphite/10 | Ball-on-disc/various counterparts/dry + water/RT | [197] (A) |
| PEEK | CF/10 wt.% | PTFE/10 wt.% graphite/10 wt.% | Ball-on-prism/various counterparts/dry + water/RT | [198] (A) |
| PEEK | CF/10 wt.% | PTFE/10 wt.% graphite/10 wt.% | Ball-on-prism/various counterparts/dry + water/RT | [31] (A) |
| PEEK | CF/10–15 wt.% | PTFE/0–25 wt.% graphite/ 10–15 wt.% | Pin-on-disc/100Cr6/ dry/RT – 220 °C | [23] (A) |
| PEEK | CF/10 wt.% | PTFE/10 wt.% graphite/10 wt.% | Pin-on-disc/100Cr6/ dry/RT – 150 °C | [181] (A) |
| PEEK | CF/15 wt.% | PTFE/0–20 wt.% | Pin-on-disc/steel/dry/ RT, 77 K | [194] (A) |
| PEEK | GF/30 | PTFE/15 | Pin-on-disc/various counterparts/dry + water/RT | [197] (A) |
| PEEK | GF/30 wt.% | PTFE/15 wt.% | Ball-on-prism/various counterparts/dry + water/RT | [31] (A) |
| PEEKK | CF | PTFE | Pin-on-disc/100Cr6/ dry/RT (?) | [23] (A) |

**Table 8.9** Hybrid Composites for Tribological Applications—Combination of Microscale Additives—*Cont'd*

| Matrix | Microscale Additives | | Wear Conditions (Setup/counterpart/ lubrication/ temperature) | Reference (+Type of hybrid) |
| | Fibrous (type/ content) | Particulate (type/ content) | | |
| --- | --- | --- | --- | --- |
| PEI | CF/15 vol.% | Graphite/5 vol.% | Pin-on-disc/steel/dry/ RT | [193] (A) |
| PEI | CF/10 vol.% | Graphite/15 vol.% | Pin-on-disc/100Cr6/ dry/RT | [200] (A) |
| PEI | CF/10 vol.% | Graphite/15 vol.% CaSiO$_3$/ TiO$_2$/5vol.% | Pin-on-disc/100Cr6/ dry/RT | [200] (D) |
| PEK | CF/10 wt.% | PTFE/10 wt.% graphite/10 wt.% | Pin-on-disc/100Cr6/ dry/RT - 220 °C | [199] (A) |
| PTFE | CF/5– 20 wt.% | Aromatic polyester/ 5–20 wt.% | Pin-on-disc/steel/dry/ RT, 77 K | [194] (B) |
| PTFE | CF/15 wt.% | PEEK/9–25 wt.% | Pin-on-disc/steel/dry/ RT, 77 K | [194] (B) |
| PEEK | CF/10 wt.% | PTFE/10 wt.% graphite/10 wt.% | Pin-on-disc/SiC abrasive paper/dry/RT | [17] (A) |
| PEEK | - | PTFE/n.a. CuS/n.a. | Pin-on-disc/steel/dry/ RT | [202] (C) |
| PEEK | - | PTFE 0–40 vol.% SiC 3.3 vol.% | | [203] (C) |
| | CF/10 wt.% | PTFE/10 wt.% graphite/10 wt.% | Pin-on-disc/SiC abrasive paper/dry/RT | [217] (A) |
| PEEK | - | PTFE/0–15 vol.% graphite/0–10 vol.% | Block-on-ring/100Cr6 dry/RT | [218] (C) |

leads to an increase of the maximum allowable product of contact pressure and sliding speed. On the other hand, lubricating additives such as graphite, or alternatively PTFE, reduce the friction coefficient and, subsequently, the shear stresses as well as the contact temperature in the mating surface are decreased.

Davim et al. [189] investigated the wear properties and friction coefficients of neat PEEK, PEEK–CF, and PEEK–GF at a constant fiber content of 30 wt% against steel surface on a long dry sliding test. They concluded that composites with CFs show the best tribological behavior and lowest friction coefficient [189]. However, orientations of the CFs (parallel, antiparallel, or normal) relative to the sliding direction play a significant role. As an example, Zhanga et al. [182] observed lower friction coefficients in SCF/PTFE/Graphite–PEEK composites, if fibers were aligned antiparallel to the sliding direction. The testing conditions are another influential factor. For instance, when changing the testing conditions from dry to wet, like in a ring–ring tribotest [190], the friction coefficients of CF–PEEK composites decrease significantly. Tanga et al. [190] relate this phenomenon to water lubrication at the interface and also to the enhanced heat transfer, the latter allowing more constant surface temperatures and thus retaining the mechanical performance. In addition factors such as the contact pressure on the sample and particularly the sliding velocity are of increased significance for wet testing conditions [190,191].

The outlined concept has been verified for PEEK as well as for other matrix materials (Tables 8.9 and 8.10). As an example, Suresha et al. [192] investigated the wear

**Table 8.10** Hybrid Composites for Tribological Applications—Combination of Microscale and Nanoscale Additives (Compositional Information as far as Mentioned in the Study)

| Matrix | Microscale and Nanoscale Additives | | Wear Conditions (Setup/counterpart/ lubrication/ temperature) | Reference (+Type of hybrid) |
|---|---|---|---|---|
| | Fibrous (type/ content) | Particulate (type/ content) | | |
| Epoxy | CF/15 | Graphite/5 TiO$_2$ (nano)/5 | Block-on-ring/100Cr6/ dry/RT | [219] (D) |
| Epoxy | CF/0–15 vol.% | Graphite/0–15 vol.% PTFE/0–15 vol.% TiO$_2$ (nano)/0–10 vol.% | Block-on-ring/100Cr6/ dry/RT | [29] (D) |
| Epoxy | CF/15 | Graphite/7 vol.% TiO$_2$ (nano)/4 vol.% | Block-on-ring/100Cr6/ dry/RT | [207] (D) |
| Epoxy | CaSiO$_3$/0 - 15 vol.% | TiO$_2$ (nano)/0–10 vol.% | Cylinder-on-flat/100Cr6/dry/ RT | [55] (B) |
| Epoxy | CF/n.a. | CNT/n.a. | n.a. | [211] (B) |
| Epoxy | CF/n.a. | Fullerene/n.a. | n.a. | [211] (A) |
| PEEK | | PTFE/0–40 vol.%, SiC (nano)/3.3 vol.% | Block-on-ring/carbon steel/dry/RT | [203] (C) |
| PEEK | | PTFE/0–10 wt.% Al$_2$O$_3$ (nano)/0–5 wt.% | Block-on-ring/carbon steel/dry/RT | [208] (C) |
| PEI | CF/10 vol.% | Graphite/15 vol.% TiO$_2$ (nano)/0–5 vol.% | Pin-on-disk/100Cr6/ dry/RT | [200] (D) |
| PEI | CF/5 - 15 vol.% | Graphite/5 vol.% TiO$_2$ (nano)/5 vol.% | Pin-on-disc/steel/dry/ RT | [193] (D) |
| PPS | CF/5 - 15 vol.% | CuO (nano)/1–4 vol.% | Pin-on-disc/hardened steel/dry/RT | [210] (B) |
| PPS | Kevlar/5 - 15 vol.% | CuO (nano)/1–4 vol.% | Pin-on-disc/hardened steel/dry/RT | [210] (B) |
| PEEK | CF/n.a. | PTFE/n.a. Graphite/n.a. Nanoadditives (ZnS, TiO$_2$)/n.a. | Pin-on-disc/100Cr6/ dry/RT | [209] (D) |

behavior of epoxy matrices (EP) in the presence of GFs and graphite, reporting a pronounced reduction in wear rate as well as in frictional coefficient. Chang et al. [193] detected a similar behavior for composites of PEI, CFs, and graphite, while the excellent performance was deteriorated at elevated contact pressures and sliding velocities. Moreover, the investigations of Theiler et al. [194] on composites consisting of PEEK, CFs, and PTFE at room temperature highlighted that the lowest frictional coefficient was obtained at 5 wt% of PTFE, while 15 wt% were essential in order to ensure the distinct minimum in wear rate. At lower temperatures of 77 K, however, the further pronounced reduction of the wear rate was attributed to the increase in polymer hardness. The beneficial use of graphite in combination with CFs has also been verified by Lu et al. [23] for PEEK matrices; particularly advantageous to reduce the frictional coefficient at elevated temperatures. Besides the well-investigated graphite and PTFE acting as solid lubricants, similar effects have recently been reported for alternative systems such as molybdenum disulfide [195]. However, for PA as a matrix material, a synergistic behavior could only be achieved in combination with CFs. Here, electron microscopic investigations of the worn surfaces revealed a thin and uniform as well a continuous transfer film under dry conditions.

In the particular case of PEEK matrices, the reinforcing effect of CFs is often combined with both graphite and PTFE, leading to synergisms in the wear behavior as demonstrated in numerous studies [23,182,194–199]. As can be seen in Fig. 8.31, both the frictional coefficient as well as the wear rate can be further reduced compared to systems containing only one of the additives. Systematic investigations of

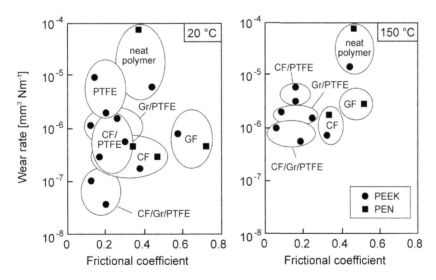

**FIGURE 8.31**

Synergism of microscale PEEK and PEN hybrids. Specific wear rate versus frictional coefficient of polymers and their composites at 20 and 150 °C

*Reproduced from Ref. [23].*

the wear behavior as well as of the transfer film formation [196] provided a detailed insight in the active wear mechanisms. Typically, the tribological system is composed of 10 wt% of CFs, graphite and PTFE, which leads to the formation of continuous and thin transfer films under dry conditions following a short running-in period. While PEEK compounds including graphite and PTFE only reveal an enhanced wear behavior, the presence of CFs provides a significantly enhanced strength and modulus and can be regarded as an optimum compromise between tribological and mechanical performance [199].

However, it should be noted that in the case of wet conditions such as under water lubrication, the formation of the transfer film is suppressed, or at least disturbed [197,198]. Subsequently, neither graphite nor PTFE can act as a solid lubricant, an effect that even leads to an increased wear rate when compared to dry conditions [198].

Besides hybrid systems revealing enhanced wear properties due to the addition of solid lubricants, Xian et al. [200] demonstrated the beneficial use of hard particles of $CaSiO_3$ to CF-reinforced PEI. The faster formation of the transfer film and the lower frictional coefficient were related to the protective effect of the particles, in particular by reducing the shear forces and the bearing of impact loads. Such hard particles are not only useful in combination with fibers, but also with solid lubricants. Prehn et al. [201] as well as Vande Voort et al. [202] investigated epoxy/graphite/SiC and PEEK/PTFE/CuS hybrids, respectively, and highlighted synergistic effects. However, the occurrence of tribochemical reactions can significantly alter the tribological performance, as in the case of PEEK/PTFE/SiC composites [203].

Finally, the complex interactions between different fillers in a PA matrix selected to enhance the wear performance were recently investigated for a combination of CFs, solid lubricants (PTFE) and reinforcing solid particles (CuO, CuS) [204,205]. Here, only the presence of both PTFE and CuO, or CuS, ensured the formation of the desired transfer film and the subsequently decreased wear rate.

As outlined in Sections 8.2 and 8.4, the wear behavior of polymeric materials can be significantly enhanced by the addition of nanoparticles. In numerous cases, lower values of the wear rate were observed when compared to microscale additives. Therefore, the question can be posed if a synergism in tribological performance exists for hybrids of microscale and nanoscale additives. This issue has been recently addressed by several studies, mostly regarding matrices of high-performance thermoplastics and epoxies (Table 8.10).

In the case of epoxy resins, for example, Zhang et al. [29] demonstrated the advantageous use of nanosized $TiO_2$ in composites consisting of CFs and graphite flakes. Although the traditional microscale fillers appeared to be more efficient to reduce the initially high wear rate, further improvements by a factor of 3 were achieved by the addition of 4–6 vol% of 300-nm-sized $TiO_2$ particles (Fig. 8.32). Similar results were observed for matrices of PA 6.6 and were ascribed to the pronounced reduction of the frictional coefficient as well as of the contact temperature [29,196]. In order to evaluate the wear mechanism, analysis of the worn surfaces indicated well-established mechanisms for fiber-reinforced systems, but much smoother surfaces in the case of the nano-$TiO_2$, even under severe operating conditions. In particular, no interfacial damage could be detected, but the fiber surfaces were finely scratched

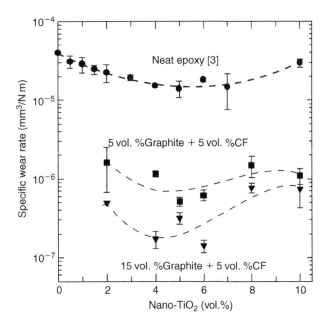

**FIGURE 8.32**

Synergism of nanoscale hybrids. Dependence of the specific wear rate of epoxy/graphite/CF composites on the content of nano-TiO$_2$

*With permission from Ref. [29].*

by the nanoparticles. Based on these results, a rolling effect of the nanoparticles has been proposed, effectively helping to reduce the friction during sliding and to lower the shear stresses and the contact temperature accordingly. Such rolling mechanisms have also been reported for other nanoparticles, for example, WS$_2$ [206].

A synergism in the wear behavior of graphite and nano-TiO$_2$ in an epoxy matrix was also observed by Xian et al. [207]. Again, the formation of a thin and continuous transfer film, as a key factor for low wear rates in the case of wear against a steel counterpart, was promoted by the presence of the two fillers. In particular, the adhesion strength of the film could be considerably enhanced as a uniform mixture of the debris was formed, while the resistance to microcracking and fatigue failure was also enhanced by the nanoparticles. At elevated sliding velocities, the subsequent temperature increase led to a partial destruction of the transfer film.

Besides the beneficial effects of nanoadditives on the wear rate under steady-state conditions, they particularly show a positive effect on the transient behavior when combined with PTFE [207]. While PTFE as a solid lubricant dominates the running-in phase by significantly reducing the frictional coefficient, the contact temperature and the time essential to reach steady-state conditions, hard nanoparticles such as TiO$_2$ advantageously contribute to the long-term and steady-state behavior (Fig. 8.33).

Regarding composites consisting of a PEEK matrix, Wang et al. [203] investigated the addition of PTFE and SiC nanoparticles (80 nm). In contrast to the

PEEK–SiC nanocomposite, the further modification with PTFE allowed neither to reduce the frictional coefficient nor the wear rate. The formation of a thick, discontinuous transfer film, plowing of the surface and, in particular, the presence of tribochemical reactions between SiC and PTFE (leading to SiFx, as analyzed by X-ray photoelectron spectroscopy analysis) contributed to the observed degradation of the tribological behavior. However, Qiao et al. [208] reported contrary results for composites of PEEK, PTFE, and nanosized $Al_2O_3$. Investigating the effect of particle size (15–500 nm), the wear performance remained similar, while the frictional coefficient was considerably reduced with decreasing particle size, but no distinct synergism was identified. In contrast, Oster et al. [209] demonstrated a synergistic behavior for various material combinations of PEEK with microscale as well as nanoscale fillers, in particular for composites made of CFs, graphite, ZnS, and $TiO_2$. Without highlighting the exact material composition in the publication, the composite revealed excellent properties even under severe loads and temperatures. It should be noted that similar concepts of material development were investigated for other high-performance thermoplastics such as PEI [193,209] and PPS [210].

Besides the aforementioned studies on hybrid materials consisting of at least one nanoscale additive, very limited information is available regarding the interaction between nanoscale carbon fillers with other additives. Only Ginzburg et al. [211] investigated the wear behavior of CF-reinforced epoxies and their hybrids with CNTs and fullerenes. For example, a wear rate strongly reduced by a factor of 2–7 was observed by the addition of 5 wt% of nanotubes, while only a minor effect on the frictional coefficient could be detected.

Based on these initial results, it appears promising to combine the previously described PEEK/CNF nanocomposites with common microscale additives, such as CFs or PTFE, in order to further optimize the tribological performance of such

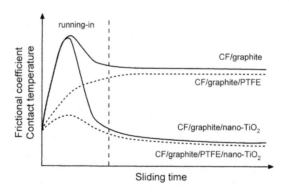

**FIGURE 8.33**

Synergistic wear performance of hybrid materials. Schematic representation of the effects of PTFE and nano-$TiO_2$ on the frictional coefficient and the contact temperature of graphite and SCF-filled polymer composites

*According to Friedrich et al. [196].*

high-performance compounds. In addition, conventional CFs should also significantly improve the mechanical properties of the pure nanocomposites and thus help to overcome the as yet limited strength and stiffness of these materials.

### 8.5.2 Tribological performance of hybrid PEEK–CNF nanocomposites

In an attempt to further characterize the potential of the CNFs as promising tribological additives, experimental hybrid compounds based on 10 wt% of nanofibers and other conventional additives were prepared. More specifically, custommade composites based on 10 wt% nanofibers and 10 wt% of the previously used macroscopic CFs, 10 wt% PTFE, and a combination of both, respectively, were manufactured. The labels and the exact compositions of all hybrid compounds evaluated in this study are listed in Table 8.11.

In order to ensure a good comparability, all manufacturing parameters were kept as outlined in Section 8.4. Following the initial twin-screw extrusion operation, CAMPUS tensile test bars were again produced from the extruded master batches by injection molding using an Arburg Allrounder 420C injection molding machine. The tensile bars were cut into $10 \times 10 \times 3$ mm$^3$ samples parallel to the injection direction. All the samples were again annealed at 200 °C for 30 min, followed by a 4 h cycle at 220 °C prior to testing to ensure maximum matrix crystallinity.

Again, in order to ensure direct comparability, all wear tests against the two previously introduced steel counterparts, namely, the 100Cr6 and the X5CrNi18-10 steel, were carried out in an identical manner as described in Section 8.4.4.

A representative scanning electron micrograph of a freeze-fractured surface of the PEEK–CF–CNF hybrid material is shown in Fig. 8.34, highlighting the morphology in the core region of a tensile bar. As can be seen, the microscale CF is strongly bonded to the nanocomposite matrix of PEEK and CNFs. Moreover, the CNFs appear to be homogeneously distributed in the matrix and around the CF, and are even embedded in the cylindrical shell of polymer coated around the microscale CF. Interestingly, the nanofibers are much less oriented compared to the CFs. Although some fiber pullouts can be detected for fibers oriented in parallel to the fracture direction, the interfacial bonding between the PEEK and the nanofibers as well as microfibers generally appeared to be good.

**Table 8.11** Overview of Hybrid Sample Designations and Compositions. Note that All Composites Contain 10 wt% of CNFs

| Sample Designation | CNFs (wt%) | Carbon fibers (wt%) | PTFE (wt%) |
|---|---|---|---|
| 10CNF | 10 | - | - |
| CNF/CF | 10 | 10 | - |
| CNF/PTFE | 10 | - | 10 |
| CF/CF/PTFE | 10 | 10 | 10 |

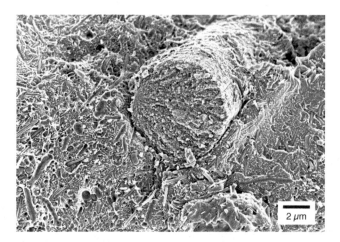

**FIGURE 8.34**

Representative SEM micrograph of a hybrid material fracture surface. Material composed of PEEK, microscale CFs and CNFs.

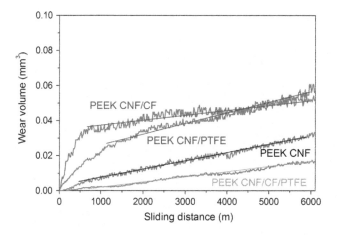

**FIGURE 8.35**

Plot of wear volume as a function of sliding distance of a 10 wt% PEEK–CNF nanocomposite as compared to modified hybrids containing an additional 10 wt% of CFs, 10 wt% of PTFE, and both [178]. (For color version of this figure, the reader is referred to the online version of this book.)

The influence of PTFE and macroscopic CFs on the wear behavior of the 10 wt% PEEK–CNF nanocomposite was explored. Wear volume versus sliding distance curves of these blends against the 100Cr6 steel are shown in Fig. 8.35, and the resulting specific wear rates are summarized in Fig. 8.36. Figure 8.36 also includes a summary of the specific wear rates observed for the CrNi steel. As can be seen in Fig. 8.35, the addition of 10 wt% of macroscopic CFs to the nanocomposite has no

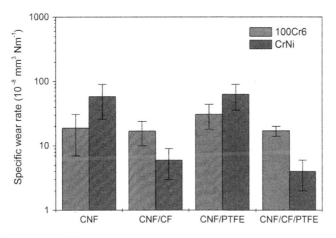

**FIGURE 8.36**

Influence of PTFE and CFs on the specific wear rate of the 10 wt% PEEK–CNF nanocomposite tested against the 100Cr6 and the X5CrNi18-10 steels [178]. (For color version of this figure, the reader is referred to the online version of this book.)

influence on the specific wear rate but does affect the initial wear behavior for the 100Cr6 steel. While the PEEK–CNF nanocomposite shows a rather constant wear rate throughout the entire test, a pronounced running-in phase can be observed for the CNF–CF composite until a preliminary steady state is reached. Furthermore, after about 450 m of sliding distance, a wear transition is observed. This transition is attributed to a tribochemical reaction even though it is not as pronounced as for the PEEK–CF composites without CNFs (compare to data discussed in Section 8.4.4) and occurs later. In contrast, the addition of the CFs significantly reduces the specific wear rate of the hybrid composite against the CrNi steel (Fig. 8.36).

The addition of 10 wt% of PTFE to the 10 wt% PEEK–CNF nanocomposite increased the wear rate against the 100Cr6 steel. This effect can be attributed most probably to a reduced shear strength of the composite. However, somewhat unexpectedly, a similar trend was not observed for the test against the CrNi steel (Fig. 8.36). Giltrow and Lancaster found that a high chromium content in the steel generally promotes the formation of a dense and smooth transfer film on its wear surfaces [212]; an effect that might explain the observed differences. In addition, a reinforcement of this transfer film by the CNFs as discussed previously might also contribute to the observed enhanced behavior.

The PEEK–CNF/CF/PTFE hybrid composite shows excellent wear behavior, verifying the general approach of combining different tribological additives. Its specific wear rate against 100Cr6 is about the same as for the CNF/CF samples, but without requiring a running-in phase. Its wear curve is linear right from the beginning of the test, and no evidence for a tribochemical reaction could be found despite the presence of the CFs. The wear rate against the CrNi steel is even lower and is superior to all other material combinations tested, and is more than three times better than the tribologically optimized, commercial CF/PTFE/Gr compound discussed in

Section 8.4.4. There is no pronounced running-in phase although the wear function is slightly curved at the beginning of the test. For both metal counterparts, the results lie within a very narrow range and hence the wear behavior of this material can be considered to be very predictable and thus reliable.

## 8.6 ADVANCED HYBRID PEEK–CNF COMPOSITES

Hybrid materials based on CNF-reinforced PEEK have been identified as promising materials for tribological applications. In the following section, recent research activities are described, which aim to exploit these excellent properties further and to develop tailored compounds for hinge bearings. As most of the work was performed in close cooperation with industrial partners, neither the exact material compositions nor the absolute values observed by the characterization are shown. The following results nevertheless highlight trends of significant scientific as well as commercial interest, which can guide the selection of advanced hybrid materials for tribological applications.

### 8.6.1 Materials strategy

For the selection of a tribological compound for a specific technical application, four decisive factors need to be considered (Fig. 8.37):

- Primary demands regarding material properties (in the present case, the tribological performance);
- Secondary demands regarding material properties (such as stiffness, strength, chemical resistance, thermal stability, impact resistance);
- Processing-related and technological requirements (such as ease-of-processing, feasibility of manufacturing complex parts, enhanced handing capabilities for the assembly process);
- Economical issues (such as low material and processing costs, increased maintenance intervals).

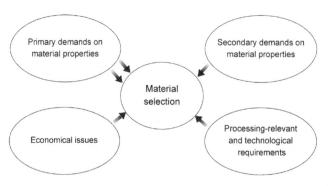

**FIGURE 8.37**

Selection of materials for tribological applications.

In order for the material to be suited for tribological applications, a sufficiently low wear rate and an adequate frictional coefficient are an absolute must-have and are, therefore, classified as primary demands. Furthermore, the material often needs to satisfy further requirements, frequently addressing the overall mechanical properties as well as the thermal stability and chemical resistance. Due to the fact that these requirements are often not as strict, they are regarded as secondary demands with a lower impact on the materials selection. Finally, factors relevant for the processing and economical efficiency need to be taken into account, while their relevance depends on the specific application and the maximum allowed costs of the final part. As an example, such issues can play a major role for the mass market, whereas their significance for high-tech applications such as in the aeronautics sector is markedly lower.

It needs to be pointed out that these various demands are often not concurrent. The discrepancy between the mechanical properties and a high flowability can guide as a first example. As most technical parts consist of thin-walled structures, typically with a thickness range of several millimeters or even less, low-viscosity materials and, thus, easy mold fillings are favored. However, in order to maintain a sufficiently high stiffness and strength of the material, reinforcing agents such as CFs need to be added which are, in turn, detrimental for the melt flowability.

Similar to the previous investigations, PEEK was selected as the polymer matrix, as it shows excellent mechanical properties in combination with a high operating temperature and good chemical resistance as it is needed for applications such as hinge bearings in automotive or aeronautical applications. In addition, PEEK provides a significantly better processability as compared to other high-performance polymers such as PI, PA imide or PBI. Due to the thin-walled structure of the bearings, a low-viscosity PEEK was chosen in order to enhance the flowability and the mold filling capability.

For all hybrid materials, CF-reinforced PEEK (PEEK/CF) was used as a reference material. As outlined before, CFs allow a considerable reduction of the wear rate by enhancing the strength and the modulus of the material at a minimum increase of the density of PEEK. Using PEEK–CF as a base material, various microscale and nanoscale additives as well as combinations thereof were added in order to systematically evaluate the optimum hybrid composition for the given technical application (Table 8.12). On the one hand, microparticles and nanoparticles potentially acting as solid lubricants (PTFE, graphite, zinc sulfide, nanographite, and molybdenum sulfide) were used; on the other hand, the addition of harder nanoscale particles (titanium dioxide, iron oxide) particularly was aimed at improving the microhardness as well as the modulus of the polymer. As shown previously, the unique wear mechanisms of the CNFs do not allow a straightforward assignment to one of these groups, as they are acting both as a reinforcing agent as well as a solid lubricant. Special emphasis was therefore placed on the interaction between the CNF and the other nanoscale additives. In all cases, the altered mechanical property profile, the enhanced formation of the transfer film as well as the initiation of other mechanisms (e.g. rolling mechanisms of the titanium dioxide particles) were anticipated to improve the wear

**Table 8.12** Composition of Hybrid Materials for Tribological Applications

| | Matrix | Fiber | Microscale Additive | | | Nanoscale Additive | | | | |
| | PEEK | CF | PTFE | Graphite | CNF | ZnS | $TiO_2$ | FeO | Nanographite | $MoS_2$ |
| --- | --- | --- | --- | --- | --- | --- | --- | --- | --- | --- |
| CF (reference) | X | X | | | | | | | | |
| CF/PTFE | X | X | X | | | | | | | |
| CF/PTFE/Gr | X | X | X | X | | | | | | |
| CF/CNF | X | X | | | X | | | | | |
| CF/CNF/ZnS | X | X | | | X | X | | | | |
| CF/CNF/TiO | X | X | | | X | | X | | | |
| CF/CNF/ZnS/TiO | X | X | | | X | X | X | | | |
| CF/nGr/MoS | X | X | | | | | | | X | X |
| CF/FeO | X | X | | | | | | X | | |

behavior, both under dry as well as under lubricated conditions. It should be pointed out that sliding wear behavior under dry conditions plays a significant role for the material selection, although the bearing is normally operated under lubrication. In the case of an unforeseen loss of the lubricant, the emergency running properties, that is, the time until the part fails under these unlubricated conditions, is of great interest.

All materials shown in Table 8.12 were compounded using a corotating twin-screw extruder equipped with a side-feed extruder. Gravimetric dosing units were used for metering exact material compositions. Subsequently, tensile test specimens according to ISO 527-2 (thickness 4 mm, width 10 mm) as well as thin-walled, rectangular plates (100 mm × 70 mm × 1.1 mm) were prepared by injection molding. For each material, the processing parameters were optimized in order to guarantee complete mold filling and a high quality of the specimens. All samples were subsequently annealed in order to ensure similar crystallinities of the matrix and planarity of the specimens. The mechanical as well as the tribological characterization techniques will be explained in detail when discussing the results.

### 8.6.2 **Tribological performance**

Friction and wear tests were performed using a modified ball-on-disc arrangement under oscillatory load at a frequency of 20 Hz with a counterbody of steel. For these tests, the injection-molded plates were used. During the wear test, the friction coefficient was continuously observed as the ratio between the measured friction force and the applied load. The tribological behavior was analyzed both under dry and lubricated conditions over a period of 6 h. For the dry wear tests, a load of 30 N was applied at room temperature and at a relative humidity below 10%. In the case of lubrication, however, both an elevated temperature of 90 °C and a higher load of 360 N were selected, as almost no wear could be detected under the conditions selected for the dry wear tests. Here, a special grease was added, which is similar to the lubricant added for the subsequent intended technical application.

The time-averaged frictional coefficient of the composites under dry and lubricated conditions is summarized in Fig. 8.38. As can be seen for the dry case, the frictional coefficient of PEEK/CF, the reference system, can only be reduced by adding either PTFE/graphite or, at least to a minor degree, the CNFs. While PTFE itself shows no significant effect on the frictional coefficient, the other hybrid systems based on the nanoparticles reveal increased values. In contrast, the materials show a completely distinct behavior under lubricated conditions. Here, the difference between all nanohybrids can be regarded as negligible. Such phenomena, as well as the generally reduced frictional coefficient have also been reported by other authors [55,196], and are attributed to the suppression or ever absence of the transfer film formation. Subsequently, solid lubricants lose their ability of reducing the frictional coefficient, as they are readily removed from the wear surface by the lubricant, leading to a similar performance as the neat PEEK/CF. Finally, the detrimental effect of adding PTFE needs to be pointed out. The continuous and significant removal of PTFE of the composite is likely to proceed due to the limited phase adhesion to the PEEK matrix. The lubricating grease appears

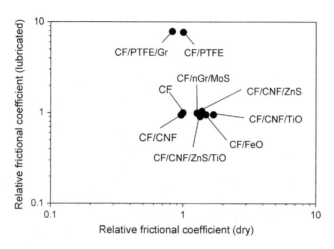

**FIGURE 8.38**

Comparison of the relative frictional coefficient of the various hybrid materials.

to promote this process. Finally, the frictional coefficient is significantly increased, both by the presence of loosely bonded or even nonbonded PTFE between the mating surfaces, and by the higher roughness of the PEEK surface.

The effect of the hybrid approach on the wear rate is shown in Fig. 8.39. Again, the values are normalized to the performance of the reference system PEEK/CF. Under dry conditions, the wear rate of PEEK/CF can be significantly reduced further by the addition of well-known systems containing solid lubricants such as PTFE as well as PTFE/graphite. Similar results are obtained for the addition of nanoparticle systems of either CNF or CNF/ZnS. In the case of all other systems, however, the wear under dry conditions is in part dramatically increased, for example, for CF/nGr/MoS.

Under lubricated conditions, the trend changes completely; in comparison to neat PEEK/CF, all hybrid systems reveal a decreased wear performance. While all systems containing nanoparticles reveal an only slightly increased wear rate, the wear rate is dramatically increased for composites containing PTFE. In turn, such materials containing either PTFE or PTFE/Gr would lead to a premature failure in the presence of lubricants. It can thus be concluded that, for the selection of an optimum material, a compromise between the wear performance under different conditions must be found. The additional material properties such as the mechanical behavior are then often used as decision guides and are discussed in the next section.

Cold remote nitrogen plasma has recently been introduced as a technique for improving the reactivity and adhesion of fibers to polymer matrices [213,214]. In the case of CFs, the method has been further modified via introducing oxygen into the plasma. As demonstrated for PEEK and other polymer matrices, the mechanical as well as tribological behavior of polymeric composites can be enhanced [215,216]. For remote nitrogen plasma mixed with 0.5% oxygen (CRNOP), the effect is related to the formation of polar groups on the fiber surface, thus allowing strong fiber–matrix

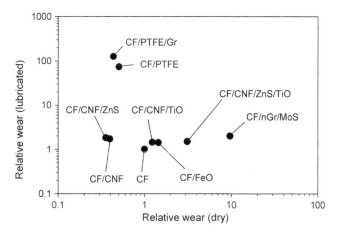

**FIGURE 8.39**

Comparison of the relative wear rate of the various hybrid materials.

interaction and adhesion. In addition, the increased surface roughness and surface area of the fibers leads to an enhanced mechanical interlocking with the matrix [214,215].

### 8.6.3 Mechanical behavior

The compression behavior can be regarded as a crucial parameter for the application of a material as a load bearing device, especially when taking into account that even a small plastic deformation may lead to a blockade and the subsequent fatal failure of the bearing. As a quantitative measure for the performance of the material, compression tests were performed using the injection-molded plates. After increasing the compressive load up to 240 MPa at 23 °C, which was applied by a cylindrical intender, the force was released to zero and the residual compressive strain was measured. The results for the various hybrid materials, summarized in Fig. 8.40, indicate that the presence of the nanoparticles mostly reduces the residual compression by a rather small factor. A considerable enhancement was only observed for the PEEK/CF system reinforced with additional CNFs only. However, a worse behavior was found for systems containing nanoscale solid lubricants only, that is, the Gr/MoS. Moreover, the residual compression was tremendously increased when adding the PTFE or PTFE/Gr, likely reflecting the soft nature of PTFE and the influence on the overall composite performance. It should be noted that a similar trend for all materials was also observed at elevated temperatures of 100 °C.

The intended technical application also demands for a high material stiffness in order to limit nondesired geometrical changes under an external load, and a good resistance to impacts; events that are numerously encountered during the lifetime of such a bearing. For both cases, characteristic values of all materials were evaluated using the dumbbell specimens and following well-known standards

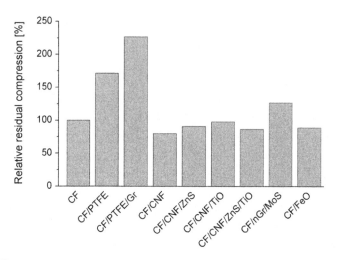

**FIGURE 8.40**

Comparative overview of relative wear rate of the various hybrid materials.

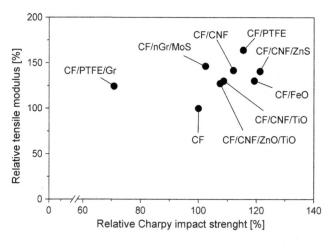

**FIGURE 8.41**

Comparison of relative tensile modulus versus relative Charpy impact strength of the various hybrid materials.

(tensile testing and Charpy impact testing according to ISO 527 and ISO 179eU, respectively). Again, the results shown in Fig. 8.41 are normalized to the PEEK/CF reference material. As can be seen, the addition of nanoparticles enhances both the impact resistance (+50%) as well as the tensile modulus (+25%). Nevertheless, distinct nanohybrids fail in showing enhanced properties, such as the CF/nGr/MoS. The materials containing PTFE, however, reveal an unexpected behavior. While the composite containing both PTFE and graphite shows a considerably

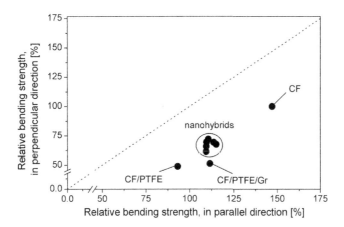

**FIGURE 8.42**

Comparative plot of relative wear rate of the various hybrid materials.

lower Charpy impact strength and only a slight stiffness enhancement, the PEEK/CF/PTFE material reveals a superior impact resistance and, moreover, the highest modulus of all materials investigated. The orientation of the CFs as a main factor influencing these properties is likely to be affected by the presence of PTFE and PTFE/graphite and, thus, alters the mechanical behavior. Regarding these results, it can be argued that all properties were observed for specimens of a relatively large thickness of 4 mm, although typically thin structures in the range of 1 mm are encountered for a bearing. In the following, this aspect will be considered in some more detail.

Finally, when discussing the property profile of multiphase composites, the orientation of the reinforcing additives in particular and the subsequent anisotropy in mechanical properties need to be taken into account. In Fig. 8.42, the bending strength of the PEEK/CF and the hybrid systems both in parallel as well as perpendicular to the injection direction can be seen (injection-molded plates as tested according to ISO 178). It should be noted that all values are normalized to the bending strength of the PEEK/CF perpendicular to the injection direction, and the diagonal line represents completely isotropic behavior. As commonly observed for fiber-reinforced systems, the strength in parallel to the injection direction is significantly higher as compared to the perpendicular direction for all systems. While the nanohybrids composed of at least one nanoscale additive show rather similar properties, the strength of the neat PEEK/CF is remarkably higher in the parallel case. Such a change in behavior is likely due to reduced orientation effects induced by the nanoparticles, which increase the viscosity and thus reduce the flow-induced orientation of the microscale fibers. Similar to the previous results, the performance of composites including PTFE is significantly reduced as compared to the other materials, in particular in the perpendicular direction. These observations are attributed to the poor interaction between the PEEK and the PTFE.

### 8.6.4 Selection guidelines of hybrid materials for tribological applications

Based on the experimental results introduced above, guidelines for the selection of the best material or, at least of the best compromise, can be deduced for a given set of technical requirements.

- Under dry sliding conditions, PEEK/CF with nanoscale solid lubricants are highly preferred, as they show excellent wear properties, if appropriately selected. In particular, CNFs as well as their combination with other nanoadditives should be pointed out, as they also allow an enhancement of the mechanical performance due to their reinforcing effect.
- Under lubricated conditions, solid lubricants show no effectiveness due to the suppression of their wear mechanism observed under dry conditions. In addition, they partly reveal a tremendously increased wear rate, as in the case of PTFE. Therefore, optimized systems composed of PEEK and CFs only are generally preferred. In particular cases, however, the use of nanoscale solid lubricants may present the better solution, as they allow an enhancement of the wear behavior in the worst case, the complete removal of the lubricant during operation. Here, the full potential of such materials needs to be considered in some more detail, for example, by running wear tests under mixed dry and lubricated conditions.
- If a particularly high mechanical performance is needed, the addition of high contents of CFs allows fulfilling these requirements in the best manner. The further addition of small contents of a solid lubricant with a reinforcing effect, CNFs in particular, can help to fine tune the sliding wear behavior and the compression as well as creep properties.

Based on the described laboratory-scale tests in order to evaluate the wear as well as mechanical performance of hybrid materials, a first preselection of promising composites can be made. Nevertheless, the structural parts and the assembly itself need to be tested under conditions similar to the intended usage, in order to allow a final conclusion. Successfully transferring the promising trends observed for the nanohybrid systems, in particular materials including CNFs, should further demonstrate the potential of such nanomaterials for technical applications and increase their commercial significance.

## 8.7 Summary

The present chapter has described the successful manufacture of a range of CNF-reinforced semicrystalline polymer composites, using an approach that should be generally applicable to other thermoplastic polymers. The commercially available CNFs were homogeneously dispersed and partly aligned in a high-performance thermoplastic PEEKs using standard polymer processing techniques such as extrusion and injection molding. The addition of low filler weight fractions led to linear increases in tensile and bending stiffness as well as tensile yield stress and strength. Furthermore, the matrix ductility was maintained up to a certain critical loading fraction, effectively leading to stronger and tougher composites. However, the nanofibers altered the macromolecular mobility of those polymer segments in their vicinity, as evidenced by an increase in the local glass transition temperature of the matrix. Similarly, the crystallization kinetics of the polymer matrices were influenced by the presence of the nanofibers, an effect that should become more pronounced as the matrix systems and processing conditions are varied and as the specific surface area of the filler increases further.

In the absence of significant morphological changes in the matrix microstructure under the chosen processing conditions, the mechanical composite performance was evaluated with respect to the effective reinforcement capability of the nanofibers by conventional short fiber composite theory. The effective nanofiber modulus provides a useful means of predicting the composite properties for future industrial applications. Although the effective nanofiber stiffness appears low, it is nevertheless comparable to that of conventional short GFs.

The CNFs should therefore be seen as a suitable reinforcement for thermoplastic matrices, especially for delicate structures such as films and fibers as well as microinjection-molded components where conventional reinforcements cannot physically be accommodated. The use of CNFs and CNTs should also add a number of other benefits to such microinjection-molded components, that is, a reduced thermal shrinkage.

More importantly, there is clear experimental evidence that the CNFs lead to an improved wear behavior of the PEEK matrix under dry sliding conditions. As demonstrated, the addition of 10 wt% of CNFs to the polymer induces a wear performance against a 100Cr6 steel comparable to that of an optimized commercial tribological compound based on 10 wt% of CFs, PTFE, and graphite. In addition, the nanofibers appear to be chemically inert, no tribochemical reaction as in the case of CFs is observed. As compared to GFs, the nanofibers are less abrasive, most likely due to their small size and graphitic nature.

The tribological performance of these PEEK–CNF nanocomposites can be further fine tuned by the addition of microscale additives such as CFs and CFs and PTFE. Such optimized nanocomposites reveal a very low wear under the selected dry sliding conditions and show a significant potential for industrial applications as neither the processing performance nor other mechanical properties are negatively affected. These compounds show an even further enhanced performance against an X5CrNi18-10 steel, in contrast to the slightly less improved behavior observed for the neat PEEK–CNF nanocomposites against this steel counterpart.

Correlating the observed improvements in wear behavior with the underlying composite microstructure, it appears likely that the nanofibers induce various wear-reducing mechanisms. On the one hand, there is clear evidence for a mechanical reinforcement of the polymer matrix without inducing additional modifications of the polymer crystal structure. It is feasible that the nanofibers also enable the formation of a uniform and reinforced transfer film. Yet the nanofibers also act as solid lubricants; an effect essentially due to the graphitic nature of the material.

In conclusion, it is clearly demonstrated that the addition of CNFs to semicrystalline PEEK can be exploited to achieve advanced tribological materials. The combination of nanofibers with conventional CFs especially leads to hybrids that can be tailored with regard to their overall mechanical and tribological properties, enabling industrial exploitation in a variety of conceivable applications. Although the tribological performance of such materials is somewhat reduced under lubricated conditions, such systems nevertheless provide a key feature for safety-critical components: in the case of a failure of the lubrication, the wear performance of such hybrids is significantly improved.

## Acknowledgments

The authors gratefully acknowledge the invaluable help and support of the various coworkers and students involved in the different aspects of the work described here. In addition, thanks are due to DaimlerChrysler, Stuttgart (U. Reimer, Dr A. Kratzsch, Dr F. Hoecker, K.-H. Röß), and Ensinger GmbH, Nufringen (A. Schmid, A. Dörper, F. Richter, K.-H. Kugele), for their interest in this study and their financial support. The authors would also like to thank Ms. Ronak Bahrami for her assistance during revising the book for the second edition.

## References

[1] P.M. Ajayan, P. Redlich, M. Ruehle, Structure of carbon nanotube-based nanocomposites, Journal of Microscopy 185 (2) (1997) 275–282.
[2] M. Meyyappan, Carbon Nanotubes: Science and Applications, CRC Press, Boca Raton, 2005.
[3] J. Coleman, U. Khan, W. Blau, Y. Gun'ko, Carbon; small but strong: a review of the mechanical properties of carbon nanotube–polymer composites, Carbon 44 (2006) 1624–1652.
[4] R.H. Baughman, A.A. Zakhidov, W.A. Heer, Carbon nanotubes—the route toward applications, Science 297 (2002) 787–792.
[5] P. Harris, Carbon Nanotubes and Related Structures, Cambridge University Press, Cambridge, 2001.
[6] V. Jamieson, in New Scientist, 2003.
[7] J.K.W. Sandler, S. Pegel, M. Cadek, F. Gojny, M. Van Es, J. Lohmar, W.J. Blau, K. Schulte, A.H. Windle, M.S.P. Shaffer, A comparative study of melt spun polyamide-12 fibres reinforced with carbon nanotubes and nanofibres, Polymer 45 (6) (2004) 2001–2015.
[8] J. Sandler, A.H. Windle, P. Werner, V. Altstädt, M.V. Es, M.S.P. Shaffer, Carbon-nanofibre-reinforced poly(ether ether ketone) fibres, Journal of Materials Science 38 (10) (2003) 2135–2141.
[9] P. Werner, R. Verdejo, F. Wöllecke, V. Altstädt, J.K.W. Sandler, M.S.P. Shaffer, Carbon nanofibers allow foaming of semicrystalline poly(ether ether ketone), Advanced Materials 17 (23) (2005) 2864–2869.

[10] K. Friedrich, Z. Lu, A.M. Hager, Overview on polymer composites for friction and wear application, Theoretical and Applied Fracture Mechanics 19 (1) (1993) 1–11.

[11] Z.P. Lu, K. Friedrich, On sliding friction and wear of peek and its composites, Wear 181–183 (2) (1995) 624–631.

[12] D. Landheer, C. Meesters, Reibung und verschleiss von ptfe-compounds (friction and wear of PTFE compounds), Kunststoffe 83 (1993) 996–1001.

[13] D. Mclean, Tribological properties of acetal, nylon and thermoplastic polyester, ANTEC Proc (1994) 3118–3121.

[14] J.H. Byett, C. Allen, Dry sliding wear behaviour of polyamide 66 and polycarbonate composites, Tribology International 25 (4) (1992) 237–246.

[15] II. Böhm, S. Betz, A. Ball, The wear resistance of polymers, Tribology International 23 (1990) 399–406.

[16] R. Sandor, Polybenzimidazole (pbi) as a matrix resin precursor for carbon/carbon composites, SAMPE Quarterly 4 (1991) 23–28.

[17] P. Mccamley, Selection guide to fiber-reinforced thermoplastics for high-temperature application, Advanced Material and Process 8 (1992) 22–23.

[18] L. Eng, Composite replaces aluminium in starter motor, Advanced Material and Process 7 (1994) 9.

[19] V. Long, PBI performance parts: a business unit of Hoechst Celanese corp, Advanced Material and Process 6 (1992) 50–51.

[20] H. Münstedt, and H. Zeiner, Polyaryletherketone—Neue Möglichkeiten für Thermoplaste, 1989. 79: p. 993–996.

[21] A. Lustiger, F.S. Uralil, G.M. Newaz, Processing and structural optimization of peek composites, Polymer Composites 11 (1) (1990) 65–75.

[22] M.F. Talbott, G.S. Springer, L.A. Berglund, The effects of crystallinity on the mechanical properties of peek polymer and graphite fiber reinforced peek, Journal of Composite Materials 21 (11) (1987) 1056–1081.

[23] Z. Lu, K. Friedrich, High temperature polymer composites for applications as sliding elements, Materialwiss. Werkst. 28 (1997) 116–123.

[24] G. Zhang, A.K. Schlarb, Correlation of the tribological behaviors with the mechanical properties of poly-ether-ether-ketones (peeks) with different molecular weights and their fiber filled composites, Wear 266 (1) (2009) 337–344.

[25] G. Erhard, 1980, University of Karlsruhe, Karlsruhe.

[26] D. Burris, W. Sawyer, A low friction and ultra low wear rate peek/ptfe composite, Wear 261 (2006) 410–418.

[27] W. Hufenbach, K. Kunze, J. Bijwe, Sliding wear behaviour of PEEK–PTFE blends, Journal of Synthetic Lubrication 20 (3) (2003) 227–240.

[28] A. Häger, M. Davies, Advances in Composite Tribology, Elsevier, Amsterdam, 1993.

[29] Z. Zhang, C. Breidt, L. Chang, F. Haupert, K. Friedrich, Enhancement of the wear resistance of epoxy: short carbon fibre, graphite, ptfe and nano-tio2, Composites Part A: Applied Science and Manufacturing 35 (12) (2004) 1385–1392.

[30] K. Friedrich, Traditional and new approaches to the development of wear-resistant polymer composites, Journal of Synthetic Lubrication 18 (4) (2002) 275–290.

[31] O. Jacobs, R. Jaskulka, C. Yan, W. Wu, On the effect of counterface material and aqueous environment on the sliding wear of various peek compounds, Tribology Letters 18 (3) (2005) 359–372.

[32] S. Bahadur, D. Gong, The role of copper compounds as fillers in the transfer and wear behavior of polyetheretherketone, Wear 154 (1) (1992) 151–165.

[33] Q. Zhao, S. Bahadur, The mechanism of filler action and the criterion of filler selection for reducing wear, Wear 225 (1) (1999) 660–668.

[34] L. Yu, S. Bahadur, An investigation of the transfer film characteristics and the tribological behaviors of polyphenylene sulfide composites in sliding against tool steel, Wear 214 (2) (1998) 245–251.

[35] L. Yu, S. Yang, H. Wang, Q. Xue, An investigation of the friction and wear behaviors of micrometer copper particle- and nanometer copper particle-filled polyoxymethylene composites, Journal of Applied Polymer Science 77 (11) (2000) 2404–2410.

[36] S. Bahadur, D. Tabor, The wear of filled polytetrafluoroethylene, Wear 98 (0) (1984) 1–13.

[37] Q. Zhao, S. Bahadur, A study of the modification of the friction and wear behavior of polyphenylene sulfide by particulate $Ag_2S$ and PBTE fillers, Wear 217 (1) (1998) 62–72.

[38] C.J. Schwartz, S. Bahadur, The role of filler deformability, filler, äìpolymer bonding, and counterface material on the tribological behavior of polyphenylene sulfide (PPS), Wear 251 (1,Äì12) (2001) 1532–1540.

[39] K. Friedrich, Friction and Wear of Polymer Composites, Elsevier, Amsterdam, 1986.

[40] J.M. Durand, M. Vardavoulias, M. Jeandin, Role of reinforcing ceramic particles in the wear behaviour of polymer-based model composites, Wear 181–183 (2) (1995) 833–839.

[41] K. Friedrich, Particulate dental composites under sliding wear conditions, Journal of Materials Science: Materials in Medicine 4 (3) (1993) 266–272.

[42] X.S. Xing, R.K.Y. Li, Wear behavior of epoxy matrix composites filled with uniform sized sub-micron spherical silica particles, Wear 256 (1) (2004) 21–26.

[43] Q. Xue, Q. Wang, Wear mechanisms of polyetheretherketone composites filled with various kinds of SiC, Wear 213 (1) (1997) 54–58.

[44] Q. Wang, Q. Xue, H. Liu, W. Shen, J. Xu, The effect of particle size of nanometer $ZrO_2$ on the tribological behaviour of peek, Wear 198 (1) (1996) 216–219.

[45] J.R. Vail, D.L. Burris, W.G. Sawyer, Multifunctionality of single-walled carbon nanotube/polytetrafluoroethylene nanocomposites, Wear 267 (1) (2009) 619–624.

[46] Q. Wang, J. Xu, W. Shen, W. Liu, An investigation of the friction and wear properties of nanometer $Si_3N_4$ filled peek, Wear 196 (1) (1996) 82–86.

[47] Q. Wang, Q. Xue, W. Shen, The friction and wear properties of nanometre $SiO_2$ filled polyetheretherketone, Tribology International 30 (3) (1997) 193–197.

[48] Q. Wang, J. Xu, W. Shen, Q. Xue, The effect of nanometer SiC filler on the tribological behavior of peek, Wear 209 (1) (1997) 316–321.

[49] Q. Wang, Q. Xue, W. Shen, J. Zhang, The friction and wear properties of nanometer $ZrO_2$-filled polyetheretherketone, Journal of Applied Polymer Science 69 (1) (1998) 135–141.

[50] C.J. Schwartz, S. Bahadur, Studies on the tribological behavior and transfer film counterface bond strength for polyphenylene sulfide filled with nanoscale alumina particles, Wear 237 (2) (2000) 261–273.

[51] S. Bahadur, C. Sunkara, Effect of transfer film structure, composition and bonding on the tribological behavior of polyphenylene sulfide filled with nano particles of $TiO_2$, ZnO, CuO and SiC, Wear 258 (9) (2005) 1411–1421.

[52] M. Zhang, M. Rong, S. Yu, B. Wetzel, K. Friedrich, Improvement of tribological performance of epoxy by the addition of irradiation grafted nano-inorganic particles, Macromolecular Materials and Engineering 287 (2) (2002) 111–115.

[53]  M. Zhang, M. Rong, S. Yu, B. Wetzel, K. Friedrich, Effect of particle surface treatment on the tribological performance of epoxy based nanocomposites, Wear 253 (9–10) (2002) 1086–1093.

[54]  G. Shi, M. Zhang, M. Rong, B. Wetzel, K. Friedrich, Friction and wear of low nanometer $Si_3N_4$ filled epoxy composites, Wear 254 (7–8) (2003) 784–796.

[55]  B. Wetzel, F. Haupert, K. Friedrich, M. Zhang, M. Rong, Impact and wear resistance of polymer nanocomposites at low filler content, Polymer Engineering & Science 42 (9) (2002) 1919–1927.

[56]  B. Wetzel, F. Haupert, M. Qiu Zhang, Epoxy nanocomposites with high mechanical and tribological performance, Composites Science and Technology 63 (14) (2003) 2055–2067.

[57]  M. Sreelka, C. Eger, Polymer Composites – from Nano- to Macroscale, Springer, Munich, 2005.

[58]  F. Li, K. Hu, J. Li, B. Zhao, The friction and wear characteristics of nanometer ZnO filled polytetrafluoroethylene, Wear 249 (10–11) (2001) 877–882.

[59]  W.G. Sawyer, K.D. Freudenberg, P. Bhimaraj, L.S. Schadler, A study on the friction and wear behavior of PTFE filled with alumina nanoparticles, Wear 254 (5–6) (2003) 573–580.

[60]  L.C. Zhang, I. Zarudi, K.Q. Xiao, Novel behaviour of friction and wear of epoxy composites reinforced by carbon nanotubes, Wear 261 (7–8) (2006) 806–811.

[61]  H. Cai, F. Yan, Q. Xue, Investigation of tribological properties of polyimide/carbon nanotube nanocomposites, Materials Science and Engineering: A 364 (2004) 94–100.

[62]  Z. Yang, B. Dong, Y. Huang, L. Liu, F. Yan, H. Li, A study on carbon nanotubes reinforced poly(methyl methacrylate) nanocomposites, Materials Letters 59 (17) (2005) 2128–2132.

[63]  Z. Yang, B. Dong, Y. Huang, L. Liu, F. Yan, H. Li, Enhanced wear resistance and microhardness of polystyrene nanocomposites by carbon nanotubes, Materials Chemistry and Physics 94 (1) (2005) 109–113.

[64]  Y. Zoo, J. An, D. Lim, D. Lim, Effect of carbon nanotube addition on tribological behavior of UHMWPE, Tribology Letters 16 (4) (2004) 305–309.

[65]  M.C. Galetz, T. Blaß, H. Ruckdäschel, J.K.W. Sandler, V. Altstädt, U. Glatzel, Carbon nanofibre-reinforced ultrahigh molecular weight polyethylene for tribological applications, Journal of Applied Polymer Science 104 (6) (2007) 4173–4181.

[66]  Y. Xue, W. Wu, O. Jacobs, B. Schadel, Tribological behaviour of UHMWPE/HDPE blends reinforced with multi-wall carbon nanotubes, Polymer Testing 25 (2) (2006) 221–229.

[67]  T. Ebbesen, Carbon Nanotubes, Preparation and Properties, CRC Press, USA, 1997.

[68]  S. Iijima, Helical microtubules of graphitic carbon, Nature 354 (6348) (1991) 56–58.

[69]  H.W. Kroto, J.R. Heath, S.C. O'brien, R.F. Curl, R.E. Smalley, C60: buckminsterfullerene, Nature 318 (6042) (1985) 162–163.

[70]  A. Oberlin, M. Endo, T. Koyama, Filamentous growth of carbon through benzene decomposition, Crystal Growth 32 (1976) 335–349.

[71]  M. Monthioux, V. Kuznetsov, Who should be given the credit for the discovery of carbon nanotubes? Carbon 44 (2006) 1621–1623.

[72]  E.V. Esquivel, L.E. Murr, A tem analysis of nanoparticulates in a polar ice core, Materials Characterization 52 (1) (2004) 15–25.

[73]  M. Reibold, P. Paufler, A.A. Levin, W. Kochmann, N. Patzke, D.C. Meyer, Materials: carbon nanotubes in an ancient damascus sabre, Nature 444 (7117) (2006) 286.

[74] R. Bacon, Growth, structure and properties of graphite whiskers, Applied Physics 31 (1960) 283–290.

[75] W. Kratschmer, L.D. Lamb, K. Fostiropoulos, D.R. Huffman, Solid C60: a new form of carbon, Nature 347 (6291) (1990) 354–358.

[76] L.X. Zheng, M.J. O'connell, S.K. Doorn, X.Z. Liao, Y.H. Zhao, E.A. Akhadov, M.A. Hoffbauer, B.J. Roop, Q.X. Jia, R.C. Dye, D.E. Peterson, S.M. Huang, J. Liu, Y.T. Zhu, Ultralong single-wall carbon nanotubes, Natuer Materials 3 (10) (2004) 673–676.

[77] M. Endo, K. Takeuchi, S. Igarashi, K. Kobori, M. Shiraishi, H.W. Kroto, The production and structure of pyrolytic carbon nanotubes (PCNTs), Journal of Physics and Chemistry of Solids 54 (12) (1993) 1841–1848.

[78] J.K.W. Sandler, J.E. Kirk, I.A. Kinloch, M.S.P. Shaffer, A.H. Windle, Ultra-low electrical percolation threshold in carbon-nanotube-epoxy composites, Polymer 44 (19) (2003) 5893–5899.

[79] D.S. Bethune, C.H. Klang, M.S. De Vries, G. Gorman, R. Savoy, J. Vazquez, R. Beyers, Cobalt-catalysed growth of carbon nanotubes with single-atomic-layer walls, Nature 363 (6430) (1993) 605–607.

[80] C. Journet, W.K. Maser, P. Bernier, A. Loiseau, M.L. De La Chapelle, S. Lefrant, P. Deniard, R. Lee, J.E. Fischer, Large-scale production of single-walled carbon nanotubes by the electric-arc technique, Nature 388 (6644) (1997) 756–758.

[81] T.W. Ebbesen, P.M. Ajayan, Large-scale synthesis of carbon nanotubes, Nature 358 (6383) (1992) 220–222.

[82] D. Colbert, J. Zhang, S. Mcclure, P. Nikolaev, Z. Chen, J. Hafner, D. Owens, P. Kotula, C. Carter, J. Weaver, A. Rinzler, R. Smalley, Growth and sintering of fullerene nanotubes, Science 266 (1994) 1218–1222.

[83] M. Cadek, R. Murphy, B. Mccarthy, A. Drury, B. Lahr, R.C. Barklie, M. In Het Panhuis, J.N. Coleman, W.J. Blau, Optimisation of the arc-discharge production of multi-walled carbon nanotubes, Carbon 40 (6) (2002) 923–928.

[84] A. Thess, R. Lee, P. Nikolaev, H. Dai, P. Petit, J. Robert, C. Xu, Y. Hee Lee, S. Gon Kim, A.G. Rinzler, D.T. Colbert, G.E. Scuseria, D. Tománek, J.E. Fischer, R.E. Smalley, Crystalline ropes of metallic carbon nanotubes, Science 273 (1996) 483–487.

[85] A.G. Rinzler, J. Liu, H. Dai, P. Nikolaev, C.B. Huffman, F.J. Rodríguez-Macías, P.J. Boul, A.H. Lu, D. Heymann, D.T. Colbert, R.S. Lee, J.E. Fischer, A.M. Rao, P.C. Eklund, and R.E. Smalley, Large-scale purification of single-wall carbon nanotubes: process, product, and characterization. 67 (1) (1998) 29–37.

[86] H.M. Cheng, F. Li, G. Su, H.Y. Pan, L.L. He, X. Sun, M.S. Dresselhaus, Large-scale and low-cost synthesis of single-walled carbon nanotubes by the catalytic pyrolysis of hydrocarbons, Applied Physics Letters 72 (25) (1998) 3282–3284.

[87] C. Rao, A. Govindaraj, R. Sen, B. Satishkumar, Synthesis of multi-walled and single-walled nanotubes, aligned- nanotube bundles and nanorods by employing organometallic precursors, Materials Research Innovations 2 (1998) 128–141.

[88] R. Andrews, D. Jacques, A. Rao, F. Derbyshire, D. Qian, X. Fan, E. Dickey, J. Chen, Continuous production of aligned carbon nanotubes: a step closer to commercial realization, Chemical Physics Letters 303 (1999) 467–474.

[89] G.G. Tibbetts, D.W. Gorkiewicz, R.L. Alig, A new reactor for growing carbon fibers from liquid- and vapor-phase hydrocarbons, Carbon 31 (5) (1993) 809–814.

[90] S. Govindjee, J.L. Sackman, On the use of continuum mechanics to estimate the properties of nanotubes, Solid State Communications 110 (4) (1999) 227–230.

[91]  M.M.J. Treacy, T.W. Ebbesen, J.M. Gibson, Exceptionally high Young's modulus observed for individual carbon nanotubes, Nature 381 (6584) (1996) 678–680.

[92]  A. Krishnan, E. Dujardin, T. Ebbesen, P. Yianilos, M. Treacy, Young's modulus of single-walled nanotubes, Physical Review B 58 (20) (1998) 14013–14019.

[93]  P. Poncharal, Z.L. Wang, D. Ugarte, W.A. De Heer, Electrostatic deflections and electromechanical resonances of carbon nanotubes, Science 283 (5407) (1999) 1513–1516.

[94]  J. Liu, Q. Zheng, Q. Jiang, Effect of a rippling mode on resonances of carbon nanotubes, Physical Review Letters 86 (21) (2001) 4843–4846.

[95]  M.R. Falvo, G.J. Clary, R.M. Taylor, V. Chi, F.P. Brooks, S. Washburn, R. Superfine, Bending and buckling of carbon nanotubes under large strain, Nature 389 (6651) (1997) 582–584.

[96]  J. Despres, E. Daguerre, K. Lafdi, Flexibility of graphene layers in carbon nanotubes, Carbon 33 (1995) 87–92.

[97]  R. Ruoff, D. Lorents, Mechanical and thermal properties of carbon nanotubes, Carbon 33 (7) (1995) 925–930.

[98]  S. Ijima, C. Brabec, A. Maiti, J. Bernholc, Structural flexibility of carbon nanotubes, Journal of Chemical Physics 104 (5) (1996) 2089–2092.

[99]  C. Ru, Effect of van der waals forces on axial buckling of a double-walled carbon nanotube, Journal of Applied Physics 87 (10) (2000) 7227–7231.

[100]  C. Ru, Elastic buckling of single-walled carbon nanotube ropes under high pressure, Physical Review B 62 (15) (2000) 10405–10408.

[101]  C. Ru, Effective bending stiffness of carbon nanotubes, Physical Review B 62 (15) (2000) 9973–9976.

[102]  C. Ru, Column buckling of multiwalled carbon nanotubes with interlayer radial displacements, Physical Review B 62 (24) (2000) 16962–16967.

[103]  C. Ru, Degraded axial buckling strain of multiwalled carbon nanotubes due to interlayer slips, Journal of Applied Physics 89 (6) (2001) 3426–3433.

[104]  C.Q. Ru, Axially compressed buckling of a doublewalled carbon nanotube embedded in an elastic medium, Journal of the Mechanics and Physics of Solids 49 (6) (2001) 1265–1279.

[105]  E.W. Wong, P.E. Sheehan, C.M. Lieber, Nanobeam mechanics: elasticity, strength, and toughness of nanorods and nanotubes, Science 277 (5334) (1997) 1971–1975.

[106]  J.-P. Salvetat, A.J. Kulik, J.-M. Bonard, G.A.D. Briggs, T. Stöckli, K. Méténier, S. Bonnamy, F. Béguin, N.A. Burnham, L. Forró, Elastic modulus of ordered and disordered multiwalled carbon nanotubes, Advanced Materials 11 (2) (1999) 161–165.

[107]  M.-F. Yu, O. Lourie, M.J. Dyer, K. Moloni, T.F. Kelly, R.S. Ruoff, Strength and breaking mechanism of multiwalled carbon nanotubes under tensile load, Science 287 (5453) (2000) 637–640.

[108]  M. Yu, B. Files, A. Arepalli, R. Ruoff, Tensile loading of ropes of single wall carbon nanotubes and their mechanical properties, Physical Review Letters 84 (24) (2000) 5552–5555.

[109]  M. Nardelli, B. Yakobson, J. Bernholc, Brittle and ductile behavior in carbon nanotubes, Physical Review Letters 81 (21) (1998) 4656–4659.

[110]  B. Yakobson, C. Brabec, J. Bernholc, Nanomechanics of carbon tubes: Instabilities beyond linear response, Physical Review Letters 76 (14) (1996) 2511–2514.

[111]  B. Yakobson, R. Smalley, Fullerene nanotubes: C1,000,000 and beyond, American Scientist 85 (4) (1997) 324–337.

[112] J. Bernholc, C. Brabec, M. Buongiorno Nardelli, A. Maiti, C. Roland, B.I. Yakobson, Theory of growth and mechanical properties of nanotubes, Applied Physics A: Materials Science & Processing 67 (1) (1998) 39–46.

[113] O. Lourie, D. Cox, H. Wagner, Buckling and collapse of embedded carbon nanotubes, Physical Review Letters 81 (8) (1998) 1638–1641.

[114] R. Ruoff, J. Tersoff, D. Lorents, S. Subramoney, B. Chan, Radial deformation of carbon nanotubes by van der waals forces, Nature 364 (1993) 514–516.

[115] J. Tersoff, R. Ruoff, Structural properties of a carbon-nanotube crystal, Physical Review Letters 73 (1994) 676–679.

[116] N. Chopra, R. Luyken, K. Cherrey, V. Crespi, M. Cohen, S. Louie, A. Zettl, Boron nitride nanotubes, Science 269 (1995) 966–967.

[117] T. Hertel, R. Walkup, P. Avouris, Deformation of carbon nanotubes by surface van der waals forces, Physical Review B 58 (20) (1998) 13870–13873.

[118] P. Avouris, T. Hertel, R. Martel, T. Schmidt, H.R. Shea, R.E. Walkup, Carbon nanotubes: Nanomechanics, manipulation, and electronic devices, Applied Surface Science 141 (3–4) (1999) 201–209.

[119] M. Yu, M. Dyer, R. Ruoff, Structure and mechanical flexibility of carbon nanotube ribbons: an atomic-force microscopy study, Journal Applied Physics 89 (2001) 4554–4557.

[120] M. Yu, M. Dyer, R. Ruoff, Structural analysis of collapsed, and twisted and collapsed, multi-wall carbon nanotubes by atomic force microscopy, Physical Review Letters 86 (1) (2001) 87–90.

[121] A. Pantano, D.M. Parks, C. Boyce Mary, Mechanics of deformation of single- and multi-wall carbon nanotubes, Journal of the Mechanics and Physics of Solids 52 (4) (2004) 789–821.

[122] D. Walters, L. Ericson, M. Casavant, J. Liu, D. Colbert, K. Smith, R. Smalley, Elastic strain of freely suspended single-wall carbon nanotube ropes, Applied Physics Letters 74 (25) (1999) 3803–3805.

[123] Z. Pan, S. Xie, L. Lu, B. Chang, L. Sun, W. Zhou, G. Wang, D. Zhang, Tensile tests of ropes of very long aligned multiwall carbon nanotubes, Applied Physics Letters 74 (1999) 3152–3154.

[124] H. Wagner, O. Lourie, Y. Feldmann, R. Tenne, Stress-induced fragmentation of multiwall carbon nanotubes in a polymer matrix, Applied Physics Letters 72 (2) (1998) 188–190.

[125] F. Li, H. Chai, S. Bai, G. Su, M. Dresselhaus, Tensile strength of single-walled carbon nanotubes directly measured from their macroscopic ropes, Applied Physics Letters 77 (2000) 3161–3163.

[126] C.M. Hsiung, M. Cakmak, Effect of processing conditions on the structural gradients developed in injection-molded poly(aryl ether ketone) (PAEK) parts. I. Characterization by microbeam X-ray diffraction technique, Journal of Applied Polymer Science 47 (1) (1993) 125–147.

[127] O.S. Carneiro, J.A. Covas, C.A. Bernardo, G. Caldeira, F.W.J. Van Hattum, J.M. Ting, R.L. Alig, M.L. Lake, Production and assessment of polycarbonate composites reinforced with vapour-grown carbon fibres, Composites Science and Technology 58 (3–4) (1998) 401–407.

[128] K. Lozano, J. Bonilla-Rios, E.V. Barrera, A study on nanofiber-reinforced thermoplastic composites (ii): investigation of the mixing rheology and conduction properties, Journal of Applied Polymer Science 80 (8) (2001) 1162–1172.

[129] K. Lozano, S. Yang, Q. Zeng, Rheological analysis of vapor-grown carbon nanofiber-reinforced polyethylene composites, Journal of Applied Polymer Science 93 (1) (2004) 155–162.

[130] P. Pötschke, M. Abdel-Goad, I. Alig, S. Dudkin, D. Lellinger, Rheological and dielectrical characterization of melt mixed polycarbonate-multiwalled carbon nanotube composites, Polymer 45 (2004) 8863–8870.

[131] P. Pötschke, T.D. Fornes, D.R. Paul, Rheological behavior of multiwalled carbon nanotube/polycarbonate composites, Polymer 43 (11) (2002) 3247–3255.

[132] T. Mcnally, P. Pötschke, P. Halley, M. Murphy, D. Martin, S.E.J. Bell, G.P. Brennan, D. Bein, P. Lemoine, J.P. Quinn, Polyethylene multiwalled carbon nanotube composites, Polymer 46 (19) (2005) 8222–8232.

[133] D. Bangarusampath, H. Ruckdäschel, V. Altstädt, J. Sandler, D. Garray, and M. Shaffer, Rheology and properties of melt-processed poly(ether ether ketone)/multi-wall carbon nanotube composites, Polymer 50 (24) 5803–5811.

[134] R. Krishnamoorti, J. Ren, A. Silva, Shear response of layered silicate nanocomposites, Journal of Chemical Physics 114 (11) (2001) 4968–4973.

[135] S. Rahatekar, K. Koziol, S. Butler, J. Elliott, M. Shaffer, M. Mackley, A. Windle, Optical microstructure and viscosity enhancement for an epoxy resin matrix containing multiwall carbon nanotubes, Journal of Rheology 50 (2006) 599–610.

[136] F. Du, R. Scogna, W. Zhou, S. Brand, J. Fischer, K. Winey, Nanotube networks in polymer nanocomposites: rheology and electrical conductivity, Macromolecules 37 (2004) 9048–9055.

[137] S. Kharchenko, J. Douglas, J. Obrzut, E. Grulke, K. Migler, Flow-induced properties of nanotube-filled polymer materials, Nature Materials 3 (2004) 564–568.

[138] H. Smallwood, Limiting law of the reinforcement of rubber, Journal Applied Physics 15 (11) (1944) 758–766.

[139] E. Guth, Theory of filler reinforcement, Journal Applied Physics 16 (1945) 20–25.

[140] I. Sanchez, R. Lacombe, Statistical thermodynamics of polymer solutions, Macromolecules 11 (6) (1978) 1145–1156.

[141] J. Brandrup, E. Immergut, E. Grulke, in: F. Edition (Ed.), Polymer Handbook, John Wiley & Sons Inc, New York, 1999.

[142] J. Sandler, P. Werner, M. Shaffer, V. Demchuk, V. Altstädt, A. Windle, Carbon-nanofibre-reinforced poly(ether ether ketone) composites, (Part A) Composites 33 (8) (2002) 1033–1039.

[143] R. Kuriger, M. Alam, D. Anderson, R. Jacobsen, Processing and characterization of aligned vapor grown carbon fiber reinforced polypropylene, (Part A) Composites 33 (2002) 53–62.

[144] A. Ciferri, I. Ward, Ultra-high Modulus Polymers, Applied Science Publishers, Barking, UK, 1979.

[145] M. Saito, S. Kukula, Y. Kataoka, Practical use of the statistically modified laminate model for injection moldings. Part 1: method and verification, Polymer Composites 19 (5) (1998) 497–505.

[146] M. Semadeni, H. Zerlik, P. Rossini, J. Meyer, E. Wintermantel, High fibre volume fraction injection moulding of carbon fibre reinforced polyetheretherketone (peek) in order to raise the mechanical properties, Polymer Composites 6 (1998) 279–286.

[147] D.J. Blundell, B.N. Osborn, The morphology of poly(aryl-ether-ether-ketone), Polymer 24 (8) (1983) 953–958.

[148] A. Waddon, M. Hill, A. Keller, D. Blundell, On the crystal texture of linear polyaryls (PEEK, PEK and PPS), Journal of Materials Science 22 (5) (1987) 1773–1784.

[149] S. Gao, J. Kim, Cooling rate influences in carbon fibre/PEEK composites. Part 1. Crystallinity and interface adhesion, Composites Part A: Applied Science and Manufacturing 31 (6) (2000) 517–530.

[150] Z. Hanmin, Z. Zhiyi, P. Weizou, P. Tiayou, Transcrystalline structure of PEEK, European Polymer Journal 30 (2) (1994) 235–237.

[151] R. Crick, D. Leach, P. Meakin, D. Moore, Interlaminar fracture morphology of carbon fibre/PEEK composites, Journal of Materials Science 22 (6) (1987) 2094–2104.

[152] T. Li, M. Zhang, H. Zeng, Strong interaction at interface of carbon fiber reinforced aromatic semicrystalline thermoplastics, Polymer 40 (15) (1999) 4307–4313.

[153] T.Q. Li, M.Q. Zhang, K. Zhang, H.M. Zeng, Long-range effects of carbon fiber on crystallization of semicrystalline thermoplastics, Polymer 41 (1) (2000) 161–168.

[154] H. Chen, R. Porter, Observation of two-stage crystallization of poly(ether ether ketone) by thermal mechanical analysis, Thermochimica Acta 243 (2) (1994) 4576–4578.

[155] K. Lozano, E. Barrera, Nanofiber-reinforced thermoplastic composites I: thermoanalytical and mechanical analyses, Journal of Applied Polymer Science 79 (2001) 125–133.

[156] E. Assouline, A. Lustiger, A.H. Barber, C.A. Cooper, E. Klein, E. Wachtel, H.D. Wagner, Nucleation ability of multiwall carbon nanotubes in polypropylene composites, Journal of Polymer Science Part B: Polymer Physics 41 (5) (2003) 520–527.

[157] M. Zhang, Y. Liu, X. Zhang, J. Gao, F. Huang, Z. Song, G. Wei, J. Qiao, The effect of elastomeric nano-particles on the mechanical properties and crystallization behavior of polypropylene, Polymer 43 (19) (2002) 5133–5138.

[158] B. Grady, F. Pompeo, R. Shambaugh, D. Resasco, Nucleation of polypropylene crystallization by single-walled carbon nanotubes, The Journal of Physical Chemistry B 106 (2002) 5852–5858.

[159] M. Mucha, J. Marszalek, A. Fidrych, Crystallization of isotactic polypropylene containing carbon black as a filler, Polymer 41 (11) (2000) 4137–4142.

[160] L. Valentini, J. Biagiotti, J.M. Kenny, S. Santucci, Morphological characterization of single-walled carbon nanotubes-PP composites, Composites Science and Technology 63 (8) (2003) 1149–1153.

[161] A. Bhattacharyya, T. Sreekumar, T. Liu, S. Kumar, L. Ericson, R. Hauge, R. Smalley, Crystallization and orientation studies in polypropylene/single wall carbon nanotube composites, Polymer 44 (2003) 2373–2377.

[162] G. Crevecoeur, G. Groenninckx, Binary blends of poly(ether ether ketone) and poly(ether imide). Miscibility, crystallization behavior, and semicrystalline morphology, Macromolecules 24 (5) (1991) 1190–1195.

[163] A. Goodwin, G. Simon, Dynamic mechanical relaxation behaviour of poly(ether ether ketone)/poly(etherimide) blends, Polymer 38 (1997) 2363–2370.

[164] A. Nogales, T. Ezquerra, Z. Denchev, I. Sics, F. Balta-Calleja, B. Hsiao, Molecular dynamics and microstructure development during cold crystallization in poly(ether-ether-ketone) as revealed by real time dielectric and X-ray methods, Journal of Chemical Physics 115 (8) (2001) 3804–3813.

[165] M.S.P. Shaffer, X. Fan, A.H. Windle, Dispersion and packing of carbon nanotubes, Carbon 36 (11) (1998) 1603–1612.

[166] G. Tsagaropoulos, A. Eisenberg, Dynamic mechanical study of the factors affecting the two glass transition behavior of filled polymers. Similarities and differences with random ionomers, Macromolecules 28 (18) (1995) 6067–6077.

[167] F. Zahradnik, Hochtemperaturthermoplaste: Aufbau—Eigenschaften—anwendungen, VDI- Verlag, Düsseldorf, 1998.

[168] J.E. Harris, L.M. Robeson, Miscible blends of poly(aryl ether ketone)s and poly-etherimides, Journal of Applied Polymer Science 35 (7) (1988) 1877–1891.

[169] L. David, S. Etienne, Molecular mobility in para-substituted polyaryls. 1. Sub-tg relaxation phenomena in poly(aryl ether ether ketone), Macromolecules 25 (1992) 4302–4308.

[170] H. Krenchel, Fibre Reinforcement, Akademisk Forlag, openhagen, 1964.

[171] F.W.J. Van Hattum, C.A. Bernardo, J.C. Finegan, G.G. Tibbetts, R.L. Alig, M.L. Lake, A study of the thermomechanical properties of carbon fiber–polypropylene composites, Polymer Composites 20 (5) (1999) 683–688.

[172] G. Tibbetts, J. Mchugh, Mechanical properties of vapor-grown carbon fiber composites with thermoplastic matrices, Journal of Materials Research 14 (1999) 2871–2880.

[173] J. Sarasua, P. Remiro, J. Pouyet, The mechanical behaviour of peek short fibre composites, Journal of Materials Research 30 (13) (1995) 3501–3508.

[174] D. Hull, An Introduction to Composite Materials, Cambridge University Press, Cambridge, 1981.

[175] M. Shaffer, A. Windle, Fabrication and characterization of carbon nanotube/poly(vinyl alcohol) composites, Advanced Materials 11 (1999) 937–941.

[176] S.-L. Gao, J.-K. Kim, Correlation among crystalline morphology of PEEK, interface bond strength, and in-plane mechanical properties of carbon/PEEK composites, Journal of Applied Polymer Science 84 (6) (2002) 1155–1167.

[177] O. Jacobs, N. Mentz, A. Poeppel, K. Schulte, Sliding wear performance of HD-PE reinforced by continuous UHMWPE fibres, Wear 244 (1–2) (2000) 20–28.

[178] P. Werner, V. Altstädt, R. Jaskulka, O. Jacobs, J.K.W. Sandler, M.S.P. Shaffer, A.H. Windle, Tribological behaviour of carbon-nanofibre-reinforced poly(ether ether ketone), Wear 257 (9–10) (2004) 1006–1014.

[179] H. Voss, K. Friedrich, On the wear behaviour of short-fibre-reinforced peek composites, Wear 116 (1987) 1–18.

[180] J. Flöck, K. Friedrich, Q. Yuan, On the friction and wear behaviour of pan- and pitch-carbon fiber reinforced peek composites, Wear 225–229 (1) (1999) 304–311.

[181] B. Mortimer, J. Lancaster, Extending the life of aerospace dry bearings by the use of hard smooth counterparts, Wear 121 (1988) 289–305.

[182] G. Zhanga, Z. Rasheva, A.K. Schlarb, Friction and wear variations of short carbon fiber (SCF)/PTFE/graphite (10vol.%) filled peek: effects of fiber orientation and nominal contact pressure, Wear 268 (7) (2009) 893.

[183] R. Jaskulka, O. Jacobs, Einfluss des gegenpartnermaterials auf den verschleiß verschiedener PEEK-compounds, Tribologie + Schmiertechnik 2 (2003) 16–20.

[184] O. Jacobs, K. Friedrich, K. Schulte, Fretting wear of continuous fibre reinforced polymer composites, ASTN Special Technical Publication 1167 (1992) 81–96.

[185] Lubricomp, A guide to LNP's internally lubricated thermoplastics, in Bulletin #254, L.E. Plastics, Editor 2001.

[186] O. Jacobs, R. Jaskulka, F. Yang, W. Wu, Sliding wear of epoxy compounds against different counterparts under dry and aqueous conditions, Wear 256 (2004) 9–15.

[187] R. Reinicke, J. Hoffmann, K. Friedrich, Einfluss unterschiedlicher füllstoffe und faserarten auf das verschleißverhalten von polyamid-46, Tribologie + Schmiertechnik 46 (1999) 29–33.

[188] M. Drubetski, A. Siegmann, M. Narkis, Hybrid particulate and fibrous injection molded composites: carbon black/carbon fiber/polypropylene systems, J. Mater. Sci. 42 (2007) 1–8.

[189] J.P. Davim, R. Cardoso, Effect of the reinforcement (carbon or glass fibres) on friction and wear behaviour of the PEEK against steel surface at long dry sliding, Wear 266 (7–8) (2009) 795–799.

[190] Q. Tanga, J. Chen, L. Liu, Tribological behaviours of carbon fibre reinforced peek sliding on silicon nitride lubricated with water, Wear 269 (2010) 541–546.

[191] G. Theiler, T. Gradt, Friction and wear of peek composites in vacuum environment, Wear 269 (2010) 278–284.

[192] B. Suresha, G. Chandramohan, N. Renukappa, Siddaramaiah, Mechanical and tribological properties of glass-epoxy composites with and without graphite particulate filler, Journal of Applied Polymer Science 103 (2007) 2472–2480.

[193] L. Chang, Z. Zhang, H. Zhang, K. Friedrich, Effect of nanoparticles on the tribological behaviour of short carbon fibre reinforced poly(etherimide) composites, Tribology International 38 (2005) 966–973.

[194] G. Theiler, W. Hübner, T. Gradt, P. Klein, Friction and wear of carbon fibre filled polymer composites at room and low temperatures, Materialwiss Werkst. 35 (2004) 683–689.

[195] J. Wang, M. Gu, S. Bai, S. Ge, Investigation of the influence of MoS$_2$ filler on the tribological properties of carbon fiber reinforced nylon1010 composites, Wear 255 (2003) 774–779.

[196] K. Friedrich, Z. Zhang, A.K. Schlarb, Effects of various fillers on the sliding wear of polymer composites, Composites Science and Technology 65 (2005) 2329–2343.

[197] R. Jaskulka, G. Schlicke, C. Tegethoff, O. Jacobs, Peek compounds in an aqueous environment, Kunststoffe-plast Europe 93 (2003) 77–80.

[198] O. Jacobs, R. Jaskulka, C. Yan, W. Wu, On the effect of counterface material and aqueous environment on the sliding wear of carbon fibre reinforced polyetheretherketone (PEEK), Tribology Letters 19 (2005) 319–329.

[199] K. Friedrich, J. Karger-Kocsis, Z. Lu, Effects of steel counterface roughness and temperature on the friction and wear of PE(E)K composites under dry sliding conditions, Wear 148 (2) (1991) 235–247.

[200] G. Xian, Z. Zhang, K. Friedrich, Tribological properties of micro- and nanoparticles-filled poly(etherimide) composites, Journal of Applied Polymer Science 101 (3) (2006) 1678–1686.

[201] R. Prehn, F. Haupert, K. Friedrich, Sliding wear performance of polymer composites under abrasive and water lubricated conditions for pump applications, Wear 259 (1–6) (2005) 693–696.

[202] J. Voort, S. Bahadur, The growth and bonding of transfer film and the role of CuS and PTFE in the tribological behavior of PEEK, Wear 181–183 (1) (1995) 212–221.

[203] Q.-H. Wang, Q.-J. Xue, W.-M. Liu, J.-M. Chen, The friction and wear characteristics of nanometer SiC and polytetrafluoroethylene filled polyetheretherketone, Wear 243 (1–2) (2000) 140–146.

[204] S. Bahadur, V.K. Polineni, Tribological studies of glass fabric-reinforced polyamide composites filled with CuO and PTFE, Wear 200 (1996) 95–104.

[205] S. Bahadur, D. Gong, J.W. Anderegg, Studies of worn surfaces and the transfer film formed in sliding by CuS-filled and carbon fiber-reinforced nylon against a steel surface, Wear 181–183 (1995) 227–235.

[206] L. Rapoport, N. Fleischer, R. Tenne, Fullerene-like WS$_2$ nanoparticles: superior lubricants for harsh conditions, Advanced Materials 15 (7–8) (2003) 651–655.

[207] G. Xian, R. Walter, F. Haupert, A synergistic effect of nano-TiO$_2$ and graphite on the tribological performance of epoxy matrix composites, Journal of Applied Polymer Science 102 (3) (2006) 2391–2400.

[208] H.-B. Qiao, Q. Guo, A.-G. Tian, G.-L. Pan, L.-B. Xu, A study on friction and wear characteristics of nanometer Al$_2$O$_3$/PEEK composites under the dry sliding condition, Tribology International 40 (1) (2007) 105–110.

[209] F. Oster, F. Haupert, K. Friedrich, M. Muller, W. Bickle, New polyetheretherketone-based coatings for severe tribological applications, Materialwiss Werkst. 35 (2004) 690–695.

[210] M.II. Cho, S. Bahadur, Study of the tribological synergistic effects in nano CuO-filled and fiber-reinforced polyphenylene sulfide composites, Wear 258 (5–6) (2005) 835–845.

[211] B. Ginzburg, D. Tochil'nikov, V. Bakhareva, A. Anisimov, O. Kireenko, Polymeric materials for water-lubricated plain bearings, Russian Journal of Applied Chemistry 79 (5) (2006) 695–706.

[212] J.P. Giltrow, J.K. Lancaster, The role of the counterface in the friction and wear of carbon fibre reinforced thermosetting resins, Wear 16 (5) (1970) 359–374.

[213] H. Cao, Y. Huang, Z. Zhang, J. Sun, Uniform modification of carbon fibers surface in 3-D fabrics using intermittent electrochemical treatment, Composites Science and Technology 65 (2005) 1655–1662.

[214] S. Tiwari, M. Sharma, S. Panier, B. Mutel, P. Mitschang, J. Bijwe, Influence of cold remote nitrogen oxygen plasma treatment on carbon fabric and its composites with specialty polymers, Journal of Materials Science (2010) 964–974.

[215] M. Sharma, J. Bijwe, P. Mitschang, Wear performance of PEEK-carbon fabric composites with strengthened fiber-matrix interface, Wear 271 (9–10) (2010) 2261–2268.

[216] M. Sharma, J. Bijwe, K. Singh, P. Mitschang, Exploring potential of micro-Raman spectroscopy for correlating graphitic distortion in carbon fibers with stresses in erosive wear studies of peek composites, Wear 270 (2011) 791–799.

[217] A.P. Harsha, U.S. Tewari, Tribo performance of polyaryletherketone composites, Polymer Testing 21 (6) (2002) 697–709.

[218] Z. Zhang, C. Breidt, L. Chang, K. Friedrich, Wear of peek composites related to their mechanical performances, Tribology International 37 (3) (2004) 271–277.

[219] H. Zhang, Z. Zhang, K. Friedrich, Effect of fiber length on the wear resistance of short carbon fiber reinforced epoxy composites, Composites Science and Technology 67 (2) (2007) 222–230.

# Wear behavior of carbon nanotube-reinforced polyethylene and epoxy composites

**Olaf Jacobs, Birgit Schädel**

*Fachhochschule Lübeck, Lübeck, Germany*

## CHAPTER OUTLINE HEAD

Tribology of Polymeric Nanocomposites. http://dx.doi.org/10.1016/B978-0-444-59455-6.00009-X

## 9.1 INTRODUCTION

Carbon nanotubes (CNTs) are a rather novel class of nanomaterials. Since their discovery by Iijima in 1991, a significant number of papers and books on structure properties and application of CNTs have been published [1,2].

Ideal CNTs are hollow cylinders with walls consisting of "rolled-up" graphite-like planes and semispherical end caps resembling halves of bucky balls (Fig. 9.1). Depending on the number of graphite planes, single-wall CNTs (SWCNTs), double-wall CNTs, and multiwall CNTs (MWCNTs) are distinguished. Their typical diameters are in the range of a few dozen nanometers, while their length is in the range of some micrometers to a few hundred micrometers.

According to their graphitic structure, CNTs possess a high thermal conductivity and an electrical conductivity that can be either semiconducting or metallike. Since the basal planes are build up of strong covalent bonds, the Young modulus of CNTs can be as high as 1000 GPa, which is approximately 5 times higher than that of steel. The tensile strength of CNTs can reach values up to 150 GPa, around 80 times higher than high-strength steel [1].

Real CNTs, however, contain a certain amount of lattice defects [1]. Irregularities in the arrangement of the atoms inside the basal planes cause strong internal stresses and a locally increased energy state. Especially the spherically shaped end caps necessarily have a high defect density. These defects are preferred sites for chemical or mechanical attack. They limit the chemical resistance of the CNTs and reduce their fracture strength. On the other hand, they provide opportunities to locally introduce chemical modifications like oxidative activation and grafting of functional groups without destroying the whole structure.

CNTs can be produced in various ways [1]. All of these processes are based on the deposition of CNTs from a vapor phase. Best qualities (in terms of few defects) can be achieved with high temperature processes like laser ablation of carbon or arc discharge. The high process temperature supports the growth of rather defect-free crystals. Chemical vapor deposition (CVD) and plasma-activated CVD allow for higher production rates though the resulting CNTs contain more defects and impurities (amorphous carbon). After production, the products contain contaminations

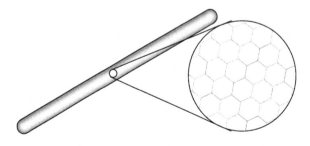

**FIGURE 9.1**

Schematic representation of a CNT.

consisting of various modifications of amorphous carbon. These contaminants have to be removed in a purification process, mostly performed by oxidative treatment in nitric acid and/or sulfur acid.

The combination of the strong covalent bonds in the axial direction together with an aspect ratio in the range of several thousands makes CNTs promising candidates as reinforcement for polymer composites. Moreover, the development of polymer composites reinforced with the highly conductive CNTs opens up new perspectives for multifunctional materials utilizing the electrical conductivity of the CNTs [3].

A number of researchers have investigated the ability of CNTs to act as a reinforcing material in metals or polymers [4–7]. A survey of these activities is presented in [8]. A CNT reinforcement of epoxy resins (EPs) can improve its glass transition temperature, tensile strength, and stiffness as well as ductility—as long as the CNTs are well dispersed in the matrix [6–8].

The dispersion (destruction of agglomerates) and homogeneous distribution of CNTs in a polymer is a basic problem. Similar to all nanofillers, CNTs have a strong tendency to form agglomerates due to their large specific surface, and the nonpolar nature of the CNTs hampers their dispersion in a polar matrix. Additionally, the extremely long CNTs (aspect ratio several thousands) get tangled up, which makes it even more difficult to singularize the CNTs. Proper dispersion of CNTs in a polymer matrix without destroying the CNTs still remains a challenge, and various approaches have been developed [9–12].

Moreover, a strong interfacial bond between CNTs and polymer matrix is required to transfer mechanical loads from the matrix to the CNT and vice versa. Otherwise, the good mechanical properties of the CNTs could not be fully exploited. This is not trivial since the graphite planes are chemically rather inert and nonpolar. The surfaces of the singularized CNTs have to be activated and/or grafted with functional groups or a coupling agent as mentioned above.

Polymers are increasingly used for components that are subjected to tribological loadings like bearing bushes, gear wheels, or low friction coatings. Some of their typical benefits are easy processing, corrosion resistance, low friction, and damping of noise and vibrations. However, the load-bearing capacity and thermal stability are inferior to metals and ceramics. Suitable fillers are added to extend the application range of polymers in tribologically loaded systems:

- The load-bearing capacity and thermal resistance are improved by adding hard particles or fibers (e.g. bronze, glass, carbon, ceramics, etc.). These hard phases additionally increase the heat conductivity so that the temperature in the sliding contact is lowered.
- Dry lubricants like graphite, $MoS_2$, or polytetrafluoroethylene (PTFE) are added in order to minimize the friction coefficient. This also lowers the shear stresses in the mating components and, thus, often reduces the wear rate.

Within the last decade, nanofillers have been increasingly used as reinforcements for tribologically loaded polymer compounds [13]. They provide several advantages over classical microreinforcements. They allow the production of micromechanical components and thin coatings, and they usually do not cause embrittlement and deterioration of tensile strength as microscopic fillers often do [7,14]. Several researchers have published studies on friction and wear of nanoparticle-reinforced thermoplastics [15–17] and thermosets [18–21]. All of them found a wear-reducing effect of the nanofillers. Our studies [18,19] revealed that nanoparticle-reinforced EP composites mostly exhibit a wear minimum at around 1–2 wt% of nanofillers for most nanoparticle types.

However, CNTs and related substances are a rather new group of nanofillers. Nevertheless, several studies on the effect of CNTs on the tribological behavior of polymers have been published so far [22–43].

In most of them, EP was used as the matrix material [22–28,41]. Further studies turned their attention on the wear of CNTs reinforcing ultrahigh-molecular-weight polyethylene (UHMWPE) [29–32] and high-density polyethylene (HDPE) [43], poly(methyl methacrylate) (PMMA) [33], PTFE [34,35], polystyrene (PS) [36], polystyrene–acrylonitrile (SAN) [40], polyamide 6 (PA6) [42], and polyimide (PI) [37,39]. The wear resistance of all these materials was significantly improved by the addition of CNTs. In all cases, the wear of the pure polymer decreased sharply when adding a small amount of CNTs. With further increasing CNT content, the wear rate decreased only slightly [25,30], reached a stable value [37], or even increased again after passing a minimum [22,24,33–36,38]. After passing through this minimum, only a slight increase was found mostly. Figure 9.2 shows the effect of CNT content on the friction and wear of in situ polymerized CNT/SAN composites as a typical example for this behavior [40]. In some cases, the wear rate was found to decrease

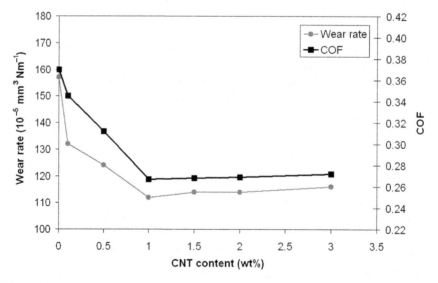

**FIGURE 9.2**

Wear rate and friction coefficient (COF) of CNT-reinforced SAN according to Ref. [40].

again when the filler content was further increased [29,41]. These small differences could be attributed to diverse test modes.

The influence of CNTs on the friction coefficient was ambiguous. The introduction of CNTs into EP [24], PMMA [33], PTFE [34], PS [36], and SAN [40] reduced the friction coefficient. However, Vail et al. [35] reported an opposite trend for CNT-filled PTFE: the friction coefficient significantly increased due to the addition of SWCNTs. The friction coefficient of UHMWPE was not affected by CNTs [29] or even increased [30]. The friction of PI/CNT composites [37–39] decreased with increasing filler content or reached a minimum depending on the applied loads and sliding speeds. An optimum of friction at low filler contents was also observed for epoxy composites reinforced with cup-stacked CNTs only at small loads a slight increase was found [28]. The same study examined the influence of ozone treatment of the filler on the tribological behavior. These differences may originate in different test conditions.

Meng et al. [42] started directly with the assumed optimum CNT content of 1 wt% and produced PA6-based nanocomposites via usual thermoplastic processing on a corotating twin-screw extruder (PA6 is a typical material for tribological components). The CNT/PA6 composites exhibited a lower wear and friction than did the pure PA6 under unlubricated and water-lubricated conditions. However, the achieved wear reduction was only small.

Johnson et al. [43] considered CNT as a possible reinforcement for a polyethylene socket of hip–joint implants. In a prestudy, they produced HDPE reinforced with various contents of CNTs. Mechanical performance and the friction and wear behavior improved with increasing CNT content up to 5 wt% CNTs.

Two studies [22,38] examined the additional effect of dry lubricants on the tribological behavior of composites containing carbon nanofillers. The addition of 10 wt% graphite in CNT and carbon nanohorn-reinforced PI could slightly reduce the friction coefficient although the wear rate increased [38].

Brake pads, too, consist of a complex mixture of additives to fulfill a wide range of diverging requirements. Hwang et al. [44] took a typical conventional brake pad material and replaced some of the conventional fillers (especially barite) by CNTs. The wear resistance increased with increasing CNT content up to 60 vol%. The friction was reduced due to the formation of a dry-lubricating interlayer consisting of undispersed CNT agglomerates. Unfortunately, no information is given about the calculation of the volume content, 60% is barely imaginable.

Chen et al. found that MWCNTs can significantly decrease the wear rate and friction coefficient of copper-based composites [45]. Lin et al., too, investigated CNT-reinforced copper composites [46]. They found minimum wear and friction at a CNT content of 10–15 vol% (unfortunately they do not explain how they calculated the volume content) and attributed it to an antiseizing effect of the CNTs and to the formation of a dry-lubricant film consisting of CNTs. This dry-lubricant film was also considered as a reason for less stick slip and for a reduced shear loading of the wear surfaces.

Lim et al. [47] investigated carbon–carbon composites coated with CNT-reinforced carbon, and they, too, found that the wear rate decreased continuously with increasing CNT content while the friction coefficient increased. Policandriotes [48] produced C–C composites with various types of nanoadditives. Again, SWCNTs were found to reduce wear as long as the filler content is not too high (0.8% in this case). The friction coefficient was significantly increased. However, the loading conditions were extremely severe, and the interface temperatures were very high in these tests.

Kim et al. [49] produced CNT-reinforced aluminum composites and characterized their tribological performance. As in the case of polymeric CNT composites, the wear rate passed a minimum at 1 wt% CNT content. The beneficial effect of the CNTs was found to depend on the dispersion method. In accordance with this investigation, Choi et al. [50] found a friction and wear minimum at 4.5 vol% (calculated assuming a density of 1.3 g cm$^{-3}$ for CNTs) for a CNT–aluminum composite.

The results presented in the present paper are based on the investigations partly published in Refs. [22,23,29]. The effect of CNT reinforcement on the sliding wear behavior of EP and UHMWPE was studied under uniform sliding against martensitic bearing steel (100Cr6) and austenitic stainless steel (X5CrNi18-10) in a ball-on-prism arrangement. The epoxy-based composites were prepared in various ways to study the effect of a pretreatment of the CNTs and of the dispersion and homogenization method on the tribological properties. The CNT content was varied systematically to find the optimal filler content. Dry lubricants were added to some compounds to optimize the wear resistance.

## 9.2 EXPERIMENTAL DETAILS

### 9.2.1 Base materials

For the first experiments, an EP was chosen as the matrix material for a number of reasons: CNTs are expensive and were only available in small quantities. Thermoplastic compounding methods would have required larger amounts of sample materials. Moreover, EPs can be compounded in the cold state. EP is therefore a good model substance to develop dispersion methods. Last but not least, EPs are often used in practical applications like repair coatings of worn surfaces and thin coatings of solid film lubricants.

The resin used in this study was an EP of the type Neukadur EP 571 based on bisphenol A, hardened with an aminic hardener Neukadur T9. Resin and hardener were produced by Altropol (Stockelsdorf, Germany).

The second part of the chapter deals with composites based on UHMWPE. In contrast to EP, UHMWPE is a thermoplastic material with nonpolar nature and low surface tension. Accordingly, pure UHMWPE has a very low friction coefficient against steel and is widely used in biomedical and technical applications because of its high wear resistance and high impact strength. A major shortcoming of UHMWPE is its extremely high melt viscosity so that it cannot be processed as usual thermoplastics and melt compounding is impossible. For these reasons, 20% HDPE were mixed into the UHMWPE. The resulting blend could be compounded and processed without any problems.

Moreover, polyethylene—in particular UHMWPE—is a viscoelastic material that possesses a relatively low creep resistance [51,52]. In the case of bearings and joints, creep deformation has a similar effect as wear: the clearance between the mating components increases with time until it exceeds the tolerances. The lifetime of such systems is therefore limited by creep and wear. It was expected that the creep resistance of the UHMWPE could be improved by blending with HDPE and reinforcement with CNTs, and the creep deformation of these composites was measured under similar conditions as the wear resistance.

The UHMWPE used was supplied from Ticona as a powder, type GUR 2122, with an average grain size of $20\,\mu m$. The average molecular weight was determined viscometrically by measuring the intrinsic viscosity of a solution of the polymer and converting the value by means of the Mark–Houwink equation. The average molecular weight was $5.0 \times 10^{-6}\,g\,mol^{-1}$. The HDPE used with an average granule diameter of $3\,mm$ was a DMD 7006A provided by QPC (Shandong, China).

The CNTs used in this study were multiwalled CNTs, produced by Namigang Co. (Shenzhen, China). The parameters of these CNTs were average diameter $10–30\,nm$ and average length $5–15\,\mu m$. The purity of CNTs was better than $95\,wt\%$, with an amorphous carbon content $<3\,wt\%$ and $<0.2\,wt\%$ catalyst residues (La, Ni). The specific surface area was between 40 and $300\,m^2\,g^{-1}$.

One set of PE composites was prepared with a different type of CNTs. These CNTs, too, were MWCNTs produced by the same company but had a larger diameter of $60–100\,nm$.

The following dry lubricants were mixed into some of the EP composites: PTFE powder with an average grain size of $6\,\mu m$, graphite powder with an average grain size of $4\,\mu m$, and $MoS_2$ with an average grain size of $3–4\,\mu m$.

## 9.2.2 Specimen preparation

### 9.2.2.1 Preparation of EP specimens

A variety of preparation methods were employed. Not all of these procedures were applied always; sometimes, some of these processing steps were skipped. Generally, the subsequent production sequence was followed:

**1.** Pretreatment (optional, some samples were produced without this pretreatment)

CNTs have a nonpolar surface that barely interacts with the epoxy matrix. The virgin MWCNTs form a kind of loosely packed strongly coherent felt (Fig. 9.3). The pores inside these nonpolar agglomerates are so small (in the range of a few hundred nanometers) that the polar EP resin can hardly infiltrate them. Some CNT samples were therefore pretreated to introduce polar surface groups and create functional groups the matrix can interact with. For this purpose, the CNTs were mixed into $HNO_3$ ($4.0\,mol\,l^{-1}$) with a CNT to $HNO_3$ weight ratio of 1:3. The mixture was continually boiled in a reflux condenser at $100\,°C$ for approximately $11\,h$ while being stirred at a rotational frequency of $300\,rpm$. After that, the mixture was washed in distilled water until the pH value approached 7 in order to eliminate the $HNO_3$.

In later experiments, the treatment time was reduced to 2 h because there was some evidence that the CNTs suffered from severe damage after 11 h etching. This is in accordance with the results of Shi et al. [53], who found that the wear resistance of carbon nanofiber (CNF)-reinforced PTFE strongly depends on the treatment time. Too short treatment times cause poor fiber–matrix interaction but too long treatment times lead to damage of the CNFs.

**2.** Application of coupling agent (optional, only few samples)

Some CNT samples were additionally modified using the coupling agent GLYMO from Degussa (Germany). GLYMO is an epoxy-functionalized silane coupling agent. First, the CNTs were dried again and pretreated in $HNO_3$ to graft functional groups to the CNT sidewalls that can react with the coupling agent. Then, the pretreated CNTs were sonicated in a solution of the coupling agent in acetone ($13.6 \, mol \, l^{-1}$) in order to avoid agglomerates of CNTs and to evaporate the solvent. The weight ratio of coupling agent to CNTs is 1:10. The sonication process was continued until all the solution was completely evaporated. After this process, the CNTs appeared like a very fine powder without tactile lumps and could easily be dispersed in the resin.

**3.** Drying of CNTs

The pretreated and untreated CNTs, respectively, were dried in an oven at 100 °C for 2 h before further processing to eliminate adsorbed moisture from the agglomerates. The dried CNTs were put into a plastic bag and stored in a desiccator.

**4.** Crushing of agglomerates

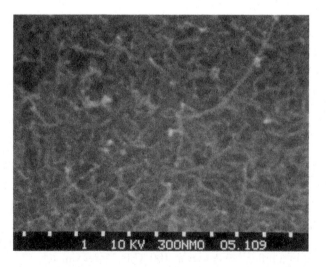

**FIGURE 9.3**

SEM micrograph of CNTs in the as-delivered state.

The CNTs were manually crushed with a mortar for approximately 10 min before being mixed into the resin. This process step was performed for all EP/CNT specimens.

**5.** Mixing of CNTs into the resin

Preceding tests pointed out that it did not make a significant difference whether the CNTs were mixed into the hardener or into the resin [23]. Therefore, this report focuses on CNTs mixed into the resin. The hardener was added after thorough homogenization of the resin/CNT mixture. Three different mixing methods were employed.

One mixing device was a four-blade stirrer (Fig. 9.4(a)) with an approximately 20 mm diameter and working at a rotational frequency of 2500 rpm. The mixture was prepared in a cup of approximately 25-mm diameter. This mixing procedure was found to introduce large amounts of air into the resin and to have a low homogenization capability.

Another mixing device was the dual asymmetric centrifuge "SpeedMixer DAC 150 FVZ" (Fig. 9.4(b)) from Hauschild (Germany). The samples (resin plus CNTs) were filled into a small pot (size: 60 ml) of Polypropylene (PP). This pot is inserted into the SpeedMixer with a 30° tilt angle. During the mixing process, the pot rotates around its $z$-axis while the base plate rotates around the vertical axis in opposite direction. The mixing process is based on a kind of high-power shaking. Due to the centrifugal forces, only little air is entrapped in the mixture. Moreover, this mixing device has some dispersion capability due to the high shear forces and impacts that occur during mixing. This mixing process lasted 2 min at 2000 and 3000 rpm, respectively.

According to Gojny et al. [7], a laboratory three-roll mill is most capable of dispersing CNTs in EP because of the extremely high shear forces between the rolls. Therefore, 1% untreated as well as 1% $HNO_3$-treated (11 h) CNTs were used for dispersion trials with a mixing mill fabricated by EXAKT (Norderstedt, Germany) (Fig. 9.5).

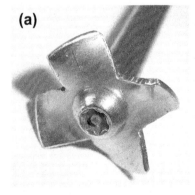

**FIGURE 9.4**

(a) Four-blade stirrer and (b) SpeedMixer.

**FIGURE 9.5**

Three-roll mixing mill from EXAKT.

The CNTs and the EP resin were manually premixed before they were dispersed in the three-roll mill. The resin passed the roll mill two times. $Al_2O_3$ was used as roll material.

**6.** Evacuating (all samples)

A large amount of air was embedded in the agglomerates when the CNTs were added to the resin. This air had to be removed after dispersion of the CNTs. After stirring, the mixture was alternately evacuated and aerated until no more bubbles were generated.

**7.** Ultrasound dispersion (optional, not all samples were sonicated)

To achieve a better deagglomeration, some samples were sonicated after the CNTs were mixed into the resin by the SpeedMixer. The parameters were as follows:

- apparatus: Dr Hielscher GmbH UP 400S (400 W max. power),
- cylindrical titanium sonotrode with a 7-mm diameter,
- full amplitude.

The process had to be interrupted every 30 s to avoid thermal softening of the polypropylene pot. The net time of this ultrasound process was 10 min, that is, 20 cycles, each lasting 30 s.

**8.** Addition of dry lubricants (optional, only a few samples)

The samples with 1 wt% untreated CNTs, sonicated plus homogenized by the Speed-Mixer, proved best. This mixture was therefore further modified by adding several dry lubricants. After the mixture of 1 wt% CNTs in EP resin had been homogenized in the SpeedMixer and sonicated, 10 wt% of a dry lubricant (PTFE, graphite, or MoS$_2$, respectively) were mixed into the resin using the SpeedMixer at 2000 rpm for 4 min. As a reference, mixtures with dry lubricants but without CNTs were produced.

**9.** Evacuation (all samples)

After thorough homogenization, the final mixture was again cyclically evacuated and repressurized until no more bubbles formed on the surface.

**10.** A hardener was stirred into the mixture.

The hardener was directly added to the mixture at a resin-to-hardener weight ratio of 4:1.

In some cases, the hardener was mixed into the EP–CNT composites by the SpeedMixer at 500 rpm for 30 s. However, these gentle mixing conditions did not lead to homogeneous mixtures. Very high mixing speeds, on the other hand, introduced too much air into the final mixture. As a compromise, the hardener was mixed into the composites at 1000 rpm for 1 min into the composites. Nevertheless, one test series was conducted with the suboptimal mixing speed of 500 rpm. Then, the ready mixture including the hardener was again evacuated for approximately 60 min.

**11.** Production of test specimens

When no more bubbles appeared, the mixtures were cast into silicone dies to produce plates of 1.5–2 mm thickness. The dies were put into an oven to cure the sample material. The cure cycle was (1) heating from 20 to 60 °C in 2 h, (2) heating from 60 to 80 °C in 2 h, (3) finally keeping at 80 °C for 4 h. The wear test specimens were cut from these plates with a handsaw into pieces of 10 mm × 10 mm × 2 mm.

Table 9.1 lists all the samples that were prepared and tested.

### 9.2.2.2 Preparation of polyethylene composites

Some specimens were produced with untreated CNTs and a second set of specimens with CNTs pretreated in nitric acid. The pretreatment was performed in the same way as for the epoxy specimens with treatment duration of 2 h. Before further processing, both CNT samples (pretreated and untreated) were dried in an oven at 100 °C for 2 h and then crushed in a mortar.

To produce pure UHMWPE specimens, the UHMWPE powder was filled into the mold of a hot press and precompacted at room temperature and a pressure of 12 MPa for 30 min. In the next step, the press was heated up to 200 °C, and the pressure was reduced to 10 MPa. 15 min later, the press was cooled to room temperature at the same pressure over a time of approximately 45 min. The square sample plates produced had an edge length of 120 mm and a thickness of approximately 2 mm.

The UHMWPE/HDPE blends were produced by mixing 20 wt% HDPE granules and 80 wt% UHMWPE powder in an HAAKE laboratory kneader at the temperature of 210 °C. The rotational speed was 10 rpm for 5 min, and then the speed was increased to 45 rpm for 10 min. In the next step, the blend was compression molded in the same die as the UHMWPE samples at the temperature of 180 °C and a pressure of 10 MPa. 20 min later, it was passively cooled under the same pressure.

**Table 9.1** List of All Sample Materials Prepared and Tested

| | Composition | Preparation Method |
|---|---|---|
| EP+0.2wt%<br>EP+0.5wt%<br>EP+1wt%<br>EP+2wt%<br>EP+4wt% | Untreated CNTs | Mixed with a four-blade stirrer |
| EP+0.2wt%<br>EP+0.5wt%<br>EP+1wt%<br>EP+2wt%<br>EP+4wt% | Pretreated CNTs, for 11h | |
| EP+0.5wt%<br>EP+1wt%<br>EP+2wt% | Untreated CNTs | Mixed with the SpeedMixer at 2000rpm hardener at 500rpm |
| EP+0.5wt%<br>EP+1wt%<br>EP+2wt% | Pretreated CNTs for 11h | |
| EP+1wt% | Untreated CNTs | Three-roll mill |
| EP+1wt% | Pretreated CNTs, for 11h | |
| EP+1wt% | Untreated CNTs | Mixed with the SpeedMixer at 2000rpm+sonication hardener at 1000rpm |
| EP+1wt% | Untreated CNTs+10wt% PTFE | Mixed with the SpeedMixer at 2000rpm+sonication hardener at 1000rpm |
| EP+1wt% | Untreated CNTs+10wt% graphite | |
| EP+1wt% | Untreated CNTs+10wt% MoS$_2$ | |
| EP | +10wt% graphite | Mixed with SpeedMixer at 2000rpm hardener at 1000rpm |
| EP | +10wt% MoS$_2$ | |
| EP+1wt% | Untreated CNTs | Mixed with the SpeedMixer at 3000rpm hardener at 1000rpm |
| EP+1wt% | Pretreated CNTs, for 2h | |
| EP+1wt% | CNTs with GLYMO treatment | |
| EP+1wt% | Untreated CNTs | Sonication, hardener at 1000rpm |
| EP+1wt% | Pretreated CNTs, for 2h | |
| EP+1wt% | CNTs with GLYMO treatment | |
| EP+1wt% | Untreated CNTs | Mixed with the SpeedMixer at 3000rpm+sonication hardener at 1000rpm |
| EP+1wt% | Pretreated CNTs, for 2h | |
| EP+1wt% | CNTs with GLYMO treatment | |

The composite specimens were produced in the same way as the UHMWPE/HDPE samples. The different weight content of CNTs (untreated or pretreated, respectively) were mixed with the UHMWPE powder and the 20wt% HDPE granulate. This mixture was again processed in the kneader and finally compression molded in the hot press to plates.

**Table 9.2** Designation of CNT-Reinforced Polyethylene Composites and Mechanical Properties of the Composites with 60–100 nm CNT Diameter.

| Designation | Composition | Yield stress (MPa) | Tensile strength (MPa) | Strain at break (%) | Young's modulus (MPa) |
|---|---|---|---|---|---|
| UHMWPE | 100% GUR 2122 | 25.0 | 24.7 | 109 | 1187 |
| PE | 80% UHMWPE+20% HDPF | 28.4 | 27.8 | 26 | 1493 |
| PE+0.2%p | PE+0.2% MWCNT (pretreated) | 29.6 | 29.7 | 34 | 1578 |
| PE+0.5%p | PE+0.5% MWCNT (pretreated) | 27.6 | 26.2 | 31 | 1610 |
| PE+1.0%p | PE+1.0% MWCNT (pretreated) | 28.8 | 25.5 | 32 | 1603 |
| PE+2.0%p | PE+2.0% MWCNT (pretreated) | 30.8 | 28.7 | 29 | 1613 |
| PE+0.2%u | PE+0.2% MWCNT (untreated) | 27.3 | 27.9 | 31 | 1508 |
| PE+0.5%u | PE+0.5% MWCNT (untreated) | 26.2 | 24.7 | 30 | 1576 |
| PE+1.0%u | PE+1.0% MWCNT (untreated) | 29.9 | 29.0 | 28 | 1594 |
| PE+2.0%u | PE+2.0% MWCNT (untreated) | 29.9 | 28.0 | 30 | 1495 |

From the sample plates, small specimens of approximately 10 mm × 10 mm were cut with a high speed saw. These specimens were glued onto the specimen holders of the wear test apparatus. For this purpose, the specimen rear side was previously treated with a special polyolefin primer. The shouldered tensile test bars were cut with a high speed saw from the sample plates to a width of approximately 5 mm. The gauge length was 25 mm.

Table 9.2 compiles the designation of the sample materials that were prepared and tested. Besides, some physical data of these sample materials are also given in the table.

## 9.2.3 **Testing procedures**

### 9.2.3.1 *Thermal and thermomechanical analysis*

The curing behavior of the CNTs–epoxy nanocomposites was determined using a differential scanning calorimetry (DSC) 204F1 from NETZSCH (Germany). Approximately 10 mg of the complete mixture including CNTs, resin, and hardener was put into aluminum pan and sealed. The sample was heated up to 80 °C at a rate of 10 K min$^{-1}$. After this curing temperature was reached, the reaction enthalpy was measured during an isothermal phase of 4 h at 80 °C.

The study of the thermomechanical behavior of some EP/CNT composites was performed by dynamic-mechanical thermal analysis (DMA) using a DMA 242 from NETZSCH (Germany). Rectangular specimens, 50 mm long, 5 mm wide, and 4 mm thick, were prepared for the measurements. The samples were exposed to an oscillating three-point bending load between 30 and 250°C at a heating rate of 5°C min$^{-1}$, the frequency was set to 1 hz, the deformation amplitude was 80 μm.

### 9.2.3.2 Wear and creep tests

All tests were performed in a ball-on-prism tribometer according to ISO 7148-2 from Dr Tillwich GmbH Werner Stehr (Germany) as shown in Fig. 9.6. The specimens were subjected to uniform and unidirectional sliding.

The polymeric specimens were glued onto the inner surfaces of the prism with 90° opening angle (Fig. 9.7(a)). The prism is fixed at one end of a lever and pressed against a ball with a diameter of $d = 12.7$ mm via a dead weight. The balls, which rotate uniformly around their vertical axis driven by a simple electro motor, can consist of different materials.

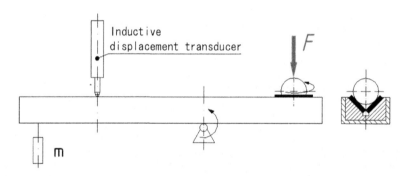

**FIGURE 9.6**

Schematic representation of a ball-on-prism tribometer. Side view (left) and front view (right).

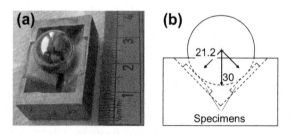

**FIGURE 9.7**

(a) Prism with specimens and ball and (b) loading situation.

All the tests were performed with a dead weight of $F_Z = 30\,\text{N}$, resulting in a normal load of

$$F_N = \frac{F_Z}{\sqrt{2}} = 21.2 \ \text{N}$$

acting on each specimen (Fig. 9.7(b)). The rotational frequency, $f$, of 1 hz resulted in a sliding speed of

$$v = \pi \cdot \frac{d}{\sqrt{2}} \cdot f = 28.2 \ \text{mm/s}$$

Each test lasted approximately 60 h, which is equivalent to approximately 6000 m sliding distance. The contact pressure, $p$, and the $pv$ value cannot be quoted because they decreased continuously during the test due to the spherical contact geometry. Yet a rough estimate can be made: the diameter of the wear mark was around 1–3 mm during steady state. The corresponding average contact pressure is 3–27 MPa and the $pv$ level below $0.76\,\text{MPa}\,\text{m}\,\text{s}^{-1}$.

One set of EP specimens was loaded statically in the test rig for approximately 60 h. Under this static load, the balls penetrated into the specimen surface only a few tenths of a micrometer due to plastic deformation and creep of the polymer. This deformation is negligible in comparison to the wear of the materials, which was in the order of some 10 µm to some hundred micrometers. The creep was therefore not further considered and skipped in the subsequent tests.

This was different for the viscoelastic polyethylene composites. In order to separate creep effects from wear, a static loading phase of 60-h duration preceded the wear tests. Under this static load, the balls penetrated into the specimen surface due to creep and plastic deformation of the polymer. The contact area increased continuously and the contact pressure decreased asymptotically to a value below which creep almost disappeared. When this situation was reached, the motor was switched on, and the wear was measured without a significant contribution caused by creep. The creep penetration was continuously recorded and the penetration depth after 60 h, when creep diminished, was taken as a measure for the creep resistance of the composites.

An inductive displacement transducer continuously measured the motion of the lever. The wear of the metallic counterparts was insignificant in all cases so that the measured system wear represented the wear of the polymer-based specimens. The displacement of the lever, $h$, was converted into a wear volume, $V$, per specimen (Fig. 9.8):

$$V = \frac{\pi}{6} \cdot \left(\frac{h}{\sqrt{2}}\right)^2 \cdot \left(3d - 2\frac{h}{\sqrt{2}}\right)$$

The wear volume was plotted versus the sliding distance, $L$, which was derived from the rotational frequency, $f$, the ball diameter, $d$, and the test duration, $t$, according to

$$L = \frac{\pi}{\sqrt{2}} \cdot d \cdot f \cdot t$$

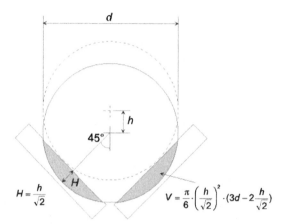

**FIGURE 9.8**

Calculation of wear volume ($\Delta V$).

After a running-in phase, all wear curves were found to become linear. A function of the type $V = V_0 + w \cdot L$ was fitted to the linear part of the curves. The slope

$$w = \frac{dV}{dL}$$

was divided by the normal load acting onto a single specimen, $F_N = 21.2\,\text{N}$, to calculate the specific wear rate, $k_s$:

$$k_s = \frac{w}{21.2\ \text{N}} = \frac{1}{F_N}\frac{dV}{dL}\left[\frac{\text{mm}^3}{\text{Nm}}\right]$$

All experiments were conducted in normal laboratory environment (~50%RH and 23 °C).

Each test was repeated at least three times. The data represented in this paper are the arithmetic mean values of the three tests. The standard deviation of the data is represented by error bars in the diagrams.

## 9.3 RESULTS AND DISCUSSION

### 9.3.1 Thermomechanical and structural characterization of EP/CNT composites

Well-dispersed CNTs have an extremely large specific surface. Since CNTs are electrically conductive, it was suspected that they may alter the curing reaction of the EP by heterogeneous catalysis. DSC measurements were carried out to clarify this question by detecting the reaction enthalpy. Figure 9.9 shows that the addition of CNTs did not have any effect on the curing behavior of the EP. The composites were therefore cured with the same parameters as the pure matrix.

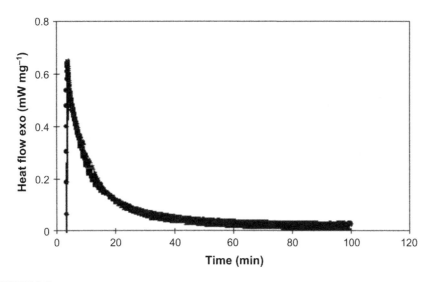

**FIGURE 9.9**

DSC curves of the curing reaction of pure epoxy/hardener mixture and epoxy composites with 1% untreated CNTs mixed by using the SpeedMixer. Parameters: heating from room temperature to 80 °C with 10 K min⁻¹, then isothermal phase for 4 h. The composites were prepared in three different ways: CNTs into resin and addition of hardener later, CNTs in hardener and addition of resin later, mixing of hardener and resin and addition of CNTs later.

Transmission electron microscopy (TEM) micrographs were taken of various EP/CNT composites. Most of the composites contained areas of large and relatively dense agglomerates as shown in Fig. 9.10(a) and areas of well-dispersed CNTs (Fig. 9.10(b)). Some singularized CNTs remained hidden inside the matrix and became visible only in randomly cracked areas (Fig. 9.11). Moreover, the arbitrary selection of the section cuts and the damage of the sample itself and the CNTs during specimen preparation has to be taken into account. Differences between the samples were therefore difficult to quantify by TEM. It must be concluded that a perfect homogenization could not be achieved with any of the dispersion methods and the differences in the dispersion states were only gradual. Finally, the effect of the dispersion state on the tribological properties is superimposed by other effects like predamage of the CNTs during pretreatment and homogenization.

The CNT reinforcement can affect the thermomechanical performance of an EP in various ways: The bending stiffness can be increased and the glass transition temperature, $T_G$, as a measure for the heat resistance can be improved as well. However, an improved $T_G$ is not necessarily accompanied by an increased stiffness. Figure 9.12 exemplifies these effects. The composite prepared with untreated CNTs via Speed-Mixer has a $T_G$ of approximately 110 °C and a Young's modulus of approximately 3000 MPa at room temperature. The pure EP showed virtually the same behavior. This means that untreated CNTs dispersed only via the SpeedMixer do not significantly

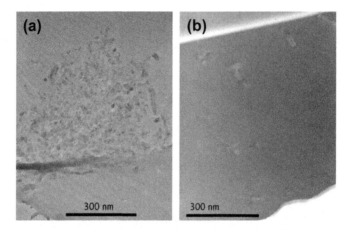

**FIGURE 9.10**

TEM micrographs of EP/CNT with 1% pretreated CNTs dispersed by the SpeedMixer without sonication. (a) Area with strong agglomeration and (b) area with well-dispersed CNTs.

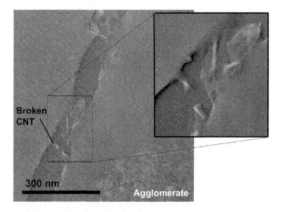

**FIGURE 9.11**

TEM micrographs of a specimen with 1% untreated CNTs plus 10% graphite in the EP, prepared with the SpeedMixer without sonication.

improve the thermomechanical performance of the EP. Ultrasound instead of the SpeedMixer increases the glass transition by approximately 30 °C but does not affect the bending stiffness. A combination of deagglomeration by homogenization by the SpeedMixer plus sonication yields the best results: $T_G$ is increased by more than 30 °C and the Young modulus at room temperature is increased by approximately 30% to 4000 MPa. We will see later on that this thermomechanical improvement does not correlate with the wear resistance of the composites.

EP/CNT composites prepared with CNTs pretreated in $HNO_3$ have an increased $T_G$, regardless of the dispersion and mixing method (Fig. 9.13). The pretreatment

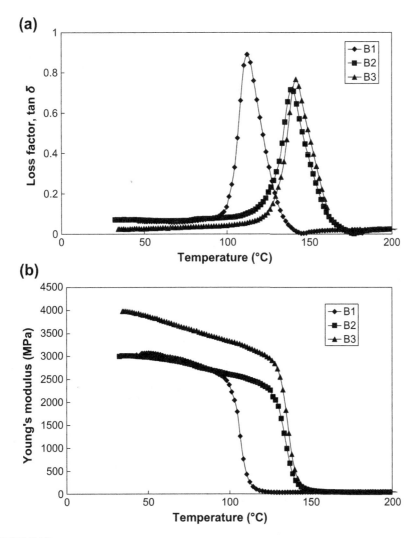

**FIGURE 9.12**

DMA measurements on EP/CNT composites prepared in various ways. (a) CNTs were untreated, CNT content = 1 wt%. (b) Curves B1 – CNTs in resin by using the SpeedMixer, curves B2 – CNTs dispersed via sonication, and curves B3 – CNTs mixed into resin via SpeedMixer and then dispersed by sonication.

generates active surface groups like –OH or –COOH, which can form covalent bonds with the epoxy molecules during curing. Those epoxy molecules that are firmly anchored on the CNT surfaces are afterward somewhat immobilized. However, a significant stiffness increase can be observed only for the composite prepared without sonication. It is suspected that sonication can cause fracture of the CNTs, and the

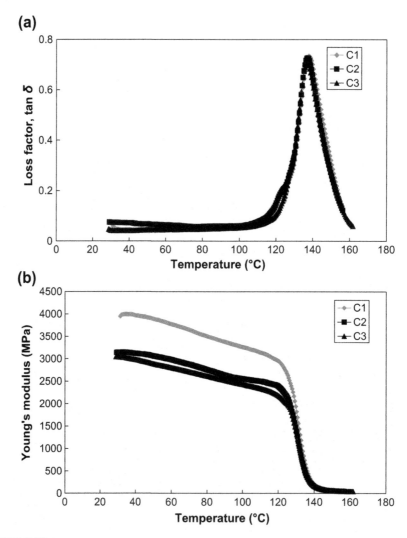

**FIGURE 9.13**

DMA result of the CNTs/epoxy composite prepared with CNTs pretreated in HNO$_3$. (a) CNT content = 1 wt%. (b) Curves C1—CNTs in resin by SpeedMixer, curves C2—CNTs dispersed via sonication, and curves C3—CNTs mixed into resin via the SpeedMixer and then dispersed by sonication.

shortened CNTs have a reduced reinforcing effect. This is apparently not true when the CNTs were not pretreated—and thus not predamaged—in HNO$_3$ (Fig. 9.12). It is only the combination of etching and introduction of high mechanical stresses in the ultrasound field that endangers the CNTs.

Applying GLYMO coupling agent leads to similar results like pretreatment in HNO$_3$ (Fig. 9.14). Both, $T_G$ and Young's modulus are increased only for the sample

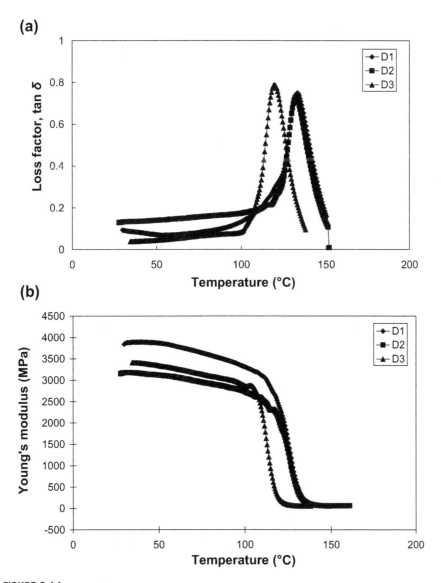

**FIGURE 9.14**

DMA result of epoxy composites including 1 wt% CNTs pretreated with coupling agent after activation in HNO₃. (a) Curves D1—CNTs in resin by the SpeedMixer, curves D2—CNTs dispersed via sonication, and (b) curves D3—CNTs mixed into resin via SpeedMixer and then dispersed by sonication.

prepared without sonication. The reason is that the CNTs had to be activated in $HNO_3$ before the application of the coupling agent. The CNTs are afterward predamaged and tend to break in the shear stress field of the ultrasound. The increase in $T_G$ is less high than in the case of pure $HNO_3$ treatment because the GLYMO-treated samples were subjected to ultrasound twice: a first time to disperse the CNTs in the GLYMO solution and a second time to disperse the pretreated CNTs in the resin.

## 9.3.2 Wear test results

### 9.3.2.1 Wear performance of CNT-reinforced epoxy composites

The first approach was to produce the samples in the simplest way: Untreated CNTs were mixed with the four-blade stirrer into the resin. In the range between 0 and 4 wt% CNTs, no significant effect of the reinforcement on the wear rate could be found (Fig. 9.15) during sliding against X5CrNi18-10. All wear rates were in the range of $10^{-5}$ mm$^3$ nm$^{-1}$. Higher CNT contents could not be prepared because the viscosity of the mixtures became too high.

A poor dispersion obviously prevented the CNTs from developing a significant reinforcing effect. The large scatter of the wear data, too, referred to an insufficient homogeneity of the mixture. Moreover, the mixtures needed >2 h of evacuation to remove the large amount of air that was entrapped inside the agglomerates.

A second attempt with pretreated CNTs led to somewhat better results (Fig. 9.16). The scatter of the data was significantly smaller, which means that the mixtures were more homogeneous. The wear rate decreased continuously as the CNT content was increased to 1 wt%. Beyond this value, the wear rate again increased slowly. This wear rate minimum in the range of a few weight percent filler content is typical for nanocomposites [18,19,21]. Nanofillers have an extremely large specific surface area where the matrix material is adsorbed. Therefore, only small amounts of nanofillers can be mixed in a polymer matrix when every particle should be properly wetted [3].

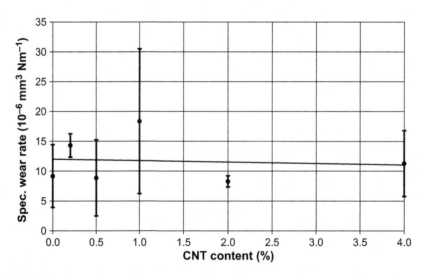

**FIGURE 9.15**

Effect of CNT content on the specific wear rate of EP/CNT composites during sliding against X5CrNi18-10. Untreated CNTs mixed with the four-blade stirrer.

Exceeding this limit usually causes embrittlement of the composites and promotes susceptibility to surface fatigue.

Figure 9.17 presents the wear rates of composites where both the CNTs (2 min at 2000 rpm) and later on the hardener (30 s at 500 rpm) were mixed into the resin with

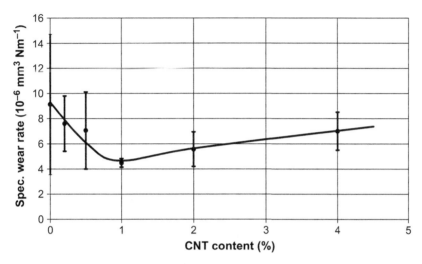

**FIGURE 9.16**

Effect of CNT content on the specific wear rate of EP/CNT composites during sliding against X5CrNi18-10. CNTs pretreated in nitric acid and mixed with the four-blade stirrer.

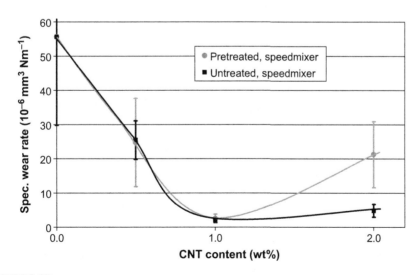

**FIGURE 9.17**

Effect of the CNT content (pretreated and untreated) on the specific wear rate of EP/CNT composites during sliding against X5CrNi18-10.

the SpeedMixer. Strikingly, the pure EP prepared in this way had a far higher specific wear rate ($\sim 55 \times 10^{-6} \, mm^3 \, nm^{-1}$) than did the EP resin mixed with the four-blade stirrer ($9 \times 10^{-6} \, mm^3 \, nm^{-1}$). It was suspected that the mild conditions during mixing of the hardener into the resin could not ensure a homogeneous hardener distribution. As a consequence, the crosslinking density may have been too low locally.

The addition of CNTs, though, had a significant effect. There was a pronounced minimum around 1 wt% CNT content. Beyond this value, the wear rate again increased slightly, in the case of the untreated CNTs less pronounced than in the case of pretreated CNTs. These results are in surprisingly good agreement with those of Dong et al. [24], who performed their tests under completely different conditions (Fig. 9.18). Around the wear minimum, there was no significant difference between untreated a pretreated CNTs. Untreated CNTs would be favorable from the practical point of view because the complicated pretreatment with the environmentally hazardous $HNO_3$ can be omitted and the composite is less sensitive to minor variations in the dosing process. The minimum wear rate of the composites prepared with untreated CNTs by the SpeedMixer is slightly smaller than the minimum wear rate of the compounds prepared with the four-blade stirrer and pretreated CNTs. The SpeedMixer seems to homogenize the CNT–epoxy mixture to a sufficiently good degree even without any pretreatment. This does not mean that a perfect dispersion was reached as the TEM micrographs had shown (Figs 9.10 and 9.11). The integrity of the CNTs—which may be disturbed by the etching process—is apparently at least as important for the wear properties as their dispersion.

Another important feature of the optimized composites with 1 wt% CNTs is the low scatter of the wear data. This is clearly visible in Fig. 9.16 as well as in Fig. 9.17.

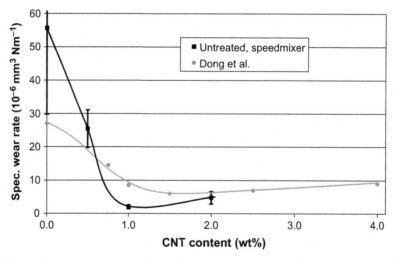

**FIGURE 9.18**

Comparison of own wear data with results of Dong et al. [24].

At lower contents, microscopic areas may suffer from CNT depletion and missing reinforcement so that the wear rate sensitively depends on the special position of the point contact; at higher contents beyond the wear minimum, more and larger agglomerates may form leading to microscopically inhomogeneous properties.

The composites with 1 wt% CNTs proved to be best with respect to the wear resistance. This composition was therefore further investigated in order to optimize the dispersion and mixing method.

According to Gojny et al. [7], the three-roll mill has the best dispersion effect for CNT composites. Two sets of samples with 1 wt% CNTs in EP were therefore prepared with this apparatus: One set of specimens with pretreated CNTs and another one with untreated CNTs. The specific wear rates measured were $14 \times 10^{-6}$ mm$^3$nm$^{-1}$ for the untreated CNTs and $5.5 \times 10^{-6}$ mm$^3$nm$^{-1}$ for the pretreated ones. The pretreatment caused a drastic improvement of the wear rate like in the case of the specimens prepared with the four-blade stirrer. As expected, the roller mill yielded better results like the four-blade stirrer for the untreated CNTs. In the case of the pretreated CNTs, no big difference could be found between roller mill and four-blade stirrer. The specimens produced with the SpeedMixer lead to better wear results in any case (Fig. 9.19). This ranking order of the preparation methods with respect to wear resistance needs not to correlate directly with the dispersion state.

Since the specimens prepared with the SpeedMixer delivered the best wear results, the corresponding process was intensified for further tests: The speed was increased from 2000 to 3000 rpm (maximum possible frequency). Moreover, the

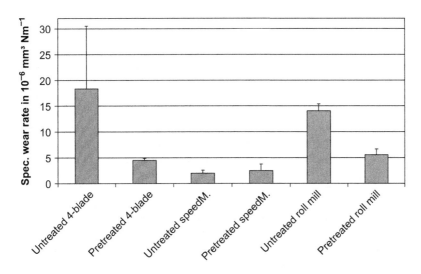

**FIGURE 9.19**

Specific wear rates of epoxy composites with 1 wt% CNTs prepared with different devices without sonication. Counterpart in the wear test was X5CrNi18-10, pretreatment: CNTs etched in HNO$_3$ for 11 h.

frequency and the processing time during mixing of the hardener into the compound were doubled from 500 to 100 rpm and from 30 to 60 s, respectively. The wear rate was in fact further reduced from $2 \times 10^{-6}$ to $1 \times 10^{-6}$ mm³ nm⁻¹.

Another approach to improve the wear resistance of the composite was sonication of the resin/CNT mixture prior to homogenization in the SpeedMixer. Simultaneously, the pretreatment time in nitric acid was reduced from 11 to 2 h to reduce predamage of the CNTs and some CNTs were treated with an epoxy-functionalized silane coupling agent (GLYMO, Degussa, Germany). The corresponding wear rates are shown in Fig. 9.20. Most data are in a small band around $1 \times 10^{-6}$ mm³ nm⁻¹. No more significant improvements could be achieved after the previously described optimizations. However, small differences are still visible: The untreated CNTs performed best. Pretreatment of the CNTs in $HNO_3$ or with coupling agent led to similar results but no further improvement as long as no ultrasound was applied. Sonication of untreated CNTs also lead to good wear results. Although, an adverse effect had the combination of pretreatment in $HNO_3$ plus sonication (the GLYMO-treated CNTs, too, were pretreated in $HNO_3$). It is suspected that etching in nitric acid damages the surfaces of the CNT, which subsequently are more prone to fracture when high stresses are applied in the ultrasound field.

This effect becomes very clear when two samples are compared that were prepared in the same way but with different etching times (Fig. 9.21). Reducing the

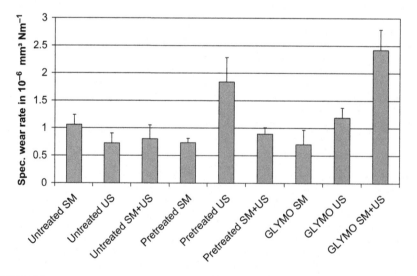

**FIGURE 9.20**

Specific wear rates of EP/CNT composites (1 wt% CNTs) prepared with different pretreatment and dispersion methods. US = dispersed by ultrasound, SM = homogenized by the SpeedMixer at 3000 rpm, pretreated = CNTs etched in $HNO_3$ for 2 h, GLYMO = coupling agent applied to CNTs. Counterpart in wear tests: X5CrNi18-10.

pretreatment time from 11 to 2 h leads to a decrease of the wear rate from almost $2.5 \times 10^{-6}\,mm^3\,nm^{-1}$ to $<1 \times 10^{-6}\,mm^3\,nm^{-1}$ for samples with 1% CNTs prepared by the SpeedMixer. This wear rate increases again very slightly, when ultrasound is applied.

One of the best EP/CNT composite so far was the mixture including 1 wt% untreated CNTs, sonicated and homogenized by the SpeedMixer. Moreover, the preparation of this composite did not require any chemical pretreatment. This compound was used as a basis for further improvements by adding 10 wt% PTFE, graphite, or $MoS_2$, respectively. Since these experiments had been performed before the final optimization of the dispersion parameters, the SpeedMixer was used at 2000 rpm. Figure 9.22 compares the wear rates of these materials. The addition of 10 wt% PTFE to the EP/CNT compound increased its wear rate slightly. This was unexpected because PTFE normally has an outstandingly beneficial effect on the wear resistance of conventional [54,55] and nanocomposites [19]. Moreover, the EP/CNT/PTFE nanocomposite was the only one where no large agglomerates could be found by TEM (Fig. 9.23), that is, where the dispersion must have been best. The reason for this unusual negative synergy between CNTs and PTFE could not be clarified so far but it seems that good dispersion is not necessarily beneficial for a good wear resistance. In contrast, $MoS_2$ and especially graphite significantly improved the wear resistance of the EP/CNT composites. EP/CNT/graphite reached a specific wear rate as low as $3.5 \times 10^{-8}\,mm^3\,nm^{-1}$ (please note the exponent −8), which was the lowest wear rate among all combinations tested.

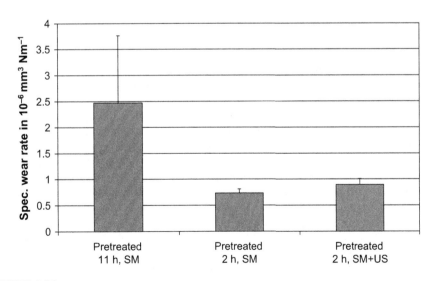

**FIGURE 9.21**

Effect of etching time and dispersion method on the wear resistance of EP/CNT with 1% pretreated CNTs. SM = speedMixer, US = ultrasound, 11 and 2 h = treatment time in hours. Counterpart in sliding test: X5CrNi18-10.

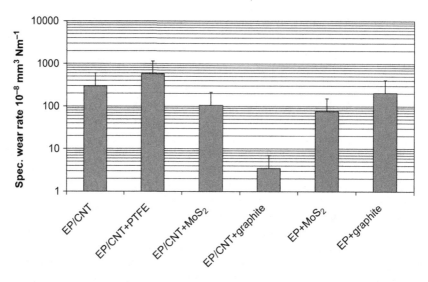

**FIGURE 9.22**

Specific wear rates of multicomponent composites. EP/CNT = EP + 1 wt% untreated CNTs (sonicated and homogenized with SpeedMixer at 2000 rpm), counterpart = X5CrNi18-10. Lubricant content was 10 wt% in all cases.

To check whether the good wear performance of the EP/CNT/graphite and the EP/CNT/MoS$_2$ composites can be traced back to the action of CNTs or to the dry lubricants, samples of pure epoxy (without CNTs) were filled with graphite and MoS$_2$, respectively. Figure 9.22 shows that these samples, too, performed better than the pure EP but worse than the multicomponent composites. Conclusively, neither the CNT reinforcement nor the dry lubrication alone can yield minimal wear rates. There is obviously a positive synergetic interaction between CNTs on the one hand and graphite or MoS$_2$, respectively, on the other hand. This mutual backing is particularly pronounced for the graphite, which chemically resembles the CNTs. Tanaka et al. [38] found an opposite trend for nanohorn- and nanotube-reinforced PI composites: additional mixing of graphite into the compounds increased the wear rate.

Previous studies [54,55] had revealed that composites containing chemically active fillers like carbon fibers, graphite, or MoS$_2$, react sensitively on the chemical composition of the counterpart material. These fillers can cause tribocorrosion of steel counterparts with a low chromium content. After this tribocorrosion commences, the composite is rapidly abraded by the corrosion products and the roughened counterpart. This is also true for the CNT composites as Fig. 9.24 points out. CNTs as well as MoS$_2$ and graphite perform worse against the unalloyed martensitic steel 100Cr6 than against the corrosion resistant austenitic steel X5CrNi18-10. The effect of the counterpart is again most pronounced for the graphite-containing sample. The MoS$_2$-filled composite has an even higher wear rate than the pure epoxy matrix during against 100Cr6.

**FIGURE 9.23**

TEM micrograph of an EP sample including 1% untreated CNTs plus 10% PTFE prepared with SpeedMixer at 2,000 rpm and sonication.

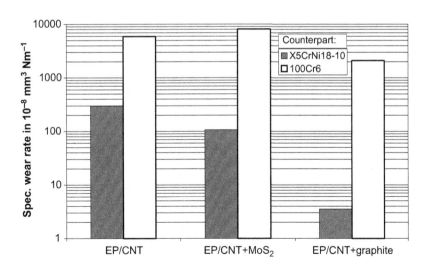

**FIGURE 9.24**

Effect of counterpart material on selected nanocomposites. "EP/CNT" means EP with 1 wt% untreated CNTs produced with the SpeedMixer at 2000 rpm and sonication. $MoS_2$- and graphite content: 10 wt%.

### 9.3.2.2 Micrographs of worn EP/CNT composites

Scanning electron microscopic (SEM) inspection of the wear marks gave only little information about the wear mechanisms and the single CNTs, whose chemical composition could not be visualized. Figure 9.25 shows some micrographs of the wear mark of EP including 1 wt% untreated CNTs mixed with the four-blade stirrer. The wear mark is rather smooth. Some loose wear debris but only little and very small scratches can be found. Adhesive wear seems to have been predominant. However, the smooth area is riddled with a lot of transverse cracks (Fig. 9.25(c)) originating from surface fatigue, which is typical for brittle materials like EP.

The wear mark of the composite consisting of 1 wt% pretreated CNTs in EP mixed with the four-blade stirrer looks different (Fig. 9.26): First of all, the wear mark has of course a far smaller diameter, which visually documents the significantly lower wear rate. However, only approximately 50% of the wear mark appears smooth, whereas the remaining part is rougher. Figure 9.26(b) shows the border between two such areas. The smooth part seems to be backtransfer, which protects the original surface from being polished. This backtransfer was missing in the case of the untreated CNTs. The backtransfer contained numerous small

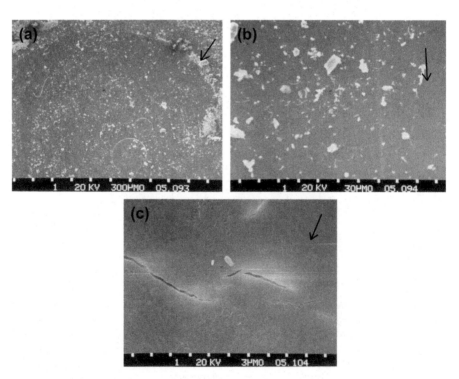

**FIGURE 9.25**

Wear mark of 1 wt% untreated CNTs in EP mixed with the four-blade stirrer. (a) Overview, (b) and (c) details. Arrows indicate the sliding direction.

transverse cracks and a few larger longitudinal cracks indicating fatigue damage of this layer followed by delamination wear. Once the backtransfer film locally detached from the specimen the original surface was again exposed to the sliding. There is less wear debris visible on the wear mark than in the case of the composite containing untreated CNTs (Figs 9.25(a) and 9.26(a)). Moreover, the wear debris has a more needlelike conformation (Fig. 9.26(d)) compared to that of the irregularly shaped debris particle in Fig. 9.25(b). These needles of some hundred-nanometer diameter must contain a high amount of CNTs because they remained very stable under the electron beam at high magnification and current. The needles may have had a kind of needle bearing effect; singularized CNTs would have been too small for such an effect.

The specimens prepared with the SpeedMixer featured similar wear marks like the specimen shown in Fig. 9.26. However, the part of the area covered by backtransfer varied slightly and was the biggest for the compound containing pretreated CNTs and prepared with the SpeedMixer (Fig. 9.27).

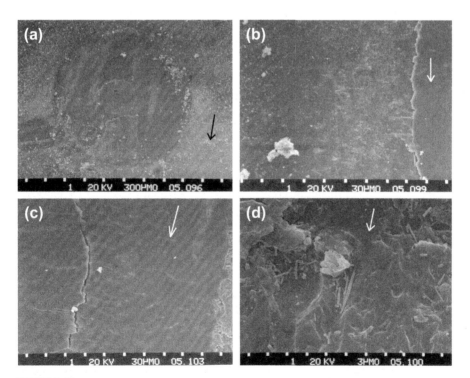

**FIGURE 9.26**

Wear mark of 1 wt% pretreated CNTs mixed with the four-blade stirrer into the EP matrix. (a) Overview, (b) border area between rough and smooth surface part, (c) magnification of smooth part, and (d) magnification of rough part. Arrows indicate the sliding direction.

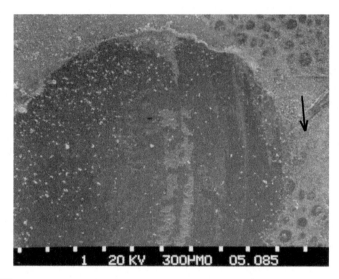

**FIGURE 9.27**

Overview of the wear mark of 1 wt% pretreated CNTs in EP, mixed with the SpeedMixer. The arrow indicates the sliding direction.

### 9.3.2.3 Wear performance of CNT-reinforced polyethylene composites
#### 9.3.2.3.1 Creep tests

Figure 9.28 shows the creep curves of the pure UHMWPE sample, the unfilled UHMWPE/HDPE blend as well as the PE blends filled with 2 wt% untreated und pretreated CNTs with the diameter of 60–100 nm. The curves represent the penetration depth of the ball into the specimen surfaces under static load without sliding. Initially, the material crept rather fast, but then the creep decelerated continuously as the contact area increased causing a reduction of the contact pressure. After approximately 50 h, the contact pressure reached a value at which the penetration motion almost diminished.

The general shape of all creep curves is identical. However, quantitatively there is a significant difference. Figure 9.29 compares the penetration depth after 60 h of creep for the unfilled materials and all composites reinforced with 60–100 nm CNTs. Pure UHMWPE has the lowest creep resistance. The penetration depth after 60 h amounts to approximately 37 μm. The composites reinforced with untreated CNTs generally have a better creep resistance. Unexpectedly, the creep resistance decreases with increasing CNT content. Concerning the composites with pretreated CNTs, there is no significant effect of the filler content. The creep resistance is only slightly better than for the pure UHMWPE. However, the best creep resistance was found for the UHMWPE/HDPE blend without CNTs. From this phenomenon, it may be concluded that the increased creep resistance of the composites was mainly caused by the addition of HDPE. Additional reinforcement with CNTs gradually compensates

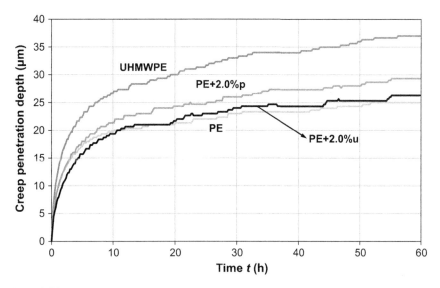

**FIGURE 9.28**

Typical creep curves UHMWPE, UHMWPE/HDPE blend (PE), and blends with 2 wt% pretreated (p) and untreated (u) CNTs.

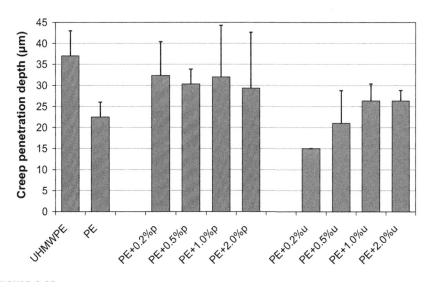

**FIGURE 9.29**

Creep penetration depth after 60 h for all the materials tested. For designations see (Table 9.2).

the creep reducing effect of the HDPE; this effect is more pronounced for the pre-treated CNTs presumably because the pretreatment induces a reduction of the average length of the nanotubes.

### 9.3.2.3.2 Wear tests

While the steel ball slid over the specimen surface, the material experienced both: creep due to the normal load and wear due to the frictional load. To separate the material wear from plastic deformation and creep, the wear tests started after the preceding creep phase. Since the creep curves became flat after 60 h, the slope of the wear curve was insignificantly affected by creep (frictional heating of the specimen was negligible under the low $pv$ values applied).

The wear volume curves of pure UHMWPE, unfilled UHMWPE/HDPE and the blends with 0.5 wt% CNTs (60–100 nm) are shown in Fig. 9.30. The intercept with the $y$-axis represents the creep volume calculated from the creep penetration depth after 60 h before the wear test started.

The pure materials exhibited a long running-in phase while that of the filled materials was much shorter. The slope of HDPE/UHMWPE slightly increased during the first 2000–3000 m. The composites and the pure UHMWPE, in contrast, showed an opposite running-in behavior with a decreasing wear rate. The slope of the CNT-filled compounds rapidly dropped down and reached a constant value after a few hundred meters while the running-in phase of the UHMWPE sample lasted approximately 2500 m. In the following steady state, the wear rate of UHMWPE and the UHMWPE/HDPE blend were almost identical. The CNT-filled composites were

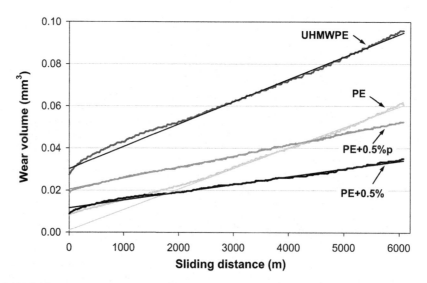

**FIGURE 9.30**

Typical wear curves of UHMWPE, UHMWPE/HDPE blend (PE), and blends containing 0.5 wt% untreated (u) or pretreated (p) CNTs with 60–100 nm diameter.

characterized by a low wear rate after a short running-in phase. The wear resistance for the samples including untreated CNTs is superior to those with the pretreated.

Figure 9.31 shows the wear rates of the samples as a function of the filler content. The composites with pretreated and untreated CNTs with 60–100 nm in diameter are

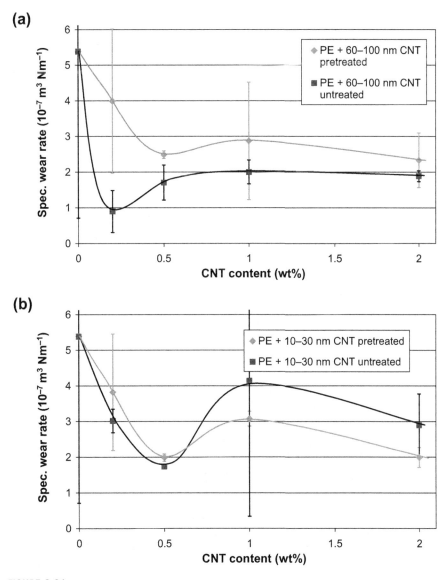

**(a)**

**(b)**

**FIGURE 9.31**

Specific wear of PE/CNT composites rate versus CNT content. (a) Thicker nanotubes (60–100 nm diameter) were used and (b) slenderer nanotubes (10–30 nm) were used.

compared in Fig. 9.31(a). Figure 9.31(b) shows the same comparison for the slenderer (10–30 nm) CNTs.

It can be seen that the wear rate of the UHMWPE/HDPE composite can be significantly reduced by adding CNTs. The wear rates of all CNT composites, both with pretreated CNTs and untreated CNTs of both sizes, decreased sharply with increasing CNT content and reached a minimum at 0.5 wt%. For the composite with 60–100 nm untreated CNT the minimum was already reached at 0.2 wt% filler content. The short increase of the wear rate after the minimum was followed by a relative stable value of wear rate for the tested filler contents up to 2 wt%. It is striking that the scatter is also smallest for the low wear rates.

The wear rate of the UHMWPE/HDPE blend was reduced >50% after the addition of only 0.5 wt% CNTs. The composites reinforced with untreated 60–100 nm CNTs had a better wear performance than did the composites with pretreated CNTs of the same diameter. This finding is in accordance with the higher creep resistance of composites with untreated CNTs even though their stiffness and tensile strength (Table 9.2) were lower than for the composites with pretreated CNTs. For the composites with the slender CNTs, the difference is less pronounced. While the wear rate of the compounds with the untreated CNT is much better for those with the bigger CNTs compared to the smaller ones, the difference between the pretreated CNTs of both diameters is almost the same. On the one hand, it seems that the pretreatment is useful for the dispersion of the smaller CNTs. On the other hand, the pretreatment may partly destroy the CNTs resulting in a deterioration of the mechanical properties and the wear resistance as it could be seen for the composites with the pretreated CNTs of the larger diameter.

Key issues for the performance of CNT-reinforced composites are interfacial bonding and the proper dispersion of the CNTs in the polymer matrix. In this study, the CNTs were mechanically dispersed into a thermoplastic polymer melt by a kneader. The central idea is to use fluid shear forces to break the nanotube agglomerates or prevent their formation. Boiling of CNTs in nitric acid before mixing in a kneader has two purposes. One purpose is the purification of the nanotubes by removal of amorphous carbon residues. For the composites discussed in this paper, this purification may have played an important role. The other purpose of this pretreatment is the grafting of functional polar groups onto the CNT surface in order to promote the filler–matrix interaction. These functional groups may have been less useful in the case of a polyethylene matrix that is nonpolar. Another unintended effect of the boiling in concentrated nitric acid is that the multiwall CNTs may be cracked and peeled.

To examine the influence of the counterpart all samples containing 2 wt% CNTs and the pure matrices were tested against 100Cr6. Figure 9.32 compares these results with the data determined against X5CrNi18-10. When sliding against the hard martensitic 100Cr6 steel, the wear resistance of pure UHMWPE improved while that of the other materials were almost unchanged or changed for the worse. But this deterioration in the wear resistance was less pronounced compared with the EP composites

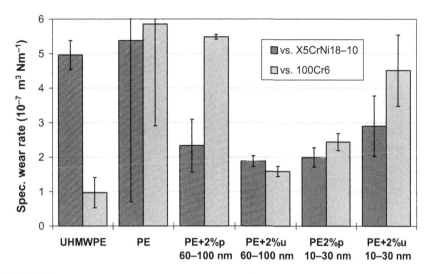

**FIGURE 9.32**

Effect of counterpart material on the specific wear rate of several polyethylene composites.

described before. In this case, the tribocorrosion was masked. This was also reported for CNF-reinforced polyetheretherketone [56].

### 9.3.2.3.3 Micrographs of worn PE/CNT composites

To study the wear mechanisms, some worn specimens were sputtered with a thin gold layer and then examined in an SEM. The steel balls were inspected by SEM without previous sputtering.

Figure 9.33 gives overviews of the wear marks on UHMWPE, on UHMWPE/HDPE, and on the composite with 2% untreated CNTs after sliding against X5CrNi18-10. Chips were formed on the exit side of all wear marks. Figure 9.34 shows details of these chips. The chips of pure UHMWPE are short and have a layered structure. On UHMWPE/HDPE and the composite with CNTs, the chips are long and look like continuous thin films. This shows that HDPE can easily be drawn mechanically than the UHMWPE, whose extremely long molecule chains are anchored by a far higher number of entanglements.

On the worn surface of all specimens (Fig. 9.35), there are scratches parallel to the sliding direction pointing out a component of abrasive wear. These scratches are the smallest on the UHMWPE surface whereas the CNT-reinforced composite exhibits wider and deeper scratches. Scratches are usually not produced by counterpart asperities, which are much smaller than the scratches, but by wear debris particles. These particles are obviously larger and harder in the case of the CNT composite than in the case of pure UHMWPE. This indicates that CNT reinforcement might be detrimental under certain conditions.

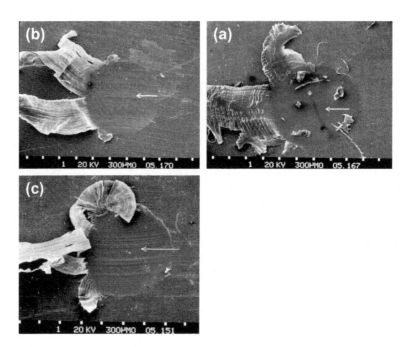

**FIGURE 9.33**

SEM micrographs of the wear marks on (a) UHMWPE, (b) UHMWPE/HDPE blend, and (c) a blend with 2 wt% untreated CNTs. Counterpart: X5CrNi18-10, the white arrows indicates the sliding direction.

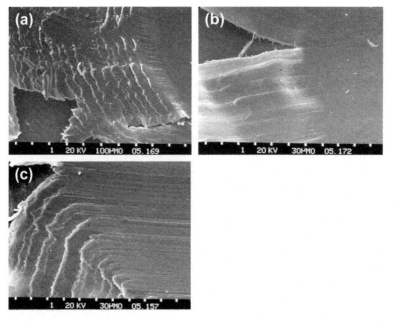

**FIGURE 9.34**

Details of the micrographs in Fig. 9.33. Chips of (a) UHMWPE, (b) UHMWPE/HDPE blend, and (c) blend with 2 wt% untreated CNTs.

The wear mark on UHMWPE exhibits the smoothest surface. The shape of the primary powder grains is still visible and the grain boundaries form narrow valleys surrounding the grains. This shows that the grains have a higher wear resistance than the grain boundaries. The UHMWPE/HDPE blend and the composite containing 2% untreated CNTs exhibited rougher surfaces with some textures transverse to the sliding direction caused by plastic flow and patchy recompacted backtransfer, which can protect the original composite surface. Obviously, larger wear particles detached from the composite surface than from the pure UHMWPE. This was more pronounced for the UHMWPE/HDPE than for the CNT-reinforced composite.

The CNT composite had small transverse cracks inside the wear mark. These transverse cracks, only found on CNT-reinforced composites, suggest a larger contribution of surface fatigue wear.

Figure 9.36 shows nanotubes in the chips formed on the exit side of the wear mark on UHMWPE/HDPE with untreated CNTs. This indicates that the nanotubes can partly be properly dispersed in the UHMWPE/HDPE compounds by the kneader used. Nevertheless, nanotube agglomerates could also be found in the debris on the counterpart surface (Fig. 9.37).

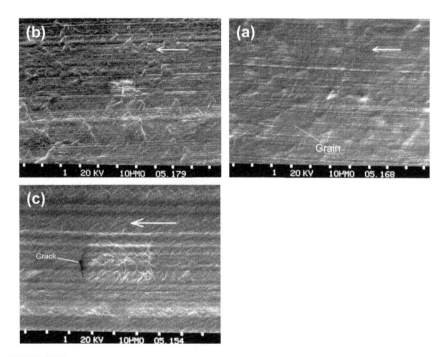

**FIGURE 9.35**

Details of the micrographs from Fig. 9.33. Worn surfaces of (a) UHMWPE, (b) blend, and (c) blend with CNTs. The white arrows indicate the sliding direction.

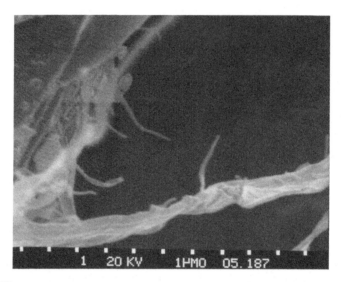

**FIGURE 9.36**

SEM micrograph: highly magnified detail of the chips formed on the exit side of the wear mark on the blend with 2 wt% untreated CNTs after sliding against X5CrNi18-10.

**FIGURE 9.37**

SEM micrograph: CNT agglomerate inside the wear debris on the surface of an X5CrNi18-10 steel ball after sliding against the blend including 2 wt% untreated CNTs.

## 9.4 CONCLUSIONS

Friction and wear are system properties and cannot be attributed to a material. The findings of this paper hold for unidirectional uniform dry sliding at low $pv$ values under normal atmosphere and should not simply be transferred to other service conditions. Under this restriction, some general conclusions may be drawn.

### 9.4.1 CNT-reinforced epoxy composites

The addition of CNTs to an EP matrix can significantly enhance its wear resistance. However, the performance of these composites sensitively depends on the pretreatment of the CNTs and the mixing procedure. Properly prepared EP/CNT samples form some sort of stable bearing needles of a few hundred nanometers diameter.

Simple stirring of untreated CNTs into the resin did not cause any improvement. Various methods were found to disperse CNTs effectively in the EP matrix:

- An activation of the CNT surface by boiling in nitric acid improves the wettability of the CNTs by the epoxy matrix. The composites produced with such pretreated CNTs yielded a better wear resistance. A short treatment time (2 h) proved better than a long treatment time (11 h) because the CNTs become simultaneously damaged and more prone to fracture during subsequent processing.
- Additional application of an epoxy-functionalized silane as coupling agent did not further improve the wear behavior of the compounds.
- An additional way to disperse the CNTs successfully in EP was the application of more sophisticated mixing methods. The worst results were achieved with samples prepared by the usual four-blade stirrer, better results were obtained with a three-roll mill, the best wear resistance was reached with samples prepared by a SpeedMixer. This SpeedMixer even allowed the preparation of samples with very high wear resistance without chemical pretreatment of the CNTs.
- Sonication also proved to be an effective way to disperse the CNTs in the matrix. CNTs pretreated in nitric acid, though, tend to break in the stress field of ultrasound. Sonication and activation in nitric acid should not be combined.
- TEM micrographs revealed that even samples with the lowest wear rates contained agglomerated CNTs as well as singularized CNTs. No clear correlation between dispersion state and wear resistance could be found. In contrast, the sample with least agglomeration (EP/CNT/PTFE) had a poor wear performance.

An optimal wear resistance was found for approximately 1 wt% CNT content, regardless of the dispersion and mixing method. The wear rate increase beyond this filler content was the smallest for the composite produced of untreated CNTs by the SpeedMixer.

The addition of dry lubricants had various effects on the wear of the CNT/EP composites. The addition of 10 wt% graphite decreased the specific wear rate significantly. It reduced the wear rate of the EP/CNT composite by a factor of almost 100 compared to the composite without graphite and >300 times compared with the pure EP. $MoS_2$ also decreased the wear rate, but to a lesser degree than the graphite. The EP/CNT filled with 10 wt% PTFE even had a slightly higher wear rate than the PTFE-free EP/CNT. CNTs on the one hand and graphite or $MoS_2$, respectively, on the other hand mutually aid one another with respect to the wear resistance of the composite. This synergetic effect is particularly pronounced for graphite.

The selection of an adequate counterpart material is even more important than the composition of the composite itself. All in all, unalloyed martensitic 100Cr6 steel caused far higher wear rates of the composites than did austenitic stainless steel X5CrNi18-10, and the differences between the composites became small. The sample containing 1 wt% CNTs and 10 wt% graphite delivered the by far best wear results ($k_s = 3.5 \times 10^{-8}$ mm$^3$ nm$^{-1}$) against the stainless steel but was extremely sensitive against the counterpart type, the wear rate against 100Cr6 was by approximately a factor of 1000 higher.

## 9.4.2 CNT-reinforced polyethylene composites

CNT-reinforced thermoplastic composites can be successfully produced by usual thermoplastic processing methods. The thermoplastic processability of UHMWPE could be sufficiently improved by adding 20% HDPE.

The composites tested had a better creep resistance than pure UHMWPE. However, it was primarily the HDPE content that improved the creep resistance of the composites. Further addition of CNTs has little effect or even lowers the creep resistance compared to UHMWPE/HDPE without CNTs, which delivered the best creep resistance. Pretreatment of the CNTs by boiling in concentrated nitric acid had a detrimental effect on the creep resistance because of some shortening of the CNTs during the etching process.

Pure UHMWPE had a slightly lower steady-state wear rate than the UHMWPE/HDPE blend without CNTs; the ranking order is opposite for the running-in phase. The UHMWPE/HDPE blend has also a better creep resistance.

The wear resistance of the UHMWPE/HDPE composite can be significantly improved by adding CNTs to the composite. As in the case of the epoxy composites, the wear rate passes a minimum and increases slightly as the CNT content further increases. Yet, the wear-reducing effect of the CNTs is less pronounced than in the case of the epoxy-based composites because the unreinforced PE matrix performs already much better than the pure EP.

Boiling in nitric acid as a pretreatment for CNTs cannot improve the wear performance of the composites. As in the case of the creep tests, the destruction of some CNTs may superimpose the beneficial effects of the pretreatment. Already the experiments on EP/CNT composites revealed that a combination of pretreatment in nitric acid and application of high stresses during dispersion may be unfavorable.

Very high shear forces must have been acting during the melt mixing of the PE composites because of the high melt viscosity of the UHMWPE/HDPE blend. However, the pretreatment had a positive effect in the case of the narrower nanotubes in PE.

As expected, the wear rates of most composites were higher against 100Cr6 than against X5CrNi18-10. In this context, CNTs obviously behave like carbon fibers or graphite, which catalyze tribocorrosion of counterparts that are not corrosion resistant. However, this effect was only small compared to the experiences with the EP/CNT composites and previously investigated microcomposites containing carbon fibers and/or graphite [54,55].

The wear rate of pure UHMWPE is lower against 100Cr6 than against X5CrNi18-10. The wear rate of pure UHMWPE against 100Cr6 was the lowest one among all combinations tested. This may be different at higher contact pressures or sliding speeds (i.e. at higher interfacial temperatures).

The CNTs were partly well dispersed but larger agglomerates still could be found. The dispersion of the CNTs in thermoplastics and the homogenization of the mixture need further improvement.

## Acknowledgments

The authors would like to thank some colleagues from the Fachhochschule Lübeck for their support and helpful discussions: Prof. B. Voß conducted the SEM investigations, Ms A.-Ch. Heidenreich and Dipl.-Ing. A. Frederich made the equipment available needed for the pretreatment in nitric acid. Altropol (Stockelsdorf, Germany) supplied the EP free of charge. The specimens produced with the three-roll mill were supplied by Dr Gojny and the TEM micrographs were made by Ms Dr Pegel, both from the Technical University Hamburg-Harburg.

The research was subsidized by the EU and the Schleswig-Holstein Ministry for science, economy, and traffic (programme Transferprojekte, Projekt "tribologisch optimierte Nanocomposites").

## References

[1]  M. Meyyappan (Ed.), Carbon Nanotubes, Science and Applications, CRC Press, Boca Raton, 2005.

[2]  S. Reich, Ch. Thomson, J. Maultzsch, Carbon Nanotubes, Basic Concepts and Physical Properties, Wiley-VCH, Weinheim, 2004.

[3]  C.A. Martin, J.K.W. Sandler, M.S.P. Shaffer, M.-K. Schwarz, W. Bauhofer, K. Schulte, A.H. Windle, Formation of percolating networks in multi-wall carbon-nanotube–epoxy composites, Composites Science and Technology 64 (2004) 2309.

[4]  R.S. Ruoff, D.C. Lorents, Mechanical and thermal properties of carbon nanotubes, Carbon 33 (1995) 925–930.

[5]  A. Allaoui, S. Bai, H.M. Cheng, J.B. Bai, Mechanical and electrical properties of a MWNT/epoxy composite, Composites Science and Technology 62 (2002) 1993–1998.

[6] F.H. Gojny, K. Schulte, Functionalisation effect on the thermo-mechanical behaviour of multi-wall carbon nanotube/epoxy-composites, Composites Science and Technology 64 (2004) 2303–2308.

[7] F.H. Gojny, M.H.G. Wichmann, U. Köpke, B. Fiedler, K. Schulte, Carbon nanotube-reinforced epoxy-composites: enhanced stiffness and fracture toughness at low nanotube content, Composites Science and Technology 64 (2004) 2363–2371.

[8] K. Schulte, F.H. Gojny, B. Fiedler, J.K.W. Sandler, W. Bauhofer, Carbon nanotube reinforced polymers: a state of the art review, in: K. Friedrich, S. Fakirov, Z. Zhang (Eds.), Polymer Composites from Nano- to Macro-Scale, Springer, New York, 2005.

[9] P. Pötschke, A.R. Bhattacharyya, A. Janke, Melt mixing of polycarbonate with multiwalled carbon nanotubes: microscopic studies on the state of dispersion, European Polymer Journal 40 (2004) 137–148.

[10] J. Sandler, M.S.P. Shaffer, T. Prasse, W. Bauhofer, K. Schulte, A.H. Windle, Development of a dispersion process for carbon nanotubes in an epoxy matrix and the resulting electrical properties, Polymer 40 (1999) 5967–5971.

[11] C. Park, Z. Ounaies, K.A. Watson, R.E. Crooks, J. Smith, S.E. Lowther, J.W. Connell, E.J. Siochi, J.S. Harrison, T.L. St. Clair, Dispersion of single wall carbon nanotubes by in situ polymerization under sonication, Chemical Physics Letters 364 (2002) 303–308.

[12] H. Xia, G. Qiu, Q. Wang, Polymer/carbon nanotube composite emulsion prepared through ultrasonically assisted in situ emulsion polymerization, Journal Applied Polymer Science 100 (2006) 3123–3130.

[13] A. Dasari, Z.-Z. Yu, Y.-W. Mai, Fundamental aspects and recent progress on wear/scratch damage in polymer nanocomposites, Materials Science and Engineering Reports 63 (2009) 31–80.

[14] Ch Roscher, J. Adam, Ch. Eger, M. Pyrlik, Novel radiation curable nanocomposites with outstanding material properties, Proceedings RadTech, 2002. April 28–May 1, 2002, Indiana.

[15] Q.-H. Wang, Q.-J. Xue, W.-M. Liu, J.-M. Chen, The friction and wear characteristics of nanometer SiC and polytetrafluorethylene filled polyetheretherketone, Wear 243 (2000) 140–146.

[16] Q. Wang, Q. Xue, W. Shen, J. Zhang, The friction and wear properties of nanometer $ZrO_2$-filled polyetheretherketone, Journal of Applied Polymer Science 69 (1998) 135–141.

[17] Q. Wang, J. Xu, W. Shen, W. Liu, An investigation of the friction and wear properties of nanometer $Si_3N_4$ filled PEEK, Wear 196 (1996) 82–86.

[18] O. Jacobs, B. Schädel, M. Cholewa, Verschleißminimierte Nano-Composites auf Epoxidharzbasis, Proceedings Tribologie-Fachtagung, Ges. f. Tribologie, 2004. September 27–29, 2004, Göttingen.

[19] O. Jacobs, B. Schädel, G. Rüdiger, Verschleißverhalten von Nano-$SiO_2$-gefülltem Epoxidharz, Tribologie und Schmierungstechnik 54 (2007) 19–24.

[20] G. Shi, M.Q. Zhang, M.Z. Rong, B. Wetzel, K. Friedrich, Sliding wear behavior of epoxy containing nano-$Al_2O_3$ particles with different pretreatments, Wear 256 (2004) 1072–1081.

[21] G. Shi, M.Q. Zhang, M.Z. Rong, B. Wetzel, K. Friedrich, Friction and wear of low nanometer $Si_3N_4$ filled epoxy composites, Wear 254 (2003) 784–796.

[22] O. Jacobs, W. Xu, B. Schädel, W. Wu, Wear behaviour of carbon nanotube reinforced epoxy resin composites, Tribology Letters 23 (2006) 65–75.

[23] H. Chen, O. Jacobs, W. Wu, G. Rüdiger, B. Schädel, Effect of dispersion method on tribological properties of carbon nanotube reinforced epoxy resin composites, Polymer Testing 26 (2007) 351–360.

[24] B. Dong, Z. Yang, Y. Huang, H.-L. Li, Study on tribological properties of multi-walled carbon nanotubes/epoxy resin nanocomposites, Tribology Letters 20 (2006) 251.

[25] A.B. Sulong, J. Park, N. Lee, J. Goak, Wear behavior of functionalized multi-walled carbon nanotube reinforced epoxy matrix composites, Journal of Composite Materials 40 (2006) 1947–1960.

[26] L.C. Zhang, I. Zarudi, K.Q. Xiao, Novel behaviour of friction and wear of epoxy composites reinforced by carbon nanotubes, Wear 261 (2006) 806–811.

[27] J.H. Lee, K.Y. Rhee, Silane treatment of carbon nanotubes and its effect on the tribological behavior of carbon nanotube/epoxy nanocomposites, Journal of Nanoscience and Nanotechnology 9 (2009) 6948–6952.

[28] Y.-K. Choi, Y. Gotoh, K. Sugimoto, S.-M. Song, T. Yanagisawa, M. Endo, Processing and characterization of epoxy nanocomposites reinforced by cup-stacked carbon nanotubes, Polymer 46 (2005) 11489–11498.

[29] Y. Xue, W. Wu, O. Jacobs, B. Schädel, Tribological behaviour of UHMWPE/HDPE blends reinforced with multi-wall carbon nanotubes, Polymer Testing 25 (2006) 221–229.

[30] Y.-S. Zoo, J.-W. An, D.-P. Lim, D.-S. Lim, Effect of carbon nanotube addition on tribological behavior of UHMWPE, Tribology Letters 16 (2004) 305–309.

[31] M.A. Samad, S.K. Sinha, Dry sliding and boundary lubrication performance of a UHMWPE/CNTs nanocomposite coating on steel substrates at elevated temperatures, Wear 270 (2011) 395–402.

[32] M.A. Samad, S.K. Sinha, Effects of counterface material and UV radiation on the tribological performance of a UHMWPE/CNT nanocomposite coating on steel substrates, Wear 271 (2011) 2759–2765.

[33] Z. Yang, B. Dong, Y. Huang, L. Liu, F.-Y. Yan, H.-L. Li, A study on carbon nanotubes reinforced poly(methyl methacrylate) nanocomposites, Materials Letters 59 (2005) 2128–2132.

[34] W.X. Chen, F. Li, G. Han, J.B. Xia, L.Y. Wang, J.P. Tu, Z.D. Xu, Tribological behavior of carbon-nanotube-filled PTFE composites, Tribology Letters 15 (2003) 275–278.

[35] J.R. Vail, D.L. Burris, W.G. Sawyer, Multifunctionality of single-walled carbon nanotube-polytetrafluoroethylene nanocomposites, Wear 267 (2009) 619–624.

[36] Z. Yang, B. Dong, Y. Huang, L. Liu, F.-Y. Yan, H.-L. Li, Enhanced wear resistance and micro-hardness of polystyrene nanocomposites by carbon nanotubes, Materials Chemistry and Physics 94 (2005) 109–113.

[37] H. Cai, F. Yan, Q. Xue, Investigation of tribological properties of polyimide/carbon nanotube nanocomposites, Materials Science and Engineering A 364 (2004) 94–100.

[38] A. Tanaka, K. Umeda, M. Yudasaka, M. Suzuki, T. Ohana, M. Yumura, S. Iijima, Friction and wear of carbon nanohorn-containing polyimide composites, Tribology Letters 19 (2005) 135–142.

[39] J. Kim, H. Im, M.H. Cho, Tribological performance of fluorinated polyimide-based nanocomposite coatings reinforced with PMMA-grafted-MWCNT, Wear 271 (2011) 1029–1038.

[40] C. Wang, T. Xue, B. Dong, Z. Wang, H.-L. Li, Polystyrene-acrylonitrile-CNTs nanocomposites preparations and tribological behavior research, Wear 265 (2008) 1923–1926.

[41] S. Wang, X. Chen, H. Song, S. Li, Y. Jin, Tribological behavior of multi-walled carbon nanotube/epoxy resin nanocomposites, Mocaxue Xuebao (Tribology) (China) 24 (2004) 387–391.

[42] H. Meng, G.X. Sui, G.Y. Xie, R. Yang, Friction and wear behavior of carbon nanotubes reinforced polyamide 6 composites under dry sliding and water lubricated condition, Composites Science and Technology 69 (2009) 606–611.

[43] B.B. Johnson, M.H. Santare, J.E. Novotny, S.G. Advani, Wear behavior of carbon nanotube/high density polyethylene composites, Mechanics of Materials 41 (2009) 1108–1115.

[44] H.J. Hwang, S.L. Jung, K.H. Cho, Y.J. Kim, H. Jang, Tribological performance of brake friction materials containing carbon nanotubes, Wear 268 (2010) 519–525.

[45] W.X. Chen, J.P. Tu, L.Y. Wang, H.Y. Gan, Z.D. Xu, X.B. Zhang, Tribological application of carbon nanotubes in a metal-based composite coating and composites, Carbon 41 (2003) 215–222.

[46] C.B. Lin, Z. Ch. Chang, Y.H. Tung, Y.-Y. Ko, Manufacturing and tribological properties of copper matrix/carbon nanotubes composites, Wear 270 (2011) 382–394.

[47] D.-S. Lim, J.-W. An, H.J. Lee, Effect of carbon nanotube addition on the tribological behavior of carbon/carbon composites, Wear 252 (2002) 512–517.

[48] T. Policandriotes, P. Filip, Effects of selected nanoadditives on the friction and wear performance of carbon–carbon aircraft brake composites, Wear 271 (2011) 2280–2289.

[49] I.-Y. Kim, J.-H. Lee, G.-S. Lee, S.-H. Baik, Y.-J. Kim, Y.-Z. Lee, Friction and wear characteristics of carbon nanotube–aluminum composites with different manufacturing methods, Wear 267 (2009) 593–598.

[50] H.J. Choi, S.M. Lee, D.H. Bae, Wear characteristic of aluminum-based composites containing multi-walled carbon nanotubes, Wear 270 (2010) 12–18.

[51] O. Jacobs, M. Kazanci, D. Cohn, G. Marom, Creep and wear behavior of ethylene–butene copolymers reinforced by ultra-high molecular weight polyethylene fibres, Wear 253 (2002) 618–625.

[52] O. Jacobs, N. Mentz, A. Poeppel, K. Schulte, Sliding wear performance of HDPE reinforced by continuous UHMWPE fibres, Wear 244 (2000) 20–28.

[53] Y. Shi, X. Feng, H. Wang, X. Lu, The effect of surface modification on the friction and wear behavior of carbon nanofiber-filled PTFE composites, Wear 264 (2008) 934–939.

[54] O. Jacobs, R. Jaskulka, F. Yang, W. Wu, Sliding wear of epoxy compounds against different counterparts under dry and aqueous conditions, Wear 256 (2004) 9–15.

[55] O. Jacobs, R. Jaskulka, Ch. Yan, W. Wu, On the effect of counterface material and aqueous environment on the sliding wear of various PEEK compounds, Tribology Letters 18 (2005) 359–372.

[56] Ph. Werner, V. Altstädt, R. Jaskulka, O. Jacobs, J.K.W. Sandler, M.S.P. Shaffer, A.H. Windle, Tribological behaviour of carbon-nanofibre-reinforced poly(ether ether ketone), Wear 257 (2004) 1006–1014.

# Tribological properties of carbon nanotube-reinforced composites

# 10

**Gong Qianming\*, Li Dan\*, Li Zhi\*, YI Xiaosu†, Liang Ji\***

*\*Department of Mechanical Engineering, Tsinghua University, Beijing, PR China, †Beijing Institute of Aeronautical Materials, Beijing, PR China*

## CHAPTER OUTLINE HEAD

**Tribology of Polymeric Nanocomposites.** http://dx.doi.org/10.1016/B978-0-444-59455-6.00010-6

## 10.1 **INTRODUCTION**

Carbon nanotubes (CNTs), which can be traced back to the discovery of the fuller-ene structure of C60 (buckyball) in 1985 [1], was first discovered by Iijima in 1991 [2]. CNTs have aroused big enthusiasm in the late years, and this could be attributed to their extraordinary mechanical, electrical and thermal properties. Many studies reported that CNTs possessed tensile modulus and strength as high as 1 TPa and 200 GPa, respectively [3]. Ruoff and Lorents estimated the tensile strength of CNTs with a wall thickness of 0.34 nm and diameter of 1 nm was about 20 GPa [4]. Yu et al. measured the tensile strength and modulus of multiwalled nanotubes, ranging from 11 to 63 GPa, and from 270 to 950 GPa, respectively [5]. In addition, thermal conductivity of CNTs is above 6000 W mK$^{-1}$ by theoretical calculation [6].

Despite of their unique properties, CNTs are limited in application in many fields due to some hindrance. For example, as enforcements in lots of composites, such as ceramic, resin and metallic composites [7–16], CNTs cannot be dispersed uniformly in composites. Moreover, in most cases, strong interfacial bonding cannot be realized in CNT-reinforced composites. However, because of their excellent self-lubricating ability like that of graphite and mechanical properties, CNTs can be used as abrasive materials in composites. In fact, whether in metal or in resin matrix composites, CNTs play an inspiring role in reducing wear.

This article is an attempt to study the friction and wear behavior of the aligned carbon nanotube (ACNT) and randomly dispersed CNT-reinforced composites and to explore the influence of the orientation and the proportion of CNTs in composites. Compared with the randomly dispersed CNT-reinforced composites, the ACNT-reinforced composites have few drawbacks since they have a more uniform dispersion and a higher proportion of CNTs. In addition, the weak interfacial bonding appearing in most metal, ceramic and resin matrix composites can be eliminated in carbon matrix composites fabricated by chemical vapor infiltration (CVI) process.

## 10.2 **MATERIALS**

### 10.2.1 **Carbon nanotube-doped carbon fiber/carbon composites**

Synthesis of CNTs was carried out in a vertical furnace with three suits of resis-tance coils as heating elements. Transition metal catalyst (Ni) was carried on diato-mite powder and then the powder was fixed in many ceramic plates. Before setting in the furnace, the plates were put on a shelf. The first step was to calcine the Ni-base catalyst at 450 °C in argon atmosphere for 30 min before being deoxidized by H$_2$ (10 l min$^{-1}$) at 500 °C for 45 min. At 650 °C, propylene was led into the furnace with a flow rate of 20 l min$^{-1}$ and the H$_2$ flow rate decreased to 5 l min$^{-1}$ at the same time.

Carbon fiber preform was prepared by stacking unidirectional long carbon fiber formed felt alternately at an angle of 45°(C/C-0), while the CNT-doped ones were prepared by two steps, that is, impregnating the felts in the suspension of pretreated CNTs first and then stacking the layers as in the first case (C/C-1 and C/C-2). CNTs accounted for about 5 wt% and 10 wt% for C/C-1 and C/C-2, respectively. The fiber used was commercial T700 PAN derived type with a diameter of 10 μm without high temperature treatment before the CVI process and its volume accounted for 35%. The initial density of the three preforms was 0.40–0.45 g cm$^{-3}$.

Subsequent densification of the three types of preforms was performed in an isothermal furnace by CVI technology at 1020 °C. Propylene, used as the carbon source in the flowing stream of nitrogen, infiltrated into the preform and was pyrolyzed on the substrate to fill the initial pores.

The morphology and microstructure of pyrocarbons in different composites were observed by polarized light microscopy (PLM) and scanning electron microscope (SEM) (LEO-1350 and S-3500N). Veckers microhardness was tested by an HMV-2 microhardness tester produced by SHIMADZU. Morphology of CNTs was investigated by a JEOL-200CX transmission electron microscopy (TEM). The graphitization degree and the crystallite size of CNTs and the resultant composites were determined by using X-ray diffractometer (XRD) (Philips, Model APD-10) using Cu Kα radiation ($\lambda = 0.15416$ nm). The interlayer spacing ($d_{002}$) was calculated from the Bragg's equation [17]. The graphitization degree (G, %) was calculated from the Maire and Merings equation [18].

Friction and wear tests for the evaluation of the effect of CNTs on tribological properties of C/C composites were carried out by using a ring-on-block M-2000 type wear and friction tester under ambient conditions. The configuration of the tester was similar to the schematic diagram in Ref. [19]. The size of the C/C composites specimen was 20 mm × 12 mm × 6 mm. A stainless steel (40Cr) ring coated with a film of Cr was used as the counterpart. Its size was Φ40 mm × 12 mm and the ring came into contact with C/C samples on the plane 20 mm × 12 mm tangentially at the beginning. Tests were carried out at a rotation rate of 200 rpm under the load of 50 N and 200 N contrastively in atmosphere.

Raman spectroscopy (Renishaw RM2000) and energy-dispersive spectroscopy (EDS) were used to investigate the wear surface besides SEM observations. The wear debris was investigated by SEM as well as the carbon fiber (Fig. 10.1(b)).

After densification by the CVI process, pyrocarbon that grew around carbon fibers or CNTs exhibits a different morphology (Fig. 10.2). Pyrocarbon in pure C/C composites is typically smooth lamella type and pyrocarbon in CNT-doped C/C composites (C/C-1 and C/C-2) is generally rough lamella type. There are no cracks between carbon fibers and pyrocarbons and few CNTs are pulled out in CNT-doped composites (Fig. 10.3). Table 10.1 shows some basic properties of C/C-0, C/C-1, and C/C-2. Corresponding to the difference in microstructure, C/C-doped composites exhibited much higher graphitization degree and rather lower hardness. The negative values of graphitization degree are meaningless in physics; nevertheless, they are valuable to distinguish the degree of perfectness of graphitic structure in the three samples.

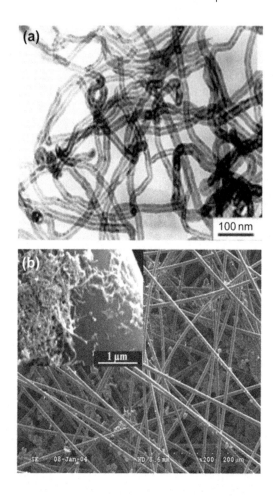

**FIGURE 10.1**

(a) TEM micrograph of CNTs and (b) SEM micrograph of CNT-doped carbon fiber preform.

## 10.2.2 Randomly dispersed carbon nanotube/epoxy composites

The epoxy resin used was bisphenol A epoxy resin (E54) and the related curing agent was dapsone (DDS, 33 phr). Following are their molecular formula:

E54 epoxy:

DDS:

**FIGURE 10.2**

(a) PLM photographs of pure C/C composites and (b) CNT-doped C/C composites.

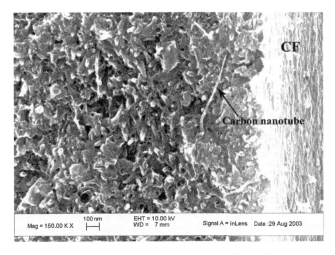

**FIGURE 10.3**

SEM micrograph of the cross-section of CNT-doped C/C composites.

E54 epoxy resin was first mixed with a certain proportion of CNTs in a three-roll mill. The black mixture was degassed at 130 °C in vacuum for 1.5 h, then cast in mold, and finally followed by a curing process at 180 °C for 3 h to produce CNT/epoxy composites.

The density of E54 epoxy was 1.21 g cm$^{-3}$ and the density of the produced composites was about 1.20–1.30 g cm$^{-3}$. Figure 10.4 shows the morphology of the fracture.

**Table 10.1** Basic Properties of Pure C/C Composites and CNT-Doped C/C Composites

| Samples | Density (g cm⁻³) | $d_{002}$ (nm) | G (%) | HV[a] |
|---------|-----------------|----------------|-------|-------|
| C/C-0 | 1.56 | 0.3481 | −47.67 | 72.8 |
| C/C-1 | 1.54 | 0.3436 | 14.65 | 52.1 |
| C/C-2 | 1.53 | 0.3422 | 20.93 | 41.5 |

[a]Vickers microhardness of pyrocarbon in different samples.

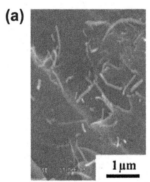

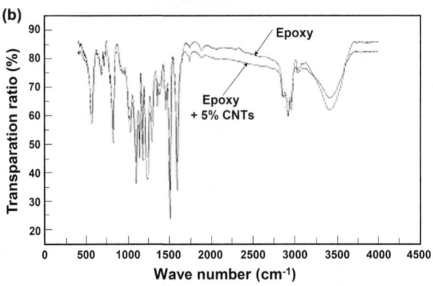

**FIGURE 10.4**

(a) SEM morphology and (b) Fourier transform infrared spectra of CNT/epoxy composites.

No big pores in the surface verified a successful curing process. While it can be noted that the CNTs have been pulled out for several hundreds of nanometers to nearly 1 μm, it inferred that interfacial bonding between CNTs and the resin was weak. In fact, there is no difference in infrared adsorption spectra of the epoxy resin and CNT/epoxy composites, which indicates that there is no chemical bonding between CNTs and the epoxy resin.

### 10.2.3 Well-ACNT/epoxy composites

The synthesis process of well aligned CNT/epoxy composites was carried out in a tubular furnace. Xylene, carried by argon and hydrogen, was pyrolyzed and deposited on quartz substrate under the catalysis of ferrocene at 700–800 °C. Figure 10.5 showed the morphology of produced ACNTs. ACNTs grew vertical to the quartz substrate (Fig. 10.1(a)) forming a preform 4–6 mm thick. The diameters of ACNTs range from 40 to 60 nm (Fig. 10.1(d)). The initial density of ACNT felt was 0.10–0.15 g cm$^{-3}$.

In ACNTs preform, the nanotubes grew closely to each other and dispersed evenly to form a regular-shaped porous block, which was difficult to put into application due to weak bonding among the nanotubes. Since the pores in ACNT felts were very small (on average 113 nm in diameter), it was obvious that resins with high viscosity were not suitable as matrix because they could not infiltrate into the felts. Epoxy resin, on the one hand, was wettable to CNTs and thus good interfacial bonding could be expected. On the other hand, ACNT/epoxy composites could be a good contrast to randomly dispersed CNT/epoxy composites because of the directionality of CNTs. As a result, epoxy was adopted to fill in the apertures of ACNTs with DDS as a curing agent.

SEM observation showed that CNTs were well aligned in the matrix and no deformation arose in the ACNTs (Fig. 10.6). The bared ends of CNTs could be discerned after Ar-ion etching of the upper side (Fig. 10.6(b)). The density of the produced ACNT/epoxy composites was 1.20–1.30 g cm$^{-3}$.

### 10.2.4 Well-ACNT/carbon composites

Different from epoxy resin, gas hydrocarbon could infiltrate into any open pores in a certain vacuum. So each CNT was coated with a layer of pyrocarbon after the CVI process without destroying the directionality of the ACNTs (Fig. 10.7). Since self-lubrication is the common characteristic of carbon materials, more excellent tribological properties could be expected for the ACNT-reinforced carbon matrix composites. Densification process was similar to that described in Section 10.2.1. After one cycle of CVI densification process (about 40 h), the resultant density was 1.30–1.50 g cm$^{-3}$.

## 10.3 EXPERIMENTS

### 10.3.1 Friction and wear tests

Friction and wear tests for CNT-doped C/C composites were carried out by using a ring-on-block M-2000 type wear and friction tester under ambient conditions.

**(a)** Macroscopic image

**Aligned carbon nanotubes (ACNTs)**

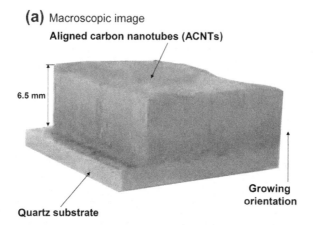

6.5 mm

**Growing orientation**

**Quartz substrate**

**(b)** Microscopic morphology

**FIGURE 10.5**

(a) Image of aligned carbon nanotubes by digital camera and (b) corresponding SEM morphology. For color version of this figure, the reader is referred to the online version of this book.

The configuration of the tester was similar to the schematic diagram in Ref. [19]. The size of the C/C composites specimen was 20 mm × 12 mm × 6 mm. A stainless steel (40Cr) ring coated with a film of Cr was used as the counterpart. Its size was $\Phi 40$ mm × 12 mm and it contacted the C/C samples on the 20 mm × 12 mm plane. The axis of the ring was vertical to carbon fiber layers. Tests were carried out at a rotation rate of 200 rpm (at the surface speed of about 0.42 m s$^{-1}$) under the load of 50 N and 200 N in normal atmosphere. The test duration was 65 min. The frictional torque was recorded every 5 min in order to calculate the coefficient of friction. At the end of the test, the wear was

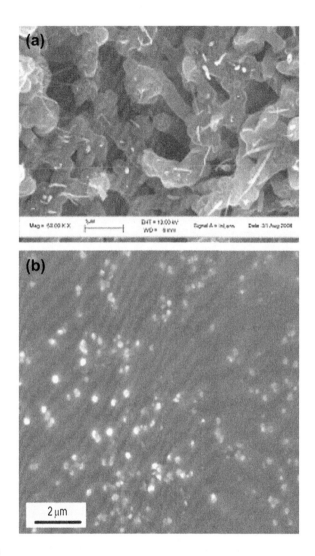

**FIGURE 10.6**

SEM morphology of ACNT/epoxy composites. (a) Intersection parallel to the axis of aligned carbon nanotube. (b) Upper surface.

measured using an analytical balance with an accuracy of 0.1 mg. Prior to the test, the surface of the specimens and the counterpart were rubbed with No.1500 abrasive paper before final polish, and then cleaned with acetone in an ultrasonic bath.

Friction and wear tests for CNT/epoxy composites were similar to that for CNT-doped C/C composites, which were carried out with MM-200 type tester. The size of CNT/epoxy composites sample was 10 mm × 6.5 mm × 8 mm and that of its counterpart, which was made of 45# steel, was Φ40 mm/Φ16–8 mm. The initial contact

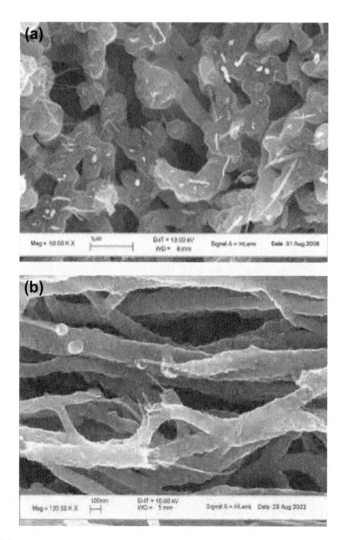

**FIGURE 10.7**

SEM morphology of aligned carbon nanotube/carbon composites. (a) The cross-section. (b) The side that is parallel to the axis of CNTs.

between the ring and block was tangentially shown in Fig. 10.8. The load was 50 N, rotation speed was 200 rpm and time duration was 30 min.

For the friction and wear tests on ACNT/C composites at high temperature, the whole test process was different. The tests were carried out with HT-500 type high-temperature friction and wear tester. Ball-disc mode was selected (Fig. 10.9). In order to maintain stable size, strength and hardness at high temperature, silicon nitride ($Si_3N_4$) ball ($\Phi4$ mm) was adopted as counterpart that was held still while

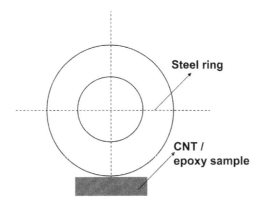

**FIGURE 10.8**

Configuration about ring-block mode for CNT/epoxy composites.

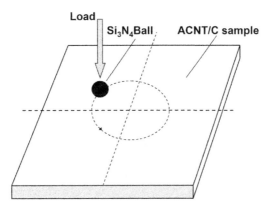

**FIGURE 10.9**

Configuration about ball-disc mode for ACNT/C composites.

ACNT/C sample was rotated (2000 rpm) by an engine. All the tests were fulfilled in ambient atmosphere and testing temperature ranged from room temperature to 500 °C. Test duration was 60 min and the load applied was 200 g.

## 10.3.2 Microstructure and related properties

The morphology of pyrocarbons in C/C composites was observed with polarized light microscope (PLM). Microstructure observations of all the composites were achieved by scanning electron microscopy (SEM) and Vickers microhardness of C/C composites was measured with the HMV-2 microhardness tester. The morphology of CNTs was investigated by TEM (same as that described in Section 10.2.1).

The graphitization degree of CNTs and the resultant C/C, CNT/C/C and ACNT/C composites was determined by using X-ray diffraction (XRD) analysis (Philips, Model APD-10) with Cu Kα radiation ($\lambda = 0.15416$ nm), which has been described in Section 10.2.1.

Raman spectroscopy and X-ray energy dispersive spectrum (EDS) were used to analyze wear surfaces of C/C and CNT/epoxy composites in addition to SEM analyses. Raman spectra were taken at room temperature under ambient conditions using a Renishaw RM2000 microscopic confocal Raman spectrometer in back scattering mode using Ar-ion laser excitation (633 nm, 5 mW; resolution 1 cm$^{-1}$). The integration time was 30 s and the spectra were averaged over three accumulations. Wear debris collected on the conducting adhesive tape was investigated by SEM as well.

## 10.4 THE FRICTION AND WEAR PROPERTIES OF CARBON NANOTUBE-REINFORCED COMPOSITES

### 10.4.1 The friction and wear properties of carbon nanotube-doped carbon/carbon composites

Figures 10.10 and 10.11 show the variations of friction coefficients of the three kinds of composites under 50 N and 200 N, respectively. Both the friction coefficients and the wear rates in a steady state sliding decrease with an increase in the ratio of CNTs in carbon fiber performs, as given in Fig. 10.10. Such change trends are in agreement with previous work. In fact, some former researchers ascribed the decrescent coefficients to the released CNTs, which not only hindered direct contact between counterparts, but also served as lubricants in the wear surface [19–21].

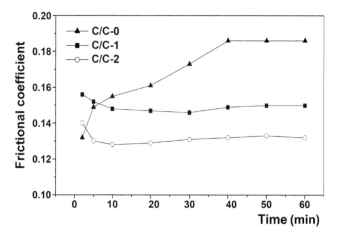

**FIGURE 10.10**

Variations of frictional coefficient as a function of time for the three kinds of C/C composites under 50 N.

However, in this study, different from that under 50 N, the change trend in frictional coefficient under 200 N displays reversal regularity while the change in the wear rates (Fig. 10.12) keeps the same (see Figs 10.10 and 10.11).

From Figs 10.10 and 10.11, it can be concluded that friction coefficients of C/C-2 remain in the range of 0.12–0.13 and keep as constant during the sliding up to a high load, while those of C/C-0 and C/C-1 decrease significantly. Thereby, the friction stability of CNT-doped C/C composites is much better. At this point, it is worth noting that there would be a different friction mechanism instead of the ones described above [19–21].

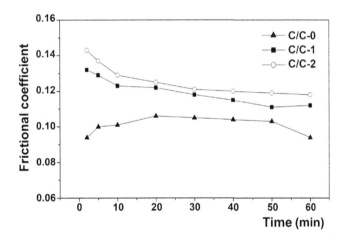

**FIGURE 10.11**

Variations of frictional coefficient as a function of time for the three kinds of C/C composites under 200 N.

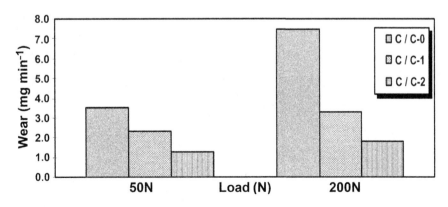

**FIGURE 10.12**

Variations of wear loss as a function load for the three kinds of composites.

### 10.4.1.1 The effect of load on frictional coefficients of C/C composites

Figure 10.13 shows the morphology of the worn surface of sample C/C-0 and C/C-2 under different loads, which would offer further information to explain the variation of the friction coefficient. Under a load of 50 N, the friction film of C/C-0 is composed of loosely agglomerated debris and bared carbon fibers show up on the worn surface. However, the friction film of C/C-2 mainly consists of compacted debris and no evident pores exist. Increased actual contact area caused by the friction film denotes improved lubrication between the counterparts due to the special graphitic structure of carbon materials [22]. Thus, the friction coefficient of C/C-2 decreases. When the load increases, some big granular or flaky debris on the wear surface would be crushed or sheared into smaller particles or thinner flakes. Concomitantly, the newly formed debris would come into being an integrated but thin friction film (Figs 10.13(b) and (d)). As a result, the small debris and integrated but thin friction film would lead to a small friction coefficient because of the decreased degree of two-body or three-body abrasive wear [23]. Since the decrease in the friction coefficients of the three samples under a high load is inevitable and the friction coefficients of C/C-2 keep constant under an increased load, it is interesting to discuss the influence of CNTs on the tribological properties of C/C composites.

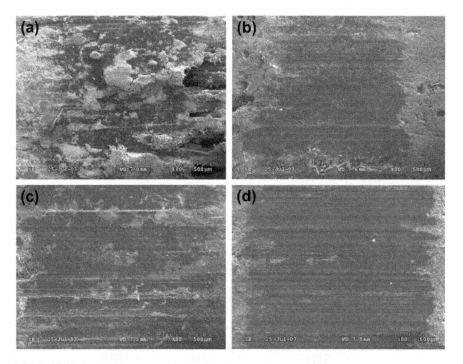

**FIGURE 10.13**

SEM micrographs of worn surface of C/C-0 and C/C-2 under different loads. (a) C/C-0 under 50 N and (b) 200 N; (c) C/C-2 under 50 N and (d) 200 N.

### 10.4.1.2 The effect of CNTs on frictional coefficients of C/C composites

Since the doped CNTs in C/C composites are covered with pyrocarbon, no bare CNTs have been observed on the worn surface except limited pyrocarbon-coated CNTs (about 200–400 nm in diameter, arrows in Fig. 10.14(b)). Correspondingly, the influence of the tribological properties on C/C composites would be in two ways. The influential mechanism is different from those mentioned in the preliminary works [19–21,24].

On the one hand, because of the high strength of CNTs and good bonding between pyrocarbon and CNTs, pryocarbon-coated CNTs that broke down to the interface can keep their shape and size under higher loads; such unaltered debris would bring about little change for friction coefficient consequentially. This reinforcing effect is similar to the effect of released CNTs in other composites [19,24].

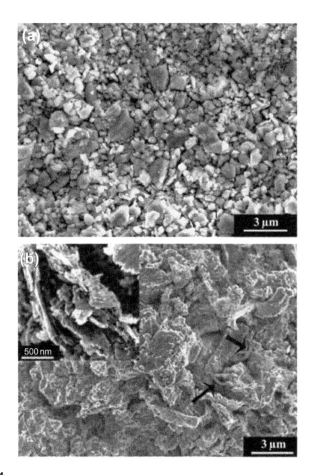

**FIGURE 10.14**

SEM observations of agglomerated wear debris of (a) C/C-0 and (b) C/C-2 under 200 N.

On the other hand, in Table 10.1, it can be drawn out that CNT-doped C/C composites possess higher graphitization degree and lower Vickers microhardness for pyrocarbons, which implies better graphitic structure for CNT-doped samples. Once the counterparts fall into contact, the microprotuberance in the wear surface of C/C-2 would be sheared into small flakes swiftly to form a layer of lubricative friction film. So the first points for CNTs-doped composites under the two loads are all just a little higher than the following points. At the same time, because of better graphitic structure, new debris can be easily and consecutively produced by shearing new protuberance once the old friction film delaminated. This mechanism might be the reason for better stability of CNT-doped samples. Based on this mechanism, higher loads that can bring about a little smaller debris and thinner friction film would only affect the friction coefficients minimally.

On the contrary, due to higher hardness and worse graphitic structure for C/C-0, the microprotuberance could not be crushed immediately and, once the counterparts meet each other, it is possible that the initial contact area is limited by the contact of the limited protuberance and in turn results in low frictional coefficients under both loads. With sliding going on, the microprotuberance broke down, but the load of 50 N could not crush them into smaller debris; the particles that formed friction media might bring about abrasive wear inevitably. High frictional coefficients are then concomitant with that kind of friction mode naturally. Even under higher loads (200 N), the debris of C/C-0 are still in particle shape (Fig. 10.14(a)), while that of C/C-2 are rolled or nonroller thin flakes with some pores (Fig. 10.14(b)). On the one hand, the porous structure would adsorb more vapor and oxygen during sliding, which can produce another lubricative film to bring about lower friction coefficients and wear loss [25]. On the other hand, debris of C/C-2 with a porous structure has larger specific surface area; it can adsorb more friction energy and result in better wear resistance.

Another interesting phenomenon was the crystallographic transformation of wear debris on the worn surface, as demonstrated in the Raman spectra (Figs 10.15 and 10.16). Two peaks are observed in the Raman spectra. The peak at 1580 cm$^{-1}$ denotes the active in-plane stretching $E_{2g}$ mode of graphite structure, known as the G-peak. The peak at 1360 cm$^{-1}$ denotes a disorder-induced mode due to a breakdown in translational symmetry by the microcrystalline structure [26]. Distinctive changes in Raman spectra are observed between unworn and worn surfaces of all the three kinds of samples, especially sample C/C-0 under 200 N. As shown in Fig. 10.15, an increase in the load on the sample C/C-0 results in an increased intensity and a decreased half width at half maximum of the G-peak and a decreased intensity of the D-peak. The two peaks in Raman spectrum are sensitive to the three-dimensional stacking of graphene sheets in the carbon materials. A decrease in the intensity of the D-peak and an increase in the $1/R$ value (where $R$ is the ratio of the integrated intensity of the G-peak to the integrated intensity of the D-peak) indicate a reduction in defects and a well-ordered graphitic structure [27]. Such a graphitization process might be brought about by the strain from the shear

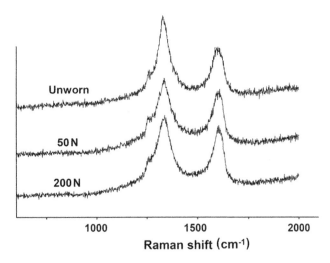

**FIGURE 10.15**

Raman spectra of unworn and worn surfaces of C/C-0 under 50 N and 200 N.

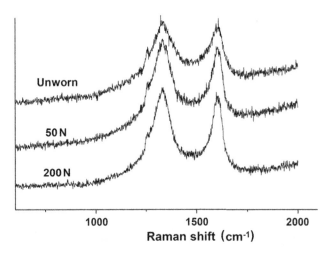

**FIGURE 10.16**

Raman spectra of unworn and worn surfaces of C/C-2 under 50 N and 200 N.

deformation on the wear surface [28]. However, compared with the aircraft braking process [28], the power density and interfacial temperatures are rather lower; hence the reason for the graphitization of wear debris deserves further research. In any case, a graphitic structure of the wear debris under a high load will lubricate counterparts efficiently, which would be a further reason for the decrease in the friction coefficient.

## 10.4.2 **Friction and Wear Properties of Randomly Dispersed Carbon Nanotube/Epoxy Composites**

Figure 10.17(a) shows the relationship between the wear loss, the friction coefficient and the ratio of CNTs added in the epoxy composites. When the ratio of CNTs is less than 5 wt%, the friction coefficients varies little and in the range of 0.45–0.50. As the ratio of CNTs increases to 10 wt% and 20 wt%, there is an obvious decrease in the friction coefficient and those values for CNT/epoxy composites drop to 0.31 and 0.25, respectively.

The change in the wear loss exhibits a different trend (Fig. 10.17(b)). There is a minimum value as the ratio of CNTs is 10 wt%, where the wear loss of CNT/epoxy

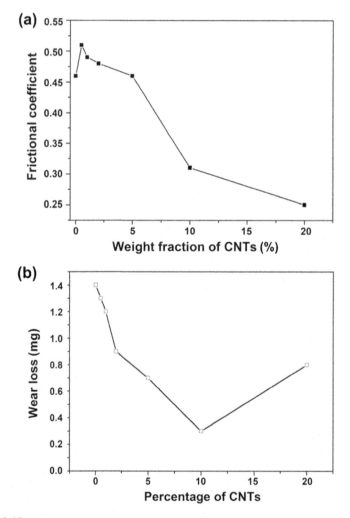

**FIGURE 10.17**

(a) Frictional coefficient and (b) wear loss of carbon nanotube/epoxy composites.

composites is about 1/5 of that of the pure epoxy resin. Till the ratio of CNTs increases to 20 wt%, the wear loss of CNT/epoxy resin is still lower than that of the pure resin.

There are discrepancies in the experimental results of the influence of CNTs on the friction coefficients and the wear loss. Microstructures of the worn surface (Fig. 10.18) might give the clues. Although the surfaces of all the samples were polished, surface topography indicates that the surfaces of the tribological contacts are covered with a certain height of asperities (Figs 10.18(a),(c),(e) and (g)). As a result, at the beginning of the contact, the summation of individual contact spots gives the real area of contact much smaller than the apparent geometrical contact area. However, after 30 min, the worn surfaces of samples with different ratios of CNTs exhibit a similar morphology.

In case that the strength and harness of steel are much higher than those of epoxy resin, once the two sliding surfaces contact, the asperites on the surface of the virgin epoxy resin would deform elastically or plastically under the given load. Some relatively higher asperities might be ruptured by the shear force between the two sliding surfaces and the ruptured debris would fall off or deform as flat film on the worn surface. As a result, the real area of contact increases significantly and a smooth worn surface is shaped (Fig. 10.18(b)). Then, it is reasonable to deduce that, with the increase in the contacting area, theoretically, the attractive interaction forces between the two contacting surfaces increase rapidly. For example, all those types of interactions that contribute to the cohesion of solids, such as metallic, covalent, and ionic, i.e. primary chemical bonds (short-range forces) as well as secondary van der Waals bonds (long-range forces), although might not appear simultaneously, might increase rapidly. Thus the increase in the friction coefficients results from a strong interaction.

Similar to the traditional carbon materials, CNTs have excellent self-lubricating ability besides their high modulus. However, the added CNTs could not form a uniform self-lubricating film on the sliding surface; in other words, the friction coefficient is largely determined by the matrix resin. Since CNTs are rather wearable, some scattered fractured CNTs protrude from the surface (Fig. 10.18(d)). Wearable properties of CNTs can be verified by EDS analyses (Fig. 10.19). The results indicate that limited CNTs (5 wt%) result in a much higher ratio of Fe element in the worn surface of CNT/epoxy composites compared with pure epoxy resin. When the ratio of CNTs is increased to 10 wt%, a smooth frictional film is formed between the two sliding surfaces (Fig. 10.18(f)) and acts as a self-lubricating media during sliding.

Comparing the unworn and worn surfaces of the sample with 20 wt% CNTs added, we could notice no big difference between the two surfaces and a layer of loose wear debris with more CNTs covering on the worn surface. A much better self-lubricating ability implies a low frictional coefficient.

In summary, the influence of CNTs on wear loss is coherent to that on the frictional coefficient. Frictional film plays a significant role in reducing the wear loss (Fig. 10.20). However, it should be noted that when the ratio of added CNTs exceeds a certain limit, the cross-linked structure of the resin matrix would be partially destroyed; thus, the shearing strength of the composites would decline remarkably. It might be the reason that 20 wt% of CNTs brought about a high wear loss. The weakened anti-delamination of the resin matrix leads to a rapid falling-off of the wear debris, and accordingly, the transferred Fe from the counterpart would escape with

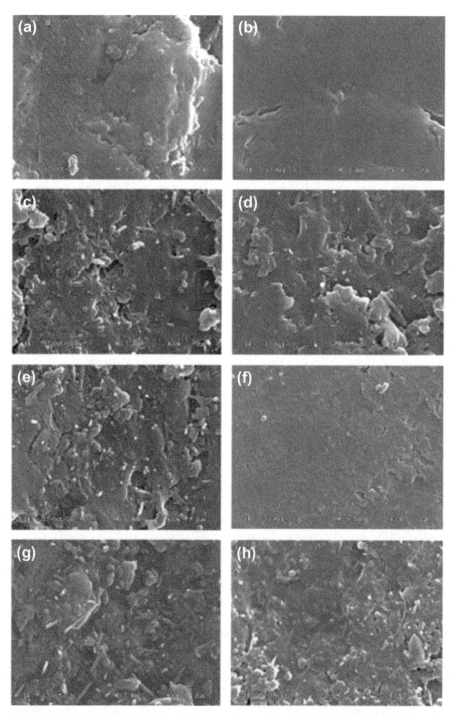

**FIGURE 10.18**

Contrastive SEM observations of unworn (a, c, e, g) and corresponding worn surfaces (b, d, f, h) of epoxy resin (a, b), CNT/epoxy composites with 5 wt% CNTs (c, d), 10 wt% (e, f) and 20 wt% (g, h).

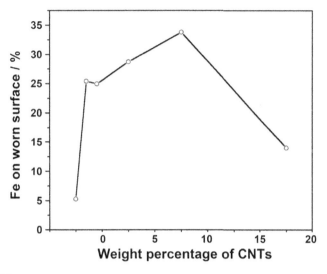

**FIGURE 10.19**

Variation of Fe content in the worn surface with different percentages of CNTs in composites.

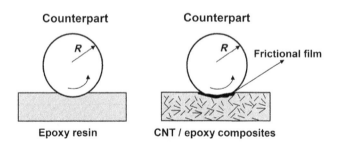

**FIGURE 10.20**

Wear mechanism of epoxy resin and CNT/epoxy composites.

the fallen debris, which is corresponding to the relatively low Fe content on the worn surface of CNT/epoxy composites with 20 wt% CNTs (Fig. 10.19).

### 10.4.3 Friction and Wear Properties of Well-ACNT/Epoxy Composites

In order to further explore the effect of CNTs on the tribological properties of epoxy resin, ACNTs were introduced as reinforcement in the composites. The friction and wear test conditions were the same as those performed in Section 10.4.2. Figure 10.21 shows the variations of the friction coefficients of different samples with time duration.

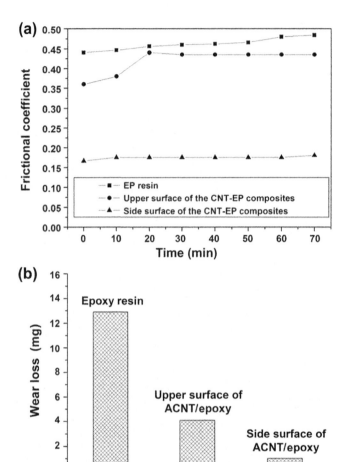

**FIGURE 10.21**

Friction and wear properties of ACNT/epoxy composites. (a) Variation of frictional coefficient with time; (b) comparison of wear loss.

As shown in Fig. 10.21, both the friction coefficients and the wear loss of the side surface are much lower than those of the pure epoxy resin and the upper surface of ACNT/epoxy resin. Calculations indicate that CNTs account for about 5 wt% in the composites and this value is close to that in randomly dispersed CNT/epoxy composites. In addition, the alignment of CNTs leads to rather different tribological properties. When the upper surface worked as the wearing surface, the real contact area between CNTs and the steel counterpart is limited. Because of the high strength, CNTs were pulled out instead of being cut into fractured short segments (Fig. 10.22(a)), and only a part of the CNTs were left on the worn surface as bared ends (inset of Fig. 10.22(a)). Consequently, the friction coefficients between the upper surface and the steel counterpart are approximate to those of the pure epoxy resin.

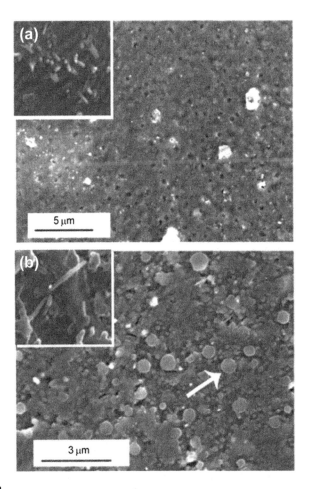

**FIGURE 10.22**

SEM observations of the worn upper surface (a) and side surface (b).

Contrastively, friction and wear between the side surface and steel are quite different, and the real area of contact between CNTs and the counterpart is obviously larger (Fig. 10.23); hence, lower frictional coefficients and wear loss can be expected, which is confirmed by experimental results. Interestingly, on the side worn surface, a lot of spherical particles appeared (Fig. 10.22(b)) and between the particles, there were some bared ends or fractured CNTs (inset of Fig. 10.22(b)) (Table 10.2).

EDS analyses show that the Fe element exists mainly in the spherical particles (Fig 10.24). We propose that CNTs are distributed in parallel on the side surface, and they could be peeled off easily under the shearing force between the sliding surfaces and then reunited as spherical agglomerations. The CNT-composed particles are much wearable and harder than epoxy resin; thereby, Fe from the counterpart mainly congregates on these particles. Further work is needed to confirm this hypothesis.

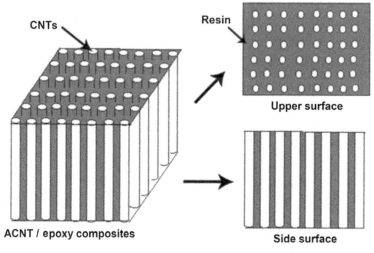

**FIGURE 10.23**

Sketch about projection of CNTs on the upper surface and side surface.

**Table 10.2** Quantitative Analyses of Different Elements in the Worn Side Surface

|  | C | O | Na | S | Fe | Total |
|---|---|---|---|---|---|---|
| Spherical particles | 59.52 | 31.41 | 0.47 | 1.18 | 7.41 | 100 |
| Area without particles | 75.22 | 22.89 | 0.00 | 1.88 | 0.00 | 100 |

### 10.4.4 Friction and Wear Properties of Well-ACNT/Carbon Composites

The traditional carbon materials have been applied in many fields as self-lubricating or sealing materials. For example, C/C composites have been developed as an international standard material for aircraft brake discs. Compared with carbon fibers, CNTs exhibit better mechanical and thermal properties. In this research, ACNT/C composites have been fabricated via the CVI process to probe their tribological properties because the CVI process densifies the preform efficiently and results in strong interfacial bonding between CNTs and matrix carbon.

Figure 10.25 shows the results of friction and wear tests for ACNT/C and C/C composites. At room temperature, frictional coefficient of ACNT/C composites is about 0.20, a little higher than that of C/C composites (~0.15). With an increase in testing temperature, the frictional coefficients fluctuate greatly. At 250 °C, the frictional coefficient of ACNT/C composites remains in the range of 0.10–0.15, while

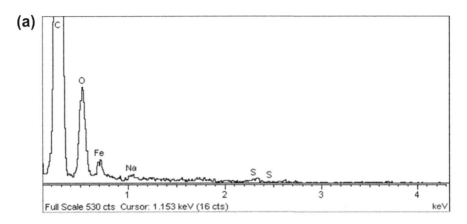

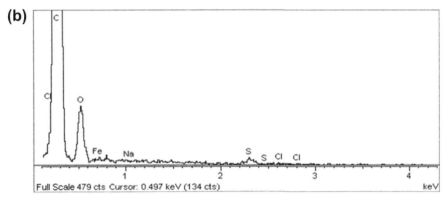

**FIGURE 10.24**

EDS analyses of the worn surface of ACNT/epoxy resin. (a) Spherical particle that the arrow indicated; (b) the area without spherical particles.

that of C/C composites increases to 0.70–0.80. At 500 °C, the friction coefficient of the two composites is in the same range of 0.45–0.60. In general, the small change in the frictional coefficients of ACNT/C composites from room temperature to 350 °C (Fig. 10.25(d)) implies more stable tribological properties of ACNT/C composites than C/C composites (Figs 10.26–Fig. 10.28).

The width of the friction track was adopted as a criterion for comparing the wear loss qualitatively since the wear loss is too small to quantify. The wear loss of ACNT/C composites is much lower than that of C/C composites at room temperature and 250 °C, but there is no big difference at 500 °C.

At room temperature, the asperities on the sliding surface were fractured under load and crushed into small pieces. In this case, "plowing friction" happened to some degree and that resulted in relatively high frictional coefficients. Comparatively, the flaky debris and smooth surface without big pores resulted in stable frictional coefficient and low wear loss for ACNT/C composites

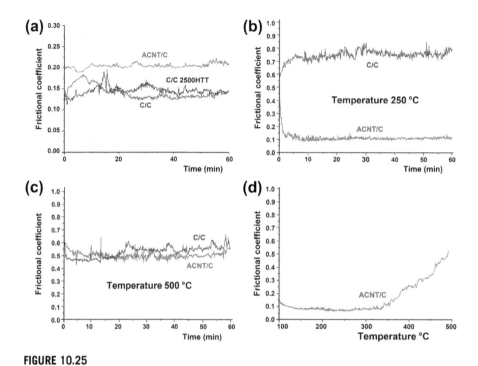

**FIGURE 10.25**

Frictional coefficients of C/C and ACNT/C composites at different temperatures. (a) At room temperature; (b) at 250 °C; (c) at 500 °C; (d) from room temperature to 500 °C. For color version of this figure, the reader is referred to the online version of this book.

(Fig. 10.29). At 250 °C, adsorbed water or vapor, which could work as lubricant during friction, was desorbed. As a result, the frictional coefficient and wear loss would increase in large scale. At the same time, the frictional film, if formed, would play a dominant role in the lubricating frictional process. On this account, perfect frictional film from ACNT/C composites leads to rather low and stable frictional coefficients (Fig. 10.30(a)). In this case of C/C composites, there was no uniform film formed to cover the worn surface, which was verified by the variation of brightness in different areas of the friction track, so the lubricating effect of the fragmentary film could not compensate for the effect brought about by the desorption of water or vapor on the worn surface. Plowing friction and adhesion friction may coexist for C/C composites. As shown in Fig. 10.30(b), the friction film on the worn surface of C/C composites was composed of layers of delaminated film and big cracks, which implied the fluctuation of the friction process. As a result, the friction coefficients for C/C composites are rather high and unstable (Fig. 10.25(b)).

At 500 °C, the oxidation occurred on the worn surface, especially on the nanosized debris on the film and interfaces between reinforcements and matrix, which was confirmed by the wider friction tracks for both samples (Fig. 10.28)

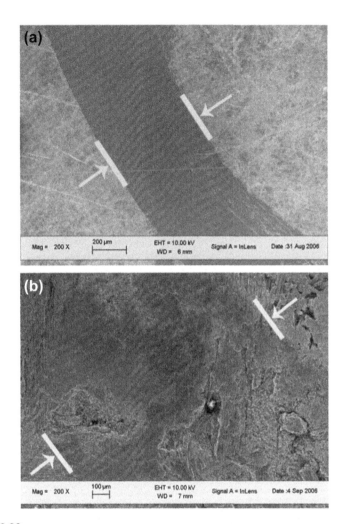

**FIGURE 10.26**

SEM images of the worn surfaces of ACNT/C and C/C composites tested at room temperature. (a) ACNT/C composites; (b) C/C composites.

and the cracks in the worn surface (Fig. 10.31). So in this case, the frictional film would be destroyed partially by oxidation and thus rather higher frictional coefficients arose for ACNT/C composites. However, for C/C composites, in a sense, oxidation might accelerate the process of film formation on the entire worn surface (Figs 10.27(b) and 10.28(b)). Accordingly, a more integrated frictional film led to lower frictional coefficients for C/C composites. Since the occurrence of oxidation, tribochemical wear mechanism should be a second contribution to the wear loss of the two samples at 500 °C besides the friction film fatigue and delamination wear mechanism.

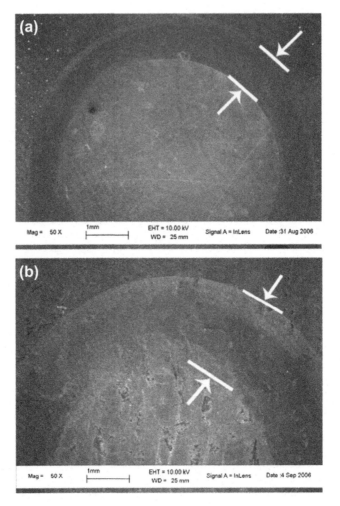

**FIGURE 10.27**

SEM images of the worn surfaces of ACNT/C and C/C composites tested at 250 °C. (a) ACNT/C composites; (b) C/C composites.

## 10.5 SOME COMMON CHARACTERISTICS ABOUT WEAR MECHANISMS OF CARBON NANOTUBE-REINFORCED COMPOSITES

CNTs, in general, conduce to maintain stable and relatively lower frictional coefficients for resin or carbon matrix composites, especially for carbon matrix composites at high temperatures. More favorably, CNTs added with a certain ratio can decrease the wear loss to a minimum value. There are some common characters in wear mechanisms for CNT-reinforced composites.

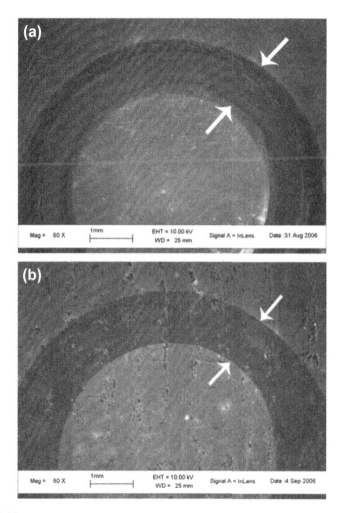

**FIGURE 10.28**

SEM images of the worn surfaces of ACNT/C and C/C composites tested at 500 °C. (a) ACNT/C composites; (b) C/C composites.

1. Although having a high modulus and hardness, CNTs do not engender abrasive wear mechanism. On the contrary, they weaken the abrasive wear mechanism and lower the wear loss. It might be attributed to their self-lubricating properties and wearable characters.
2. CNTs form a uniform self-lubricating film on the friction track or worn surfaces. As a result, they weaken the adhesive wear mechanism for composites and lead to the development of periodic surface (or film) fatigue and delamination wear mechanisms.

**FIGURE 10.29**

SEM images of the worn surfaces of ACNT/C and C/C composites tested at room temperature. (a) ACNT/C composites; (b) C/C composites.

## 10.6 CONCLUSIONS AND SUGGESTIONS

This chapter focused on the research of the influence of CNTs (randomly dispersed or well aligned) on the tribological properties of epoxy resin matrix and carbon matrix composites. The fundamental behavior of CNTs on the frictional coefficients and wear loss were briefly studied on the basis of morphology and microstructure of CNT-reinforced composites. In addition, some hypotheses about the influence of CNTs on wear mechanisms were discussed qualitatively

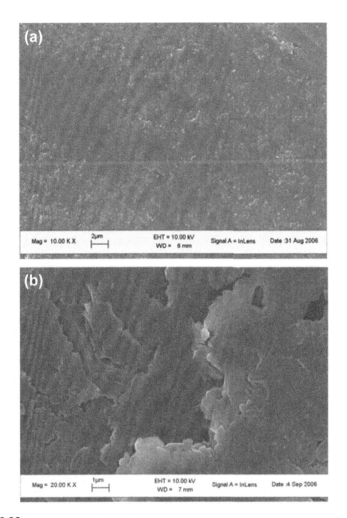

**FIGURE 10.30**

SEM images of the worn surfaces of ACNT/C and C/C composites tested at 250 °C.
(a) ACNT/C composites; (b) C/C composites.

based on the analyses of wear loss, worn surface and debris. All the experimental results suggest that CNTs can maintain stable low frictional coefficients and wear loss, especially at high temperatures for carbon matrix composites. It should be attributed to the wearable and self-lubricating properties of CNTs and their effect on weakening abrasive wear mechanism or adhesive wear mechanism. In summary, besides some theoretical research about CNTs in composites, we should be accelerating the process of CNTs application in composites, especially in friction field.

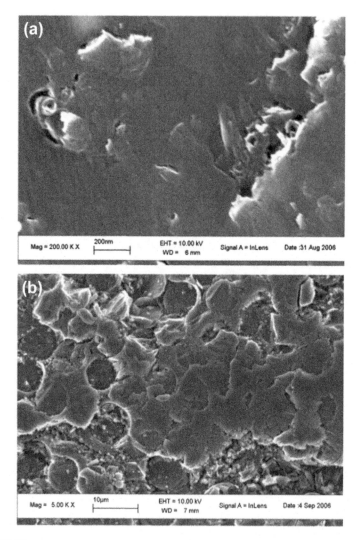

**FIGURE 10.31**

SEM images of the worn surfaces of ACNT/C and C/C composites tested at 500 °C. (a) ACNT/C composites; (b) C/C composites.

## Acknowledgment

The financial support by the National Science Council of the People's Republic of China and Aviation Research Fund of China is gratefully acknowledged.

# References

[1] M.S. Dresselhaus, G. Dresselhaus, P.C. Eklund (Eds.), Science of Fullerenes and Carbon Nanotubes, Academic Press, London, 1996.

[2] S. Iijima, Helical microtubules of graphitic carbon, Nature 354 (1991) 56–58.

[3] Kin-Tak Lau, David Hui, The revolutionary creation of new advanced materials—carbon nanotube composites, Composites: Part B 33 (2002) 263–277.

[4] R.S. Ruoff, D.C. Lorents, Mechanical and thermal properties of carbon nanotubes, Carbon 33 (1995) 925–930.

[5] M.F. Yu, O. Lourie, M. Dyer, K. Moloni, T.F. Kelly, R.S. Ruoff, Strength and breaking mechanism of multi-walled carbon nanotubes under tensile load, Science 287 (2000) 637–639.

[6] M.S. Dressel, G. Dresselhaus, P. Avouris (Eds.), Carbon Nanotubes: Synthesis, Structure, Properties and Applications, Springer, New York, 2000.

[7] R. Haggenmueller, H.H. Gommans, A.G. Rinzler, et al., Aligned single-wall carbon nanotubes in composites by melt process methods, Chemical Physics Letters 330 (2000) 219–225.

[8] Zhaoxia Jin, K.P. Pramoda, Guoqin Xu, et al., Dynamic mechanical behavior of melt-processed multi-walled carbon nanotube/poly(methyl methacrylate) composites, Chemical Physics Letters 337 (2001) 43–47.

[9] Kin-Tak Lau, David Hui, Effectiveness of using carbon nanotubes as nano-reinforcements for advanced composite structures, Carbon 40 (2002) 1605–1606.

[10] A. Allaoui, S. Bai, H.M. Cheng, et al., Mechanical and electrical properties of a MWNT/epoxy composite, Composites Science and Technology 62 (2002) 1993–1998.

[11] J. Sandler, M.S.P. Shaffer, T. Prasse, et al., Development of a dispersion process for carbon nanotubes in an epoxy matrix and the resulting electrical properties, Polymer 40 (1999) 5967–5971.

[12] D. Qian, E.C. Dickey, R. Andrews, et al., Load transfer and deformation mechanism in carbon nanotube-polystyrene composites, Applied Physics Letter 76 (20) (2000) 2868–2870.

[13] C. Bower, R. Rosen, L. Jin, et al., Deformation of carbon nanotubes in nanotube-polymer composites, Applied Physics Letter 74 (22) (1999) 3317–3319.

[14] Kin-tak Lau, San-qiang Shi, Failure mechanism of carbon nanotube/epoxy composites pretreated in different temperature environments, Carbon 40 (2002) 2965–2966.

[15] S.R. Dong, J.P. Tu, X.B. Zhang, et al., An investigation of the sliding wear behavior of Cu-matrix composite reinforced by carbon nanotubes, Materials Science and Engineering: A 313 (2001) 83–87.

[16] A. Peigney, Ch. Laurent, E. Flahaut, et al., Carbon nanotubes in novel ceramic matrix nanocomposites, Ceramic International 26 (2000) 677–683.

[17] B.D. Cullity, Elements of X-ray Diffraction, Addison Wesley, New York, 1978.

[18] K. Dasgupta, D. Sathiyamoorthy, Disordered carbon – its preparation, structure, and characterization, Materials Science and Technology 19 (8) (2003) 995–1002.

[19] W.X. Chen, F. Li, G. Han, J.B. Xia, L.Y. Wang, J.P. Tu, Z.D. Xu, Tribological behavior of carbon-nanotube-filled PTFE composites, Tribology Letters 15 (3) (2003) 275–278.

[20] D.S. Lim, J.W. An, H.J. Lee, Effect of carbon nanotube addition on the tribological behavior of carbon–carbon composites, Wear 252 (5–6) (2002) 512–517.

[21] W.X. Chen, J.P. Tu, L.Y. Wang, H.Y. Gan, Z.D. Xu, X.B. Zhang, Tribological application of carbon nanotubes in a metal-based composite coating and composites, Carbon 41 (2) (2003) 215–222.

[22] R.H. Savage, Graphite lubrication, Journal of Applied Physics 19 (1) (1948) 1–10.

[23] B.K. Yen, T. Ishihara, On temperature-dependent tribological regimes and oxidation of carbon–carbon composites up to 1800 °C, Wear 196 (1–2) (1996) 254–262.

[24] K.T. Lau, H. David, Effectiveness of using carbon nanotubes as nano-reinforcements for advanced composite structures, Carbon 40 (9) (2002) 1605–1606.

[25] B.K. Yen, Influence of water vapor and oxygen on tribology of carbon materials with $sp^2$ valence configuration, Wear 192 (1–2) (1996) 208–215.

[26] R.J. Nemanich, S.A. Solin, First- and second-order Raman scattering from infinite sized crystals of graphite, Physics Review B 20 (1) (1979) 92.

[27] M. Endo, Y.A. Kim, Y. Fukai, T. Hayashi, M. Terrones, H. Terrones, et al., Comparison study of semi-crystalline and highly crystalline multiwalled carbon nanotubes, Applied Physics Letters 79 (10) (2001) 1531–1533.

[28] Gong QM, Study on Manufacture Process and Properties of C/C Composites Used for Aircraft Brakes, Central South University, PR China, PhD thesis, 2002.

# Wear and wear maps of hard coatings

# 11

**Dongfeng Diao\*, Koji Kato\*\***

*\*Nanosurface Science and Engineering Research Institute, Shenzhen University, Shenzhen, P.R.China, Key Laboratory of Education Ministry for Modern Design and Rotor-Bearing System, Xi'an Jiaotong University, Xi'an, P.R.China, \*\*Department of Mechanical Engineering, College of Engineering, Nihon University, Koriyama, Japan*

## CHAPTER OUTLINE HEAD

## 11.1 INTRODUCTION

The coating of triboelement is mainly made to protect the surface of the substrate from wear, and a material harder than the substrate is generally chosen for the coating [1]. Although there are commonly known wear modes of abrasive, adhesive and corrosive for metals [2], they are all included in the term of "progressive wear" in this chapter and the wear mode of "delamination" is highlighted in the following as the most harmful type of wear of coating by generating flakelike or ribbonlike wear particles [3].

## 11.2 DELAMINATION OF HARD COATINGS BY INDENTATION

Figure 11.1 shows the indentation mark and crack pattern as functions of normal load $W$ and coating thickness with three regions [4]. In Fig. 11.1(a), a SiC ball of 1.5 mm radius is indented into the $Al_2O_3$ coating, and three patterns of no crack, ring crack and partial delamination between ring cracks are observed around the indentation mark. It can be seen from the figure in the thickness range of $t = 1\sim13$ μm that an indentation mark without ring cracks[1] (Hertz crack) is observed with the load below $W = 500$ N and one or two ring cracks are observed with the load $W = 800 \sim 1900$ N. When $W$ is greater than about 2000 N, more than three ring cracks are generated and partial delamination takes place between neighboring rings.

Tribology of Polymeric Nanocomposites. http://dx.doi.org/10.1016/B978-0-444-59455-6.00011-8

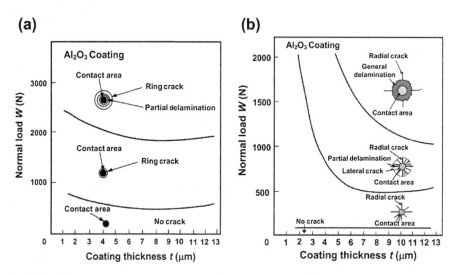

**FIGURE 11.1**

The indentation mark and crack pattern as a function of normal load $W$ and coating thickness $t$ at the surface of $Al_2O_3$ coating on WC-Co substrate. (a) Indentation of a SiC ball of 1.5 mm radius; (b) indentation of a diamond pin of 0.2 mm tip radius [4].

In Fig. 11.1(b), a spherical diamond pin of 0.2 mm tip radius is indented into the same $Al_2O_3$ coating as in Fig. 11.1(a), where the contact pressure is much higher with a smaller contact area than that of a SiC ball of 1.5 mm radius under the same load. Observed crack patterns are classified into four types with four regions: no crack, radial cracks[2] only around the indentation mark under relatively small load, radial and lateral cracks together with partial delamination of coating under medium load and general delamination of coating by the general propagation of lateral cracks[3] under a relatively large load.

1. The ring crack or Hertz crack is generated during the loading process along the periphery of the Hertz contact region when the maximum tensile stress there exceeds a critical value of the brittle material with the preexistent surface cracks and is propagated downwards to form the shape of a cone. Hertz contact is defined as the elastic contact between balls or a ball and a flat [5].
2. The radial crack is generated during the loading process at the central part of the plastically deformed zone in the contact region of the indenter when the maximum tensile stress there exceeds a critical value of the brittle material and is propagated radially from the center of contact [6].
3. The lateral crack is generated during the unloading process at the central part of the plastically deformed zone in the contact region of the indenter when the maximum tensile stress generated by the residual stress there exceeds a critical value of the brittle material, and is propagated in parallel to the surface [7].

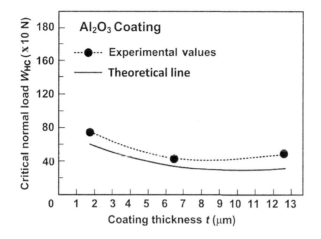

**FIGURE 11.2**

Experimental and theoretical critical loads for ring crack generation in Al$_2$O$_3$ coating on WC-Co substrate by indentation of a SiC ball of 1.5 mm radius [4].

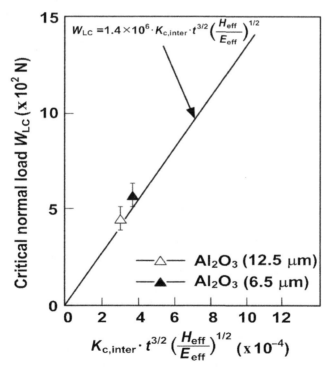

**FIGURE 11.3**

Experimental and theoretical critical loads for lateral crack generation in Al$_2$O$_3$ coating on WC-Co substrate by indentation of a diamond pin of 0.2 mm tip radius [4].

The critical load for generating partial delamination of coating in Fig. 11.1(b) is almost constant at around $W = 500$ N for the coating thickness $t$ above 6 μm; however, it is increased rapidly if the thickness $t$ is reduced from 6 to 1 μm. The critical load for generating general delamination of coating behaves similarly against the coating thickness at a higher level of the load.

The critical load $W_{HC}$ for generating a ring or Hertz crack and the critical load $W_{LC}$ for a lateral crack in the coating are described theoretically by D. F. Diao et al. [4] as follows:

$$W_{HC} = 200K_{IC,C}^3 R^2 E_{eff}^{-2} C^{-3/2} \tag{11.1}$$

$$W_{LC} = 1.4 \times 10^6 K_{IC,inter} t^{3/2} \left( \frac{H_{eff}}{E_{eff}} \right)^{1/2} \tag{11.2}$$

where $K_{IC,C}$ is the fracture toughness of coating, $R$ the radius of indenter, $E_{eff}$ the effective elastic modulus of coating and substrate, $C$ the length of preexistent surface crack, $K_{IC,inter}$ the fracture toughness of bonding interface, $t$ the coating thickness and $H_{eff}$ the effective hardness of coating and substrate.

Figure 11.2 shows the agreement between the experimental and theoretical values of $W_{HC}$ for $Al_2O_3$ coating on WC-Co substrate in the coating thickness range from 1 to 13 μm. Figure 11.3 shows a good agreement between the experimental and theoretical values of $W_{LC}$ for the same coating.

## 11.3 PROGRESSIVE WEAR AND DELAMINATION OF HARD COATINGS IN REPEATED SLIDING

In repeated sliding of a diamond pin of 33 μm tip radius against an $Al_2O_3$ coating of 1.0 μm thickness on WC-Co substrate, only a small amount of plastic deformation of asperities on the coating surface is generated as shown in Fig. 11.4(a) and (b) when the contact is mild with a relatively light load (0.1 N), where the wear mode is plowing [3]. Fine powderlike wear particles are generated after a certain number of repeated sliding cycles, as shown in Fig. 11.4(c) and (d). A large amount of powderlike wear particles are stacked on the tip surface and scattered along the wear track after 100 sliding cycles as shown in Fig. 11.4(e) and (f).

Under more severe contact with a relatively large load (1.5 N), cracks are introduced at either or both sides of the wear track, and flakelike wear particles are generated by delamination of coating on the outside of the wear track as shown in Fig. 11.5(a)–(c). After progressive wear and thinning of coating, long ribbonlike wear particles are generated by delamination of coating on the inside of the wear track as shown in Fig. 11.5(d) and (e) [8].

Under very severe contact with a large load above 1.5 N, cracks are developed far below the surface of the wear track and thick flakelike wear particles are generated by delamination. Figure 11.6 shows the magnified view of a flakelike wear particle, which shows substrate material bonding to the flake and indicates that the crack path was in the substrate [8]. Figure 11.7 shows the regimes of wear modes observed in Figs 11.4 and 11.5 as a function of normal load $W$ and the number of sliding cycles

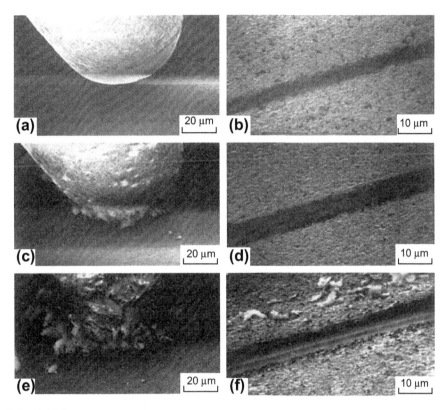

**FIGURE 11.4**

Scanning electron micrographic images of progressive wear of Al$_2$O$_3$ coating of 1 µm thickness observed in repeated sliding of a diamond pin of 33 µm tip radius under a 0.1-N load. (a) and (b) At the 1st cycle; (c) and (d) at the 50th cycle; (e) and (f) at the 100th cycle [3].

$N$. Figure 11.8 gives the values of the wear rate (mm$^3$ Nm) observed in the load range from 0.1 to 1.8 N, where the load of 0.1 N only gives the practically acceptable value of $w_s$ below 10$^{-6}$ mm$^3$ Nm.

The possible mechanism of coating delamination on the outside of the wear track by sliding of a hard pin is considered as illustrated in Fig. 11.9 [9]. The preexistent microcracks at the bonding interface are propagated to the vertical direction against sliding in Fig. 11.9(a) and a large crack is formed along the bonding interface in Fig. 11.9(b) due to the localized plastic deformation in the contact region. The compressive stress $\sigma_c$ generated by the plastic deformation finally induces delamination of coating as shown in Fig. 11.9(c) by releasing the compressive stress. A load that is large enough to introduce the plastic deformation in the contact region is therefore necessary for this mechanism.

The possible mechanism of coating delamination on the inside of the wear track by repeated sliding is considered as illustrated in Fig. 11.10. The preexistent microcracks at the bonding interface are propagated to the direction of sliding in

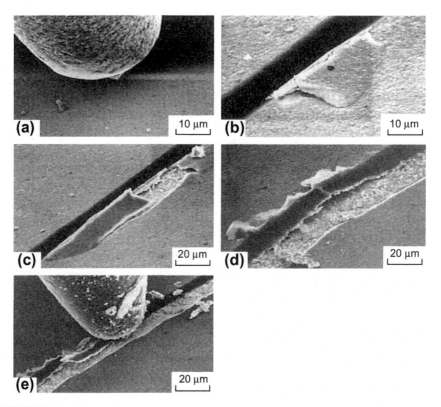

**FIGURE 11.5**

Scanning electron micrographic images of delamination of $Al_2O_3$ coating of 1 μm thickness observed on repeated sliding of a diamond pin of 33 μm tip radius under a 1.5-N load. (a) At the 1st cycle; (b) at the 3rd cycle; (c) at the 4th cycle; (d) at the 9th cycle; (e) at the 10th cycle [8].

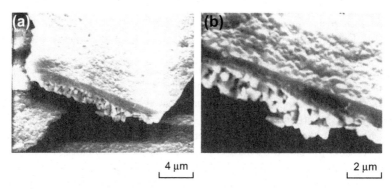

**FIGURE 11.6**

Scanning electron micrographic images of a flakelike wear particle of $Al_2O_3$ coating delaminated by crack propagation in the substrate. (a) An overview; (b) side view showing the substrate material [8].

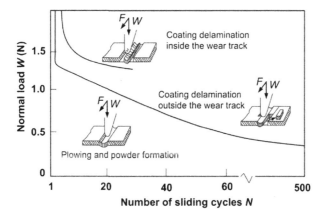

**FIGURE 11.7**

Wear modes of $Al_2O_3$ coating of 1 μm thickness observed in repeated sliding of a diamond pin of 33 μm tip radius under various loads and sliding cycles. $W$, load; $F$, friction [3].

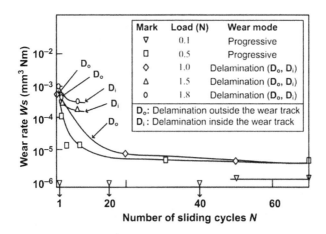

**FIGURE 11.8**

Wear rates and wear modes of $Al_2O_3$ coating of 1 μm thickness observed in repeated sliding of a diamond pin of 33 μm tip radius under various loads and sliding cycles. ▽ Under 0.1-N and ☐ under 0.5-N load, progressive wear; ◇ under 1.0-N load, coating delamination outside the wear track; △ under 1.5-N and ○ under 1.8-N load, coating delamination outside and inside the wear track [3].

Fig. 11.10(a), and a large crack is formed by repeated sliding along the bonding interface in Fig. 11.10 (b). When the coating thickness is reduced to a critical value by progressive wear and/or when the large crack is grown to a critical size, delamination of coating is finally induced as shown in Fig. 11.10(c) by releasing the residual compressive stress $\sigma_r$ in the coating.

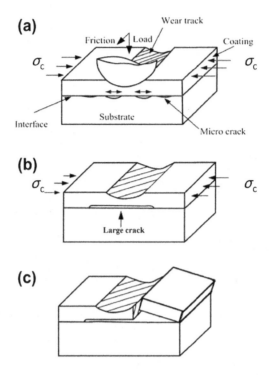

**FIGURE 11.9**

A model of delamination of coating at the outside of the wear track induced by the compressive stress $\sigma_c$ generated by the localized plastic deformation in the contact region. (a) Propagation of microcracks; (b) formation of a large crack; (c) delamination of coating.

## 11.4 LOW CYCLE FATIGUE LAW OF DELAMINATION

Figure 11.11 shows the wear track (1 μm × 1 μm) of $CN_x$ coating of 100 nm thickness on Si wafer after 40 transverse repeated sliding cycles by a diamond pin of 150 nm tip radius under a 14-μN load. The wear track of the $CN_x$ coating is formed by wear as shown in Fig. 11.12 where the progressive and sudden increase in depth of the wear track is described as a function of the number of repeated sliding cycles [10]. Similar results are observed in [11,12].

It is clearly shown in the figure that the depth of the wear track is drastically increased by the repeated sliding cycle from 13th to 14th and from 29th to 30th.

This kind of sudden increase in the depth of the wear track is caused by the sudden delamination of a thin surface layer of the coating in the wear track as shown in Fig. 11.13, where the wavy surface of the wear track at the 18th sliding cycle in Fig. 11.13(a) suddenly disappears by surface delamination at the 19th sliding cycle in Fig. 11.13(b). The critical number of sliding cycles for the generation of delamination is decreased by increasing the contact load from 14 to 45 μN as shown in Fig. 11.14.

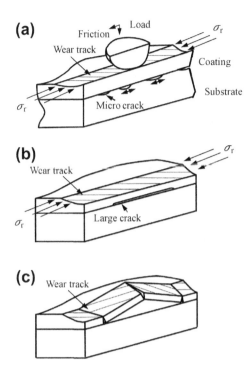

**FIGURE 11.10**

A model of delamination of coating at the inside of the wear track induced by progressive wear and the residual compressive stress $\sigma_r$ in the coating. (a) Propagation of microcracks; (b) formation of a large crack; (c) delamination of coating.

If a simple model of contact is described as shown in Fig. 11.15 where $a$ is the contact radius and $R$ the radius of the pin tip, the representative plastic strain $\Delta\varepsilon_p$ at the contact region of the diamond pin is given as follows [13]:

$$\Delta\varepsilon_p = 0.18a/R \tag{11.3}$$

The value of $a$ is calculated by the following equation:

$$a^2 = R^2 - (R - \Delta h_a)^2 \tag{11.4}$$

where $\Delta h_a$ is the average increase in wear depth per sliding cycle and is defined as follows:

$$\Delta h_a = \sum_{n=r}^{N_d} \Delta h_n / N_d \tag{11.5}$$

In Eqn (11.5), $\Delta h_n$ is the increase in wear depth per sliding cycle, which is obtained from the depth of the wear track in Fig. 11.16, and $N_d$ the number of sliding cycles for delamination.

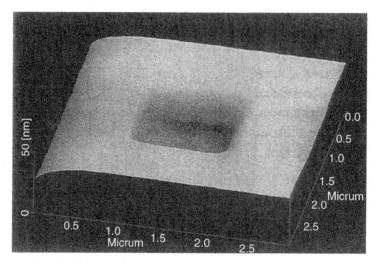

**FIGURE 11.11**

Atomic force microscopic image of wear track of $CN_x$ coating on Si wafer formed by 40 transverse sliding cycles with a diamond pin of 150 nm tip radius [10].

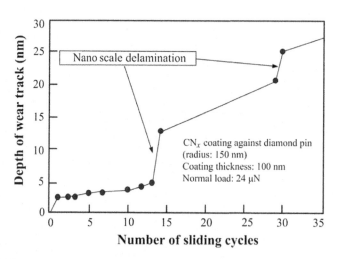

**FIGURE 11.12**

The sudden increase in the depth of wear track caused by nanoscale delamination of $CN_x$ coating after progressive wear in repeated sliding with a diamond pin of 150 nm tip radius [10].

**(a)**

18 sliding cycles

**(b)**

19 sliding cycles

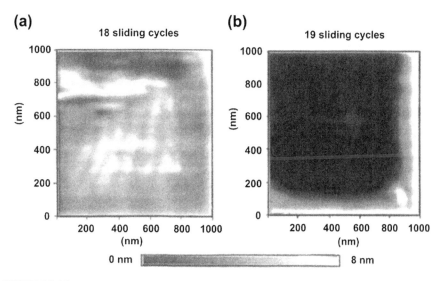

0 nm  8 nm

**FIGURE 11.13**

Atomic force microscopic images of wear track of $CN_x$ coating on Si wafer formed by repeated sliding with a diamond pin of 150 nm tip radius. (a) Wavy surface of wear track after 18 cycles; (b) dark image of wear track after surface delamination at 19th cycle [10].

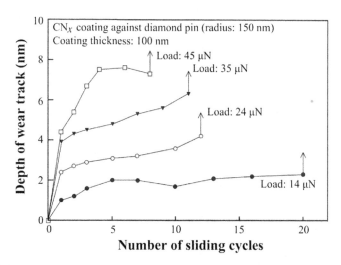

**FIGURE 11.14**

The load dependency of the critical number of sliding cycles for the sudden increase in depth of wear track by surface delamination of $CN_x$ coating of 100 nm thickness on Si wafer by repeated sliding with a diamond pin of 150 nm radius [10].

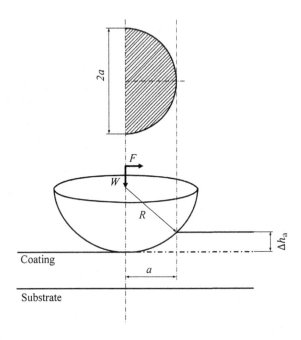

**FIGURE 11.15**

Schematic image of sliding contact between the coating and a hard spherical pin. $R$, radius of pin tip; $a$, contact radius; $\Delta h_a$, wear track depth at the $n$th sliding cycle; $W$, load; $F$, friction.

Figure 11.16 shows the relationship between the experimental values of $\Delta\varepsilon_p$ and $N_d$ obtained by changing the pin radius and contact load [11,12], where it is described by the following equation:

$$N_d^\beta \Delta\varepsilon_p = c \qquad (11.6)$$

where $\beta$ and $c$ are experimental constants.

The observed relationship in Eqn (11.6) clearly tells that the wear mode of delamination follows the power law of low cycle fatigue proposed by Coffin and Manson [14,15].

## 11.5 WEAR MAPS OF HARD COATINGS

It is shown in Sections 11.3 and 11.4 that a hard coating of brittle material is worn out by delamination under repeated sliding of a hard pin. The thickness of wear particles can be similar to the thickness of the coating, or smaller or larger than that of the coating depending on the severity of contact. In any case, the mechanism of delamination seems to be low cycle fatigue associated with the repeated plastic strain cycle in the localized zone. Because of this understanding about the mechanism of delamination of coating, the stress distribution of equivalent stress of von Mises, $\sigma_{VM}$, on the $x$–$z$ plane at $y = 0$ in Fig. 11.17 is calculated with the semianalytical

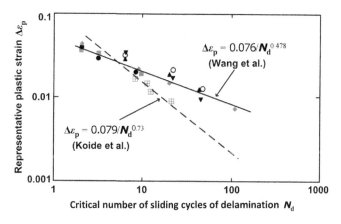

**FIGURE 11.16**

The power law observed between the representative plastic strain $\Delta\varepsilon_p$ and the critical friction cycle $N_d$ for surface delamination: broken line, by the diamond pin of 150 nm tip radius under a 15–45 μN load; solid line, by the diamond pin of 10 μm tip radius under a 10–250 μN load [11].

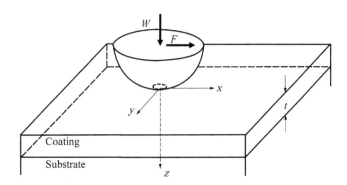

**FIGURE 11.17**

The model of sliding contact between a ball and the coating. W, load; F, friction; t, coating thickness.

method, which is based on the conjugate gradient method, discrete convolution and the fast Fourier transform technique, for the sliding contact between a ball and the coating of thickness $t$ under the load $W$ and friction $F$ [16].

Figure 11.18 shows the contours of $\sigma_{VM}$ normalized by the maximum Herzian contact pressure, $P_{max}$, on the diagram of $x/a$ and $z/a$ where $a$ is the contact radius [16]. The values of $\sigma_{VM}/P_{max}$ in the figure are for the case of the friction coefficient $\mu = 0.25$ and $E_c/E_b = 2$, where $E_c$ is the elastic modulus of the coating and $E_b$ that of

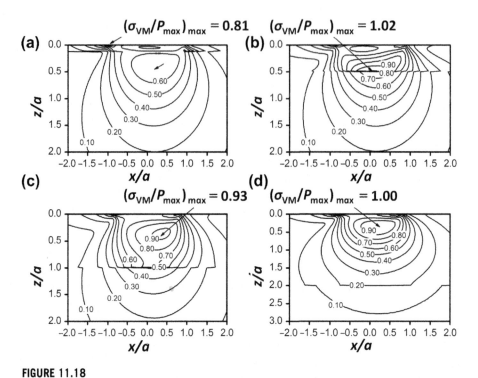

**FIGURE 11.18**

Contours of the normalized stress of von Mises stress, $\sigma_{VM}/P_{max}$, in the $x$–$z$ plane at $y = 0$ for $\mu = 0.25$ and $E_c/E_b = 2$. (a) $t = 0.125a$, (b) $t = 0.5a$, (c) $t = a$, (d) $t = 2a$ [16].

the substance. In Fig. 11.18(a) for $t = 0.125a$, the regions of high value of $\sigma_{VM}/P_{max}$ are observed on the coating surface, in the coating and in the substance, where the maximum value of $(\sigma_{VM}/P_{max})_{max} = 0.81$ is generated on the coating surface. In Fig. 11.18(b) for $t = 0.5a$, the regions of high values of $\sigma_{VM}/P_{max}$ are observed in the coating and around the bonding interface, where $(\sigma_{VM}/P_{max})_{max} = 1.02$ is generated at the bonding interface. In both Fig. 11.18(c) for $t = a$ and Fig. 11.18(d) for $t = 2a$, the region of the highest value of $\sigma_{VM}/P_{max}$ is generated in the coating, where $(\sigma_{VM}/P_{max})_{max} = 0.93$ in (c) and $(\sigma_{VM}/P_{max})_{max} = 1.00$ in (d).

Figure 11.19 shows the contours of $\sigma_{VM}/P_{max}$ obtained for the case of $\mu = 0.5$ and $E_c/E_b = 2$ by the calculation similar to that of Fig. 11.18 [16]. In Fig. 11.19(a) for $t = 0.125a$, the regions of high value of $\sigma_{VM}/P_{max}$ are observed on the coating surface and in the coating, where $(\sigma_{VM}/P_{max})_{max} = 1.33$ is generated on the coating surface. In Fig. 11.19 (b) for $t = 0.5a$, (c) for $t = a$ and (d) for $t = 2a$, the region of the highest value of $\sigma_{VM}/P_{max}$ is generated on the coating surface, where $(\sigma_{VM}/P_{max})_{max} = 1.34$ in (b), $(\sigma_{VM}/P_{max})_{max} = 1.42$ in (c) and $(\sigma_{VM}/P_{max})_{max} = 1.50$ in (d).

After having similar distributions of contours of $\sigma_{VM}/P_{max}$ for $\mu = 0$ and $\mu = 0.70$ in the range from $t = 0$ to $t = 4a$, the site of yield initiation, where the maximum value of $\sigma_{VM}$ exceeds the yield stress of the material, was confirmed in the contact stress

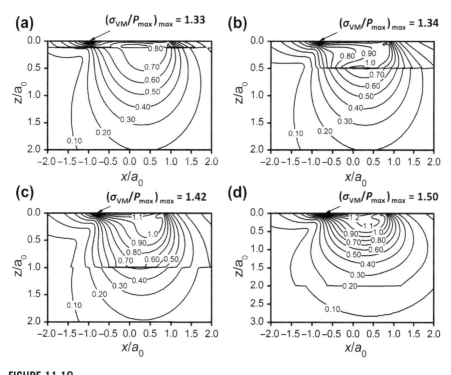

**FIGURE 11.19**

Contours of the normalized von Mises stress, $\sigma_{VM}/P_{max}$, in the $x$–$z$ plane at $y = 0$ for $\mu = 0.5$ and $E_c/E_b = 2$. (a) $t = 0.125a$, (b) $t = 0.5a$, (c) $t = a$, (d) $t = 2a$ [16].

region for various combinations of $t/a$ and $Y_c/Y_b$, where $Y_c$ is the yield strength of coating and $Y_b$ of substrate.

The four regions I, II, III and IV, which indicate the sites of local yield initiation in the contact stress region of the coating and substrate, are shown on the diagram of $t/a$ and $Y_c/Y_b$ in Fig. 11.20, where the yield strength of the bonding interface is assumed to be the same as that of the substrate and $E_c/E_b = 2.0$ [16].

Figure 11.20(a) for $\mu = 0$ shows that the yield initiation takes place in the substrate on region I, at the bonding interface on region II and in the coating on region III. Figure 11.20(b) for $\mu = 0.25$ and $E_c/E_b = 2.0$ shows that the small region IV appears for the yield initiation on the coating surface in addition to the regions I, II and III. When the friction coefficient $\mu$ is further increased to 0.50 in Fig. 11.20(c) and to 0.70 in Fig. 11.20(d), only two regions are observed: region II for the yield initiation at the bonding interface and region IV for the yield initiation at the coating surface.

When the yield is initiated in the substrate on region I, wear particles similar to those in Fig. 11.6 can be generated, and when the yield is initiated at the bonding interface on region II, wear particles similar to those in Fig. 11.5 can be generated. The delamination observed in Fig. 11.13 can be generated by the yield initiation

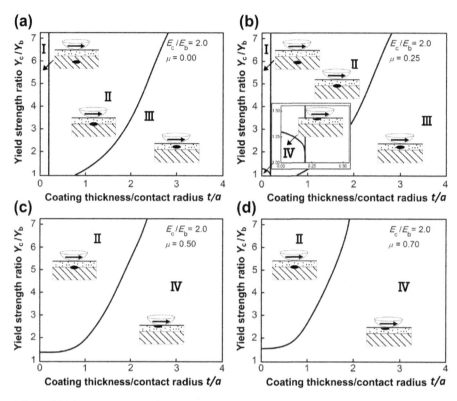

**FIGURE 11.20**

Wear maps showing the crack initiation sites by local yield for delamination of coating. $Y_c/Y_b$ = yield strength of coating/yield strength of substrate; $t/a$ = coating thickness/contact radius. (a) $\mu = 0.25$, (b) $\mu = 0.25$, (c) $\mu = 0.50$, and (d) $\mu = 0.70$ [16].

within the coating on region III. When the yield initiates at the coating surface on region IV, the progressive wear observed in Fig. 11.4 can be generated in general. Figure 11.20(a)–(d) is useful in this way to predict the possible wear mode for the given condition of $t/a$, $Y_c/Y_b$, $E_c/E_b$ and $\mu$. Therefore, the diagram like the ones in Fig. 11.20 may be called as the "wear map of coating."

## 11.6 SUMMARY

The mechanisms of delamination wear of hard coatings are described by the theoretical models based on experimental results in nano- and micrometer scales. Equations (11.1) and (11.2) of critical loads introduced for predicting the generation of ring (Hertz) and lateral cracks in indentation of a hard spherical pin tip are practically useful.

The wear map shown in Fig. 11.20 is also useful for predicting wear modes corresponding to the coating thickness and material properties of coating and substrate together with the friction coefficient at the contact interface. Wear maps of this kind are useful for the optimum design of coating by considering the combination of materials, coating thickness, friction coefficient, contact load and shape of contact elements.

It must be noted that the thickness of the coating becomes smaller by progressive wear under repeated sliding and as a result the value of $t/a$ (coating thickness/contact radius) is shifted to the side of the smaller value of $t/a$ on the maps of Fig. 11.20 introducing the transition of wear mode from one to another. It is also important to notice that the contact radius is a function of the pin tip radius, elastic moduli of the coating and substrate, load and shape of the wear track formed by the progressive wear.

The wear map shown in Fig. 11.20 must be modified when the yield strength at the bonding interface is very different from that of the coating or substrate. It can be further modified for better practical use if any new criterion of crack initiation and/or propagation would be introduced in addition to the local yield criterion in the coating, at the bonding interface and in the substrate.

## References

[1] K. Holmberg, A. Mattkews, in: D. Dowson (Ed.), Coatings Tribology, Tribology Series, vol. 28, Elsevier, The Netherlands, 1994.

[2] K. Kato, Micro-mechanisms of wear – wear models, Wear 153 (1992) 277–295.

[3] K. Kato, Microwear mechanics of coating, Surface and Coatings Technology 76–77 (1995) 469–474.

[4] D.F. Diao, K. Kato, K. Hokkirigawa, Fracture mechanisms of ceramic coatings in indentation, Transactions ASME Journal of Tribology 116 (October, 1994) 860–869.

[5] B.R. Lawn, Fracture of Brittle Solids, Cambridge University Press, New York, 1993 251–256.

[6] B.R. Lawn, A.G. Evans, D.B. Marshall, Elastic/plastic indentation damage in ceramics: the medium/radial crack system, Journal of American Ceramic Society 63 (1980) 574–581.

[7] D.B. Marshall, B.R. Lawn, A.G. Erans, Elastic/plastic indentation damage in ceramics: the lateral crack system, Journal of American Society 65 (1982) 561–566.

[8] K. Kato, D.F. Diao, M. Tsutsumi, The wear mechanism of ceramic coating film in repeated sliding friction, ASME Wear of Materials (1991) 243–248.

[9] D.F. Diao, Y. Sawaki, Fracture mechanisms of ceramic coating during wear, Thin Solid Films 270 (1–2) (1995) 362–366.

[10] K. Kato, H. Koide, N. Umehara, Micro-wear mechanisms of thin hard coatings sliding against diamond tip of AFM, ASME, Advances in Information Storage Systems 9 (1998) 289–302.

[11] D.F. Wang, K. Kato, Nano-scale fatigue wear of carbon nitride coatings: part I – wear properties, Transactions ASME Journal of Tribology 125 (2003) 430–436.

[12] D.F. Wang, K. Kato, Nano-scale fatigue wear of carbon nitride coatings: part II – wear mechanisms, Transactions ASME Journal of Tribology 125 (2003) 437–444.

[13] D. Tabor, The Hardness of Metals, Clarendon Press, 1951.

[14] S.S. Manson, Fatigue: a complex subject – some simple approximations, Experimental Mechanics 5 (1965) 193–226.

[15] L.F. Coffin Jr., N.Y. Schenectady, A study of the effects of cyclic thermal stresses on a ductile metal, Transactions ASME 76 (1954) 931–950.

[16] P.Y. Zhang, D.F. Diao, Z.J. Wang, Three-dimensional local yield maps of hard coating under sliding contact, ASME Transaction Journal of Tribology 134, (2012.) 011301-1–8.

# Hybridized carbon nanocomposite thin films: synthesis, structures and tribological properties

# 12

**Eiji Iwamura**

*R&D Center, Arakawa Chemical Industries, Ltd, Tsurumi-ku, Osaka, Japan*

## CHAPTER OUTLINE HEAD

## 12.1 INTRODUCTION

Nanocomposite thin films have been intensely investigated because the combined structures, which contain two or more components with different physical properties, make it possible to achieve better performance and functionality than do those of

Tribology of Polymeric Nanocomposites. http://dx.doi.org/10.1016/B978-0-444-59455-6.00012-X

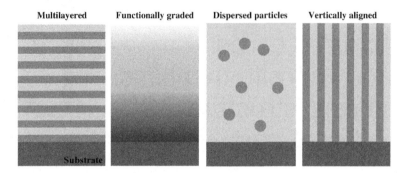

**FIGURE 12.1**

Configurations of nanocomposite thin films.

each individual component. Various three-dimensional composite configurations as shown in Fig. 12.1 have been developed, particularly in bulk materials. Multilayered and functionally graded structures have been frequently employed in thin films due to the process feasibility in industrial applications. On the other hand, there are technological difficulties in fabricating dispersed and vertically aligned arrangements in normal thin film deposition procedures.

In the field of hard and wear-resistant coatings, microstructural design of individual features, such as grain size, defect density, and phase arrangements in a one-, two- or three-dimensional manner, plays a key role in modern development [1]. Nanocomposite films are expected to be a promising solution to achieve strength and toughness at the same time. It is generally believed that the harder materials are favorable for achieving higher wear resistance. However, the intrinsically hard coating materials are usually brittle and have poor toughness, which cause a serious reliability problem in practical applications. Nanocomposite films with various buffer components have been investigated to improve robustness of the harder coating films. There are numerous studies relating to the material designs. The concepts for designs can be summarized as superhard coatings in which the film hardness significantly exceeds the value expected from the rule of mixture and a high strength phase, which is incorporated with a deformable phase or structures. Particularly, the generic concept utilizing a thermodynamically driven spinodal phase separation results in superhard and thermally stable nanocomposite coatings [2–4]. The ideas concerning multilayered configurations including heterostructures have been widely utilized in a great variety of combinations of elements and components. Typical combinations of materials are the following:

- Elemental combinations of (C, B, N) and (Ti, Si, Al)
- Structural combinations of (Ni, Cu) and (TiN, ZrN, CrN)
- Structural combinations of (a-C, a-BN, a-CN$_x$, a-Si$_3$N$_4$) and (nitride, carbide, and boride)

## 12.2 HYBRIDIZATION OF CARBON ALLOTROPES IN THIN FILM CONFIGURATIONS

Carbon-based materials, not only graphite and diamonds but also nanotubes (CNTs), fullerenes, graphenes and amorphous carbon (a-C), have been regarded as one of the most important materials in nanotechnology. They can be used for various applications due to their superior and diverse physical properties, which arise from unique carbon structures. From the point of view of tribological coating applications, diamond is one of the intrinsically hardest materials due to the strong $sp^3$ bonding, and polycrystalline diamond films have been industrially used in protective coatings on cutting tools. Graphite is known to be a solid self-lubricant due to the layered crystal structure of $sp^2$ bonding in which weak van der Waals bonds connect the graphite planes. Thin films, which impregnate concentric graphite-like shells known as carbon onions, also exhibit superb low friction as seen in the films including fullerene-like $MoS_2$, $WS_2$ and $CN_x$ nanoparticles [5–10]. a-C films have been practically employed in low friction and wear-resistant coatings for years [11,12].

The advantage of a-C coatings is that it is relatively easy to control film qualities over a wide range, namely, the properties of a-C films can be changed from mechanically soft to hard, from electrically conductive to insulate, and from optically transparent to opaque. It is generally accepted that the structural and functional properties of a-C are critically influenced by the ratio between the numbers of $sp^2$ and $sp^3$ bonds. Depending on the atomistic bonding configurations and the film density ranging from 1.4 to 3.5 Mg m$^{-3}$, the mechanical properties of the films can be changed from relatively soft like graphite to superhard close to diamond [13,14]. The wide controllability of film properties suggests that a combination of appropriate amorphous and graphitic structures may lead to a novel or multifunctional mechanical property, which cannot be achieved by a single-phase film.

In thin film synthesis, the formation of structurally ordered carbon strongly relies on complex plasma or physical film growth processes. The structural ordering of carbon atoms is generally realized in a high-energy state. Special forming conditions, such as an intense electron beam (EB), high temperature, high pressure or ultra high vacuum, are required in most fabrication processes. Various synthesis techniques have been reported for synthesizing the ordered carbon structures in a thin film form as summarized in Fig. 12.2 [15–29]. The processes involve the following key factors: (1) energetic particle irradiation, (2) highly dense film formation, (3) carbon cluster source, (4) catalytic preferential growth, and (5) incorporation of nitrogen in the carbon network. Those factors have been combined together in addition to annealing.

The formation of carbon onion structures in ultra high vacuum in transmission electron microscopy (TEM) by subjecting them to high energy and a high dose of electrons has stimulated synthesizing specific graphitic structures in an amorphous state [15]. The phenomenology of the formation of graphitic structures, namely, breaking and rebuilding of carbon networks concurrently taking place during electron irradiation, has been clearly revealed by in situ TEM observations [16,30–32]. Local stress and highly metastable conditions encourage the reconstruction of carbon

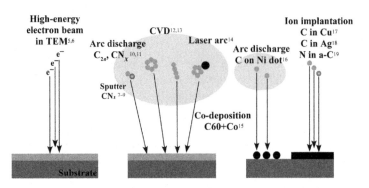

**FIGURE 12.2**

Synthesis techniques for graphitic structures in thin film form.

networks. However, the high-energy convergent EB techniques do not seem feasible for industrial applications.

The ideas of energetic particle irradiation have been widely utilized within various deposition techniques [33–36]. Arc-discharge plasma deposition and ion beam deposition that can fabricate relatively high-density carbon films have been frequently used together with ion bombardment in growing films. Combining cathodic vacuum arc deposition with ion bombardment, nanocrystallites of diamond were obtained in tetrahedral a-C films. The nanocrystallites are reported to form either within the arc plasma or due to kinetic processes involved in rapid thermal quenching of $C^+/C^{++}$ ions, and a graphitic particle source is assumed to be necessary for preseeding of the diamond nanocrystals [33].

Introduction of a graphitic cluster source is another approach. It is reported that a hybridized carbon nanocomposite with a high density of graphitic components of CNTs, onions and graphites was dispersedly formed in the thin films deposited by a cathodic arc technique. The generation of carbon clusters ($C_{2n}$) in the presence of high helium gas pressure appears to promote the formation of graphitic structures [20]. $C_{60}$ has also been used as a graphitic particle source [25].

It is well recognized that nitrogen doping into carbon networks encourages the formation of the graphitic networks. The incorporation of a pentagon bonding configuration in the graphite basal plane bends the networks to form cross-linking [17]. Nitrogen atoms are introduced into carbon films by means of a gas source, or ion implantation during and postdeposition. Because this approach is quite simple, there have been a number of research studies relating to $CN_x$ thin films with graphitic components.

Catalytic elements, such as Fe, Co and Ni, have been frequently used for graphitization in a-C films [37–42]. A relatively high temperature at least more than 1273 K is usually required to graphitize a-C structures [43,44], whereas catalysts reduce the temperature in the range between 573 and 1073 K [37–39]. Graphitization on the surface of catalysts appears to couple with diffusion-induced local stresses [45,46]. The Kirkendall effect likely generates strong compressive stress at the interface between the catalysts and the carbon networks. The ordered carbon structures that are derived from carbide precursors are also known as carbide-driven carbon [47].

On the other hand, synthesis of vertically aligned nanocomposite configurations seems difficult in most of the above-mentioned techniques. Using Ni dots as graphitization sites, it is possible to fabricate a graphite column selectively on the Ni dots surrounded with tetrahedral a-C [26]. This technique is dissimilar to other techniques with respect to both the vertical and intentional arrangement of graphitization sites. However, the Ni dots have to be formed prior to carbon deposition by means of annealing at 1073 K.

Considering practical applications, it is favorable to synthesize hybrid carbon films via a process as simple as and at a temperature as low as possible. Postdeposition treatment using low-energy electron irradiation after a-C formation followed by Fe atom implantation synthesizes many patterns containing graphitic regions in amorphous matrices without elevating the temperature [48,49]. Using this technique, it is possible to fabricate vertically aligned nanocomposite configurations of hybridized carbon films.

Some of those films exhibit excellent tribological properties. Superhard and highly elastic properties are common features in those hybridized carbon nanocomposite films [17,20,50]. Those properties are attributed to the curved graphitic structures [17,24]. It is expected that the hybridized carbon films may enhance utilization of the materials in the practical applications used under harsh circumstances. However, the way to control functionally hybridized order/disorder structures has not been established yet. This can be achieved through the development of a new concept in both material and related process design, utilizing the features of carbon that form various structures with different dimensions.

In the following sections, structures and tribological properties are depicted in detail for two different types of hybridized carbon films with nanocomposite configurations consisting of amorphous and graphitic regions. One configuration is the multilayered structure, stacking amorphous and graphitic layers with a thickness of a nanometer-length scale, alternately. In another, amorphous and graphitic regions are aligned normal to the film plane in the form of column/intercolumn structures. Synthesis techniques are important to achieve the configurations at a low temperature. Two kinds of electron irradiation techniques were employed to provide energy to the films, in situ during film growth and ex situ after film deposition. The formation processes are also described in detail.

## 12.3 HYBRIDIZED CARBON NANOCOMPOSITE FILMS WITH MULTILAYERED CONFIGURATION

In this section, hybridized carbon nanocomposite films with multilayered configuration will be described. Using electron and ion irradiation onto the growing films, alternate stacking structures composed of amorphous and graphitic layers were fabricated. Unbalanced magnetron (UBM) sputtering deposition is a key process in realizing the configurations in a feasible way at low temperature.

### 12.3.1 Synthesis

#### 12.3.1.1 Characteristics of UBM sputtering

Sputtering deposition techniques generally have the disadvantage of a low deposition rate. Magnetron sputtering systems, which comprise magnets behind sputtering targets, were invented in order to overcome the low deposition rate problem by generating higher plasma density in the vicinity of the surface of sputtering targets. Usually magnets with the same magnetic properties are arranged to form a magnetic field as confined as possible in which electrons are trapped in the field to enhance the discharge. This is the so-called balanced magnetron (BM). On the other hand, a UBM sputtering system in which magnets with dissimilar magnetic properties are intentionally arranged to form an open magnetic field is employed to utilize the diffusion of ions and electrons along with the magnetic flux leaked from the vicinity of the sputtering targets. Therefore, in UBM sputtering, energetic electrons can effectively irradiate the growing films.

Figure 12.3 shows a schematic drawing of the UBM sputtering configuration with a ferrite as the central magnet and SmCo as the annular magnet. Due to the difference in magnetic force between the ferrite and SmCo magnets, some magnetic flux originating from the annular magnets reached the substrate, without passing into the cylindrical central magnet.

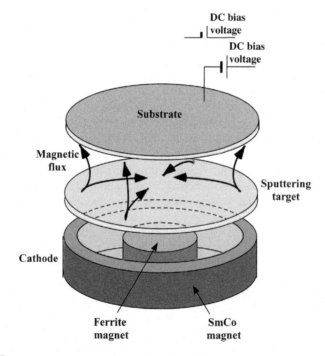

**FIGURE 12.3**

Schematic drawing of the UBM sputtering system.

The features of UBM sputtering are attributed to the higher energy and influx of electrons or a higher influx of ions induced by the unbalanced magnetic arrangement [51,52]. Figure 12.4 shows the single probe (I–V) characteristics of UBM sputtering in comparison to BM sputtering using a ferrite as the annular magnet instead of the SmCo magnet in Fig. 12.3. The influx of electrons reached 8.31 mA at the ground potential as opposed to 1.95 mA in the BM configuration. Figure 12.5 shows a second deviation coefficient deduced from the I–V characteristics. The plasma space potential in the UBM configuration was estimated to be about −7 V with respect to

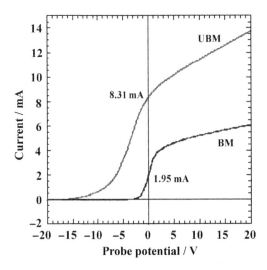

**FIGURE 12.4**

Probe characteristic curve in the UBM and the BM.

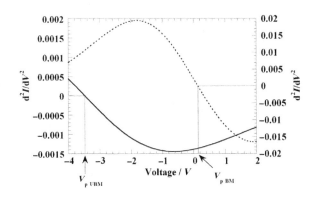

**FIGURE 12.5**

Second deviation coefficient deduced from single probe characteristics of the UBM and the BM.

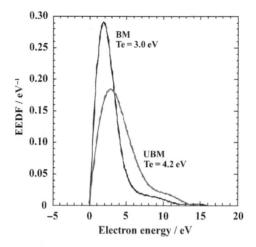

**FIGURE 12.6**

Electron energy distribution function of the UBM and the BM.

the chamber wall at the ground potential, while a slightly positive potential existed in the BM. As shown in Fig. 12.6, electron energy distribution functions deviated via the Druyvesteyn method were shifted to the higher energy side in the UBM, and the electron temperature deduced from the I–V profile was about 4.2 eV as opposed to 3.0 eV in the BM. These facts indicate that high-energy electrons with a four times larger influx struck the growing films at ground potential in the UBM process.

The magnetic field strongly affects these conditions of the electrons. The mobility of electrons is usually much greater than that of ions; therefore, the plasma potential tends to become positive with respect to the chamber wall in the conventional BM sputtering. At the substrate position in the UBM configuration, the magnetic influx incident on the substrate plane-normal is higher than in the BM configuration due to the unbalanced magnetic arrangement. Electrons diffuse along the magnetic and potential field. The high-energy electrons generated in the high-density plasma over the target can diffuse to the surface of the substrate along the magnetic flux without passing into the central magnet. When the component of the electron mobility perpendicular to the magnetic flux is significantly reduced by Lorentz force while the effect is weaker for ions, there can be a magnetic field where the total outflux of electrons from plasma is smaller than that of ions. Maintaining a quasineutral condition, the plasma space potential spontaneously becomes negative to the chamber wall. Consequently, the amount of electron influx to the substrate at ground potential increases.

On the other hand, the strong magnetic field induced by the annular SmCo magnets results in high ion density in the UBM plasma as well as high electron density. Figure 12.7 shows the specimen electric current density in the UBM and the BM as a function of DC discharge current with the parameter of substrate DC bias voltages. Applying a negative bias voltage up to −200 V to the substrate, the ions were incident

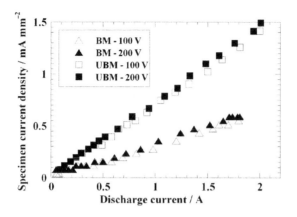

**FIGURE 12.7**

Specimen current density as a function of discharge current in UBM and BM sputtering.

to the substrate at 1.5 mA at the discharge power of 1 kW, and the arrival rate ratio of Ar$^+$/C was estimated to be about 1.4. As well as electrons, the influx of Ar ions in the UBM was about 2.5 times that with the BM, nearly independent of the substrate bias voltage. Using the bombardment conditions of the energetic particles of electrons and ions, the film microstructures can be effectively controlled.

### 12.3.1.2 Synthesis of multilayered films

The a-C films were deposited onto high-speed steel substrates (Hv = 800, diameter = 50 mm × thickness = 8 mm). Tungsten/carbon gradient content adhesive layers were formed prior to the a-C deposition [53]. Using an anisotropic graphite target of 6 inches in diameter, a-C films at up to 1 μm in thickness were synthesized in pure Ar gas at a pressure of $3.0 \times 10^{-1}$ Pa. The base chamber pressure was below $2 \times 10^{-4}$ Pa. The sputtering DC power density was 0.057 W mm$^{-2}$, and the resultant deposition rate was nominally about 0.5 nm s$^{-1}$. The substrate was DC biased with −200 V or grounded. Multilayered structures composed of alternate stacks of amorphous and graphitic layers were formed by periodically changing the applied bias voltages to the substrates.

## 12.3.2 Microstructures and tribological properties

### 12.3.2.1 Microstructures of multilayered film

Figure 12.8 shows a cross-sectional TEM microstructure of an a-C multilayered film stacking with a nominal modulation period of 10 nm. Image contrast in the micrograph is caused by the difference in microstructures. The dark layers of about 8 nm thickness are graphitic layers formed under the conditions of electron irradiation by keeping the substrate at ground potential, while the bright ones of about 2 nm thickness are amorphous layers formed by ion bombardment induced by the negative substrate bias. Figure 12.9(a) and (b) shows plan-view TEM microstructures of single-layer films corresponding to each component layer shown in Fig. 12.8. A number of graphitic clusters were observed in the films deposited under the condition of ground potential (Fig. 12.9(a)). Some of the clusters appeared to have onion-like features. It is presumed that the influx of relatively high-energy electrons to the growing film in the UBM sputtering deposition encouraged evolution of the structural ordering [54]. The film density measured via Rutherford back scattering was estimated to be 2.0–2.1 Mg m$^{-3}$. The highly disordered amorphous structures were formed by applying a negative substrate bias voltage of −200 V due to the peening effects of energetic Ar ion bombardment (Fig. 12.9(b)). The film density was estimated to be 2.6–2.7 Mg m$^{-3}$.

### 12.3.2.2 Tribological properties

The microstructures strongly affected the mechanical properties of the a-C films. Film hardness evaluated by a nanoindentation technique by applying a load of 4.9 mN linearly increased, as the film density increased from about 2.0 to 2.7 Mg m$^{-3}$. The

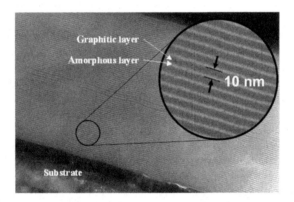

**FIGURE 12.8**

Cross-sectional TEM microstructure of a hybridized carbon multilayered film.

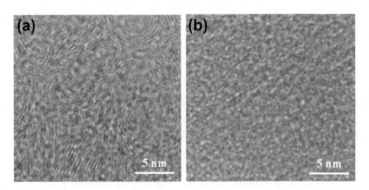

**FIGURE 12.9**

High-resolution TEM micrographs of single-layered films: (a) graphitic film; (b) amorphous film.

multilayered films exhibited an intermediate hardness that follows the so-called rule of mixture. The film hardness increased monotonically from that of the soft layer to a hard one depending on the volume fraction ratio of each layer. The films with a 3 nm/3 nm stacking sequence showed just half the hardness of that of each single layer.

Wear resistance was significantly improved on the multilayered films. Figure 12.10(a) and (b) shows wear resistance of the multilayered film with the stacking sequence of 3 nm/3 nm compared to each single-layered film. The sliding wear was measured in a ball-on-disc test against an $Al_2O_3$ ball of 6 mm diameter. The applied vertical load was 4.9 N. The sliding speed was 100 mm s$^{-1}$. The wear volume after sliding for 2 km and the abrasive wear tracks were evaluated using scanning probe microscopy. The wear rate of the amorphous film was about one-half slower than that of the graphitic film. In the multilayered film, it improved about 30% more. Abrasive wear resistance was evaluated by means of shooting $Al_2O_3$ particles of 30 μm diameter against the film surface at an incident angle of 30°. The shooting pressure was

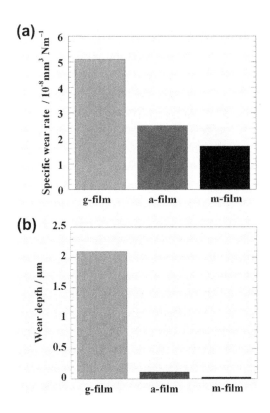

**FIGURE 12.10**

(a) Sliding wear resistance and (b) abrasive wear resistance for three kinds of a-C film: (g-film) a single-layered graphitic film, (a-film) highly disordered amorphous film, and (m-film) hybridized carbon multilayered film.

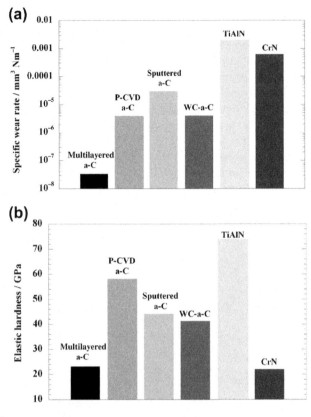

**FIGURE 12.11**

Specific wear volume (a) and elastic hardness (b) of hard coatings.

5 kg cm$^{-2}$, and the flow of the particles was 2.5 l s$^{-1}$. In the abrasive wear test, wear depth in the multilayered film was distinctively decreased up to 0.03 μm as were the amorphous films of 0.11 μm.

The specific wear volume and the microhardness of the hybridized carbon multilayers are given in Fig. 12.11 together with values for the typical wear protection coatings commercially available. The wear volume was measured in a ball-on-disc test against an SUS440C ball in ambient air. The applied vertical load was 1.8 GPa and the sliding speed was 1.6 m s$^{-1}$. The film hardness was evaluated by a microhardness tester with an applied load of 20 mN. The multilayered films exhibited superior wear resistance even though they were not so hard. The specific correlation between wear resistance and film hardness was indistinctive.

Figure 12.12 shows the change in friction coefficient of the multilayered film with a stacking sequence of 3 nm/3 nm compared with each single-layered film. A TiAlN film deposited by an arc ion plating technique served as a control. The friction coefficients against SUJ2 balls in ambient air were measured in a pin-on-disc test.

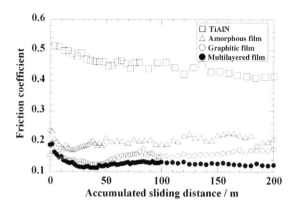

**FIGURE 12.12**

Friction coefficient as a function of sliding distance.

A fixed SUJ2 steel ball 10 mm in diameter was slid back and forth on the a-C films with an applied load of 4.9 N. The reciprocation distance for one passage was 20 mm and the relative sliding speed was 20 mm s$^{-1}$. The friction coefficient during the test was measured up to 200 m of accumulated sliding distance. The depth of the wear track was measured using a stylus surface profiler. The most stable and lowest friction coefficient of 0.11 was obtained with the multilayered film. A comparatively soft film containing graphitic clusters was in the range of 0.14–0.17, while it was about 0.2 for the hard amorphous film.

The intrinsically hard films generally exhibit better wear resistance as shown in Fig. 12.10; however, the harder coatings usually cause a serious reliability problem in practical applications. Figure 12.13 shows a typical failure observed on a sliding wear track of a relatively hard homogeneous a-C film commercially used. Even though the coatings are hard enough to tolerate the applied loads, fragmentation on micrometer-length scale frequently occurs during the tests followed by propagation of delamination leading to severe wear loss. The multilayered structures suppress the generation and propagation of failures due to discontinuity of the film structures [55]. Another reason for the wear resistance is that the multilayered nanocomposite system is expected to reduce the internal stress in the films [56]. In the synthesis of a multilayered system, every new layer is started to form on the surface of previous layers with sufficiently high nucleation sites. This implies that there is a smaller stress generation of curvature effect acting on the surface of a small island in the initial film formation stage occurring in the deposition on a fresh substrate surface [57]. Moreover, the thickness of each layer is quite low. There is no accumulation of strain energy due to the incorporation of excess adatoms during film growth.

The graphitic clusters will play a role as deformable phases to reduce the wear resistance of a-C coatings, while they act as a lubricant. On the other hand, the highly disordered amorphous structures without crystallographic sliding planes of graphite show increased film hardness and generate comparatively high friction, while they

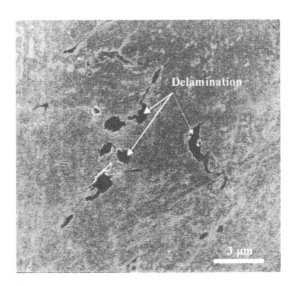

**FIGURE 12.13**

SEM micrograph of a typical microfragmentation occurred on wear tracks in a hard a-C film.

act as a load-supporting component. Thus, the excellent tribological performance of the hybridized carbon multilayered films can be obtained due to segmentation on a nanometer-length scale and the functional combination of different carbon structures, even though the films have intermediate film hardness of each individual component.

## 12.4 HYBRIDIZED CARBON NANOCOMPOSITE FILM WITH VERTICALLY ALIGNED CONFIGURATION

The columnar structures commonly found in vapor-deposited thin films have been classified by what have been termed structure zone models [58]. It is reported that amorphous films as well as polycrystalline films form zone 1 and zone T structures under deposition conditions of low adatom mobility [59,60]. The microstructure in amorphous thin films can be characterized as columns with relatively high density surrounded by lower density regions. The lower density regions usually exhibit network structures, which have been called void networks, crack networks or honeycomb-like networks [61]. Although the characteristics of the lower density region are not fully understood, they have found various unique applications due to their inhomogeneous but quasiperiodic and anisotropic structure.

In this section, a new class of hybridized carbon nanocomposite films will be described. Using column/intercolumn structures that are spontaneously formed in vapor-deposited thin films as a template, vertically aligned configurations were fabricated. The postdeposition treatment with low-energy electron irradiation is a key process in realizing hybridization converting amorphous to graphitic structures at low temperatures.

**FIGURE 12.14**

Low-energy electron radiation system. (For color version of this figure, the reader is referred to the online version of this book.)

### 12.4.1 Synthesis

#### 12.4.1.1 Deposition of column/intercolumn films

The process was divided into two parts, namely, film deposition and postdeposition structural modification. The frame of the vertically aligned anisotropic structures was fabricated by the column growth of a-C films. Hydrogenated a-C thin films having a thickness of 500 nm were deposited onto Si substrates by DC magnetron sputtering in $Ar + CH_4$ gas mixtures at a relatively high pressure up to 4.0 Pa. Evolution of the column/intercolumn structures was observed normal to the film plane. Subsequently, the preformed a-C films were exposed to an electron shower in a vacuum of $1 \times 10^{-3}$ Pa using a low-energy electron radiation system.

#### 12.4.1.2 Low-energy EB irradiation process

Hybridization was accomplished via preferable graphitization of the intercolumn region by low-energy EB irradiation. Figures 12.14 and 12.15, respectively, show an outlook and a schematic diagram of a low-energy EB radiation system (Min-EB manufactured by USHIO, Inc.). The most important feature in this process is that the energy of the electrons can be efficiently applied to a thin layer of the sample surface, and consequently, it is possible to keep the specimen and the system at a relatively low temperature.

The electron emission unit consisting of a thermal cathode and a grid is encapsulated in a glass tube sealed under high vacuum. Electrons from the hot cathode are accelerated by the potential difference between the cathode and the EB transmission window, which is made of a 3-μm Si film. The typical accelerating voltage and current are 60 kV and 0.3 mA, respectively. The electrons are emitted into the

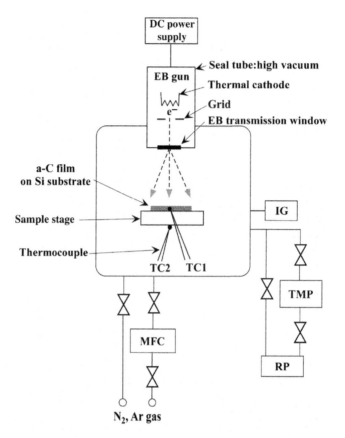

**FIGURE 12.15**

Schematic drawing of low-energy electron radiation system.

radiation chamber and irradiated onto the specimens. When the electrons are transmitted through the Si window, the energy of the electrons is reduced and the beam is scattered. Therefore, irradiation effects strongly depend on the gun voltage and current or the gap between the Si window and the specimen. Transmission depth and the energy absorption rate are generally determined by the electron energy, which is attributed to the accelerating voltage and the material density. The electron path can be approximated by the following equation [62].

$$S = 0.0667 V^{5/3} / \rho$$

where $S$ is the electron path vertical to the material surface, $V$ is the accelerating voltage and $\rho$ is the mass density through which the electrons are transmitted. When the electrons are accelerated by 60 and 30 kV, the energy absorbed in the Si window is approximated to be 7 and 11 keV, respectively. The resultant low electron energy less than 50 keV on the specimen surface suppresses secondary radiation to a sufficiently low level.

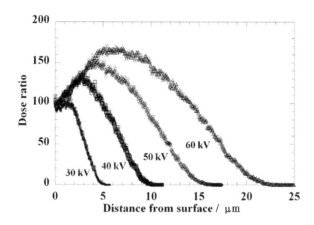

**FIGURE 12.16**

Achievable distance of electrons in Si as a function of accelerated voltage.

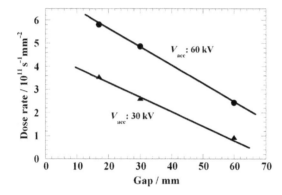

**FIGURE 12.17**

Electron dose rate as a function of accelerated voltage.

Figure 12.16 shows the achievable depth profile of a Si wafer estimated by Monte Carlo simulation. Most of the electron energy is lost within the region up to 10 μm from the top surface, and the maximum achievable distance is about 20 μm at most. Thus, most of the electrons transpierce the carbon films, and the energy is absorbed in the thin surface region of the specimens.

It is possible to conceive three effects as a result of electron irradiation: knock-on, the heating effect via energy transfer and electron–electron interaction. The number of knock-on atoms strongly depends on the primary knock-on atom cross-section and the electron fluence. Figure 12.17 shows the electron dose rate as a function of accelerating voltage and the gap between the Si window and the specimen. Dose rate increased linearly as the gap decreased. Under the conditions where the gun voltage and current were 60 kV and 0.3 mA, respectively, and the gap was 15 mm, the dose rate became $6.0 \times 10^{11}$ s$^{-1}$ mm$^{-2}$. Compared to the convergent intense EB used in TEM for onion formation, which is reported to be in the range of $10^{17}$–$10^{18}$ s$^{-1}$ mm$^{-2}$, the electron dose rate used in this treatment is extremely low. The low energy of the radiated electrons after scattering by the Si window results in a small knock-on cross-section of the carbon atoms. In addition, the low dose rate yields a low fluence even under irradiation for several tens of kilosecond. Consequently, the displacement per atom is approximated to be much less than $10^{-6}$ in the irradiation process. This value is too small to cause a macroscopic knock-on effect.

The heating effect via energy absorption at the surface of the Si substrate is presumed to play a dominant role in the formation of graphitic structures. The specimen temperature was not controlled, but the temperature measured at about 50 μm underneath the film surface did not exceed 450 K during electron irradiation. Figure 12.18 shows the change in temperature of the specimen and the stage during EB irradiation. The specimen temperature quickly increased as the EB irradiation started and became saturated at about 450 K in 2 ksec. The temperature of the sample stage gradually increased up to about 310 K. Hydrogen atoms that chemically bond to a-C networks significantly desorbed at an elevated temperature of more than 850 K. When 40at%H-containing a-C films were subjected to electron irradiation, a change in only a couple of atomic percentage of hydrogen content was obtained by SIMS (secondary ion mass spectrometry) analysis, which was in the range of experimental error. Therefore, it was estimated that the heating effect of the electron irradiation corresponds to about 800 K at most [63].

Electron–electron interaction may affect structural modification in a short-scale range. In a Raman spectrum, graphitization was recognized in the film with a relatively short-time EB irradiation of 1–2 min. Moreover, with EB irradiation for 10 s,

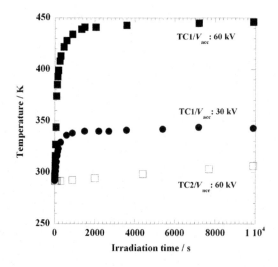

**FIGURE 12.18**

Change in specimen and stage temperature during electron irradiation.

a change in transparency probably due to graphitization was observed in the range of the visible and infra red spectrum.

## 12.4.2 Structures and tribological properties of hybridized column/intercolumn film

### 12.4.2.1 Structural modification induced by low-energy electron irradiation

Figure 12.19(a) shows a plan-view TEM microstructure of an as-deposited film of 40at%H:a-C. The microstructures correspond to the columnar structures commonly found in vapor phase-synthesized thin films as shown in an insert in the schematic illustration in Fig. 12.19. The specific unit sizes, such as size of the column regions and width of the intercolumn regions, are roughly determined by film thickness and deposition conditions [64]. The dominant film structures are reported to evolve with film thickness in accordance with a power low relation [65]. Each unit size increased linearly with the 0.8 power of the film thickness, independent of gas pressures. The column sizes were 1/3 of those of the dominant surface morphology. This relationship between the dominant structural units size and film thickness is the same as that previously noted in a-Ge and a-Si:H by Messier et al. [66].

Figure 12.19(b) shows an energy-filtered TEM micrograph at 0 eV after electron irradiation for 3.6 ksec. Comparing the irradiated film to the as-deposited film, the reversal of image contrasts between the column and intercolumn regions was distinctive [55,63]. Analyzing the corresponding zero-loss spectrum, the intercolumn region was found to transform from 12% lower density to a 5% larger density relative to the column region, which remained at 1.8 Mg m$^{-3}$. That is, the density of the intercolumn region increased from 1.6 to 1.9 Mg m$^{-3}$. In addition, the plasmon energy of

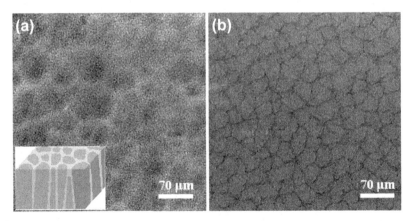

**FIGURE 12.19**

Energy-filtered TEM micrographs of a-C films with column/intercolumn structures (a) before and (b) after electron irradiation.

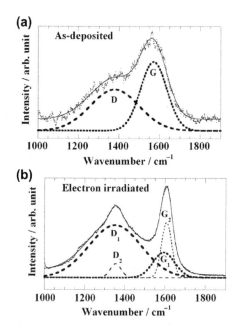

**FIGURE 12.20**

Raman spectra before (a) and after (b) electron irradiation.

$\pi + \sigma$ excitation in this region distinctively shifted from 16 to 24 eV, which is about 1 eV more than that of the column regions. Using the Drude model, the valence electron density in the intercolumn region can be approximated to be at least about 32% less than that in the column region.

Figure 12.20(a) and (b) shows Raman spectra using an Ar (488 nm) laser obtained before and after electron irradiation for 120 s. The spectra can be fitted by sets of a pair of a D peak at around 1350 cm$^{-1}$ and a G peak at around 1500–1600 cm$^{-1}$, which are attributed to the $A_{1g}$ and $E_{2g}$ symmetry mode of graphite, respectively. A D-G peak pair fit the profile of the as-deposited film. Applying the Tuinstra–Koenig relationship [13], the in-plane correlation length of the graphitic structure (La) was estimated to be less than 1 nm. After electron irradiation, the Raman spectrum was fitted by two D-G peak pairs: a broad peak set at La = 2 nm and a relatively sharp peak set at La = 16 nm.

The observed TEM image contrast presumably comes from a quantitative difference in structural components among the regions. The intercolumn structure is formed as a result of the effects of shadowing and surface diffusion during film deposition. Such intercolumn regions usually include many defects and antibonds. The defective structures are thought to be unstable and can be easily modified by EB irradiation. Because distinctive structural changes were not recognized in the column regions, structural modification involving graphitization primarily occurred in the intercolumn regions. Considering the significant change in TEM image contrast and

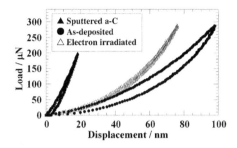

**FIGURE 12.21**

Load-displacement curves obtained during nanoindentation testing.

the low process temperature at which this transformation occurred, ordered graphitic structures were preferably formed in the intercolumn region probably due to unstable loose bonding of carbon atoms inferred from the substantially lower density.

The width of the intercolumn region appeared to be shrunk after electron irradiation. The width of around 10 nm is smaller than the correlation length of graphitic structures. In addition, the shrinkage likely brings about internal stress normal to the column boundaries. The c-plane of graphite tends to arrange normal to the direction of the stress due to the significant difference in stiffness between the in-plane and out-of-plane directions. Therefore, the c-plane of the graphitic structures likely formed along the column structures orienting parallel to the normal axis of the film plane.

### 12.4.2.2 Novel tribological properties induced by vertically hybridized structures

Figure 12.21 shows load-displacement curves for the nanocomposite films and a typical sputter-deposited a-C film used in hard protective coatings. The elastic modulus and plastic deformation hardness of the as-deposited film were determined to be about 15 GPa and about 1.6 GPa, respectively. The mechanical properties of the irradiated film improved only slightly to 20.9 GPa in elastic modulus and 2.3 GPa in hardness. The curve of the irradiated film showed a minimal plastic deformation regime. The elastic recovery increased to about 50% from about 35% in the as-deposited film. On the other hand, the typical sputtered a-C film showed 138.1 GPa in elastic modulus and 13.6 GPa in hardness. Thus, the mechanical properties of the column/intercolumn film are at least 1/6–1/7 of the typical a-C film. The properties are equivalent to so-called glassy carbon that is generally known as a very soft material.

In order to study the mechanical properties on a smaller scale, the films were examined by scanning probe microscopy. Figure 12.22 shows topographic and elastic image profiles for the as-deposited film and the modified film. Higher intensity in the elastic profiles means that the regions are relatively hard. The concave regions (indicated by arrows in Fig. 12.22(a)) correspond to the intercolumn regions. Surface roughness of the modified films was almost the same as that of the as-deposited films. In the as-deposited film, the intercolumn region showed

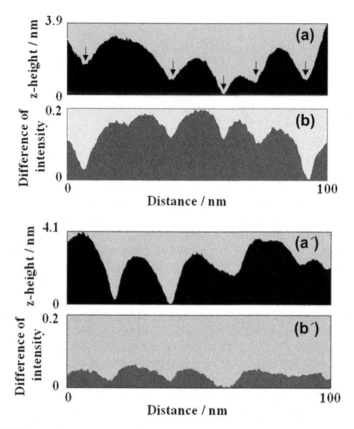

**FIGURE 12.22**

Topographic (a, a′) and elastic image (b, b′) profiles measured by SPM: (upper images) as-deposited film, (lower images) after electron irradiation.

lower mechanical properties as expected from its lower density than the column regions. It is noted that the difference in relative intensity in the elastic image profiles between the column and intercolumn regions was reduced significantly. The results imply that the mechanical properties in the intercolumn region were significantly improved probably due to the structural modification induced by electron irradiation.

Figure 12.23 shows wear development curves for each kind of film. The multiple scratching tests were performed by a Triboindentor® (manufactured by Hysitron, Inc.) at a constant load of 10 μN on an area of 2 μm by 2 μm. The as-deposited film exhibited severe wear of about 27 nm in 10 cycles of scanning, because the film contains low-density regions and is substantially soft. On the other hand, the modified film exhibited superior wear resistance with almost no wear in 20 cycles and with only 1.5 nm in wear depth after 50 cycles. This performance means that only one atomic layer was removed in about 20 scanning cycles. The superb wear resistance is

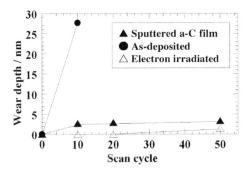

**FIGURE 12.23**

Wear depth as a function of reciprocating scratch cycles.

equivalent to the reported performance of a superhard cubic-boron-nitride film [67]. The specific wear factor $Ws = V/WL$, where $V$ is the wear volume, $W$ is the load, and $L$ is the total scan length, was $10^{-5}$ mm$^3$ N$^{-1}$ m$^{-1}$ for the modified film. The value was two orders of magnitude smaller than that of the as-deposited film and one order superior to the typical sputtered film.

Even in macroscopic tribological properties evaluated by a ball-on-disc test with a load of 50 N and a sliding speed of 31 mm s$^{-1}$, the irradiated film exhibited significant improvement in wear resistance and a stable friction coefficient in the range of 0.15–0.18 [68].

The superior wear resistance can be attributed to the selective structural modification via the electron irradiation. The formation of graphitic structures in the intercolumn regions brings out hardening of the nanoscaled regions. The phenomena occurring in wear processes appear to be dissimilar to those of the multilayered films with intermediate hardness described in Section 3. The differences are distinctive in the correlation between film hardness and wear resistance induced by the mechanical properties and spatial arrangements of the components. In the films before electron irradiation, both the components are low density, and the soft film naturally exhibits poor wear resistance. In the irradiated films, an increase in density resulting from graphitization at the intercolumn regions improved the local strength, but the regions still remained rather low density. The column regions preserve the same structures and low density as before electron irradiation. The resultant film hardness is too deficient to attain the obtained high wear resistance. Those results imply that there is another factor that acts to improve wear resistance in the films with low hardness. In microtribology, the physical and chemical properties of film surface are generally the dominant factors rather than the mechanical properties of the bulk or of each component of the nanocomposite structures. However, a high wear resistance was obtained in a modified film with low film hardness. This fact indicates that not only the surface structures exposing the dual phases but also the anisotropic inner structures play an important role in the excellent wear resistance.

### 12.4.2.3 Wear resistance in dry microstamping

With increasing drive toward further size reduction of electrical devices and products, manufacturing of electrical parts has faced requirements of miniaturization with the geometry of millimeters/submillimeters with strict accuracy in tolerance. Lubrication is essential in machining and forming, but lubricating oils are unfavorable for the small-sized electrical parts. It is not only because reduction of lubricating oil consumption becomes a critical issue to improve the environmental friendliness in manufacturing processes, but also because complete removal of oils on the work surface is difficult or nearly impossible even by using a large amount of cleansing agents. In the challenge of highly precise production of miniaturized parts and components coupled with green manufacturing, a dry precise stamping using a self-lubricant protective coating is a unique environmentally benign processing.

A microstamping test was designed and performed to demonstrate that the structurally modified column/intercolumnar coating has superb wear resistance even in severe bending and ironing under dry conditions [69,70]. The progressive stamping sequence was composed of shearing and bending as shown in Fig. 12.24. One sequence advanced in 1 s. In the step of shearing process, holes were punched out in a sheet of AISI-304 austinitic stainless steel with the thickness of 0.211 mm without any significant burrs only by using the cemented carbide tools and dies. The cemented carbide punch was coated with the nanocomposite film with the thickness of 1 μm. Figure 12.25 shows a successfully formed part after one progressive stamping. The work sheet is first bended and ironed to fix the bended configuration with the angle of 90°. In the bending and ironing step, severe wear might occur on the ironed surface between punch and work sheet since the total clearance between die and punch is strictly fixed by the tolerance of much less than 5% of the sheet material thickness.

Figure 12.26(a) and (b) compares the ironed surface of coated punch between two tools with the as-deposited film, namely, without electron irradiation, and with

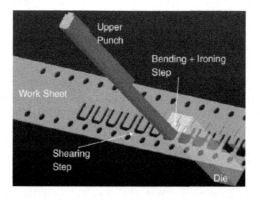

**FIGURE 12.24**

Schematic drawing of dry microstamping test with progressive stamping sequence. (For color version of this figure, the reader is referred to the online version of this book.)

the modified film after bending and ironing a 100 parts. Without electron irradiation, the as-deposited film broke away and separated partially from the cemented carbide tool in only a few steps of the progressive stamping sequence. On the contrary, the irradiated film was still effective without significant delamination even after stamping 2000 parts. Although a little scratched trace was observed on the surface, no increase in stamping load was experienced in the process. The above-mentioned difference in dry stamping performance can be attributed to the highly elastic deformability and ductility coupled with superior wear resistance induced by the structural modification.

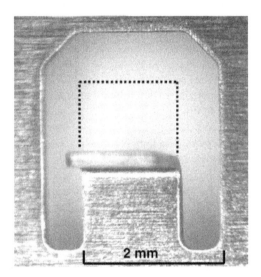

**FIGURE 12.25**

Sample part after bending and ironing steps in progressive microstamping. (For color version of this figure, the reader is referred to the online version of this book.)

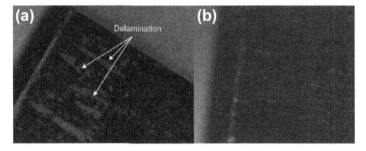

**FIGURE 12.26**

Contact surface of coated punch after bending and ironing for 100 parts: (a) without electron irradiation; (b) with electron irradiation. (For color version of this figure, the reader is referred to the online version of this book.)

### *12.4.2.4 Pliability of hybridized a-C films on polyimide film substrates*

a-C films generally have a problem of adhesion to other materials due to their strong carbon–carbon bonding in the amorphous networks. Hard a-C coatings are usually fabricated with the assistance of ion bombardment during film growth in order to achieve higher density. They usually show extremely high internal compressive stress due to peening effects. The high compressive stresses frequently cause buckling and wrinkle instability, or complete delaminations, particularly when there is an imperfection on the surface of the substrates [71,72]. The hard a-C coatings have been applied as a protective coating on diverse flexible polymeric substrates, and various practical applications were found for sealing and sliding parts, such as rubber O-rings for a compact camera [73]. Although these materials exhibit excellent wear resistance and a sufficiently low friction coefficient, they still include buckles and wrinkles, or cracks on the film surface, when they are subjected to significant bending or a severe load.

One of the unique properties of the hybridized a-C thin films with vertically aligned components is that they are pliable to deformation of the soft substrates. Figure 12.27 shows the surface of an a-C film with the structurally modified column/intercolumn structures after a bending test of 100 cycles. The a-C film of 500-nm thickness was deposited onto a polyimide film (Kapton® H) substrate and exposed to a low-energy electron shower for 5 min. The top surface structures were observed by field emission scanning electron microscope (SUPRA™ 35 manufactured by Carl Zeiss, Inc.) with a low accelerating voltage of 0.4 kV. The bending test was carried out by sliding the films along the periphery of a cylinder of 5 mm diameter as shown in the insert in Fig. 12.24. No delamination failure, such as buckles, wrinkles or cracks, was found on a macro- and microscopic scale even after harsh deformation of the soft substrates.

## 12.5 CONCLUSIONS

It has been demonstrated that excellent tribological performance can be obtained in hybridized carbon nanocomposite films consisting of amorphous and graphitic segments. Intermediate or even low film hardness of the hybridized carbon films resulted

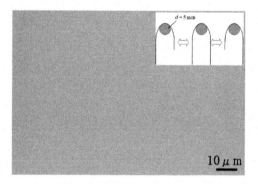

**FIGURE 12.27**

Surface of the hybridized a-C film on polyimide substrate after bending test for 100 cycles.

in superior wear resistance. The performances are dissimilar to those of the ordinal superhard nanocomposite films that significantly exceed the hardness given by the rule of mixture. These facts imply that by employing an appropriate arrangement on a nanometer-length scale and a combination of tribological properties, superhardness is not necessary to achieve distinctive wear resistance. It is particularly noted that the superb wear resistance equivalent to a superhard film was obtained in vertically aligned nanocomposite configurations fabricated by structural modification of column/intercolumn structures. It is suggested that the approach using three-dimensionally arranged carbon nanocomposite structures might provide a solution to the conventional loss of reliability in the hard coating industry.

## Acknowledgments

Thanks are due to Kiyoshi Yamamoto, Hiroko Okano and Haruyoshi Niki of Kobelco Research Institute, Inc., for their technical assistance in the energy-filtered TEM and FESEM experiments. Thanks are also due to Masanori Yamaguchi of USIO, Inc., for his help in the low-energy electron irradiation experiments.

## References

[1] P.H. Mayrhofer, C. Mitterer, L. Hultman, H. Clemens, Microstructural design of hard coatings, Progress in Materials Science 51 (2006) 1032–1114.

[2] S. Veprek, M.G.J. Veprek-Heijman, P. Karvanva, J. Prochazka, Different approaches to superhard coatings and nanocomposites, Thin Solid Films 476 (2005) 1–29.

[3] S. Veprek, R.F. Zhang, M.G.J. Veprek-Heijman, S. Sheng, A.S. Argon, Superhard nanocomposites: origin of hardness enhancement, properties and applications, Surface and Coatings Technology 204 (2010) 1898–1906.

[4] S. Veprek, Recent attempts to design new super- and ultrahard solid leads to nano-sized and nano-structured materials and coatings, Journal of Nanoscience and Nanotechnology 11 (2011) 14–35.

[5] M. Chhowalla, G.A.J. Amaratunga, Thin films of fullerene-like $MoS_2$ nanoparticles with ultra-low friction and wear, Nature 407 (2000) 164–167.

[6] L. Hultman, S. Stanfstrom, Z. Czigany, J. Neidhardt, N. Hellgren, I.F. Brunell, K. Suenaga, C. Colliex, Cross-linked nano-onions of carbon nitride in the solid phase: existence of a novel $C_{48}N_{12}$ Aza-fullerene, Physical Review Letter 87 (2001) 225503/1–225503/4.

[7] J. Neidhardt, L. Hultman, Beyond β-$C_3N_4$-Fullerene-like carbon nitride; A promising coating material, Journal of Vacuum Science Technology A25 (2007) 633–644.

[8] C. Muratore, A.A. Voevodin, Chameleon coatings: adaptive surfaces to reduce friction and wear in extreme environments, Annual Review of Materials Research 39 (2009) 297–324.

[9] H. Friedman, O. Eidelman, Y. Feldman, A. Moshkovich, V. Perfilier, L. Rapoport, H. Cohen, A. Yoffe, R. Tenne, Fabrication of self-lubricating cobalt coatings on metal surfaces, Nanotechnology 18 (2007) 115703/1–115703/8.

[10] O. Eidelman, H. Friedman, R. Rosentsveig, A. Moshkovith, V. Perfilier, S.R. Cohen, Y. Feldman, L. Rapoport, A. Yoffe, R. Tenne, Chromium-rich coatings with $WS_2$ nanoparticles coating fullerene-like structure, NANO 6 (2011) 313–324.

[11] A. Erdemir, C. Donnet, Tribology of diamond-like carbon films: recent progress and future prospects, Journal of Physics D 39 (2006) R311–R327.

[12] S.V. Hainsworth, N.J. Uhure, Diamond like carbon coatings for tribology: production techniques, characterization methods and applications, International Materials Reviews 52 (2007) 153–174.

[13] J. Robertson, Diamond-like amorphous carbon, Materials Science and Engineering R37 (2002) 129–281.

[14] E. Iwamura, Characterization of amorphous carbon hard coating films and tribology, Materia Japan 41 (2002) 635–643.

[15] D. Ugarte, Curling and closure of graphitic networks under electron-beam irradiation, Nature 359 (1992) 707–709.

[16] F. Banhart, The role of lattice defects in the formation of new carbon structures under electron irradiation, Journal of Electron Microscopy 51 (2002) S189–S194.

[17] H. Sjostrom, S. Stafstrom, M. Boman, J.-E. Sundgren, Superhard and elastic carbon nitride thin films having fullerenelike microstructure, Physical Review Letter 75 (1995) 1336–1339.

[18] D. Roy, M. Chhowalla, N. Hellgen, T.W. Clyne, G.A.J. Amaratunga, Probing carbon nanoparticles in $CN_x$ thin films using Raman spectroscopy, Physical Review B 70, 2004. 035406/1–6.

[19] R. Gago, G. Abrasonis, A. Muechlich, W. Moeller, Zs. Czigany, G. Radnoczi, Fullerenelike arrangement in carbon nitride thin films grown by direct ion beam sputtering, Applied Physics Letter 87, 2005. 071901/1–3.

[20] M. Chhowalla, R.A. Ahrronov, C.J. Kiely, I. Alexandrou, G.A.J. Amaratunga, Generation and deposition of fullerene- and nanotube-rich carbon thin films, Philosophical Magazine Letter 75 (5) (1997) 329–335.

[21] I. Alexandrou, C.H. Kelly, A.J. Papworth, G.A.J. Amaratunga, Formation and subsequent inclusion of fullerene-like nanoparticles in nanocomposite carbon thin films, Carbon 42 (2004) 1651–1656.

[22] A.P. Burden, S.R.P. Silva, Fullerene-like carbon nanoparticles generated by radio-frequency plasma-enhanced chemical vapor deposition, Philosophical Magazine Letter 78 (1998) 15–19.

[23] M. Veres, M. Fuele, M. Koos, I. Pocsik, J. Kokavecz, Z. Toth, G. Randnoczi, Simultaneous preparation of amorphous solid carbon films, and their cluster building blocks, Journal of Non-Crystalline Solids 351 (2005) 981–986.

[24] I. Alexandrou, H.-J. Scheibe, C.J. Kiely, A.J. Papworth, G.A.J. Amaratunga, B. Schultrich, Carbon films with an $sp^2$ network structure, Physics Review B 60 (1999) 10903–10907.

[25] V. Lavrentiev, H. Abe, S. Yamamoto, H. Naramoto, K. Narumi, Formation of carbon nanotubes under conditions of Co + C60 film, Physica B 323 (2002) 303–305.

[26] C.S. Lee, T.Y. Kim, K.R. Lee, J.P. Ahn, K.H. Yoon, Nanocomposite ta-C films prepared by the filtered vacuum arc process using nanosized Ni dots on a Si substrate, Chemical Physics Letter 380 (2003) 774–779.

[27] H. Abe, S. Yamamoto, A. Miyashita, K. Sickafus, Formation mechanisms for carbon onions and nanocapsules in C+-ion implanted copper, Journal of Applied Physics 90 (2001) 3353–3358.

[28] T. Cabioc'h, E. Thune, J.P. Riviere, S. Camelio, J.C. Girard, P. Guerin, L. Henrard, Ph. Lambin, Structure and properties of carbon onion layers deposited onto various substrates, Journal of Applied Physics 91 (2002) 1560–1567.

[29] D.L. Baptista, F.C. Zawislak, Hard and $sp^2$-rich amorphous carbon structure formed by ion beam irradiation of fullerene, a-C and polymeric a-C:H films, Diamond and Related Materials 13 (2004) 1791–1801.

[30] T.J. Konno, R. Sinclair, Crystallization of amorphous carbon in carbon-cobalt layered thin films, Acta Metallurgica Et Materialia 43 (1995) 471–484.

[31] B.S. Xu, S.-I. Tanaka, Formation of giant onion-like fullerenes under Al nanoparticles by electron irradiation, Acta Materialia 46 (1998) 5249–5257.

[32] T. Nishijima, R. Ueki, Y. Miyazawa, J. Fujita, Artificial tailoring of carbon nanotube and its electrical properties under high-resolution transmission electron microscope, Microelectronic Engineering 88 (2011) 2519–2523.

[33] S. Ravi, P. Silva, X. Xu, B.X. Tay, H.S. Tan, W.I. Milne, Nanocrystallites in tetrahedral amorphous carbon films, Applied Physics Letter 69 (1996) 491–493.

[34] P. Patsalas, S. Logothetidis, P. Douka, M. Gioti, G. Stergioudis, Ph. Komninou, G. Nouet, Th. Karakostas, Polycrystalline diamond formation by post-growth ion bombardment of sputter-deposited amorphous carbon films, Carbon 37 (1999) 865–869.

[35] W. Kulisch, C. Popov, On the growth mechanisms of nanocrystalline diamond films, Physica Status Solidi 203 (2006) 203–219.

[36] O.S. Panwar, S. Sushil, Ishpal, A.K. Srivastava, A. Chouksey, R.K. Tripathi, A. Basu, Effect of substrate bias in nitrogen incorporated amorphous carbon films with embedded nanoparticles deposited by filtered cathodic jet technique, Materials Chemistry and Physics 132 (2012) 659–666.

[37] A.G. Ramirez, T. Itoh, R. Sinclair, Crystallization of amorphous carbon thin films in the presence of magnetic media, Journal Applied Physics 85 (1999) 1508–1513.

[38] F. Banhart, N. Grobert, M. Terrones, J.C. Charlier, P.M. Ajayan, Metal atoms in carbon nanotubes and related nanoparticles, International Journal of Modern Physics B15 (2001) 4037–4069.

[39] J. Fujita, M. Ishida, T. Ichihashi, Y. Ochiai, T. Kaito, S. Matsui, Graphitization of Fe-doped amorphous carbon pillars grown by focused-ion-beam-induced chemical-vapor deposition, Journal of Vacuum Science and Technology B20 (2002) 2686–2689.

[40] E. Iwamura, Structural ordering of metal-containing amorphous carbon thin films induced by low-energy electron beam projection, Review Advanced Material Science 5 (2003) 34–40.

[41] K. Higashi, M. Ishida, S. Matsui, J. Fujita, Driving force of an iron particle's movement in solid-phase graphitization, Japanese Journal of Applied Physics 46 (2007) 6282–6285.

[42] S. Aikawa, T. Kizu, E. Nishikawa, Catalytic graphitization of an amorphous carbon film under focused electron beam irradiation due to the presence of sputtered nickel metal particles, Carbon 48 (2010) 2997–2999.

[43] W.A. de Heer, D. Ugarte, Carbon onions produced by heat treatment of carbon soot and their relation to the 217.5 nm interstellar absorption feature, Chemical Physics Letter 207 (1993) 480–486.

[44] A. Oberlin, Pyrocarbons, Carbon 40 (2002) 7–24.

[45] F. Banhart, T. Fuller, Ph. Redlich, P.M. Ajayan, The formation, annealing and self-compression of carbon onions under electron irradiation, Chemical Physics Letter 269 (1997) 349–355.

[46] F. Banhart, P. Redlich, P.M. Ajayan, The migration of metal atoms through carbon onions, Chemical Physics Letter 292 (1998) 554–560.

[47] V. Presser, M. Heon, Y. Gogotsi, Carbon-derived carbons-from porous networks to nanotubes and graphene, Advanced Functional Materials 21 (2011) 810–833.

[48] E. Iwamura, T. Aizawa, Fabrication of patterned domains with graphitic clusters in amorphous carbon using a combination of ion implantation and electron irradiation techniques, Materials Research Society Symposia Proceedings 908E, (2006) OO11.5.1.

[49] E. Iwamura, T. Aizawa, Nano-graphitization in amorphous carbon films via electron beam irradiation and the iron implantation, Materials Research Society Symposia Proceedings 960E, (2007) N12–04.

[50] G.A.J. Amaratunga, M. Chhowalla, C.J. Kiely, I. Alexandrou, R. Ahrronov, R.M. Devenish, Hard elastic carbon thin films from linking of carbon nanoparticles, Nature 383 (1996) 321–323.

[51] B. Window, N. Savvides, Charged particle fluxes from planar magnetron sputtering sources, Journal of Vacuum Science and Technology A4 (1986) 196–202.

[52] B. Window, G.L. Harding, Ion-assisting magnetron sources: principles and uses, Journal of Vacuum Science Technology A8 (1990) 1277–1282.

[53] E. Iwamura, N. Matsuoka, M. Takeda, T. Miyamoto, K. Akari, Adhesion strength improvement of a-C films by employing W-C amorphous intermediate layers, in: Proc. 2nd Inter. Conf. on Advanced Mater. Development and Performance, Tokushima, 1999, pp. 83–88.

[54] E. Iwamura, Friction and wear performance of unbalanced magnetron sputter-deposited diamond-like carbon multiplayer coatings, in: Proc. 2nd Inter. Conf. Processing Mater. for Properties, San Francisco, 2000, pp. 263–266.

[55] T. Aizawa, E. Iwamura, T. Uematsu, Formation of nano-columnar amorphous carbon films via electron beam irradiation, Journal of Material Science 43 (2008) 6159–6166.

[56] M. Gioti, S. Logothetidis, C. Charitidis, Stress relaxation and stability in thick amorphous carbon films deposited in layer structure, Applied Physics Letter 73 (1998) 184–186.

[57] F. Spaepen, Interface and stresses in thin films, Acta Materialia 48 (2000) 31–42.

[58] J.A. Thornton, High rate thick film growth, Annual Review of Materials Science 7 (1977) 239–260.

[59] R. Messier, A.P. Giri, R.A. Roy, Revised structure zone model for thin film physical structure, Journal of Vacuum Science and Technology A2 (1984) 500–503.

[60] G.S. Bales, A. Zangwill, Macroscopic model for columnar growth of amorphous films by sputter deposition, Journal of Vacuum Science and Technology A9 (1991) 145–149.

[61] A. Staudinger, S. Nakahara, The structure of the crack network in amorphous films, Thin Solid Films 45 (1977) 125–133.

[62] S. Schiller, Electron Beam Technology, John Wiley & Sons, New York, 1983.

[63] E. Iwamura, M. Yamaguchi, Nano-structural modification of amorphous carbon thin films by low-energy electron beam irradiation, Transaction Material and Heat Treatment 25 (2004) 1247–1252.

[64] E. Iwamura, Characterization of nanometer-scale columnar and low-density boundary network structures in hydrogenated amorphous carbon films, Ceramic Transaction 148 (2003) 139–146.

[65] R. Messier, Toward quantification of thin film morphology, Journal of Vacuum Science and Technology A4 (1986) 490–495.

[66] R. Messier, J.E. Yehoda, Geometry of thin-film morphology, Journal of Applied Physics 58 (1985) 3739–3746.

[67] S. Miyake, S. Watanabe, M. Murakawa, R. Kaneko, T. Miyato, Tribological study of cubic boron nitride film, Thin Solid Films 212 (1992) 262–266.

[68] T. Aizawa, E. Iwamura, unpublished data.

[69] T. Aizawa, E. Iwamura, K. Itoh, Development of nano-columnar carbon coating for dry micro-stamping, Surface and Coatings Technology 202 (2007) 1177–1181.

[70] T. Aizawa, E. Iwamura, K. Itoh, Nano-lamination in amorphous carbon for tailored coating in micro-dry stamping of AISI-304 stainless steel sheets, Surface and Coatings Technology 203 (2008) 794–798.

[71] M.-W. Moon, J.-W. Chung, K.-R. Lee, K.H. Oh, R. Wang, A.G. Evans, An experimental study of the influence of imperfections on the buckling of compressed thin films, Acta Materialia 50 (2002) 1219–1227.

[72] S. Faulhaber, C. Mercer, M.-W. Moon, J.W. Hutchinson, A.G. Evans, Bucking depamination in compressed multilayers on curved substrates with accompanying ridge cracks, Journal of Mechanics and Physics Solids 56 (2006) 1004–1028.

[73] T. Nakahigashi, Y. Tanaka, K. Miyake, H. Oohara, Properties of flexible DLC film deposited by amplitude-modulated RF P-CVD, Tribology International 37 (2004) 907–912.

# Sliding friction and wear of "nanomodified" and coated rubbers

# 13

**József Karger-Kocsis***, **Dávid Felhős**[†]

*MTA–BME Research Group for Composite Science and Technology and Department of Polymer Engineering, Faculty of Mechanical Engineering, Budapest University of Technology and Economics, Budapest, Hungary, [†]Department of Polymer Engineering, Faculty of Materials Science and Engineering, University of Miskolc, Miskolc, Hungary*

## CHAPTER OUTLINE HEAD

## 13.1 INTRODUCTION

The practical importance of rubbers as constructional materials does not need detailed elaboration. The most spectacular properties of rubbers are their high strain to break (several hundred percent), exceptional resilience and low modulus. Based on these properties the dimensional tolerance of rubber is relatively uncritical, which is in favor of the mass production of different seals [1]. The high dry sliding friction of rubbers is a clear disadvantage for seals but one of the major benefits for tire tread compounds. Because a fundamental understanding of the friction and wear with rubbers is still lacking, the related material development is of empirical feature. Fueled by tire applications, a large body of work was devoted to assess the abrasion behavior of rubbers (e.g. [2–7]. and references

**Tribology of Polymeric Nanocomposites. http://dx.doi.org/10.1016/B978-0-444-59455-6.00013-1**

therein). These studies resulted in suitable predictions for the abrasion, however, only for certain arrangements and conditions. The wear of polymers, including rubbers, is differently classified (e.g. [8,9]), emphasizing various aspects. This already hints for the complexity of rubber tribology. The friction and wear of rubbers differ from those of most of the other materials as apart from cohesive and wear components considerable energy dissipation may occur by hysteresis. Hysteretic heating is characteristic for all viscoelastic materials (such as rubbers). This comes from the viscoelastic deformation ("flow") of the rubber over the surface topography (asperities) of the "hard" counterpart. Fundamental works on the friction and wear of rubbers have been performed until the end of the 1980s (e.g. [2,5,7,10]). The interest renewal for this topic is likely due to the following aspects:

1. Appearance of novel modifiers and formulation methods for rubbers,
2. Industrial need to model (i.e. "to predict") the behavior of the rubber parts under the foreseen loading collectives, and
3. Novel possibilities for the surface texturing of the rubber and the counterparts (e.g. laser patterning).

Nowadays, there is a clear shift in the R&D activities with rubbers from abrasion toward sliding wear. Sliding friction and wear are key issues for technical rubber goods such as seals, bearings and crankshafts. They are often low-cost products but the consequence of their failure can be very high – just remember the explosion of the Challenger shuttle in 1986, which was attributed to the failure of joints sealed by O-rings [1]. A similar scenario can be transferred to everyday life by considering, for example, the brake system of automobiles. Accordingly, recent targets of material development are improved reliability, prolonged lifetime, and maintenance-free service. Needless to say, this demand can only be met when material development is accompanied by proper design and suitable manufacturing process.

This chapter is aimed at showing the potential of recent formulation and coating strategies for rubbers in which "nanomodification" and "nanostructuring" are the key issues. In order to determine and collate the tribological characteristics of different rubber recipes, tests were performed under dry sliding conditions against smooth counterparts (mostly steel). It should be kept in mind that during dry sliding the contribution of the interfacial wear should be markedly larger than that of the hysteretic contribution [8]. The configurations of the test rigs used are depicted schematically in Fig. 13.1. The test conditions selected agree fairly with the service conditions of rubber seals in reciprocating sliding and in high-frequency oscillation (fretting) duties. Accordingly, the sliding speeds and apparent contact pressures were usually below 0.25 m s$^{-1}$ and 2.5 MPa, respectively (Fig. 13.1).

## 13.2 RUBBER (NANO) COMPOSITES

### 13.2.1 Fillers

It is intuitive that the coefficient of friction (COF) regulates the stresses transmitted between the sliding surfaces and thereby the COF controls the deformation and

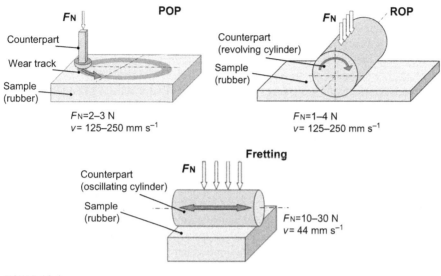

**FIGURE 13.1**

Schemes of the tribotesting devices used. POP, pin (steel) on plate (rubber); ROP, roller (steel) on plate (rubber). This figure indicates also the usual testing conditions.

thus wear behavior. Lubricants allow the loaded counterfaces to slide over each other with a minimum COF corresponding to a minimum tangential resistance [11].

Traditional rubber mixes and thermoplastic rubbers often contain processing oils. Although their primary function is to support homogenization and guarantee easy processing, the oil may work as a lubricant, too. However, controlled release of the oil to the contact area can hardly be achieved. Nowadays, trials are in progress to make use of the microcapsulation technique. However, liquid lubrication is at odds with the strategy of maintenance-free engineering design. The latter is the driving force to check the potential of solid lubricants. There is a large variety of solid-phase lubricants (e.g. [11]); however, only a few are used for polymeric systems: stearic acid, waxes, $MoS_2$, graphite, ultrahigh molecular weight polyethylene and especially poly(tetrafluoro ethylene) (PTFE) (e.g. [12,13]). It has to be borne in mind that rubber recipes contain stearic acid and even waxes, however, for reasons other than lubrication. Stearic acid is a steady ingredient in sulfur-curing recipes. It helps the activation process via a zinc-based transition complex by the "organophilization" of the zinc compound (viz. ZnO). Waxes are used as antiozonants. It has been earlier recognized that PTFE is an excellent solid lubricant. PTFE became recently available in micron- and even nanoscaled powders. Micronized poly(ether ether ketone) (PEEK) is now also commercialized. PEEK and related high temperature-resistant thermoplastics are outstanding wear-resistant materials both under sliding and abrasive conditions (e.g. [12,14,15]). They form the matrices of injection moldable short glass and carbon fiber-reinforced composites, which usually contain solid lubricants (PTFE, $MoS_2$) as well [12,15]. Therefore, it was of interest to investigate whether the incorporation of micronized PTFE and PEEK powders reduces the COF and wear. Figure 13.2 displays the COF values of hydrogenated nitrile rubber (HNBR) mixes containing various amounts of carbon black (CB) with and without additional PTFE or PEEK powder. One can recognize that there is practically no effect of these powders under pin-on-plate (POP) configuration testing (cf. Fig. 13.2(a)). On the other hand, some reduction in the COF was found in fretting, at least when one compares the COF values of compounds at the same overall filler content (cf. Fig. 13.2(b)).

The specific wear rate was enhanced (approximately doubled) by the incorporation of both PTFE and PEEK powders. Scanning electron microscopic (SEM) pictures taken from the wear track show the reason for the enhanced wear (cf. Fig. 13.3). Comparing Fig. 13.3(a) and (b), one can notice the development of a Schallamach wave pattern [16]. Note that the Schallamach wavy pattern becomes more intense with decreasing CB content. Additional incorporation of micronized PTFE in the HNBR increased wear instead of reducing it. Figure 13.3(c) shows that the Schallamach waves are disrupted especially where high contact pressure prevails (midsection of the wear track in Fig. 13.3(c)). This is due to the missing adhesion between the PTFE particles and HNBR matrix. The holes after removal of the particles (cf. Fig. 13.3(d)) act as "flaws" initiating the premature failure of the rubber being subjected locally to a combined tensile/shear stress field.

The detrimental effect of such holes (acting as stress concentrators) was recently confirmed by a detailed SEM study [17]. When the interfacial bonding is so crucial, well-bonded PTFE particles should have a beneficial effect on the friction and wear performance. This was shown by Haberstroh et al. [18]. The cited authors functionalized the PTFE microparticles making use of electron beam irradiation ($\gamma$ irradiation with a dose of less than 500 kGy). Irradiation produces peroxy radicals of long life, which may work as grafting sites toward both unsaturated monomers and polymers. As a consequence PTFE can be grafted by unsaturated monomers, eventually bearing a second functional group (e.g. glycidyl methacrylate), or with suitable rubbers (oligobutadienes) during compounding. It was established that the COF can be markedly reduced via this concept (reduction by about 40%). However, the wear rate, depending on numerous factors, was not always reduced. Nonetheless, Sohail Khan et al. [19] recently demonstrated that incorporation of PTFE micropowder in ethylene propylene diene rubber (EPDM) can markedly reduce both the COF and specific wear rate.

Incorporation of solid lubricants is especially straightforward for thermoplastic rubbers. They are traditionally incorporated through melt compounding in twin-screw extruders. The basic problem with the incorporation of solid lubricant is that quite a large amount is needed (>20 wt%) to reach tribological improvements. This high loading is usually accompanied with a steep reduction in the most relevant mechanical properties. Nanoscale dispersion of solid lubricants would reduce the necessary amount of the solid phase lubricant. Therefore, this is a preferred research topic at present.

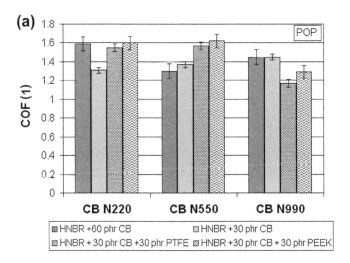

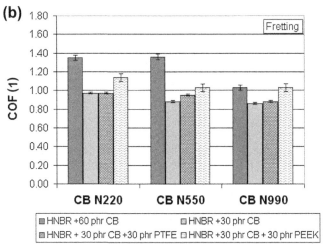

**FIGURE 13.2**

COF of various HNBR mixes determined in POP (a) and fretting (b) conditions.
The activity of the CB increases with decreasing number according to ASTM D 1765;
60 and 30 parts per hundred rubber (phr) correspond to 34 and 21 wt%, respectively, in
the related recipes. Algoflon® L206 (Solvay Solexis, Bollate, Italy) and Vicote® 704
(Victrex Europe GmbH, Hofheim, Germany) served as PTFE and PEEK powders. Testing
conditions for POP: normal load, 10 N; $v = 0.1$ m s$^{-1}$; pin diameter, 10 mm; its
arithmetical roughness, $R_a = 0.9$ µm; testing time, 6 h. Testing condition for fretting:
normal load, 100 N; stroke, 2 mm; frequency, 10 Hz; $R_a = 1$ µm; testing
time, 3 h.

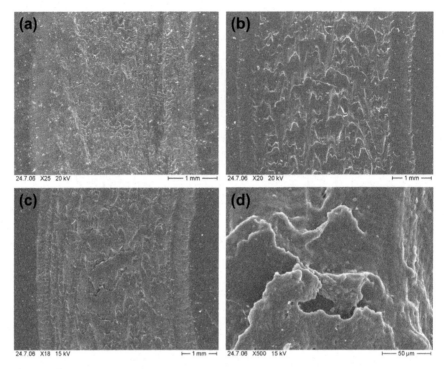

**FIGURE 13.3**

SEM pictures of the wear traces of HNBR + 60 phr CB N550 (a), HNBR + 30 phr CB N550 (b), HNBR + 30 phr CB N550 + 30 phr PTFE (c and d) after the POP tests. For the corresponding notes cf. Fig. 13.2. Sliding direction is downward.

## 13.2.2 **Reinforcements**

Polymeric nanocomposites (organic/inorganic hybrid materials) have attracted considerable interest in both academia and industry. This is due to the outstanding properties achieved at low "nanofiller" content (usually less than 5 wt%). The nano-composites contain inorganic fillers the size of which, at least in one dimension, is in the nanometer range. Albeit the term "nanocomposites" being introduced recently, their industrial history is dated to almost a century back. Recall that the widely used CB itself is a nanoscaled reinforcing additive. Like other fillers and reinforcements, the nanoparticles can be grouped with respect to their aspect ratio (ratio of length to thickness). So, one can distinguish three- (e.g. zeolites and mesoporous glasses), two- (e.g. layered silicates and graphene), and one-dimensional (e.g. carbon nanofibers (CNFs) and carbon nanotubes (CNTs)) versions. In the following sub-sections the sliding wear behavior of rubbers containing layered silicates, CNFs and CNTs, and layered graphite (graphene) will be discussed.

### 13.2.2.1 Organophilic clays

It was demonstrated in several papers that the property profile achieved by adding 30–40 wt% traditional filler (spherical chalk, platy talk) or 10–20 wt% reinforcement (discontinuous glass fiber) can also be set by dispersing 4–5 wt% layered silicate ([20–23] and references therein). Apart from the mechanical properties, other properties like barrier and flame retardancy can also be improved by adding layered silicates [21,23].

Layered silicates of natural (in that case termed as clays) and synthetic origins possess a layer thickness of about 1 nm. The lateral dimension of this platy (disc-shaped) reinforcement may reach the micrometer range. However, the aspect ratio of clays (bentonite and montmorillonite) is usually less than 200. Isomorphic substitution of higher valence cations in the silicate framework by lower valence ones renders the layers negatively charged. This negative surface charge of the layers is counterbalanced by alkali cations located between the clay galleries. They can be replaced by large cationic organic surfactants (termed intercalant) by ion-exchange reactions. The initially hydrophilic silicate becomes organophilic by this ion-exchange process. At the same time the interlayer spacing widens. To set an interlayer distance more than 1.5 nm seems to be necessary to facilitate the intercalation with the polymer molecules. Intercalation means that the initial interlayer spacing increases; however, by X-ray scattering well-detectable long-range order remains between the layers. In case of exfoliation, the individual layers are dispersed in the matrix without any structural ordering. Intercalation is governed by thermodynamical and kinetical parameters [20–25]. All these parameters should be considered when selecting the right "organophilization" of the layered silicate for a given polymer. Karger-Kocsis and Wu [26] argued that rubbers are best suited for organoclay modification due to the following:

1. Possible interaction between the usual amine-type intercalant of the organophilic clay and sulfur curatives;
2. High melt viscosity supporting the delamination of the clay stacks;
3. Relatively low temperature of curing, which avoids the thermal degradation of the clay intercalant; and
4. Swelling in aqueous and organic solutions, which allows us to adopt latex and solution compounding processes.

Irrespective of the latter option, rubber nanocomposites are mostly produced by melt compounding. Accordingly, many works were already devoted to different aspects with rubber/clay nanocomposites ([26–29] and references therein). Opposed to clay-modified thermosets and thermoplastics, the wear behavior of the related rubbers was scarcely studied [30–32].

Gatos et al. [30] reported that organoclay modification of rubbers may improve or deteriorate the resistance to wear when determined by POP testing. It was argued that the in-plane orientation (i.e. in the sliding direction) of the clay layers (especially when exfoliated) may enhance the wear. The clay layers are removed from the contact area via a mechanism termed "can opening" due to similarity with the related process. The authors also demonstrated that the rubber exhibits improved resistance to sliding wear when it contains intercalated stacks randomly oriented across the thickness. In a follow-up study a peroxide-cured HNBR with and without 10 parts per hundred rubber (phr) reinforcement loading (corresponds to about 8 wt% based on the related recipe) was subjected to various tribotests. As reinforcement the following materials were selected: organoclay (Cloisite® 30B of Southern Clay, Gonzales, TX, USA), multiwall carbon nanotube (MWCNT, Baytubes® C 150 P from Bayer MaterialScience Leverkusen, Germany), fumed silica (Cab-O-Sil® M-5 of Cabot, Leuven, Belgium) and silica (Ultrasil® VN2 of Degussa, Frankfurt, Germany). Note that the Cloisite® 30B organoclay was best intercalated/exfoliated by HNBR due to its intercalant, viz. methyl-tallow-bis(2-hydroxyethyl) quaternary ammonium salt [33]. The basic mechanical properties of the above HNBR rubbers are collated in Table 13.1.

**Table 13.1** Basic Mechanical and Sliding Wear Properties of Peroxide-Cured HNBR with and without Various Reinforcements

| Property, Unit | | HNBR | HNBR + Organoclay | HNBR + MWCNT | HNBR + Fumed Silica | HNBR + Silica (Ultrasil) |
|---|---|---|---|---|---|---|
| | | | | Composition (Filler Content: 10 phr) | | |
| Density (g cm$^{-3}$) | | 1.057 | 1.058 | 1.057 | 1.054 | 1.066 |
| Shore A (°) | | 42 | 46 | 56 | 47 | 49 |
| Tensile strength (MPa) | | 4.4 | 3.4 | 4.9 | 19.5 | 4.7 |
| Tensile strain (%) | | 280 | 247 | 179 | 279 | 265 |
| Tear strength (kN m$^{-1}$) | | 4.2 | 5.3 | 8.1 | 5.3 | 5.3 |
| POP | COF steady state (1) | 1 | 1.46 | 1.13 | 1.61 | 1.28 |
| | Specific wear rate (mm$^3$ Nm$^{-1}$) | 0.113 | 0.151 | 0.004 | 0.053 | 0.056 |
| ROP | COF steady state (1) | 2.95 | 2.23 | 1.86 | 2.3 | 3.44 |
| | Specific wear rate (mm$^3$ Nm$^{-1}$) | 0.038 | 0.005 | 0.00078 | 0.005 | 0.006 |
| Fretting | COF steady state (1) | 0.9 | 0.8 | 1.3 | 1.1 | 1.6 |
| | Specific wear rate (mm$^3$ Nm$^{-1}$) | 0.0015 | 0.00076 | 0.00135 | 0.00056 | 0.00307 |

*The reinforcement content was 10 phr (≈8 wt%). Mechanical data determined according to the usual standards at ambient temperature using 500 mm min$^{-1}$ crosshead speed. Wear conditions for POP: $F_N$ = 3 N; v = 0.25 m s$^{-1}$; steel pin diameter, 10 mm; pin roughness $R_a$ = 1 μm; test duration, 90 min. Wear conditions for ROP: $F_N$ = 4 N; v = 0.25 m s$^{-1}$; steel roller diameter, 10 mm; its roughness $R_a$ = 1 μm; test duration, 90 min. Fretting conditions: $F_N$ = 30 N; $v_{(mean)}$ = 0.228 m s$^{-1}$; steel cylinder diameter, 15 mm; its roughness $R_a$ = 0.9 μm; test duration, 90 min.*

The tribological results in Table 13.1 demonstrate how different they may be in function of the test setup. The COF increased for all reinforcements under POP and fretting, but decreased under roller-on-plate (ROP) configuration. Deviation from the related tendencies was only observed for the silica filler. Note that the COFs are very similar to each other under POP and fretting, whereas the ROP-related ones are much higher. This is related to the effects of the corresponding test rigs. The steady "line" contact in ROP produces a severe heat buildup due to which the COF increases. In addition, the revolving roll may be covered by a rubber layer yielding a mixed type of sliding steel/rubber on rubber. The specific wear rate under POP increased slightly for the organoclay, decreased moderately for both silica and markedly for MWCNT. Under ROP testing the specific wear rate followed the order: silica > fumed silica ≈ organoclay > MWCNT. No such ranking could be established for the fretting results. The organoclay and fumed silica reduced the specific wear markedly, MWCNT only slightly, whereas silica enhanced (doubled) the wear. It is noteworthy that the related changes are quite small when taking into account the usual scatter of such tests. In a further work [34] the dry sliding wear performance of HNBR containing CB, silica and MWCNT was studied and compared. The above reinforcements, added in 20 phr, reduced the specific wear rate compared to plain rubber. On the other hand, the COF increased or decreased depending on the test setup.

The sliding friction and wear of organoclay-modified thermoplastic polyurethane (TPU) rubbers were also investigated [31]. The specific wear rate of a low-hardness, polyetherdiol-based TPU increased with organoclay loading under POP and fretting testing. On the other hand, a high-hardness polyesterdiol-based TPU exhibited some slight improvement in wear resistance. Figure 13.4 shows that under fretting the high-hardness TPU failed by gouging, which is characteristic neither for rubbers nor

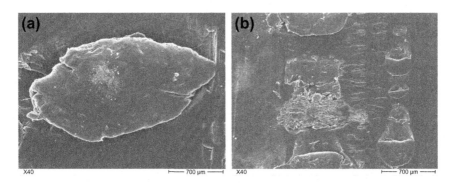

**FIGURE 13.4**

Characteristic SEM pictures showing the failure due to fretting ($F_N = 100$ N) in a TPU of Shore $D = 65°$ without (a) and with (b) organoclay. The organoclay type was an octodecylamine-modified montmorillonite, added in an amount of 5 wt% [31]. Oscillation direction is downward.

for fretting against a smooth steel surface. In the presence of organoclay, gouging occurred mostly in the surface layer and less in the bulk opposed to the plain TPU (cf. Fig. 13.4).

The difference in the failure mode (cf. Fig. 13.4) suggests that the organoclay might have altered the morphology of the TPU. Note that the thermoplastic character of the TPU is given by crystalline domains acting as nodes of the corresponding "physical network" structure [24]. Finnigan et al. [35] pointed out that the morphology of TPUs is strongly affected by the organoclay dispersed therein. Accordingly, the observed effects cannot be traced solely to the organoclay, as its effects on the TPU morphology should be considered, as well. This note is also valid when interpreting the wear behavior of organoclay-modified semicrystalline thermoplastics the morphology of which may be highly influenced by the organoclay [22]. The organoclay modification of cross-linked (thermoset) and thermoplastic rubbers is less promising as supposed. Our recent results achieved on polypropylene (PP)-based thermoplastic dynamic vulcanizates (TPVs) seem to support the above conclusion [36].

It is the right place to mention that anionic clays, i.e. those with positive surface charge and thus anions (–OH prior to anion exchange) in the interlayer galleries, are scarcely used for the modification of rubbers [37].

### 13.2.2.2 Carbon nanofibers and nanotubes

The use of CNTs and CNFs of different types became the subject of numerous studies after their discovery by Iijima in 1991. Based on their unique stiffness (Young's modulus <1 TPa) and strength (<200 GPa), for which different values are reported (e.g. [38–42] and references therein), emphasis was put on their reinforcing effect in organic polymers. Apart from their reinforcing potential (which often remained below the expectation), efforts were made to exploit further characteristics of CNTs, like electric conductivity and sensing properties. CNT was already incorporated in rubbers [43–49] and the related works summarized [36,49].

Table 13.2 lists the basic mechanical and tribological properties of HNBR rubbers containing 10 and 30 phr MWCNT [50]. For comparison purpose the related data for the plain HNBR and its 30 phr silica-containing version are also indicated. The tensile strengths of the 30 phr MWCNT- and silica-reinforced HNBR are practically the same. On the other hand, MWCNT outperforms silica with respect to tear strength, which has been almost doubled. However, the ultimate tensile strain of the MWCNT-modified HNBR was markedly below the silica-containing one. This suggests a clear reinforcing effect usually caused by anisometric particles having a high aspect ratio. The tribological data indicate again that the COFs and specific wear rates strongly depend on the type of loading in the corresponding test rigs. The steady state COF increases with increasing MWCNT content under POP, whereas its increment for fretting stabilizes at about 1.2. On the other hand, the COF is highly reduced and leveled off as a function of MWCNT in ROP. The specific wear rate in POP decreased by two orders of magnitudes compared to neat HNBR when 30 phr MWCNT was incorporated. A similar large reduction, but without the effect of the MWCNT content, was established under ROP testing. A prominent upgrade in the

**Table 13.2** Basic Mechanical and Sliding Wear Properties of Peroxide-Cured HNBR with and without Various Reinforcements

| Property, Unit | | HNBR | HNBR + 10 phr MWCNT | HNBR + 30 phr MWCNT | HNBR + 30 phr Silica (Ultrasil) |
|---|---|---|---|---|---|
| Density (g cm$^{-3}$) | | 1.057 | 1.057 | 1.135 | 1.141 |
| Shore A (°) | | 42 | 56 | 82 | 61 |
| Tensile strength (MPa) | | 4.4 | 4.9 | 16.5 | 17.2 |
| Tensile strain (%) | | 280 | 179 | 132 | 476 |
| Tear strength (kN m$^{-1}$) | | 4.2 | 8.0 | 22.0 | 12.5 |
| POP | COF steady state (1) | 1 | 1.13 | 1.4 | 1.89 |
| | Specific wear rate (mm$^3$ Nm$^{-1}$) | 0.113 | 0.004 | 0.001 | 0.033 |
| ROP | COF steady state (1) | 2.95 | 1.86 | 2.13 | 3.08 |
| | Specific wear rate (mm$^3$ Nm$^{-1}$) | 0.038 | 0.00078 | 0.00066 | 0.00135 |
| Fretting | COF steady state (1) | 0.9 | 1.3 | 1.2 | 1.15 |
| | Specific wear rate (mm$^3$ Nm$^{-1}$) | 0.0015 | 0.00135 | $3.624 \times 10^{-5}$ | 0.001436 |

*The 10- and 30-phr reinforcements correspond to ≈8 and ≈21 wt%, respectively. For further notes cf. those in the legend of Table 13.1.*

resistance to wear was found under fretting conditions for the HNBR with increasing MWCNT content. The improvement is well highlighted when one takes into consideration the related values of the plain HNBR and its 30 phr silica-containing variant (cf. Table 13.2). SEM pictures taken from the worn surfaces of HNBR and HNBR with 30 phr MWCNT after various tribotests help us to interpret the sliding wear results in Table 13.2 [50].

The worn surface of the HNBR after POP is rough; the related wear mechanism is mostly pitting (cf. Fig. 13.5(a)). Reinforcement with MWCNT results in a smoothed, "ironed" surface. At high magnification microroll formation can be resolved. The reason for the appearance of the bandlike surface pattern is unclear (cf. Fig. 13.5(b)). The HNBR fails by pitting along with massive fragmentation in ROP (cf. Fig. 13.5(c)). This suggests high COF and high wear that were found (cf. Table 13.2). By contrast, the worn surface of the 30 phr MWCNT-containing HNBR suggests that the MWCNT worked as a solid-phase lubricant (cf. Fig. 13.5(d)). The fretting worn surface of the plain HNBR (cf. Fig. 13.5(e)) is similar to those produced by POP and ROP (cf. Fig. 13.5(a) and (c)). The fretting surface of the HNBR with 30 phr MWCNT supports again the lubrication effect of the MWCNT, which is well embedded in the matrix (cf. Fig. 13.5(f)). As a consequence the wear rate is very low. Interestingly, the wear resistance under fretting was accompanied with a rather high COF (cf. Table 13.2). This suggests the formation of a transfer layer

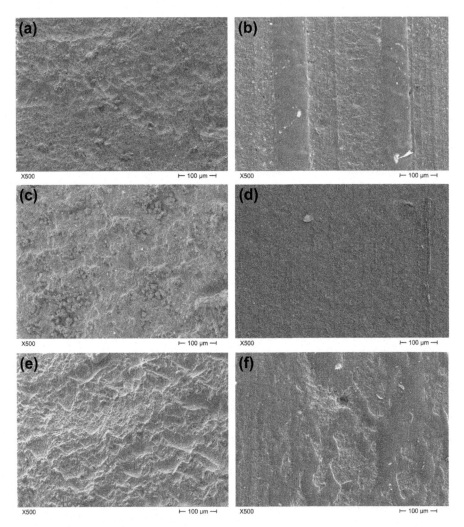

**FIGURE 13.5**

SEM pictures of the worn plain HNBR (a,c,e) and its 30 phr (=21 wt%) MWCNT version (b,d,f) after the POP (a,b), ROP (c,d) and fretting tests (e,f). Sliding/moving direction is downward. The MWCNT type agrees with that disclosed in the text. The wear conditions are in accordance with those given in Fig. 13.1.

and/or considerable temperature rise. It is noteworthy that the MWCNT particles could not be fully disentangled to their constituents by mixing in a two-roll mill (cf. Fig. 13.6).

It is well known that the dispersion stage of the CNTs strongly depends on the compounding technique and its parameters. This aspect was addressed by us

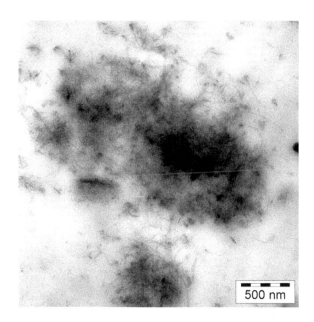

500 nm

**FIGURE 13.6**

Characteristic TEM image showing the MWCNT dispersion in the HNBR mix with 10 phr (=8 wt%) MWCNT.

using an aqueous dispersion of CNF (containing 15 wt% CNF) produced by Grupo Antolin (Burgos, Spain). CNF was deposited on EPDM particles of porous structure (Buna® AP447 of Bayer, Leverkusen, Germany) by evaporating the dispersing fluid. Afterwards it was incorporated in a PP-based TPV (Santoprene® 8000 TPV 8201-80HT of Advanced Elastomer Systems, Akron, OH, USA) by melt blending [51]. Note that this kind of thermoplastic elastomer was already the subject of sliding wear studies [52]. The CNF content was set for 1 and 5 wt%, respectively, in the studied TPV [51]. The transmission electron microscopic (TEM) picture in Fig. 13.7 clearly shows that the CNF was finely dispersed by this way. The TEM investigations also showed that the CNF was preferentially embedded in the EPDM rubber phase instead of the PP.

Data in Table 13.3 demonstrate that the COF of the TPV/EPDM/CNF systems was reduced with increasing CNF content under POP, whereas an adverse trend was found for ROP. Similar changes were noticed for the specific wear rates of the related systems (cf. Table 13.3) [51]. This finding underlines the widespread claim that wear is a "system-dependent" property. It means that the tribological data strongly depend on the configuration and parameters of the tribotests used, even for the same material. SEM pictures in Fig. 13.8 display that the CNF reinforcement increases the resistance to sliding wear under POP by reducing the pitting of the thermoplastic rubber. Accordingly, the overall appearance of the worn surface is smoother ("ironed") in presence of CNF compared to the plain version (Fig. 13.8).

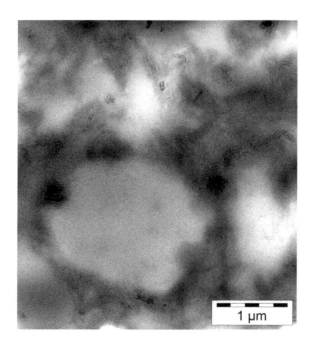

**FIGURE 13.7**

TEM picture of the TPV/EPDM blend (90/10 parts) containing 1 wt% CNF. Due to staining the rubber phase appears dark, while the PP appears white.

**Table 13.3** Effects of CNF Content on the COF and Specific Wear Rate of a Thermoplastic Elastomer Composed of TPV/EPDM (90/10)

| Composition TPV/EPDM/ CNF (parts) | COF (1) | | Specific Wear Rate (mm³ Nm⁻¹) | |
|---|---|---|---|---|
| | POP | ROP | POP | ROP |
| 90/10 | 1.55 | 1.33 | 0.0029 | $1.2 \times 10^{-6}$ |
| 90/10/1 | 0.9 | 1.42 | 0.0029 | $6.46 \times 10^{-5}$ |
| 90/10/5 | 0.79 | 1.57 | 0.0017 | $1.35 \times 10^{-4}$ |

*For notes cf. those given in Table 13.1.*

### 13.2.2.3 Expanded graphites, graphenes

Graphene is an atomically thick sheet composed of $sp^2$ hybridized carbon atoms forming a honeycomb structure. Graphene can be treated as the building unit of other graphitic carbon allotropes, such as fullerenes, CNT and nanoribbons. The single-layer graphene is the strongest material ever measured having a Young's modulus of 1 TPa and strength of 130 GPa. Its thermal and electric conductivities are 5000 W mK⁻¹ and 6000 S cm⁻¹, respectively. Graphene can be produced by top-down and bottom-up

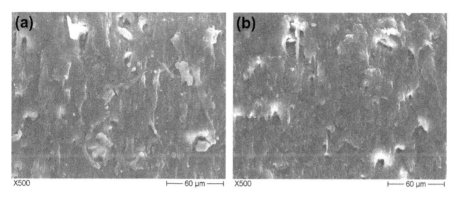

**FIGURE 13.8**

SEM pictures of the worn surfaces of a thermoplastic rubber (TPV/EPDM = 90/10 parts, a) and its 5 phr CNF-reinforced version (b). Sliding direction is downward.

approaches. The former means cleavage/exfoliation of graphite and related derivatives, whereas the latter methods comprise various synthetic routes ([53] and references therein). The difference in the production methods is likely the reason for the different designations used including expanded graphite, layered graphite, graphite nanosheet, graphene etc. Graphene is an ideal nanofiller, also for rubbers, due to its high aspect ratio and high surface area. Moreover, it can be functionalized (whereby oxidation is preferred) to tune the surface tension and reactivity. The modification of rubbers with graphene is an emerging research field because apart from improvements in structural performance functional properties can also be generated [54]. Wang et al. [55] studied the sliding friction and wear of expanded graphite-modified nitrile rubber. The graphite (content up to 10 phr) was present either in micro- or in nanoscale dispersed form, which was achieved by mechanical blending and latex compounding, respectively. The COF decreased for both composites as the load and sliding velocity increased. The wear rate increased with increasing load, but it went through a maximum as a function of the sliding velocity. The nanocomposite outperformed the microcomposite with respect to the reduction in the wear rate and COF. This was attributed to the formation of a continuous, uniform and stable lubricant film under the sliding conditions used.

### 13.2.3 "Hybrid" rubbers

Blending of different rubbers is a well-established method to meet given property requirements in the rubber industry. However, there are many other routes via which the rubber as matrix can be modified. Hybrid rubber/thermoplastic and rubber/thermoset systems are very promising, especially with respect to the development of wear-resistant materials. Note that such systems were not yet topics of tribological investigations. Hence, the brief description below reports on some preliminary results and outlines also some promising concepts.

### 13.2.3.1 Thermoplastics

Rubber/thermoplastic hybrids can be produced by melt mixing of rubbers with thermoplastics prior to curing the former. A further option is to combine with the curable rubber monomers and oligomers that can be polymerized under the same conditions in which the rubber vulcanization is performed. Exactly, this concept is followed when using cyclic butylene terephthalate (CBT) oligomers. CBT oligomers melt at about 140 °C above which they polymerize into poly(butylene terephthalate) (PBT) ([56–59] and references therein). The beauty of this material is that it polymerizes and also crystallizes below the melting of the resulting PBT (melting temperature $\approx$ 225 °C). For that purpose isothermal treatment between 170 and 190 °C is usually preferred [57]. Recall that this temperature agrees fairly with that of rubber vulcanization. In addition, the polymerization time of CBT can also be adjusted to that of rubber curing by selecting suitable catalysts. To check the potential of this hybridization, HNBR mixes with CBT® 160 (Cyclics, Schwarzheide, Germany) were produced by mixing on a two-roll mill. In a subsequent step the mixes were hot pressed at $T = 190$ °C for 25 min. This was enough to polymerize the CBT and cure the HNBR, respectively. Differential scanning calorimetric traces confirmed that the CBT was polymerized to PBT but not well crystallized. Hampered crystallization is a peculiar feature of the polymerized CBT [58]. The COF and specific wear data of the HNBR/PBT hybrid and the reference HNBR are displayed in Fig. 13.9.

Note that the COF is practically not influenced by the PBT, except the fretting, where some increase was found. Combination with PBT reduced the specific wear rate under both POP and ROP but slightly increased for fretting. It has to be added that this modification doubled the tensile strength and even the ultimate tensile strain, whereas in the tear strength an eightfold increase was achieved [60]. The related hybrid should possess an interpenetrating network (IPN, more correctly a semi-IPN) structure, i.e. both vulcanized rubber and thermoplastic PBT phases are continuous [61]. This was supported apart from the failure mode of the tensile loaded specimens (showing "trellis" effect) also by the appearance of the worn surface (cf. Fig. 13.10). It has to be emphasized that in an IPN the size of the constituent "bands" is in the range of some hundreds of nanometers.

In a follow-up work the effects of CBT amount and its polymerization degree on the sliding wear performance of HNBR-based systems were addressed [62]. It was established that the recrystallized, unpolymerized CBT acts as efficient reinforcement whereby reducing the specific wear rate, as well. Romhány et al. [59] emphasized that the solid-phase high-energy ball milling of CBT with graphitic nanofillers (graphene, CNT, etc.) is the most straightforward method for the disintegration of the filler agglomerates. The related CBT/nanofiller powder is well suited for the incorporation in rubber mixes.

Apart from the ring-opening polymerization, other polymerization techniques can also be adopted to produce rubber/thermoplastic hybrids. For example, the cross-linked rubber can be swollen in suitable liquid monomers and polymerized into the corresponding thermoplastic further on. The structure of the resulting hybrid is a semi-IPN. The most promising candidates for hybridization are, however,

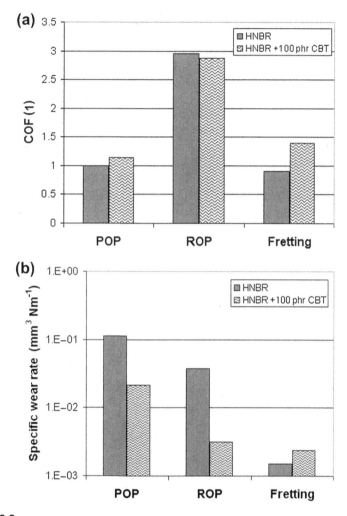

**FIGURE 13.9**

COFs (a) and specific wear rates (b) of the HNBR and HNBR + 100 phr CBT (=45 wt%) as assessed in different tribotests. The testing conditions agree with those indicated in Fig. 13.1.

polyurethanes (PUs). This note is based on the fact that PUs exhibit outstanding wear resistance (e.g. [63]). Moreover, their polyaddition reactions are less affected by those of rubber curing. The potential of rubber/PU hybridization will be shown in the next Section 13.2.3.2 on the example of reactive (i.e. cross-linkable) PU versions.

### 13.2.3.2 Thermosets

The routes to produce rubber/thermoset hybrids are basically the same as outlined in Section 13.2.3.1. The only difference is that the functionality of the applied monomers, oligomers or other precursors (at least one of them) should be higher than two.

The type of the cross-linking reactions of the rubber and thermoset may be identical or different. In the former case the rubber and the thermoset phase are chemically interlinked, "grafted". The resulting morphology of such hybrids is very versatile. However, in many cases the target is to generate an IPN structure (in that case termed as full-IPN because both continuous phases are cross-linked) as this provides usually the best combination of properties of the hybrid components.

Metal salts of unsaturated carboxylic acids (zinc and magnesium methacrylates and dimethacrylates) are successfully used to modify the rubber networks (e.g. [64–67]). During peroxide curing of the rubber these "coagents" undergo homopolymerization and graft copolymerization at the same time. Moreover, in their presence additional cross-linking sites of ionic nature appear. Adding a multifunctional grade in higher amount and/or cross-linking the rubber by curatives other than peroxide, IPN-structured systems can also be produced. Incorporation of zinc diacrylate (Saret 633 of Sartomer Inc., Exton, PA, USA) in 30 phr (about 21 wt%) in a peroxide-cured HNBR resulted in mechanical properties that were comparable with 30 phr silica rubber (cf. Table 13.2). The COF of the zinc diacrylate-modified rubber was lower than the silica-filled one at the same loading for all the tribotests used. The specific wear rates of those compounds were also comparable for ROP and fretting, whereas under ROP the acrylate-modified HNBR was far more resistant to wear than the silica filled grade [68].

Our interest was next focused on HNBR/PU hybridization. The PU "precursors" (i.e. polyol and blocked polyisocyanate) were introduced in the HNBR mix in powder form in an open mill. The deblocking of the polyisocyanate followed by the polyaddition reaction of PU formation finished at $T = 140$ °C. So curing the HNBR with peroxide at $T = 175$ °C for 15 min was accompanied with the development of

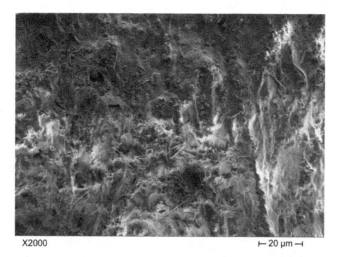

X2000                                      ⊢ 20 µm ⊣

**FIGURE 13.10**

SEM picture of the worn surface of the HNBR containing 100 phr (=45 wt%) polymerized CBT after the POP test. Sliding direction is downward.

cross-linked PU. PU was introduced in 50 and 100 phr (about 30 and 45 wt%, respectively) in the HNBR mix. This modification improved the tensile strength, strain and tear strength by <100%, <60% and <200%, respectively [69]. The related COFs and specific wear rates, compared to those of the plain HNBR, are shown in Fig. 13.11.

The COF values did not change much as a function of PU content and also did not differ significantly from that of the plain HNBR, except the fretting conditions. Slight and substantial improvement in the specific wear rate was found, however, for POP and ROP, respectively. By contrast, the resistance to sliding wear was reduced due to fretting [69]. In a recent work some peculiarities of such HNBR/PU hybrids

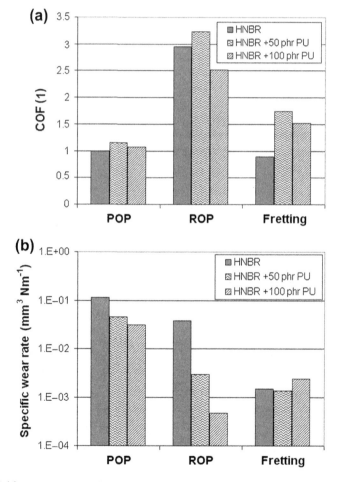

**FIGURE 13.11**

COFs (a) and specific wear rates (b) of the HNBR and its 50- (=30 wt%) and 100 phr (=45 wt%) PU-containing versions as assessed in different tribotests. The testing conditions agree with those indicated in Fig. 13.1.

related to the morphology and CB location have been clarified [70]. The same PU hybridization principle was adapted for an EPDM rubber [71]. Unfortunately, this EPDM/PU did not show the expected improvement in the tribological properties. This was attributed to the missing interfacial adhesion between the EPDM and PU phases. However, the authors are convinced that by proper selection of compatibilizers the adhesion between various rubbers and in situ curable PU can be improved and thus hybrids with enhanced wear resistance can be produced. Moreover, many options exist to tailor the microstructure, curing and properties of the PU hybrid components (e.g. [72]).

## 13.3 RUBBER COATINGS

As the resistance of rubbers to sliding wear is much smaller than that of thermoplastics and thermosets, efforts were always undertaken to coat the rubbers by them accordingly. New impetus to the related efforts was given by the appearance of novel coating methods such as laser-based technologies, plasma spraying, ion implantation, and microwave-assisted methods. Alternative rubber coatings than those with thermoplastics and thermosets are beyond the scope of this chapter. It should be kept in mind that rubber coatings for tribological application are still in their embryonic stage of development.

### 13.3.1 Thermoplastics

High temperature-resistant thermoplastics, such as PEEK and poly(phenylene sulfide) (PPS) are preferred materials exhibiting outstanding resistance to sliding wear [12,14,15]. Unfortunately, their melting temperature is too high compared to those of vulcanization and short- and long-term temperature resistance of rubbers. This limits their application as coating materials for rubbers when they are deposited by thermal processes. In addition, feasibility tests by bonding films of high-temperature plastics to rubbers (during or after vulcanization) showed that acceptable adherence to the rubber substrate can only be achieved by using suitable primers. It is therefore of paramount importance to match the thermal properties of the coating materials with those of the related rubber. This means, for example, that the conditions of thermal spraying of a suitable powder should be below the short-term temperature resistance of the rubber, or comparable with the vulcanization condition of the latter when the surface coating is finished during vulcanization. Combination of soft (rubber)/hard (thermoplastics) systems through injection molding teaches us that semicrystalline polymers, such as polyamides (PAs), linear polyesters, and PUs (TPU), are most suited potential coating materials. Note that their melting temperature is markedly below those of PEEK, PPS and the like, which is beneficial for thermal coating processes, such as flame spraying, plasma deposition and laser cladding. In addition, the ductility and toughness of PAs and linear polyesters are usually higher than those of high temperature-resistant thermoplastics, which mean a better "property match"

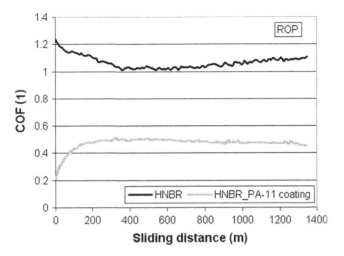

**FIGURE 13.12**

Change of the COF as a function of the sliding distance under ROP for the HNBR with and without flame-sprayed PA-11 coating. The HNBR contained 60 phr (=34 wt%) N550 type CB. The testing conditions agree with those indicated in Fig. 13.1.

with rubbers. Figure 13.12 shows that the COF of a PA-11 coating on HNBR under ROP is much smaller than that of the HNBR substrate, as expected. SEM pictures taken from the worn surfaces after ROP test highlight the basic difference in the wear between HNBR and the PA-11 coating (Fig. 13.13). However, the coating disappears with time (wearing out). The wear protection by PA-11 proved to depend on the tribotests and their conditions and followed the ranking: ROP > POP > fretting [68].

"Nanoreinforcement" of the PA-11 coating by adding MWCNT resulted in a further improvement in the COF, at least for a given time interval. This was demonstrated on coatings deposited by laser cladding on TPU. Figure 13.14 compares the traces of the COFs as a function of sliding distance for the TPU with and without PA-11 coating. One of the PA-11 coatings, produced by laser cladding, contained additionally 1 wt% MWCNT. It is noteworthy that the initial roughness of the PA-11 with MWCNT was much higher than that of the PA-11 reference. The related SEM pictures (cf. Fig. 13.15) indicate that the PA-11 coating on TPU fails by surface cracking and gouging. The latter are hampered by the presence of MWCNT reinforcement.

Benefits of PA-11 coatings, with and without further additives (MoS$_2$, PTFE particles), deposited on the surface of TPU by laser cladding were demonstrated also by Verheyde et al. [73].

Thermal coating processes [74] may be practiced with in situ polymerizable CBT and TPU powders (the latter may be both reactive and ready-to-use thermoplastic versions). It is worth noting that modification of CBT, PBT and TPU by nanofillers (organoclay, CNF, MWCNT, and graphene) may further improve the wear characteristics assuming that their change runs parallel with that of the mechanical properties.

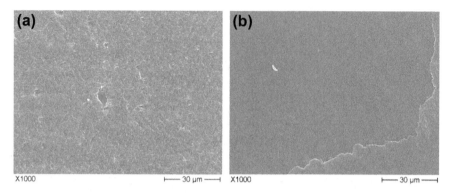

**FIGURE 13.13**

SEM pictures of the worn surfaces of HNBR and its PA-11 coated version after ROP. The HNBR grade agrees with that indicated in Fig. 13.12. The sliding direction is downward.

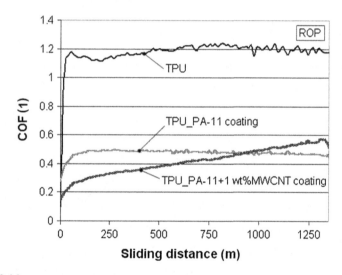

**FIGURE 13.14**

Change of the COF as a function of the sliding distance under ROP for the TPU with and without laser-deposited PA-11 coatings. Note that one of the PA-11 coating contained 1 wt% MWCNT. The testing conditions agree with those indicated in Fig. 13.1.

"Cold" coating processes, such as cold or atmospheric polymerization may be a further interesting coating option that is now of academic interest [75].

## 13.3.2 Thermosets

Little information is available on coating with thermosets. Immersion and powder coating with cross-linkable PUs and thermal spraying with epoxy-based systems

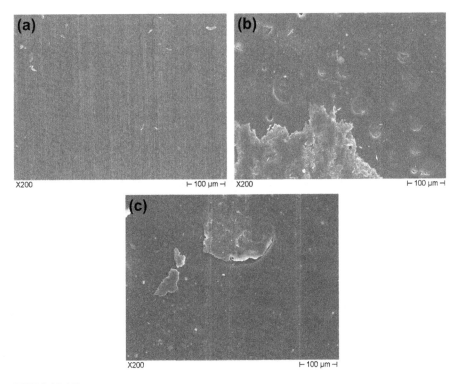

**FIGURE 13.15**

SEM pictures of the worn surfaces of TPU (a) and its PA-11-coated versions (b,c) with (c) and without (b) 1 wt% additional MWCNT reinforcement after ROP. The coatings were produced by laser cladding. The sliding direction is downward.

may be interesting techniques. Plasma (especially "cold" plasma) polymerization for coatings is very promising, too. Verheyde et al. [73] produced a plasma-polymerized polysiloxane coating on the surface of HNBR, which yielded an almost 80% reduction in the COF.

However, the basic problems with all kind of coatings on rubber substrates are poor adhesion and a large difference in the mechanical properties (especially in stiffness) between the coating material and rubber. Adherence between the rubber and coating can be improved by various techniques (use of primers, IPN structuring, incorporation of fillers, and modification of the rubber with polymers the chemical buildup of which is similar to that of the polymer coating). Avoiding stress concentration due to stiffness mismatch is more problematic. To generate a gradient structure in the interphase between rubber and coating is very challenging, especially with respect to industrial realization. Being very variable in their preparation and properties, PUs remain the first choice coating materials [76]. For their property modification surface-treated nanoparticles will be used in increasingly in the future.

## 13.4 SUMMARY AND OUTLOOK

Modification of rubbers by novel nanofillers may improve their resistance to sliding wear. CNT-layered graphites, when well dispersed and well adhering to the rubber matrix, are more promising additives than organoclays. Moreover, the sol/gel route is an interesting technique to produce nanomodified rubbers [77], also when targeting wear resistance [78,79]. This technique has also a fair chance for rubber coatings.

As thermoplastic rubbers are gaining importance against thermoset ones, the modification of the former will be pushed forward. This will be solved, however, in the reactor via in situ polymerization techniques rather than by additional compounding processes. Increasing amounts of rubbers' recipe ingredients will be obtained from renewable resources. The related products are not restricted for processing aids, but will cover "natural" reinforcements, such as cellulose whiskers, rice husk ash, too [80]. Development of rubbers with semi- and full-IPN structures may be an alternative route to tailor the sliding wear behavior upon request.

The future belongs to rubbers and their coatings of self-assembled (nano)structure the driving force of which is of thermodynamical origin (e.g. [81]). Their realization will concur with surface pattering techniques for which photolithographic, laser-assisted techniques are favored at present (e.g. [82,83]). Further on, considerable efforts will be devoted to model the sliding wear behavior of rubbers (e.g. [84–91]), to get a deeper insight into the processes taking place in the rubber/substrate contact zone [92], and to assess tribochemical aspects of sliding wear [93].

## Acknowledgments

This work was performed in the framework of the KRISTAL project of the EU (Contract Nr.: NMP3-CT-2005-515837; www.kristal-project.org) and K100294 of the Hungarian Scientific Research Fund (OTKA), respectively.

## References

[1] B.S. Nau, The state of the art of rubber-seal technology, Rubber Chemistry and Technology 60 (1987) 381–416.

[2] D.I. James (Ed.), Abrasion of Rubber, Maclaren and Sons, London, 1967.

[3] A.D. Sarkar, Friction and Wear, Academic Press, New York, 1980.

[4] G.M. Bartenev, V.V. Lavrentev, Friction and Wear of Polymers, Elsevier, Amsterdam, 1981.

[5] A.N. Gent, C.T.R. Pulford, Mechanisms of rubber abrasion, Journal of Applied Polymer Science 28 (1983) 943–960.

[6] J-Å Schweitz, L. Åhman, Mild wear of rubber-based compounds, in: K. Friedrich (Ed.), Friction and Wear of Polymeric Composites, Elsevier, Amsterdam, 1986, pp. 289–327. (Chapter 9).

[7] S.-W. Zhang, Tribology of Elastomers, Elsevier, Amsterdam, 2004.

[8] B.J. Briscoe, S.K. Shina, Tribology of polymeric solids and their composites, in: G.W. Stachowiak (Ed.), Wear – Materials, Mechanisms and Practice, Wiley, New York, 2005, pp. 223–267. (Chapter 10).

[9] N.K. Myshkin, M.I. Petrokovets, A.V. Kovalev, Tribology of polymers: adhesion, friction, wear and mass-transfer, Tribology International 38 (2005) 910–921.

[10] M. Barquins, Adherence, friction and wear of rubber-like materials, Wear 158 (1992) 87–117.

[11] I.L. Singer, in: I.L. Singer, H.M. Pollock (Eds.), Solid Lubrication Processes in Fundamentals of Friction: Macroscopic and Microscopic Processes, Kluwer Academic, Dordrecht, 1992, pp. 237–261.

[12] K. Friedrich, Z. Zhang, A.K. Schlarb, Effects of various fillers on the sliding wear of polymer composites, Composites Science and Technology 65 (2005) 2329–2343.

[13] W. Brostow, M. Keselman, I. Mironi-Harpaz, M. Narkis, R. Peirce, Effects of carbon black on tribology of blends of poly(vinylidene fluoride) with irradiated and non-irradiated ultrahigh molecular weight polyethylene, Polymer 46 (2005) 5058–5064.

[14] A.P. Harsha, U.S. Tewari, The effect of fibre reinforcement and solid lubricants on abrasive wear behavior of polyetheretherketone composites, Journal of Reinforced Plastics and Composites 22 (2003) 751–767.

[15] K. Friedrich, Z. Zhang, P. Klein, Wear of polymer composites, in: G.W. Stachowiak (Ed.), Wear – Materials, Mechanisms and Practice, Wiley, New York, 2005, pp. 269–290. (Chapter 11).

[16] A. Schallamach, How does rubber slide? Wear 17 (1971) 301–312.

[17] S.V. Hainsworth, An environmental scanning electron microscopy investigation of fatigue crack initiation and propagation in elastomers, Polymer Testing 26 (2007) 60–70.

[18] E. Haberstroh, C. Linhart, K. Epping, T. Schmitz, Verbesserte tribologische eigenschaften von elastomeren durch PTFE pulver, Kautschuk, Gummi, Kunststoffe 59 (2006) 447–453.

[19] M. Sohail Khan, R. Franke, D. Lehmann, G. Heinrich, Physical and tribological properties of PTFE micropowder-filled EPDM rubber, Tribology International 42 (2009) 890–896.

[20] T.J. Pinnavaia, G.W. Beall (Eds.), Polymer-Clay Nanocomposites, Wiley, Chichester United Kingdom, 2000.

[21] S. Sinha Ray, M. Okamoto, Polymer/layered silicate nanocomposites: a review from preparation to processing, Progress in Polymer Science 28 (2003) 1539–1641.

[22] J. Karger-Kocsis, Z. Zhang, Structure-property relationships in nanoparticle/semicrystalline themoplastic composites, in: G.H. Michler, F.J. Baltá Calleja (Eds.), Mechanical Properties of Polymers Based on Nanostructure and Morphology, CRC Press, Boca Raton, Florida, 2005, pp. 553–602. (Chapter 13).

[23] L.A. Utracki, Clay-Containing Polymeric Nanocomposites, Rapra Technology, Shawbury, United Kingdom, 2004.

[24] J. Karger-Kocsis, Nanoreinforcement of thermoplastic elastomers, in: S. Fakirov (Ed.), Handbook of Condensation Thermoplastic Elastomers, Wiley-VCH, Weinheim, 2005, pp. 473–488. (Chapter 16).

[25] S. Varghese, J. Karger-Kocsis, Layered silicate/rubber nanocomposites via latex and solution intercalations, in: K. Friedrich, S. Fakirov, Z. Zhang (Eds.), Polymer Composites from Nano- to Macroscale, Springer, Berlin, 2005, pp. 77–90. (Chapter 5).

[26] J. Karger-Kocsis, C.M. Wu, Thermoset rubber/layered silicate nanocomposites. Status and future trends, Polymer Engineering and Science 44 (2004) 1083–1093.

[27] S. Sadhu, A.K. Bhowmick, Morphology study of rubber based nanocomposites by transmission electron microscopy and atomic force microscopy, Journal of Materials Science 40 (2005) 1633–1642.

[28] K.G. Gatos, J. Karger-Kocsis, Effects of primary and quaternary amine intercalants on the organoclay dispersion in a sulphur-cured EPDM rubber, Polymer 46 (2005) 3069–3076.

[29] K.G. Gatos, J. Karger-Kocsis, Effect of the aspect ratio of silicate platelets on the mechanical and barrier properties of hydrogenated acrylonitrile butadiene rubber (HNBR)/layered silicate nanocomposites, European Polymer Journal 43 (2007) 1097–1104.

[30] K.G. Gatos, K. Kameo, J. Karger-Kocsis, On the friction and sliding wear of rubber/layered silicate nanocomposites, eXPRESS Polymer Letters 1 (2007) 27–31.

[31] J. Karger-Kocsis, Dry friction and sliding behavior of organoclay reinforced thermoplastic polyurethane rubbers, Kautschuk, Gummi, Kunststoffe 59 (2006) 537–543.

[32] D. Xu, J. Karger-Kocsis, Dry rolling and sliding friction and wear of organophilic layered silicate/hydrogenated nitrile rubber nanocomposite, Journal of Materials Science 45 (2010) 1293–1298.

[33] K.G. Gatos, N.S. Sawanis, A.A. Apostolov, R. Thomann, J. Karger-Kocsis, Nanocomposite formation of hydrogenated nitrile rubber (HNBR)/organo-montmorillonite as function of the intercalant type, Macromolecular Materials and Engineering 289 (2004) 1079–1086.

[34] D. Xu, J. Karger-Kocsis, A.K. Schlarb, Friction and wear of HNBR with different fillers under dry rolling and sliding conditions, eXPRESS Polymer Letters 3 (2009) 126–136.

[35] B. Finnigan, D. Martin, P. Halley, R. Truss, K. Campbell, Morphology and properties of thermoplastic polyurethane nanocomposites incorporating hydrophilic layered silicates, Polymer 45 (2004) 2249–2260.

[36] D. Felhös, J. Karger-Kocsis, Friction and wear of rubber nanocomposites containing layered silicates and carbon nanotubes, in: V. Mittal, J.K. Kim, K. Pal (Eds.), Recent Advances in Elastomeric Nanocomposites, Springer, Berlin, 2011, pp. 343–379. (Chapter 13).

[37] M. Galimberti (Ed.), Rubber-Clay Nanocomposites, Wiley, Hoboken, New Jersey, USA, 2011.

[38] V.N. Popov, Carbon nanotubes: properties and application, Materials Science and Engineering R 43 (2004) 61–102.

[39] O. Breuer, U. Sundararaj, Big returns from small fibers: a review of polymer/carbon nanotube composites, Polymer Composites 25 (2004) 630–645.

[40] M. Moniruzzaman, K.I. Winey, Polymer nanocomposites containing carbon nanotubes, Macromolecules 39 (2006) 5194–5205.

[41] J.-H. Du, J. Bai, H.-M. Cheng, The present status and key problems of carbon nanotube based polymer composites, eXPRESS Polymer Letters 1 (2007) 253–273.

[42] H.C. Zheng, Carbon nanotube-based nanocomposites, in: H.S. Nalwa (Ed.), Handbook of Organic–Inorganic Hybrid Materials and Nanocomposites, American Scientific Publ., Los Angeles, 2003, pp. 151–180. (Chapter 4).

[43] J.D. Wang, Y.F. Zhu, X.W. Zhou, G. Sui, J. Liang, Preparation and mechanical properties of natural rubber powder modified by carbon nanotubes, Journal of Applied Polymer Science 100 (2006) 4697–4702.

[44] D. Yue, Y. Liu, Z. Shen, L. Zhang, Study on preparation and properties of carbon nanotubes/rubber composites, Journal of Material Science 41 (2006) 2541–2544.

[45] A. Fakhru'l-Razi, M.A. Atieh, N. Girun, T.G. Chuah, M. El-Sadig, D.R.A. Biak, Effect of multi-wall carbon nanotubes on the mechanical properties of natural rubber, Composite Structures 75 (2006) 496–500.

[46] M.A. López-Manchado, J. Biagiotti, L. Valentini, J.M. Kenny, Dynamic mechanical and Raman spectroscopy studies on interaction between single-walled carbon nanotubes and natural rubber, Journal of Applied Polymer Science 92 (2004) 3394–3400.

[47] A.M. Shanmugharaj, J.H. Bae, K.Y. Lee, W.H. Noh, S.H. Lee, S.H. Ryu, Physical and chemical characteristics of multiwalled carbon nanotubes functionalized with aminosilane and its influence on the properties of natural rubber composites, Composites Science and Technology 67 (2007) 1813–1822.

[48] L. Bokobza, Multiwall carbon nanotube-filled natural rubber: electrical and mechanical properties, eXPRESS Polymer Letters 6 (2012) 213–223.

[49] R. Verdejo, M.A. Lopez-Manchado, L. Valentini, J.M. Kenny, Carbon nanotube reinforced rubber composites, in: S. Thomas, R. Stephen (Eds.), Rubber Nanocomposites, Wiley, Singapore, 2010, pp. 147–168. (Chapter 6).

[50] D. Felhös, J. Karger-Kocsis, D. Xu, Tribological testing of peroxide cured HNBR with different MWCNT and silica content under dry sliding and rolling conditions against steel, Journal of Applied Polymer Science 108 (2008) 2840–2851.

[51] J. Karger-Kocsis, D. Felhös, R. Thomann, Tribological behavior of a carbon-nanofiber-modified Santoprene thermoplastic elastomer under dry sliding and fretting conditions against steel, Journal of Applied Polymer Science 108 (2008) 724–730.

[52] J. Karger-Kocsis, D. Felhös, D. Xu, A.K. Schlarb, Unlubricted sliding and rolling wear of thermoplastic dynamic vulcanizates (Santoprene®) against steel, Wear 265 (2008) 292–300.

[53] H. Kim, A.A. Abdala, C.W. Macosko, Graphene/polymer nanocomposites, Macromolecules 43 (2010) 6515–6530.

[54] G. Chen, W. Zhao, Rubber/graphite nanocomposites, in: S. Thomas, R. Stephen (Eds.), Rubber Nanocomposites, Wiley, Singapore, 2010, pp. 527–550. (Chapter 19).

[55] L. Wang, L. Zhang, M. Tian, Effects of expanded graphite (EG) dispersion on the mechanical and tribological properties of nitrile rubber/EG composites, Wear 276–277 (2012) 85–93.

[56] D.J. Brunelle, Synthesis and polymerization of cyclic polyester oligomers, in: J. Scheirs, T.E. Long (Eds.), Modern Polyesters: Chemistry and Technology of Polyesters and Copolyesters, Wiley, New York, 2003, pp. 117–142. (Chapter 3).

[57] Z.A. Mohd Ishak, Y.W. Leong, M. Steeg, J. Karger-Kocsis, Mechanical properties of woven glass fabric reinforced in situ polymerized poly(butylene terephthalate) composites, Composites Science and Technology 67 (2007) 390–398.

[58] J. Karger-Kocsis, P.P. Shang, Z.A. Mohd Ishak, M. Rösch, Melting and crystallization of in-situ polymerized cyclic butylene terephthalates with and without organoclay: a modulated DSC study, eXPRESS Polymer Letters 1 (2007) 60–68.

[59] G. Romhány, J. Vígh, R. Thomann, J. Karger-Kocsis, I.E. Sajó, pCBT/MWCNT nanocomposites prepared by in situ polymerization of CBT after solid-phase high-energy ball milling, Macromolecular Materials Engineering 296 (2011) 544–550.

[60] J. Karger-Kocsis, D. Felhös, T. Bárány, T. Czigány, Hybrids of HNBR and in situ polymerizable cyclic butylene terephthalate (CBT) oligomers: properties and dry sliding behaviour, eXPRESS Polymer Letters 2 (2008) 520–527.

[61] D. Xu, J. Karger-Kocsis, A.A. Apostolov, Hybrids from HNBR and in situ polymerizable cyclic butylene terephthalate (CBT): structure and rolling wear properties, European Polymer Journal 45 (2009) 1270–1281.

[62] D. Xu, J. Karger-Kocsis, Rolling and sliding wear properties of hybrid systems composed of uncured/cured HNBR and partly polymerized cyclic butylene terephthalate (CBT), Tribology International 43 (2010) 289–298.

[63] N.M. Barkoula, J. Karger-Kocsis, Process and influencing parameters of the solid particle erosion of polymers and their composites, Journal of Materials Science 37 (2002) 3807–3820.

[64] Y. Lu, L. Liu, C. Yang, M. Tian, L. Zhang, The morphology of zinc dimethacrylate reinforced elastomers investigated by SEM and TEM, European Polymer Journal 41 (2005) 577–588.

[65] Y. Lu, L. Liu, C. Yang, M. Tian, H. Geng, L. Zhang, Study on mechanical properties of elastomers reinforced by zinc dimethacrylate, European Polymer Journal 41 (2005) 589–598.

[66] Z. Peng, J. Qian, D. Yin, Y. Zhang, Y. Zhang, Reinforcement of elastomers by in situ prepared aluminium methacrylate, Kautschuk, Gummi, Kunststoffe 55 (2002) 94–99.

[67] Z. Wei, Y. Lu, Y. Meng, L. Zhang, Study on wear, cutting and chipping behaviours of hydrogenated nitrile butadiene rubber reinforced by carbon black and in-situ prepared zinc dimethacrylate, Journal of Applied Polymer Science 124 (2012) 4564–4571.

[68] J. Karger-Kocsis, D. Felhös, unpublished results.

[69] J. Karger-Kocsis, D. Felhös, D. Xu, Mechanical and tribological properties of rubber blends composed of HNBR and in situ produced polyurethane, Wear 268 (2010) 464–472.

[70] U. Šebenik, J. Karger-Kocsis, M. Krajnc, R. Thomann, Dynamic mechanical properties and structure of in situ cured polyurethane/hydrogenated nitrile rubber compounds: effects of carbon black type, Journal of Applied Polymer Science 125 (2012) E41–E48.

[71] D. Xu, J. Karger-Kocsis, Unlubricated rolling and sliding wear against steel of carbon-black-reinforced and in situ cured polyurethane containing ethylene/propylene/diene rubber compounds, Journal of Applied Polymer Science 115 (2010) 1651–1662.

[72] J.H. Tan, X.P. Wang, J.J. Tai1, Y.F. Luo, D.M. Jia, Novel blends of acrylonitrile butadiene rubber and polyurethane-silica hybrid networks, eXPRESS Polymer Letters 6 (2012) 588–600.

[73] B. Verheyde, M. Rombouts, A. Vanhulsel, D. Havermans, J. Meneve, M. Wangenheim, Influence of surface treatment of elastomers on their frictional behavior in sliding contact, Wear 266 (2009) 468–475.

[74] V. Viswanathan, T. Laha, K. Balani, A. Agarwal, S. Seal, Challenges and advances in nanocomposite processing techniques, Materials Science and Engineering Reports 54 (2006) 121–285.

[75] A. Wildberger, H. Geisler, R.H. Schuster, Atmosphärendruckplasmaverfahren, Kautschuk, Gummi, Kunststoffe 60 (2007) 24–31.

[76] D.K. Chattopadhyay, K.V.S.N. Raju, Structural engineering of polyurethane coatings for high performance applications, Progress in Polymer Science 32 (2007) 352–418.

[77] L. Bokobza, A.L. Diop, Reinforcement of silicone rubbers by sol–gel in situ generated filler particles, in: S. Thomas, R. Stephen (Eds.), Rubber Nanocomposites, Wiley, Singapore, 2010, pp. 63–85. (Chapter 3).

[78] C. Nah, D.H. Kim, W.D. Kim, W.-B. Im, S. Kaang, Friction and abrasion properties of in-situ silica-filled natural rubber nanocomposites using sol–gel process, Kautschuk, Gummi, Kunststoffe 57 (2004) 224–226.

[79] L. Busse, K. Peter, C.W. Karl, H. Geisler, M. Klüppel, Reducing friction with $Al_2O_3$/$SiO_2$-nanoparticles in NBR, Wear 271 (2011) 1066–1071.

[80] S. Kamel, Nanotechnology and its applications in lignocellulosic composites, eXPRESS Polymer Letters 1 (2007) 546–575.

[81] V.V. Tsukruk, Nanocomposite polymer layers for molecular tribology, Tribology Letters 10 (2001) 127–132.

[82] H.-W. Fang, Y.-C. Su, C.-H. Huang, C.-B. Yang, Influence of biological lubricant on the morphology of UHMWPE wear particles generated with microfabricated surface textures, Materials Chemistry and Physics 95 (2006) 280–288.

[83] Y. Martelé, K. Naessens, P. Van Daele, R. Baets, K. Callawaert, E. Schacht, Micropatterning polyurethane surfaces with lasers, Polymer International 52 (2003) 1641–1646.

[84] A. Le Gal, X. Yang, M. Klüppel, Evaluation of sliding friction and contact mechanics of elastomers based on dynamic-mechanical analysis, Journal of Chemical Physics 123, 2005. 014704.

[85] A. Le Gal, M. Klüppel, Investigation and modelling of adhesion friction on rough surfaces, Kautschuk, Gummi, Kunststoffe 59 (2006) 308–315.

[86] B.N.J. Persson, O. Albohr, U. Tartaglino, A.I. Volokitin, E. Tosatti, On the nature of surface roughness with application to contact mechanics, sealing, rubber friction and adhesion, Journal of Physics: Condensed Matter 17 (2005) R1–R62.

[87] M. Thomine, J.-M. Degrange, G. Vigier, L. Chazeau, J.-M. Pelletier, P. Kapsa, L. Guerbé, G. Dudragne, Study of relations between viscoelasticity and tribological behaviour of filled elastomer for lip seal application, Tribology International 40 (2007) 405–411.

[88] L. Pálfi, T. Goda, K. Váradi, Theoretical prediction of hysteretic rubber friction in ball on plate configuration by finite element method, eXPRESS Polymer Letters 3 (2009) 713–723.

[89] P. Gabriel, A.G. Thomas, J.J.C. Busfield, Influence of interface geometry on rubber friction, Wear 268 (2010) 747–750.

[90] J.M. Bielsa, M. Canales, F.J. Martínez, M.A. Jiménez, Application of finite element simulations for data reduction of experimental friction tests on rubber-metal contact, Tribology International 43 (2010) 785–795.

[91] F.J. Martínez, M. Canales, S. Izquierdo, M.A. Jiménez, F.J. Martínez, Finite element implementation and validation of wear modeling in sliding polymer-metal contacts, Wear 284–285 (2012) 52–64.

[92] F. Grün, W. Sailer, I. Gódor, Visualization of the process taking place in the contact zone with in-situ tribometry, Tribology International 48 (2012) 44–53.

[93] L. Martínez, R. Nevshupa, D. Felhös, J.L. de Segovia, E. Román, Influence of friction on the surface characteristics of EPDM elastomers with different carbon black contents, Tribology International 44 (2011) 996–1003.

# Scratch damage resistance of silica-based sol–gel coatings on polymeric substrates

**Zhong Chen\*, Linda Y.L. Wu†**

*\*School of Materials Science and Engineering, Nanyang Technological University, Singapore,*
*†Singapore Institute of Manufacturing Technology, Singapore*

## CHAPTER OUTLINE HEAD

## 14.1 INTRODUCTION

Sol–gel coating is a low-cost and low-temperature process, which can be applied to almost any type of substrates of any shapes. While the coatings are designed for a wide range of functionality from electrical, electromagnetic, optical, and chemical to mechanical applications, their mechanical integrity is always an important consideration for device reliability and durability. To achieve optimum mechanical integrity, the coating composition, microstructure and processing parameters have to be adjusted for different types of substrates under different functional constraints. An additional advantage of sol–gel coatings is that their properties can be easily modified through chemical formulation and hybridization with organic or inorganic compounds. Comprehensive reviews on sol–gel coatings were provided by Sakka and Yoko [1]. Malzbender et al. [2] have given an extensive coverage on the mechanical property measurement of sol–gel coatings. New development for various types of sol–gel formulation, the resulting properties and applications can be found in many recent publications [3–20].

The focus of this chapter is on the scratch behavior of relatively hard and stiff coatings produced by sol–gel processes on relatively soft and compliant substrates. In particular, since sol–gel processing temperature can be as low as 100 °C, it has been widely used as scratch-resistant coatings on a range of polymeric substrates that typically have very low glass transition temperatures. Application examples include plastic lenses, automobile topcoat, safety windows, display panels, etc. Many more applications are expected with the growing market for personal portable electronic gadgets, and flexible electronic devices and displays. Without doubt, the resistance to accidental scratch damage depends on the mechanical properties and other nonmaterial factors, including the size and shape of the scratching object, the coating thickness, and the test environment and speed. While it is important to understand the effect of all these factors on scratch failure, the current work will focus mainly on the mechanical properties and the thickness of the coatings. The mechanical properties of consideration include Young's modulus, hardness, fracture toughness and adhesion toughness to the substrate. The polymeric substrate in consideration is compliant and easy to deform plastically.

Scratch is a physical process during which a sharp object is pressed onto, and drawn over the surface of the coating simultaneously. The normal load is either kept constant or progressively increased, depending on the purpose of the test and machine availability. Usually progressively increased scratch testing aims to induce a critical point of damage such as coating delamination, coating cracking (in the case of brittle coatings) or whitening (in the case of polymeric coatings) by performing only one test. The critical load or its derivative (e.g. scratch hardness, defined as the load divided by indented area) will then be used to compare the performance of different coatings. In constant-load scratch testing, the normal force on the scratch stylus is maintained at a constant level during the test. Multiple tests at increased constant load levels can be used to determine the critical scratch load or scratch hardness. The constant-load scratch requires more tests to find the critical point of damage, but it can be carried out in a less costly testing rig compared to the progressive loading scratch test. It can also be used to detect nonuniformity of the coating over the entire surface. The American Society for Testing and Materials (ASTM) International has published quite a number of testing standards related to scratch tests over the years. For example, ASTM C 1624 [21] describes a standard test method for hard (Vickers hardness HV ≥ 5 GPa) ceramic coatings, while ASTM D 7027 [22] is for the evaluation of polymeric coatings and plastics. Nanoscratch of soft coatings is described by ASTM D 7187 [23]. For sol–gel coatings on polymeric substrates, the industry-preferred standard seems to be the pencil scratch test by ISO 15184 [24], or its counterpart by ASTM D 3363 [25]. There are some minor differences between the two standards. ISO 15184 has designated the normal load to be (750 ± 10) g at the pencil tip, while ASTM D 3363 only specifies a constant pressure to be exerted by the operator. In addition, the range of pencil lead grades recommended by ISO 15,184 is from 9B to 9H, while in ASTM D 3363 it varies from 6B to 6H. Apparently, between the two standards, ISO 15184 is able to provide more consistent results since it uses a controlled load. Both standards were intended for the so-called film hardness assessment of soft coatings like paints and varnishes, but in practice, they were popularly applied to scratch tests on protective hard coatings by the industry. The pencil scratch test presents an interesting variation from other scratch tests. It is a constant-load scratch test; however, it does not require increased constant load to reach the critical point of failure. Instead, it uses pencil leads of different hardness grades as the scratch stylus, compared to the diamond stylus usually used in most other test standards. By applying the same normal load with indenters of different hardness, a critical pencil lead grade is cited as the reference point for coating damage tolerance. The standard specifies that the highest pencil lead that does not cause damage to the coating is assigned to be the pencil hardness of the coating. Details about the pencil scratch test will be given later.

In the following sections, a brief review of the mechanics involved in a scratch test will be given. Potential failure modes, their mechanisms, and the relation with materials properties and loading conditions are discussed. Mechanical characterization of the coating fracture toughness and interfacial toughness by a controlled buckling experiment will be introduced. Finally, results of the pencil scratch test on sol–gel hybrid coatings are presented and discussed. General discussion will be made of the key factors that affect the scratch resistance performance of hard coatings.

## 14.2 MECHANICS OF SCRATCH

### 14.2.1 Stresses due to normal contact on a monolithic material

Scratch action exerts a combined normal (contact) and tangential load to the coating surface. For the convenience of understanding the mechanical analysis, we begin with the summary of contact load on a monolithic material first. The contact problem was initially studied by Hertz [26], Boussinesq [27], and Sneddon [28]. Comprehensive solutions for various contact problems have been provided by Johnson [29] in his book entitled *Contact Mechanics*. For a spherical indenter, the contact area is circular, and the pressure distribution under the indented area is given by

$$p\left(r\right) = p_0 \left[1 - \left(\frac{r}{a}\right)^2\right]^{1/2}$$

(14.1)

where $p_0 = \dfrac{3P_n}{2\pi a^2}$ is the maximum pressure at the center of the contact zone, $P_n$ is the normal load, $a$ is the radius of the contact, and $r$ is the distance from the center of symmetry. The radial ($\sigma_r$), circumferential ($\sigma\theta$), and axial ($\sigma_z$, along the indentation direction) stresses inside and outside the contact zone can be analytically solved [29] but they are not detailed here. Instead, two important aspects related to scratch damage are highlighted.

First, there is a transition from compressive to tensile in the radial stress at the contact surface. The radial stress is given by

$$\frac{\sigma_r}{p_0} = \begin{cases} \dfrac{(1-2v)}{3}\left(\dfrac{a}{r}\right)^2\left[1 - \left(1 - \dfrac{r^2}{a^2}\right)^{3/2}\right] - 2v\left(1 - \dfrac{r^2}{a^2}\right)^{1/2}, & r \le a \\[3mm] \dfrac{(1-2v)}{3}\left(\dfrac{a}{r}\right)^2, & r > a \end{cases}$$

(14.2)

From Eqn. (14.2), the maximum stress at the contact edge ($r = \pm a$) is tensile and its magnitude depends on Poisson's ratio, $v$:

$$\frac{\sigma_r^{max}}{p_0} = \frac{(1-2v)}{3}$$

(14.3)

For materials with $v = 0.25 \sim 0.3$, $\sigma_r^{max} = 0.17 \sim 0.13 p_0$. Figure 14.1 shows a finite element result by Djabella and Arnell [30] together with the analytical solution [29] for a spherical indentation. Similar stress distribution may be found when a coating is added.

Second, analysis on the stresses beneath the contact zone reveals that plastic deformation will begin at about $0.5a$ below the center of the contact circle [31]. Figure 14.2 shows a reproduction of the Tresca stress distribution under a spherical indenter based on the analysis by Davies [31]. Following Tresca yield criterion, Tabor [32] found that the initial yielding occurs when

$$p_m \approx 1.1\sigma_y$$

(14.4)

where $p_m$ is the mean pressure, which is $2/3p_0$ for a spherical indentation, and $\sigma_y$ is the yield strength of the material. As the load is increased further, the plastic zone increases in size and ultimately spreads to the surface. Further increase in load will see the plastic deformation filling up the whole contact area beneath the indenter, and the resistance to indentation is solely determined by plastic properties of the material. Tabor [32] suggested that at this stage, the hardness (defined as the mean pressure of an indentation) can be related to the material's yielding by a simple formula

$$H = C\sigma_{\text{flow}} \tag{14.5}$$

where $C$ is a constant that depends on indenter geometry only, and $\sigma_{\text{flow}}$ is the flow stress at certain representative plastic strains [32,33]. For fully yielded indentation of

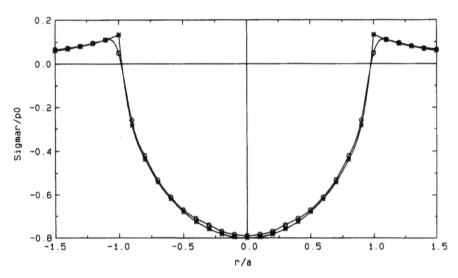

**FIGURE 14.1**

Comparison between analytical and finite element modeling results of the radial stress on a spherical contact [30]. The vertical axis represents the radial stress normalized by the maximum pressure ($\sigma_r/p_0$). Symbol * indicates the analytical, and symbol o the finite element solutions.

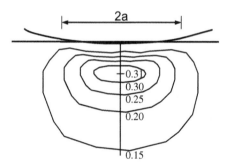

**FIGURE 14.2**

Elastic deformation of a spherical indenter on a flat surface (after [31]). The stress contour is normalized with the maximum pressure, $p_0$. The maximum Tresca stress is $0.31p_0$ (when $v = 0.3$), located at about $0.5a$ below the surface.

ductile materials, $C$ typically lies in the range of 2.8–3.0. Notice that $\sigma_{flow}$ could be taken at the plastic strain $\varepsilon = 0.2\%$, at which the point of yielding is defined.

Clearly, the stress–strain response depends on the level of loading in relation to the material's properties. Fischer–Cripps [34] divided the response into three regimes based on Tabor's analysis [32]:

Regime I: when $p_m < 1.1\sigma_y$, deformation is fully elastic and recoverable. There is no plastic deformation in the specimen during and after the indentation.

Regime II: when $1.1\sigma_y < p_m < C\sigma_{flow}$, the plastic deformation zone is constrained by the surrounding elastic material.

Regime III: when $p_m = C\sigma_{flow}$, there is complete plastic deformation. A notable feature is that in this regime the mean pressure (or hardness) hardly changes with increasing load.

Figure 14.3 shows a schematic of the mean pressure (or the measured hardness)—applied load relation for the above classification. The classification gives a good indication of the level of the applied load and deformation with respect to the indented material.

## 14.2.2 Influence of a tangential load on the normal contact

In a scratch test as a result of friction, a tangential load is added to the normal load. This friction traction superimposes a compressive stress at the front edge of the contact and a tensile stress to the trailing edge. Figure 14.4 shows the radial stress inside and outside a spherical stylus contact zone with different friction coefficients (after [35,36]). It is clear that the tensile stress at the trailing edge is intensified, while at the front edge the tensile stress is reduced even to compressive state, depending on the friction coefficient. It is also noticeable that at some location under the contact zone, the radial stress on the trailing side may change from compressive to tensile. The maximum tensile stress at the trailing edge is given by Hamilton [35,36]:

$$\frac{\sigma_r^{max}}{p_0} = \left( \frac{1-2\nu}{3} + \frac{4+\nu}{8} \pi\mu \right) \tag{14.6}$$

in which $\mu$ is the friction coefficient.

With the addition of the tangential friction force, the point of first yielding shifts forwards and upwards to the surface [37,38]. When the friction coefficient exceeds 0.3, the first yield point reaches the surface, so that the plastic deformation zone begins to form at the surface instead of at ~0.5$a$ below the surface as in the case of the pure contact shown in Fig. 14.2. In summary, the existence of a tangential force leads to intensified tensile stress at the trailing edge and moves the yield zone closer to the surface.

## 14.2.3 Stresses in coated systems under scratch loading

Djabella and Arnell [30,38] used finite element analysis to study the elastic contact stresses of coating/substrate systems with and without tangential loading. In their studies, coating modulus was varied from one to four times the modulus of

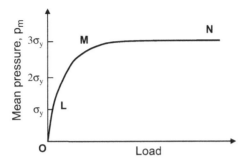

**FIGURE 14.3**

Mean pressure–load characteristics of an ideally plastic metal (after [32]). OL: regime I; LM: regime II; MN: regime III.

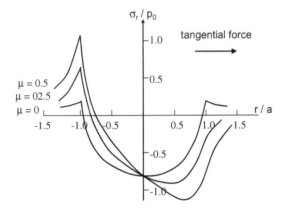

**FIGURE 14.4**

Radial stress caused by a spherical stylus moving from left to right (after [35,36]). The vertical axis shows the normalized radial stress by the maximum pressure, $\sigma_r/p_0$. The horizontal axis indicates the radial distance, $r/a$. When friction coefficient $\mu = 0$, the curve is the same as the one in Fig. 14.1.

the substrate. The ratio of the coating thickness over the contact radius, $t/a$, varied from zero (no coating) to unity. In general, it was found that in a frictionless contact problem [30], the modulus ratio (as defined by the quotient of coating modulus over substrate modulus) has little effect on the substrate stress but a marked effect on the coating stress. The maximum radial surface stress at the contact edge is greater than the uncoated case (see Fig. 14.1) when the coating is relatively thin ($t/a = 0$–0.35), but it becomes less after $t/a \approx 0.35$ (Fig. 14.5). For thin coatings, the magnitude of the tensile stress increases as the modulus ratio increases, while for thick coatings, the magnitude of the stress decreases when the modulus ratio increases. Further increasing the $t/a$ ratio beyond 1, it was found that the effect from the modulus ratio on the

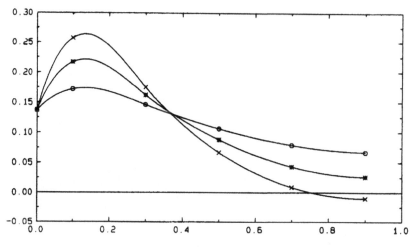

**FIGURE 14.5**

Variation of the coating surface stress at the contact edge ($r/a = 1$) with coating thickness [30]. The vertical axis shows the normalized radial stress, $\sigma_r/p_0$, and the horizontal axis the relative coating thickness, $t/a$. Different symbols represent different coating/substrate modulus ratio: o, modulus ratio = 2; *, modulus ratio = 3; ×, modulus ratio = 4.

maximum radial surface stress starts to become very little for $t/a \geq 5$ and virtually has no effect when $t/a \geq 8$. This is understandable since when the coating is very thick in comparison to the contact radius, the response can be treated as one of indentation on a monolithic material—the coating material only.

The radial stress $\sigma_r$ at the interface between the coating and the substrate shows a strong dependence on the thickness, as shown in Fig. 14.6. When the coating is thin ($t/a < 0.1$), the stress at the contact edge is tensile, similar to the case with no coating. When the thickness ratio is greater than 0.1, the stress becomes compressive, and the magnitude increases with increasing modulus ratio. A similar trend has been observed for the circumferential stress $\sigma\theta$ [30]. In general, it was concluded that the stresses of both the outer surface of the coating and the coating–substrate interface are complicated functions of the modulus ratio and the thickness ratio [30]. Therefore, it may not be surprising to find that in some coating thicknesses the in-plane stress is tensile while for others it is compressive. The corresponding failure modes will be quite different as a result of such transition.

When extending their analysis to include friction traction, Djabella and Arnell [38] have found that the in-plane surface stress at the trailing edge is affected by the coating thickness in a similar way as in the frictionless contact case (Fig. 14.5). The stress is greater than the uncoated stress when the coating is relatively thin, but it becomes less when the coating is thick. The critical thickness (at which the transition occurs) depends on the friction coefficient, and varies from $t/a \approx 0.2$ ($\mu = 0.15$) to $t/a \approx 0.5$ ($\mu = 0.5$). For thin coating thicknesses, the magnitude of the tensile stress increases as the modulus ratio increases. Similarly, for thick coating thicknesses, the

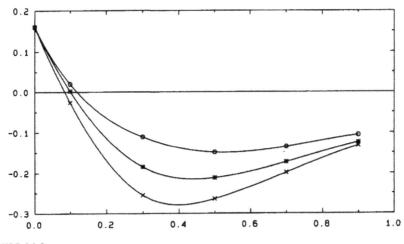

**FIGURE 14.6**

Variation of the normalized radial stress ($\sigma_r/p_0$) at the interface with coating thickness [30]. The point of interest is right beneath the contact edge ($r/a = 1$). The vertical axis shows the normalized radial stress, $\sigma_r/p_0$, and the horizontal axis the relative coating thickness, $t/a$. Different symbols represent different coating/substrate modulus ratio: o-modulus ratio = 2; *-modulus ratio = 3; ×-modulus ratio = 4.

magnitude of the stress decreases when the modulus ratio increases. Again the substrate effect is almost negligible when $t/a \geq 8$. At the front edge, the in-plane stress is always more compressive than for the uncoated substrate. The absolute magnitude increases with increasing modulus ratio and increasing friction coefficient [38].

The variation in the in-plane stress at the coating interface beneath the trailing edge shares some similarities with the frictionless case shown in Fig. 14.6. Take an example with $\mu = 0.15$, for very thin coating $t/a < 0.05$, the stress is tensile; for $0.05 < t/a < 0.6$, the stress is compressive and the absolute magnitude increases as the modulus increases. For $t/a > 0.6$, the absolute magnitude of the compressive stress decreases with increasing modulus ratio and may go back to tensile stress region [38]. For higher friction coefficients, similar trends exist, but the transition from tensile to compressive is much delayed till much thicker coating. As a result, when $\mu > 0.5$, most of the thickness range is in tensile stress [38]. The in-plane stress at the coating interface beneath the front edge is compressive when there is no coating (Fig. 14.4) due to the existence of the friction traction. The compressive stress increases in magnitude (more compressive) when $t/a$ reaches 0.1; after $t/a = 0.1$, the magnitude begins to decrease (less compressive). However, within the range of $\mu = 0.15$–0.5, this stress remains compressive. The magnitude of the stress always depends on the modulus ratio.

To conclude this section, the stress in a scratch test is a complex function of the thickness ratio, the modulus ratio, and the friction coefficient. The nature of stress in the coating can vary from tensile to compressive, making the prediction of coating failure a challenging task. Nevertheless, several general trends can be summarized here. First, the in-plane surface stress at the trailing edge is tensile, and its magnitude increases when the coating is relatively thin. When the coating is thick, the magnitude of the tensile stress decreases but it largely remains in the tensile range. On the other hand, the in-plane surface stress at the front edge is largely compressive due to the addition of the friction traction. Second, at the coating interface, the in-plane stress at the trailing edge is tensile for thin coatings, and it becomes compressive with thick coatings. Further increase in coating thickness reduces the magnitude of the compressive stress and may even bring it to the tensile stress range. Similar to the surface stress, the interface stress at the front edge is always compressive. Third, the tangential frictional force brings the maximum shear stress zone upwards to the surface and forward in the direction of scratch. When the coating is thin, the highest maximum shear stress zone is at the surface of the coating. For a relatively thick coating, the surface remains to be a maximum stress region, and meanwhile at the coating/substrate interface another region of high shear stress may also form. This summary is largely based on the finite element work by Djabella and Arnell [30,38]. Many other researches on the modeling and simulation of a scratch-loaded coating on substrate systems have also been reported [39–46].

## 14.3 SCRATCH FAILURE MODES

Two major groups of coating failure may occur during a scratch test: coating cracking and delamination. Coating cracking is mainly caused by tensile in-plane stress, while delamination is usually caused by in-plane compressive stress. However, due to the complexity in the stress state as discussed above, the exact failure pattern, location and sequence can be quite different even within the general category of coating cracking or delamination. In general, delamination-related failures happen when the interface is relatively weak and when the substrate experiences large deformation, corresponding to regimes II and III shown in Fig. 14.3. Coating cracking may occur in all three regimes.

ASTM C 1624-05 [21] has provided a scratch atlas for hard coatings on steel substrates. However, it does not provide the mechanisms for these different failure modes. Many studies have been dedicated to the investigation into the mechanisms behind coating failures [2,3,47–53]. Here we will provide a discussion of the failure modes that are likely to be encountered in the systems of brittle coatings on relatively compliant substrates. According to the schematic map of scratch failure modes by Bull [53], when the substrate hardness is low and the coating hardness is high, the most likely failure will be the through-thickness coating cracking. This is true only when the interface is strong enough. Otherwise, coating delamination related failure modes should never be overlooked.

Figure 14.7 depicts an exaggerated deformation sketch caused by a sliding sphere over a coating on the substrate [44]. Combined with the overview of the coating stresses in the previous section, it is understandable that the main driving forces come from in-plane tensile stress or compressive stress. The stresses are induced by the joint action of contact-induced bending, the friction traction from sliding, and the pileup of the substrate due to plastic deformation. Table 14.1 lists the failure modes and their schematics, which will be discussed individually later. Spherical sliding contact is assumed. We have left out the mode of chipping since it typically occurs in brittle bulk specimens or within hard coatings on hard substrates [53]. The present

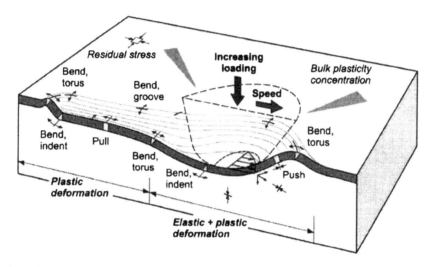

**FIGURE 14.7**

Schematics of loads and deformation caused by a sliding sphere over a coating on substrate [44].

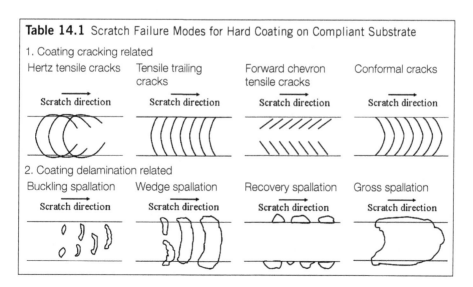

work is limited to hard coatings on compliant substrates; thus this may not occur unless the coating thickness is very large.

## 14.3.1 Hertz tensile cracks

Hertz tensile cracks typically form as a result of tensile radial stress (Fig. 14.1) caused by normal contact force on brittle materials. Cone cracks in bulk glass specimens were initially observed and studied by Hertz [26]. When the indenter is sliding, the friction traction tilts the cone crack axis by changing the symmetric radial stress in the nonfrictional case [26,35]. As a result, only partial cone cracks can be observed [54,55]. Similar to the bulk behavior, a brittle coating on a ductile substrate will also show partial rings on the surface under sliding spherical contact [51,53], but the partial cone cracks may just stop at the interface if the substrate is much tougher. Depending on the interface toughness, underneath delamination under the contact may or may not occur. As the indenter moves along, several partial ring cracks are formed along the track and these rings may intercept with each other to cause a network of cracks. This mode can occur even with only elastic deformation of the substrate, and the cracks closely follow the shape of the indenter. The fact that the cracks on the surface can form more than half-full a ring indicates that the contact-induced stress is the dominant factor for the cracking; the tensile traction force only modifies the magnitude of the stress and its distribution.

Chai and Lawn [56] carried out a thorough analysis about the effect of coating thickness on the stress distribution, failure initiation, and the corresponding critical loads in pure contact loading. When the coating is relatively thick, $t/a \gg 1$, the coatings may be treated as a monolithic material where cone crack initiates on the top surface around the contact in the classical Hertz mechanics [26,54,55]. With intermediate coating thickness, $t/a \sim 1$, the coating flexural stress begins to dominate, which will produce a maximum stress at the bottom surface leading to a radial crack, plus secondary ring cracks on the top surface. In such a case, the critical load, $P_{cri}$, depends on the coating fracture strength, $\sigma_{cf}$, and the thickness of the coating as in

$$P_{cri} \propto \sigma_{cf} t^2 \tag{14.7a}$$

When the coating is thin, $t/a \ll 1$, through-thickness ring cracks initiate on the top surface again, and the critical load is related to the indenter radius, $R_i$, the coating fracture strength, as well as the modulus of both the coating and the substrate in the following relation:

$$P_{cri} \propto (E_s/E_c)(\sigma_{cf}/E_c)^2 \sigma_{cf} R_i^2 \tag{14.7b}$$

where $E_s$ stands for substrate modulus and $E_c$ for coating modulus. The thickness-dependent behavior was derived based on the pure contact analysis [56], but it is qualitatively applicable to analyzing sliding contacts as well. Hsueh [57] derived an analytical "master curve" solution for the indentation displacement for different modulus ratios and coating thicknesses. The solution would enable the prediction of coating/substrate mechanical properties based on the indentation response. Contact cracking of brittle bilayers on compliant substrate was also investigated by Hsueh and Miranda [58].

## 14.3.2 **Tensile trailing cracks**

Besides the stresses caused by frictional pulling, coating bending due to the contact action also contributes to the tensile stress that opens up the crack perpendicular to the sliding direction behind the moving stylus. The contribution from the bending depends on the relative thickness of the coating, and can become significant when $t/a \leq 1$ [56]. When the contact load is large (e.g. regime III in Fig. 14.3), severe grooving in the substrate will also cause cracking in the coating alongside the groove edge. In addition, plastic pileup in front of the indenter can make a considerable change to the load distribution. When the pileup occurs, the load on the moving stylus is mainly borne by the front half of the indenter so the effective friction coefficient increases drastically. The increase in the friction traction in turn raises the frictional stress at the trailing end (refer to Eqn. (14.6)), which favors the formation of the curved parallel cracks behind the advancing indenter.

Although both the trailing tensile cracks and Hertz cracks may show curved, rather than straight lines, there are some differences that set them apart. First, the tensile cracks are formed under the dominant effect of frictional traction stress; therefore, a large friction coefficient favors the tensile trailing cracks. Hertz contact rings may be observed with very small traction stress, and when there is little or no friction, the partial ring cracks will revert back to full rings, where the radius is at the minimum. Second, in the case of a trailing tensile crack, the crack lines do not overlap with each other as they may in the Hertz tensile cracking. These lines may still be curved but the radius is typically greater than the contact radius (i.e. flatter). This provides a good physical feature differentiating the two modes.

Tensile trailing cracking closely resembles the parallel channeling cracks under uniaxial stress that have been extensively studied [59–66]. Thouless [64] derived a simple relation between the energy release rate, $G_{cra}$, of a thin coating and the minimum crack spacing between the parallel cracks:

$$G_{cra} = \begin{cases} \dfrac{1.98\sigma_c^2 \left(1-v_c^2\right)t}{E_c} & l \geq 8t \\[4mm] \dfrac{\left[0.5(l_c/t)-0.0316(l_c/t)^2\right]\sigma_c\left(1-v_c^2\right)t}{E_c} & l \leq 8t \end{cases}$$

(14.8)

where $\sigma_c$ is the coating stress, $l_c$ is the crack spacing and $v_c$ is the Poisson's ratio for the coating materials. At the point of cracking, $G_{cra} = \Gamma_c$, the fracture toughness of the coating. Equation (14.8) shows that the fracture toughness of the coating can be estimated if the stress causing the cracking is known. However, this expression was derived by assuming that the substrate has the same elastic modulus with the coating. This may introduce considerable error when the substrate is more compliant than the coating [59–63]. For ceramic-based sol–gel coatings on polymeric substrates, the substrate is indeed much more compliant than the coating; therefore, the accuracy of applying Eqn. (14.8) remains uncertain. More work is needed to take into account the substrate compliance effect.

Detailed analysis of the parallel cracking problem was given by Hsueh et al. [65,66] with the possibility of residual stress included. Based on their work, if crack density—applied strain function—is known for one film thickness, it can be used to predict those for other thicknesses. When the residual stress is unknown, the measured crack density—applied strain relation—can be used to calculate the residual stress.

### 14.3.3 Forward chevron tensile cracks

This mode of failure also occurs at the trailing end, but the fracture initiates near the two edges of the contact groove, forming a slanted angle to the sliding direction. Experimental observation found that the forward chevron cracking occurred at a restively low indenting depth after substantial groove formation [67]. With further increase in the load, both the density and the length of the cracks increase. Eventually this mode of failure transits to tensile trailing cracking, where cracks extend across the full width of the groove. Three-dimensional finite element analysis by Holmberg et al. [44,67] revealed that at the chevron crack formation depth, the tension from movement of the indentation tip and the deformations near the contact edge interact with each other. The joint action induces peak stress at the tail end to form at an approximate angle of 45° from the plane of symmetry. When the loading depth is further increased, there is a transition in the state of stress to be dominated by the tension, and the tensile stress peak is located behind the contact region right before the part of the substrate that has experienced elastic unloading [67]. Corresponding to this stress, the full-width tensile trailing cracks form. Figure 14.8 summarizes the failure mode transition illustrated by Holmberg et al. [44].

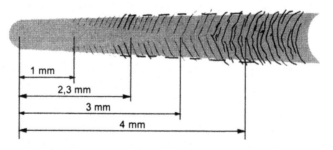

**FIGURE 14.8**

Schematics showing cracks observed with progressively increased scratch loading test [44]. The actual experiment was done on a 2-μm-thick TiN coating on a high-speed steel substrate. The sliding stylus was Rockwell C type with a spherical diamond tip of 200 μm radius.

### 14.3.4 **Conformal cracks**

Conformal cracking occurs under the similar driving force as the one in buckling spallation, which will be discussed later. The reason for coating cracking without delamination is clearly the strong interface adhesion. The cracks follow semicircular trajectories parallel to the leading edge of the spherical indenter [68] when substantial substrate pileup has occurred (Fig. 14.7). The tensile stress that causes the cracking mainly comes from the bending in the pileup, and is sensitive to the normal loading and the plastic properties of the substrate. Calculation of the exact magnitude of such stress can be very complex, and may require the help from finite element tools. If the conformal cracks have formed in front of the moving stylus, the trailing cracks in the wake will be less likely to occur. This is because the existing cracks can prevent stress buildup when the same piece of coating moves to the trailing end of the indenter.

### 14.3.5 **Buckling spallation**

Buckling spallation is caused by the compressive stress ahead of the moving indenter. Apart from the elastic compressive stresses as analyzed by Djabella and Arnell [38], additional stresses come from the plastic deformation induced pileups (Fig. 14.7). Relatively weak locations at the coating interface in front of the moving indenter start to delaminate so that the coatings in these locations buckle up. Coating spallation results when the buckles are bent to a critical curvature. Bull [53] found that this failure mode occurs for relatively thin ceramic coatings on steel substrates, and the critical load for buckling spallation increases as the coating thickness increases. The estimated compressive stress causing the buckling spallation also increases with coating thickness. Evans [69] expressed the critical buckling stress as

$$\sigma_{bk} = \frac{1.22E_c}{1 - \nu_c^2}\left(\frac{t}{R_d}\right)^2 \tag{14.9}$$

where $R_d$ is the radius of the circular area of interfacial delamination. This equation agrees well with the observed trend.

Xie and Hawthorne [45] analyzed the compressive stress in the coating and found that the stress is affected mainly by the relative contact radius (as indicated by the ratio of contact radius over the indenter radius), the yield strength of the coating, Young's modulus of the coating, and the friction coefficient. They have derived a simple empirical expression for the mean compressive stress as

$$\sigma_{mc} = 0.15\left(\frac{P_n H_c}{H_s}\right)^{0.5}\frac{E_c^{0.3}E_s^{0.2}}{R_i} \tag{14.10}$$

where $H_c$ and $E_c$ are the hardness and Young's modulus of the coating, and $H_s$ and $E_s$ are the hardness and Young's modulus of the substrate. By comparing the stress dependence on the indenter geometry, it was concluded that a large indenter radius and high normal load favors the compressive coating stress as compared to the tensile bending stress [46]. As a result, it is easier to induce delamination under such conditions.

Thouless [52] analyzed the case of a detached piece of coating in a trapezoidal shape as in Fig. 14.9(a). Two assumptions were made: (1) there is no pileup ahead of the indenter so the driving force comes from the sliding indenter only; (2) coating cracking occurs readily through the thickness of the coating. Following a general solution by Evans and Hutchinson [70], the energy release rate for delaminating the interface is given by

$$G_{\text{dela}} = \frac{P_{\text{lat}}}{8E_c (L\tan\beta + d)^2 t} \tag{14.11}$$

where $P_{\text{lat}}$ is the lateral force acting on the edge of the spall. $L_d$, $d$ and $\beta$ are the spall dimensions as shown in Fig. 14.9. At the point of delamination, $G_{\text{dela}} = \Gamma_{\text{int}}$, the interfacial fracture toughness. By analyzing the required bulking force, Thouless [52] found that the interfacial toughness can be directly estimated from the geometry of the spall by

$$\Gamma_{\text{int}} = 0.35 \frac{E_c t^5}{L_d^4} \left( \frac{\tan\beta + 2d/L_d}{\tan\beta + d/L_d} \right)^2 \tag{14.12a}$$

If considering the presence of residual stress, $\sigma_r$, then the interface toughness becomes

$$\Gamma_{\text{int}} = \frac{E_c t}{2} \left[ 0.5 \left( \frac{\sigma_r}{E_c} \right)^2 + 0.42 \left( \frac{\sigma_r}{E_c} \right) \left( \frac{t}{L_d} \right)^2 + 0.353 \left( \frac{t}{L_d} \right)^4 \right] \tag{14.12b}$$

Modifying Thouless's solution for a curved front edge, which is more realistic in experimental observation as shown in Fig. 14.9(b), Malzbender, den Toonder et al. [2,71] proposed that

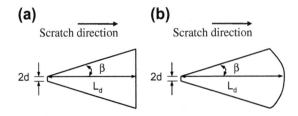

**(a)** Scratch direction    **(b)** Scratch direction

**FIGURE 14.9**

Geometry of a spalled region. (a) Trapezoidal shape as considered by Thouless [52]; (b) modified version by Malzbender, den Toonder et al. considering a curved front edge [2,71].

$$\Gamma_{\text{int}} = \frac{1.42 E_c t^5}{L_d^4} \left( \frac{\frac{d}{L_d} + \frac{\beta \pi}{2}}{\frac{d}{L_d} + \beta \pi} \right)^2 + \frac{t (1 - \nu_c) \sigma_r^2}{E_c} + \frac{3.36 (1 - \nu_c) \sigma_r t^3}{L_d^2} \left( \frac{\frac{d}{L_d} + \frac{\beta \pi}{2}}{\frac{d}{L_d} + \beta \pi} \right)$$

$$(14.13)$$

Caution should be exercised when applying various types of formulas for the interface delamination energy release rate. Most of them were derived based on the elastic energy stored in the coating only (e.g. [52,60,71–74]), which will significantly underestimate the energy release rate when the substrate is more compliant than the coating. Cotterell and Chen [75,76] have shown that for indium-tin-oxide film on polyethylene terephthalate substrate, which has the modulus ratio $(E_c/E_s)$ around 60, the calculated interface toughness would only be 1/7 of the actual value if the substrate effect is neglected. The finding was confirmed and extended by Yu and Hutchinson [77]. Nevertheless, these empirical formulas can be very useful in parametric analysis into the key factors leading to delamination.

### 14.3.6 **Wedge spallation**

Wedge spallation is also caused by the compressive stress ahead of the moving indenter, as well as along the scratch groove. This mode of failure occurs on relatively thick coatings compared to the ones in buckling spallation. When the coating is thick, buckling is much more difficult (since buckling stress scales with $t^2$ as shown in Eqn. (14.9)). Wedge spallation begins with compressive shear fracture of the coating, and then interface delamination [50,51,53]. The difference in the fracture patterns between buckle spallation and wedge spallation can be found by observing the edge of the spalled coating. In buckle spallation, the angle of the edge is perpendicular to the coating–substrate interface since it is fractured by the in-plane tensile stress; but in wedge spallation the edge forms a slanted angle to the interface due to the shear [50]. Another distinct feature for wedge spallation is that the spalled area tends to be larger and semicircular. The large delaminated area is due to the greater amount of energy stored in a thicker coating, and for the same reason, the spall may propagate into trackside if the adhesion is relatively weak. On the other hand, buckling spallation happens in relatively thin coatings so it is more likely to produce relatively small spalled areas within the scratch track.

Bull [53] had observed that this failure mode occurs for relatively thick ceramic coatings (typically >10 μm) on steel substrates, and the critical load for wedge spallation decreases as the coating thickness increases. The estimated compressive stress causing the buckling spallation also decreases with coating thickness. Evans [69] estimated that the biaxial stress in the coating necessary to cause wedge crack should be

$$\sigma_{\text{wd}} = \left[ \frac{4 E_c \Gamma_c}{(1 - \nu_c) \lambda} \right]^{\frac{1}{2}}$$

$$(14.14)$$

where $\lambda$ is the width of the wedge spalled area. The stress to produce the spall after the shear cracking is given by

$$\sigma_{sp} = \left[ \frac{E_c \Gamma_{int}}{(1 - \nu_c) t} \right]^{\frac{1}{2}} \tag{14.15}$$

This stress is usually greater than the stress for wedge cracking. Therefore, the observed wedge spallation stress dependence on the coating thickness [53] can be well explained. In addition, based on Eqn. (14.15), if the spallation stress is known, by plotting $\sigma_{sp}$ against the reciprocal of the square root of thickness, $t^{1/2}$, the interfacial toughness can be estimated.

The source for the stresses leading to wedge delamination and spallation could mainly come from the friction traction [35,36] when there is little plastic deformation in the substrate, or from the substrate dilation when there is a severe substrate plastic deformation. The plastic deformation of the substrate drives the coating away from the center of symmetry causing delamination with or without spallation. To induce a large plastic deformation in the substrate, the applied load should be quite high. For example, up to 1.5 kN load was applied through a Rockwell "C" indenter to induce delamination of ~1-μm diamond film on Ti-6Al-4V substrate [78,79]. Assuming lateral displacement is the only driving force for coating delamination, the compressive stress will increase with increasing volume of plastic deformation [70]. Based on the same principle, indentation induced delamination by a wedge indenter was investigated by Vlassak et al. [80].

## 14.3.7 Recovery spallation

This mode of failure occurs after the sliding indenter has passed over the loaded region. Delaminated areas are alongside the edges of the scratched groove. It is caused by the difference in the amount of the elastically recovered strain between the coating and the substrate. The mismatch in the strain imposes a shear stress at the coating interface leading to delamination on the sides of the scratch track. There is a good similarity in appearance between wedge spallation and recovery spallation since both can occur on the sides of the scratch track, but the driving forces are very different. The former occurs when the load is applied, and the latter only appears after the load is released. It is not unusual to find both types of failure to operate in the same test piece, and when this happens, it is difficult to differentiate the two. Bull [51] suggested that the amount of residual stress could be used to tell the difference between the two: when compressive residual stress increases, the wedge spallation becomes easier, while the residual stress does not affect the recovery spallation.

## 14.3.8 Gross spallation

Gross spallation is a sign of (1) extremely poor adhesion or (2) the presence of extremely large residual stress. A small interface delaminated spot can propagate a considerable distance ahead and on either side of the track before coming to a stop, causing a large area of spallation. The initial interface defects may either nucleate at a weak interface under buckling load, or derive from through-thickness crack. Therefore, its mechanism is similar to buckling spallation or wedge spallation, and the corresponding mechanics applies. When the residual stress exceeds the critical value $\sigma_r > \sqrt{2E_c\Gamma_{int}/t}$, spontaneous delamination will occur [52].

## 14.4 FRACTURE TOUGHNESS AND INTERFACE TOUGHNESS MEASUREMENT

From the above discussion, it is clear that two important material properties that could determine the transition from coating fracture-related modes to delamination-related modes are the fracture toughness of the coating and the interface toughness. When the coating adhesion is weak or when the driving force for delamination is relatively greater (for example, high indenter radius and high normal load increase the delamination stress more than coating cracking stress [46]), delamination-related failures are more likely to occur. On the other hand, when the interface adhesion is strong or when the loading configuration provides a relatively weaker driving force for delamination, the modes of coating cracking become dominant. Sufficient evidence has been shown by many researchers that pretreatment of the coating surface could significantly improve the interface adhesion, changing the failure mode from interface related to coating related, resulting in higher critical loads for scratch failure [43,47]. It is obvious that both properties play important roles in scratch resistance. However, unlike other properties like Young's modulus and hardness, which can be reliably measured on the coating itself by nanoindentation, fracture toughness and interface fracture toughness values have been mostly calculated based on empirical models (for example, [2,52,64,69,71]). In this section, we will present a scheme using controlled buckling test for the measurement of the fracture toughness and interface toughness. This technique was developed by Chen, Cotterell et al. [61–63,75,76], which shows very consistent and reliable results [81,82].

### 14.4.1 The controlled buckling experiment

The test samples are prepared by applying a layer of coating, typically 0.1–5 μm thick, on one side of the substrate that is a few hundred micrometers thick. The samples are in the form of a slender thin plate (beam). Polymer substrate is usually chosen since it has the advantage of a large elastic strain limit and low modulus compared to metals. It is easy to buckle a polymeric beam specimen, and when the coating fails in a brittle manner, the substrate remains elastic. As will be seen later, the mechanical analysis of the system is much simpler when both the coating and the substrate remain elastic up to the point of coating failure. If high-temperature coating has to be conducted, polymer substrates become unsuitable; then metallic substrates may have to be used. In such a case, plastic deformation in the substrate may occur. The procedure to analyze plastically deformed substrate will also be given in this section.

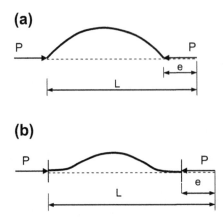

**FIGURE 14.10**

The controlled buckling test. (a) Simple support (free ends); (b) built-in support (clamped ends). The film is coated on one side of the substrate only.

The controlled buckling test is illustrated in Fig. 14.10. The ends of the composite (coating on substrate) beam are either free (simple support) or clamped (built in). The experiment is carried out with progressive displacement applied, either using a mechanical testing machine, or manually by an operator using a purpose built fixture. Under the axial loading, the beam buckles, and the thin coating on one side of the substrate experiences uniform stress (or strain).

The test fixture shown in Fig. 14.11 can be used for the measurement of both the coating fracture toughness and the interface fracture toughness between the coating and the substrate [61–63,75,76]. When the coating is placed on the tensile side of the plate, channeling crack will form when the energy release rate reaches a critical point. Similarly, when the coating is placed on the compressive side of the plate, steady state delamination will occur when the critical energy release rate reaches its critical value. The onset of cracking or delamination can be observed using an optical microscope (Fig. 14.11). The only parameter that needs to be taken after the experiment is the lateral displacement at which the coating cracking or delamination occurs. The corresponding critical strain can be calculated by large displacement beam bending theory. Detailed calculation will be given later.

Referring to Fig. 14.10, when both the coating and the substrate remain elastic to the point of fracture, the coating on substrate plate under controlled buckling test can be analyzed as a plane strain beam loaded along its axis. Large deformation buckling theory of beams correlates the maximum curvature of bending to the displacement by the following equations [83]:

$$\delta = 2 \left[ 1 - \frac{E(k)}{K(k)} \right]; \quad \frac{l}{R} = 4K(k)\,k \tag{14.16}$$

**FIGURE 14.11**

Controlled buckling test fixture operated manually under an optical microscope. This fixture can take both the clamped or simple supported slender thin plate samples. (For color version of this figure, the reader is referred to the online version of this book.)

where $K(k)$ and $E(k)$ are complete elliptic integrals of the first and second, $L$ is the original length of the beam, $R$ is the minimum radius of curvature corresponding to the maximum bending, and $\delta = e/L$ is the contraction ratio. $k$ is an intermediate variable in the above equation. For the two schemes in Fig. 14.10, $l = L$ for simple support and $l = L/2$ for built-in ends. By the input of the displacement, $e$, measured from the experiment, the radius of the maximum bending curvature, $1/R$, can be calculated.

Since the ratio of the coating thickness to the substrate thickness is much less than one, the neutral axis is close to the geometric center of the composite beam. Thus the strain in the coating, $\varepsilon_c$, is approximately uniform through the coating thickness and is given by

$$\varepsilon_c = \frac{(t + t_s)}{2R} \tag{14.17}$$

where $t$ and $t_s$ are the thicknesses of the coating and substrate, respectively.

In the case where the substrate exceeds the elastic limit when coating failure occurs, an elastic–plastic stress–strain relation for the substrate material has to be known first. A detailed buckling analysis for an elastic–linear work hardening plate has been provided by Liu et al. [84]. Following the approach, similar analysis for any other type of work hardening materials can be carried out. Again from the lateral displacement, $e$, the maximum bending curvature at the point of coating failure can be obtained.

## 14.4.2 Fracture toughness calculation

When the coating is placed under the tensile side of the buckling beam, parallel channeling cracks will form through the width of the plate when the critical bending strain (stress) is reached. They are an indication of steady state propagation of through-film-thickness cracking. The steady state energy release rate, $G_{cra}$, is given by [59,60]

$$G_{cra} = \frac{1}{2}\bar{E}_c \varepsilon_c^2 \pi t g \left(\alpha, \beta\right)$$

(14.18)

in which $\bar{E}_c$ is the plane strain modulus of the coating. The factor $g(\alpha,\beta)$ is a function of the Dundurs' parameters, $\alpha$ and $\beta$, which for plane strain condition are given by

$$\alpha = \frac{\bar{E}_c - \bar{E}_s}{\bar{E}_c + \bar{E}_s}; \quad \beta = \frac{\bar{E}_c \left(\frac{1 - 2\nu_s}{1 - \nu_s}\right) - \bar{E}_s \left(\frac{1 - 2\nu_c}{1 - \nu_c}\right)}{2\left(\bar{E}_c + \bar{E}_s\right)}$$

(14.19)

where $\bar{E}_s$ is the plane strain modulus of the substrate, and $\nu_s$ is the Poisson's ratio of the substrate. Of these two parameters that determine the value of $g(\alpha,\beta)$, $\alpha$ is far more influential, while $\beta$ has nearly negligible effect on $g(\alpha,\beta)$. The finite element scheme that calculates $g(\alpha,\beta)$ has been given in [59]. The critical energy release rate, or the fracture toughness, $\Gamma_c$, of the coating can be obtained by Eqn. (14.18) once the critical fracture strain, $\varepsilon_{cf}$, is known from Eqns (14.16) and (14.17).

When the substrate yielding occurs, the energy release rate can still be expressed by the general form of Eqn. (14.18), but the $g(\alpha,\beta)$ now has to be replaced by a function of not only $\alpha$ and $\beta$, but also the applied stress level $(\sigma_c/\sigma_y)$ and the work hardening coefficient, $n$:

$$G_{cra} = \frac{1}{2}\bar{E}_c \varepsilon_c^2 \pi t g \left(\alpha, \beta, \frac{\sigma_c}{\sigma_y}, n\right)$$

(14.20)

Finite element scheme to calculate $g\left(\alpha, \beta, \frac{\sigma_c}{\sigma_y}, n\right)$ has been outlined by Beuth and Klingbeil [85]. Substrate yielding is seldom encountered with brittle coatings even when metallic substrates are used, since the coating usually fractures at lower strain compared to the substrate. If it does happen, the deformation in the substrate promotes crack opening and thus increases the energy release rate. It was found [63] that the energy release rate is affected by the plastic deformation more significantly for substrate materials of low work hardenability. In practice, this means that the same coating will crack more easily with more compliant substrate, especially when the work hardening coefficient is lower.

### 14.4.3 Interface toughness calculation

When the coating is placed on the compressive side of the bent beam, delamination of the coating occurs when the critical energy release rate is reached. Since the coating is under uniaxial stress, a steady state channeling delamination perpendicular to the loading direction will be observed [75,76]. Depending on the maximum tensile stress in the buckle, coating cracking may or may not occur along with the channeled delamination. Superficially under the optical microscope, a buckled and cracked delamination failure looks very similar to the channeling tensile cracking, but closer inspection with scanning electron microscopy (SEM) or atomic force microscopy can reveal the difference. In the former case, the cracks grow behind the delaminated buckles [61,62,75]. The calculation for the energy release rate of the delamination with and without coating cracking was initially provided by Hutchinson and Suo [60], and Thouless [72], respectively. However, it was later found that both these analyses that consider the energy stored in the coating itself can introduce significant amount of error when the substrate is more compliant than the coating [75–77]. This is indeed the case of sol–gel coating on polymer substrate. Thus, consideration of the energy stored in the substrate has to be made. Such analyses were provided by Cotterell and Chen [75,76] and Yu and Hutchinson [77]. Interested readers should refer to the papers for details.

# 14.5 SCRATCH PERFORMANCE OF SOL–GEL COATINGS ON POLYMERIC SUBSTRATES

Based on the above analysis on scratch failure modes, it is obvious that it is critical to improve both the coating properties and its adhesion to the substrate. In this section, several examples are illustrated through coating improvement as well as interface adhesion enhancement based mainly on our previous work. The first example uses organically modified silicate (Ormosil) coatings prepared via a sol–gel route. The sol–gel coatings contain 3-glycidoxypropyl trimethoxysilane (GLYMO), tetraethylorthosilicate (TEOS) and silica nanofillers. The second example is an ultraviolet (UV)-curable sol–gel coating with incorporated functional polyhedral oligomeric silsesquioxanes (POSS) nanoparticles, which further enhance the scratch resistance and durability of the coating. The third example demonstrates interfacial adhesion enhancement through thermal impregnation of a chemical layer onto the surface of the polymer substrate, providing a hardened interface and chemical linking between the substrate and the sol–gel coating. Together with other findings by other researchers, general discussion will also be made on the key factors involved in the scratch resistance analysis.

## 14.5.1 GLYMO-TEOS coatings with colloidal silica fillers [86,87]

### 14.5.1.1 Sample preparation and characterization

A stock solution of GLYMO–TEOS was prepared by hydrolyzing them in ethanol (EtOH) and water ($H_2O$) in an acidic solution (HIt, itaconic acid). The molar ratios of the components were GLYMO:TEOS:EtOH:$H_2O$:HIt = 1.0:1.63:2.19:5.0:0.26. The GLYMO and TEOS were hydrolyzed separately and then mixed together. The mixture was stirred for 24 h and used as the base solution for coatings. To this base solution (A), a colloidal silica solution (Ludox AS-40) was first acidified by HIt to pH 3, and then added as hard filler in different molar ratios of 0.7, 2.08, 4.27, and 5.48. The colloidal silica was first coated with a monolayer of the sol–gel by adding a small amount of solution A (15 wt%). The purpose was to stabilize the colloidal particles avoiding flocculation when added to the sol. After measuring the density of the cured unfilled coating, which was 1.3 g cm$^{-3}$, the volume percent of the filler in the coating matrix was calculated. The above molar ratios of silica correspond to 6.7, 17.6, 30.5 and 36.0 vol% in the cured coatings. Since the particle size of the silica is about 20 nm, the coatings remained transparent. This is important for many applications that require optical transparency. Just before the dip coating process, a small amount (0.05 wt%) of ethylenediamine (ED) was added to the coating solution as the cross-linking agent of the epoxy ring in GLYMO.

Polycarbonate (PC) substrates, measured 100 mm × 50 mm × 3 mm, were supplied with a layer of protective film. The film was peeled off before the oxygen plasma treatment of the PC substrate. The treatment was carried out just before the dip coating for all specimens used in the current work. The purpose for such a treatment is to remove organic contaminations on the PC surface and activate the surface for better wetting and adhesion between the coating and the substrate [88]. The treatment was done at the following conditions: RF power 400 W; pressure 100 Torr; oxygen flow rate 400 sccm, and treatment time 5 min.

The pretreated PC substrates were dip coated with the above solutions in different withdrawal speeds so that the effect of layer thickness on the coating's hardness and scratch resistance could be studied. After the dip coating, specimens were placed in a bench top furnace for drying and curing. The drying was done at 80 °C for 40 min and curing at 110 °C for 90 min. To achieve thicker coatings (>10°m), varying the coating speed was proven to be inadequate. Thus, multiple coatings were applied. After each coating and curing step, a plasma treatment was carried out before the subsequent layer was applied. This is to avoid mixing of the two layers and to eliminate potential risk of cracking upon curing.

The thickness of the coating was measured using a profilometer (Talysurf Series 2 Stylus Profilometer) across the uncoated and coated areas on the same specimen. The scratch resistance of the coating was characterized by a commercial pencil hardness tester (Scratch Hardness Tester Model 291, Erichsen Testing Equipment). The test conformed to the ISO standard 15184 [24], where a vertical force of 7.5 ± 0.1 N was applied at the tip of the pencil. The pencil was fixed at 45° angle to the horizontal coating surface as the pencil was moved over the coated specimen. The pencil lead was flattened before the test as specified in the standard. The highest pencil grade that does not cause damage to the coated specimen will be termed as the pencil hardness of the coating. Meanwhile the intrinsic hardness (referred to as indentation hardness thereafter) and Young's modulus of the coatings were measured using a nanoindenter (NanoTest™). The depth of the indentation was controlled to be less than 1/10 of the coating thickness in order to minimize the effect from the substrate. To measure the film fracture toughness, coatings were applied on thinner PC substrates 200 μm thick and tested by the controlled buckling test as detailed above. At least eight samples were tested for each condition. The residual stress caused by curing shrinkage was calibrated in the toughness calculation. The residual stress is measured by the curvature method [89].

### 14.5.1.2 Results and analysis

Table 14.2 shows the indentation hardness, Young's modulus, and the scratch pencil hardness grade of coatings with different colloidal silica contents. In order to minimize the potential influence of coating thickness on the pencil hardness results, all the measurements were made on specimens with approximately the same coating thickness (within the range of $5 \pm 0.5$ μm). The results show that indentation hardness and modulus increase with the silica content. The increase in hardness and elastic modulus is easily understandable since the added silica is harder and stiffer than the matrix. However, it may be premature to conclude that the pencil hardness increase was due to the indentation hardness increase. The discussion will only be possible after the examination of the scratch failure mode, which will be made in the following subsections.

To enable a better understanding of the protection provided by the sol–gel coating, separate measurement was carried out on the PC substrate without coating. The pencil hardness was found to be 6B, and the indentation hardness 0.19 GPa. It is clear that even with no silica added, the current sol–gel coating has provided substantial protection to scratch damage. In Table 14.2, the pencil hardness grade is B for the basic sol–gel coating without addition of the colloidal silica.

There was a small increase in coating fracture toughness with silica content increase. With more colloidal silica added, the coating becomes harder and stiffer, and usually the ability to absorb energy should decrease. However, in our sol–gel coatings, the beneficial effect with increased colloidal silica content could be explained by the difference in porosity. Sol–gel coatings tend to possess lots of pores after curing. Coatings with more colloidal silica added have a lower residual porosity due to the filling of the pores and the chemical bonding of the silica nanoparticles with the sol–gel matrix. Therefore, the pores in these coatings will be less in number and smaller in size than the ones in the coatings with less colloidal silica. The pores in brittle coatings act as flaws causing stress concentration; the larger the flaw size, the lower the coating fracture resistance.

**Table 14.2** Pencil Hardness, Indentation Hardness, Young's Modulus, and Fracture Toughness of Coatings with Different Silica Contents

| Colloidal silica content (vol%) | 0.0 | 6.7 | 17.6 | 30.5 | 36.0 |
|---|---|---|---|---|---|
| Pencil hardness grade | B | HB | F | H | H |
| Indentation hardness (GPa) | $0.59 \pm 0.003$ | $0.62 \pm 0.013$ | $0.66 \pm 0.022$ | $0.71 \pm 0.029$ | $0.88 \pm 0.032$ |
| Young's modulus (GPa) | $2.65 \pm 0.02$ | $5.26 \pm 0.05$ | $5.43 \pm 0.06$ | $8.30 \pm 0.15$ | $9.99 \pm 0.19$ |
| Coating thickness (μm) | 4.9 | 5.2 | 4.6 | 5.5 | 5.2 |
| Fracture toughness (J/m$^2$) | $8.5 \pm 3.3$ | $8.5 \pm 1.7$ | $8.7 \pm 3.1$ | $9.4 \pm 0.6$ | $10.2 \pm 2.7$ |

*Coating thickness is within the range of $5 \pm 0.5$ μm.*

With the increase in coating thickness, while the colloidal silica content was fixed at 30.5 vol%, the pencil hardness grade improved significantly, as shown in Table 14.3. A similar trend was also observed in specimens with other colloidal silica content. From these observations, it is clear that the improvement in scratch resistance by thickening the coating is an effective way of improving the scratch resistance. This finding is of practical interest since the amount of silica that can be added to reinforce the matrix is limited, while thickening the coating is relatively easy and unlimited.

It is also interesting to note that the scratch damage occurs when the pencil lead is generally much softer than the coating. Figure 14.12 shows the hardness of different pencil grades measured by nanoindentation. Despite the scatter, which is understandable due to the inhomogeneity of the mixture of clay and graphite in the pencil lead, the general trend of increasing hardness with the pencil grade is clear. The highest pencil grade, 9H, has an average hardness of 0.53 GPa, which is still lower than the 0.59 GPa from the softest sol–gel coating (Table 14.2). Nevertheless, coating damage occurred at much lower pencil grades than 9H.

Figure 14.13 shows typical scratch marks after the pencil scratch test. Despite the discrepancy among test runs, coating cracking was found to be the dominant mode of failure. With softer pencil grades, there was no scratch damage to the coating.

**Table 14.3** Pencil Hardness Increase with Coating Thickness

| Coating thickness (μm) | 5.5 | 7.9 | 10.5 | 13.9 | 20.5 | 25.1 |
|---|---|---|---|---|---|---|
| Pencil hardness grade | H | 2H | 2H | 3H | 3H | 5H |

*The colloidal silica content for all these specimens is 30.5 vol%.*

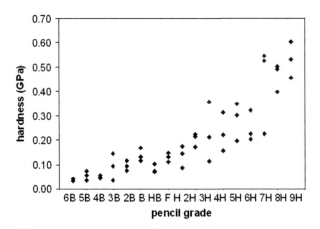

**FIGURE 14.12**

Nanoindentation hardness of the whole series of pencil leads.

Some crumbs from the pencil lead were observed on the surface (Fig. 14.13(a)). The fracture patterns show that coating damage was caused by tensile cracking behind the indenter. There was little or no coating detachment at either the leading edge or the trailing end of the scratch, indicating that the adhesion between the coating and the substrate was very strong under the test conditions (Figs 14.13(b) and (c)). When the pencil lead grade increased further, there was partial delamination of the severely cracked coatings (Fig. 14.13(d)). The general good adhesion was attributed to the oxygen plasma treatment before the dip coating. Blees et al. [49] also found that the adhesion of their sol–gel coatings on polypropylene substrates was much improved after microwave oxygen plasma treatment. Ong et al. [88] reported improvement in adhesion, as well as the mechanical properties of amorphous carbon films on PC plastics after the oxygen plasma treatment. Surface treatment before coating has been a popular choice for adhesion improvement, and its limitation and effect on other properties have been widely reported [90–94]. Another example of interface enhancement with thermal impregnation treatment will be demonstrated in a later section.

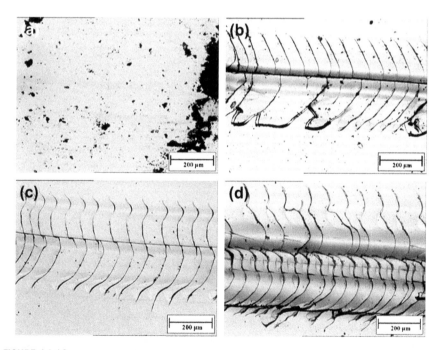

**FIGURE 14.13**

Micrographs of pencil scratched coating surfaces. The coating is 10.2 μm thick and without silica filler. The scratch direction is from right to left for all pictures. Pencil grade was assigned to be 2H in this case. (a) Scratched by pencil 2H, μ = 0.13. The dark spots are the pencil crumbs at the starting point of the scratch; (b) scratched by pencil 3H, μ = 0.57; (c) scratched by pencil 4H, μ = 0.28; (d) scratched by pencil 5H, μ = 0.73. (For color version of this figure, the reader is referred to the online version of this book.)

In all the sol–gel coatings tested, the effective friction coefficient ranged from 0.27 to 0.73 when coating cracking occurred. The magnitude generally increased with increasing pencil grade. When there was no scratch damage, the friction coefficient stayed around 0.13 for all coatings. Figure 14.14 shows an example of the friction coefficient of the same coating used in Fig. 14.13, but has included three sets of independent measurement. The sudden increase in the friction coefficient provides a good indication for the beginning of scratch damage.

The most severe damage took place when the pencil lead gouged into, and plowed along the substrate, as shown in Fig. 14.15 as an example. This type of damage happened when (1) the hardest pencil leads were used, (2) the coating was relatively thin, and (3) the coating was without silica filler. The last condition,

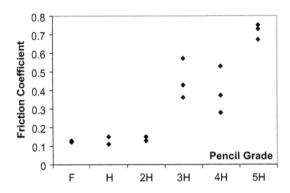

**FIGURE 14.14**

Effective friction coefficient measured on the 10.2-μm-thick coating without silica filler. Three sets of measurements were carried out. When coating failure starts to occur, the friction coefficient increases.

**FIGURE 14.15**

Substrate gouge in an 8.2-μm-thick sol–gel coating without silica filler, but with 0.05 wt% of ED added. The pencil used in the scratch test was 5H. (For color version of this figure, the reader is referred to the online version of this book.)

again, leaves a question mark on whether it is the modulus or the indentation hardness, or other properties and factors, that are behind the scratch resistance. This is discussed later.

### 14.5.2 UV-curable sol–gel coating with incorporated functional polyhedral oligomeric silsesquioxanes nanoparticles

POSS is a novel family of functional inorganic nanofillers with unique structural features and superior properties. They are often used as nanoscale building blocks for the construction of hybrid polymer nanocomposite materials. A typical POSS nanoparticle has a cubic inorganic $Si_8O_{12}$ core surrounded by eight organic corner groups as shown in Fig. 14.16, and has a particle size of about 1.5 nm. The eight organic corner groups can be functionalized with a variety of organic substituents. Commonly, one of them is functionalized with a polymerizable or reactive group (e.g. acrylate group), while the others are designed to have identical nonreactive groups (e.g. isobutyl group). With this unique structure, POSS molecules can be applied as macromonomers in various copolymerization systems to prepare a novel family of nanoscale-structured hybrid copolymers covalently tethered with POSS nanoparticles. It has been found in a great number of studies that such covalent POSS incorporation can significantly enhance thermal and mechanical properties of polymer materials [95–98]. Since the sol–gel chemical formulation process allows the freedom in selecting precursors and fillers with potential stoichiometry chemical reaction to combine the organic and inorganic components at the nanoscale, suitable POSS materials can be selected for both thermal and UV-curable sol–gel coatings. In the following case, the POSS side groups are covalently connected to the suitable functional groups of the sol–gel matrix for enhanced mechanical and surface properties.

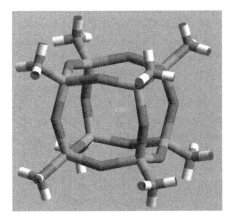

**FIGURE 14.16**

Chemical structure of cubic silsesquioxane ($R_8Si_8O_{12}$). (For color version of this figure, the reader is referred to the online version of this book.)

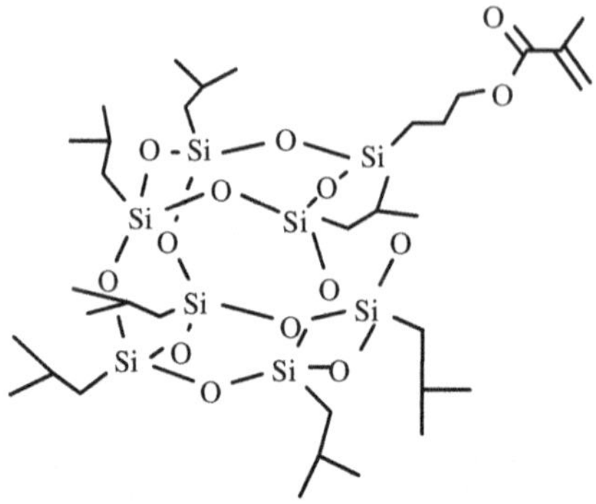

**FIGURE 14.17**

Chemical structures of isobutyl methacrylate POSS (UV-POSS).

With the objective of improving the mechanical properties of the sol–gel coating without scarifying the light transmittance of the coating, several types of POSS materials were synthesized with different chemical side groups for the thermal curable sol–gel coatings. Here we describe a UV-curable sol–gel coating solution formulated from methacrylate silane precursor, in which a hard POSS with an acrylic function group was incorporated. As a result, the mechanical properties and pencil scratch resistance of the coating were greatly improved.

The hard POSS was synthesized with one or more acrylic side group(s) and isobutyl groups at the other corners as shown in Fig. 14.17. The methacrylate group(s) in the POSS material will be polymerized with the UV sol–gel upon UV curing to form an integrated network. The UV-curable sol–gel coating materials were formulated using methacrylate silane precursor (3-(trimethoxysilyl) propyl methacrylate, MEMO) and alkoxy silane (TEOS) with optional colloidal silica nanoparticles. MEMO and TEOS were hydrolyzed in water and acetic acid and ethanol in molar ratios of MEMO:TEOS:EtOH:$H_2O$:HAc = 0.25:1.0:1.4:2.4:1.3. After stirring overnight, colloidal silica was added drop-wise and stirred overnight. UV-POSS was dissolved in a small amount of ethanol, added drop-wise into the mixture, and stirred for 4 h. Photoinitiator in 0.5 wt% of the total solution was added right before the dip coating.

Poly(methyl methacrylate) (PMMA) substrates measuring 100 mm × 100 mm × 3 mm were cleaned with isopropanol (IPA) and blown dried by $N_2$ gas. Dip coating was carried out by a dipping machine at withdrawal speed selected from 10 to 25 mm $s^{-1}$. The samples were dried in the oven at 60 °C for 10 min and then cured under a UV fusion lamp with a total energy of 1–2.5 J $cm^{-2}$. The UV-POSS containing UV-cured sol–gel coating provides transparent protective coatings for improved mechanical properties. Table 14.4 lists the properties of the POSS-containing coating on PMMA compared to the sol–gel coating without POSS. For comparison, values for bare PMMA surface and thermally cured sol–gel coating are also included. It is seen that both elasticity modulus and hardness have been increased when POSS is added. The pencil hardness increased from 3–4H to 8H due to the addition of POSS.

**Table 14.4** Properties of the POSS-Containing Coating on PMMA Compared to PMMA Substrate Without Coating and Sol–Gel Coatings (UV and Thermal Cured) Without POSS

| Sample | Coating Thickness (μm) | Young's Modulus (GPa) | Hardness (GPa) | Pencil Hardness Grade |
|---|---|---|---|---|
| PMMA without coating | / | 4.5–5.0* | 0.20–0.35* | 1H |
| Thermal sol–gel on PMMA | 4.0 | 6.46 ± 0.15 | 0.72 ± 0.013 | 2H |
| UV sol–gel on PMMA | 4.0 | 11.67 ± 1.57 | 0.95 ± 0.130 | 3–4H |
| UV sol–gel + UV-POSS on PMMA | 4.5 | 20.06 ± 2.35 | 1.023 ± 0.134 | 8H |

*Reference values from B. J. Briscoe, L. Fiori, E. Pelillo, Nano-indentation of polymeric surfaces, Journal of Physics D: Applied Physics, 31 (1998), 2395–2405.

### 14.5.3 Interface adhesion enhancement by thermal impregnation treatment [99]

Surface treatments for adhesion improvement have been reported by many researchers [86–88,90–94]. $O_2$ plasma treatment is probably the most popular method. The plasma treatment changes surface polarity and activates the surface for better wetting and adhesion [100–102]. Other approaches include surface roughening (morphology change only) and chemically modifying the surface structure by etching or impregnation. Comparatively, surface chemistry change is often more effective than surface morphology change alone. For example, hydrogen peroxide–sulfuric acid ($H_2O_2$–$H_2SO_4$) solution was applied to introduce hydroxyl groups and to increase the oxygen content on the polymer surface for bonding with partially hydrolyzed TEOS, resulting in improved wettability and adhesion with the formation of Si–O–Si bond within the top 100 Å of polymer [103]. An aminolysis mechanism was reported to be operative when a primer solution of 3-aminopropyl triethoxysilane (APS) in IPA solution is used to modify the surfaces of PC and PMMA [104].

We developed a heated impregnation treatment process [99] using a homemade device, which contains a pressure vessel with bottom heating, vacuum, $N_2$ gas input and chemical solution dosing control. A typical process flow chart for temperature, pressure and time control is shown in Fig. 14.18.

A hybrid precursor solution was formulated from methyl trimethoxylsilane (MTMS) and APS in a mixed solvent of ethanol and methyl ethyl ketone (MEK) by the following components (in molar ratio): MTMS:APS:EtOH:MEK:$H_2O$:HAc = 1.0:1.0:14.0: 4.0:6.0:0.4. MTMS was hydrolyzed in water, ethanol and acetic acid. APS was also hydrolyzed first in water, then MEK was added. Then the two solutions

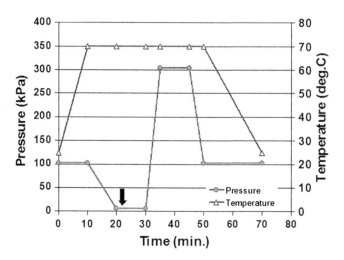

**FIGURE 14.18**

A typical process flow chart for temperature, pressure, and time of the heated impregnation treatment process. (For color version of this figure, the reader is referred to the online version of this book.)

were mixed and stirred for at least 2 h. Further dilution by ethanol was carried out to ensure the solution is not too viscous, which could result in nonuniform layer thickness. This hybrid solution was dosed into the heated impregnation process and spread onto the substrate surface uniformly and subjected to heating and pressure, therefore, was impregnated into the top layer of the polymer substrate and provided bonding sites for subsequent sol–gel coating. The sol–gel formulation follows the same one in Section 14.5.1 [86,87].

The treatment goes through the following steps: (1) load cleaned sample on stage that has a declined angle of 15° to the horizon; (2) close the chamber; (3) adjust the temperature to 60, 70 or 80 °C and wait till the temperature reaches the set temperature; (4) switch on the vacuum and hold for 10–20 min; (5) turn on the dosing switch to inject the hybrid precursor solution onto the sample surface till the entire surface has been covered; (6) turn on the $N_2$ gas switch and open the valve till the pressure increases to 3 bar; (7) hold for 5, 10, 15 or 30 min to force the solution to impregnate into the substrate; (8) switch off the $N_2$ gas, and open the dosing knob to force out the remaining solution in the tubing so as to prevent clotting; (9) lower the temperature to room temperature and open the chamber for cooling for 10 min; and (10) clean the chamber and switch off the power. The treated samples were further dried in the oven at 80 °C for 1 h and were ready for sol–gel coating.

Figure 14.19 shows a SEM image of the cross-section of a sample treated by impregnation followed by sol–gel coating. It is seen that a uniform impregnated layer was formed directly on the PC surface. There was an interfacial layer between the impregnated layer and the sol–gel coating. This indicates that a merge

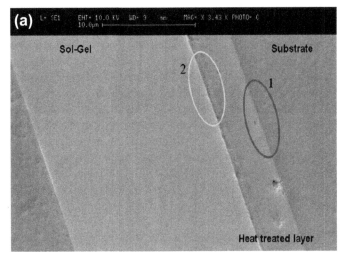

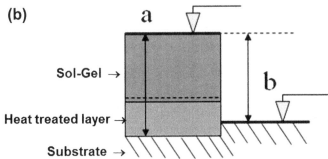

**FIGURE 14.19**

(a) SEM image of the cross-section of impregnated PC surface with sol–gel coatings (b) and schematic diagram of the cross-section layer structures. Around 5-μm-thick impregnated layer and penetration of about 1.5 μm are identified from the cross-section image. (For color version of this figure, the reader is referred to the online version of this book.)

or interpenetration between the coating material and the impregnated layer had occurred. This is because the impregnated layer was purposely not fully cured (only dried at 80 °C for 1 h), so that the available hydroxyl groups could react and condense together with the dip-coated sol–gel material to form strong chemical bonding at interface 2. Interface 1 between the polymer substrate and the impregnated layer looks seamless. After measuring the layer thickness and the profile depth of the coated surface relative to the original polymer surface, the penetration depth of the impregnation into the substrate can be calculated as described in Fig. 14.19(b). It was found that the impregnated layer was about 5 μm thick and the penetration into polymer top surface was about 1.5 μm. This depth is larger than most of the reported pretreatment depths, which are typically below 100 nm [103].

Table 14.5 shows the pencil scratch grades at different treatment conditions, after sol–gel coating by 1-dip, 2-dip and 3-dip layers. From the results obtained, it can be inferred that the dip coating material formed strong adhesion with the impregnated surface and provided a hard coating for the PC substrate. The best parameters of the impregnation process are 70 °C, pressure 3 bar, and duration 10–30 min. With multilayer thicker coating, the scratch resistance was improved, in line with the results in Section 14.5.1. Pencil hardness of 8H has been achieved on a two-layer-coated sample. The second dip coating layer also showed smoother surface as compared to the first layer indicating higher cross-linking of the inorganic and organic components. However, after coating with the third layer, cracks were observed due to increased energy release rate (e.g. Eqn. (14.18)). When compared to the commercial lenses (with pencil scratch grade of HB), the pencil scratch grade of heat-treated and dip-coated PC is higher even after only the first dip coating.

Hardness and Young's modulus of the impregnated surface and the sol–gel coated surface were measured by nanoindentation. The results are shown in Table 14.6. The improvements on both Young's modulus and hardness of the surface are obvious after impregnation and further after the sol–gel coating. The impregnated layer provided a gradient layer between the sol–gel coating and the PC substrate. Due to this

**Table 14.5** Pencil Scratch Grade and Coating Thickness of Samples After Surface Treatment

| Treatment Temperature (°C) | Pencil Hardness Grade/Total Coating Thickness (μm) | | | |
|---|---|---|---|---|
| | Heat treatment | 1-Dip coat layer | 2-Dip coat layer | 3-Dip coat layer |
| 60 | 2B/8.4 | F/19.8 | 2H/20.4 | - |
| 70 | B/7.1 | 2H/10.7 | 8H/20.2 | 9H/31.3 with cracks |
| 80 | B/5.1 | F/9.6 | 6H/14.05 | - |
| Commercial lenses | | HB | - | - |
| Plasma-treated PC | | H/12.7 | 4H/21.3 | - |

**Table 14.6** Hardness and Young's Modulus of Original PC, Impregnated Layer and Sol–Gel Coated Layer

| Sample | Indentation Hardness (GPa) | Young's Modulus (GPa) |
|---|---|---|
| Original PC | 0.19 | 2.38** |
| Impregnated PC | 0.35 ± 0.04 | 4.98 ± 0.21 |
| Sol-gel coated surface | 0.95 ± 0.13 | 11.67 ± 1.6 |

**Reference value from Materials Science and Engineering: An Introduction, W. D. Callister, Jr., John Wiley, 7th Edition, 2006.*

stiffened/hardened substrate and the chemical bonds formed, the resistance to gouge failure is increased; therefore, plastic deformation in substrate is delayed resulting in higher scratch resistance of the coated surface.

### 14.5.4 General discussion on coating scratch resistance

In the pencil scratch tests described above, the contact load at the tip of the pencil was fixed at $(750 \pm 10)$ g. Therefore, the maximum contact stress is mainly determined by the contact area. When a flattened pencil lead was tilted at $45°$ to the coating surface (as illustrated in Fig. 14.20), the sharp tip may induce very high pressure at the point of contact. However, if the pencil lead is fragile (e.g. lower pencil grade), then the fracture of the pencil lead will result in a drastic increase in the contact area, and thus reduction in the pressure. The lead crumbs (Fig. 14.13(a)) may also serve as a solid lubricant so that when there is no damage in the coating, the friction coefficient stays at a low level.

The tendency of crumb formation in the pencil lead of harder grades is reduced due to the increase in the clay content in the mix. Despite some wear, the pencil tip remains sharp so that the contact pressure is kept at a high level. In such a case, there is an increased chance of substrate deformation after the coating cracks. Once the pencil tip plows into the substrate, the increase in the friction traction exacerbates the substrate deformation [37,38]. The hard coating can help alleviate the pressure so that substrate plowing is suppressed even after coating fracture. However, this ability might be hampered if the coating is too thin. Therefore, the bending stiffness of the coating plays an important role in preventing substrate gouge. Once substrate gouging occurs, the analysis will be similar to the one for monolithic polymer scratch [105,106], which is beyond the scope of the current work.

The schematic in Fig. 14.20 explains how the stresses are generated in the coating during a typical pencil scratch test. As discussed before, cracking at the trailing edge was caused by the contact action and the friction traction. Careful examination of Fig. 14.13(c) and (d) reveals that the cracks are more curved along certain scratch subtracks, indicating a strong influence of the contact stress. Delamination could also occur under these highly indented regions (Fig. 14.13(d)). The less curved portion of the cracks was more influenced by the tangential friction stress. In theory, if only

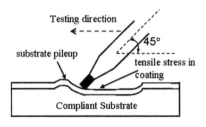

**FIGURE 14.20**

Schematics of a scratched sol–gel coating on a polymer substrate.

tangential friction stress exists, the cracks should form straight lines perpendicular to the scratch direction.

Now we can estimate the tendency of coating cracking from the energy point of view. The energy release rate of a coating under tensile stress is given by

$$G_{cra} = \frac{\sigma_c^2 t}{2E_c} g\left(\alpha, \beta, \frac{\sigma_c}{\sigma_y}, n\right) \qquad (14.21)$$

This is the same as Eqn. (14.20), which considers substrate plastic deformation, but expressed as a function of the coating stress. The tensile stress is contributed from both the contact bending and the friction traction, as in

$$\sigma_c = C_1 \frac{P_n}{t^2} + C_2 \frac{\mu P_n}{Wt} \qquad (14.22)$$

where $C_1$ and $C_2$ are constants, and $W$ the scratch width.

Substituting Eqn. (14.22) into (14.21), the critical load for coating cracking failure is given by

$$P_{cri} = \sqrt{2tE_c\Gamma_c}\left(\frac{tW}{C_1 W + C_2 t\mu}\right) g\left(\alpha, \beta, \frac{\sigma_c}{\sigma_y}, n\right)^{-\frac{1}{2}} \qquad (14.23)$$

Equation (14.23) reveals that the resistance to coating cracking is adversely affected by the friction coefficient. Meanwhile, it also shows that increasing the elasticity modulus and the thickness of the coating helps to reduce the tendency of cracking failure. Mechanically, an increase in the coating modulus and thickness translates into an increase in the stiffness of the coating; therefore, the stress is reduced when other factors remain unchanged. Increasing the coating toughness surely is beneficial; however, there is a limited room for the improvement for sol–gel coatings, as illustrated in Table 14.2. Increasing the coating thickness has resulted in a remarkable improvement of four grades in the pencil hardness test of the sol–gel coatings (Table 14.3). Increasing silica addition improves the modulus quite significantly, and has resulted in an improvement of three pencil grades (Table 14.2). Similarly, the addition of POSS has shown significant impact on the coating modulus and scratch resistance (Table 14.4).

A closer look into Eqn. (14.23) finds that if the tensile trailing stress is dominated by the friction traction, $C_2 \gg C_1$, the critical load will be inversely proportional to the friction coefficient and square root of the thickness ($t^{1/2}$). In such a case, increasing the thickness and reducing the friction coefficient will be very effective in raising the critical scratch load for the tensile trailing or forward chevron cracking (Table 14.1). When contact loading is the main cause for coating cracking, $C_1 \gg C_2$, the thickness has a greater influence ($t^{3/2}$) on the critical load. Increasing the thickness will help preventing the Hertz tensile cracking.

Delamination-driven scratch failure is another focus of this discussion. Blees et al. [49] have compared several models predicting the critical scratch load and found that for a sol–gel coating on polypropylene substrate, failure could be predicted by a model similar to the one derived from the model by Evans [69] for wedge spallation in Eqn. (14.15). The critical load was inversely proportional to the friction coefficient as in

$$P_{cri} = \frac{W}{v_c \mu} \sqrt{2tE_c \Gamma_{int}} \tag{14.24}$$

As discussed previously, this equation has not considered the effect of substrate on the energy release rate. So error might arise if it is to be used to calculate the precise value of interface fracture toughness. For parametric analysis, this relation clearly shows that the critical scratch load increases with the square root of the coating thickness, elasticity modulus, and the interface fracture toughness. Therefore, to improve the scratch failure load, one should reduce the friction coefficient, and increase the coating thickness, elasticity modulus and interface fracture toughness. The example shown in Section 14.5.3 is an attempt to improve all these key factors.

The influence of friction coefficient on the scratch resistance has been the focus of many studies [49,107–111]. From the above analyses, it is clear that it has an adverse effect on the critical scratch load whether the scratch failure is by coating cracking or delamination. Experimentally, various ways of achieving different friction coefficients were employed in the studies; they include the use of different indenter materials, shapes and diameters; deposition of a thin top layer; and applying chemical cleaning, plasma treatment, and lubrication on the coating surface. Details are available in these references.

## 14.6 CONCLUSION

In this work, the mechanics of scratch and the mechanisms of various modes of scratch failure have been reviewed first. Particular attention was paid to hard and brittle coatings on compliant and ductile substrates. In general, scratch resistance may depend on material properties, such as the hardness, Young's modulus, fracture toughness, and interface toughness, of the coating, the substrate, and the indenter itself. Other factors include the coating surface conditions, thickness, indenter shape, testing speed and testing environment. The importance is highlighted to quantify the coating fracture toughness and the delamination toughness between the coating and the substrate. These two parameters are not as easily obtainable as other conventional materials properties. A test method using controlled beam buckling has been introduced. Other methods have also been covered in early sections.

Due to the complexity in diverse application background, it is not our intension, nor is it possible, to exhaust various examples that have been attempted by various research groups. Rather, we focused on the semiquantitative parametric analysis, which could provide good insights into the key factors behind scratch failures. Such understanding is of great help in the design of coatings with improved scratch resistance.

There are two main areas that should be improved when designing scratch resistant brittle coatings on compliant substrate. The first is the adhesion between the coating and the substrate. Weak interface induces premature failure by delamination and spallation. Surface treatment such as plasma bombardment on the substrate has been proven effective to improve the adhesion of coatings. Interlayers have also shown a positive effect for the enhancement of interface toughness. The second is the cracking resistance. With sufficient interfacial toughness, coating cracking inevitably will occur with increasing load. To delay the occurrence of cracking, several possible approaches can be contemplated. (1) Since cracking is induced by the tensile stress, keeping a compressive residual stress in the coating will help counteract the tensile stress. However, the creation of compressive stress in sol–gel coating is unlikely since the coating usually shrinks after curing, leaving residual tensile stress in the coating. (This may partially explain why mainly cracking, not delamination, occurred in our sol–gel coatings.) (2) Effective friction coefficient can be reduced by appropriate surface treatment. Reduction in the surface asperity will also be effective. Lubrication, if allowed, would certainly reduce the traction stress. (3) Coating thickness should be increased within the limit of its functional and processing constraints. For a fixed magnitude of load, a thick coating increases the stiffness and reduces the maximum stress of the coating, and thus the risk of cracking.

Several examples of improving scratch resistance of sol–gel coatings are given. The first case shows GLAMO-TEOS coatings on PC substrate containing different amounts of colloidal silica particles. The failure mode of this type of coating was brittle cracking at the trail end. When harder pencils were used, substrate gouge started to take place. Key factors influencing the scratch failure were found to be the elastic modulus, fracture toughness, and thickness of the coating material. By increasing the silica content, which leads to marked increase in the modulus, and the thickness of the coating, the pencil hardness grade was increased from B to 5H. The second example incorporated functional POSS nanoparticles into the sol–gel coating to further enhance the coating stiffness. Such coating has shown higher scratch-resistant pencil grade. Finally, it is demonstrated that interfacial adhesion enhancement through chemical impregnation, together with the coating stiffness reinforcement by thickness increase, is a very effective way to further enhance the pencil scratch resistance of the same GLAMO-TEOS sol–gel coating. More recently published results have largely supported our analyses [112–124].

In a scratch test, it is always desirable to reduce the effective friction traction since it contributes to both delamination-induced and cracking-induced failures. However, it should be noted that when the indenting object has plowed into the substrate during the scratch action, the effective friction coefficient can increase significantly and this increase has little to do with the coating surface morphology. In other words, when the normal load is too large, reducing the friction traction is almost impossible. To prevent substrate plowing, coating thickness and elasticity modulus have to be increased.

## Acknowledgments

Experimental work on the hybrid sol–gel coatings on polycarbonate substrates by Edmund Chwa, Tham Otto and Lisa Boon is gratefully acknowledged. The authors also wish to express their appreciation for the helpful discussion on the first edition of the chapter with Tony Atkins, Brian Cotterell, Gordon Williams, Yiu-Wing Mai, Alexander Korsunsky, Michel Ignat, Mark Hoffman, Tongyi Zhang, Zbigniew Stachurski, Jianhui Liu, Xingfu Chen, Yong Sun, Hanshan Dong, and Chun-Hway Hsueh. Continued encouragement by Prof. Klaus Friedrich, one of the editors of this book, to update the work for this second edition is very much appreciated. Zhong Chen particularly acknowledges the guidance and help he has received over the years from his PhD supervisor, Prof. Tony Atkins, a great mechanical engineer and mentor. Prof. Atkins' energetic approach to deformation and fracture has inspired many researchers in the field of mechanics of materials.

## References

[1] S. Sakka, T. Yoko, Sol-gel-derived coating films and applications, in book series: structure and bonding, in: R. Reisfeld, C.K. Jørgensen (Eds.), Chemistry, Spectroscopy and Applications of Sol-gel Glasses, vol. 72, Springer-Verlag, 1992, pp. 89–118.

[2] J. Malzbender, J.M.J. den Toonder, A.R. Balkenende, G. de With, Measuring mechanical properties of coatings: a methodology applied to nano-particle-filled sol-gel coatings on glass, Materials Science and Engineering R – Report 36 (2002) 47–103.

[3] G. Schottner, K. Rose, U. Posset, Scratch and abrasion resistant coatings on plastic lenses – state of the art, current developments and perspectives, Journal of Sol-gel Science and Technology 27 (2003) 71–79.

[4] K.H. Haas, S. Amberg-Schwab, K. Rose, Functionalized coating materials based on inorganic-organic polymers, Thin Solid Films 351 (1999) 198–203.

[5] J.D. Blizzard, S.V. Perz, P.J. Popa, J.L. Stasser, J.S. Tonge, Silicon-acrylic sol-gel modified automotive clear topcoats for etch and scratch resistance, Surface Coatings International Part B – Coatings Transactions 84 (2001) 205–211.

[6] P. Innocenzi, M. Esposto, A. Maddalena, Mechanical properties of 3-glycidoxypropyl-trimethoxysilane based hybrid organic-inorganic materials, Journal of Sol-gel Science and Technology 20 (2001) 293–301.

[7] A. Fidalgo, L.M. Ilharco, The defect structure of sol-gel-derived silica/polytetrahydro-furan hybrid films by FTIR, Journal of Non-crystalline Solids 283 (2001) 144–154.

[8] K.-C. Song, J.-K. Park, H.-U. Kang, S.-H. Kim, Synthesis of hydrophilic coating solution for polymer substrate using glycidoxypropytrimethoxysilane, Journal of Sol-gel Science and Technology 27 (2003) 53–59.

[9] S.R. Davis, A.R. Brough, A. Atkinson, Formation of silica/epoxy hybrid network polymers, Journal of Non-crystalline Solids 315 (2003) 197–205.

[10] J.H. Harreld, A. Esaki, G.D. Stucky, Low-shrinkage, high-hardness, and transparent hybrid coatings: poly(methyl methacrylate) cross-linked with silsesquioxane, Chemistry of Materials 15 (2003) 3481–3489.

[11] G.M. Wu, J. Shen, T.H. Yang, B. Zhou, J. Wang, Preparation of scratch-resistant nano-porous silica films derived by sol-gel process and their anti-reflective properties, Journal of Materials Science and Technology 19 (2003) 299–302.

[12] D.K. Hwang, J.H. Moon, Y.G. Shul, K.T. Jung, D.H. Kim, D.W. Lee, Scratch resistant and transparent UV-protective coating on polycarbonate, Journal of Sol-gel Science and Technology 26 (2003) 783–787.

[13] C. Charitidis, A. Laskarakis, S. Kassavetis, C. Gravalidis, S. Logothetidis, Optical and nanomechanical study of anti-scratch layers on polycarbonate lenses, Superlattices and Microstructures 36 (2004) 171–179.

[14] F.S. Li, S.X. Zhou, L.M. Zhou, Effects of preparation method on microstructure and properties of UV-curable nanocomposite coatings, Journal of Applied Polymer Science 98 (2005) 1119–1124.

[15] W. Tanglumlert, P. Prasassarakich, P. Supaphol, S. Wongkasemjit, Hard-coating materials for poly(methyl methacrylate) from glycidoxypyltrimethoxysilane-modified silatrane via a sol-gel process, Surface and Coatings Technology 200 (2006) 2784–2790.

[16] L.Y.L. Wu, G.H. Tan, X.T. Zeng, T.H. Li, Z. Chen, Synthesis and characterization of transparent hydrophobic sol-gel hard coatings, Journal of Sol-gel Science and Technology 38 (2006) 85–89.

[17] E. Chwa, L. Wu, Z. Chen, Factors towards pencil scratch resistance of protective sol-gel coatings on polycarbonate substrate, Key Engineering Materials 312 (2006) 339–344.

[18] T. Takaki, K. Nishiura, Y. Mizuta, Y. Itou, Advanced polymer-inorganic hybrid hard coatings utilizing in situ polymerization method, Journal of Nanoscience and Nanotechnology 6 (2006) 3965–3968.

[19] M. Spirkova, M. Slouf, O. Blahova, T. Farkacova, J. Benesova, Submicrometer characterization of surfaces of epoxy-based organic-inorganic nanocomposite coatings. A comparison of AFM study with currently used testing techniques, Journal of Applied Polymer Science 102 (2006) 5763–5774.

[20] P. Fabbri, B. Singh, Y. Leterrier, J.A.E. Manson, M. Messori, F. Polita, Cohesive and adhesive properties of polycaprolactone/silica hybrid coatings on poly(methyl methacrylate) substrate, Surface and Coatings Technology 200 (2006) 6706–6712.

[21] ASTM C 1624 – 05, Standard Test Method for Adhesion Strength and Mechanical Failure Modes of Ceramic Coatings by Quantitative Single Point Scratch Testing, ASTM International, 2005.

[22] ASTM D 7027 – 05, Standard Test Method for Evaluation of Scratch Resistance of Polymeric Coatings and Plastics Using an Instrumented Scratch Machine, ASTM International, 2005.

[23] ASTM D 7187 – 05, Standard Test Method for Measuring Mechanistic Aspects of Scratch/Mar Behavior of Paint Coatings by Nanoscratching, ASTM International, 2005.

[24] ISO 15184:1998, Paints and Varnishes – Determination of Film Hardness by pencil test, ISO, 1998.

[25] ASTM D3363 – 00, Standard Test Method for Film Hardness by Pencil Test, ASTM International, 2000.

[26] H. Hertz, On contact between elastic solids, Journal für die reine und angewandte Mathematik 92 (1882) 156–171, and a full English translation in: Miscellaneous Papers, Translated by D. E. Jones and G. A. Schott (1896), Macmillan, London.

[27] J. Boussinesq, Application des Potentials à l'etude de l'équilibre et du Mouvement des solides élastiques, Gauthier-Villars, Paris, 1885.

[28] I.N. Sneddon, Fourier Transforms, McGraw-Hill, New York, 1951.

[29] K.L. Johnson, Contact Mechanics, Cambridge University Press, Cambridge, UK, 1985.

[30] H. Djabella, R.D. Arnell, Finite element analysis of the contact stresses in an elastic coating on an elastic substrate, Thin Solid Films 213 (1992) 205–219.

[31] R.M. Davies, The determination of static and dynamic yield stresses using a steel ball, Proceedings of the Royal Society of London 197A (1949) 416–432.

[32] D. Tabor, Hardness of Metals, Oxford University Press, Oxford, 1951.

[33] A.G. Atkins, D. Tabor, Plastic indentation in metals with cones, Journal of the Mechanics and Physics of Solids 13 (1965) 149.

[34] A.C. Fischer-Cripps, Nanoindentation, Springer-Verlag, New York, 2002.

[35] G.M. Hamilton, L.E. Goodman, The stress field created by a sliding circular contact, Journal of Applied Mechanics 33 (1966) 371–376.

[36] G.M. Hamilton, Explicit equations for the stress beneath a sliding spherical contact, Proceedings of the Institution of Mechanical Engineers Part C – Journal of Mechanical Engineering Science 197 (1983) 53–59.

[37] R.D. Arnell, The mechanics of the tribology of thin-film systems, Surface and Coatings Technology 43-44 (1990) 674–687.

[38] H. Djabella, R.D. Arnell, Finite-element analysis of the contact stresses in elastic coating substrate under normal and tangential load, Thin Solid Films 223 (1993) 87–97.

[39] K. Mao, Y. Sun, T. Bell, A numerical model for the dry sliding contact of layered elastic bodies with rough surfaces, Tribology Transactions 39 (1996) 416–424.

[40] L. Zheng, S. Ramalingam, Stresses in a coated solid due to shear and normal boundary tractions, Journal of Vacuum Science and Technology A 13 (1995) 2390–2398.

[41] E. Kral, K. Komvopoulos, Three-dimensional finite element analysis of surface deformation and stresses in an elastic-plastic layered medium subjected to indentation and sliding contact loading, Journal of Applied Mechanics 63 (1996) 365–375.

[42] L.S. Stephens, Y. Liu, E. Meletis, Finite element analysis of the initial yielding behavior of a hard coating/substrate system with functionally graded interface under indentation and friction, Journal of Tribology 122 (2000) 381–387.

[43] N. Panich, Y. Sun, Mechanical characterization of nanostructured $TiB_2$ coatings using microscratch techniques, Tribology International 39 (2006) 138–145.

[44] K. Holmberg, A. Laukkanen, H. Ronkainen, K. Wallin, Tribological analysis of fracture conditions in thin surface coatings by 3D FEM modelling and stress simulations, Tribology International 38 (2005) 1035–1049.

[45] Y. Xie, H.M. Hawthorne, A model for compressive coating stresses in the scratch adhesion test, Surface and Coatings Technology 141 (2001) 15–25.

[46] Y. Xie, H.M. Hawthorne, Effect of contact geometry on the failure modes of thin coatings in the scratch adhesion test, Surface and Coatings Technology 155 (2002) 121–129.

[47] Y. Liu, L. Li, X. Cai, Q. Chen, M. Xu, Y. Hu, T.-L. Cheung, C.H. Shek, P.K. Chu, Effects of pretreatment by ion implantation and interlayer on adhesion between aluminum substrate and TiN film, Thin Solid Films 493 (2005) 152–159.

[48] P. Bertrand-Lambotte, J.L. Loubet, C. Verpy, S. Pavan, Understanding of automotive clearcoats scratch resistance, Thin Solid Films 420-421 (2002) 281–286.

[49] M.H. Blees, G.B. Winkelman, A.R. Balkenende, J.M.J. den Toonder, The effect of friction on scratch adhesion testing: application to a sol-gel coating on polypropylene, Thin Solid Films 359 (2000) 1–13.

[50] S.J. Bull, E.G. Berasetegui, An overview of the potential of quantitative coating adhesion measurement by scratch testing, Tribology International 39 (2006) 99–114.

[51] S.J. Bull, Failure modes in scratch adhesion testing, Surface and Coatings Technology 50 (1991) 25–32.

[52] M.D. Thouless, An analysis of spalling in the microscratch test, Engineering Fracture Mechanics 61 (1998) 75–81.

[53] S.J. Bull, Failure mode maps in the thin film scratch adhesion test, Tribology International 30 (1997) 491–498.

[54] B.R. Lawn, Partial cone crack formation in a brittle material loaded with a sliding spherical indenter, Proceedings of the Royal Society of London 299A (1967) 307–316.

[55] D.A. Hills, R.L. Munisamy, D. Nowell, A.G. Atkins, Brittle fracture from a sliding Hertzian contact, Proceedings of the Institution of Mechanical Engineers Part C – Journal of Mechanical Engineering Science 208 (1994) 409–415.

[56] H. Chai, B.R. Lawn, Fracture mode transitions in brittle coatings on compliant substrates as a function of thickness, Journal of Materials Research 19 (2004) 1752–1761.

[57] C.H. Hsueh, P. Miranda, Master curves for Hertzian indentation on coating/substrate systems, Journal of Materials Research 19 (2004) 94–100.

[58] C.H. Hsueh, P. Miranda, Modeling of contact-induced radial cracking in ceramic bilayer coatings on compliant substrates, Journal of Materials Research 18 (2003) 1275–1283.

[59] J.L. Beuth, Cracking of thin bonded films in residual tension, International Journal of Solids and Structures 29 (1992) 1657–1675.

[60] J.W. Hutchinson, Z. Suo, Mixed mode cracking in layered materials, Advances in Applied Mechanics 29 (1992) 63–191.

[61] Z. Chen, B. Cotterell, W. Wang, E. Guenther, S.J. Chua, A mechanical assessment of flexible opto-electronic devices, Thin Solid Films 394 (2001) 202–206.

[62] Z. Chen, B. Cotterell, W. Wang, The fracture of brittle thin films on compliant substrate in flexible displays, Engineering Fracture Mechanics 69 (2002) 597–603.

[63] Z. Chen, Z. Gan, Fracture toughness measurement of thin films on compliant substrate using controlled buckling test, Thin Solid Films 515 (2007) 3305–3309.

[64] M.D. Thouless, Crack spacing in brittle films on elastic substrates, Journal of the American Ceramic Society 73 (1990) 2144–2146.

[65] C.H. Hsueh, M. Yanaka, Multiple film cracking in film/substrate systems with residual stresses and unidirectional loading, Journal of Materials Science 38 (2003) 1809–1817.

[66] C.H. Hsueh, A.A. Wereszczak, Multiple cracking of brittle coatings on strained substrates, Journal of Applied Physics 96 (2004) 3501–3506.

[67] K. Holmberg, A. Laukkanen, H. Ronkainen, K. Wallin, S. Varjus, A model for stresses, crack generation and fracture toughness calculation in scratched TiN-coated steel surfaces, Wear 254 (2003) 278–291.

[68] P.J. Burnett, D.S. Rickerby, The relationship between hardness and scratch adhesion, Thin Solid Films 154 (1987) 403–416.

[69] H.E. Evans, Modeling oxide spallation, Materials at High Temperatures 12 (1994) 219–227.

[70] A.G. Evans, J.W. Hutchinson, On the mechanics of delamination and spalling in compressed films, International Journal of Solids and Structures 20 (1984) 455–466.

[71] J. den Toonder, J. Malzbender, G. de With, R. Balkenende, Fracture toughness and adhesion energy of sol-gel coatings on glass, Journal of Materials Research 17 (2002) 224–233.

[72] M.D. Thouless, Combined buckling and cracking of films, Journal of the American Ceramic Society 76 (1993) 2936–2938.

[73] M.D. Thouless, J.W. Hutchinson, E.G. Liniger, Plane-strain, buckling-driven delamination of thin-films - model experiments and mode-II fracture, Acta Metallurgica et Materialia 40 (1992) 2639–2649.

[74] M.D. Thouless, H.M. Jensen, E.G. Liniger, Delamination from edge flaws, Proceedings of the Royal Society of London 447A (1994) 271–279.

[75] B. Cotterell, Z. Chen, Buckling and cracking of thin films on compliant substrate under compression, International Journal of Fracture 104 (2000) 169–179.

[76] B. Cotterell, Z. Chen, Buckling and fracture of thin films under compression, Key Engineering Materials 183 (2000) 187–192.

[77] H.H. Yu, J.W. Hutchinson, Influence of substrate compliance on buckling delamination of thin films, International Journal of Fracture 113 (2002) 39–55.

[78] M.D. Drory, J.W. Hutchinson, Diamond coating of titanium alloys, Science 263 (1994) 1753–1755.

[79] M.D. Drory, J.W. Hutchinson, Measurement of the adhesion of a brittle film on a ductile substrate by indentation, Proceedings of the Royal Society of London 452A (1996) 2319–2341.

[80] J.J. Vlassak, M.D. Drory, W.D. Nix, A simple technique for measuring the adhesion of brittle films to ductile substrates with application to diamond-coated titanium, Journal of Materials Research 12 (1997) 1900–1910.

[81] B. Balakrisnan, C.C. Chum, M. Li, Z. Chen, T. Cahyadi, Fracture toughness of Cu-Sn intermetallic thin films, Journal of Electronic Materials 32 (2003) 166–171.

[82] Z. Chen, M. He, B. Balakrisnan, C.C. Chum, Elasticity modulus, hardness and fracture toughness of $Ni_3Sn_4$ intermetallic thin films, Materials Science and Engineering 423A (2006) 107–110.

[83] S.J. Britvec, The Stability of Elastic Systems, Pergamon Press, New York, 1973.

[84] J.H. Liu, A.G. Atkins, A.J. Pretlove, The effect of inclined loads on the large deflection behaviour of elastoplastic work-hardening straight and pre-bent cantilevers, Journal of Mechanical Engineering Science 209 (1995) 87–96.

[85] J.L. Beuth, N.W. Klingbeil, Cracking of thin films bonded to elastic-plastic substrates, Journal of the Mechanics and Physics of Solids 44 (1996) 1411–1428.

[86] L.Y.L. Wu, E. Chwa, Z. Chen, X.T. Zeng, A study towards improving mechanical properties of sol-gel coatings for polycarbonate, Thin Solid Films 516 (2008) 1056–1062.

[87] Z. Chen, L.Y.L. Wu, E. Chwa, O. Tham, Scratch resistance of brittle thin films on compliant substrates, Materials Science and Engineering A 493 (2008) 292–298.

[88] H.C. Ong, R.P.H. Chang, N. Baker, W.C. Oliver, Improvement of mechanical properties of amorphous carbon films deposited on polycarbonate plastics, Surface and Coatings Technology 89 (1997) 38–46.

[89] Z. Chen, X. Xu, C.C. Wong, S. Mhaisalkar, Effect of plating parameters on the intrinsic stress in electroless nickel plating, Surface and Coatings Technology 167 (2003) 170–176.

[90] J. Gilberts, A.H.A. Tinnemans, M.P. Hogerheide, T.P.M. Koster, UV curable hard transparent hybrid coating materials on polycarbonate prepared by the sol-gel method, Journal of Sol-gel Science and Technology 11 (1998) 153–159.

[91] J. Wen, G.L. Wilkes, Synthesis and characterization of abrasion resistant coating materials prepared by the sol-gel approach. 1. Coatings based on functionalized aliphatic diols and diethylenetriamine, Journal of Inorganic and Organometallic Polymers 5 (1995) 343–375.

[92] S.Y. Lee, J.D. Lee, S.M. Yang, Preparation of silica-based hybrid materials coated on polypropylene film, Journal of Materials Science 34 (1999) 1233–1241.

[93] Z. Gan, Z. Chen, S.G. Mhaisalkar, M. Damayanti, Z. Chen, K. Prasad, S. Zhang, J. Ning, Effect of electron beam treatment on adhesion of Ta/polymeric low-k interface, Applied Physics Letters 88, 2006. Article number: 233510.

[94] K. Jang, H. Kim, The gas barrier coating of 3-aminopropyltriethoxysilane on polypropylene film, Journal of Sol-gel Science and Technology 41 (2007) 19–24.

[95] J.D. Jeyaprakash, S. Samuel, J. Rühe, A facile photochemical surface modification technique for the generation of microstructured fluorinated surfaces, Langmuir 20 (2004) 10080–10085.

[96] B. Park, C.D. Lorenz, M. Chandross, M.J. Stevens, G.S. Grest, O.A. Borodin, Frictional dynamics of fluorine-terminated alkanethiol self-assembled monolayers, Langmuir 20 (2004) 10007–10014.

[97] S. Veeramasuneni, J. Drelich, J.D. Miller, G. Yamauchi, Hydrophobicity of ion-plated PTFE coatings, Progress in Organic Coatings 31 (1997) 265–270.

[98] K.Y. Mya, C.B. He, J.W. Xu, S. Lu, X.H. Lu, Y.L. Wu, Y.X. Wang, Star-like polyurethane hybrids with functional cubic silsesquioxanes: preparation, morphology and thermomechanical properties, Journal of Polymer Science Part A: Polymer Chemistry 47 (2009) 4602–4616.

[99] L.Y.L. Wu, L. Boon, Z. Chen, X.T. Zeng, Adhesion enhancement of sol–gel coating on polycarbonate by heated impregnation treatment, Thin Solid Films 517 (2009) 4850–4856.

[100] A. Kaminska, H. Kaczmarek, J. Kowalonek, The influence of side groups and polarity of polymers on the kind and effectiveness of their surface modification by air plasma action, European Polymer Journal 38 (2002) 1915–1919.

[101] P. Munzert, U. Schulz, N. Kaiser, Transparent thermoplastic polymers in plasma-assisted coating processes, Surface and Coatings Technology 174-175 (2003) 1048–1052.

[102] S. Vallon, B. Drevillon, C. Senemaud, A. Gheorghiu, V. Yakovlev, Adhesion of a thin silicon oxide film on a polycarbonate substrate, Journal of Electron Spectroscopy and Related Phenomena 64-65 (1993) 849–856.

[103] T.P. Chou, G.Z. Cao, Adhesion of sol-gel-derived organic-inorganic hybrid coatings on polyester, Journal of Sol-gel Science and Technology 27 (2003) 31–41.

[104] C.H. Li, G.L. Wilkes, The Mechanism for 3-aminopropyltriethoxysilane to strengthen the interface of polycarbonate substrates with hybrid organic-inorganic sol-gel coatings, Journal of Inorganic Organometallic Polymers 8 (1998) 33–45.

[105] B.J. Briscoe, P.D. Evans, E. Pelillo, S.K. Sinha, Scratch maps for polymers, Wear 200 (1996) 137–147.

[106] B.J. Briscoe, E. Pelillo, S.K. Sinha, Scratch hardness and deformation maps for polycarbonate and polyethylene, Polymer Engineering and Science 36 (1996) 2996–3005.

[107] P.A. Steinmann, Y. Tandy, H.E. Hintermann, Adhesion testing by the scratch test method: the influence of intrinsic and extrinsic parameters on the critical load, Thin Solid Films 154 (1987) 333–349.

[108] J. Valli, A review of adhesion test methods for thin hard coatings, Journal of Vacuum Science and Technology 4A (1986) 3007–3014.

[109] S.J. Bull, D.S. Rickerby, A. Matthews, A. Leyland, A.R. Pace, J. Valli, The use of scratch adhesion testing for the determination of interfacial adhesion – the importance of frictional drag, Surface and Coatings Technology 36 (1988) 503–517.

[110] MD.E. Coghill, D.H. StJohn, Scratch adhesion testing of soft metallic coatings on glass, Surface and Coatings Technology 41 (1990) 135–146.

[111] F. Attar, T. Johannesson, Adhesion evaluation of thin ceramic coatings on tool steel using the scratch testing technique, Surface and Coatings Technology 78 (1996) 87–102.

[112] S.W. Chen, B. You, S.X. Zhou, L.M. Wu, Preparation and characterization of scratch and mar resistant waterborne epoxy/silica nanocomposite clearcoat, Journal of Applied Polymer Science 112 (2009) 3634–3639.

[113] H. Zhang, L.C. Tang, Z. Zhang, L. Gu, Y.Z. Xu, C. Eger, Wear-resistant and transparent acrylate-based coating with highly filled nanosilica particles, Tribology International 43 (2010) 83–91.

[114] Z.Z. Wang, P. Gu, Z. Zhang, Indentation and scratch behavior of nano-$SiO_2$/polycarbonate composite coating at the micro/nano-scale, Wear 269 (2011) 21–25.

[115] H. Zhang, H. Zhang, L.C. Tang, L.Y. Zhou, C. Eger, Z. Zhang, Comparative study on the optical, surface mechanical and wear resistant properties of transparent coatings filled with pyrogenic and colloidal silica nanoparticles, Composites Science and Technology 71 (2011) 471–479.

[116] Y.S. Lin, M.S. Weng, T.W. Chung, C.M. Huang, Enhanced surface hardness of flexible polycarbonate substrates using plasma-polymerized organosilicon oxynitride films by air plasma jet under atmospheric pressure, Surface and Coatings Technology 205 (2011) 3856–3864.

[117] M. Dinelli, E. Fabbri, F. Bondioli, $TiO_2$–$SiO_2$ hard coating on polycarbonate substrate by microwave assisted sol-gel technique, Journal of Sol-Gel Science and Technology 58 (2011) 463–469.

[118] T. Gururaj, R. Subasri, K.R.C.S. Raju, G. Padmanabham, Effect of plasma pretreatment on adhesion and mechanical properties of UV-curable coatings on plastics, Applied Surface Science 257 (2011) 4360–4364.

[119] B.D. Beake, V.M. Vishnyakov, A.J. Harris, Relationship between mechanical properties of thin nitride-based films and their behaviour in nano-scratch tests, Tribology International 44 (2011) 468–475.

[120] H.T. Chiu, C.Y. Chang, C.L. Chen, T.Y. Chiang, M.T. Guo, Preparation and characterization of UV-curable organic/inorganic hybrid composites for NIR cutoff and antistatic coatings, Journal of Applied Polymer Science 120 (2011) 202–211.

[121] M. Barletta, A. Gisario, G. Rubino, Scratch response of high-performance thermoset and thermoplastic powders deposited by the electrostatic spray and 'hot dipping' fluidised bed coating methods: the role of the contact condition, Surface and Coatings Technology 205 (2011) 5186–5198.

[122] C.C. Chang, T.Y. Oyang, F.H. Hwang, C.C. Chen, L.P. Cheng, Preparation of polymer/silica hybrid hard coatings with enhanced hydrophobicity on plastic substrates, Journal of Non-crystalline Solids 358 (2012) 72–76.

[123] T.O. Kaarlainen, P.J. Kelly, D.C. Cameron, B. Beake, H.Q. Li, P.M. Barker, C.F. Struller, Nanoscratch testing of atomic layer deposition and magnetron sputtered $TiO_2$ and $Al_2O_3$ coatings on polymeric substrates, Journal of Vacuum Science & Technology A 30, 2012. Article number: 01A132.

[124] L. Sowntharya, S. Lavanya, G. Ravi Chandra, N.Y. Hebalkar, R. Subasri, Investigations on the mechanical properties of hybrid nanocomposite hard coatings on polycarbonate, Ceramics International 38 (2012) 4221–4228.

# Scratch behavior of polymeric materials

# 15

**Robert L. Browning\*, Han Jiang†, Hung-Jue Sue‡**

*\*Polymer Technology Center, Texas A&M University, USA, †Polymer Technology Center, Department of Mechanical Engineering, Texas A&M University, College Station, TX, USA, ‡Polymer Technology Center, Texas A&M University, TAMU, College Station, USA*

## CHAPTER OUTLINE HEAD

## 15.1 INTRODUCTION

In the past, a scratch has been envisioned as a sliding indentation, either under a constant or increasing load. Unfortunately, as is the case with most aspects of polymer science, scratch behavior is not that simple. Addressing this issue is complicated not only due to the fact that polymers have complex material and mechanical behavior characteristics but also because the quantitative evaluation of scratch resistance requires the elimination of ambiguity and subjectivity.

Diamond is the hardest material in existence and is thus impenetrable while an ideal rubber is soft and will recover completely if deformed. Therefore, in principle, these two materials possess the highest possible scratch resistance against surface

*Tribology of Polymeric Nanocomposites.* http://dx.doi.org/10.1016/B978-0-444-59455-6.00015-5

deformation and damage upon removal of an applied load. Most polymers lie at the other extreme and are extremely susceptible to scratch damage. As polymers have come to be utilized in a plethora of applications over the years, the demand for reliable scratch testing methods and objective analysis techniques has increased substantially. Until recently, no such testing method or analysis technique has existed.

In studying polymer scratch behavior, there are three main areas of concern: esthetics, structural integrity, and durability. With regard to estheticism, one can easily find its relevance in many products such as car dashboards and cellular phones where visible scratches reduce the original product's superficial quality while the intended functionality remains. Poor durability means that surface scratches can lead to damage of the underlying substrate. This is especially important in transport applications where metal pipes are often coated with polymeric materials to prevent corrosion while they are exposed to the atmosphere or buried underground. Surface quality is extremely critical in upholding structural integrity for packaging applications. The thin layer(s) of a food packaging film, such as those employed for military Meals Ready to Eat, are particularly susceptible to scratches that can lead to failure of its structural integrity, thus spoiling the contents. In data storage, scratches on hard disks and optical storage devices can cause permanent loss of data.

Often, scratch resistance is merely described with vague terms like "good" and "bad" or it is given a "pass" or "fail" designation. These terms do nothing to describe the nature of the material or why that particular material possesses "good" or "bad" scratch resistance. Employing a reliable testing and analysis methodology that is founded upon the principles of material science can make the objective understanding of polymer scratch behavior possible. Ideally, once the properties of the material are known and its behavior fairly well understood, mechanical models can be constructed to predict the outcome when a polymer is subjected to a particular testing scenario.

The aim of this chapter is to present the reader with an evolution of polymer scratch testing and analysis from the most basic means to the establishment of a standardized testing methodology and the development of predictive finite element method (FEM) modeling. A review of past test methods will be presented with the intention of identifying shortcomings that have been improved upon. The standardized testing and analysis methodology will be discussed in detail, and the results of successful experimental case studies will be presented. With the validation of the experimental procedure, great progress has been made with respect to obtaining predictive models through FEM, which will also be discussed.

## 15.2 REVIEW OF PAST WORK

### 15.2.1 Scratch testing and evaluation

Evaluation related to scratch resistance has been going on for more than a century. In 1824, Friedrich Mohs introduced his scratch resistance scale based on the relative hardness of 10 minerals [1]. Although this scale is rather ambiguous and somewhat arbitrary in nature, it is still used to this day as a quick reference when assessing the hardness of a material. More meaningful work was done with metals by Tabor in the

1950s [2,3] where the scratch hardness was related to the indentation hardness. This involved the construction of one of the first scratch testing devices. The apparatus consisted of a lever arm rigidly pivoted on one end and fitted with a sharp tool for scratching on the other end. The scratch tip geometry was the intersection of three flat, ground surfaces. A scratch is applied by imposing a dead weight to the arm and then moves the specimen under the scratching tip [2]. Other measurements of hardness can be obtained through indentation tests. The Rockwell, Brinell, and Knoop hardness tests measure the contact area as a function of the applied normal force for indenters of different geometries.

Literature on polymer scratch testing can be found dating back to 1955 where modified polystyrenes were explored as a material for telephone housings [4]. Since then, much work has been done to more effectively tie together material properties with scratch behavior [5–46]. The important idea of polymer scratch maps came from Briscoe et al. in 1996 [10,13]. These maps provided generalized information regarding scratch damage as a function of several scratch testing and material parameters such as testing speed, modulus, and ductility. Numerous other researchers have also tried novel ways of relating scratch behavior to material properties. However, even after more than one and a half centuries since Mohs proposed his mineral hardness scale, reliable scratch testing and analysis methodologies for polymers based on fundamental principles are scant.

Needless to say, numerous scratch test devices and evaluation methods have been designed in recent years. A comprehensive review of devices is given by Wong et al. [43]. Among those test methods highlighted are the pencil test [47], the single-pass pendulum test [48,49], the pin-on-disc test [50], the Revetest scratch device [51], the Taber scratcher [52], and the Ford Five-Finger scratch tester [23,53]. The pencil test is similar in practice to the Mohs hardness test in that pencils of various hardness values are used to scratch the polymer surface. The hardness of the polymer is taken to be comparable to the pencil that can cause surface scratch damage. However, this method is highly dependent on user-influenced factors like how hard the pencil is pressed on the surface and how fast the surface is scratched. The other test methods mentioned involve more mechanical means of applying scratches and are not as reliant on the competency of the user. Even though these methods can produce scratches in a more or less repeatable manner, there are still factors that affect their reliability.

In addition to developing appropriate testing methods, the evaluation of polymer scratch behavior presents a great challenge, as well. Hutchings [54] conducted a study to rank polymer scratch visibility using a panel of human observers. The results from the human panel showed reasonable agreement with their digital image analysis, which employed differences in the gray level between the damaged and undamaged surface. However, natural differences between the eyes of human observers can make objective evaluation troublesome. More meaningful analysis can be conducted with methods similar to the one described by Rangarajan [55] and Kody [52]. These methods use cameras and uniform lighting at different angles to measure the scattering of incident light caused by surface features. Scratch visibility can be addressed in an even more robust manner with the method developed by Wong et al. [44]. This method uses a progressive load scratch test and digital image analysis software to more precisely quantify where the onset of scratch visibility occurs. This methodology will be described in detail in Section 15.4.2.

## 15.2.2 Numerical analysis of scratch behavior

Early studies of scratch mechanics relied on the knowledge gained from indentation research. With the help of contact mechanics, the Hertzian indentation problem was extended to the action of sliding [56–60]. However, these analyses are only applied to materials that are essentially linearly elastic in nature and hence cannot sufficiently describe the complicated deformation experienced by the polymer during a scratch. Moreover, static analysis is only valid up to the instant when the scratch tip is about to slide.

There are research efforts on dynamic aspects of the scratch problem [61–63], but they are also limited to isotropic linearly elastic materials. For a viscoelastic material, the Hertzian indentation problem of a rigid spherical punch was studied [64,65]. During a scratch process, the substrate may deform plastically under extensive straining and such plastic flow should be considered in the analysis. Due to the nonlinearity of polymeric materials, there has been no analytical work on the complex material responses under indentation or sliding. It is evident from these works that the material description and scope of analysis limit a comprehensive study of polymer scratch mechanics.

For numerical techniques, although the molecular dynamics method has been attempted to simulate behavior of polymer surfaces during scratch testing [66,67], it is not suitable to address neither the macroscopic penetration and recovery nor the microscopic deformation and damage due to its molecular scale of simulation. The most common approach for scratch simulation is the use of FEM [68]. Research efforts using FEM in scratch applications remain scant, and most works are restricted to the study of indentation [69]. The elastic–plastic and plane-strain behavior of a layered substrate of bilinear materials under normal and tangential contact stresses was investigated [70]. However, the dynamic effects of the moving indenter and the contact between the indenter and substrate were not modeled. Lee et al. modeled a steel ball scratching a rotating polycarbonate (PC) [71]. While a more realistic material law was adopted for the PC, the three-dimensional (3D) problem was oversimplified to a two-dimensional plane-strain case that was not applicable to the original problem. Other efforts included the 3D simulations of a smooth, rigid conical indenter scratching on an elastic–perfectly plastic and bilinear material [72,73]. Since the material rheology in those works do not address the strain softening and strain hardening nature of polymers, they are not suitable to study the scratch response of polymers.

## 15.3 THE NEED FOR RESEARCH AND DEVELOPMENT

The main drive for research in the field of polymer scratch behavior involves obtaining a set of parameters that can quickly and easily provide preemptive guidelines as to how a material will behave. Material parameters such as Young's modulus, tensile strength, and ductility all contribute valuable insights as to how the polymer will mechanically respond under most loading cases, but the same cannot be said about scratch behavior. The first step to rectifying this shortcoming is to address the issue of polymer scratch behavior from a fundamental material science standpoint with the purpose of developing meaningful experimental methodologies. Through rigorous research and application of the scientific method, it is hoped that polymer scratch behavior can become just as important a concept as tensile or fracture behavior can.

There are some aspects of scratch behavior, however, that cannot be addressed in a practical manner using experimental means. For this reason, numerical simulation is more suitable to perform the scratch analysis than an analytical approach due to the complexity of material behavior and scratch damage mechanisms. Among different numerical techniques, FEM is preferred for its versatility to accommodate complicated phenomena such as surface contact, frictional interaction, and atypical material responses.

The main concern of scratch behavior study is to investigate the mechanical response of polymeric materials under the scratching action of a sliding asperity. It is critical that the applied numerical modeling approach possesses the following important attributes [74]:

- Capacity for 3-D dynamic analysis—In the actual scratch experimental setup, the material response is too complex to be simplified to a 2-D formulation. Also, dynamic analysis should be performed to account for the transient response of scratch problem.
- Consideration of contact and frictional interactions—The surface contact between the scratch tip and substrate has to be featured. The simulation needs to provide an accurate understanding of its mechanical response. Verified by experiments, the frictional interaction between a substrate and the scratch tip has been shown to be significant in the scratch process.
- Geometric nonlinearity—During the scratch process, the substrate material under the scratch tip undergoes extremely large and localized deformation. Such geometrical nonlinear deformation could severely distort the initial FEM mesh and lead to convergence problems. Good mesh quality should be preserved.
- A descriptive material model for polymers—An appropriate constitutive model for the polymeric material to be studied is critical for successful numerical analysis. To capture the true representation of the material response, the constitutive law should have the ability to describe strain-softening, strain-hardening, and rate-dependent characteristics that are typical for polymeric materials.
- Damage criteria for polymers—To account for the complicated damage acquired during the scratch, appropriate damage criteria should be implemented in the FEM numerical modeling to predict their initiation and development.

## 15.4  RECENT PROGRESS

### 15.4.1  Test methods

In utmost simplicity, a scratch involves applying a load to an asperity and sliding that asperity over the surface of a material. The usual testing parameters (temperature, strain rate, testing speed, indenter geometry, etc.) will have an effect and will need to be considered, but more important is the need to apply consistent scratches that reliably determine the material's scratch behavior.

As stated previously, there have been a number of attempts at achieving this goal. The methods that involve modifying the setup of an atomic force microscopy (AFM) or a nanoindenter to conduct scratches on the near- and submicrometer level are appropriate for understanding small-scale fundamental material behavior under a slow rate of testing. In fact, these setups are quite convenient in their scanning capabilities. This allows for immediate observation of the applied scratch after the test has been conducted without having to use posttest analysis equipment such as a scanning electron microscope (SEM).

However, when it comes to real-life scenarios where the applied load, the size of the indenter, and the rate of testing are orders of magnitude larger than that of an AFM or nanoindenter, the capability of these setups makes it practically impossible to obtain relevant data of interest. Some larger-scale apparatus have been developed over the years, but the most widely used are five-finger scratch testers like the Multi-Finger Scratch/Mar Tester (Model 710) by Taber Industries, Inc. (Fig. 15.1) [75]. This scratch tester is used quite often in the automotive field for evaluating the scratch resistance of interior and exterior polymeric automobile surface.

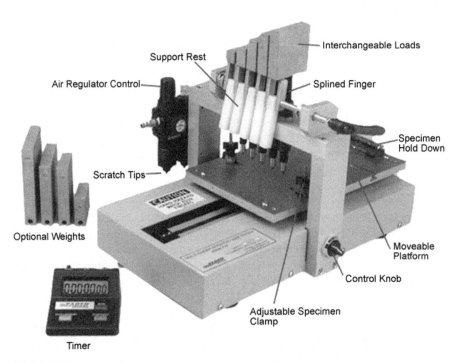

**FIGURE 15.1**

The Multifinger Scratch/Mar Tester (Model 710) manufactured by Taber Industries, Inc.

The five-finger test applies a "dead" load to each stylus. Here, the term "dead" refers only to the fact that the weight placed on the stylus is of a fixed value. As a result, the applied load will fluctuate due to the discontinuous scratch-induced deformation and damage incurred on the polymer surface. The viscoelasticity of polymers will also have an effect on the stability of the applied testing rate. When comparing a soft and a hard polymer, the softer polymer will resist tip movement more due to a deeper penetration of the scratch tip into the polymer subsurface. As a result, the tip movement will become slower in the beginning. Also, the compressible nature of the air in the pneumatic system of a typical five-finger apparatus will amplify rate instability.

Bearing the above drawbacks in mind, a new testing apparatus was developed. It employs a gear-driven servo motor for scratch tip displacement and a servopneumatic system for load application. The use of a gear-driven servomotor for displacement will result in a precisely controlled velocity, regardless of the material. Since air pressure is used rather than dead weight, the applied normal load can be linearly increased in a single pass. This leads to a continuous progression of deformation and damage on the scratch path, making the results more robust and meaningful for analysis. This is the main improvement over the five-finger test method. Furthermore, the apparatus can be equipped with sensors that detect the applied normal force and tangential load as well as the vertical and horizontal tip displacement and store the information for later analysis.

The results of this new methodology are so consistent and repeatable that it has been adopted as an industrial standard under the designation ASTM D7027-05 [76] and ISO 19252:08. It has gained a great deal of attention not only from industries in automotive, coating, packaging, etc., but also from the military.

## 15.4.2 Evaluation

The main challenge in analyzing scratch behavior lies in the fact that the perceived quality or severity of the scratch will vary from observer to observer. Often, a scratched specimen is observed by a panel and given a rating of "pass" or "fail" by a panel of observers. This provides little or no information regarding what makes the material good or bad in terms of scratch resistance.

In most applications, the main concern with scratch behavior is visibility. Scratch visibility is largely a result of the scattering of incident light due to rough or nonuniform surface features. These features can include cracks, crazes, and localized molecular orientation induced during scratching. In order for scratch damage to become visible, the size of rough or nonuniform surface features must be above the wavelength of visible light. Therefore, at low applied normal loads, the scratch damage will appear to be quite subtle and invisible. The contrast between the scratch damaged area and the virgin surface will increase as the normal load increases. In coatings, sometimes visibility can occur as a result of a transparent coating becoming adhesively delaminated from its substrate. If there are fillers present in the material that naturally scatter light (i.e. talc, clay), they will contribute to scratch visibility if any particles are near the surface and become exposed.

The standardized progressive load scratch test method previously mentioned has the ability to relate the applied normal load to the scratch distance, thus making it possible to locate at which load scratch visibility occurs [44,76]. The difficulty in obtaining the onset point of scratch visibility is due to the natural differences between the eyes of human observers. To address this need, several digital techniques are employed to objectively determine the onset of scratch visibility.

First, the scratched sample is scanned with a flatbed desktop scanner at an appropriate resolution of choice. To ensure that the scanner captures the image with levels of brightness and contrast that are a true representation of what is seen with the naked eye, a card known as a "color checker" is scanned along with the image. The colors on the "color checker" are standardized and represent shades and hues of objects found in nature as well as a range of shades from black to white. The red, green, and blue (RGB) values of each color on the card are known, so the RGB levels of the digital image can be balanced against the standard levels to more closely replicate the true object as possible.

Second, the digital image is processed using a software package called ASV©. The software package uses an algorithm to balance the levels of RGB in the image according to how the human eye perceives RGB and then converts the digital image to grayscale. The software then scans along the scratch path and calculates the contrast of the pixels associated with the scratch relative to the background. The point where the contrast reaches a level of 3% while also satisfying continuity and geometrical criteria is considered to be the onset of visibility [77]. An example of a scratched polymer sample that has been processed by the grayscale threshold method is shown in Fig. 15.2.

**(a)**

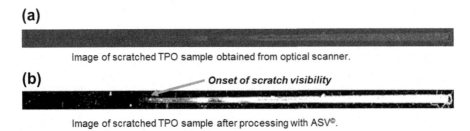

Image of scratched TPO sample obtained from optical scanner.

**(b)**                                    *Onset of scratch visibility*

Image of scratched TPO sample after processing with ASV©.

**FIGURE 15.2**

Illustration of scratch visibility image analysis using digital software. The white pixels of b) correspond to the visible portion of the scratch based on the ≥3% contrast criterion. (Please note that white pixels will be red in online version of this book.)

Finally, the onset of visibility can be obtained by relating the applied normal load to the scratch distance with the following relationship:

$$F_c = \frac{(L-x)}{L} * (F_f - F_0) + F_0$$

where $F_c$ is the critical applied normal load at which visibility occurs, $L$ is the scratch length, $x$ is the length of the visible portion of the scratch, and $F_f$ and $F_0$ are the final and initial applied normal loads, respectively. In this manner, five or more specimens can be tested and analyzed, thus yielding an average value for the onset of scratch visibility. Typically, a standard deviation of <10% is obtained.

In a recent study, the mechanistic origins for scratch visibility were investigated [77]. A schematic of a cross-section of a typical scratch is shown in Fig. 15.3. The difference in H1 and H2 is taken as the "Shoulder Height" and is a measure of how much material is displaced by the scratch tip and shored up on the sides of the scratch.

The study looked at scratched samples of thermoplastic olefin (TPOs) with a direct light source shining either parallel or perpendicular to the scratch path. It was concluded that light could be scattered by the features of a scratch in two different ways: either by a change in the roughness associated with the scratch path or by the height of the shoulders on either side of the scratch (Fig. 15.4).

The study was able to establish a criterion for the measurement of scratch visibility that is directly related to the physical deformation of the polymer surface. By measuring both the roughness along the scratch path and the height of the shoulder on either side of the scratch, it is now possible to make correlations among scratch deformation, visibility, and mechanical properties.

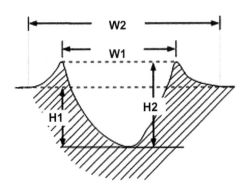

**FIGURE 15.3**

Cross-sectional definition of scratch geometry used to measure groove profile dimensions.

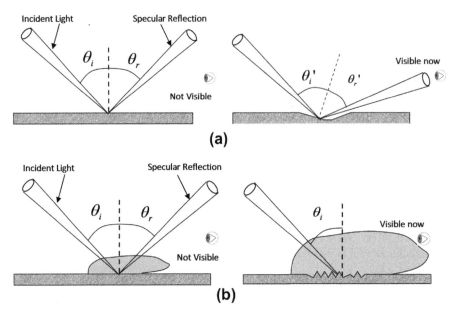

**FIGURE 15.4**

Scratch visibility induced by alteration of surface reflection with different scratch damage mechanisms: a) groove profile change and b) surface roughness increase. (For color version of this figure, the reader is referred to the online version of this book.)

### 15.4.3 Experimental demonstration of methodology

The following section presents the reader with examples of how the new ASTM methodology has been used effectively in various applications. It will be shown that in addition to bulk polymer systems, polymer coatings and nanocomposites can be evaluated just as effectively.

#### 15.4.3.1 Effect of additives on the scratch behavior of TPOs

TPOs are polypropylene (PP)-based polymers that are used widely in the automobile industry [78]. Since TPOs are highly susceptible to scratches, additives, such as erucamide slip agents, can be used to increase their scratch resistance.

For this study, four TPO systems were used. Systems A–C contained 20 wt% untreated talc and various loadings of slip agent (0, 0.3, and 0.6 wt%). System D contained 20 wt% surface-treated talc and 0.3 wt% erucamide. All systems contained 78 wt% TPO (PP/EPR) and approximately 2% carbon black filler to provide scratch visibility contrast. The four systems were tested under ASTM conditions.

The results of the posttest analysis can be seen in Fig. 15.5. It was observed that in Systems A–C, the critical load for the onset of scratch visibility increased only slightly as the erucamide loading increased. However, System D showed a marked improvement over Systems A–C. To investigate the behavior at the surface,

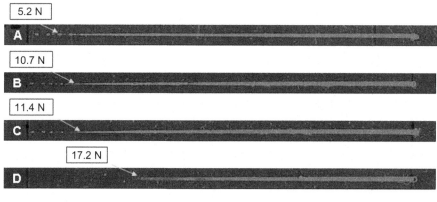

**FIGURE 15.5**

Optical scans of scratched TPO specimens containing talc and various loadings of erucamide [21].

*Reprinted with permission of John Wiley & Sons, Inc., ©2006.*

the scratch coefficient of friction (SCOF) was calculated by taking the ratio of the tangential force to the applied normal load at each data point collected.

Figure 15.6 shows a plot of the SCOF as a function of scratch length for all four systems. It can be seen that within a certain region (Zone 1), the SCOF of System D is much lower than those for Systems A–C. The reason for the convergence of the SCOF curves in Zone 2 is believed to be due to the penetration of the scratch tip through the erucamide layer on the surface.

The surface friction is expected to be directly proportional to the amount of slip agent that reaches the surface upon migration. To investigate this, Fourier transform infrared spectroscopy in attenuated total reflectance mode (FTIR ATR) was employed. The results in Fig. 15.7 indicate that System D has the highest peak intensity corresponding to the C=O stretch mode of the amide group in erucamide (1645 cm$^{-1}$). This implies that the surface concentration of erucamide is the highest in System D. Upon elemental analysis with backscattering SEM on cross-sections of all four materials, aggregations of talc were found in Systems B and C, while none were observed for System D (Fig. 15.8). This implies that there exists an interaction between untreated talc and erucamide that inhibits the migration of the slip agent to the surface of the TPO.

### 15.4.3.2 Scratch behavior of acrylate coatings: thickness effect

Polymer coatings play an important role in providing both protection and esthetic appeal for material surfaces. Thus, it is important to understand their scratch behavior. As with bulk materials, visible scratches on a coating can lower the perceived quality of a part. But even more importantly, the fact that coatings are usually quite thin presents a challenge in regard to structural integrity. Microcracks can expose the substrate to harmful conditions, thus deleting the protective functionality of the coating. Therefore, a coating of an appropriate thickness must be applied to alleviate these concerns.

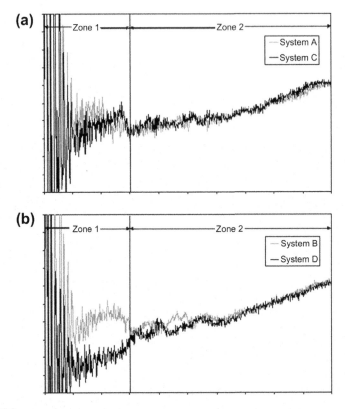

**FIGURE 15.6**

SCOF of scratched TPO systems containing talc and erucamide [21].

*Reprinted with permission of John Wiley & Sons, Inc., ©2006.*

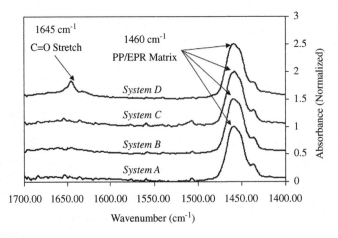

**FIGURE 15.7**

FTIR-ATR curves for surfaces of TPO systems containing talc and erucamide. The peak at 1645 cm⁻¹ corresponds to erucamide [21].

*Reprinted with permission of John Wiley & Sons, Inc., ©2006.*

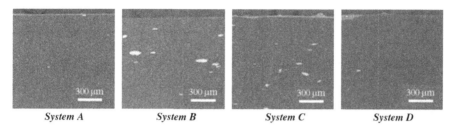

| System A | System B | System C | System D |

**FIGURE 15.8**

SEM micrographs (backscattering mode) of TPO systems containing talc and erucamide [21].

*Reprinted with permission of John Wiley & Sons, Inc., ©2006.*

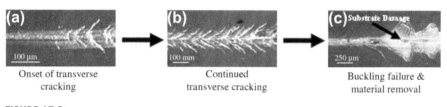

| Onset of transverse cracking | Continued transverse cracking | Buckling failure & material removal |

**FIGURE 15.9**

SEM micrographs showing a typical scratch damage evolution for a thin acrylate coating applied to a phosphatized steel substrate [22]. (For color version of this figure, the reader is referred to the online version of this book.)

*Reprinted with permission of Elsevier, ©2006.*

This particular study focused on how the thickness of an acrylate coating applied to a steel substrate affected the coating's scratch behavior. UV-cured coatings ranged in thickness from 10 to approximately 60 μm were applied to 800-μm-thick phosphatized steel plate substrates. The coatings were tested with an applied normal load range of 1–50 N, a scratch speed of 100 mm s$^{-1}$ and a scratch length of 150 mm. The longer scratch length allowed for observation of damage transitions with a greater resolution.

Three coating specimens, C1 (thickness = 10 μm), C2 (thickness = 36 μm), and C3 (thickness = 61 μm) were chosen to illustrate the thickness effect. In all specimens, a pattern of damage evolution was observed as shown in Fig. 15.9. As with visibility, there will come a point where the coating will begin to form small cracks, and then at higher loads, the coatings will actually fracture and break away. This exposes the substrate to scratch damage (shown in Fig. 15.9(c)). Although the damage pattern was the same for all specimens, the magnitude was affected by the coating thickness.

Figure 15.10 shows the optically scanned images of specimens C1, C2, and C3. SEM micrographs of the damage transitions can be seen in Fig. 15.11. It was observed that after the onset of buckling failure, the maximum width of the fractured region remained constant throughout the duration of the scratch.

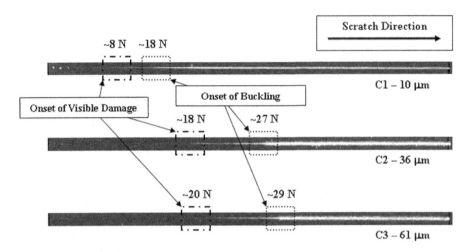

**FIGURE 15.10**

Comparison of scratch damage of acrylate coatings of different thicknesses [22].

*Reprinted with permission of Elsevier, ©2006.*

When C1 is compared to C2 and C3, both the onset for transverse cracking and buckling failure occur at a lower applied normal load. However, when C2 and C3 are compared, there appears to be little change in the magnitude of the scratch damage. Likewise, the maximum width of the fractured region after the onset of buckling failure was the smallest in C1, while C2 and C3 showed little difference between them. This implies that, with increasing thickness, these particular damage features reach a plateau and the coating begins to behave more like a bulk material. To illustrate this point more clearly, Fig. 15.12 shows plots of the discussed damage features as a function of coating thickness. Each curve seems to exhibit a plateau that appears at a particular coating thickness. This implies that a critical thickness may exist for polymer coatings where the scratch behavior represents that of the bulk material.

### 15.4.3.3 Investigation of epoxy nanocomposite scratch behavior

The mechanical performance of epoxy-based materials containing nanofillers has been studied extensively in recent years [79–86]. With good dispersion, toughening particles like core–shell rubber (CSR) [79–83] can resist crack propagation while α-zirconium phosphate (ZrP) can provide improvements in modulus and strength [84–86]. While the addition of nanofillers brings improvement to the bulk properties, the corresponding benefits on surface scratch properties, if any, still remain uncertain.

To explore this issue, two epoxy nanocomposite systems as well as a neat epoxy system were prepared; one containing CSR with a size of 100 nm and the other containing well-exfoliated α-zirconium phosphate with nanoplatelet thickness of 1 nm (formulations are listed in Table 15.1). Micrographs of the morphology of the systems obtained from a transmission electron microscope (TEM) can be viewed in Fig. 15.13. Mechanical properties of these systems were evaluated and are listed in Table 15.2.

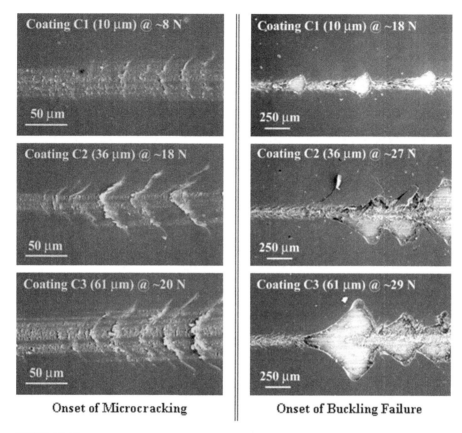

| | |
|---|---|
| Coating C1 (10 µm) @ ~8 N | Coating C1 (10 µm) @ ~18 N |
| 50 µm | 250 µm |
| Coating C2 (36 µm) @ ~18 N | Coating C2 (36 µm) @ ~27 N |
| 50 µm | 250 µm |
| Coating C3 (61 µm) @ ~20 N | Coating C3 (61 µm) @ ~29 N |
| 50 µm | 250 µm |
| **Onset of Microcracking** | **Onset of Buckling Failure** |

**FIGURE 15.11**

SEM micrographs depicting the onset of damage features for acrylate coatings of different thicknesses [22].

*Reprinted with permission of Elsevier, ©2006.*

The epoxy nanocomposites were tested using a linearly increasing applied normal load of 1–90 N, a scratch velocity of 10 mm s$^{-1}$, and a scratch length of 100 mm with a stainless steel spherical tip of a 1-mm diameter. The resulting scratch behavior of the nanocomposites was compared to that of a neat epoxy resin.

Scratch damage appears in the form of microcracks in the nanocomposites systems, while no cracking (except for periodic shear lips) was observed for the neat system under the stated testing conditions. Micrographs of the epoxy/CSR and epoxy/ZrP sample surfaces obtained from an optical microscope (OM) can be seen in Fig. 15.14. The micrographs were taken at the point where microcracks were first observed and the raw data were used to extract the applied normal load. Detailed SEM micrographs of the scratch-induced damages are shown in Figs 15.15 and 15.16. In addition to the surface damage, the subsurface scratch damage was viewed and is shown in Fig. 15.17.

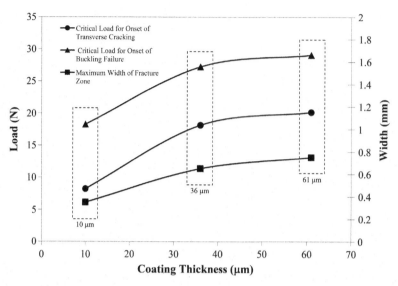

**FIGURE 15.12**

Plots showing the critical load for the onset of transverse cracking, critical load for onset of buckling, and the maximum width of the fracture zone after initiation of buckling as a function of coating thickness [22].

*Reprinted with permission of Elsevier, ©2006.*

**Table 15.1** Formulations of Epoxy Nanocomposites for Scratch Study

|  | Concentration of the Nanofiller |
| --- | --- |
| Neat epoxy | 0 |
| Epoxy/ZrP | 2.0 vol% |
| Epoxy/CSR | 3.0 wt% |

As shown, in comparison with epoxy/ZrP system, the CSR appears to delay the onset of microcracking, which suggests that it is better at providing scratch resistance than ZrP is. The scratch-induced microcracks in the epoxy/ZrP system have a higher frequency than in the epoxy/CSR system. Both the delay in microcracking and the lower frequency of microcracks upon initiation can likely be attributed to the higher ductility and fracture toughness ($K_{IC}$) of the epoxy/CSR system. However, it should be noted that the neat epoxy system with no nanofillers showed no signs of microcracking throughout the duration of the scratch test. Therefore, these results suggest that improvements in mechanical properties like ductility and fracture toughness alone do not mean that scratch resistance will be improved.

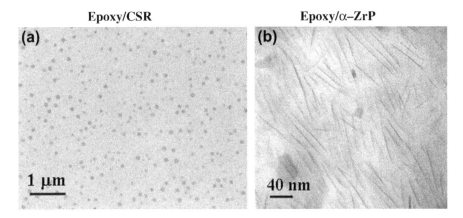

**FIGURE 15.13**

TEM micrographs showing morphology of two epoxy nanocomposites containing (a) CSR and (b) ZrP.

**Table 15.2** Mechanical Property Summary of Epoxy Nanocomposite Systems Used in the Scratch Study

|  | Neat Epoxy | Epoxy/CSR | Epoxy/ZrP |
|---|---|---|---|
| Young's modulus (GPa) | 2.90 ± 0.05 | 2.56 ± 0.06 | 3.72 ± 0.21 |
| Elongation at break (%) | 4.1 ± 0.4 | 6.5 ± 0.5 | 3.9 ± 0.3 |
| $K_{IC}$ (MPa m$^{1/2}$) | 0.72 ± 0.02 | 0.92 ± 0.08 | 0.79 ± 0.04 |
| Residual scratch depth at onset of microcrack formation (mm) | N/A | 8.2 ± 0.1 | 3.5 ± 0.2 |

It was stated earlier that lowering the tangential force is likely to be the key to improving scratch performance in polymers. To that end, both the tangential force ($F_t$) and SCOF of the polymer nanocomposites studied here are shown in Fig. 15.18. For both $F_t$ and the SCOF, it can be observed that both nanocomposites show a higher $F_t$ than the neat epoxy at some point within the scratch test, while the SCOF shows that the Epoxy/ZrP system can be more clearly differentiated from the other two systems.

The above findings imply that nanoparticles will not necessarily help in improving the scratch resistance of polymers. Improvements in tensile modulus/strength (i.e. epoxy/ZrP) [84–86] and in ductility/toughness (i.e. epoxy/CSR) [79–83] over those of a neat epoxy do not result in a more scratch resistant polymer. A similar outcome was also observed in a few other polymer nanocomposites. This, again, indicates the complexity of the polymer scratch behavior. Modeling tools are needed to help address the structure–property relationship regarding polymer scratch.

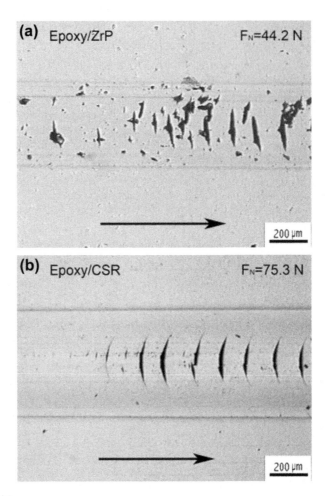

**FIGURE 15.14**

OM micrographs of scratched sample surfaces of (a) epoxy/ZrP and (b) epoxy/CSR nano-composites at the load where microcracks are first noticed. (Solid black arrows indicate the scratch direction.)

## 15.4.4 Implementation of finite element modeling

With the help of simulation by numerical modeling and the validation of experimental results, fundamental scratch deformation and the associated damage mechanisms can be comprehensively understood. In the following section, modeling the effects of different scratch conditions on polymer scratch behavior will be addressed through the use of FEM. Using this approach, the influence of material and surface properties can be assessed and meaningful guidelines to improve the scratch performance of polymeric

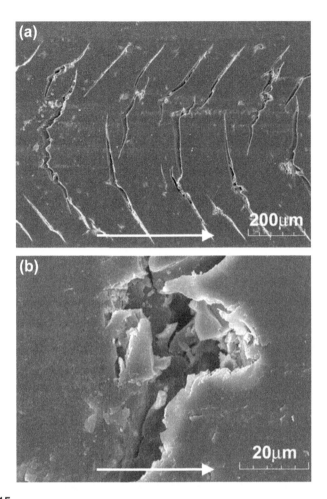

**FIGURE 15.15**

SEM micrographs showing details of scratch-induced microcracks in epoxy/ZrP nanocomposite at an applied normal load of approximately 58.6 N (a = low magnification; b = high magnification; solid white arrows indicate the scratch direction).

materials can be obtained for engineering applications. It is also possible to predict the damage initiation during a scratch process with proper choices of damage criteria. Owing to its user-friendly layout, the FEM software ABAQUS® [87] is chosen as the numerical tool to handle the complicated aspects of modeling the scratch problem.

### 15.4.4.1 Scratch deformation and damage mechanisms

For a successful numerical simulation, it is important to understand the essential processes occurring in an actual scratch experiment. A scratch process can be separated

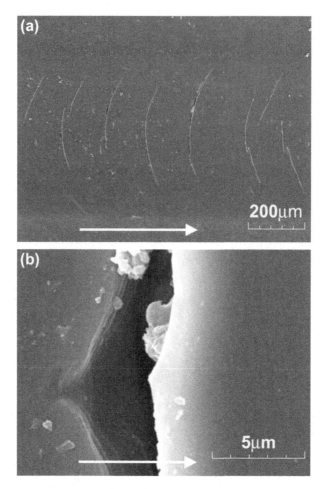

**FIGURE 15.16**

SEM micrographs showing details of scratch-induced microcracks in epoxy/CSR nanocomposite at an applied normal load of approximately 81.4 N (a = low magnification; b = high magnification; solid white arrows indicate the scratch direction).

into three mechanical steps: indentation, scratch and spring-back (Fig. 15.19). Indentation is where the tip makes initial contact with the substrate. The scratch step is where the tip moves forward and plows through the surface of the material. For the final step, the tip stops moving and is lifted up to allow for elastic recovery of the material.

An example of a representative mesh and boundary conditions of an FEM model used to simulate actual experimental conditions are illustrated in Fig. 15.20. An explicit scheme is employed to describe the time evolution of the dynamic stress/strain analysis during the steps. Three-dimensional linear eight-node solid elements (C3D8R) with selective reduced integration and hourglass mode control are utilized.

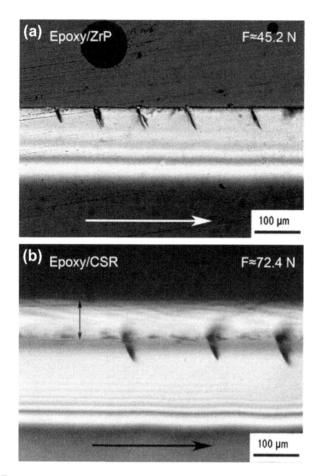

**FIGURE 15.17**

OM micrographs of longitudinal-sections of epoxy/ZrP and epoxy/CSR. (Arrows indicate the scratch direction.)

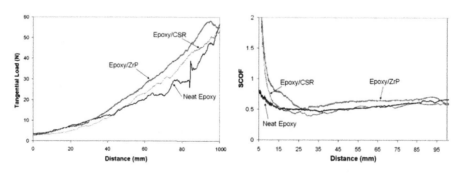

**FIGURE 15.18**

Tangential load and SCOF of neat epoxy, Epoxy/ZrP and Epoxy/CSR nanocomposites.

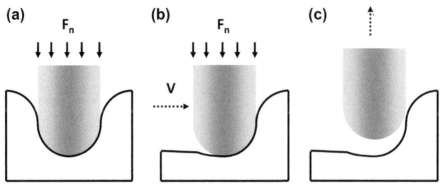

**FIGURE 15.19**

Various steps in scratch process: (a) indentation step; (b) scratch step; (c) spring-back step.

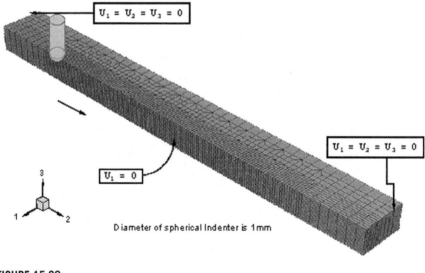

**FIGURE 15.20**

FE mesh and boundary conditions [74].

The true stress-strain curve of PP at various strain rates was adopted [88,89] and von Mises yielding criterion was employed. The Coulomb friction model was incorporated for the frictional interaction between the tip and substrate. Convergence study was performed to determine the optimal mesh so that sufficient accuracy is obtained with acceptable computational cost.

Without an expensive high-speed video camera, the scratch deformation and damage processes will occur too rapidly to obtain the detailed dynamic and mechanistic

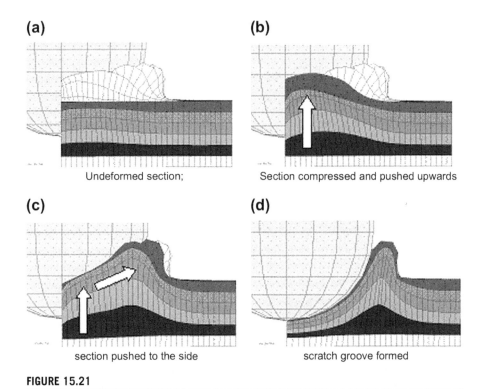

**(a)**

Undeformed section;

**(b)**

Section compressed and pushed upwards

**(c)**

section pushed to the side

**(d)**

scratch groove formed

**FIGURE 15.21**

The sequential formation of a scratch groove [74].

information during scratch. FEM simulation becomes useful as it can capture the time evolution of the scratch process that can be visualized to illustrate the sequence of material deformation and damage events. For example, as shown in Fig. 15.21, as the scratch tip is moving outwards toward the reader, a barely deformed material (gray- and black-colored elements) is in front of the approaching tip (Fig. 15.21(a)). The material is undergoing compression and is pushed upwards as the tip moves further ahead (Fig. 15.21(b)). Then, the moving tip keeps compressing the material and pushes it sideways at the same time (Fig. 15.21(c)). Finally, the tip plows through the material ahead of it and forms a scratch groove (Fig. 15.21(d)). The above information cannot be easily obtained experimentally. This illustrates the usefulness of FEM to understand complex deformation mechanisms during scratching.

Other than showing the sequential formation of scratch grooves, the examination of the material mechanical response close to the tip allows one to make further predictions of possible dominant deformation and damage mechanisms. From the numerical simulation, one can find that the material beneath the front of the tip remains under compression, the material behind the tip is in tensile stress state. Tensile stresses can also be observed for material that is further ahead of the indenter. Figure 15.22(a) shows the direction of maximum principal stress in the elements

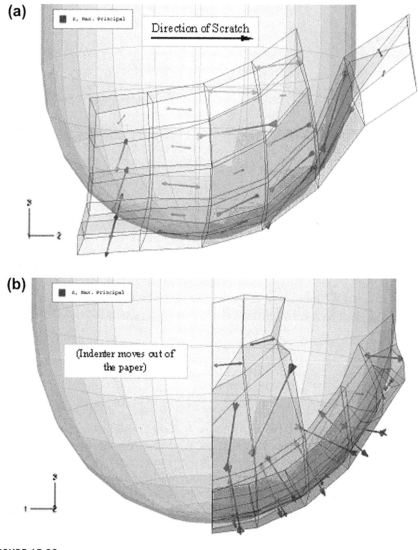

**FIGURE 15.22**

Direction of the maximum principal stresses in the elements around the indenter for (a) perspective in 2–3 plane and (b) perspective in 1–3 plane [36].

around the scratch tip. The tensile stress for the elements right behind the tip is generally in the 2-direction (i.e. the scratch direction) while the stress in the left-most rows of elements are in the vertical 3-direction and with slight bias to the 2-direction. This suggests that as the scratch tip moves, the material directly behind it is stretched in the direction of the scratch. If fracture does occur, it should be perpendicular to the scratch direction. As the tip keeps moving, the same material is now pulled in the vertical direction, possibly leading to the material being spalled off. The outer-most row of elements ahead of the tip are stretched outwardly in the 1-direction, suggesting that the material in that region tears apart as the indenter plows through, forming cracks parallel to the scratch direction (Fig. 15.22(b)). Thus, the phenomenological occurrence of the scratch damage mechanism can be deduced based on mechanical analysis of the FEM results.

### 15.4.4.2 Parametric study via FE modeling

The challenges for polymer scratch research are objective evaluation of material scratch performance and the identification of essential material parameters that will have a significant effect on scratch behavior. An earlier attempt was made to assess the effect of Young's modulus, $E$, and the radius of stylus tips on the scratch performance in which an elastic material type was adopted [45]. The findings from the modeling were limited to its ability to realistically predict the scratch damage of polymers. To address how the key material parameters and surface properties would affect polymer scratch performance, parametric studies using FEM were carried out to give useful guidelines toward designing scratch resistant polymers. While an experimental parametric study is difficult due to the extreme care and effort required to effectively control polymer material properties, it is convenient to conduct this task with numerical modeling. All that is required is to change the material parameters of the numerical simulation.

Without losing essential material information, an elastic–perfectly plastic material constitutive relationship was adopted in an earlier parametric study [90] to examine the effect of selected material and surface properties, that is, Young's modulus, Poisson's ratio, yield stress, and coefficient of adhesive friction, on the scratch performance of PP. The 3D FEM modeling scheme similar to that in the previous section was adopted here. The residual scratch depth, defined as the difference between the instantaneous scratch depth and the amount of recovered depth, was adopted as a useful quantity to rank scratch visibility.

It was found that the tendency of the effect of the four parameters on the scratch is independent of the load range studied (Fig. 15.23). Poisson's ratio has a minimal effect. On the other hand, yield stress and coefficient of adhesive friction are critical in influencing the residual scratch depth of PP.

Part of the parametric study using numerical modeling was validated by a set of model systems study on PP blends [90] and by the experimental work in which the slip agent was added to the matrix to reduce the friction coefficient while other material properties remained unaltered [21]. This work illustrates the capability of numerical modeling for providing guidelines to improve scratch performance of polymeric materials.

While the utilization of an elastic–perfectly plastic model is sufficient to understand the influence of preyield mechanical behavior, it is unable to capture the effect of postyield mechanical behavior. Furthermore, it should be noted that the compressive and tensile behaviors of polymers are practically never symmetric. To abet

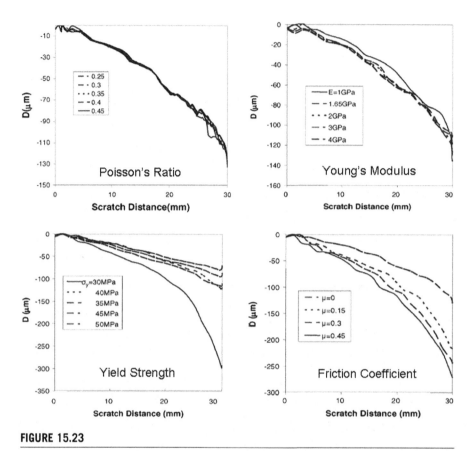

**FIGURE 15.23**

Parametric study of scratch residual depth [90].

*Reprinted with permission of John Wiley & Sons, Inc., ©2007.*

the understanding gained from the first study, more FEM parametric studies were conducted using piecewise linear true stress–strain plots like the one represented in Fig. 15.24 to more closely simulate the actual stress–strain behavior of polymers in numerical simulation.

The nomenclature for the variables in Fig. 15.24 is as follows: $\sigma_y$ Yield stress; $s$ Strain softening slope; $h$ Strain hardening slope; $\varepsilon_r$ Strain at yield stress recovery; and $\varepsilon_{ES}$ prehardening strain.

A subscript of "T" on any of the above variables denotes tension while a subscript of "C" denotes compression.

One study involved varying the five main parameters listed above [91,92]. The FEM simulation used the scratch depth as well as the shoulder height as metrics for comparing different cases with variation in yield and postyield material parameters. It was shown that, out of all the parameters investigated, $\sigma_y$, $\varepsilon_r$, and the $h$ beyond the

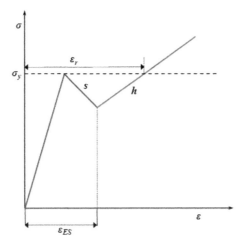

**FIGURE 15.24**

Piecewise linear true stress–strain plot.

strain at stress recovery are the most important factors to consider when designing polymeric materials with superior scratch resistance [91,92].

The next study was similar to the previous one in that it incorporated piecewise linear true stress–strain plots. However, this time the asymmetry between compressive and tensile behavior was considered [93]. Furthermore, several styrene–acrylonitrile (SAN) copolymers with differing acrylonitrile (AN) contents and a PC system were used to experimentally validate the findings of the numerical modeling. The reasons for choosing SAN and PC as model systems for the experimental study are as follows:

1. SAN is brittle in tension and ductile in compression, while PC shows ductile behavior under both scenarios.
2. A range of tensile strengths can be achieved with SAN by varying the AN content and/or the molecular weight.

The uniaxial tension and compression properties of the SAN and PC model systems are listed in Table 15.3. The various systems were chosen in an effort to experimentally parameterize the properties of interest.

The cases employed for the asymmetric compressive/tensile stress–strain FEM simulation were as follows:

Case 1: Compressive and tensile behavior are the same;
Case 2: Same as Case 1, but tensile behavior is elastic–perfectly plastic;
Case 3: Compressive yield stress > tensile yield stress;
Case 4: Same as Case 3, but tensile behavior is elastic–perfectly plastic;
Case 5: Compressive yield stress < tensile yield stress;
Case 6: Same as Case 5, but tensile behavior is elastic–perfectly plastic.

**Table 15.3** Uniaxial Tensile and Compressive Properties of SAN and PC Model Systems

|  | SAN 19 | SAN 27A | SAN 27B | SAN 27C | SAN 35 | PC |
|---|---|---|---|---|---|---|
| Tensile modulus (GPa) | 3.4 ± 0.0 | 3.6 ± 0.1 | 3.7 ± 0.1 | 3.7 ± 0.1 | 3.7 ± 0.0 | 2.3 ± 0.0 |
| Tensile strength (MPa) | 68.9 ± 1.5 | 63.7 ± 2.3 | 75.1 ± 3.0 | 79.0 ± 1.0 | 81.9 ± 0.7 | 65.2 ± 0.0 |
| Compressive modulus (GPa) | 3.5 ± 0.1 | 3.6 ± 0.1 | 3.6 ± 0.3 | 3.5 ± 0.2 | 3.4 ± 0.2 | 2.1 ± 0.1 |
| Compressive yield stress (MPa) | 117.6 ± 0.8 | 115.2 ± 0.5 | 117.2 ± 0.4 | 117.2 ± 0.2 | 113.7 ± 1.4 | 75.3 ± 0.7 |

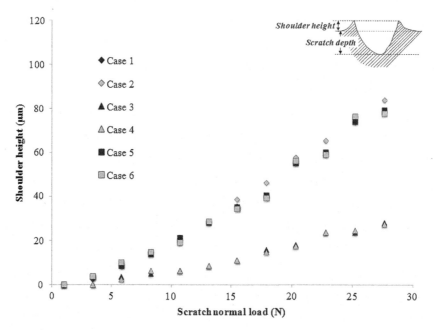

**FIGURE 15.25**

Plot of shoulder height as a function of scratch normal load (FEM simulation) [93].

When the experimental observations are compared with the numerical simulation, a good agreement is found between the two. These results show that compressive behavior (i.e. $\sigma_{y,C}$, $\varepsilon_{r,C}$, and $h_C$ beyond $\varepsilon_{r,C}$) is strongly linked to the development of the scratch depth and shoulder height of the groove formed during the scratch process.

To illustrate, the curves in Figs 15.25 and 15.26 show the results of the FEM modeling with regard to the formation of shoulder height and scratch depth for the

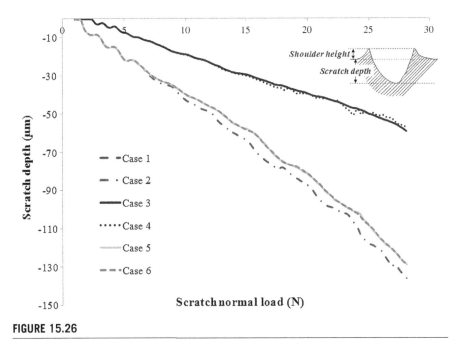

**FIGURE 15.26**

Scratch depth profile (FEM simulation) [93].

six cases listed above. As shown, a higher compressive yield stress, $\sigma_{y,C}$, and a lower strain at stress recovery in compression, $\varepsilon_{r,C}$ (Cases 3 and 4) induces lower shoulder height and shallower scratch depth. In other words, modification in compressive behavior is expected to alter the scratch depth and shoulder height formation during the scratch process, whereas modification of tensile behavior has little influence. To experimentally validate the findings above, shoulder height and scratch depth for a grade of SAN and PC were measured and shown in Figs 15.27 and 15.28, respectively. Despite the similarity in tensile strength, the lower compressive yield stress of PC induces higher shoulder height and deeper scratch depth. Likewise, it was also shown that, despite the differences in tensile strengths of different grades of SAN, similar compressive behavior induces similar scratch depth and shoulder height along the scratch path.

Since the development of the maximum tensile stress component occurs at the rear of the scratch tip, the tensile behavior is responsible for the formation of cracking/crazing that brings about the scratch-induced change in surface roughness that is a source of scratch visibility in brittle polymers like SAN (Fig. 15.29(a)). Polymers that exhibit ductile tensile behavior will either resist cracking and crazing completely and show little change in scratch path roughness like PC (Fig. 15.29(b)), or they will be susceptible to formation of fish scales as in TPOs (Fig. 15.29(c)).

Physical measurements of surface roughness were taken with a laser-scanning confocal microscope, and the results are shown in Figs 15.30 and 15.31. Figure 15.30 compares different grades of SAN, both brittle in tension. SAN 35 has a higher

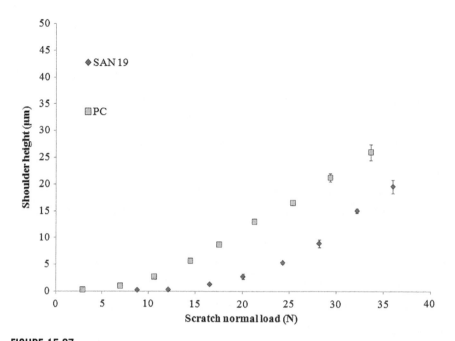

**FIGURE 15.27**

Comparison of shoulder height in SAN 19 and PC [93].

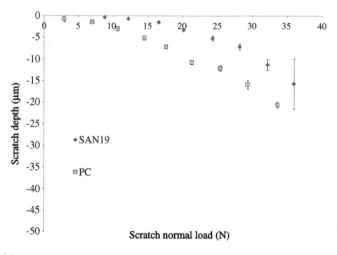

**FIGURE 15.28**

Comparison of scratch depth in SAN 19 and PC [93].

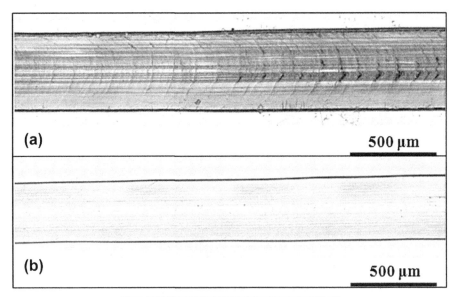

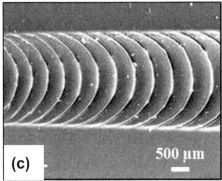

**FIGURE 15.29**

Micrographs of scratch paths showing (a) microcracking in SAN, (b) resistance to surface roughness development in PC, and (c) fish-scale formation in TPO. (Note—SAN and PC are shown at the same load range to illustrate PC's higher resistance to the formation of a rough scratch path. The TPO micrograph is simply a representation of the fish-scale feature.)

tensile strength than SAN 27A, and, thus, shows a delayed development of surface roughness. Figure 15.31 shows that a material that exhibits brittle behavior in tension (SAN 19) will be highly susceptible to the development of surface roughness. In short, tensile behavior is important for delaying or preventing the development of scratch-induced surface roughness in polymers.

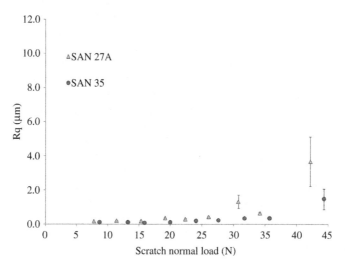

**FIGURE 15.30**

Comparison of surface roughness in SAN 27A and SAN 35 [93].

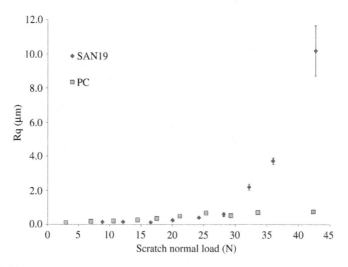

**FIGURE 15.31**

Comparison of surface roughness in SAN 19 and PC [93].

## 15.5 SUMMARY AND CONCLUSIONS

To close, the study of polymer scratch behavior has come a long way in the past 50 years. Based on the groundwork laid out by others, there now exists a reliable testing method to produce repeatable scratches that allow for unbiased evaluation of the material response to scratch. It is also now possible to assess the resistance against scratch visibility with little ambiguity. Numerical simulation using FEM has become vital in characterizing the complicated mechanistic aspects of polymer scratch behavior that is otherwise virtually impossible to achieve experimentally. However, the simulation is effective only if correct conditions are applied and appropriate constitutive models are chosen. Coupling the experimental methodology with the numerical simulation has resulted in a substantiated fundamental means to address polymer scratch behavior.

The ultimate goal for the above efforts is to obtain models that are able to predict polymer scratch behavior in a preemptive fashion. In this way, the coupled ASTM and FEM approach becomes much more effective, saving time, money, and improving material performance. At the present, the experimental evaluation is carried out first, so as to provide the necessary components for an effective numerical simulation. Then, the developed model is used in another scenario to predict the material scratch behavior. Thus, the process becomes reiterative and the numerical analysis and material design can be progressively fine tuned.

A great number of obstacles still lie ahead in achieving the above objective. The foremost is the evaluation of scratch visibility. Even though it has been shown here that polymer scratch visibility can be assessed in a quantitative way, there still remains disagreement in whether or not digitally enhanced images of scratches truly represent what a common observer would see. Furthermore, at least two other main hindrances exist in the realm of numerical simulation. First is the availability of an appropriate constitutive model that can represent the true mechanical response of polymeric materials under scratch conditions. Viscoelasticity, viscoplasticity, skin-core morphology, and other factors can all contribute to the complexity of the polymer constitutive model. Currently, the problem is so complex that there is no single model that can fully predict the mechanical response of a polymer. Second, in order to capture the representative physical damage mechanisms during the scratch process such as fracture, microcracks and crazes, simplifications must be made to ensure that numerical simulation is practical for current computational capabilities. Attempting to model even a single physical damage mechanism could potentially require a large amount of time and effort.

Overcoming the barriers stated above seems like a daunting task at first glance, but a strong foundation has been built thanks to the recent developments in FEM and the establishment of ASTM D7027-05 and ISO 19,252:08. The key to advancing this research lies not only in employing the new ASTM and ISO scratch test in a fundamental fashion but also in correlating these results with FEM analyses. This will pave the way for the development of models that can possibly one day fully predict the scratch behavior of polymeric materials.

## Acknowledgments

The authors would like to thank the Texas A&M Scratch Behavior Consortium (Advanced Composites, Atlas-MTS, Cabot, Ciba Specialty Chemical, Clorox, Dow Chemical, Imerys, Innovene, Japan Polypropylene, Kaneka, Kraton, Phillips-Sumika, Solvay Engineered Polymers, Sumitomo Chemical, Surface Machine Systems, and

Visteon) for providing the financial support in this research endeavor. The authors would also like to acknowledge the financial support from the State of Texas (ARP #32191-73,130) and Defense Logistic Agency (SP0103-02-D-0003). Special thanks are also due to the Society of Plastics Engineers—South Texas Section, for their generous donation of equipment for this research. Allan Moyse, Goy Teck Lim, Mohammad Motaher Hossain, and Ehsan Moghbelli are recognized for their contributions to this body of research work.

# References

[1] F. Mohs, Grundriss der mineralogie, 1824. (English translation by W. Haidinger: Treatise on Minerology, Constable, Edinburgh, 1825).

[2] H. Hogberg, Methods of testing modified polystyrenes for use in telephone housings, Modern Plastics 33, (1955) 3, 150, 152, 157 and 259.

[3] D. Tabor, Mohs hardness scale – a physical interpretation, Proceedings of the London Physical Society, Section B 67 (411) (1954) 249–257.

[4] D. Tabor, The physical meaning of indentation and scratch hardness, British Journal of Applied Physics 7 (5) (1956) 159–166.

[5] B.J. Briscoe, A.C. Smith, Influence of dynamic loading on sliding friction, Nature 278 (5706) (1979) 725–726.

[6] B.J. Briscoe, T.A. Stolarski, Wear of polymers in the pin-on-disk configuration, ACS Symposium Series 287 (1985) 303–313.

[7] B.J. Briscoe, P.D. Evans, Scratch hardness as an evaluation of cure temperature for glass-fiber reinforced polyester, Composites Science and Technology 34 (1) (1989) 73–90.

[8] B.J. Briscoe, Materials aspects of polymer wear, Scripta Metallurgica et Materialia 24 (5) (1990) 839–844.

[9] B.J. Briscoe, S.K. Biswas, S.K. Sinha, S.S. Panesar, The scratch hardness and friction of a soft rigid-plastic solid, Tribology International 26 (3) (1993) 183–193.

[10] B.J. Briscoe, E. Pelillo, S.K. Sinha, Scratch hardness and deformation maps for polycarbonate and polyethylene, Polymer Engineering and Science 36 (24) (1996) 2996–3005.

[11] B.J. Briscoe, E. Pelillo, S.K. Sinha, Characterisation of the scratch deformation mechanisms for poly(methylmethacrylate) using surface optical reflectivity, Polymer International 43 (4) (1997) 359–367.

[12] B.J. Briscoe, S.K. Sinha, Density distributions characteristics of green ceramic compacts using scratch hardness, Tribology International 30 (7) (1997) 475–482.

[13] B.J. Briscoe, P.D. Evans, E. Pelillo, S.K. Sinha, Scratching maps for polymers, Wear 200 (1–2) (1996) 137–147.

[14] B.J. Briscoe, Isolated contact stress deformations of polymers: the basis for interpreting polymer tribology, Tribology International 31 (1–3) (1998) 121–126.

[15] B.J. Briscoe, A. Chateauminois, T.C. Lindley, D. Parsonage, Fretting wear behaviour of polymethylmethacrylate under linear motions and torsional contact conditions, Tribology International 31 (11) (1998) 701–711.

[16] B.J. Briscoe, E. Pelillo, F. Ragazzi, S.K. Sinha, Scratch deformation of methanol plasticized poly(methylmethacrylate) surfaces, Polymer 39 (11) (1998) 2161–2168.

[17] B.J. Briscoe, A. Delfino, E. Pelillo, Single-pass pendulum scratching of poly(styrene) and poly(methylmethacrylate), Wear 229 (1999) 319–328.

[18] B.J. Briscoe, A. Chateauminois, Measurements of friction-induced surface strains in a steel/polymer contact, Tribology International 35 (4) (2002) 245–254.

[19] B.J. Briscoe, S.K. Sinha, Scratch resistance and localised damage characteristics of polymer surfaces - a review, Materialwiss Werkst 34 (10–11) (2003) 989–1002.

[20] D. Britz, R.A. Ryntz, V. Jardret, P.V. Yaneff, Understanding the effect of processing conditions on the scratch and mechanical behavior of automotive plastic coating on TPO, Journal of Coatings Technology 3 (6) (2006) 40–46.

[21] R. Browning, G.T. Lim, A. Moyse, L.Y. Sun, H.J. Sue, Effects of slip agent and talc surface-treatment on the scratch behavior of thermoplastic olefins, Polymer Engineering and Science 46 (5) (2006) 601–608.

[22] R.L. Browning, G.T. Lim, A. Moyse, H.J. Sue, H. Chen, J.D. Earls, Quantitative evaluation of scratch resistance of polymeric coatings based on a standardized progressive load scratch test, Surface and Coatings Technology 201 (6) (2006) 2970–2976.

[23] J. Chu, C. Xiang, H.J. Sue, R.D. Hollis, Scratch resistance of mineral-filled polypropylene materials, Polymer Engineering and Science 40 (4) (2000) 944–955.

[24] A. Dasari, S.J. Duncan, R.D.K. Misra, Atomic force microscopy of scratch damage in polypropylene, Materials Science and Technology-London 18 (10) (2002) 1227–1234.

[25] A. Dasari, S.J. Duncan, R.D.K. Misra, Micro- and nano-scale deformation processes during scratch damage in high density polyethylene, Materials Science and Technology-London 19 (2) (2003) 239–243.

[26] A. Dasari, J. Rohrmann, R.D.K. Misra, Atomic force microscopy characterisation of scratch deformation in long and short chain isotactic polypropylenes and ethylene–propylene copolymers, Materials Science and Technology-London 19 (9) (2003) 1298–1308.

[27] A. Dasari, J. Rohrmann, R.D.K. Misra, Atomic force microscopy characterisation of mechanically induced surface damage in ethylene–propylene diblock copolymeric materials, Materials Science and Technology-London 19 (10) (2003) 1458–1466.

[28] A. Dasari, R.D.K. Misra, J. Rohrmann, Scratch deformation characteristics of micrometric wollastonite-reinforced ethylene-propylene copolymer composites, Polymer Engineering and Science 44 (9) (2004) 1738–1748.

[29] A. Dasari, J. Rohrmann, R.D.K. Misra, On the scratch deformation of micrometric wollastonite reinforced polypropylene composites, Materials Science and Engineering A, Structural Materials 364 (1–2) (2004) 357–369.

[30] R. Hadal, A. Dasari, J. Rohrmann, R.D.K. Misra, Susceptibility to scratch surface damage of wollastonite- and talc-containing polypropylene micrometric composites, Materials Science and Engineering A, Structural Materials 380 (1–2) (2004) 326–339.

[31] R.S. Hadal, R.D.K. Misra, Scratch deformation behavior of thermoplastic materials with significant differences in ductility, Materials Science and Engineering A, Structural Materials 398 (1–2) (2005) 252–261.

[32] R. Hadal, Q. Yuan, J.P. Jog, R.D.K. Misra, On stress whitening during surface deformation in clay-containing polymer nanocomposites: a microstructural approach, Materials Science and Engineering A, Structural Materials 418 (1–2) (2006) 268–281.

[33] V. Jardret, R. Ryntz, Visco-elastic visco-plastic analysis of scratch resistance of organic coatings, Journal of Coatings Technology and Research 2 (8) (2005) 591–598.

[34] F.N. Jones, W. Shen, S.M. Smith, Z.H. Huang, R.A. Ryntz, Studies of microhardness and mar resistance using a scanning probe microscope, Progress in Organic Coatings 34 (1–4) (1998) 119–129.

[35] G.T. Lim, J.N. Reddy, H.J. Sue, Finite element modeling for scratch damage of polymers, Stimuli-Responsive Polymeric Films and Coatings 912 (2005) 166–180.

[36] G.T. Lim, M.H. Wong, J.N. Reddy, H.J. Sue, An integrated approach towards the study of scratch damage of polymer, Journal of Coatings Technology and Research 2 (5) (2005) 361–369.

[37] L. Mi, H. Ling, W.D. Shen, R. Ryntz, B. Wichterman, A. Scholten, Some complementary scratch resistance characterization methods, Journal of Coatings Technology and Research 3 (4) (2006) 249–255.

[38] R.D.K. Misra, R. Hadal, S.J. Duncan, Surface damage behavior during scratch deformation of mineral reinforced polymer composites, Acta Materialia 52 (14) (2004) 4363–4376.

[39] R.A. Ryntz, B.D. Abell, G.M. Pollano, L.H. Nguyen, W.C. Shen, Scratch resistance behavior of model coating systems, Journal of Coating Technology 72 (904) (2000) 47–53.

[40] R.A. Ryntz, D. Britz, Scratch resistance behavior of automotive plastic coatings, Journal of Coating Technology 74 (925) (2002) 77–81.

[41] S.K. Sinha, 180 years of scratch testing, Tribology International 39 (2) (2006) 61.

[42] J.S.S. Wong, H.J. Sue, K.Y. Zeng, R.K.Y. Li, Y.W. Mai, Scratch damage of polymers in nanoscale, Acta Materialia 52 (2) (2004) 431–443.

[43] M. Wong, G.T. Lim, A. Moyse, J.N. Reddy, H.J. Sue, A new test methodology for evaluating scratch resistance of polymers, Wear 256 (11–12) (2004) 1214–1227.

[44] M. Wong, A. Moyse, F. Lee, H.J. Sue, Study of surface damage of polypropylene under progressive loading, Journal of Materials Science 39 (10) (2004) 3293–3308.

[45] C. Xiang, H.J. Sue, J. Chu, B. Coleman, Scratch behavior and material property relationship in polymers, Journal of Polymer Science Part B: Polymer Physics 39 (1) (2001) 47–59.

[46] C. Xiang, H.J. Sue, J. Chu, K. Masuda, Roles of additives in scratch resistance of high crystallinity polypropylene copolymers, Polymer Engineering and Science 41 (1) (2001) 23–31.

[47] P.R. Guevin, State-of-the-art instruments to measure coating hardness, Journal of Coating Technology 67 (840) (1995) 61–65.

[48] O. Vingsbo, S. Hogmark, Single-pass pendulum grooving – a technique for abrasive testing, Wear 100 (1–3) (1984) 489–502.

[49] Y.N. Liang, S.Z. Li, D.F. Li, S. Li, Some developments for single-pass pendulum scratching, Wear 199 (1) (1996) 66–73.

[50] A. Chanda, D. Basu, A. Dasgupta, S. Chattopadhyay, A.K. Mukhopadhyay, A new parameter for measuring wear of materials, Journal of Materials Science Letters 16 (20) (1997) 1647–1651.

[51] A. Krupicka, M. Johansson, A. Hult, Use and interpretation of scratch tests on ductile polymer coatings, Progress in Organic Coatings 46 (1) (2003) 32–48.

[52] R.S. Kody, D.C. Martin, Quantitative characterization of surface deformation in polymer composites using digital image analysis, Polymer Engineering and Science 36 (2) (1996) 298–304.

[53] J. Chu, L. Rumao, B. Coleman, Scratch and mar resistance of filled polypropylene materials, Polymer Engineering and Science 38 (11) (1998) 1906–1914.

[54] I.M. Hutchings, P.Z. Wang, G.C. Parry, An optical method for assessing scratch damage in bulk materials and coatings, Surface and Coatings Technology 165 (2) (2003) 186–193.

[55] P. Rangarajan, M. Sinha, V. Watkin, K. Harding, J. Sparks, Scratch visibility of polymers measured using optical imaging, Polymer Engineering and Science 43 (3) (2003) 749–758.

[56] R.D. Mindlin, Compliance of elastic bodies in contact, Journal of Applied Mechanics 16 (1949) 259–268.

[57] G.M. Hamilton, L.E. Goodman, Stress field created by a circular sliding contact, Journal of Applied Mechanics 33 (2) (1966) 371–376.

[58] W.T. Chen, Stresses in some anisotropic materials due to indentation and sliding, International Journal of Solids and Structures 5 (1969) 191–214.

[59] W. Lin, C.H. Kuo, L.M. Keer, Analysis of a transversely isotropic half space under normal and tangential loadings, Journal of Tribology 112 (2) (1991) 335–338.

[60] M.T. Hanson, I.W. Puja, The elastic field resulting from elliptical Hertzian contact of transversely isotropic bodies: closed-form solutions for normal and shear loading, Journal of Applied Mechanics 64 (3) (1997) 457–465.

[61] V.A. Churilov, Action of an elliptic stamp moving at a constant speed on an elastic half-space, Journal of Applied Mathematics and Mechanics 42 (6) (1978) 1176–1182.

[62] M. Rahman, Hertz problem for a rigid punch moving across the surface of a semi-infinite elastic solid, Zeitschrift für angewandte Mathematik und Physik 47 (4) (1996) 601–615.

[63] L.M. Brock, Exact analysis of dynamic sliding indentation at any constant speed on an orthotropic or transversely isotropic half-space, Journal of Applied Mechanics 69 (3) (2002) 340–345.

[64] E.H. Lee, J.R.M. Radok, The contact problem for viscoelastic bodies, Journal of Applied Mechanics 27 (1960) 438–444.

[65] S.C. Hunter, The Hertz problem for a rigid spherical indenter and a viscoelastic half-space, Journal of the Mechanics and Physics of Solids 8 (1960) 219–234.

[66] W. Brostow, J.A. Hinze, R. Simoes, Simulations of scratch resistance and recovery in one and two-phase polymers, ANTEC Conference Proceedings (2003) 3613–3617.

[67] W. Brostow, J.A. Hinze, R. Simoes, Tribological behavior of polymers simulated by molecular dynamics, Journal of Materials Research 19 (3) (2004) 851–856.

[68] J.N. Reddy, An Introduction to the Finite Element Method, second ed., McGraw-Hill, New York, 1993.

[69] J. Mackerle, Finite element and boundary element simulations of indentation problems – a bibliography (1997–2000), Finite Elements in Analysis and Design 37 (10) (2001) 811–819.

[70] H. Tian, N. Saka, Finite-element analysis of an elastic-plastic 2-layer half space sliding contact, Wear 148 (2) (1991) 261–285.

[71] J.H. Lee, G.H. Xu, H. Liang, Experimental and numerical analysis of friction and wear behavior of polycarbonate, Wear 251 (2) (2001) 1541–1556.

[72] J.L. Bucaille, E. Felder, G. Hochstetter, Mechanical analysis of the scratch test on elastic and perfectly plastic materials with the three-dimensional finite element modeling, Wear 249 (5–6) (2001) 422–432.

[73] G. Subhash, W. Zhang, Investigation of the overall friction coefficient in single-pass scratch test, Wear 252 (1–2) (2002) 123–134.

[74] G.T. Lim, Scratch Behavior of Polymers, Ph.D. Thesis, Texas A&M University, College Station, TX, 2005.

[75] http://www.taberindustries.com/PDFs/scratch_mar_710.pdf

[76] ASTM D7027–05, Standard test method for evaluation of scratch resistance of polymeric coatings and plastics using an instrumented scratch machine, Annual Book of ASTM Standards, 2005. (ASTM International).

[77] H. Jiang, R.L. Browning, M.M. Hossain, H.J. Sue, M. Fujiwara, Quantitative evaluation of scratch visibility resistance of polymers, Applied Surface Science 256 (21) (2010) 6324–6329.

[78] C. Xiang, H.J. Sue, Iosipescu shear deformation and fracture in model thermoplastic olefins, Journal of Applied Polymer Science 82 (13) (2001) 3201–3214.

[79] K.T. Gam, M. Miyamoto, R. Nishimura, H.J. Sue, Fracture behavior of core-shell rubber-modified clay-epoxy nanocomposites, Polymer Engineering and Science 43 (10) (2003) 1635–1645.

[80] H.J. Sue, Craze-like damage in a core-shell rubber-modified epoxy system, Journal of Materials Science 27 (11) (1992) 3098–3107.

[81] H.J. Sue, E.I. Garcia-Meitin, D.M. Pickelman, P.C. Yang, Optimization of mode-1 fracture-toughness of high-performance epoxies by using designed core-shell rubber particles, Advances in Chemistry 233 (1993) 259–291.

[82] H.J. Sue, Study of rubber-modified brittle epoxy systems .1. Fracture-toughness measurements using the double-notch 4-point-bend method, Polymer Engineering and Science 31 (4) (1991) 270–274.

[83] H.J. Sue, Study of rubber-modified brittle epoxy systems 2. Toughening mechanisms under mode-I fracture, Polymer Engineering and Science 31 (4) (1991) 275–288.

[84] W.J. Boo, L.Y. Sun, J. Liu, A. Clearfield, H.J. Sue, M.J. Mullins, H. Pham, Morphology and mechanical behavior of exfoliated epoxy/alpha-zirconium phosphate nanocomposites, Composites Science and Technology 67 (2) (2007) 262–269.

[85] H.J. Sue, K.T. Gam, N. Bestaoui, A. Clearfield, M. Miyamoto, N. Miyatake, Fracture behavior of alpha-zirconium phosphate-based epoxy nanocomposites, Acta Materialia 52 (8) (2004) 2239–2250.

[86] H.J. Sue, K.T. Gam, N. Bestaoui, N. Spurr, A. Clearfield, Epoxy nanocomposites based on the synthetic alpha-zirconium phosphate layer structure, Chemistry of Materials 16 (2) (2004) 242–249.

[87] ABAQUS® User's Manual, Ver. 6.4. ABAQUS® Inc., Pawtucket, Rhode Island, 2003.

[88] E.M. Arruda, S. Azhi, Y. Li, A. Ganesan, Rate dependant deformation of semi-crystalline polypropylene near room temperature, Journal of Engineering Materials and Technology 119 (1997) 216–222.

[89] H.J. Sue, R.A. Pearson, A.F. Yee, Mechanical modeling of initiation of localized yielding under plane stress conditions in rigid-rigid polymer alloys, Polymer Engineering and Science 31 (11) (1991) 793–802.

[90] H. Jiang, G.T. Lim, J.N. Reddy, J.D. Whitcomb, H.J. Sue, Finite element method parametric study on scratch behavior of polymers, Journal of Polymer Science Part B: Polymer Physics 45 (12) (2007) 1435–1447.

[91] M.M. Hossain, H. Jiang, H.J. Sue, Effect of constitutive behavior on scratch visibility resistance of polymers – a finite element method parametric study, Wear 270 (11–12) (2011) 751–759.

[92] M.M. Hossain, H. Jiang, H.J. Sue, Correlation between Constitutive Behavior and Scratch Visibility Resistance of Polymers – A Finite Element Method Parametric Study, SPE Automotive TPO Conference, 2011, Troy, Michigan.

[93] M.M. Hossain, R. Browning, R. Minkwitz, H.J. Sue, Effect of asymmetric constitutive behavior on scratch-induced deformation of polymers, Tribology Letters 47 (2012) 113–122.

# Wear and scratch damage in polymer nanocomposites

# 16

**Aravind Dasari\*, Zhong-Zhen Yu[†], Yiu-Wing Mai[‡]**

*\*School of Materials Science & Engineering, Nanyang Technological University, Singapore,
[†]State Key Laboratory of Organic-Inorganic Composites, Department of Polymer Engineering,
College of Materials Science and Engineering, Beijing University of Chemical Technology,
Beijing, China, [‡]Centre for Advanced Materials Technology (CAMT), School of Aerospace,
Mechanical and Mechatronic Engineering, The University of Sydney, Sydney, NSW, Australia*

## CHAPTER OUTLINE HEAD

## 16.1 BACKGROUND

In our previous chapter on "Wear and Scratch Damage in Polymer Nanocomposites" (Chapter 16 of first edition), a review of wear/scratch damage processes occurring in various polymer nanocomposite systems was presented and the parameters responsible for controlling the surface integrity and material removal were deduced and described [1]. In short, we have pointed to the inherent tribological complexities with polymer nanocomposites and the qualitative nature of specified/identified mechanisms. Results varied widely from study to study with only subtle changes of testing conditions, material/filler or even characterization techniques. Considering the complexities involved in quantitatively evaluating the tribological response of polymer nanocomposites, limited efforts have been made to model stress fields induced by different slider geometries, and no explicit correlations between material parameters and wear/scratch damage, particularly for polymer nanocomposites at nanoscale, were available.

Tribology of Polymeric Nanocomposites. http://dx.doi.org/10.1016/B978-0-444-59455-6.00016-7

Further, it was suggested that the presence of nanoparticles by themselves or the improved material mechanical properties do not always result in improvements in wear performance [1,2]. However, the formation of uniform and stable transferred materials adhered to the counterface as a result of tribochemical reactions or by some other means between the fillers and the slider counterface was a governing wear response factor. Other reasons were also given including the generation of free radicals during the sliding process, polishing of the counterface slider by fillers, and wear debris acting like a lubricant or third-body in reducing the wear rate by a rolling action. Specifically, during scratching of polymeric materials, the yield stress was considered as the main parameter influencing the scratch depth [3–5]. Moreover, the coefficient of adhesive friction and strain at stress recovery were found to affect the scratch depth and shoulder height. Besides, all these reasons were directly or indirectly influenced by size, volume fraction, geometry, orientation and dispersion of nanoparticles in the polymer matrix. In this chapter, we will update the status of the advancements/knowledge in the field in the past 5 years. For prior knowledge and background on this subject area, interested readers are encouraged to consult our previous contributions [1,2].

## 16.2 WEAR/SCRATCH DAMAGE IN POLYMER NANOCOMPOSITES

As mentioned in our earlier review, added to the coupling of inherent tribological complexities, poor characterization of nanocomposites and their properties often made interpretations of experimental results more difficult. This was even the case with some of the recent studies. However, there are many others that revealed some interesting results with novel approaches. Below, we review both types of studies and delineate the issues of importance.

### 16.2.1 Effect of coefficient of friction

Analogous to the approach of self-healing of epoxy-based composites by reinforcing with resin-filled capsules/fibers, Zhang and coworkers [6] adopted an approach by adding lubricant oil-loaded microcapsules (8 phr) into hybrid epoxy composites (containing 1 phr short carbon fibers and 5 phr silica ($SiO_2$) nanoparticles). It was expected that during wear/scratch conditions, the capsules would break, releasing the oil in the contact area, and thereby reducing the frictional coefficient and material loss. As shown in Fig. 16.1, the presence of oil-loaded microcapsules alone has a significant effect on wear resistance (increased by 60 times) and coefficient of friction (reduced by ~75%) compared to neat epoxy; additional presence of short carbon fiber and/or silica nanoparticles has little effect on tribological properties. However, flexural strength and modulus of these hybrid composites improved considerably compared to epoxy with oil-loaded microcapsules. But, these improvements are negligible when compared to neat epoxy. Although this approach seems to yield positive results on tribological properties, the probability of fracture of oil-loaded

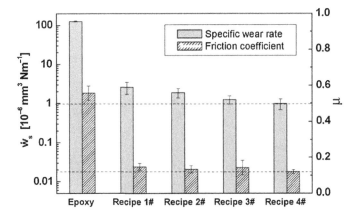

**FIGURE 16.1**

Specific wear rate and friction coefficient of epoxy and its composites. Sliding wear test conditions: block-on-ring apparatus, carbon-steel ring, constant velocity ~0.42 m s⁻¹ and a constant pressure ~3 MPa. Recipe #1, 8 phr oil-loaded capsules; recipe #2, 8 phr oil-loaded capsules and 5 phr $SiO_2$; recipe #3, 8 phr oil-loaded capsules and 1 phr short carbon fiber; and recipe #4, 8 phr oil-loaded capsules along with 5 phr $SiO_2$ and 1 phr short carbon fiber.

microcapsules during the manufacturing process of the composites is large. This in turn results in localized pockets of oil in the matrix, which might affect not only the processability of these materials but also some other properties.

Similarly, some studies have also noted that to improve wear/scratch resistance, addition of slip agents (mainly derived from fatty acids) in polymers is the best approach. These compounds generally "bloom" or migrate to the surface of the polymer (during the processing step) [7,8] forming a waxy layer that lowers the coefficient of adhesive friction. However, migration will be hindered in the additional presence of other fillers, as slip agents tend to adsorb on the surface of the former. Browning et al. [9] noted that the incorporation of slip agents like erucamide (Fig. 16.2) to thermoplastic olefin (TPO) resulted in a significant decrease in the sliding friction coefficient during scratching, thereby improving the scratch resistance. We have also used stearic acid (SA) (octadecanoic acid, $CH_3(CH_2)_{16}COOH$, Fig. 16.2), a saturated fatty acid with a long hydrophobic (aliphatic) tail $(CH_3(CH_2)_{16})$ and a hydrophilic head (–COOH), as a slip agent in polypropylene (PP) and PP/(calcium carbonate) $CaCO_3$ systems [10]. In neat PP, the coefficient of friction was similar at all the wear testing conditions studied. This was attributed to similar wear mechanisms operating under all testing conditions (owing to frictional heating). When 1.5 phr of SA was incorporated, friction was reduced (and also specific wear rate) irrespective of the presence of $CaCO_3$ or its content. For example, at a sliding speed of 0.1 m s⁻¹ and a load of 7 N, the friction coefficient for neat PP was 0.296 ± 0.01, while for PP with unmodified 15 wt%

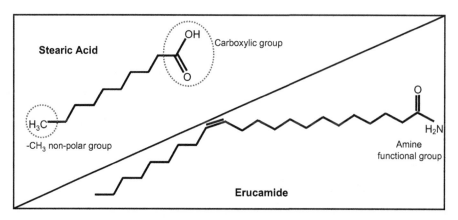

**FIGURE 16.2**

Chemical structures of stearic acid and erucamide. For color version of this figure, the reader is referred to the online version of this book.

CaCO$_3$, PP/SA blend (1.5 phr SA), and PP with 1.5 phr SA-treated CaCO$_3$ are 0.329 ± 0.01, 0.237 ± 0.007, and 0.241 ± 0.005, respectively.

Ha et al. [11] noted that friction coefficients and wear volume losses of epoxy/clay nanocomposites at low loadings of 2–6 wt% were higher than that of neat epoxy. When the loading increased to 10 wt%, significant reductions in both frictional coefficient and wear loss were noted (Fig. 16.3). They believed that the clay layers acted as a lubricating material with increase in clay concentration. This result cannot be clarified as even the clay layers were not (organically) modified in this study and actually negates the idea of a surfactant acting as a lubricating/slip agent. Besides, why this phenomenon occurs only at 10 wt% is unclear.

In another study, the concept of (debonded) nanoparticles acting as rolling balls in retarding the wear and coefficient of friction was extended to even scratching of a polymer nanocomposite system [12]. A polycarbonate (PC)/SiO$_2$ system was scratched with loads of 500, 1000, 2000, and 3000 μN to a distance of 10 μm. As frictional coefficients and scratch penetration depths of the composites are lower than those of neat PC under all testing conditions investigated (Fig. 16.4), the authors suggested that spherical-like SiO$_2$ nanoparticles detached from the matrix during the scratching process and acted as rolling balls between the counterface slider and surface. It is hard to imagine that this process can occur even during a one-step scratching process, as generally, this was suggested to occur during the wearing process after the surface material disintegrates over a period of time. It is also important to note that although several studies on wearing of polymer nanocomposites have used this concept to explain their results, no clear evidence was provided to date to substantiate this. For instance, in a study on the tribological properties of epoxy nanocomposites, Chang et al. [13] proposed a positive rolling effect of the nanoparticles to interpret the remarkable reduction in the frictional coefficient after the addition of nano-TiO$_2$.

Pendleton et al. [14] used fullerene as an additive to biofluids and investigated the lubrication response of total joint replacements (TJRs). This is to develop an alternative

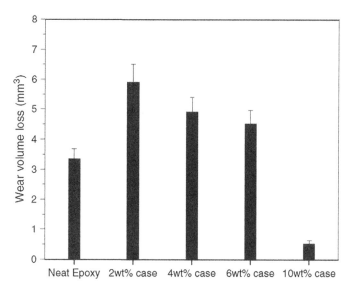

**FIGURE 16.3**

Wear volume loss of epoxy and its nanocomposites at different clay loadings. Wear testing conditions: counterface slider, carbon-steel ball (~12.7 mm in diameter); sliding distance, 3251 m; applied load, 10 N; and sliding speed, 0.12 m s⁻¹.

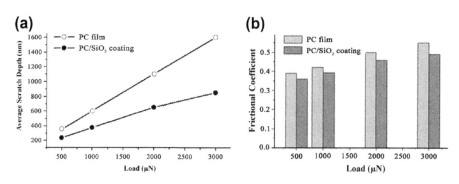

**FIGURE 16.4**

(a) Average scratch depth and (b) coefficient of friction of neat PC and PC/SiO₂ nanocomposite at different applied loads.

to the commonly used in vivo lubrication with periprosthetic synovial fluid, which has to be done for many years after the surgery to restrict TJRs deterioration via wear. In the study, the authors also compared the wear performance of fullerene with a ringlike molecule, crown ether. These additives were simply added to deionized water to make ~37.6 wt%, or 27.2 vol% and 24.9 vol% of crown ether and fullerene, respectively. Tribological testing was done in a reciprocal motion using a ball-on-disc tribometer, with a Ti-6Al-4V ball against a pure titanium disc. During the tribological experiment, 1 ml of fullerene or crown ether-containing fluid was applied to the substrate using a pipette. Results indicated that fullerene provided the lowest coefficient of friction under the testing conditions (sliding speeds varying from 0.1 to 0.5 cm s$^{-1}$, track length 1 cm and applied loads 1, 4, and 5 N) (see Fig. 16.5(a)). They further performed transmission electron microscopic (TEM) investigations of crown ether and fullerene particles before and after wear tests to understand the reasons behind the results and found that crown

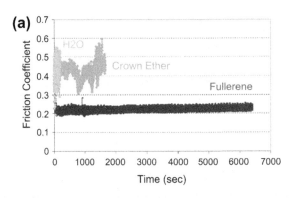

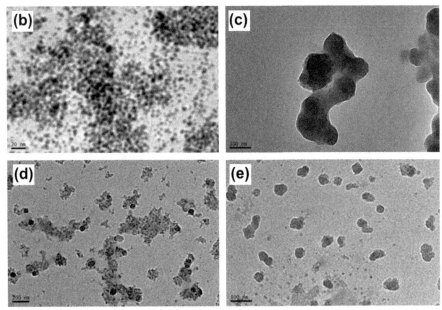

**FIGURE 16.5**

(a) Coefficient of friction for the three fluids tested over a period of time. (b–e) TEM images of additives before (b, d) and after (c, e) wear testing; (b, c) crown ether and (d, e) fullerene. For color version of this figure, the reader is referred to the online version of this book.

ether molecules were adhered together during the wearing process, plausibly due to friction (Fig. 16.5(b), (c)), while fullerene particles were separated (Fig. 16.5(d), (e)). However, it is not clear how crown ether particles aggregated and why this affects the tribological properties as they are still in the size range of 100–400 nm compared to 50–200 nm sizes of fullerene. Moreover, the slider diameter was 6 mm.

## 16.2.2 **Evaluation of scratch/wear resistance**

Liu et al. [15] correlated the dielectric properties of high-density polyethylene (HDPE)/carbon nanofiber (CNF) nanocomposites to the magnitude of wearing, that is, using dielectric response as a reflection of dipole movements (in insulating materials), which is very sensitive to the polar groups in polymers. Evidently, this is system dependent and only for qualitative comparison; any attempt to quantitatively evaluate the performance will result in misleading conclusions. For instance, it is hard to envelop the effects of frictional heat, ups/downs in the wear coefficient with time, and lubrication among other factors. Nevertheless, to enhance the interaction of CNF with HDPE, silanization was used, which also provides a nonpolar hydrocarbon layer on top of the nanofiber surface. During the wear process, it was believed that cutting some polymer chains and separating nanofibers from the polymer matrix, as well as causing damage to the surface of the nanofiber, would create polar groups. Therefore, with longer wear time, the dielectric response could be stronger since more wear would result in more polar groups on or near the surface. To clarify this, the authors also varied the thickness of silane coating on nanofibers and obtained reasonable correlation with permittivity (see Table 16.1). Figure 16.6 also indicates that the permittivity increases linearly with increasing wear coefficient. In another approach, plasticity index obtained via indentation was used as an indicator of scratch penetration depth. This index is defined as $\psi = W_{ir}/(W_{ir} + W_r)$, where $W_{ir}$ and $W_r$ represent the irreversible work done during the indentation and the reversible work recovered by viscoelastic processes during the unloading, respectively (see Fig. 16.7(a)). Therefore, the lower the plasticity index, the higher the elasticity recovery. To illustrate this, epoxy/silica nanocomposite was used as an example, for which the plasticity index values at different loadings are shown in Fig. 16.7(b) along

**Table 16.1** Effect of Silane Coating Thickness on CNFs versus Permittivity of the Nanocomposites Before and After Wear Testing

| Samples | Thickness of Silane Coating on CNF Surface (nm) | Permittivity Before Wearing | Permittivity After 120 h of Wear Process |
|---|---|---|---|
| Nanocomp-ox | 0 | 5.2 | 6.3 |
| Nanocomp-A | 1.2 | 4.8 | 5.3 |
| Nanocomp-B | 2.8 | 4.6 | 4.9 |
| Nanocomp-C | 46 | 4.1 | 4.4 |

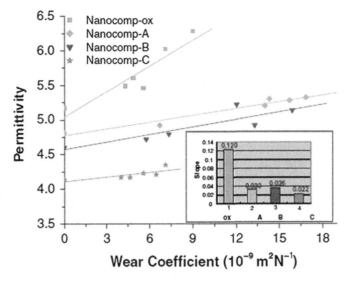

**FIGURE 16.6**

Permittivity trends at $1 \times 10^3$ Hz versus corresponding wear coefficient of HDPE/CNF nanocomposites with untreated and treated CNFs. Compositions and designations of materials are identified in Table 16.1. For color version of this figure, the reader is referred to the online version of this book.

with the scratch penetration depths (Fig. 16.7(c)). About 8% of $SiO_2$/epoxy samples exhibited the lowest plasticity index and correspondingly better scratch resistance. Clearly, this approach cannot accommodate the frictional heat (tangential force) and distinctive changes in scratch damage mechanisms.

Devaprakasam et al. [16] compared the tribological properties (i.e. friction and wear) at the nanoscale for a nanocomposite and a microcomposite based on a resin matrix consisting of various monomeric dimethacrylates, bisphenol-A-glycidyldi-methacrylate, triethylene glycoldimethacrylate, and urethane dimethacrylate. Both nano- and microscale $SiO_2$ particles at a volume fraction of 56% were used. They noted that the average friction force was always higher for the microcomposite than the nanocomposite for a given applied load (for a small scan area $1 \times 1$ $\mu m^2$ using the lateral force microscopy mode of atomic force microscopy, AFM). The wear depths of both composites increased with increasing number of scans (Fig. 16.8(a)); a significant increase was noted with the microcomposite compared to the nanocomposite (wear depths were measured after scanning $10 \times 10$ $\mu m^2$ area with a load of 15 $\mu N$). In addition, based on the AFM phase images, they noted that energy dissipation of the microcomposite during the wear process was higher due to its high surface heterogeneity (Fig. 16.8(b) and (c)); that is, inelastic collisions of an AFM tip with sharp edges on the surface resulted in higher phase differences than the nanocomposite. However, correlating the phase differences to the energy dissipation is not as direct

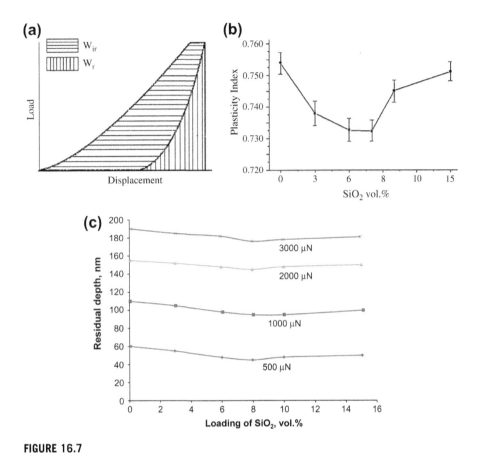

**FIGURE 16.7**

(a) Schematic showing the calculation process of plasticity index, (b) plasticity index and (c) residual depth of epoxy/silica nanocomposites at different loadings of silica. For color version of this figure, the reader is referred to the online version of this book.

as claimed by the authors. Phase differences are always expected even for a simple polymer blend whether it is ductile or brittle. Even the protrusions of the micropar- ticles (as they are not uniformly covered with resin) from the composite surface could result in significant phase differences. For the nanocomposite, considering the size, it is expected that the nanoparticles (even on the surface) were well covered by the resin and might not result in such large phase differences.

Chang et al. [17] used poly(2-hydroxyethyl methacrylate) as a matrix and silica nanoparticles modified by a silane, 3-trimethoxysilyl propyl methacrylate, to prepare hard hybrid coatings. To reduce particle–particle aggregation, before incorporation of silane-modified nanoparticles in the matrix, they were further modified by a capping agent, trimethyethoxylsilane (TMES) to consume the Si-OH groups on their surface. They used polyethylene terepthalate as a substrate for these ultra violet-curable coatings.

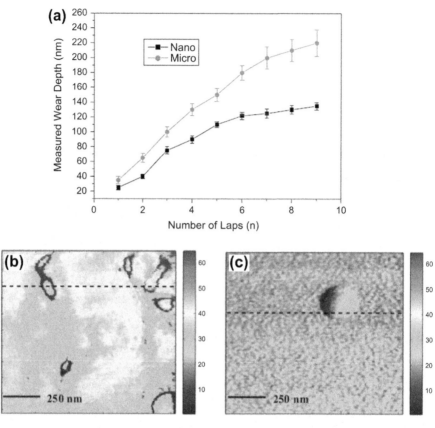

**FIGURE 16.8**

(a) Maximum wear depth versus the number of scanning laps for micro- and nanocomposites; (b and c) phase images of micro- and nanocomposites, respectively. For color version of this figure, the reader is referred to the online version of this book.

However, it is surprising to note that the pencil hardness for all hybrid coatings is the same irrespective of the loadings of silica and TMES (a value of 4H). Although the authors suggested increased cross-linking as a possible reason for this result, the substrate effect could not be excluded. As the coating thickness is only 15 μm, the application of pencil hardness for characterization is questionable. They have also conducted the Taber abrasion test, but only tested the transmittance after the test and revealed that similar transmission results under all conditions (varied from 66 to 70%) including the neat matrix. Similarly, Yao et al. [18] prepared water-soluble polyurethane-based $CaCO_3$ coatings on glass slides, and based on their pencil hardness tests (all the values are essentially between 2H and 3H), they concluded that reinforced coatings are scratch resistant. This again reiterates the need for appropriate application of testing techniques to delineate the differences among the materials and/or rank candidate materials. Seo and Han [19] also reported that the pencil hardness of acrylate coatings increased with the percentage loading of silane coupling agent-modified silica. They attributed this to the chemical bonding between the particles and the matrix. However, their resistance to penetration decreased with more than 30 phr of silica nanoparticles.

Similarly, Lin and Kim [20] noted that the friction coefficient of 10-µm-thick polymethyl methacrylate (PMMA) and polystyrene (PS)-based silica composite coatings on silicon substrate increased by more than 60% with silica loading (up to 0.3 wt%) in all composites independent of the matrix. It was also noted that the depth of the worn groove reduced in PMMA/silica composites and increased in PS/silica composites compared to their corresponding neat polymers. However, no specific details on the depth or width of the worn groove were listed, which makes it even harder to compare between samples. Besides, these differences were attributed to the compatibility and interfacial bonding variation of silica with PMMA and silica with PS; however, no supporting information was presented to validate the proposition.

### 16.2.3 Surface roughness

Surface roughness is another parameter that affects the surface stresses and frictional properties and thereby the scratch/wear response of polymeric materials. When two surfaces slide against each other, the actual contact is between the asperities of the two surfaces. Junctions will be formed at these contact regions due to physical or chemical interactions possibly caused by heating at the interface [21]. Depending on the contact surface area of the two interacting surfaces, the stress state and the resulting deformation can change substantially during testing. Several attempts were made to model the effect of surface roughness on the contact stresses during sliding. Comprehensive theories and models of surface roughness effects on friction and wear, such as Greenwood–Williamson (G–W) model [22], Majumdar–Bhushan (M–B) model [23], and Cantor set contact models [24,25], have been proposed. The G–W model is based on the assumption that all the asperities have the same radius of curvature, while the M–B model is fractal based on elastic and plastic contact between rough surfaces. Contact models (based on Cantor sets) are also developed on fractal characterization of surfaces aiming to quantify.

In general, the deformation undergoes a transition from an elastic regime to a plastic regime as the applied load increases [26]. To account for this transition, a parameter scratch coefficient of friction (SCOF) was introduced. It is the ratio of the tangential force experienced by the scratch tip during scratching to the applied normal force similar to the conventional coefficient of sliding friction. SCOF is divided into two parts [9,27,28] such that SCOF = $(\mu_s + \mu_r)$, where $\mu_s$ is the surface sliding coefficient of friction (dependent on surface adhesion and contacts) and $\mu_r$ accounts for material deformation mechanisms, such as ploughing, cracking and crazing. As expected, at lower applied loads, the magnitude of scratch-induced deformation will be subtle and $\mu_s$ dominates. At higher loads, an additional contribution from $\mu_r$ should be accounted for due to the inherent material resistance to deformation (Fig. 16.9).

Sue and his coworkers [26] correlated surface roughness, applied loads and scratch visibility for a model TPO. They noted that surface roughness was less important at higher

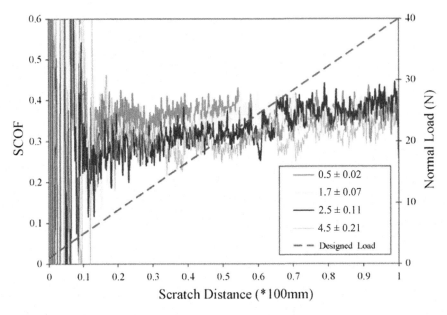

**FIGURE 16.9**

SCOF versus applied load for a TPO system with differing surface roughness. For color version of this figure, the reader is referred to the online version of this book.

loads where $\mu_r$ dominates. More importantly, the onset of scratch visibility became convoluted as the background average roughness increased. For a given surface background, scratch visibility is an indication of the first position at which the contrast is sufficient to be observed. Therefore, if the surface scatters more visible light and introduces noise, it will mask the onset point. This was also confirmed by studying a textured surface, like the "random animal skin", which hid the true onset of visible surface damage.

## 16.2.4 Anisotropic response

Considering the asymmetric response of some of the polymers in compression and tension, many of the tribological properties were found to vary with the direction of sliding. Sue and coworkers [3] reported that this behavior will significantly influence the scratching response of polymers as the stress field evolves with the indenter movement during a progressive load scratch test. For this purpose, different model systems were considered with variations in tensile and compressive behaviors (see Table 16.2). Figure 16.10 shows comparisons of shoulder height and scratch depth as a function of the scratch normal load. In the first case, three systems were compared with similar compressive properties, but differ in their tensile properties. Irrespective of these differences, as shown in Fig. 16.10(a) and (b), the shoulder height and scratch depth of the materials remain essentially the same. In the second scenario, two systems were compared with similar tensile strength values and different compressive yield strengths. Clearly, the lower compressive yield strength of PC compared to styrene acrylonitrile (SAN) 19 makes it more susceptible to

**Table 16.2** Tensile and Compressive Properties of Selected Model Systems

| Property | SAN 19 | SAN 27B | SAN 27C | Neat PC |
|---|---|---|---|---|
| Tensile modulus, GPa | 3.4 ± 0.0 | 3.7 ± 0.1 | 3.7 ± 0.1 | 2.3 ± 0.0 |
| Tensile strength, MPa | 68.9 ± 1.5 | 75.1 ± 3.0 | 79.0 ± 1.0 | 65.2 ± 0.0 |
| Compressive modulus, GPa | 3.5 ± 0.1 | 3.6 ± 0.3 | 3.5 ± 0.2 | 2.1 ± 0.1 |
| Compressive yield strength, MPa | 117.6 ± 0.8 | 117.2 ± 0.4 | 117.2 ± 0.2 | 75.3 ± 0.7 |

*The numbers after SAN represent the weight percentage of AN; 27B and 27C differ in molecular weight, 119 and 134 kg mol, respectively.*

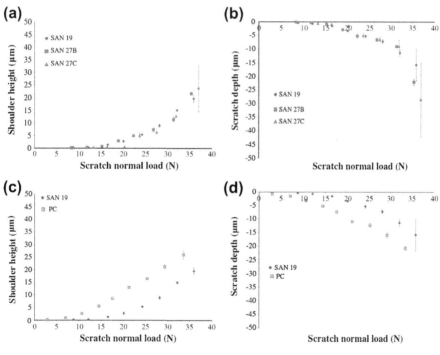

**FIGURE 16.10**

Comparisons of (a, c) shoulder height and (b, d) scratch penetration depth of neat PC and other model systems as a function of applied normal load illustrating the importance of compressive properties. For color version of this figure, the reader is referred to the online version of this book.

scratch damage resulting in higher shoulder height and deeper scratch depth. This suggests that shoulder height and scratch depth are predominately affected by compression loading properties. However, it should also be noted that SAN 19 exhibited brittle behavior and neat PC yielded a ductile behavior under uniaxial tensile testing conditions.

In the case of nanocomposites, recently, we have shown that even the orientation and extent of intercalation of clay layers are important parameters influencing the magnitude of scratch damage [29]. Residual depths were lower for scratches performed on the cross-sections (normal to the flow direction) of the nanocomposites compared to those on the surface (parallel to the flow direction), and greater scratch penetration resistance was noted for the nanocomposite with higher intercalation extent of organoclay (Fig. 16.11). This clearly suggests that the orientation and dispersion

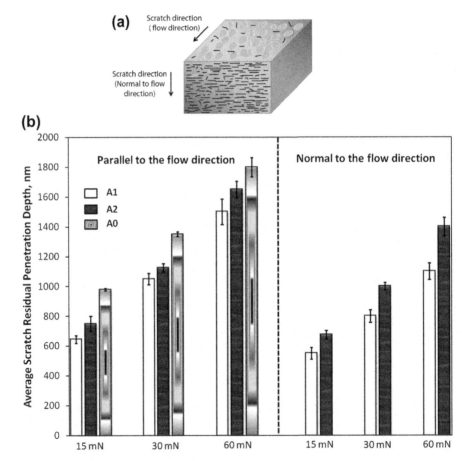

**FIGURE 16.11**

(a) Schematic of two different locations on an injection-molded sample (parallel and normal to flow direction) where scratch tests are performed showing the orientation of clay platelets at both locations and (b) average scratch residual penetration depth for neat polyamide (PA) 6 (A0), exfoliated PA 6/organoclay (90/10) nanocomposite (A1) and intercalated PA 6/organoclay (90/10) composite (A2) at different loads parallel and normal to the injection-molding direction using a Berkovich indenter. For color version of this figure, the reader is referred to the online version of this book.

**Table 16.3** Instantaneous and Residual Penetration Depths and Viscoelastic Recovery of PC and PC/ZnO Composite Along Longitudinal and Transverse (Sliding) Directions

| Material | Sliding Direction: Longitudinal | | | Sliding Direction: Transverse | | |
|---|---|---|---|---|---|---|
| | Penetration Depth ($\mu$m) | Residual Depth ($\mu$m) | Viscoelastic Recovery (%) | Penetration Depth ($\mu$m) | Residual Depth ($\mu$m) | Viscoelastic Recovery (%) |
| PC | 292.8 | 37.5 | 87.7 | 130.2 | 48.8 | 62.5 |
| PC/ZnO | 157.3 | 45.9 | 70.8 | 143.1 | 59.4 | 58.5 |

*Viscoelastic recovery* = $[1-(R_h/R_p)] \times 100$; where $R_p$ instantaneous penetration depth and $R_h$ is the residual depth (measured after 2 min of testing).

of clay layers must be simultaneously considered in determining the effective structural reinforcement in polymer/clay nanocomposites. Otherwise, the results can be misleading. In another similar study, Bermudez et al. [30] also showed that neat PC exhibited an anisotropic behavior during scratching; that is, the scratch response was dependent on the direction of scratching. It was observed that the instantaneous penetration depth for neat PC increased by almost 55% in the longitudinal direction compared to the transverse direction under progressive loading (0.03–29 N) scratch test at the highest applied load (Table 16.3). However, in contrast to the above-discussed study [29], with the incorporation of ZnO nanoparticles, the anisotropic behavior was minimal, and a mere 9.9% increase was noted under similar conditions. Even the residual penetration depth varied with the sliding direction, but in this case, both PC and PC/ZnO materials showed a similar variation of ~23% increase in the transverse direction compared to the longitudinal direction. These differences were attributed to the variations in viscoelastic recovery of PC and PC/ZnO composite in the longitudinal and transverse directions. However, this reasoning cannot answer the differences in the instantaneous penetration depths ($R_p$) between longitudinal and transverse directions, which are significant.

## 16.3 COATINGS

Generally, in a metal–metal contact (such as drilling and cutting tools) and in dry sliding systems, to prevent the abrasive and adhesive wear of the softer material by the harder material, application of functional coatings is one of the common techniques utilized. With much advancement in coatings technology, deposition of films (or multilayered films) with tailored properties is achievable. Many such coatings were developed depending on the specific purpose; soft coatings were chosen particularly for lubricating purposes and hard (and superhard) coatings for load bearing as well as sacrificial layers [31–34]. Examples of hard coating materials include TiN, MoS$_2$, TiC, Al$_2$O$_3$, diamond, and diamond-like carbon (DLC) as single or multilayer combinations. Particularly, DLC coatings have found tremendous applications as they decreased the coefficient of friction and the wear rate by more than an order of magnitude. The presence of hydrogen up to 40 at% in these materials strongly influences their mechanical and tribological behaviors [35]. Depending on the requirement of these coatings, other elements like nitrogen, silver, silicon, and tungsten were also incorporated. In some specific cases, very low coefficients of friction 0.003–0.008 in dry sliding were also achieved [36]. For some applications like magnetic recording discs, a combination of hard DLC and soft polymeric coatings were used.

Cavallin et al. [37] used cold plasma enhanced chemical vapor deposition to treat poly(vinyl chloride) (PVC) and PC surfaces with TiO$_2$ coatings with an objective to increase surface hardness and induce superhydrophilic characteristics. Although hardness was increased (an average value of 0.57 GPa for PVC with TiO$_2$ compared to 0.15 GPa for neat PVC and 0.78 GPa for PC with TiO$_2$ compared to 0.23 GPa for neat PC) and contact angles were reduced to <10° from 65 to 80° for neat materials, there were huge cracks in the coatings (see Fig. 16.12 for TiO$_2$ on PC). The island-like structures clearly suggest the differences in the thermal expansion coefficients of the polymeric materials and TiO$_2$ coating. Besides, the interaction (adhesion) between the substrate and coating is not considered in the study. In another investigation, polyetherimide (PEI), polyether-ether-ketone (PEEK) and PA 12 materials were deposited on low-carbon steel by low-velocity flame spray technology with an optimized spray distance of 85 mm and a preheating temperature of 230 °C [32]. The

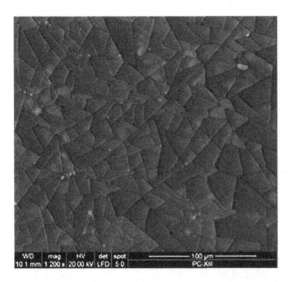

**FIGURE 16.12**

SEM micrograph of an 8-μm-thick TiO$_2$ coating on PC substrate demonstrating the island-like structures.

average coating thickness was ~300 μm. Three-body abrasion tests (abrasive ~250-μm silica dry sand, applied load ~45 N, 6000 revolutions and a sand flow ~400 g min⁻¹) were performed on the coatings, which revealed that the performance of PA 12 coating was far better than the other two. The mass loss value of PA 12 was ~50% of that of PEEK and ~20% less than PEI values. This is despite minor differences in the (tensile) adhesion values and microhardness of the coatings. However, the authors neither provided any reasons for the differences nor studied the mechanisms of wear, which makes it tough to deduce any specific reason for this behavior.

Despite the many advancements in this field, particularly deposition technologies, there are many issues to be addressed. Some studies have already been directed to understand the significance of properties of coating materials such as elastic modulus, hardness or shear strength and fracture toughness as well as other parameters like critical load in scratching [38]. However, there are many other parameters that should be completely evaluated. For inorganic hard coatings (on inorganic substrates), intrinsic stresses are critical in many studies resulting in failure. These are a combination of stresses induced during the deposition/growth process of coatings and due to the mismatch of coating–substrate thermomechanical properties [39]. For polymeric substrates, parameters like chemical inertness, thermal resistance and structural weakness to ultraviolet radiation and ion bombardment were also important. As they change with the inherent properties of the polymer, the effects of these parameters vary from system to system. Another important facet is the adhesion of coatings and substrates. For polymer–polymer systems, many approaches were already established like the chemical functionalization through the use of reactive groups or the creation of a broad interface region that enables the realization of covalent bonds. However, for polymer-inorganic systems, as mentioned before, there are still many questions to be answered.

## 16.4 **CONCLUDING REMARKS**

As discussed in this chapter, in some of the contributions, novel approaches were adopted to tackle the issues of friction and wear, while some dealt with enhancing our understanding of the existing tribological knowledge of polymer nanocomposites. Unfortunately these contributions are limited in number, and the majority of the studies, although considered issues of serious concern, failed either at the approach level or in providing a deeper insight into the wear/scratch/friction mechanisms. This in turn has indirect impacts, particularly when dealing with biomedical applications, as more research is diverted into bioceramics and biocompatible polymer nanocomposites that experience tribological contacts at different length scales. Another new area of promise is the usage of polymer-based nanocomposites as shoe soles, particularly in the sports field. The potential of nanoparticles to enhance adhesion and friction between the materials and surface would be critical in these applications. Obviously the usage of polymer nanocomposites in these different fields under different contact conditions requires the overcoming of several hurdles discussed earlier.

## Acknowledgments

We thank Singapore's Ministry of Education through AcRF Tier 1 grant (RG45/11), the Start-up Grant from Nanyang Technological University, Australian Research Council and the National Natural Science Foundation of China (50873006 and 51125010) for financially supporting our research on various issues of polymer nanocomposites.

## References

[1] A. Dasari, Z.Z. Yu, Y.-W. Mai, Wear and scratch damage in polymer nanocomposites, . (Chapter 16), in: K. Friedrich, A.K. Schlarb (Eds.), Tribology of Polymeric Nanocomposites, vol. 55, Elsevier, Oxford, 2008, pp. 374–399.

[2] A. Dasari, Z.Z. Yu, Y.-W. Mai, Fundamental aspects and recent progress on wear/scratch damage in polymer nanocomposites, Materials Science and Engineering R: Reports 63 (2009) 31–80.

[3] M.M. Hossain, R. Browning, R. Minkwitz, H.-J. Sue, Effect of asymmetric constitutive behavior on scratch-induced deformation of polymers, Tribology Letters 47 (2012) 113–122.

[4] M.M. Hossain, H. Jiang, H.-J. Sue, Effect of constitutive behavior on scratch visibility resistance of polymers: a finite element method parametric study, Wear 270 (2011) 751–759.

[5] H. Jiang, G.T. Lim, J.N. Reddy, J.D. Whitcomb, H.J. Sue, Finite element method parametric study on scratch behavior of polymers, Journal of Polymer Science Part B: Polymer Physics 45 (2007) 1435–1447.

[6] Q.B. Guo, K.T. Lau, M.Z. Rong, M.Q. Zhang, Optimization of tribological and mechanical properties of epoxy through hybrid filling, Wear 269 (2010) 13–20.

[7] A.S. Rawls, D.E. Hirt, M.R. Havens, W.P. Roberts, Evaluation of surface concentration of erucamide in LLDPE films, Journal of Vinyl and Additive Technology 8 (2002) 130–138.

[8] J. Edenbaum, Plastics Additives and Modifiers Handbook, vol. 107, Chapman & Hall, London, 1996.

[9] R. Browning, G.T. Lim, A. Moyse, L. Sun, H.-J. Sue, Effects of slip agent and talc surface-treatment on the scratch behavior of thermoplastic olefins, Polymer Engineering & Science 46 (2006) 601–608.

[10] A. Dasari, On Toughening and Wear/scratch Damage in Polymer Nanocomposites, PhD School of Aerospace, Mechanical and Mechatronic Engineering, The University of Sydney, Sydney, 2007, http://hdl.handle.net/2123/1911.

[11] S.-R. Ha, K.-Y. Rhee, H. Shin, Effect of MMT concentration on tribological behavior of MMT/Epoxy nanocomposite, Journal of Nanoscience and Nanotechnology 8 (2008) 4869–4872.

[12] Z.Z. Wang, P. Gu, Z. Zhang, Indentation and scratch behavior of nano-$SiO_2$/polycarbonate composite coating at the micro/nano-scale, Wear 269 (2010) 21–25.

[13] L. Chang, Z. Zhang, C. Breidt, K. Friedrich, Tribological properties of epoxy nanocomposites I. Enhancement of the wear resistance by nano-$TiO_2$ particles, Wear 258 (2005) 141–148.

[14] A. Pendleton, P. Kar, S. Kundu, S. Houssamy, H. Liang, Effects of nanostructured additives on boundary lubrication for potential artificial joint applications, Journal of Tribology 132 (2010) 031201.

[15] T. Liu, W. Wood, W.-H. Zhong, Sensitivity of dielectric properties to wear process on carbon nanofiber/high-density polyethylene composites, Nanoscale Research Letters, 2010.

[16] D. Devaprakasam, P.V. Hatton, G. Möbus, B.J. Inkson, Nanoscale tribology, energy dissipation and failure mechanisms of nano- and micro-silica particle-filled polymer composites, Tribology Letters 34 (2008) 11–19.

[17] C.-C. Chang, T.-Y. Oyang, F.-H. Hwang, C.-C. Chen, L.-P. Cheng, Preparation of polymer/silica hybrid hard coatings with enhanced hydrophobicity on plastic substrates, Journal of Non-crystalline Solids 358 (2012) 72–76.

[18] L. Yao, J. Yang, J. Sun, L. Cai, L. He, H. Huang, R. Song, Y. Hao, Hard and transparent hybrid polyurethane coatings using in situ incorporation of calcium carbonate nanoparticles, Materials Chemistry and Physics 129 (2011) 523–528.

[19] J.Y. Seo, M. Han, Multi-functional hybrid coatings containing silica nanoparticles and anti-corrosive acrylate monomer for scratch and corrosion resistance, Nanotechnology 22 (Jan 14 2011) 025601.

[20] L.-Y. Lin, D.-E. Kim, Tribological properties of polymer/silica composite coatings for microsystems applications, Tribology International 44 (2011) 1926–1931.

[21] B.J. Briscoe, B.H. Stuart, S. Sebastian, P.J. Tweedale, The failure of poly (ether ether ketone) in high-speed contacts, Wear 162 (Apr 1993) 407–417.

[22] J.A. Greenwood, J.B.P. Williamson, Contact of nominally flat surfaces, Proceedings of the Royal Society of London: Series A – Mathematical Physical and Engineering Sciences 295 (1966) 300–319.

[23] A. Majumdar, B. Bhushan, Fractal model of elastic-plastic contact between rough surfaces, Journal of Tribology-transactions of the ASME 113 (1991) 1–11.

[24] T.L. Warren, D. Krajcinovic, Random Cantor set models for the elastic-perfectly plastic contact of rough surfaces, Wear 196 (1996) 1–15.

[25] G. Liu, Q.J. Wang, C. Lin, A survey of current models for simulating the contact between rough surfaces, Tribology Transactions 42 (Jul 1999) 581–591.

[26] H. Jiang, R. Browning, J. Fincher, A. Gasbarro, S. Jones, H.-J. Sue, Influence of surface roughness and contact load on friction coefficient and scratch behavior of thermoplastic olefins, Applied Surface Science 254 (2008) 4494–4499.

[27] H. Jiang, R.L. Browning, M.M. Hossain, H.-J. Sue, M. Fujiwara, Quantitative evaluation of scratch visibility resistance of polymers, Applied Surface Science 256 (2010) 6324–6329.

[28] P. Liu, R. Lee Browning, H.-J. Sue, J. Li, S. Jones, Quantitative scratch visibility assessment of polymers based on Erichsen and ASTM/ISO scratch testing methodologies, Polymer Testing 30 (2011) 633–640.

[29] A. Dasari, Z.-Z. Yu, Y.-W. Mai, J.-K. Kim, Orientation and the extent of exfoliation of clay on scratch damage in polyamide 6 nanocomposites, Nanotechnology 19 (2008) 055708.

[30] M.D. Bermúdez, W. Brostow, F.J. Carrión-Vilches, J. Sanes, Scratch resistance of polycarbonate containing ZnO nanoparticles: effects of sliding direction, Journal of Nanoscience and Nanotechnology 10 (2010) 6683–6689.

[31] A. Rempp, A. Killinger, R. Gadow, New approach to ceramic/metal-polymer multilayered coatings for high performance dry sliding applications, Journal of Thermal Spray Technology 21 (2012) 659–667.

[32] C.R.C. Lima, N.F.C. de Souza, F. Camargo, Study of wear and corrosion performance of thermal sprayed engineering polymers, Surface and Coatings Technology, 2012.

[33] X. Guan, L. Wang, The tribological performances of multilayer graphite-like carbon (GLC) coatings sliding against polymers for mechanical seals in water environments, Tribology Letters 47 (2012/07/01) 67–78.

[34] S.K. Field, M. Jarratt, D.G. Teer, Tribological properties of graphite-like and diamond-like carbon coatings, Tribology International 37 (2004) 949–956.

[35] G. Dearnaley, J.H. Arps, Biomedical applications of diamond-like carbon (DLC) coatings: a review, Surface and Coatings Technology 200 (2005) 2518–2524.

[36] J.A. Heimberg, K.J. Wahl, I.L. Singer, A. Erdemir, Superlow friction behavior of diamond-like carbon coatings: time and speed effects, Applied Physics Letters 78 (2001) 2449.

[37] T. Cavallin, N. El Habra, M. Casarin, F. Bordin, A. Sartori, M. Favaro, R. Gerbasi, G. Rossetto, Superhydrophilic and tribological improvements of polymeric surfaces via plasma enhanced chemical vapor deposition Ceramic coatings, Journal of Nanoscience and Nanotechnology 11 (2011) 8079–8082.

[38] D.L. Burris, B. Boesl, G.R. Bourne, W.G. Sawyer, Polymeric nanocomposites for tribological applications, Macromolecular Materials and Engineering 292 (2007) 387–402.

[39] K.H. Lau, K.Y. Li, Y.-W. Mai, Influence of hardness ratio on scratch failure of coatings, International Journal of Surface Science and Engineering 1 (2007) 3–21.

# Polytetrafluoroethylene matrix nanocomposites for tribological applications

# 17

**David L. Burris\*, Katherine Santos\*, Sarah L. Lewis[†], Xinxing Liu[†], Scott S. Perry\*, Thierry A. Blanchet[†], Linda S. Schadler[†], W. Gregory Sawyer\***

*\*University of Florida, FL, USA, [†]Rensselaer Polytechnic Institute, NY, USA*

## CHAPTER OUTLINE HEAD

## 17.1 INTRODUCTION

### 17.1.1 Motivation and organization

Proper lubrication enables the smooth operation of nearly all moving devices we use on a daily basis, from cabinet drawers to automobiles. Traditional fluid and grease

Tribology of Polymeric Nanocomposites. http://dx.doi.org/10.1016/B978-0-444-59455-6.00017-9

lubricants are ideal for many applications and provide characteristically low friction coefficients and wear rates, but in an increasing number of applications, these lubrication techniques may be impractical or even precluded. The reservoirs, pumps and filters required in fluid lubricated systems add cost, size and weight, and grease reapplication may be prohibitive. Fluid and grease lubricants also have inherent environmental limitations. They can be contaminated by dirt and debris and can themselves lead to contamination of the product or the environment. Furthermore, they outgas in low pressure and vacuum applications and function only over limited temperature ranges.

Solid lubricants are often used in applications where fluid and grease lubricants do not provide the required performance. Solid lubricants provide a number of advantages including low cost, simplicity, structural integrity and environmental insensitivity. Most importantly, however, solid lubricants provide lubrication internally and do not require external lubricants for successful operation. Because the low friction mechanism in solid lubricants often involves failure at weak internal interfaces, self-lubricated systems generally exhibit higher friction and wear than fluid- and grease-lubricated systems. Significant research efforts are currently dedicated to reducing friction and wear at self-lubricated interfaces and deducing the mechanisms by which solid lubricants operate.

Typically self-lubricated bearings entail either bulk bushings or slideways of polymers or polymer composites, or thin coatings of lamellar solids such as molybdenum disulphide or graphite. Polytetrafluoroethylene (PTFE) is a common solid lubricant used for its unique combination of low friction, low chemical reactivity and large operational temperature range. Its applications include nonstick frying pan coatings, low-friction seals, oil additives and vascular stents. Despite its promising combination of unique physical properties and widespread use in engineering, the high wear rate of PTFE has greatly limited its use as a solid lubricant. The goal of the work summarized here is to identify mechanisms of wear and wear resistance in PTFE composites in order to guide the future design of low-wear PTFE-based solid lubricants. The remainder of this chapter is organized as follows.

In the introduction, we examine and discuss the PTFE tribology literature of unfilled, filled and nanofilled PTFE-based tribosystems. In the following section, initial findings of a multiuniversity collaboration focusing on the fundamental mechanisms in PTFE-based tribosystems are discussed. The chapter concludes by offering a hypothesized model of PTFE wear resistance that is based on a combination of the literature with findings from the current research. Properties and wear resistance mechanisms of PTFE nanocomposites are discussed and the importance of synergies is highlighted; it is suggested that 100× improvements in wear resistance can be achieved with any of several wear-resistance mechanisms, but 10,000× improvements are only achieved through synergies unique to nanocomposites.

## 17.1.2 Polytetrafluoroethylene as a solid lubricant

PTFE is a linear chain polymer of smooth molecular profile consisting of 20,000–200,000 repeating units of tetrafluoroethylene ($C_2F_4$). The fluorine encasement of the carbon backbone provides high chemical inertness while its smooth profile provides low-friction sliding. PTFE also has a large useful temperature range (4–500 K) and a very low vapor pressure (low outgassing) making it a viable material for solid lubrication in space.

PTFE is viscoelastic in nature and as a result, its tribological properties are strong functions of both sliding speed and temperature. As temperature is reduced or speed increased, the friction coefficient increases [1–6]. In low-speed (<10 mm s$^{-1}$) applications PTFE has a low friction coefficient (between $\mu = 0.03$ and $\mu = 0.1$) and moderate wear resistance ($10^{-5}$ mm$^3$ Nm$^{-1}$). Makinson and Tabor [3] found that as the sliding speed increased to above 10 mm s$^{-1}$ at room temperature, a transition from mild to severe wear (from $10^{-5}$ to $10^{-3}$ mm$^3$ Nm$^{-1}$) accompanied increased friction. They combined the early structural work of Bunn, Cobold and Palmer [7] and Speerschnieder and Li [8] with their own tribological results and electron diffraction work to relate the tribological behavior of PTFE to deformation of its crystalline structure. They hypothesized that at conditions of both low speed (<10 mm s$^{-1}$) and high temperature (>30 °C), shearing occurs in the rate-sensitive amorphous regions giving rise to a lamellar-type response of crystals or molecules. As speed is increased or temperature decreased from an original condition of low friction and moderate wear, the stress required to shear amorphous regions exceeds the stress required to cause failure at boundaries between crystalline regions in the sintered material. They concluded that this leads to larger debris and increased wear rates. Tanaka proposed a similar model with failure occurring at boundaries of the characteristic "banded" structure of PTFE [6]. Blanchet and Kennedy [1] studied this severe wear transition at several temperatures and found an increase in the transition speed to accompany increased temperature. When the wear rate, $k$, was plotted versus the friction coefficient, $\mu$, the transition to severe wear occurred at $\mu = 0.1$ in each case. These results are consistent with the proposed transition mechanism of Makinson and Tabor [3] and suggest that the severe wear transition is a response to the stress state and thus the friction coefficient, while the friction coefficient is a function of both speed and temperature. Several samples were microtomed perpendicular to the wear surface in the direction of sliding after mild and severe wear had taken place. Cracks were found to propagate in the direction of sliding beneath a layer of worked material at subsurface depths consistent with observed debris thicknesses for severe wear samples. No such cracks were found in mild wear samples. They conjectured that defects in the sintered material act as initiated cracks. When speeds are low, the kinetic friction coefficient at the tribological interface is low, and the static friction coefficient between internal crack faces is sufficient to fully support the surface tractions. However, when the kinetic coefficient of friction at the tribological interface increases with increased sliding speed and exceeds the static coefficient of friction ($\mu \sim 0.1$) at internal PTFE/PTFE interfaces, the crack tips must support shear. This leads to a progressive delamination wear process similar to that described in Suh's delamination theory of wear [9]. The severe wear of PTFE at speeds greater than the temperature-dependent transition speed has precluded its use in many applications, and motivates the use of fillers to abate the onset of severe wear.

### 17.1.3 **PTFE-based tribological composites**

For decades, fillers have been successfully used to reduce the wear of PTFE. In Fig. 17.1, wear rate is plotted versus filler weight percentage for testing of some representative PTFE composites found in the literature [10–15]. Despite being tested with varying configurations, testers, methods, pressures, speeds and fillers, there is a systematic trend of decreased wear rate with increased filler fraction up to 50 wt%.

The wear-reducing mechanism of fillers in PTFE-based composites remains a topic of debate. Lancaster [16] proposed that hard wear-resistant fillers, especially those with a high aspect ratio, preferentially support the load, reducing the wear of PTFE in the composite. Sung and Suh [17] found that vertically oriented fibers were most effective in reducing wear, but suggested that the critical role of the filler was to arrest crack propagation, rather than to support the load. Tanaka et al. [6,18] suggested that the filler prevented the initial transfer of the PTFE to the counterface, and thus prevented transfer wear. Briscoe [19] noted the formation of a thin, well-adhered transfer film for a high-density polyethylene system and hypothesized that fillers provide augmented transfer film adhesion, and thus reduced transfer wear by slowing transfer film removal and the requisite replenishment. Using X-ray photoelectron spectroscopy (XPS), Gong et al. [20,21] found that the wear rate of PTFE was independent of chemical bonding with the counterface, and concluded that cohesive failure within the PTFE must govern its wear rate. Blanchet et al. had similar findings with XPS analysis of PTFE and PTFE composites in dry sliding, and concluded

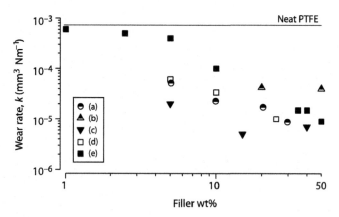

**FIGURE 17.1**

Wear rate versus filler fraction for some of the PTFE-based microcomposite systems found in the tribology literature: (a) Li et al., 2000, graphite; (b) Bahadur and Gong, 1992, graphite; (c) Lu et al., 1995, PEEK; (d) Burroughs et al., 1999, $B_2O_3$; (e) Menzel and Blanchet, 2002, irradiated FEP. Neat PTFE has a high wear rate ($k \sim 10^{-3}$ mm³ Nm⁻¹) at speeds above 10 mm s⁻¹, while composites approach a moderate wear rate ($k \sim 10^{-5}$ mm³ Nm⁻¹) as filler fraction increases above 10 wt%.

that the wear-reducing role of the filler is to slow primary removal of material from the bulk by arresting crack propagation rather than slowing secondary removal of material from the counterface via increased transfer film adhesion [22]. Bahadur and Tabor [23] and Blanchet and Kennedy [1] saw direct relationships between wear rate, debris size and the ease with which debris are expelled from the contact, and concluded that the fillers interrupt the formation of the larger debris that form during severe wear of PTFE.

### 17.1.4 PTFE-based nanocomposites

In practice, PTFE microcomposites are loaded with more than 20% (by volume) filler and because the fillers are inherently hard and wear resistant, abrasion to the counterface can be problematic. Nanoparticles are of the same size scale as counterface asperities and therefore have potential as wear-resistant, nonabrasive fillers. In addition, nanocomposites can have tremendous particle number densities and surface areas with low filler fractions (<10%) of nanoparticles. In various other polymeric systems, low filler fractions of nanoparticles have resulted in impressive improvements in mechanical properties such as strength, modulus and strain to failure [24–27].

In 1996, Wang et al. demonstrated the potential benefits of nanoparticles in tribology using a polyetheretherketone (PEEK) matrix [28–32]. PEEK nanocomposites had optimal wear rates with roughly 2–4% filler fractions of nanoscale $ZrO_2$, $Si_3N_4$ and $SiO_2$ [28–30]. Later experiments with nanoscale SiC showed optimal wear performance at 1% filler fraction [31]. They also demonstrated the superior wear and friction performance of a 10-wt% SiC nanocomposite over 10-wt% SiC micrometer particle and whisker-filled PEEK composites [32]. In all cases, the improvements in friction and wear of the nanocomposite samples were attributed to the tenacity and uniformity of the protective transfer films.

Despite the success of microfillers in abating severe wear of PTFE and the demonstrated benefits of nanoparticles on the properties of other polymer matrices, conventional wisdom suggested that nanoscopic fillers would be ineffective in reducing the wear of PTFE. Tanaka and Kawakami [18] found evidence in support of this in a study varying the filler in a PTFE matrix. They found that the submicrometer-hard particle-filled composite had poor wear resistance compared to the larger hard particle-filled composites in the study, and concluded that fillers must be of sufficient size to be effective. They suggested that submicrometer fillers were unable to prevent transfer and large-scale destruction because they were simply swept away within the matrix during plowing by relatively large counterface asperities.

In 2001, Li et al. [33] filled PTFE with nanoscale ZnO, and found 15 wt% to be the optimal filler fraction for wear resistance while retaining a low coefficient of friction. Chen et al. [34] created a PTFE nanocomposite with single-walled carbon nanotubes and found that the friction coefficient was reduced slightly and wear resistance was improved by more than two orders of magnitude over unfilled PTFE. Sawyer et al. [35] made nanocomposites of PTFE with 38 nm $Al_2O_3$ and found a 600×

reduction in wear with 20 wt% filler concentration. Wear was reduced monotonically as filler concentration was increased to 20 wt%. While similar improvements in wear resistance were found by Burris and Sawyer using 44-nm alumina nanoparticles of the same phase (70:30 Δ:Γ) [36], the use of 80-nm α phase alumina was found to reduce wear by an additional 100× [37]; it was hypothesized that an irregular particle shape was responsible for the vast improvement in tribological performance.

Figure 17.2 shows wear rate plotted versus filler fraction for the PTFE nanocomposites in the literature. In general, microcomposites are optimized at higher filler fractions, while nanocomposites are optimized at lower filler fractions. With as little as 0.4 wt% filler fraction of nanoparticles, wear rate can be reduced by a factor of 10, while negligible reductions in wear are observed with less than 5 wt% filler fraction of microparticles. In addition, the wear rate of PTFE with low filler fractions of nanoparticles can have improved wear resistance over even highly reinforced microcomposites ($10^{-7}$ mm$^3$ Nm$^{-1}$ at 1 wt% versus $10^{-5}$ mm$^3$ Nm$^{-1}$ at 50 wt%).

Initial rules of mixtures and preferential load support models of wear resistance have been inadequate to predict the success of nanofillers at low filler fractions. Early investigations of the dominant wear reduction mechanisms in PTFE nanocomposites focused on strengthening and toughening of the matrix and the transfer films. Li et al. [33] used scanning electron microscopy (SEM) to study cross-sections of unfilled and nanofilled PTFE. The neat PTFE had many fibers drawn from the bulk, while the nanocomposite did not. They suggested that the nanoparticles effectively prevented the destruction of the banded structure. They also found thick, patchy transfer films formed by unfilled PTFE,

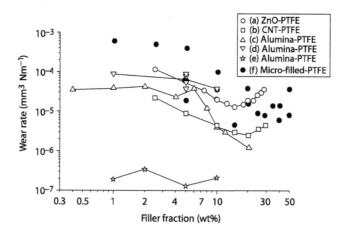

**FIGURE 17.2**

Wear rate plotted versus filler fraction for the PTFE nanocomposites in the tribology literature. The microcomposites from Fig. 17.1 are shown collectively in Fig. 17.2 as filled circles. Data sets are listed as (a–f): (a) Li et al., 2001; (b) Chen et al., 2003; (c) Sawyer et al., 2003; (d) Burris and Sawyer, 2005; (e) Burris and Sawyer, 2006; (f) see references in Figure 17.1. Nanocomposites can have substantially improved wear rates at filler fractions below 10 wt%, but show a large degree of variability between studies.

while thin, tenacious transfer films were formed by the wear-resistant nanocomposite. It was offered that the nanoparticles help bond the transfer film to the counterface, which promotes low wear by protecting the soft composite from direct asperity damage. Chen et al. [34] also found evidence to suggest that the nanotubes prevented destruction of the crystalline structure of the PTFE. The high aspect ratio fillers were thought to reinforce the matrix by intertwining with PTFE crystals. In addition, they hypothesized that the nanotubes may provide additional self-lubrication after breaking off from the composite during wear. In the study by Sawyer et al. [35], SEM revealed that the PTFE particles were decorated by the nanoscopic alumina during a powder blending process that preceded compression molding. The resulting structure after molding was cellular with thin regions of highly concentrated alumina-rich material surrounding micrometer-sized domains of nominally unfilled PTFE. These concentrated regions were hypothesized to act as barriers to crack propagation, reducing the delamination wear of PTFE. Further, it was offered that with increasing filler concentration, the number, size and possibly strength of the compartmentalizing regions increased.

In general, the results from the polymer nanocomposites tribology literature are striking. Contrary to early suggestions, the use of nanoparticle fillers in PTFE has been very successful with 1000× improvements in wear resistance occurring with as little as 1 wt% nanoscale filler. Unfortunately, the literature is seemingly inconsistent with wear rates between studies varying by as much as 1000× for a given filler fraction. With the large number of variables between studies and the qualitative descriptions of transfer films, debris morphology, mechanical properties and most importantly, nanoparticle dispersion, it is difficult to identify the sources of the inconsistency that makes global statements about wear resistance mechanisms difficult. Previous studies suggest that the wear rates of these systems are complex and coupled, possibly involving crack deflection, filler/matrix interactions, regulation of debris size and debris/counterface interactions, but there is a current need for more quantitative measurements to enable application-specific system design. The following studies represent the beginning of an effort to clarify some of the underlying mechanisms that govern the tribology of PTFE nanocomposite systems. The first studies directly examine the effects of filler particle size in PTFE composites. The following studies quantitatively examine the morphological, tribological, compositional and chemical properties of the transfer films and address their influences on the tribosystem. The next series of studies investigates the nature of the matrix/filler interface and its effect on the tribology of the system. The final studies discussed focus on the phase and morphology changes in the PTFE that occur as a result of nanoparticle inclusion, and the effects of these changes on the tribosystem.

## 17.2 CURRENT AND ONGOING STUDIES

### 17.2.1 Investigations of particle size

The hypothesis that filler particles must be of sufficient size to effectively provide wear resistance to PTFE was founded on the observation of wear rates of submicrometer (0.3 μm) $TiO_2$-filled composites exceeding those using fillers at sizes of several micrometers or more of materials such as chopped glass fibers, bronze and $ZrO_2$ [18]. In that study, however, each filler material was tested at a single particle size; thus, the effect of filler particle size was not isolated from any possible effect of filler material type. Despite wearing more rapidly than the glass, bronze and $ZrO_2$-filled composites, the submicrometer $TiO_2$ fillers in several instances did in fact provide wear rate reductions of two orders of magnitude compared to unfilled PTFE and no clear evidence is provided that larger $TiO_2$ fillers would have augmented this wear resistance any further. Li et al. [33] demonstrated the ability of nanofillers to provide wear resistance to PTFE with 50-nm ZnO particles reducing wear rate by nearly 100-fold in some cases. However, no micrometer-scale ZnO was included as a control to quantify this effectiveness relative to larger filler particle sizes. In their demonstration of the ability of an 80-nm alumina filler to provide PTFE wear resistance, Burris and Sawyer [36] also included a 0.5-μm filler for comparison; however, this study still lacked a control within the range from several micrometers to about 30 μm conventionally deemed suitable for PTFE wear resistance [18].

To more clearly assess the effects of filler size on PTFE wear resistance, a study [38] was performed using two alumina nanofillers of size 40 nm (27–43 nm) and 80 nm, as well as four alumina microfillers of size 0.5 μm (0.35–0.49 μm), 1 μm, 2 μm (0.9–2.2 μm), and 20 μm, all being provided by the same commercial manufacturer and being of the same α phase. In some cases the manufacturer specified a particle size range, indicated parenthetically. Each alumina particle type was blended at 5 wt% into commercially available PTFE powder of ~30 μm particle size using a Hauschild mixer, with the powder mixture being cold pressed at 40 MPa into a cylindrical preform, which was sintered in nitrogen at 360 °C. Composite pin specimens (4 mm × 4 mm cross-section) were machined from the sintered preforms and tested in sliding contact with 304 stainless steel counterfaces ($R_a = 0.05$ μm) in air at room temperature. Intermittent pin mass loss measurements were used to quantify composite wear.

The evolution of wear with increasing sliding distance for composite pins produced using a commercial G580 PTFE resin is plotted in Fig. 17.3. These tests were conducted in unidirectional sliding contact on a three-pin-on-disc tribometer at a sliding speed of 10 mm s$^{-1}$ and a contact pressure of 3.1 MPa. As expected, unfilled PTFE wore so rapidly, losing nearly 100 mm$^3$ of wear volume within 1 km of sliding distance, that it falls very tight to the vertical axis. In contrast, the two nanocomposites wore so slowly, requiring sliding distances in excess of 100 km to reach wear volumes on the order of a couple cubic millimeters, that their wear records fall very tight to the horizontal axis. An inset is provided expanding this portion of the plot, so that the nanocomposite wear behaviors may be more clearly seen. The four microcomposites fall within a diagonal band across the plot, clearly providing unfilled PTFE an increase in the wear resistance, although far less than the wear resistance provided by the superior nanocomposites.

At steady state under constant load, the volume loss increases in proportion to the sliding distance such that for each composite a steady state wear rate is quantified and plotted as a function of filler particle size in Fig. 17.4. As compared to the unfilled PTFE value near $0.7 \times 10^{-3}$ mm$^3$ Nm$^{-1}$, the microcomposites each provided

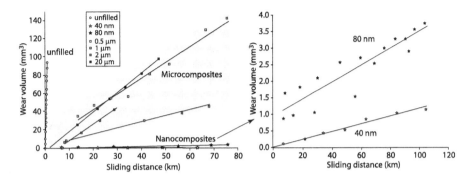

**FIGURE 17.3**

Wear records for unfilled PTFE as well as for PTFE micro- and nanocomposites incorporating α phase alumina filler particles at 5 wt%. While the microcomposites fall within a diagonal band, an inset is provided with a magnified wear volume axis so that the low levels of nanocomposite wear may be seen.

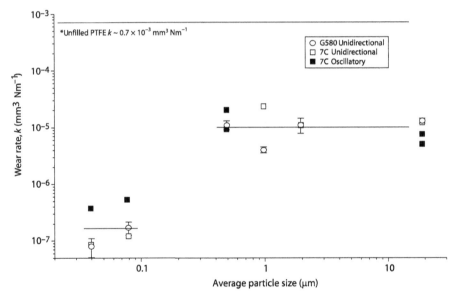

**FIGURE 17.4**

Steady state wear rate of alumina-filled PTFE composites as a function of filler particle size. Composites formed with either G580 or 7C resin and tested in either unidirectional or oscillatory (45 mm stroke length) sliding contact against 304 stainless steel. There is a two order of magnitude reduction in wear rate from moderate wear to low wear as filler size is reduced from 0.5 μm to 80 nm.

wear reductions of nearly two orders of magnitude, with wear rates falling closer to $10^{-5}$ mm$^3$ Nm$^{-1}$. The two nanocomposites provided an additional two orders of magnitude decrease in wear rate, with wear rates near $10^{-7}$ mm$^3$ Nm$^{-1}$. The wear behavior as a function of filler particle size was also duplicated in a series of composites produced using an alternate (7C) commercial PTFE grade, not only in the same unidirectional three-pin-on-disc tribometer but also under oscillatory sliding contact against the polished 304 stainless steel at the same contact pressure and speed using a stroke length of 45 mm [39]. Despite drastically altering the wear behavior of PTFE, the alumina filler particles had little effect on the friction coefficient as the filled and unfilled PTFE all had measured friction coefficients of approximately $\mu = 0.18$ at the 10 mm s$^{-1}$ test speed.

Following less than a kilometer of unidirectional sliding of unfilled PTFE, the countersurface was covered with abundant platelike debris, while the microcomposites required 50 km or more to generate a comparable volume of characteristically finer debris. Despite being given more than 100 km of sliding distance, the nanocomposite wear debris were observed to be both fine and sparsely distributed about the wear track edges. SEM within the wear tracks (not shown) revealed that the microcomposites left abraded grooves and loosely bound debris on the stainless steel counterfaces. The nanocomposites were found not to cause such abrasion but instead left thin transfer films, and even though cracking did appear in the thickest regions of these films, they appeared to remain coherent and well adhered without liberating numerous transfer wear debris.

Clear differences in the transfer morphologies of unfilled, microfilled and nanofilled PTFE after oscillatory sliding can be observed in the mapping stylus profilometric transfer film measurements shown in Fig. 17.5. The unfilled PTFE transferred plates of material with characteristic in-plane dimensions of 1–2 mm and thicknesses of 5–10 μm. The microcomposite transferred substantially smaller platelets with in-plane dimensions of 100–200 μm and thicknesses of 0.5–2 μm. A comparable reduction in transfer size is obtained by replacing microparticles with nanoparticles. The transferred particles of nanocomposite material have in-plane dimensions of 20–50 μm and thicknesses of 50–500 nm.

The SEM image from the nanocomposite pin specimen in Fig. 17.6 shows "mud-flat" cracking throughout an otherwise smooth and coherent surface layer covering the worn pin. Higher magnification imaging (not shown) revealed that fibrillated PTFE spanned these cracks and appeared to stabilize the surface layer against breakdown and wear debris formation. In contrast, the worn microcomposite shows an incomplete flowing surface layer that appears to be flaky and less adherent, transforming into wear debris of larger size. Energy-dispersive X-ray spectra taken from these worn surfaces reveal Kα peaks at energies of 0.85, 1.49 and 6.40 keV, respectively, for F from the PTFE matrix, Al from the alumina filler and Fe from steel particles abraded from the countersurface. Ratios of the heights of these Al and Fe peaks to the F matrix peak can serve as indicators of the relative amounts of alumina filler and steel wear debris upon the composite surface. For unworn composites at 5 wt% alumina, the Al/F ratio is observed to be approximately 0.2. After sliding, this ratio may increase as the filler

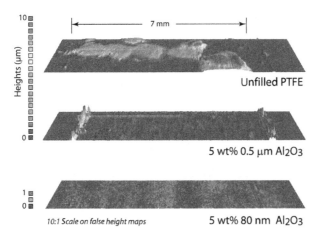

**FIGURE 17.5**

Mapping stylus profilometry measurements of transfer films on polished counterfaces. (Top) Transfer film of unfilled PTFE; (center) 5-wt% 500-nm alpha alumina-filled PTFE transfer film; (bottom) 5-wt% 80-nm alpha alumina-filled PTFE. The stylus has a 12.5 μm diameter tip and measurements were made using a contacting force of 100 μN. (For color version of this figure, the reader is referred to the online version of this book.)

accumulates at the near surface region due to preferential removal of the less wear-resistant PTFE matrix as previously reported for PTFE microcomposites [40]. Following testing, microcomposites were found to accumulate more filler at the surface with Al/F ratios ranging from 1 to 2 while the Al/F ratios of the nanocomposites never exceeded 0.5. Consistent with SEM observations of counterface abrasion, the Fe/F ratios of the microcomposites were greater than 0.2 (often well beyond 0.2), while those of the nanocomposites were less than 0.2. This difference is even more striking when one considers the extended sliding distances of the nanocomposite tests and the fact that the near-surface material of the microcomposites, including the embedded steel from counterface abrasion, is removed at a rate 100 times greater than the near-surface material of the nanocomposites.

An additional test program produced composites having a mixture of nanoparticles (40 nm) and microparticles (20 μm), so that comparison of the wear behaviors to those of composites filled at 5 wt% with only nanoparticles or microparticles might elucidate predominating wear mechanisms. As shown in Fig. 17.7, these mixed-filler composites were produced either with 5 wt% of each filler or with 2.5 wt% of each. In either case, the mixed-filler composites displayed wear rates more near to the $10^{-5}$ mm$^3$ Nm$^{-1}$ microcomposite value than the $10^{-7}$ mm$^3$ Nm$^{-1}$ nanocomposite value. It can be concluded that while microparticles interfere with the severe wear mechanisms that result in large debris and rapid wear of PTFE, they supplant the wear resistance offered by the nanoparticles by providing an otherwise unavailable wear pathway. Presuming the nanocomposite wear mechanism is one of transfer wear, where additional removal of nanocomposite material is activated to replenish

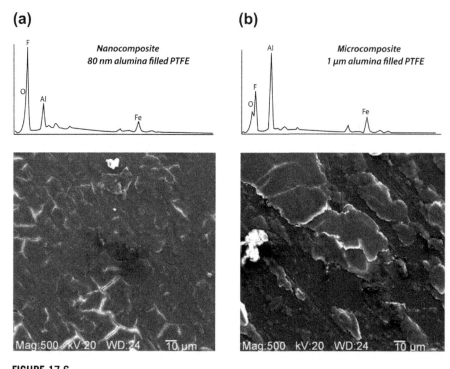

**(a)** **(b)**

**FIGURE 17.6**

Secondary electron images from example wear surfaces of a (a) nanocomposite (80 nm filled) and (b) microcomposite (1 μm filled). EDXS (energy dispersive X-ray spectroscopy) spectra from each are also provided, indicating F from the composite matrix, Al from the composite filler, and Fe from debris abraded off the mating steel countersurface.

detached and discarded transfer films, countersurface abrasion induced by the microparticles would increase wear rate by reducing the residence time of the transfer material. Such a hypothesis would benefit from more focused transfer film investigations on microcomposites and nanocomposites.

## 17.2.2 Investigations of transfer films

Contrary to early suggestions that nanoparticles could not appreciably improve the wear resistance of PTFE, it has been shown that nanofillers can be far superior to microfillers with a transition in the dominant wear reduction mechanism occurring at a particle size approaching 100 nm. Preliminary evidence suggests that reduced counterface abrasion, third body wear and retention of protective transfer films are primarily responsible for the improvements in wear resistance. Thin, uniform transfer films consistently accompany wear resistance in the tribological nanocomposites literature [28–30,33,36,37,41,42], but quantitative measurements of these films are lacking. Some authors suggest that wear resistance of the nanocomposite is due to

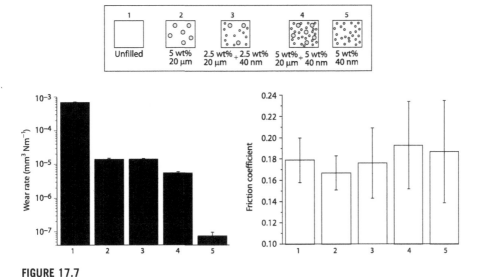

**FIGURE 17.7**

Wear and friction behavior of five different PTFE composite materials sliding against 304 stainless steel, indicating the effect of inclusion of alumina microparticles (20 μm) and nanoparticles (40 nm), as well as mixtures of micro- and nanoparticles. The microparticles disrupt low wear by providing a wear pathway that is not otherwise available in the nanocomposite.

protection from the transfer film while others offer that the films are formed as a consequence of low wear. It is currently unclear why and how these films form; how they facilitate wear resistance; if they are composed primarily of the PTFE, the filler or the composite; and if chemical reactions are involved.

Burris and Sawyer [36] conducted a study with 5-wt% α phase and $\Delta$:$\Gamma$ phase alumina-PTFE nanocomposites against various rough counterfaces to study the effect of asperity size on the transfer and wear of different PTFE nanocomposites. The surfaces were made using different standard finishing techniques and interferometry measurements of these surfaces are shown in Fig. 17.8.

Wear rates for these composites are plotted versus counterface roughness $R_q$ in Fig. 17.9. The different phases of alumina were found to result in widely different tribological properties despite identical processing and testing. Wear rates of $\Delta$:$\Gamma$ alumina nanocomposites increased monotonically from 50 to $300 \times 10^{-6}$ mm$^3$ Nm$^{-1}$ with increased surface roughness. In addition, wear debris were relatively large and transfer films thick and discontinuous. Wear rates of α alumina nanocomposites did not correlate with roughness and were significantly lower than those of the $\Delta$:$\Gamma$ nanocomposites ranging from $0.8^{-10} \times 10^{-6}$ mm$^3$ Nm$^{-1}$. Wear rates from tests conducted on counterfaces without predominant orientation were equivalently low despite roughness ranging from 80 to 580 nm $R_q$. Transfer films on these surfaces were all thin and uniform. Testing against the oriented wet-sanded surface on the

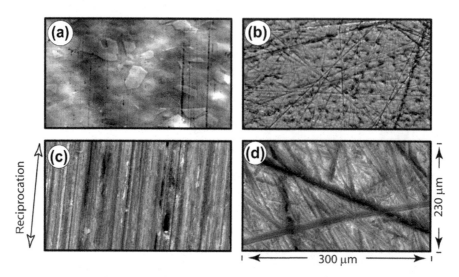

**FIGURE 17.8**

Surfaces used to study roughness effects on PTFE nanocomposite transfer and wear: (a) electropolished, $R_q$ (root mean squared roughness) = 80 nm; (b) lapped, $R_q$ = 160 nm; (c) wet-sanded, $R_q$ = 390 nm; (d) dry-sanded, $R_q$ = 580 nm. Note that the lay of the wet-sanded surface is oriented in the direction of sliding; it is smoother in the direction of sliding than against it.

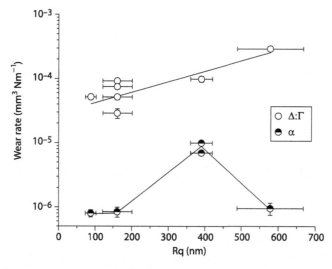

**FIGURE 17.9**

Wear rate plotted versus counterface surface roughness, $R_q$ for 5-wt% 40-nm $\Delta$:$\Gamma$ phase and 80-nm $\alpha$ phase alumina-PTFE nanocomposites. The two phases of alumina filler produce wear rates that differ by 100× on average with different surface sensitivities.

other hand increased the wear rate of the nanocomposites by an order of magnitude. A repeat at this condition confirmed the validity of the result. Transfer films on the wet-sanded surface were incomplete, thick and banded in the direction of sliding [36].

### 17.2.2.1 Transfer film morphology

An examination of the data collected throughout the test reveals an additional key difference between $\Delta{:}\Gamma$ and $\alpha$ phase alumina nanocomposites. The $\Delta{:}\Gamma$ nanocomposites reached steady state almost immediately, while the $\alpha$ nanocomposites had a significant transient period of moderate wear followed by a transition to a lower steady state wear rate. Despite the relative insensitivity of steady state wear rates to counterface roughness for $\alpha$ nanocomposites, the transient wear rate during transfer film development increased monotonically with increased roughness. This is intuitive since larger asperities tend to remove more material during wear events. In addition, the total volume removed during the transient portion of the test increases with increased roughness. These results suggest that as material is removed from the sample and deposited onto the counterface, more of the asperities become covered and the wear rate is reduced. Larger asperities require more material to transfer before steady state is reached, but at steady state, abrasion is insignificant and wear rate is independent of roughness. The orientation of the wet-sanded surface likely disrupted the formation of a stable transfer film, resulting in comparable transient and steady state wear rates. It can be concluded that the presence of a protective transfer film is necessary for low wear of PTFE nanocomposites. It is also interesting to note that when neither composite was sufficiently protected by transfer films, either during the transient region or against the wet-sanded surface, the $\alpha$ alumina nanocomposites outperformed the $\Delta{:}\Gamma$ nanocomposites. This suggests a difference in the wear mechanisms, which likely governs the ability of the composite to form protective films during sliding. Qualitatively, transfer films were found to increase in thickness and discontinuity with increasing wear rate.

Thin, uniform transfer films and fine debris consistently accompany wear resistance in these studies and in the nanocomposites tribology literature. Global relationships between wear rates and transfer films were studied by quantitatively measuring transfer films of widely varying PTFE-based tribosystems using either scanning white-light interferometry or mapping stylus profilometry. These systems include 5-wt% alumina-PTFE composites with $\alpha$ and $\Delta{:}\Gamma$ particle phases; 40 nm, 80 nm and 0.5 µm particle sizes; and counterfaces of polished, lapped, wet-sanded and dry-sanded surface finishes. Wear rate is plotted as a function of maximum transfer film thickness in Fig. 17.10. Despite varying particle phase, size and surface finish, wear rate is approximately proportional to the maximum thickness of the transfer film cubed. Not only do thicker films imply larger debris, but it is also suggested that thick transfer films are more easily removed by the passing pin, and as a consequence, need more rapid replenishment.

It is well known that under certain low-speed sliding conditions, PTFE deposits very thin and oriented transfer films [3,43–47]. The orientation is thought to produce

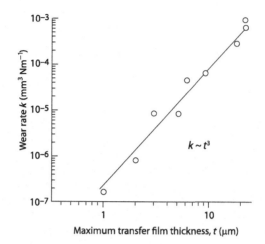

**FIGURE 17.10**

Wear rate plotted versus the maximum transfer film thickness as measured with optical interferometry. This data includes results of 5-wt% 44-nm Δ:Γ, 80-nm α and 0.5-μm α composites against the best and worst performing counterfaces, and 5-wt% 80-nm and 0.5-μm composites and unfilled PTFE against polished surfaces. Wear rate is proportional to the maximum transfer film thickness cubed.

a model sliding condition where chain entanglement is minimized and pure axial sliding of PTFE chains past one another results in the very low friction coefficients observed under these conditions ($\mu = 0.03$–$0.07$). It is hypothesized that the role of the filler is to reduce gross damage to PTFE under severe sliding conditions, facilitating the formation of thin and aligned PTFE films, and enabling low wear of the nanocomposite.

The friction and wear properties of the films themselves were measured using microtribometry to test the hypothesis that thin, aligned films of unfilled PTFE are wear resistant. Model films of PTFE were deposited onto a thin steel foil with a sliding velocity of 254 μm s$^{-1}$ for 1000 reciprocation cycles at 25 °C under 6.3 MPa of normal pressure. Atomic force microscopy (AFM) was used to estimate an average film thickness of 50 nm. After creation, the film-covered foils were cut into rectangular samples for testing. Custom designed sample mounts fixed opposing foils into a crossed-cylinder geometry. This geometry reduces misalignment sensitivity, minimizes edge effects and helps reduce pressures to values more typical of those found in macroscale testing. Parallel (chains oriented in the direction of sliding) and perpendicular (chains oriented against the direction of sliding) aligned films were tested to study the hypothesized tribological anisotropy of aligned PTFE films. Normal and friction forces were continuously measured at the stationary pin, while a 600-μm reciprocation displacement was imposed on the counterface. Tests with an average sliding speed of 100 μm s$^{-1}$ and a normal load of 500 mN were conducted over 250 sliding cycles. The contact patch was estimated with ex situ optical observation to be

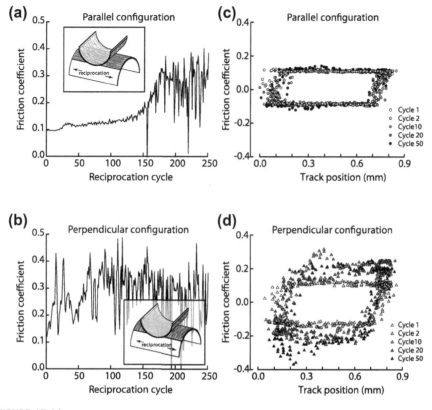

**FIGURE 17.11**

Microtribometry friction results for the crossed cylinder-oriented PTFE transfer film tests. Friction coefficient is examined versus reciprocation cycle for (a) parallel and (b) perpendicular configurations. The evolution of friction coefficient along the reciprocation track is also plotted for both the (c) parallel and (d) perpendicular configurations. The perpendicular alignment of the films leads to rapid failure of the films due to a tendency to reorient during unidirectional sliding.

200 µm in diameter; this translates into an average pressure of 15 MPa. The results of the microtribometry experiments are shown in Fig. 17.11.

In line with the hypothesis that orientation in the sliding direction facilitates low friction and wear, perpendicular alignment of the films led to complete failure of the film (denoted by $\mu > 0.2$) in about 10 cycles, while parallel aligned films were at least 10× more wear resistant. Despite having similar average values of friction coefficient for the first few passes, differences can be seen in the positionally resolved friction data on the right of Fig. 17.11. Examining the first pass, the parallel sample has a steady friction loop, while the perpendicular friction loop has significant scatter. The mechanism of motion accommodation appears more damaging in the case of the perpendicularly aligned films, and the tendency of these films to reorient into the direction of sliding is likely responsible for the erratic friction and wear behavior. The parallel aligned films have much lower wear presumably as a result of the stable orientation.

Since protective transfer films are necessary to reduce wear of these nanocomposites against counterface asperities, the wear rate of the transfer film places a lower limit on the wear rate of the composite. An estimate of the wear rate for the parallel aligned films was calculated to determine whether model films of unfilled PTFE could possibly support low-wear sliding. The contact areas on the top foil (pin) and bottom foil (counterface) are 0.033 and 0.12 mm$^2$, respectively, so failure of the top film should occur first, followed by direct asperity contact and rapid deterioration of the bottom film. Failure of the top film in parallel alignment occurred after approximately 150 cycles. The wear rate in this case is calculated as

$$k = \frac{50 \times 10^{-6} \text{ mm}}{15 \text{ N mm}^{-2} \cdot 150 \text{ cycles} \cdot 1.2 \times 10^{-3} \text{ m cycle}^{-1}} = 2 \times 10^{-5} \text{mm}^3 \text{ Nm}^{-1}$$

(17.1)

Despite the superiority of the parallel aligned transfer film, estimation of the wear rate reveals that rates are still orders of magnitude higher than those found for many low-wear PTFE nanocomposites ($\sim 10^{-5}$ versus $10^{-7}$ mm$^3$ Nm$^{-1}$). It can be concluded that even model thin and aligned transfer films of PTFE are incapable of supporting low-wear sliding. The films formed by sliding of a low-wear nanocomposite must therefore comprise composite material or some more wear-resistant variant of PTFE.

Similar experiments were conducted for transfer films of a low-wear 10-wt% PEEK-filled PTFE composite. In each configuration, the composite film had lower and more stable friction coefficients for the duration of 1000 cycle tests with no obvious signs of wear in posttest analysis. Clearly, the compositions and chemistries of these films are additional factors that require quantification for a more complete understanding of these nanocomposite systems.

### 17.2.2.2 Transfer film composition

Although XPS chemical and compositional analyses of unfilled polymer and microcomposite transfer films have been conducted by several investigators [20–22], it is unclear how composition and chemistry evolve in the transfer films of PTFE nanocomposites or how this evolution influences tribological phenomena observed during testing. XPS was used to test the hypothesis that transfer films consist of composite material. PTFE nanocomposites were created with ⅛%, ½% and 1% filler fractions of 40-nm α alumina nanoparticles. In a control set, the particles were untreated and in a second set of samples, the nanoparticles were treated with a fluorinated silane. The treatment was hypothesized to improve dispersibility and compatibility with the matrix. Treated (fluorinated) and untreated samples were compression molded, machined and tested. The tribological experiments were conducted on a linear reciprocating tribometer; the pin passes back and forth across a linear wear track. Photoelectron spectra of the C, O, F, Al and Si regions were collected using a PHI 5700 X-ray Photoelectron spectrometer equipped with a monochromatic Al Kα X-ray source ($h\nu = 1486.7$ eV) incident at 90° relative to the axis of a hemispherical analyzer. The spectrometer was operated at high resolution with a pass energy of 23.5 eV, a photoelectron takeoff angle of 45° from the surface normal and an analyzer spot diameter of 1.1 mm. All spectra were collected at room temperature with a base pressure of $1 \times 10^{-9}$ Torr. Electron binding energies were calibrated with respect to the C 1s line at 291.6 eV (C–F).

Even at very low filler fractions, alumina was found to transfer to the counterface with the PTFE (trace amounts of Si were also found in fluorinated samples). The atomic fraction of aluminum in the transfer films is plotted versus the aluminum content in the bulk in Fig. 17.12. Aluminum in the transfer films of fluorinated samples (filled circles) was found in direct proportion to the filler fraction in the bulk sample. The untreated ⅛% sample had an unexpectedly high amount of aluminum in the film, possibly due to poor dispersion or agglomeration during processing. Alternatively, aluminum could have accumulated in the transfer film with wear of the sample— this sample had 50× higher wear than any of the other samples—Blanchet and Han [40,48] previously described this mechanism as one of preferential removal of PTFE from the system, which leaves the interface rich with the filler.

Similar XPS analyses of transfer films formed from both types of alumina particles revealed a higher proportion of oxygen than would otherwise be predicted based on the aluminum present, suggesting tribochemical oxidation of the PTFE during extended sliding. Furthermore, these measurements demonstrate a correlation between friction coefficient and the oxygen content of the transfer film (Fig. 17.13). It is unclear whether an increase in friction of the fully formed transfer film drives the oxidation or if oxidation itself, occurring in the creation of the transfer films, leads to the higher friction coefficients.

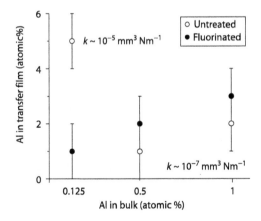

**FIGURE 17.12**

Aluminum atomic content (percentage) in the transfer films of virgin and fluorinated nanocomposites plotted versus the atomic content as prepared in the bulk. Measurements were made in the center of each wear track. In low-wear samples the aluminum content in the transfer film is linearly proportional to the aluminum content in the bulk sample. Evidence of filler accumulation was observed for the less wear-resistant sample.

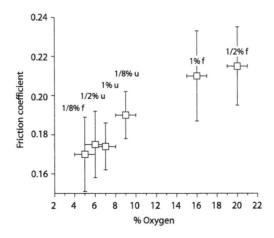

**FIGURE 17.13**

Friction coefficient plotted versus oxygen content as measured using X-ray photoelectron spectroscopy. The symbol "u" denotes untreated alumina and "f" denotes fluorinated alumina.

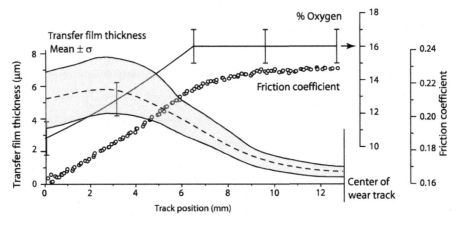

**FIGURE 17.14**

Transfer film thickness, oxygen content and friction coefficient plotted versus track position over half of the wear track for a 1% fluorinated sample. The transfer film thickness envelope represents the mean plus and minus one standard deviation. The position-dependent friction coefficient correlates well with both oxygen content and thickness.

The transfer films formed from PTFE nanocomposites containing fluorinated alumina exhibited frictional behavior uncharacteristic of PTFE; friction coefficients were lowest at the reversals (ends of the wear track) and highest in the center. Based on this behavior and the oxygen–friction coefficient correlation presented above, it was hypothesized that the oxygen content should decrease toward the ends of the wear track where friction was found to be the lowest. Figure 17.14 shows the friction coefficient, transfer film thickness and oxygen content as functions of the pin center track position over half of the track for a 2-wt% fluorinated 40-nm α phase alumina-PTFE nanocomposite. XPS was used to determine the oxygen content in the films and three dimensional mapping stylus profilometry was used to map film thickness; the thickness envelope shown in Fig. 17.14 reflects averaged results of mapping stylus profilometric measurements of the transfer film plus and minus one standard deviation. Both oxygen content and thickness correlate well with the friction coefficient along the wear track for the final cycle of sliding.

### 17.2.2.3 Transfer film chemistry

The oxidation of PTFE is initially surprising given its known chemical inertness. However, very wear-resistant materials produce very thin transfer films that are exposed to prolonged frictional energy dissipation at the interface. Over extended sliding distances, sufficient energy can be absorbed by these thin layers to initiate even low probability chemical events. XPS analysis of a 5-wt% 80-nm α alumina transfer film after sliding with a wear rate of $10^{-7}$ mm$^3$ Nm$^{-1}$ further probed the tribochemical degradation of low-wear PTFE. The comparison of the core level C 1s spectra of unfilled PTFE to that of the nanocomposite transfer film shown in Fig. 17.15 demonstrates clear evidence of a chemical transformation of PTFE in the process of transfer film deposition and wear. The data confirm a reduction in C–F intensity at 292 eV, consistent with defluorination of the transfer film and the measured reduction in the F 1s integrated intensity (data not shown). This change is accompanied by the appearance of new C species giving rise to intensity at 288

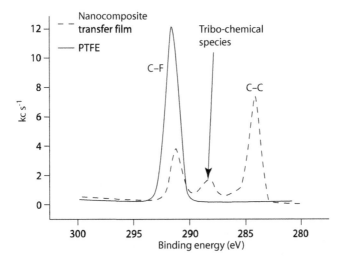

**FIGURE 17.15**

Comparison of the core level C 1s spectra of unfilled PTFE to that of the nanocomposite transfer film. The appearance of a new peak at 288 eV provides evidence of a chemical transformation in the PTFE that occurs during wear.

and 284 eV. While the relative changes in the 292 eV and 284 regions could be rationalized through the adsorption/deposition of adventitious carbon during sliding or sample preparation, the 288 eV feature can be correlated with the relative wear resistance of filled PTFE transfer films. As such, we conclude that the formation of unique chemical species in these wear-resistant transfer films accompanies the relative changes in atomic concentration.

Several investigators have intentionally degraded PTFE using various techniques including gamma ray, electron beam, and ultraviolet irradiation and have correlated chemical degradation with changes in mechanical properties [49–52]. For example, Zhao et al. [53] noted a rapid fluorine loss on the surface of PTFE when exposed to vacuum ultraviolet radiation as well as increased optical absorbance due to carbon exposure, which resulted in a brownish appearance of the sample. Similar brown discolorations are often found in low-wear transfer films. Lappan et al. [54–56] and Oshima et al. [57] used infrared spectroscopy to identify various reaction products from the degradation, noting processes involving defluorination and chain scission, which produce terminal and branched $CF_3$ groups, C–C double bonds and branched (cross-linked) carbon structures. COF and COOH have also been observed. High-speed magic angle spinning (HS MAS) nuclear magnetic resonance (NMR) spectroscopy measurements of degraded PTFE by Katoh et al. [58] and Fuchs et al. [59] provided evidence of cross-linking in degraded PTFE. Oshima et al. [57] used the HS MAS NMR data to identify chemical structures of $CF_2$, CF (cross-linking) and $=CF$ (double bonded carbon) in XPS spectra.

Blanchet et al. [11,15,60] previously conducted tribology experiments on irradiated PTFE and FEP (fluorinated ethylene propylene) composites and demonstrated that the wear resistance of each was improved by several orders of magnitude with 30 Mrad doses of electron irradiation. More recently, we conducted an analogous study to further explore the influence of degradation on the tribological properties of PTFE. A commercial chemical etch was used to emulate tribochemical degradation by stripping fluorine from the surface of an unfilled PTFE wear sample. Intensity at 288 eV was verified in the XPS spectrum of the etched surface. The tribological properties of the degraded PTFE surface coating were measured for linear reciprocating sliding at 6.3 MPa and 50 mm s$^{-1}$. The degraded PTFE was found to be 100× more wear resistant than unfilled PTFE, verifying that conjugated PTFE offers the possibility of increased wear resistance. However, a period of moderate wear was followed by a sharp increase in wear and a 10% reduction in friction coefficient as the more wear-resistant degraded surface wore through to the virgin PTFE beneath. The rate of material consumption due to wear here was greater than the rate of tribochemical degradation, and it is likely that in these nanocomposite systems, the degradation mechanism supplements the other more dominant mechanisms that enable degradation to occur.

### 17.2.2.4 Role of transfer films

It has been shown that transfer film morphology, composition and chemistry all play important roles in determining the wear rate of the films and thus the tribosystem. It has also been shown that certain surface characteristics can destabilize the transfer film, which disables low-wear sliding. It is clear that the presence of a high-quality transfer film is a necessary condition for low-wear sliding of PTFE nanocomposites; it is unclear whether it is a sufficient condition for low wear. To test this hypothesis, an alumina-PTFE nanocomposite known to ordinarily produce poor-quality transfer films (5-wt% 44-nm Δ:Γ alumina) was tested upon a predeposited transfer film formed under ultralow-wear sliding conditions ($k < 10^{-8}$ mm$^3$ Nm$^{-1}$ [61]). It was found that the composite had the same wear rate ($7 \times 10^{-5}$ mm$^3$ Nm$^{-1}$) whether it was tested against a wear-resistant transfer film or a fresh counterface. Despite the presence of an ultralow-wear transfer film, thick platelets indicative of delamination wear were deposited on top of the preexisting transfer film. The abrasive wear that is reduced by the transfer film appears to have a negligible influence on the wear rate of this nanocomposite due to the severity of its delamination wear. In situ optical microscopy of the wear track revealed that the transferred material is very unstable, moving appreciably after each cycle. Clearly, the mechanics of the nanocomposite itself dominate the properties of this system and govern the development of the transfer film. These results suggest that while thin transfer films are required for low wear, they form as a result of the low-wear debris morphology and are not themselves the source of wear resistance.

### 17.2.3 Investigations of internal interfaces

The critical role of transfer films in enabling low wear of PTFE nanocomposites has been demonstrated, but bulk properties of the composite seem to dictate the initiation and development of the films as well as the ability of the composite to achieve low wear against high-quality transfer films. Bahadur and Tabor [23] and Blanchet and Kennedy [1] noted a trend of decreased wear debris size with decreased wear rate and suggested that the primary role of the filler was to reduce the size of the wear debris. Because the wear rate is proportional to volume, which is proportional to the cube of a characteristic diameter of the debris, reducing debris size inherently reduces the wear rate, promotes engagement of debris with the surface and improves transfer film stability. Burris and Sawyer [37] hypothesized that the size and shape of the debris during run-in were critical in the development of these transfer films, and that the film morphology observed during low wear was a consequence of low wear rather than the cause. They envisioned a wear model proposed by Blanchet and Kennedy [1], where the cracks that lead to the destructive delamination in PTFE were effectively arrested by the filler, resulting in reduced debris size and stable transfer film formation. The strength at the filler/matrix interface would have a critical influence on such a system.

Often, nanoparticles and polymer matrices are inert by design to limit environmental sensitivity of the tribological response. This inertness limits chemical interaction at the filler/matrix interface and can lead to inherent weakness. Wagner and Vaia [62] articulated the importance of the bonding at the interface for a nanotube-reinforced polymer. They calculated interfacial shear strength to be 3 MPa with only van der Waals interactions present, and in excess of 100 MPa with only 1% covalent bonding of the carbon atoms. Many investigators have successfully improved interfacial bonding in composites with surface coatings that compatibilize the filler with the matrix. He et al. [63] found improved mechanical properties and dispersion when the nanoparticles were plasma modified, and Eitan et al. [25] found improved load transfer via strain-dependent Raman spectroscopy and improved bulk mechanical properties of a treated multiwalled carbon nanotube-filled polycarbonate over the nanocomposite with untreated nanotubes. While the studies of Burris and Sawyer [36,37] showed a strong dependence of wear rate on alumina phase suggesting the potential importance of these internal interfaces in tribology, it was unclear whether observed differences were due to particle phase or size since both phase and size varied in the experiments. Recently, experiments were conducted on PTFE nanocomposites with smaller $\alpha$ phase alumina to evaluate the potential size effect. Figure 17.16 shows transmission electron micrographic (TEM) images of the 44-nm $\Delta$:$\Gamma$, 40-nm $\alpha$ and 80-nm $\alpha$ phase nanoparticles. The $\alpha$ phase particles have an irregular platelike morphology, while the $\Delta$:$\Gamma$ particles are spherical.

Tribological experiments were conducted on nanocomposites of these particles at various filler fractions against lapped 304 stainless steel counterfaces in standard laboratory conditions at 50 mm s$^{-1}$ and 6.3 MPa of normal pressure. Wear rate is plotted versus alumina filler fraction in Fig. 17.17. The wear rates of the 40- and 80-nm $\alpha$ alumina-PTFE nanocomposites are insensitive to size and filler fraction in the range from $\frac{1}{2}$–5% filler and these particles provide additional 100–1000× improvements in wear resistance over the $\Delta$:$\Gamma$ phase alumina-PTFE nanocomposites. It can be concluded that the differences in wear rates are not attributable to particle size, and although these

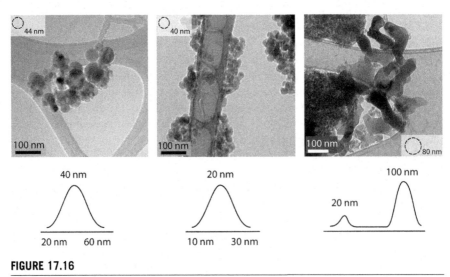

**FIGURE 17.16**

Transmission electron images of 44-nm $\Delta$:$\Gamma$ (left), 40-nm $\alpha$ (center) and 80-nm $\alpha$ (right) particles used in this study with size distributions estimated from TEM surveying.

dispersions have not yet been characterized, the difference is thought not to be due to dispersibility since the 40-nm α phase particles have higher specific surface area and were observed to agglomerate substantially more than the other particles during nanoparticle imaging. Although the wear reduction mechanism of the α phase particles remains unclear, it appears to be related to the nature of the interface.

It was hypothesized that additional gains in tribological performance could be achieved by compatibilizing the nanoparticles with the matrix. A nanoparticle surface fluorination was thought to provide compatibility with the matrix, and the decrease in nanoparticle surface energy due to the fluorination was thought to aid dispersion; poor dispersibility was suspected as the source of the unusual scatter in the 40-nm α nanocomposites. The 40-nm α phase particles were chemically treated with 3,3, 3-trifluoropropyl trimethoxysilane. Infrared absorption spectroscopy confirmed the presence of the fluorinated groups and thermal gravimetric analysis was used to estimate the mass fraction at 3%. Nanocomposites with filler fractions of ⅛%, ½% and 1% untreated and fluorinated nanoparticles were tested to investigate the effects of the interface treatment on the tribological properties of the nanocomposite. Wear volume is plotted versus the sliding distance on the left hand side of Fig. 17.18. In general, the fluorinated samples were very well behaved in comparison to the untreated samples. Transients are steeper and longer lasting with more material removed for samples of lower nanoparticle filler fractions. The untreated alumina nanocomposites behaved erratically by comparison. Steady state wear rates are

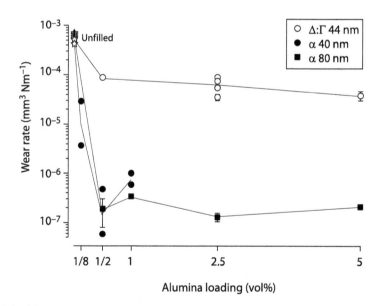

**FIGURE 17.17**

Wear rate plotted versus filler fraction for alumina-PTFE nanocomposites with varying nanoparticle phase against a lapped counterface. Error bars represent the experimental standard uncertainty in the measurement of wear rate.

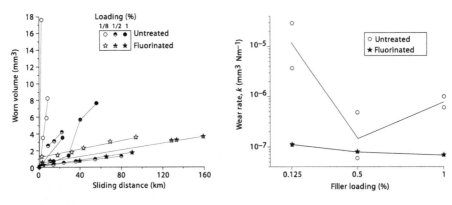

**FIGURE 17.18**

(Left) Worn volume plotted versus sliding distance for untreated and fluorinated 40-nm α phase alumina-filled PTFE. (Right) Wear rate plotted versus filler fraction for untreated and fluorinated 40-nm α phase alumina-filled PTFE. Linear reciprocation experiments were conducted against lapped 304 stainless steel counterfaces under standard laboratory conditions with a normal load of 250 N.

plotted versus filler fraction on the right of Fig. 17.18. The large sample-to-sample variation of the untreated nanocomposites is clear with the low weight percentage samples having nearly an order of magnitude difference in wear rates of identically prepared samples. The wear rates of the functional nanocomposites differ by less than 2× from ⅛ to 1%, and decrease linearly with increasing filler fraction; variations in wear rates of 2× or less for PTFE nanocomposites are rare even for sample–to–sample variations. With the addition of ⅛% functional nanoparticles, the wear resistance of PTFE was improved by over 5000×. It is unclear whether the variations in wear rates were dominated by dispersion, interface strength or both. Efforts to interrogate and quantify dispersions are currently underway.

## 17.2.4 Investigations of matrix phase and morphology

Previous studies have shown that subtle changes, such as nanoparticle shape, phase and surface chemistry, can have dramatic effects on the wear rate of the PTFE nanocomposite. A previous hypothesis was that these factors increased the particle/matrix interface strength, but filler particle surfaces can also affect the local phase, morphology and mobility of the polymer chains at the interface. This layer of affected polymer in the vicinity of the particle is known as the interfacial region, interaction zone and interphase. We will refer to it as an interfacial region here. Due to its characteristic nanometer size scale, it is often appropriately neglected in microcomposites. However, the interfacial region can often be comparable in thickness to nanoparticle fillers and can therefore dominate the properties of a nanocomposite. In amorphous polymers, the primary impact is on the chain morphology and mobility. Sternstein et al. [27,64] did a systematic study of the particle/matrix interface strength effects on the rheology of nanocomposites and proposed a theory for reinforcement and nonlinear viscoelasticity of polymer nanocomposites that is based on the trapped entanglements of chains near the polymer/matrix interface and the consequent far-field effects on other polymer chains. Maiti and Bhowmick [65] used AFM to measure interfacial thickness, and found the thickness of the region to increase with increased filler/matrix compatibility. Eitan et al. [25] observed a similar relationship for treated and untreated nanotubes in a polycarbonate matrix. In addition, it was found that fracture occurred preferentially within the matrix itself rather than at the filler/matrix interface, suggesting that the interface and interfacial region can both be stronger than the unfilled polymer. The manifestations of interface effects in amorphous polymers have been observed by numerous researchers and include changes in glass transition temperature, modulus, toughness and rheology [24,25,27,64, 66–69]. In semicrystalline polymers, nanoparticles can alter the crystalline phase [70], morphology [71,72] and the degree of crystallinity [73,74], which can extend the influence of the nanoparticles. Such changes in crystalline morphology can have significant influences on mechanical behavior, but it is unclear whether such effects are present in PTFE nanocomposites or to what extent tribological properties are altered due to these effects. The role of crystallinity and crystal phase is explored in the following paragraphs.

PTFE is a semicrystalline polymer known to have a complex molecular organization with three phases existing near room temperature at ambient pressure. Phase II [75–80] is typically stable below 19 °C and is characterized by a triclinic unit cell with $a = b = 0.559$ nm and $\gamma = 119.3°$. The molecular conformation is described by Clark [75] as a noncommensurable 13/6 helix. As the temperature increases above 19 °C, phase IV becomes stable, the molecules untwist to a possibly commensurable 15/7 helix and the unit cell becomes hexagonal with a 1% increase in the lattice parameter ($a = b = 0.566$ nm) [75]; macroscopically, this results in an increase in volume [80]. In addition, Kimmig et al. [81] noted a rapid increase in the density of coherent helical reversal defects possibly associated with the onset of helix commensurability. As the temperature increases above 30 °C, the hexagonal unit cell becomes distorted [82], the Bragg reflection at $2\theta = 42°$ broadens [83] and molecular disorder increases [77,81,84] as intramolecular forces begin to dominate intermolecular forces [81]. It has been suggested that molecular disorder increases with increased temperature because axial and rotational oscillations of molecules become more pronounced and helical reversal defects increase in length and become incoherent.

While it has been shown that the degree of crystallinity has a minimal role on the wear rate of PTFE [6,85], phase and temperature have both been found to have dramatic influences on the mechanical and tribological properties of PTFE. Flom and Porile noted a dramatic effect of the phase of PTFE on its tribological properties [2]. They performed sliding experiments with self-mated PTFE at speeds of 11 and 1890 mm s$^{-1}$ and found an abrupt and reversible increase in the friction coefficient as the background temperature increased above a threshold value near room temperature in both cases. They hypothesized that the increase was associated with the phase transition from II to IV at 19 °C. Steijn [5] also found increased friction coefficients as the temperature increased from 19 to 30 °C despite the more global trend of reduced friction coefficient with increased temperature. Makinson and Tabor [3] found evidence of strong adhesion to the counterface and proposed a lamellar effect of intercrystalline shear governed by van der Waals interactions between polymer chains. Studies by Steijn [5], Tanaka et al. [6], Blanchet and Kennedy [1], and McCook et al. [86] supported this hypothesis finding frictional responses of PTFE to be thermally activated and consistent with van der Waals interactions. More generally, Joyce et al. [87], Brown and Dattelbaum [88], Brown et al. [89], Rae and Dattelbaum [90] and Rae and Brown [91] found similar dependencies of various mechanical properties to temperature in the range from 200 to 400 K. The measured property in each of these studies has been normalized by the room temperature measurement and is plotted versus temperature in Fig. 17.19.

Tanaka et al. [6,18] focused on the characteristic bands of PTFE that contain both crystalline and amorphous material. In a study investigating the wear mechanisms of PTFE, they found insensitivity of wear rate to crystallinity, but found that with certain processing conditions, the characteristic size of the banded structure could be reduced with a subsequent reduction in the wear rate [6]. In varying the background temperature and sliding speed of PTFE during testing, a transition to low wear rate was found to occur as temperature increased past a critical temperature. At the 50-mm s$^{-1}$ sliding

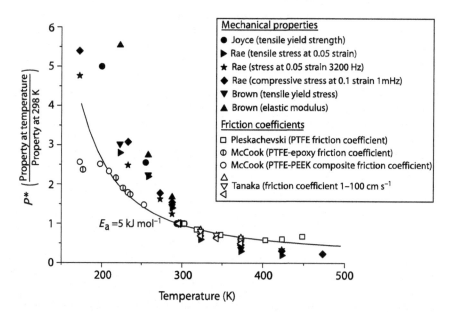

**FIGURE 17.19**

Normalized properties plotted versus temperature for variable temperature studies available in the literature. The normalized property is defined as the ratio of the value at temperature to the room temperature value. Various tribological and mechanical experiments suggest that deformation of PTFE is a thermally activated process.

speed used here, the data from Tanaka et al. [6] indicates a transition temperature near the IV to I phase transition of PTFE (30 °C). A recent study of the effects of PTFE phase on toughness by Brown and Dattelbaum [88] helps explain this result. Increased fracture toughness was found for phase I ($T > 30$ °C) over phases II ($T < 19$ °C) and IV ($19$ °C $< T < 30$°), especially at high strain rates. Phase II was characterized by brittle fracture, while ductility and fibrillation were observed in the high toughness phase I. The fibrils were thought to bridge cracks and reduce the stress concentrations at crack tips. These results have important implications to tribology because both fracture and wear require energy to create new surfaces. These results also suggested superiority of phase I over phases II and IV in wear applications, since the events that occur in tribological contacts typically occur with high strain rates.

One constant observation in low- and ultralow-wear PTFE-based composites is that of fibrillation under stress at room temperature [38,61]. SEM imaging was used to study the wear surface of a 5% 80-nm $\alpha$ phase alumina-PTFE nanocomposite with a wear rate of $k \sim 10^{-7}$ mm$^3$ Nm$^{-1}$. Figure 17.20 shows the results of these observations at two magnifications. In the low-magnification image, "mudflat" cracking is observed on the wear surface. The mudflats are on the order of 10 μm in diameter and while they appear to be poorly connected to the bulk, vacancies from debris liberation are not observed. Higher magnification imaging reveals fibrils spanning the cracks and appearing to prevent the liberation of the cracked material as debris. The same alumina-PTFE nanocomposite was fractured in bending at 25 °C and the resulting crack imaged. This crack is shown in Fig. 17.21. Fibrils are observed to span the entire length of the 150-μm crack. This degree of fibrillation is extraordinary for PTFE under

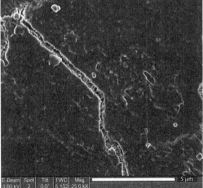

**FIGURE 17.20**

SEM images at different magnifications of the worn surface of a 5% 80-nm $\alpha$ phase alumina-PTFE nanocomposite. The "mudflat" cracking is a characteristic that is repeatedly observed for these wear-resistant PTFE nanocomposites. Wear debris appears to be on the order of 1 μm, while the cracking patterns encompass tens of micrometers of material. The liberation of large wear debris appears to be inhibited by fibrils spanning the cracks.

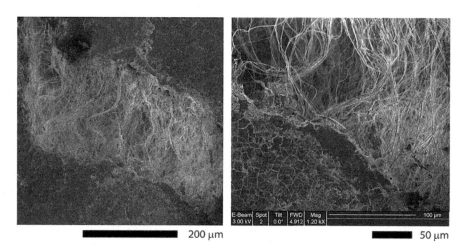

200 μm          50 μm

**FIGURE 17.21**

SEM images at different magnifications of the worn surface of a 5% 80-nm α phase alumina-PTFE nanocomposite after being fractured at room temperature. Fibrils completely span a 150-μm crack.

these conditions and suggests that the nanoparticles influence the crystalline morphology of the PTFE, and possibly stabilize the tougher phase I.

It is clear from the literature that the phase and morphology of PTFE are important in determining toughness and ease of fibrillation [88,89]. It is also well known that nanoparticles can stabilize metastable crystalline phases in polymers and change the crystalline morphology [72]. Therefore, one of the mechanisms that could explain the dramatic changes in wear rate at such small filler fractions of nanoparticles is an effect of the particles on the polymer phase and morphology. The following studies were conducted to characterize the effects of the nanoparticles on the matrix to gain insights into how these morphological changes affect the tribology of PTFE.

Melting behaviors of neat and filled samples were monitored using differential scanning calorimetry (DSC) and structures were examined using X-ray diffraction (XRD) to determine the role of matrix morphology on tribological properties. The XRD patterns near the main diffraction peak (18°) in Fig. 17.22 reveal that the nanocomposite has a larger full width half maximum (FWHM) and a larger amorphous background than the unfilled PTFE. This implies that the nanoparticles have interrupted the lamellar crystalline structure. There is also a slight shift in the peak value suggesting an interruption of the unit cell. The high-temperature DSC shown in Fig. 17.23 supports this result. The DSC shows that the degree of crystallinity of the neat matrix is significantly higher than that of the 1-wt% nanocomposite. DSC also shows a melt peak for each material near the traditional 327 °C melt temperature of PTFE. However, the nanocomposite has a larger melt peak at a higher temperature of about 345 °C consistent with the melt temperature of virgin PTFE before melt processing. The higher temperature is often attributed to the larger,

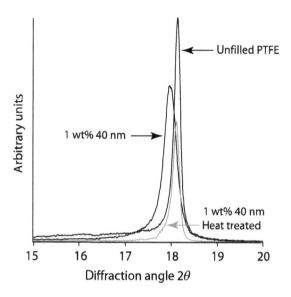

**FIGURE 17.22**

X-ray diffraction for the main reflections of neat PTFE, a low-wear nanocomposite and the same nanocomposite after a heat treatment to 400 °C.

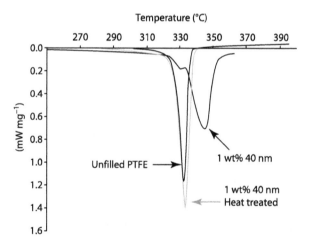

**FIGURE 17.23**

High-temperature differential scanning calorimetry of neat PTFE, a low-wear nanocomposite and the same nanocomposite after a heat treatment.

more perfect crystals of the virgin resin. Despite being processed under identical conditions (360 °C, 4 MPa), the unfilled PTFE has melt characteristics indicative of melt processed PTFE, while the nanocomposite retains the melt characteristics of virgin resin. The nanoparticles may impede the mobility of PTFE molecules, requiring greater amounts of thermal energy before reorganization of the PTFE to the lower order "gel" state can occur.

One hypothesis for wear resistance of these systems is that the virgin structure facilitates fibrillation of the nanocomposites during wear. Another is that the mechanical destruction that occurs during processing results in a fibrillated structure that is stabilized by the nanoparticles to temperatures above the process temperature; the processing involves jet-mill grinding and mixing of the powder constituents, which has been shown to result in elongation and 5× size reductions of the PTFE particles [35]. In either case, higher temperatures are needed to remove any stabilized structures. To test this hypothesis, the nanofilled sample was heat treated at 400 °C and characterized. Following heat treatment, the XRD peak at 18° has sharpened and returned to the same position as the unfilled PTFE.

Figure 17.24 shows that in the region most sensitive to the phases present, the 1-wt% nanofilled resin has the lowest peak height to amorphous background ratio (1.5 to 1) compared to the neat resin (3 to 1) and the heat treated 1-wt% nanofilled resin (about 2 to 1). From the literature we know that as the proportion of phase I increases, the peaks (especially the peak at 42°) decrease in intensity compared to the background. Because these samples are mixtures of at least 2 phases, it is difficult to ascertain exactly what structure is present. These results, however, imply that the 1-wt% nanofilled resin has the most phase I and the least phase IV of the materials shown and that the heat treatment to 400 °C leads to the most phase IV.

This is also supported by the DSC data at low temperature shown in Fig. 17.25. The data show that the 1-wt% nanofilled sample has much less of a phase I to IV transition than the unfilled PTFE, while the heat-treated composite shows a significant I to IV transition. Therefore, we would expect more phase I in the 1-wt% nanofilled composite and the least in the heat-treated 1-wt% nanofilled composite. In addition,

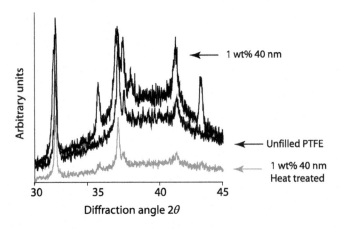

**FIGURE 17.24**

X-ray diffraction of neat PTFE, a low-wear nanocomposite and the same nanocomposite after a heat treatment in the region sensitive to the phases present.

the transformation temperature from I to IV is lower for the 1-wt% nanofilled sample further supporting the argument that it has the most phase I present. Finally, at room temperature the IV to II transition has begun in the heat-treated sample. This would explain the significantly higher ordering in this sample seen during XRD measurement. Both XRD and DSC provide evidence that more of the tough phase I is present in the untreated nanocomposite than either the heat-treated nanocomposite or the unfilled PTFE at room temperature. Therefore, we conclude from these results that the addition of untreated α phase nanoparticles results in a stabilization of phase I and of the local deformation that occurs during jet-milling. Heat treatment at 400 °C removes these benefits and should therefore result in increased wear.

AFM was used to probe the PTFE morphologies in the nanocomposite before and after heat treatment.

The AFM measurements shown in Fig. 17.26 support the features observed in XRD and DSC. Clear differences can be seen before and after heat treatment with the nanocomposite before having much thinner and more organized lamellae than it had after. Before heat treatment, the lamellae are well aligned and appear to have the folded-ribbon morphology that is often cited for virgin PTFE [83,92,93], while after heat treatment, lamellae are thick, tangled and packed into substantially larger characteristic regions. Marega et al. suggest that shifts in the phase transition temperature to lower temperatures (DSC) are indicative of increased molecular disorder and thinning of lamellae [83]. XRD and AFM measurements of the control nanocomposite show evidence of increased molecular disorder, and thinner lamellae, respectively. These characteristics may explain the stability of phase I for this sample at lower temperatures, and the high

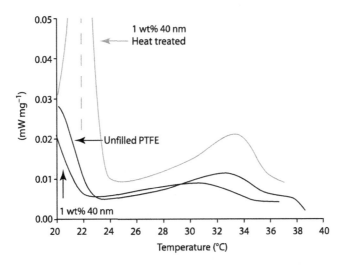

**FIGURE 17.25**

Low-temperature differential scanning calorimetry measurements of neat PTFE, a low-wear nanocomposite and the same nanocomposite after a heat treatment. Stabilization of the phases to lower temperatures no longer exists after heat treatment.

Untreated     $k = 5.9 \times 10^{-8}$ mm$^3$ Nm$^{-1}$     Heat treated     $k = 640 \times 10^{-8}$ mm$^3$ Nm$^{-1}$

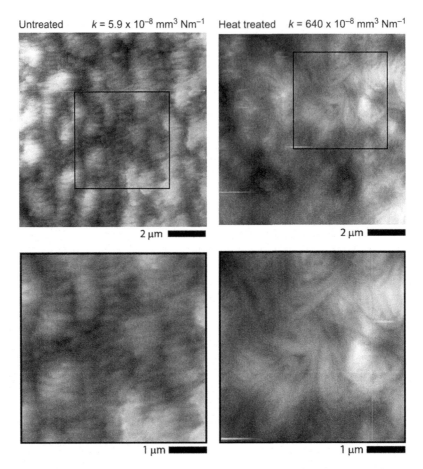

2 μm      2 μm

1 μm      1 μm

**FIGURE 17.26**

Atomic force microscopy of the crystalline morphology of 1-wt% 40-nm alumina-PTFE before (left) and after (right) heat treatment. As prepared, the nanocomposite has very small lamellae with finer scale, more ordered packing.

degree of lamellar alignment over the field of view supports the hypothesis that the nanoparticles stabilize higher melt temperatures by impeding molecular mobility.

To test the hypothesis that the effects of the nanoparticles on the morphology of PTFE reduce wear, tribological experiments were conducted on the same nanocomposite before and after heat treatment. The control experiment was conducted on the 1-wt% 40-nm α alumina-filled PTFE sample in linear reciprocation with 50% relative humidity at a temperature of 25 °C. The sliding speed and normal pressure were 50.8 mm s$^{-1}$ and 6.3 MPa, respectively. After standard processing, the nanocomposite was tested for approximately 80 km. The sample was then heated to 400 °C to anneal the crystalline structure induced during the original processing. A new counterface was used, the sample was faced, removing only the material necessary to make it flat, and the experiment was continued. After the heat treatment, the friction coefficient was reduced by about 10%, wear rate increased by 100× and transfer film thickness and discontinuity increased. The tribological results are shown in Fig. 17.27. The nanoparticle phase, chemistry, size, filler fraction and dispersion are essentially constant before and after the heat treatment, so the increase in wear rate is attributed only to the morphological effects of the nanoparticles on the PTFE.

DSC and XRD measurements of a 5-wt% alumina-PTFE microcomposite were made to determine if similar morphological effects could be detected. It can be seen in Fig. 17.28 that the micrometer-scale fillers do not cause a broadening of the XRD peak at 18° or a shift in the peak position. This is not surprising because the scale of the particle is much larger than the lamellae size. On the other hand, the DSC and the XRD do show a lower crystallinity in the micrometer filled sample. In addition, the XRD peaks between 30 and 60° in Fig. 17.29 indicate that the micrometer filled sample has a large amorphous peak, but the peak to amorphous background ratio is 2 to 1 compared to that in the 1-wt% nanofilled sample indicating a higher percentage of phase I in the nanofilled sample.

Similar experiments were conducted on fluorinated alumina-PTFE nanocomposites to investigate the roles of phase and morphology on the wear rates of these systems. Unlike the control nanocomposite, the fluorinated nanocomposite had a melt temperature near 327 °C, the low temperature peak. The XRD peak at 18° shown in Fig. 17.28 has an FWHM similar to that of the unfilled PTFE with no peak shift. This implies that the nanoparticles have not interrupted the lamellar structure or the unit cell. In addition, for the 30°–60° region, the peak at 42° is absent in this sample. This

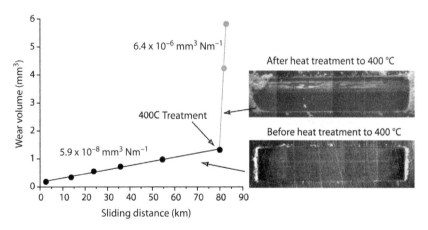

**FIGURE 17.27**

Wear volume plotted versus sliding distance for a 1-wt% 40-nm alumina-PTFE nanocomposite before and after heat treatment to 400 °C. The microstructural benefits of the nanoparticles are lost after heat treatment and the wear rate increases by 100×. Optical images of the transfer film before and after heat treatment are shown on the right.

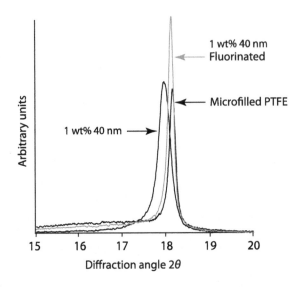

**FIGURE 17.28**

X-ray diffraction of a microcomposite and fluorinated and virgin alumina nanocomposites in the region of the primary reflection.

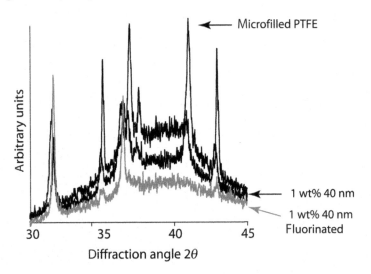

**FIGURE 17.29**

X-ray diffraction of a microcomposite and fluorinated and virgin alumina nanocomposites in the region sensitive to the phases present.

indicates that the sample is primarily phase I. The room temperature DSC data shown in Fig. 17.30 is less clear. The phase I to IV transformation is small in the fluorinated material, but no smaller than in the as received 1-wt% nanocomposite. While the temperature shift is less extreme for the fluorinated sample compared to the control

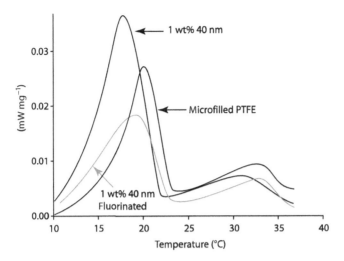

**FIGURE 17.30**

Low-temperature differential scanning calorimetry measurements of a microcomposite and fluorinated and untreated alumina nanocomposites.

sample, there is a longer tail of material transitioning at temperatures well below the transition temperature.

The fluorinated sample had comparable wear resistance to the control nanocomposite sample, but it lacked some of the distinct morphological differences, such as interruption of the lamellar structure and unit cell, and high melt temperature, which were hypothesized to be responsible for wear resistance. It did, however, show a strong phase I character in XRD and a shift in phase transitions to lower temperatures in DSC. It was therefore hypothesized that one of the critical roles of the nanoparticles was the stabilization of the high toughness phase I at the test temperature. The previous DSC data indicate that the phase II transition of the control nanocomposite occurs at a temperature of 18 °C. Thus if stabilization of phase I or IV leads to low wear at room temperature, then the wear rate of this sample should increase as temperature is reduced below some critical temperature near 18 °C. To isolate the effects of thermal history from phase stabilization, an additional nanocomposite test specimen was cut from the original low-wear 1-wt% 40-nm α phase alumina-PTFE puck described earlier. The sample was run-in on a fresh counterface for 2500 m of sliding with 6.3 MPa pressure at 40 °C. The wear rate during run-in was $6 \times 10^{-7}$ mm$^3$ Nm$^{-1}$, consistent with the run-in observed in the original experiment. Forecasting from the initial experiment, the wear rate after a 2500-m run-in is near steady state and on the order of $k \sim 10^{-7}$ mm$^3$ Nm$^{-1}$. After an additional 25 m of sliding, the temperature was continuously decreased over the next 30 m to a target of 10 °C; the dew point temperature was approximately 7–8 °C.

The results of the varied temperature experiment are shown in Fig. 17.31 with wear rate and counterface temperature plotted as functions of sliding distance. The wear-rate curve was estimated from mass loss measurements and in situ observation of the transition. The data shown here were recorded after the initial run-in period.

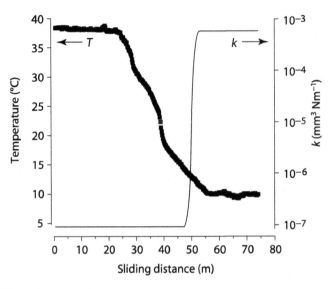

**FIGURE 17.31**

Wear rate and counterface temperature versus sliding distance for variable temperature tribology testing of a wear-resistant PTFE nanocomposite. Wear rate is an estimate based on mass measurements and in situ observation of the transition. Normal pressure and sliding speed were 6.3 MPa and 50 mm s$^{-1}$, respectively. An abrupt and drastic change in the wear mechanism occurs as the temperature is reduced below about 14 °C.

For the first 47 m of sliding, no observable debris was liberated from the contact indicating retention of the low-wear state. At 47 m, the temperature was 14 °C and the first wear fragment appeared. This was followed by rapid deterioration of performance until a steady state wear rate of $k \sim 5 \times 10^{-4}$ mm$^3$ Nm$^{-1}$ was reached at 52 m and about 12 °C.

It was hypothesized that the wear resistance of the nanocomposite was due to a disruption of the crystalline structure by the nanoparticles that resulted in increased disorder, stabilization of the higher toughness phase to lower temperatures and fibrillation under stress. The results from DSC of this sample are plotted versus temperature in Fig. 17.32.

The transition to high wear occurs when most of the sample has transitioned to the low toughness phase II. In addition to being more brittle, Brown et al. [89] found phase II to be stronger and stiffer than phases I and IV. It is possible that as the temperature is dropped and the material begins to transition to phase II, deformation preferentially occurs in the remaining phase IV material. At a critical point, enough of the material has been converted such that phase II must contribute to motion accommodation. Eventually, crack initiation, propagation and gross failure occur. The fact that unfilled PTFE does not exhibit such high wear resistance at comparable locations on the phase diagram suggests that facilitation of fibril formation is an additional critical wear-resistance mechanism in PTFE nanocomposites. The breakdown of wear

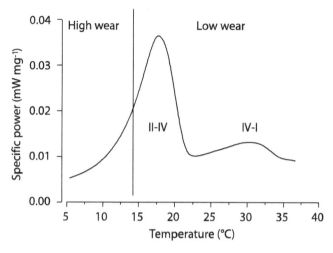

**FIGURE 17.32**

Normalized DSC power plotted versus temperature. The transition to high wear occurs at a critical point after the low toughness phase II transition temperature.

performance coinciding with transition to phase II suggests that the beneficial fibril-related toughening mechanism that results from the nanoparticle filler, is disabled in phase II.

## 17.3 HYPOTHESIZED MODEL OF WEAR RESISTANCE MECHANISMS IN PTFE SOLID LUBRICANTS

### 17.3.1 Introduction

Although many future studies are still needed to complete our understanding of these systems, the material science characterization and tribology literature combined with the recent studies just discussed have helped shape a more complete understanding of PTFE nanocomposites' tribology. The previously discussed research has shown that many factors have direct effects on the tribology of PTFE and suggests the degree of complexity that arises as these factors couple together to determine global tribosystem characteristics. We will now discuss a hypothesis for wear and wear-resistance mechanisms of PTFE-based systems.

### 17.3.2 Wear of unfilled PTFE

In unfilled PTFE, low toughness and easy slip of internal interfaces dominates the wear rate. Below the critical speed, the friction coefficient at the tribological interface is low and statically loaded internal interfaces of self-mated PTFE can fully support the traction. In this situation, adhesive and abrasive wear dominate and lead to wear rates on the order of $k \sim 10^{-5}$ mm$^3$ Nm$^{-1}$. Above the critical speed, the friction coefficient at the tribological interface exceeds that at the internal interfaces, cracks rapidly propagate and severe wear occurs ($k \sim 10^{-3}$ mm$^3$ Nm$^{-1}$) [1]. The thick, plate-like debris deposited onto the counterface are easily ejected from the contact by the passing pin.

### 17.3.3 **Wear of PTFE microcomposites**

Microscale fillers have been successfully used to arrest the crack propagation that leads to severe delamination wear in PTFE. As a crack encounters a particle, it is deflected or arrested and debris size is reduced. Because cracks can travel readily through unreinforced PTFE, the interparticle spacing is a critical parameter. Often, high filler fractions of these particles are required to limit this spacing, to provide load support against asperity contact and, ultimately, to improve wear performance. These fillers must themselves be hard and wear resistant to promote wear resistance of the system and often abrade transfer films and counterfaces. We have shown that as particle size is reduced from the micrometer scale to the nanoscale, a transition from moderate wear ($k \sim 10^{-5}$ mm$^3$ Nm$^{-1}$) to low wear ($k \sim 10^{-7}$ mm$^3$ Nm$^{-1}$) can occur. It was also shown that the introduction of micrometer fillers to a low-wear nanocomposite disrupts the wear-resistance mechanisms of the nanocomposite and the system transitions to moderate wear. Observations of gross destruction of the composite and abraded counterfaces and transfer films suggest that micrometer fillers prevent low wear by abrading the transfer films needed for protection against direct asperity contact.

### 17.3.4 **Wear of PTFE nanocomposites**

In order to facilitate discussions of the various coupled interactions that occur in PTFE nanocomposites, we will describe wear-resistance mechanisms as being either primary or secondary. A wear-resistance mechanism is defined here as a property or effect that has been shown to have a direct impact upon the wear rate. A primary wear-resistance mechanism is defined here as one that requires no pre-existing condition of wear resistance to occur, while a secondary wear-resistance mechanism occurs only as a result of wear resistance. Primary wear-resistance mechanisms that have been identified include (1) bonding and strength at the filler/matrix interface, (2) dispersion and mechanical effects of load support and crack deflection, (3) morphological effects of nanoparticles on the matrix and (4) fibrillation and toughening. Secondary wear-resistance mechanisms that have been identified include (1) transfer film protection, (2) transfer film orientation and (3) chemical degradation. The hypothetical system shown in Fig. 17.33 illustrates each of these mechanisms, highlighting the interactions of these different primary and secondary wear-resistance mechanisms.

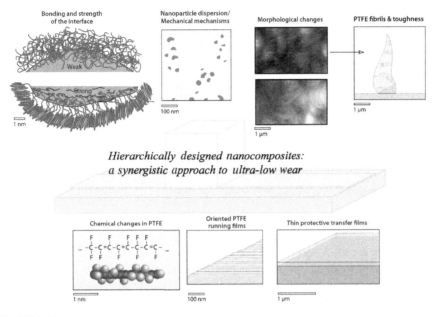

**FIGURE 17.33**

Hypothetical model of a PTFE-based tribological nanocomposite. Arrows point from cause to effect. Thin, aligned and degraded transfer films are thought to develop as a result of low-wear debris morphology and once initiated, provide low-wear feedback into the system.

Bonding at the filler/matrix interface occurs over the smallest length scales, involving both entanglement of individual chains and chemical bonds between atoms. Because particle densities and specific areas are inherently large in nanocomposites, subtle differences in the bonding state between a matrix chain and filler particle can translate into substantial improvements to bulk properties. Such differences are thought to be responsible for the superior wear resistance of $\alpha$ phase alumina-filled PTFE over $\Delta{:}\Gamma$ phase alumina-filled PTFE. There is also evidence to suggest that activation and effectiveness of the other mechanisms depends on the nature of the filler/matrix interface.

At a larger size scale, nanoparticle dispersion provides an additional mechanism of wear resistance by interfering with crack propagation and by shielding the matrix from direct asperity contact. Because low filler fractions are desirable, the direct mechanical effects of the nanoparticles are thought to be minimal, e.g. $\Delta{:}\Gamma$ nanocomposites. Both interfacial strength and nanoparticle dispersion determine the effectiveness of crack arrestment. When the filler/matrix interface is weak, particles arrest cracks less effectively and may actually nucleate additional cracks. When nanoparticle dispersion is poor, interparticle distance is large, and cracks may freely propagate through unfilled material. In addition, nanocomposites with agglomerations may be brittle because agglomerations are poorly bound by matrix material.

If the nanoparticles are well dispersed, strong interactions at the nanoparticle surface can influence the crystalline phase and morphology on a micrometer-size scale. Smaller lamellar thickness and increased molecular disorder can stabilize a higher toughness phase. In addition, these morphological effects are thought to determine the composite's ability to fibrillate. When fibrillation is activated, extensive deformation is required before material removal can occur.

When the debris size is sufficiently reduced via primary wear-resistance mechanisms, secondary mechanisms are activated. During sliding, the highly deformed and fibrillated matrix is easily drawn into thin oriented films, which protect the composite from asperity damage and facilitate low shear sliding. As wear rate is reduced, material is resident at the interface longer, more frictional energy is absorbed and degradation produces more wear-resistant, conjugated PTFE.

### 17.3.5 Summary

Results of these studies support many of the consistent observations found in the PTFE nanocomposites tribology literature, and more generally, the polymer nanocomposite tribology literature. In addition, these results provide quantitative evidence to support previous qualitative observations from the literature. The contributions of the nanoparticles span well beyond the traditional rules of mixtures models that adequately describe many of the commercial PTFE microcomposites, with nanoparticle dispersion, surface morphology and chemistry likely driving the fundamental wear-resistance mechanisms that synergistically couple to produce low wear. Currently, it is difficult to make a global interpretation of the state of the art of PTFE nanocomposite tribology because of the inherently large number of variables and the limited number of quantitative measurements of these different variables; critical components such as nanoparticle dispersion are almost completely absent from discussion in the literature (this work included). We have gathered results and offered a hypothesis based on a limited but ongoing research initiative to promote thought and discussion, provide new insights into overlooked aspects of the literature and motivate future, directed studies of these systems. The authors believe that with independent and quantitative materials and tribological characterization, great strides will soon be made in constructing a more accurate and complete description of these systems to enable novel composite design for current and future needs.

## Acknowledgments

This material is based upon an AFOSR-MURI grant FA9550-04-1-0367. Any opinions, findings, and conclusions or recommendations expressed in this material are those of the authors and do not necessarily reflect the views of the Air Force Office of Scientific Research. We thank Will Heward for the X-ray diffraction work, Su Zhao for particle surface fluorination and Sashi Kandanur and Steven McElwain for tribological experimentation.

# References

[1] T. Blanchet, F. Kennedy, Sliding wear mechanism of polytetrafluoroethylene (PTFE) and PTFE composites, Wear 153 (1) (1992) 229–243.

[2] D. Flom, N. Porile, Friction of Teflon sliding on Teflon, Journal of Applied Physics 26 (9) (1955) 1088–1092.

[3] K. Makinson, D. Tabor, Friction + transfer of polytetrafluoroethylene, Nature 201 (491) (1964) 464.

[4] K. Mclaren, D. Tabor, Visco-elastic properties and friction of solids – friction of polymers – influence of speed and temperature, Nature 197 (487) (1963) 856.

[5] R. Steijn, Sliding experiments with polytetrafluoroethylene, ASLE Transactions 11 (3) (1968) 235.

[6] K. Tanaka, Y. Uchiyama, S. Toyooka, Mechanism of wear of polytetrafluoroethylene, Wear 23 (2) (1973) 153–172.

[7] C. Bunn, A. Cobbold, R. Palmer, The fine structure of polytetrafluoroethylene, Journal of Polymer Science 28 (117) (1958) 365–376.

[8] C. Speerschneider, C. Li, Some observations on structure of polytetrafluoroethylene, Journal of Applied Physics 33 (5) (1962) 1871.

[9] N. Suh, Delamination theory of wear, Wear 25 (1) (1973) 111–124.

[10] S. Bahadur, D. Gong, The action of fillers in the modification of the tribological behavior of polymers, Wear 158 (1–2) (1992) 41–59.

[11] T. Blanchet, Y. Peng, Wear resistant irradiated FEP unirradiated PTFE composites, Wear 214 (2) (1998) 186–191.

[12] B. Burroughs, J. Kim, T. Blanchet, Boric acid self-lubrication of $B_2O_3$-filled polymer composites, Tribology Transactions 42 (3) (1999) 592–600.

[13] F. Li, et al., The tribological behaviors of copper-coated graphite filled PTFE composites, Wear 237 (1) (2000) 33–38.

[14] Z. Lu, K. Friedrich, On sliding friction and wear of PEEK and its composites, Wear 181 (1995) 624–631.

[15] B. Menzel, T. Blanchet, Effect of particle size and volume fraction of irradiated FEP filler on the transfer wear of PTFE, Lubrication Engineering 58 (9) (2002) 29–35.

[16] J.K. Lancaster, Polymer-based bearing materials – role of fillers and fiber reinforcement in wear, Wear 22 (3) (1972) 412.

[17] N. Sung, N. Suh, Effect of fiber orientation on friction and wear of fiber reinforced polymeric composites, Wear 53 (1) (1979) 129–141.

[18] K. Tanaka, S. Kawakami, Effect of various fillers on the friction and wear of polytetrafluoroethylene-based composites, Wear 79 (2) (1982) 221–234.

[19] B. Briscoe, A. Pogosian, D. Tabor, Friction and wear of high-density polythene – action of lead oxide and copper oxide fillers, Wear 27 (1) (1974) 19–34.

[20] D. Gong, Q. Xue, H. Wang, ESCA study on tribochemical characteristics of filled PTFE, Wear 148 (1) (1991) 161–169.

[21] D. Gong, et al., Effect of tribochemical reaction of polytetrafluoroethylene transferred film with substrates on its wear behavior, Wear 137 (2) (1990) 267–273.

[22] T. Blanchet, F. Kennedy, D. Jayne, XPS analysis of the effect of fillers on PTFE transfer film development in sliding contacts, Tribology Transactions 36 (4) (1993) 535–544.

[23] S. Bahadur, D. Tabor, The wear of filled polytetrafluoroethylene, Wear 98 (1–3) (1984) 1–13.

[24] B. Ash, L. Schadler, R. Siegel, Glass transition behavior of alumina/polymethylmethacrylate nanocomposites, Materials Letters 55 (1–2) (2002) 83–87.

[25] A. Eitan, et al., Reinforcement mechanisms in MWCNT-filled polycarbonate, Composites Science and Technology 66 (9) (2006) 1162–1173.

[26] C. Ng, L. Schadler, R. Siegel, Synthesis and mechanical properties of TiO$_2$-epoxy nanocomposites, Nanostructured Materials 12 (1–4) (1999) 507–510.

[27] A. Zhu, S. Sternstein, Nonlinear viscoelasticity of nanofilled polymers: interfaces, chain statistics and properties recovery kinetics, Composites Science and Technology 63 (8) (2003) 1113–1126.

[28] O. Wang, Q. Xue, W. Shen, The friction and wear properties of nanometre SiO$_2$ filled polyetheretherketone, Tribology International 30 (3) (1997) 193–197.

[29] Q. Wang, et al., An investigation of the friction and wear properties of nanometer Si$_3$N$_4$ filled PEEK, Wear 196 (1–2) (1996) 82–86.

[30] Q. Wang, et al., The effect of particle size of nanometer ZrO$_2$ on the tribological behaviour of PEEK, Wear 198 (1–2) (1996) 216–219.

[31] Q. Wang, et al., The friction and wear characteristics of nanometer SiC and polytetrafluoroethylene filled polyetheretherketone, Wear 243 (1–2) (2000) 140–146.

[32] Q. Xue, Q. Wang, Wear mechanisms of polyetheretherketone composites filled with various kinds of SiC, Wear 213 (1–2) (1997) 54–58.

[33] F. Li, et al., The friction and wear characteristics of nanometer ZnO filled polytetrafluoroethylene, Wear 249 (10–11) (2001) 877–882.

[34] W. Chen, et al., Tribological behavior of carbon-nanotube-filled PTFE composites, Tribology Letters 15 (3) (2003) 275–278.

[35] W. Sawyer, et al., A study on the friction and wear behavior of PTFE filled with alumina nanoparticles, Wear 254 (5–6) (2003) 573–580.

[36] D. Burris, W. Sawyer, Tribological sensitivity of PTFE/alumina nanocomposites to a range of traditional surface finishes, Tribology Transactions 48 (2) (2005) 147–153.

[37] D. Burris, W. Sawyer, Improved wear resistance in alumina-PTFE nanocomposites with irregular shaped nanoparticles, Wear 260 (7–8) (2006) 915–918.

[38] S. Mcelwain, Wear Resistant PTFE Composites via Nano-Scale Particles, Master's Thesis, Rensselaer Polytechnic Institute, Troy, New York, 2006.

[39] S. Kandanur, Master of Science Thesis, Rensselaer Polytechnic Institute, Tory, New York, 2007.

[40] S. Han, T. Blanchet, Experimental evaluation of a steady-state model for the wear of particle-filled polymer composite materials, Journal of Tribology-Transactions of the ASME 119 (4) (1997) 694–699.

[41] P. Bhimaraj, et al., Effect of matrix morphology on the wear and friction behavior of alumina nanoparticle/poly(ethylene) terephthalate composites, Wear 258 (9) (2005) 1437–1443.

[42] C. Schwartz, S. Bahadur, Studies on the tribological behavior and transfer film-counterface bond strength for polyphenylene sulfide filled with nanoscale alumina particles, Wear 237 (2) (2000) 261–273.

[43] G. Beamson, et al., Characterization of PTFE on silicon wafer tribological transfer films by XPS, imaging XPS and AFM, Surface and Interface Analysis 24 (3) (1996) 204.

[44] D. Breiby, et al., Structural surprises in friction-deposited films of poly(tetrafluoroethylene), Macromolecules 38 (6) (2005) 2383–2390.

[45] C. Pooley, D. Tabor, Friction and molecular structure – behavior of some thermoplastics, Proceedings of the Royal Society of London Series A-Mathematical and Physical Sciences 329 (1578) (1972) 251.

[46] D. Wheeler, The transfer of polytetrafluoroethylene studied by X-ray photoelectron-spectroscopy, Wear 66 (3) (1981) 355–365.

[47] J. Wittmann, P. Smith, Highly oriented thin-films of poly(tetrafluoroethylene) as a substrate for oriented growth of materials, Nature 352 (6334) (1991) 414–417.

[48] T. Blanchet, A model for polymer composite wear behavior including preferential load support and surface accumulation of filler particulates, Tribology Transactions 38 (4) (1995) 821–828.

[49] S. Abdou, R. Mohamed, Characterization of structural modifications in poly-tetra-fluoroethylene induced by electron beam irradiation, Journal of Physics and Chemistry of Solids 63 (3) (2002) 393–398.

[50] B. Briscoe, H. Mahgerefteh, S. Suga, The effect of gamma irradiation on the pressure dependence of the room temperature transition in PTFE, Polymer 44 (3) (2003) 783–791.

[51] B. Fayolle, L. Audouin, J. Verdu, Radiation induced embrittlement of PTFE, Polymer 44 (9) (2003) 2773–2780.

[52] O. Harling, G. Kohse, K. Riley, Irradiation performance of polytetrafluoroethylene (Teflon(R)) in a mixed fast neutron and gamma radiation field, Journal of Nuclear Materials 304 (1) (2002) 83–85.

[53] X. Zhao, et al., An experimental study of low earth orbit atomic oxygen and ultraviolet radiation effects on a spacecraft material – polytetrafluoroethylene, Polymer Degradation and Stability 88 (2) (2005) 275–285.

[54] H. Dorschner, U. Lappan, K. Lunkwitz, Electron beam facility in polymer research: radiation induced functionalization of polytetrafluoroethylene, Nuclear Instruments & Methods in Physics Research Section B-Beam Interactions with Materials and Atoms 139 (1–4) (1998) 495–501.

[55] U. Lappan, et al., Number-average molecular weight of radiation-degraded poly(tetrafluoroethylene). An end group analysis based on solid-state NMR and IR spectroscopy, Polymer 43 (16) (2002) 4325–4330.

[56] U. Lappan, et al., Radiation-induced branching and crosslinking of poly (tetrafluoroethylene) (PTFE), Nuclear Instruments & Methods in Physics Research Section B-Beam Interactions with Materials and Atoms 185 (2001) 178–183.

[57] A. Oshima, et al., Chemical structure and physical properties of radiation-induced cross-linking of polytetrafluoroethylene, Radiation Physics and Chemistry 62 (1) (2001) 39–45.

[58] E. Katoh, et al., Evidence for radiation induced crosslinking in polytetrafluoroethylene by means of high-resolution solid-state F-19 high-speed MAS NMR, Radiation Physics and Chemistry 54 (2) (1999) 165–171.

[59] B. Fuchs, U. Scheler, Branching and cross-linking in radiation-modified poly(tetrafluoroethylene): a solid-state NMR investigation, Macromolecules 33 (1) (2000) 120–124.

[60] T. Blanchet, Y. Peng, Wear-resistant polytetrafluoroethylene via electron irradiation, Lubrication Engineering 52 (6) (1996) 489–495.

[61] D. Burris, W. Sawyer, A low friction and ultra low wear rate PEEK/PTFE composite, Wear 261 (3–4) (2006) 410–418.

[62] D. Wagner, R. Vaia, Nanocomposites: issues at the interface, Materials Today (2004) 38–42.

[63] P. He, et al., Surface modification and ultrasonication effect on the mechanical properties of carbon nanofiber/polycarbonate composites, Composites Part A-Applied Science and Manufacturing 37 (9) (2006) 1270–1275.

[64] S. Sternstein, et al., Reinforcement and nonlinear viscoelasticity of polymer melts containing mixtures of nanofillers, Rubber Chemistry and Technology 78 (2) (2005) 258–270.

[65] M. Maiti, A. Bhowmick, New insights into rubber–clay nanocomposites by AFM imaging, Polymer 47 (17) (2006) 6156–6166.

[66] H. Koerner, et al., Deformation–morphology correlations in electrically conductive carbon nanotube thermoplastic polyurethane nanocomposites, Polymer 46 (12) (2005) 4405–4420.

[67] D. Ratna, et al., Poly(ethylene oxide)/clay nanocomposite: thermomechanical properties and morphology, Polymer 47 (11) (2006) 4068–4074.

[68] K. Yang, et al., Morphology and mechanical properties of polypropylene/calcium carbonate nanocomposites, Materials Letters 60 (6) (2006) 805–809.

[69] A. Yasmin, et al., Mechanical and thermal behavior of clay/epoxy nanocomposites, Composites Science and Technology 66 (2006) 2415–2422.

[70] Y. Kojima, et al., Mechanical-properties of nylon 6-clay hybrid, Journal of Materials Research 8 (5) (1993) 1185–1189.

[71] H. Yang, et al., Crystal growth in alumina/poly(ethylene terephthalate) nanocomposite films, Journal of Polymer Science, Part B: Polymer Physics 45 (7) (2007 April 1) 747–757.

[72] Z. Xiao, et al., Probing the use of small-angle light scattering for characterizing structure of titanium dioxide/low-density polyethylene nanocomposites, Journal of Polymer Science Part B-Polymer Physics 44 (7) (2006) 1084–1095.

[73] E. Petrovicova, et al., Nylon 11/silica nanocomposite coatings applied by the HVOF process. I. Microstructure and morphology, Journal of Applied Polymer Science 77 (8) (2000) 1684–1699.

[74] F. Kuchta, et al., Materials with improved properties from polymer–ceramic-nanocomposites, Materials Research Society Symposium Proceedings 576 (1999) 363–368.

[75] E. Clark, The molecular conformations of polytetrafluoroethylene: forms II and IV, Polymer 40 (16) (1999) 4659–4665.

[76] J. Weeks, E. Clark, R. Eby, Crystal-structure of the low-temperature phase(II) of polytetrafluoroethylene, Polymer 22 (11) (1981) 1480–1486.

[77] J. Weeks, R. Eby, E. Clark, Disorder in the crystal-structures of phase-1 and phase-2 of copolymers of tetrafluoroethylene and hexafluoropropylene, Polymer 22 (11) (1981) 1496–1499.

[78] B. Farmer, R. Eby, Energy calculations of the crystal-structure of the low-temperature phase (II) of polytetrafluoroethylene, Polymer 22 (11) (1981) 1487–1495.

[79] C. Bunn, E. Howells, Structures of molecules and crystals of fluorocarbons, Nature 174 (4429) (1954) 549–551.

[80] H. Rigby, C. Bunn, A room-temperature transition in polytetrafluoroethylene, Nature 164 (4170), 1949. 583–583.

[81] M. Kimmig, G. Strobl, B. Stuhn, Chain reorientation in poly(tetrafluoroethylene) by mobile twin-helix reversal defects, Macromolecules 27 (9) (1994) 2481–2495.

[82] F. Tieyuan, et al., Study of factors affecting room temperature transition of polytetrafluoroethylene, Chinese Journal of Polymer Science 2 (1986) 170–179.

[83] C. Marega, et al., Relationship between the size of the latex beads and the solid–solid phase transitions in emulsion polymerized poly(tetrafluoroethylene), Macromolecules 37 (15) (2004) 5630–5637.

[84] M. D'amore, et al., Disordered chain conformations of poly(tetrafluoroethylene) in the high-temperature crystalline form I, Macromolecules 37 (25) (2004) 9473–9480.

[85] H. Ting-Yung, N. Eiss, Effects of molecular weight and crystallinity on wear of polytetra-fluoroethylene, in: Wear of Materials: International Conference on Wear of Materials, 1983, pp. 636–642.

[86] N. McCook, et al., Cryogenic friction behavior of PTFE based solid lubricant composites, Tribology Letters 20 (2) (2005) 109–113.

[87] J. Joyce, Fracture toughness evaluation of polytetrafluoroethylene, Polymer Engineering and Science 43 (10) (2003) 1702–1714.

[88] E. Brown, D. Dattelbaum, The role of crystalline phase on fracture and microstructure evolution of polytetrafluoroethylene (PTFE), Polymer 46 (9) (2005) 3056–3068.

[89] E. Brown, et al., The effect of crystallinity on the fracture of polytetrafluoroethylene (PTFE), Materials Science & Engineering C-Biomimetic and Supramolecular Systems 26 (8) (2006) 1338–1343.

[90] P. Rae, D. Dattelbaum, The properties of poly (tetrafluoroethylene) (PTFE) in compression, Polymer 45 (22) (2004) 7615–7625.

[91] P. Rae, E. Brown, The properties of poly(tetrafluoroethylene) (PTFE) in tension, Polymer 46 (19) (2005) 8128–8140.

[92] R. Pucciariello, V. Villani, Phase behavior at low temperature of poly(tetrafluoroethylene) by temperature-modulated calorimetry, Journal of Fluorine Chemistry 125 (2) (2004) 293–302.

[93] R. Pucciariello, V. Villani, Melting and crystallization behavior of poly (tetrafluoroethylene) by temperature modulated calorimetry, Polymer 45 (6) (2004) 2031–2039.

# Development of nanostructured slide coatings for automotive components

# 18

**Andreas Gebhard\*, Frank Haupert†, Alois K. Schlarb‡**

*\*NanoProfile GmbH, Kaiserslautern, Germany,*
*†Hochschule Hamm-Lippstadt, Hamm, Germany,*
*‡Lehrstuhl für Verbundwerkstoffe, Composite Engineering cCe,*
*University of Kaiserslautern, Kaiserslautern, Germany*

## CHAPTER OUTLINE HEAD

## 18.1 INTRODUCTION

One of the major drawbacks of the internal combustion engine is its dependence on limited fossil fuel and, coupled to this, its emission of greenhouse gases like carbon dioxide ($CO_2$) or nitric oxides ($NO_x$) as well as the emission of harmful particulates [1]. In 2004, the overall $CO_2$ emission in the US was estimated to be 6.0 billion tons. Thirty percent of this, that is, 1.8 billion tons, is caused by the burning of fossil fuels in transportation [2]. It has been estimated that a 10% reduction of mechanical losses would lead to a 1.5% reduction in fuel consumption [3], which would correspond to an annual

Tribology of Polymeric Nanocomposites. http://dx.doi.org/10.1016/B978-0-444-59455-6.00018-0

**Table 18.1** Commercially Available Slide Coatings (Selection)

| Manufacturer | Brand | Solvent | Solid Lubricant(s) | Binder | Properties |
|---|---|---|---|---|---|
| 3M | Scotchkote-series | Powder coating | Unknown | EP | Corrosion protection |
| Acheson industries | DAG 213 | Unknown | Graphite | EP | Good adhesion, good lubrication at elevated temperatures, good wear protection |
| Acheson industries | DAG 154 | Isopropanol | Graphite | Cellulose | Max. operation temperature 400°C, for use in internal combustion engines, dries at room temperature |
| Acheson industries | Molydag 250 | Organic | MoS$_2$, Graphite | Thermoset | Max. operation temperature 350°C, good wear protection |
| SermaGard | SermaLube | Unknown | Unknown | Unknown | – |
| Everlube | Everlube 9002 | Water | MoS$_2$ | EP | High resistance to sliding and abrasive wear, highly resistant to chemicals, for high loading, water based |
| Klüber lubrication | Klübertop 06-111 | Unknown | MoS$_2$ | Organic | Maximum operation temperature 220°C, highly resistant to chemicals, loading >10MPa |
| KS Paul | Moly-Paul ITC-bond | Methylacetate, isopropanol | MoS$_2$ | Resin | High corrosion and wear resistance, maximum operation temperature 300°C, used for piston skirts |
| Lubrication engineers | Almasol 9200 | Unknown | Unknown | Resin | Operation temperature: −73 bis 343°C, loading up to 100.000 psi (670MPa) |
| OKS | OKS 536 | Water | Graphite | Unknown | Highly loadable, temperature resistant, water based |
| Solvay adv. polymers | AI-10 | Unknown | Unknown | Polyamide imide | Maximum operation temperature 260°C, high chemical resistance (e.g. jet fuel, motor oil, diesel fuel) |
| Whitford worldwide | Xylan 5250 | Water | PTFE | Resin | Corrosion and wear protection, water based |

saving of 27 million tons of $CO_2$ in the US alone. The annual benefit of the reduction of the internal combustion engine's friction and wear to the US economy are estimated to be US$ 120 billion. Overall, the combined effect of the increasing environmental awareness of customers and of the more stringent global competition and governmental regulations on greenhouse gas emission has made the increase of fuel economy a highly important field of tribological research during the past decades [4–8].

In a medium-sized passenger vehicle, the major portion of energy loss occurs in the engine (41% of total energy consumption), followed by vehicle weight (29%) and rolling resistance (15%) [9,10]. Overall, the engine accounts for 66% of the total frictional loss, most of which is caused by the sliding of the piston rings (19%) and the piston skirt (25%) against the cylinder wall and by the engine's bearings (22%). In two of these tribological systems, namely, slide bearings and piston skirt coatings, polymeric slide coatings constitute the state of the art and effectively abate friction and increase wear resistance. Modern slide coatings consist of three basic components: a binder, solid lubricants, and a solvent. Since they are basically varnishes with color pigments exchanged by friction and wear-reducing components, they can be applied using standard procedures, for example, spraying followed by heating in an oven [11,12]. Table 18.1 shows a selection of the wide range of the commercially available slide coatings. As can be seen, thermosets are the most popular polymeric binders for slide coatings, but there are also coatings based on high-performance thermoplastics like polyamide imide (PAI).

There are also a wide variety of functional fillers available, for example, solid lubricants for reducing friction, reinforcing fibers for high mechanical strength or hard particles for abrasion resistance. Recently, all these fillers have become commercially available as nanoscale fillers, which leads to significant research in the field of nanocomposites and especially in the field of nanocomposite tribology. In automotive applications, nanocomposites are considered to be the key to reduced frictional loss and increased wear resistance. The possible benefits are the reduction of the oil and fuel consumption, the prolongation of tribological components' lifetimes their ability to operate at higher $pv$ products and at higher temperatures. Furthermore, slide coatings based on nanocomposites offer the usual advantages of polymeric anti friction coatings, namely, increased corrosion resistance compared to metallic tribosystems, reliable long-term protection from fretting and galling, low cost, wide applicability and superior dry running capabilities, which is important in the case of oil lack or during cold starting.

## 18.2 MECHANICAL PROPERTIES OF SLIDE COATINGS

In automotive applications, some of the main requirements for slide coatings are high operation temperature, high mechanical properties, low friction, and a high wear resistance. Once the slide coating is applied to its substrate and cured, all material properties are in principle influenced by both matrix and fillers (Fig. 18.1). Nevertheless, there are some properties that are predominantly determined by the matrix, for example, thermal stability, damage tolerance, toughness, and chemical resistance.

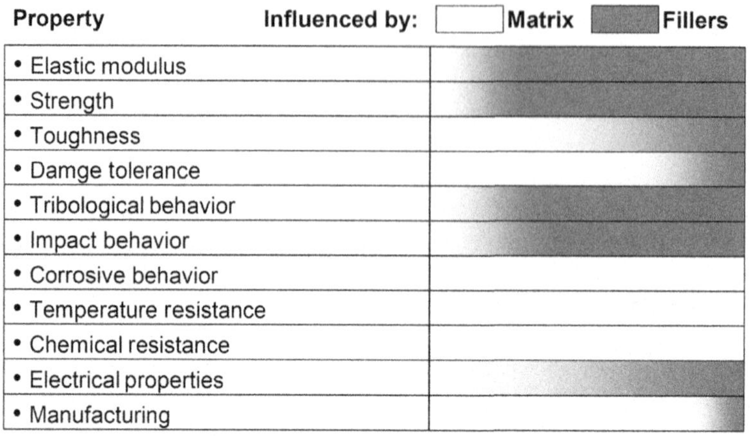

| Property | Influenced by: | Matrix | Fillers |
|---|---|---|---|
| • Elastic modulus | | | |
| • Strength | | | |
| • Toughness | | | |
| • Damge tolerance | | | |
| • Tribological behavior | | | |
| • Impact behavior | | | |
| • Corrosive behavior | | | |
| • Temperature resistance | | | |
| • Chemical resistance | | | |
| • Electrical properties | | | |
| • Manufacturing | | | |

**FIGURE 18.1**

Relative influence of matrix and fillers on selected properties of a slide coating.

On the other hand, there are properties that mainly depend on the fillers, like strength, elastic modulus, wear resistance, and coefficient of friction.

In order to obtain high mechanical properties, fibers are the most important common fillers for slide coatings. They increase the coating's load carrying capacity, its maximum operation temperature, and they reduce its thermal coefficient of expansion and thereby the thermal stress between the coating and the metallic substrate.

In general, the improvement of the wear resistance and the lowering of the coefficient of friction of bulk polymeric composites by functional fillers also hold true for thin films used as slide coatings. The action of nanoscale and microscale fillers on bulk materials is extensively reviewed in a previous chapter by L. Chang et al. Therefore, this section focuses on the description of selected structure–property relationships that allow for the optimization of a coating's operation temperature and compression loading capability by the careful selection of a proper matrix and optimum filler contents.

## 18.2.1  Operation temperature

The temperature threshold up to which polymers can be used as matrices for coatings is limited by the thermal stability of the matrix itself, the retention of its mechanical properties and the deterioration of the composite's tribological performance at elevated temperatures [13,14]. As will be demonstrated in Section 16.4.2.1 nanosized antiwear modifiers have proven to be especially useful in obtaining a high wear resistance and a low coefficient of friction at elevated temperatures.

Basically, the retention of the composite's mechanical properties at elevated temperatures can be facilitated by choosing thermoplastics with a high and stable stiffness and hardness up to the glass transition temperature (which should be high too)

**Table 18.2** Residual Mechanical Properties at 150 °C of a Bisphenol A-Resin Cured with Two Different Amine Hardeners

|  | Bisphenol A/DDM | | Bisphenol A/DDS | |
| --- | --- | --- | --- | --- |
|  | At 20 °C | At 150 °C | At 20 °C | At 150 °C |
| Tensile strength (MPa) | 53 | 19 | 59 | 37 |
| Tensile modulus (GPa) | 2.75 | 1.54 | 3.07 | 1.47 |
| Compressive strength (MPa) | 111 | 29 | 107 | 63 |
| Compressive modulus (GPa) | 2.67 | 0.72 | 2.0 | 1.28 |
| Flexural strength (MPa) | 116 | 41 | 102 | 49 |
| Flexural modulus (GPa) | 2.73 | 1.68 | 2.79 | 1.16 |

*Data from Refs. [37,75].*

or thermosets with a high softening temperature, respectively. While the choice of a matrix defines the basic mechanical and thermal performance of the composite the maximum operation temperature of thermoplastics can be increased by the addition of functional fillers, for example, glass or carbon fibers. This increase in the thermal resistance by the incorporation of fibers is also feasible for thermosets. While for thermoplastics the composite's basic performance concerning thermal resistance and retention of mechanical properties is set by the matrix, a distinct influence of the specific chemical nature of the resin/hardener combination is observed for thermosets. Table 18.2 demonstrates this by comparing the thermal retention of selected mechanical properties of a Bisphenol A resin that has been cured with 4,4′-diamino diphenyl methane (DDM) and 4,4′-diamino diphenyl sulfone (DDS), respectively. The higher performance of DDS- over DDM-cured resins is attributed to a more rigidly crosslinking due to the higher rigidity of the sulfone moiety compared to the methylene moiety [15].

Furthermore, does the chemical structure of the resin and the hardener significantly influence the cured material's resistance to thermal degradation? In contrast to the maximum operation temperature and the retention of mechanical properties, the thermal degradation of the matrix is hardly improved by functional fillers. While most fillers like hard ceramic particles, layered silicates, fibers, solid lubricants, or carbon nanotubes are unaffected by temperatures up to 400 °C, polymeric matrices suffer severe thermal degradation. The main mechanisms therefore are chemical reactions that break up the polymer chains to form smaller and more stable molecules. One of the first unambiguous demonstrations of this effect was demonstrated for Bisphenol A that has been cured with DDM and with a Methyl nadic anhydride (MNA), an acidic anhydride hardener [16]. In Thermogravimetric Analysis under nitrogen atmosphere, the diglycidyl ether of Bisphenol A (DGEBA) cured with DDM does not exhibit any weight loss up to 300 °C. For MNA, however, thermal degradation occurs in a two-step process with the first significant weight loss taking place at only 150 °C. This is caused by the decarboxylation of esters formed during the curing.

Furthermore, this decarboxylation is considered to generate free radicals, which are highly reactive chemical species that are capable of damaging the resin's molecular structure beyond the effect of the decarboxylation alone. Since amine hardeners form thermally more stable β-hydroxypropylamines instead of esters, resins cured with polyamines generally exhibit a higher thermal stability.

Information concerning the chemical resistance of polymers is frequently given for standard conditions. Very often this does mean exposure at 20–25 °C for some hours to a few days. When selecting a polymeric binder for a high-temperature slide coating that will be exposed to chemicals, it must be kept in mind that the chemical resistance deteriorates with high increasing temperature. For certain forms of chemical attack like oxidation or hydrolysis, this deterioration is disproportionately severe with increasing temperature. PAI, for example, is significantly impaired by hot water. This is especially important when choosing PAI for slide coatings that are exposed to oil and/or fuel. During short trip cold start conditions, these two liquids may contain up to 5% water [4].

## 18.2.2 Compression loading capability

Since slide coatings substitute metallic surfaces like the bronze layer in journal bearings, they need a high compressive strength and a high compression modulus. As can be seen from Fig. 18.1, the mechanical properties of a composite materials mainly depend upon the action of the fillers. The compressive properties of polymeric composites are most commonly improved by the addition of short fibers in contents between 10 and 30 vol.%. While clear relationships between the tensile properties of a fiber-reinforced composite and the fundamental tensile properties of the fibers themselves have been established, only very little is known about the analogous relationship for compression loading. Mainly, this is due to the lack of an experimental procedure that allows for the direct determination of the compressive properties of fibers with diameters of only a few micrometers. Therefore, indirect methods like the subjecting of unidirectional reinforced polymers to macroscopic standard compression tests are used. From these, the fibers' fundamental compressive properties are then estimated by a rule of mixtures [17,18]. For a detailed review of testing methods for the compressive properties of fibers, see the review by Kumar [19].

The predominant failure modes of polymeric fiber-reinforced composites under compression loading include fiber kinking, longitudinal splitting of the fiber, and pure compression or shear failure [20]. Most of the currently available microbuckling models [21,22] assume that the kinking of the fiber is fostered by insufficient lateral support from the matrix. This lateral support originates from the shearing and compression of the matrix due to the radial expansion of an axially compressed fiber. This mechanism has been investigated experimentally by an indirect test method that subjects a composite to a compressive strain higher than the fibers' ultimate strain [23]. The fibers' mechanical properties are then estimated from the mean fragment length of fibrillar failed fibers. Using this method, the compressive strength of different Polyacrylnitrile (PAN)- and pitch-based fibers in an epoxy resin (EP) matrix

**Table 18.3** Compressive Strengths of Different Carbon Fibers

|  | Torayca T-300 (HT-PAN Fiber) | Torayca M-40 (HM-PAN Fiber) | Tonen HTX (HT-Pitch Fiber) | Tonen HMX (HM-Pitch Fiber) |
|---|---|---|---|---|
| Tensile strength (GPa) | 3.50 | 2.88 | 3.34 | 4.33 |
| Estimated compressive strength (GPa) | 2.06 | 0.78 | 1.25 | 0.54 |
| Estimated compressive strength (% of tensile strength) | 59 | 27 | 37 | 12 |

*Data from Ref. [23].*

were studied. In order to obtain different lateral support the modulus of the matrix was varying the sample's temperature from 20 to 100 °C. For all fibers studied, a steady decrease of the composite's apparent compressive strength was found with increasing temperature. Since the real compressive strength of carbon fibers is considered to be constant in this temperature interval, the observed effect must be due to a change in the fiber–matrix interaction. One of these interactions is a thermal stress between the fiber and the surrounding matrix, which acts perpendicularly onto the fiber surface. This stress originates from the different coefficients of thermal expansion of the matrix and fiber during the curing of a resin or during the cooling down of a thermoplastic. For the investigated EP/SCF systems, a change in the composite's microscopic failure mode was observed: while at temperatures, <100 °C compressive failure of the fiber itself occurs, buckling is found to be the predominant failure mode at ≥100 °C. The thermal stress due to different thermal expansion was estimated to be about 6 MPa at 20 °C. At 100 °C, however, this stress is relaxed to only about 0.01 MPa, and at the same time, the stiffness of the EP matrix is reduced to 1% of its corresponding stiffness at 20 °C. Therefore, the observed change of the failure mode is reasonably attributed to the loss of lateral support by thermal deterioration of the fiber/matrix prestrain and of the matrix stiffness.

The "real" fiber compressive strengths were obtained from the observed temperature dependence of the embedded fibers strength by extrapolation to zero thermal stress and zero elastic modulus of the matrix (Table 18.3).

Most interesting is the fact that there are significant differences between HT- and HM- as well as between PAN- and pitch-based fibers. While the compressive strength of the HT-PAN fiber Torayca T-300 amounts to nearly 60% of its tensile strength (3.5 GPa), the compressive strength of the HM-fiber Torayca M-40 amounts to only 27% of its tensile strength (2.9 GPa). The same result, namely, the HT-fibers having a higher absolute and relative compressive strength than HM-fibers is also found for pitch-based fibers. Overall, PAN-based fibers possess higher absolute compressive strengths than do their pitch-based counterparts. Additionally, their compressive strength (normalized to their tensile strength) is significantly higher than for pitch-based fibers.

| Table 18.4 Compressive Strengths of Various Fibers | |
| --- | --- |
| | **Compressive Strength (GPa)** |
| UHMWPE | 0.17 |
| Kevlar | 0.34–0.48 |
| PBO | 0.2–0.4 |
| PBT | 0.26–0.41 |
| Pitch CF | 0.48 |
| S-2 Glass | 1.1 |
| PIPD | 1.0–1.7 |
| PAN CF | 1.05–2.75 |
| *Data from Refs. [20,24].* | |

Besides pure carbon fibers, there are also fibers spun from polymers based on defined organic species. The most common fibers are based on Ultrahigh Molecular Weight Polyethylene (UHMWPE), Polyethylene Terephthalate and Polybutylene Terephthalate (PBT), organic Aramides (Kevlar, Nomex), and Polyphenylene Benzobisoxazole (PBO). PBO is currently the polymeric fiber with the best tensile properties, that is, it has the highest tensile modulus and tensile strength of all polymeric fibers. Additionally, PBO is among the most thermally stable of all commercially available polymeric fibers. However, it exhibits a rather poor axial compressive strength of only 200–400 MPa, which corresponds to only 4–7% of its tensile strength [24]. The axial compressive strength being only a fraction of the corresponding tensile strength is a common flaw of most polymeric fibers as can be seen from Table 18.4, which compares the compressive strengths of several fiber types. Currently, the polymeric fiber with the highest axial compressive strength is based on Polydi(imidazo) pyridinylene phenylene (PIPD). This polymer forms strong intermolecular hydrogen bonds in two dimensions, which translate into a fiber compressive strength of 1.017 GPa. This is the highest value found for polymeric fibers to this date.

Nevertheless, the compressive strength of PAN-based carbon fibers is unmatched by any other fiber type, and concerning the maximum achievable improvement, it is therefore the first choice for enhancing the compressive strength of slide coatings. However, do carbon fibers with a high axial compressive strength tend to have a lower tensile modulus [25]? For a detailed review on the compressive properties of carbon and polymeric fibers, see the reviews of Kumar [19,24–26] and the references given therein.

Overall, the matrix plays an important role in achieving high compressive loading composites by fiber reinforcement. While it is true that the fiber's compressive strength sets the upper limit that a composite's strength can reach, a proper matrix that is able to laterally support the fiber is required in order to prevent early failure due to microbuckling. Since the shear modulus of the matrix determines the maximum radial load that can be transferred from the fiber to the matrix at any given axial fiber strain, a matrix with a high modulus of elasticity is favorable. According to the combined influence of fiber stiffness and lateral support by the matrix buckling occurs for low modulus fibers (e.g. aramid fibers) and low modulus matrices ($E < 3$ GPa), whereas stiff PAN- or pitch-based fibers ($E > 350$ GPa) rather exhibit compressive failure of the fiber itself. Again, this is even more important at elevated temperatures, since the general decrease of mechanical properties and the relaxation of residual thermal stresses between the matrix reduce the matrix' capability to support the fiber.

Compressive failure based on microbuckling can be reduced by higher matrix shear strength and yield strength, by increasing the fiber–matrix adhesion and by reducing fiber waviness and defect content [20,27]. While fiber waviness is rather a problem of laminates, an optimization of the fiber orientation into loading direction should be aimed at for short fiber-reinforced composites so that a bigger fiber fraction is loaded axially. Furthermore, can microbuckling be prevented increasing the resin's shear modulus for increasing the lateral load that is transferred to the matrix at a given strain or by introducing a prestrain between the fiber and the matrix during fabrication. The enhancement of the fiber–matrix bonding strength can reduce fiber slippage and increases the transferable axial load. If pure compressive failure of the fibers is the microscopic reason for macroscopic failure, the only way to achieve better loading capability is to replace the current fiber with a stiffer fiber.

## 18.3 MATRICES FOR SLIDE COATINGS

When selecting a polymeric binder for a slide coating, a choice must be made between thermosets and thermoplastics, each of which possesses inherent advantages and disadvantages. Generally, thermosets have a higher mechanical strength and stiffness, but at the same time, they are more brittle than thermoplastics are. The latter are the preferred option for dynamically loaded parts since they have a better damage tolerance, toughness, and damping performance, which greatly help in reducing noise emission.

This section focuses on the detailed discussion of EPs since they are the most common matrices for commercial slide coatings and on PAI since it is a high-performance thermoplastic that is also very frequently used in commercial slide coatings, especially for piston skirts [28,29]. For comparison, two other bulk thermoplastics (as possible candidates for composite coatings) are also briefly discussed, that is, a poly-para-phenylene (PPP) copolymer and polyetheretherketone (PEEK).

### 18.3.1 Polyamide imide

Although being an amorphous thermoplastic PAI exhibits a high mechanical strength and stiffness, a high creep resistance as well as excellent thermal and tribological properties. This is attributed to the stiff polymer backbone, the pronounced noncovalent interaction between the polymer chains and a crosslinking that takes places during a thermal curing after fabrication. The neat PAI's

**Table 18.5** Comparison of the Compressive Properties and the Impact Strengths of Selected High-Temperature Thermoplastics

|  | Compressive Strength (MPa) | Compressive Modulus (GPa) | Impact Strength (Izod, Notched) (J cm$^{-1}$) |
|---|---|---|---|
| Torlon 4203 (PAI) | 166 | 3.3 | 1.06 |
| Tempalux PEI | 151 | 3.3 | 0.53 |
| Techtron PPS | 148 | 3.0 | 0.23 |
| Ketron PEEK | 138 | 3.5 | 0.53 |

compressive strength of 166 MPa and its impact strength are higher than those of other high-performance polymers (Table 18.5). Furthermore, PAI's elastic modulus is >3 GPa, which has been formulated as a minimum value to effectively prevent fiber buckling. By the addition of 30% carbon fibers, PAI's compressive strength at yield is increased to 256 MPa. Another advantage of PAI over other thermoplastics is its high impact strength.

Although PAI is an amorphous thermoplastic, it requires a postcure after application to surfaces in order to increase its molecular weight, which increases the chemical and thermal stabilities of PAI. If correctly tempered, PAI exhibits an excellent temperature resistance and excellent residual mechanical properties at high temperatures (Table 18.6).

The neat PAI-matrix is chemically degraded by hot steam, high pH alkalines like a 30% aqueous solution of sodium hydroxide or by strong organic acids like benzene sulfonic acid or formic acid (88%). The mechanism of the chemical attack is considered to be the disruption of hydrogen bonding and/or the chemical decomposition of the amide bond in the polymer chain backbone. Otherwise, the neat PAI matrix exhibits an excellent chemical resistance against a wide range of weak acids and caustics, aqueous saline solutions, alcohols, aldehydes and ketones, chlorinated organic compounds, esters, ethers, unpolar hydrocarbons, nitriles, nitro compounds and, most important concerning automotive applications, PAI is highly resistant against diesel fuel, gasoline, hydraulic oil, and motor oil [30].

The glass transition temperature of neat PAI is 285 °C, which constitutes its thermal limit since PAI is completely amorphous. Without mechanical loading, no thermal degradation of the neat PAI matrix is observed below 300 °C. At 250 °C, it takes several thousand hours in order to deteriorate PAI's tensile strength noticeably [31]. At a loading of 1.82 MPa, PAI filled with 3 wt.% $TiO_2$ and 0.5 wt.% polytetrafluoroethylene (PTFE) exhibits a heat deflection temperature of 274 °C, which is one of the highest known for polymeric materials. Due to its excellent thermal resistance and its very good mechanical properties up to 230 °C, PAI is highly favorable for the design of tribological coatings for use in the engine compartment of a vehicle or even in an internal combustion engine itself.

Besides its very high-temperature resistance, PAI has a very low coefficient of thermal expansion. With $25 \cdot 10^{-6}$ K$^{-1}$, the thermal expansion of the PAI composite

**Table 18.6** Mechanical and Thermal Properties of Selected PAI Composites [76,30]

| Composition (Wt.-%) | SCF | 0 | 0 | 0 | 30 | 0 | 0 |
|---|---|---|---|---|---|---|---|
| | SGF | 0 | 0 | 30 | 0 | 0 | 0 |
| | Graphite | 20 | 12 | 0 | 0 | 0 | 12 |
| | PTFE | 3 | 3 | 0 | 0 | 0.5 | 8 |
| | $TiO_2$ | 0 | 0 | 0 | 0 | 3 | 0 |
| Density (g cm$^{-3}$) | | 1.51 | 1.46 | 1.61 | 1.48 | 1.42 | 1.50 |
| Tensile strength (MPa) | −196 °C | 130 | – | 204 | 158 | 218 | – |
| | 23 °C | 131 | 164 | 205 | 203 | 192 | 123 |
| | 135 °C | 116 | 113 | 160 | 158 | 117 | 104 |
| | 232 °C | 56 | 73 | 113 | 108 | 66 | 54 |
| Tensile modulus (GPa) | 23 °C | 7.8 | 6.6 | 10.8 | 22.3 | 4.9 | 6.0 |
| Strain at break (%) | −196 °C | 3 | – | 4 | 3 | 6 | – |
| | 23 °C | 7 | 7 | 7 | 6 | 15 | 9 |
| | 135 °C | 15 | 20 | 15 | 14 | 21 | 21 |
| | 232 °C | 17 | 17 | 12 | 11 | 22 | 15 |
| Flexural strength (MPa) | −196 °C | 203 | – | 381 | 315 | 287 | – |
| | 23 °C | 212 | 219 | 338 | 355 | 244 | 189 |
| | 135 °C | 157 | 165 | 251 | 263 | 174 | 144 |
| | 232 °C | 111 | 113 | 184 | 177 | 120 | 100 |
| Flexural modulus (GPa) | −196 °C | 9.6 | – | 14.1 | 24.6 | 7.9 | – |
| | 23 °C | 7.3 | 6.9 | 11.7 | 16.5 | 5.0 | 6.3 |
| | 135 °C | 5.6 | 5.5 | 10.7 | 15.6 | 3.9 | 4.4 |
| | 232 °C | 5.1 | 4.5 | 9.9 | 13.1 | 3.6 | 4.3 |
| Compressive strength (MPa) | 23 °C | 124 | 166 | 264 | 254 | 221 | 130 |
| Compressive modulus (GPa) | 23 °C | 4.0 | 5.3 | 7.9 | 9.9 | 4.0 | – |
| Impact strength (Izod) (J m$^{-1}$) | Notched | 85 | 60 | 80 | 50 | 140 | 69 |
| | Unnotched | 250 | 405 | 505 | 340 | 1070 | – |
| Water absorption (24 h) (%) | | 0.33 | 0.28 | 0.24 | 0.26 | 0.33 | 0.17 |
| Shrinkage (%) | | 0.4 | 0.5 | 0.2 | <0.2 | 0.7 | – |

containing 20 vol.% graphite and 3 vol.% PTFE is very similar to the one of aluminum ($24 \cdot 10^{-6}$ K$^{-1}$). The material is designated as "bearing grade" and is therefore an interesting candidate for the coating of the piston skirts of aluminum pistons (Section 16.4.1). Other common automotive applications of all PAI materials are thrust washers, seal rings, sliding vanes, bushings, clutch rollers, gears, valve plates, and valve seats. Figure 18.2 shows the specific wear rates of selected PAI-bearing grade composites in dry sliding against steel [30]. With one exception, all values

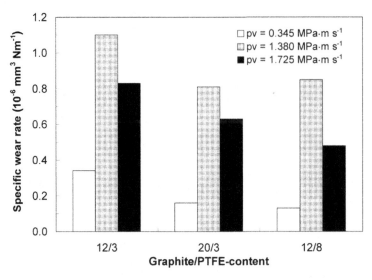

**FIGURE 18.2**

Specific wear rates of the commercially available bearing grade PAI materials.

*Data from Ref. [30].*

lie in the range of $10^{-7}\,mm^3\,Nm^{-1}$, which is among the lowest wear rates known for polymeric composites under the given loading parameters.

## 18.3.2 Epoxy resins

In 1962, US patent No. 3,293,203 filed by G. F. Paulus of Acheson Industries describes a series of thermosettable resin compositions for forming low friction surface coatings [32]. Among other matrices like a thermosettable silicone resin, the epoxy resin Epi-Rez 201 was filled with low-molecular-weight PTFE to form a highly wear resistant coating with good adhesion to steel and a coefficient of friction of only 0.044. Today, EPs are still the most popular matrices for slide coatings and many commercial state-of-the-art products utilize the EP matrix. Examples are the Scotchkote Series (3M), DAG 213 (Acheson Industries), or Everlube 9002 (Everlube Products). The main reasons for this are the beneficial high mechanical strength and stiffness of EPs, their high chemical resistance, their easy processability, and their superior performance at temperatures that occur in automotive applications.

The most distinct feature of EP resins however is the extraordinary wide range of the monomers, which allows for tailoring of coatings to suit individual applications' demands beyond the mere choice of a single matrix and fillers. Epoxy resins consist of two components, one of them being a polyfunctional epoxide that is chemically reacted with a curing agent [33–35]. Today >90% of the world's overall EP production is based on DGEBA. However, composites for high-temperature applications are mainly based on novolac resins since their glass transition temperatures is about

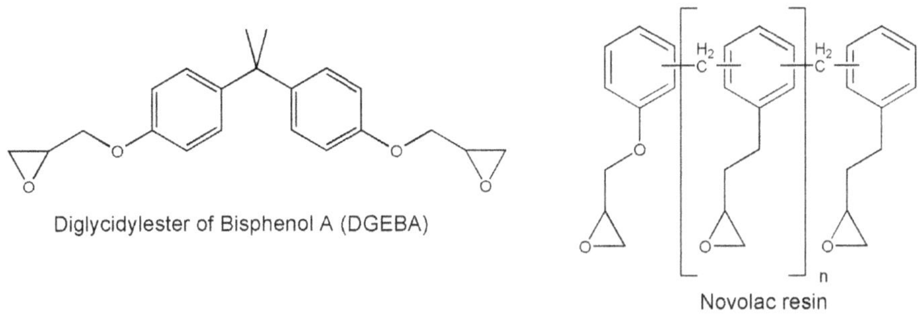

**FIGURE 18.3**

Chemical structure of DGEBA (left) and novolac resins (right).

50 °C higher than that for Bisphenol A resins. Novolacs are oligomers of two to five monomers formed by the condensation of formaldehyde with phenol and the subsequent epoxidation with epichlorohydrin (Fig. 18.3).

Novolacs have superior mechanical and thermal properties as well as higher chemical resistance than DGEBA resins [36]. However, their adhesion to metallic substrates does not completely reach the high level of DGEBA resins. Current commercial novolac resins include for example the Dow Epoxy Novolac series from The Dow Chemical Company [37].

As curing agents for EPs organic polyamines or anhydrides of organic acids are available. Since amine hardeners generally give higher mechanical properties than do anhydride hardeners, they are commonly used to cure resins for high-temperature applications. Amine hardeners can contain aliphatic and aromatic species. Since aromatic groups have a higher intrinsic mechanical stability and a higher resistance to solvents, resins cured with aromatic amine hardeners have a higher mechanical strength, higher $T_g$, and higher chemical resistance. Thus, a Bisphenol A resin cured with Diethylene triamine, an aliphatic hardener, a $T_g$ of 112 °C is obtained. Curing the same resin with DDS, a $T_g$ of 150 °C is obtained.

Furthermore, the size of the monomers has a great influence on the properties of the cured resin. Compact resin molecules as well as compact hardeners generally result in a higher density of covalent bonds. The bonding energies of covalent bonds are about one to two orders of magnitude higher than those of noncovalent bonds, and hence, the strength and stiffness are higher than a network formed by large molecular species with large distances between the epoxy functionalities. Additionally, compact species tend to have a higher content of aromatic groups in their backbone, which are rigidized by π-electron interactions. Aromatic species therefore exhibit a higher thermal stability and a better retention of mechanical properties at elevated temperatures.

Following the structure–property relationships stated above as guiding principles, the achievement of high-performance properties by the careful choice of monomers was demonstrated [27]. Therefore, the trifunctional resin DGEBA resin TDE-85 was cured with different amine hardeners (Fig. 18.4).

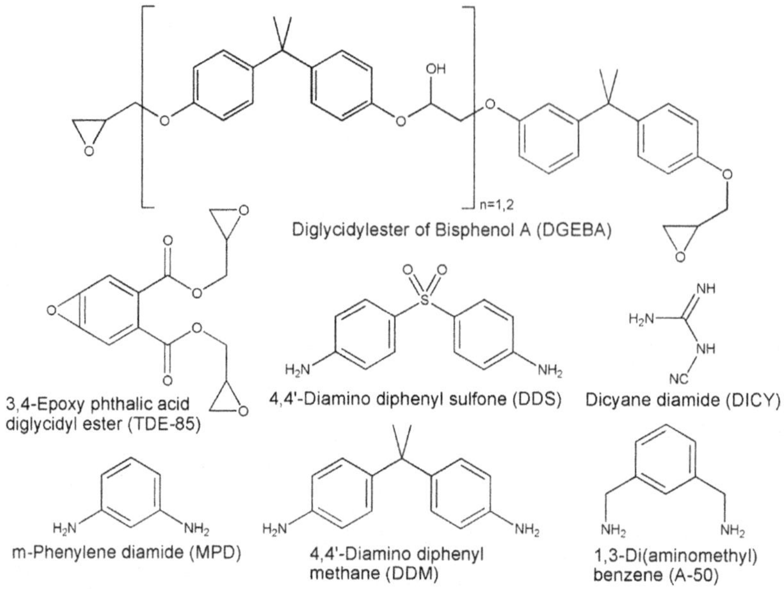

**FIGURE 18.4**

Chemical structure formulae of the resins and hardeners used to study structure–property relationships of EPs.

The series of hardeners has been designed so that the amines are getting more compact from DDM to MPA (meta-phenylene diamine) and that therefore a denser three-dimensional network is achieved and the free volume in the cured material is minimized. As can be seen from Table 18.7, this results in an increase in the bulk density, which is accompanied by an increase of strength and stiffness while reducing the materials impact strength.

The compressive properties, the flexural modulus and the heat distortion temperature, which are essential for the use in highly loaded slide coatings, increase significantly when curing TDE-85 with the smallest hardener (MPA). Compressive strength, compressive modulus, and heat distortion temperature are 15–30% higher than for DDS, which is commonly considered to be one of the most efficient amine hardeners. When curing CYD-128 with the highly compact DICY (Dicyane diamide) hardener, much lower mechanical and thermal properties are obtained. This is attributed to CYD-128 being considerably larger than TDE-85 and having a more flexible molecular backbone, which leads to the formation of a less dense network of covalent bonds with larger, free volumes.

## 18.3.3 PPP copolymer and PEEK

PEEK as a high-temperature thermoplastic polymer is recognized for its excellent mechanical and tribological properties. Therefore, the material is used in many

**Table 18.7** Mechanical and Thermal Properties of Different Cured Model Resins

| Epoxy Resin | TDE-85 | | | | DGEBA |
|---|---|---|---|---|---|
| Hardener | MPD | A-50 | DDS | DDM | DICY |
| Density (g cm$^{-3}$) | 1.369 | 1.318 | 1.311 | 1.309 | 1.215 |
| Tensile strength (MPa) | 85.7 | 86.6 | 75.0 | 72.2 | 73.2 |
| Tensile modulus (GPa) | 5.3 | 5.1 | 4.3 | 4.1 | 2.96 |
| Strain at break (%) | 2.5 | 2.31 | 2.3 | 1.6 | 1.36 |
| Flexural strength (MPa) | 215 | 197 | 133 | 113 | 112 |
| Compressive strength (MPa) | 233 | 228 | 194 | 203 | 125 |
| Compressive modulus (GPa) | 7.3 | 6.67 | 5.59 | 5.32 | – |
| Impact strength (J m$^{-2}$) | 11.9 | 10.4 | 17.1 | 14.6 | 8.6 |
| Heat distortion temperature (°C) | >250 | >250 | 214 | 202 | – |
| Glass transition temperature (°C) | 230 | 228 | 222 | 214 | – |
| Coefficient of thermal expansion (10$^{-6}$ K$^{-1}$) | 44 | 47 | 50 | 44 | 62 |

Data from Ref. [27].

challenging applications especially slide bearings and frictional coatings for high pressures and elevated ambient temperatures [38–42].

PPP copolymer (Teca-max SRP, Ensinger, Germany; or PrimoSpire self-reinforced polyphenylene SRP, Solvay, USA) is a new, ultrahigh-performance-thermoplastic material, which is based on very stiff poly(1,4-phenylene)-structured polymer chains [43–45]. At each phenylene ring, special substituents are attached, which allow it to become more flexible so it can be processed by extrusion or compression molding. In this way, the stiff molecular backbone provides high strength and stiffness to the system, while the substituents lead to better processability. At present, this self-reinforcing polymer can be considered (at room temperature) as the stiffest, strongest, and hardest polymer worldwide (i.e. without an additional type of reinforcement, such as short glass or carbon fibers) [46].

Preliminary studies with PPP have demonstrated a tensile strength of 207 MPa (i.e. higher than the yield strength of general purpose constructional steels, e.g. St 33,DIN 17100 [47]), a Young's modulus of 8300 MPa (much higher than PEEK, with 3650 MPa), a compressive strength of >620 MPa, and an impact toughness of 1200 kJ m$^{-2}$, all at a density of 1.21 g cm$^{-3}$ [46].

The comparison of the PPP's excellent mechanical properties under tensile conditions in comparison to other polymers used in tribological applications is demonstrated in Fig. 18.5. Both modulus and strength are more than two times higher than the corresponding values as measured for thermoplastic PEEK, polyphenylenesulfide (PPS), polyetherimide (PEI), or thermosetting polyimide.

Concerning the flexural and compressive properties, the compressive strength value of PPP especially exceeds those of PEEK, polycarbonate, PPS by a factor of 5 [46].

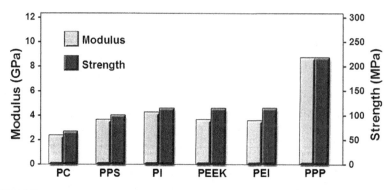

**FIGURE 18.5**

Modulus and strength of different polymeric materials.

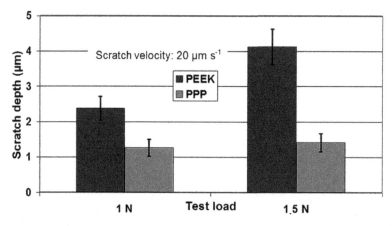

**FIGURE 18.6**

Results of scratch resistance tests of PPP versus PEEK.

The scratch resistance was determined, using a diamond indenter under various loading conditions. The relative resistance of the two different polymers under consideration (PPP vs. PEEK) can be estimated from the penetration scratch depth as a function of normal load, applied to the diamond scratch indenter (Fig. 18.6). Especially at the higher load applied (1.5 N), PPP exhibits a resistance against scratch formation, being 3 times better than the PEEK material. This was evaluated from laser profilometer scans [46].

Numerous studies on the wear of PEEK and its composites against 100 Cr6-steel confirmed that the specific wear rate of the neat polymer varies between $9.3 \cdot 10^{-6}$ and $7.61^{-7} mm^3 Nm^{-1}$, depending on the particular load collective (nominal pressure $p$ and sliding speed $v$) applied [48]. The coefficient of friction measured under these conditions, varied around $\mu = 0.39$ [49]. A comparison of these values to the data measured for PPP under room temperature sliding conditions with a pin-on-disc device clearly showed a higher friction and wear level of the PPP [46].

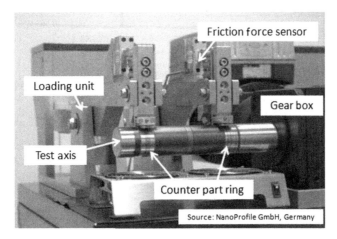

**FIGURE 18.7**

Block-on-ring testing machine of NanoProfile GmbH, Kaiserslautern, Germany. (For color version of this figure, the reader is referred to the online version of this book.)

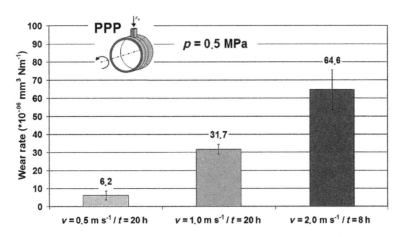

**FIGURE 18.8**

Wear rate of PPP as a function of sliding speed [46].

A direct comparison of the PPP material with PEEK was carried out with a newly built wear testing machine at the company NanoProfile GmbH [50]. In this case, the machine operates at a very low vibration, with high precision and over a wider load/velocity range, following the block-on-ring principle (Fig. 18.7).

The wear rate of PPP increased with increasing sliding speed up to higher values. This trend was found with the block-on-ring configuration (Fig. 18.8). The discrepancies in the absolute values are simply a matter of the differences in the testing machines, various running-in conditions and the typical scatter known for wear studies in general.

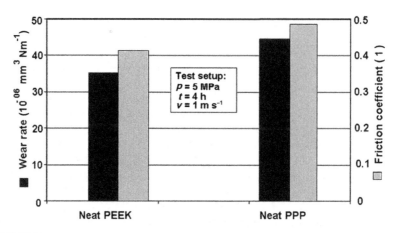

**FIGURE 18.9**

Wear rate and coefficient of friction of PPP versus PEEK.

Using a load collective of $1\,ms^{-1}$ and $5\,MPa$, PEEK resulted in slightly better results than did the PPP material (Fig. 18.9). This fact could be explained by the slightly better temperature stability of the semicrystalline structure of the PEEK compared to the amorphous structure of PPP.

According to these previous results, it can be expected that this material also has exceptional properties under other loading conditions. The very high strength and modulus values advert to a high usability in tribological applications. Due to the higher coefficients of friction and wear rates in comparison to the PEEK polymer [46], it must be stated that an additional reinforcement with particles like graphite or molybdenum disulfide are important to introduce PPS into friction and wear applications. In combination with these functional filler, the PPP seems to have high potential.

## 18.4 NANOPARTICLE-FILLED SLIDE COATINGS FOR AUTOMOTIVE APPLICATIONS

This section describes three automotive applications in which coatings with a combined microreinforcement and nanoreinforcement have proven to be superior to conventional microparticle or fiber-reinforced slide coatings: A PAI-based piston skirt coating and tribological coatings for polymer/metal-slide bearings for high-temperature use and for use under boundary or mixed lubrication.

### 18.4.1 Slide coatings for piston skirts

Concerning slide coatings for pistons, at least three tribological topics must be discerned. The first is the piston ring pack (Fig. 18.10). In most cases, it consists of three metallic rings [51] that are from top to bottom: the top and the second compression

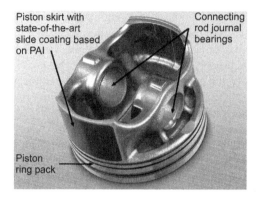

Piston skirt with state-of-the-art slide coating based on PAI

Connecting rod journal bearings

Piston ring pack

**FIGURE 18.10**

Tribosystems of an aluminum diesel engine piston where slide coatings can be used to achieve low wear and friction and thus contribute to the improvement of fuel economy. (For color version of this figure, the reader is referred to the online version of this book.)

*Picture courtesy KS Kolbenschmidt GmbH, modified.*

ring whose function is to effectively seal the combustion chamber off the crankcase, and the oil control ring. The oil control ring is designed to leave an oil film of only a few micrometers thickness on the cylinder wall as the piston descends. All three rings experience high mechanical, tribological and thermal stress and are usually made of hard-coated metals or of high-performance polymers [52].

The second area of interest in the field of pistons is the potential future substitute of connecting rod journal bearings by slide coatings. Here, the coatings will be subjected to a compression loading of up to 2000-bar loading and to sliding speeds of up to $20\,m\,s^{-1}$. Currently, metal/polymer-slide bearings are inserted into the piston during manufacturing. Here, the direct application of a slide coating, which could happen simultaneously with the coating of the piston skirts, has the potential to reduce the number of parts, shorten the manufacturing time and hence reduce manufacturing costs.

The third area of tribological interest is the piston skirt, which lies directly below the ring pack and upon which we will focus in the remainder of this section. The piston skirt transmits transverse loads onto the cylinder wall. The secondary motion of the piston, which causes these transverse forces, includes transverse motion and tilting about the main piston axis [53,54]. This leads to noise generation due to collisions of the piston with the cylinder wall and to severe frictional loss and high wear. The piston skirt–cylinder wall interaction is estimated to contribute 30% to the piston assembly's frictional loss and is therefore one of the largest single portion of a vehicle's total frictional loss [6]. Hence, the reduction of piston skirt friction is a very important task in automotive tribology.

Although not foremost dedicated to the reduction of friction but rather aimed at the increase of the wear resistance, one of the earliest patents related to piston coatings was issued in 1934 [55]. At that time, light-weight pistons from aluminum alloys were becoming increasingly popular. Although these were light weighted, easily manufactured, and had high thermal conductivities, their wear resistance was not as good as for pistons made from heavy metals. Therefore, a layered coating with a hard nickel layer and the subsequent addition of a tin layer that greatly reduced wear and scuffing is described in the patent. In 1957, one of the earliest patents dealing with the reduction of piston skirt/cylinder wall friction by the use of polymers describes the usage of PTFE as a slide coating for the piston skirts [56].

The main requirements for modern piston coatings, wherever they are to be applied, are thermal stability up to 200°C, high wear resistance, a low coefficient of friction, and a lifetime of at least 2000 h. Since the actual operation temperature strongly depends upon the heat deduction from the piston into the engine block, a high thermal conductivity and a thickness of the coating of only 5–25 μm are beneficial. One of the high-performance polymers that fulfills these requirements is PAI filled with PTFE, $MoS_2$, and graphite [28].

An important step in the development of a slide coating is the optimization of filler contents. Usually, this is done by the tribological testing of a series of materials in which the content of one filler is varied while the content of all the other fillers is kept constant. A very interesting variation of the classical single variant method proposed by McCook utilizes a simplex optimization procedure in order to find compositions for optimized wear resistance and low friction [57].

Other methods utilize only one specimen that has a spatial variation in properties [58] for determining the influence of single components. Recently, the use of artificial neural networks (ANNs) has become very popular in predicting tribological properties and their dependence on singular variants like filler content, loading or sliding speed [59]. The key feature of ANNs is that they predict a wear surface from an experimental database without the need of establishing explicit functional relationships between filler content and wear rate or coefficient of friction [60,61].

Figure 18.11 shows the result of a two-step optimization of the graphite content of a nanoparticle-filled PAI-based slide coating. The coating also contains SCF and graphite as microscale fillers with an initial content of 7 vol.% graphite and 9 vol.% SCF, respectively. While its nanoparticle contents are considered to be already at their respective optimum values concerning the wear rate, the SCF and the graphite content of a PAI-based slide coating was optimized.

In a first step, a series of coatings was made in which the SCF content was kept constant at 9 vol.% and the graphite content was varied. In a series of ring-on-plate dry sliding experiments, an optimum graphite content of 7 vol.% was determined. Therefore, the initial material is already composed at an optimum graphite content.

In the second step of the optimization procedure, the SCF content was changed while maintaining the graphite content at 7 vol.%. As can be seen from Fig. 18.6, the specific wear rate monotonically decreased with increasing SCF content. Within the scope of this study, the optimum amounts for the coating's microfillers are 9 vol.% graphite and 13 vol.% short carbon fibers. Overall, the specific wear rate was improved by 25% by increasing the SCF content from 9 to 13 vol.% while keeping the graphite content at its initial level.

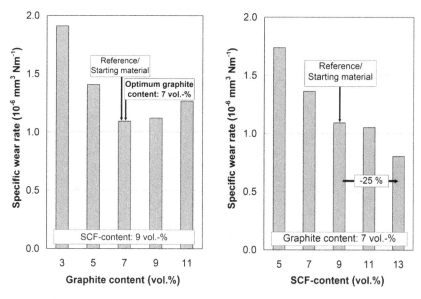

**FIGURE 18.11**

Reduction of the specific wear rate of a slide coating by a two-stage single variant optimization of the filler composition.

## 18.4.2 Polymer/metal-slide bearings

Polymer/metal-slide bearings consist of a steel back coated with a tribologically optimized polymer layer. They therefore combine the beneficial mechanical stability, creep resistance, and high thermal conductivity of steel with the superior tribological performance of polymeric composites. Today, slide bearings are important components in automobile manufacturing. Among many other applications like door hinges, they are used in high-temperature applications like connecting rods, motor bearings, and fuel injection pumps. The main requirements for such applications are a low coefficient of friction, a low wear rate, and high operation temperatures.

### 18.4.2.1 Use in high-temperature applications

In this section, we focus on the comparison of the high-temperature tribological performance of the commercial PEEK-based microcomposite Lubricomp-LTW and a similar nanoparticle-filled composite [62,63]. While the commercial reference contains 10 vol.% SCF, graphite and PTFE, respectively, the nanocomposite contains 10 vol.% SCF, graphite, nano-ZnS and nano-$TiO_2$, respectively. Both materials were applied to a steel plate with a thickness of 100 μm and subjected to sliding wear against 100Cr6 in the temperature range of 23–225 °C using the plate-on-ring geometry. Figure 18.12 depicts the coefficient of friction and Fig. 18.13 the specific wear rate of both PEEK composites as a function of temperature.

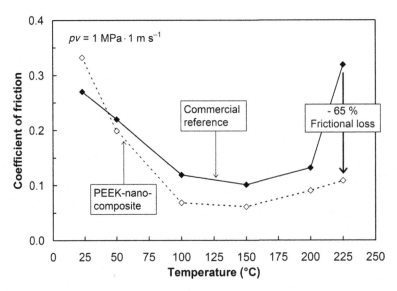

**FIGURE 18.12**

Dependence of the coefficient of friction of both Micro-PEEK and Nano-PEEK at an ambient temperature of 23 °C.

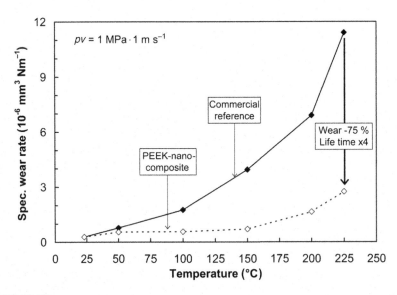

**FIGURE 18.13**

Dependence of the specific wear rate of Micro-PEEK and Nano-PEEK at an ambient temperature of 23 °C.

At 23 °C, both materials exhibit similar specific wear rates of about $0.6 \cdot 10^{-6}$ mm$^3$ Nm$^{-1}$ while the nanoparticle-filled composite has a slightly higher coefficient of friction. At higher temperatures, however, the nanocomposite exhibits significant advantages over the commercial microcomposite. The coefficient of friction of both materials is reduced with rising temperature. At 150 °C, which is the $T_g$ of both materials, it reaches a minimum of 0.06 for the nanocomposite and 0.10 for Luvocom LTW, respectively. Above 150 °C, friction increases continuously with temperature, but in all cases, the nanocomposite remains at a significantly lower coefficient of frictions. At 225 °C, the coefficient of friction of the nanoparticle-filled composite is still only 0.11, which is 65% less than the micromaterials coefficient of friction of 0.32. Starting at an ambient temperature of 100 °C, the nanofillers cause a significant reduction of the specific wear rate related to the microcomposite's wear rate. At a temperature of 225 °C, this amounts to a 75% advantage, which corresponds to a threefold lifetime prolongation.

### 18.4.2.2 Use under mixed/boundary lubrication

Acceleration resistance contributes up to 35% of a vehicle's total energy consumption. If the car is used in city traffic, it is especially important to reduce the mass to be accelerated and to be decelerated since the vehicle's mass contributes approximately 80% to the acceleration resistance. It is assumed that the reduction of the vehicle's weight by 100 kg reduces the fuel consumption by 0.51 per 100 km. An important method in the design of fuel efficient vehicles is the reduction of vehicle weight by the downsizing of components. However, this also forces tribological components to operate under higher $pv$ products, which can shift the operation conditions into the mixed lubrication regime or into boundary lubrication that are usually associated with high frictional loss and a high wear rate. In the boundary friction regime, for example, there is no bearing oil film like in the hydrodynamic regime, and hence, asperity contact carries the load. However, in contrast to dry sliding, an adsorbed layer of lubricant molecules that are bound to the surfaces by physisorption protects the asperities.

Under boundary lubrication, polymeric slide coatings have many advantages over their metallic counterparts. They have superior antiseizure and antigalling properties, a high corrosion resistance and very good dry running properties. Furthermore, polymeric slide coatings are able to embed foreign particles like oil contaminants or wear debris. One component that operates under very high tribological loading is the fuel injection pump. Current state-of-the-art diesel fuel injection pumps rely completely on the intrinsic lubrication of the delivered fuel. Due to the simplicity of design and manufacturing as well as for reducing costs, there is no separate oil based lubrication for the continuously stressed journal bearing/camshaft-pairing (Fig. 18.14).

Since efficient fuel combustion requires high fuel injection speeds, conveying pressures of up to 2000 bar are used and 3000 bar are aimed at. Even if the hydrodynamic velocity within the slide bearing is never zero during operation, the formation of a bearing hydrodynamic film can locally be disrupted at such severe loading due to the low viscosity and the poor intrinsic lubricity of the diesel fuel [64]. Hydrocarbons,

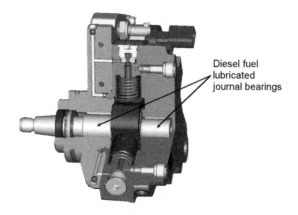

**FIGURE 18.14**

Diesel fuel injection pump. (For color version of this figure, the reader is referred to the online version of this book.)

*Courtesy of Robert Bosch GmbH, modified.*

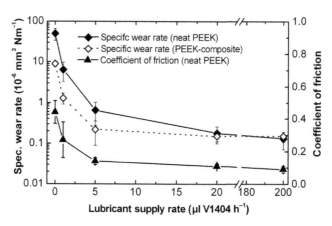

**FIGURE 18.15**

Decrease of friction and wear with increasing lubricant rate.

which constitute the largest fraction of diesel fuel, poorly physisorb to metals and are poor lubricants and consequently significant portions of boundary friction may occur in the camshaft bearing. While polymer/metal bearings counter this by tribologically optimizing the slide coating, there are also current attempts to enhance the lubricity of diesel fuels by adding organic acids, organic esters, aliphatic amines [65], or bio-diesel [66] to low sulfur petrodiesel.

Since PEEK is a common slide coating in the journal bearings of diesel fuel injection pumps, its tribological behavior and mixed and boundary lubrication have been studied recently [67]. Additionally, a fully optimized tribological nanocomposite based on PEEK has been studied. This tribological composite contains 10 vol.% SCF graphite

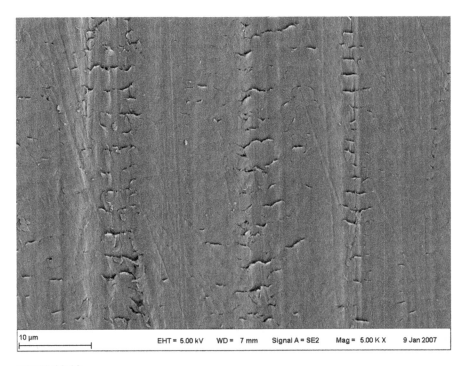

**FIGURE 18.16**

Formation of fatigue cracks perpendicular to the sliding direction (bottom to top) due to lubricant sorption into the neat PEEK matrix [67].

*Source: Solvay.*

nano-ZnS and nano-TiO$_2$, respectively. Using the calibrating fluid V1404 [68] as a lubricant and using the coefficient of friction as indicator for the different lubrication regimes, boundary lubrication and mixed lubrication were selectively engaged in sliding wear tests by applying different, but constant lubricant rates in the low microliter per hour range. It is found that friction and wear decreased steadily when applying a higher lubricant rate. Compared to dry lubrication, the specific wear rate decreases by one order of magnitude in boundary lubrication and by two orders of magnitude in mixed lubrication (Fig. 18.15). Under boundary lubrication, the wear rate of the PEEK composite is only 25% of the neat matrix. Since this advantage of the composite over the neat matrix vanishes at higher lubricant supply rates, it is considered to be due to the solid lubricants of the composite. Since the latest results indicate that there is a separate contribution from the classical microfiller (graphite) and from the nanofillers, the deconvolution of the individual fillers' contributions is the subject of ongoing investigations.

Concerning the wear mechanism of polymeric slide coatings under boundary lubrication, there are significant differences in metallic slide bearings. For PEEK [67–69], Nylon 11 [69], and PTFE [70], it has been observed that hydrocarbon lubricants (diesel fuel, paraffin oil, decanoic acid, dodecylamine, and cresol) are absorbed into the first 5–10 µm of the matrix. This leads to a plasticization of the polymer matrix that reduces surface hardness and elasticity. Consequently, fatigue failure and crack formation perpendicular to the sliding direction are observed under the shear stress induced by sliding (Fig. 18.16).

## 18.5 CONCLUSIONS AND OUTLOOK

Slide coatings combine the high mechanical and thermal stability of metallic sliding surfaces with the superior tribological performance of modern polymeric composites. Therefore, they have become indispensable components in the manufacturing of light-weight, high-power, and safe vehicles. With the advent of high-performance composites, slide coatings are increasingly used in high-temperature and high-pressure applications, for example, in the engine compartment.

The future development of polymeric slide coatings is mainly determined by the current necessity to increase fuel economy. Besides avoiding mechanical losses, one important way is to reduce the vehicle's weight. Since modern vehicles already utilize aluminum and polymeric composites to a high extent without increasing the costs beyond a sensible limit, the most important way to achieve a further weight reduction is the reduction of component size. This will require the corresponding tribological systems to operate at higher sliding speeds and at higher pressures in order to maintain the same power output. The future development of slide coatings will therefore aim at satisfying this need for higher $pv$ products without reducing the coating's service lifetime. Concerning polymer/metal-slide bearings, the direct application of the slide coating to a metallic bulk instead of joining a separate bearing into the bulk is desirable. This would, for example, economize the need to join two bearings into each piston (Fig. 18.5) and would therefore reduce assembly time and costs.

A new polymeric material, PPP copolymer, was introduced with regard to mechanical and tribological properties. It turned out that this material has a high potential for tribological applications, as long as the temperature range does not exceed 140 °C. The material possesses also a very high hardness and scratch resistance, which can be very helpful if good surface properties are requested. However, the sliding wear resistance was not as good as the one measured for the neat PEEK, but it is assumed that this can be improved when incorporating special fillers into a PPP matrix [46].

Besides this procedural advantage and the pure enhancement of the mechanical and tribological properties of composites, nanosized fillers can also provide 'auxillary' functions. One of them, which is in our opinion not yet fully utilized, is the reduction of a composite's thermal resistance by the incorporation of a nanofiller with high thermal conductivity (e.g. single walled carbon nanotubes) [71–74]. Concerning the piston example the nanofiller improves the thermal deduction even more over the direct contact due to the direct application. In effect the temperature rise in the coating due to the heat flux can be greatly reduced. Thus the nanofillers can contribute significantly to a coating's overall tribological performance by preventing the latter's thermal deterioration.

# References

[1] U.S. Environmental Protection Agency, Air Quality Criteria for Particulate Matter (October 2004), U.S. Environmental Protection Agency, Washington, DC, 2004.

[2] U.S. Environmental Protection Agency, Inventory of U.S. Greenhouse Gas Emissions and Sinks: 1990–2004, U.S. Environmental Protection Agency, Washington, DC, 2006.

[3] B.S. Anderson, Company perspectives in vehicle tribology—Volvo, The 17th Leeds-Lyon Symposium on Tribology—Vehicle Tribology, Tribological Series, 18, UK Elsevier Ltd, Oxford, 1991, pp. 503–506.

[4] S.C. Tung, L.M. McMillan, Automotive tribology overview of current advances and challanges for the future, Tribology International 37 (2004) 517–536.

[5] S. Korcek, Automotive lubricants for the next millennium, Industrial Lubrication and Tribology 52 (2000) 209–220.

[6] S. Boyde, Green lubricants. Environmental benefits and impacts of lubrication, Green Chemistry 4 (2002) 293–307.

[7] T.I. Taylor, Improved fuel efficiency by lubricant design: a review. Proceedings of the Institution of Mechanical Engineers Part J:, Journal of Engineering Tribology 214 (2000) 1–15.

[8] Y. Enomoto, T. Yamamoto, New materials in automotive tribology, Tribology Letters 5 (1998) 13–24.

[9] M. Nakasa. Engine friction overview, in: Proceedings of International Tribology Conference, Japan, Yokohama, (1995).

[10] J.A. Spearot. Friction, wear, health, and environmental impacts – Tribology in the new millenium. A keynote lecture at the STLE Annual Meeting, Nashville, Tennessee, May, 2000.

[11] K. Doren, D. Stoye, W. Freitag, Water-Borne Coatings: The Environmentally-Friendly Alternative, Hanser Gardner Publications, 1994 ISBN–13: 978-1569901397.

[12] V. Durga, N. Rao, D.M. Kabat, B.W. Brian Lizotte. U.S. Patent 5482637-Anti-friction coating composition containing solid lubricants.

[13] K. Tanaka, Y. Yamada, Effect of temperature on the friction and wear of some heat-resistant polymers, ACS Symposium Series 287, 1985 Washington (DC), pp. 103–128.

[14] J.P. Critchley, G.J. Knight, W.W. Wright, Heat-Resistant Polymers, Plenum Press, New York, 1983.

[15] H.K. Soni, R.G. Patel, V.S. Patel, Structural, physical and mechanical properties of carbon fibre-reinforced composites of diglycidyl ether of bisphenol-A and bisphenol-C, Angewandte Makromolekulare Chemie 211 (1) (1993) 1–8.

[16] H.C. Anderson, Thermal degradation of epoxide polymers, Journal of Applied Polymer Science 6 (22) (1962) 484–488.

[17] S. Kumar, Structure and properties of high performance polymeric and carbon fibers—an overview, SAMPE Quart 20 (2) (1989) 3–8.

[18] J.H. Sinclair, C.C. Chamis, Compression Testing of Homogeneous Materials and Composites, ASTM STP 808 ASTM, Philadelphia, 1983 p. 155.

[19] V.V. Kozev, H. Jiang, V.R. Mehta, S. Kumar, Compressive behavior of materials: part II. High performance fibers, Journal of Materials Research 10 (4) (1995) 1044–1061.

[20] L.A. Pilato, M.J. Michno, Advanced Composite Materials, Spinger, 1994 ISBN: 0-387-57563-4.

[21] M.J. Shuart, Failure of compression loaded multidirectional composite laminates, AIAA Journal 27 (9) (1989) 1274–1279.

[22] A.A. Caiazzo, B.J. Sullivan, B.W. Rosen, Analysis of micromechanical and microstructural effects on compression behaviour of unidirectional composites, Journal of Reinforced Plastics and Composites 12 (4) (1993) 457–469.

[23] T. Ohsawa, M. Miwa, M. Kawade, E. Tsushima, Axial compressive strength of carbon fiber, Journal of Applied Polymer Science 39 (8) (1975) 1733–1743.

[24] H.G. Chae, S. Kumar, J. Rigid rod polymeric fibers, Journal of Applied Polymer Science 100 (1) (2006) 791–802.

[25] M.L. Minus, S. Kumar, The processing, properties and structure of carbon fibers, JOM Journal of the Minerals Metals and Materials Society 57 (2) (2005) 52–58.

[26] X.-D. Hu, S.E. Jenkins, B.G. Min, M.B. Polk, S. Kumar, Rigid-rod polymers: synthesis, processing, simulation, structure, and properties, Macromolecular Materials and Engineering 288 (2003) 823–843.

[27] Y. Zheng, R. Ning, Glass fiber-reinforced polymer composites of high compressive strength, Journal of Reinforced Plastics and Composites 23 (16) (2004) 1729–1740.

[28] K. Saito, Y. Fuwa, M. Sugiyama, M. Murakami, Trans. Soc. Autom. Engrs. Japan (September, 1995) 101.

[29] W. Bickle, F. Haupert, W. Schubert, G. Bürkle. Kolben für Brennkraftmaschinen. European Patent EP 1729003A2.

[30] J.E. Fitzpatrick, Polyamide-imides, in: C.A. Dostal (Ed.), Engineered Materials Handbook, Engineering Plastics, vol. 2, CRC-Press, 1988.

[31] J.D. Felberg, Polyamidimide (german), in: G.W. Becker, D. Braun (Eds.), Kunststoffhandbuch, Band 3/3. Kapitel 7, Hanser, München, 1994.

[32] G. F. Paulus. US Patent 3,293,203-Thermosettable resin compositions and method of forming low friction surface coatings.

[33] J. O'Toole, Epoxy Resin Reference Manual, Akzo Chemie America, 1985.

[34] C.A. May (Ed.), Epoxy Resins, second revised and expanded ed., Marcel Dekker, 1987.

[35] B. Ellis (Ed.), Chemistry and Technology of Epoxy Resins, Kluwer Academic Publishers, 1993.

[36] L. Soos, New epoxy composites and their applications, European Polymers Paint Colour Journal 183 (4338) (1993) 490.

[37] The Dow Chemical Company, Dow Epoxy Novolac Resins—High Temperature, High-Performance Epoxy Resins, 1998 Midland.

[38] Z. Zhang, C. Breidt, L. Chang, K. Friedrich, Wear of PEEK composites related to their mechanical performances, Tribology International 37 (2004) 271–277.

[39] R. Reinicke, F. Haupert, K. Friedrich, On the tribological behavior of selected, injection molded thermoplastic composites, Composites – A 29 (7) (1998) 763–777.

[40] J. Flöck, K. Friedrich, Q. Yuan, K. Takano, On the friction and wear behavior of PAN- and pitch-carbon fiber reinforced PEEK-composites, Wear 225–229 (1999) 304–311.

[41] X.-Q. Pei, K. Friedrich, Sliding wear properties of PEEK, PBI and PPP, Wear 274–275 (2012) 452–455.

[42] L. Chang, K. Friedrich, Enhancement effect of nanoparticles on the sliding wear of short fiber reinforced polymer composites: a critical discussion of wear mechanisms, Tribology International 43 (2010) 2355–2364.

[43] N. Malkovich, R. Gane, Parmax1 Self Reinforced Polymers; the World's Strongest, Stiffest, Hardest Polymers, Mississippi Polymer Technologies Inc., Presentation to NDIA, May 17, 2005.

[44] N.N: Tecamax: SRP-Extreme Festigkeit ohne Faserverstärkung. Ensinger GmbH, Germany, 2005 Produktinformation.

[45] N.N: PrimoSpire1: Self-Reinforced Polyphenylene. Solvay Advanced Polymers, USA, 2007 Product Information.

[46] K. Friedrich, T. Burkhart, A.A. Almajid, F. Haupert, Mechanical properties and scratch/wear behaviour of a new poly-para-phenylene-copolymer (PPP), International Journal of Polymeric Materials 59 (2010) 680–692.

[47] M. Merkel, K.-H. Thomas, Taschenbuch der Werkstoffe, Hanser, Leipzig, 2000.

[48] A.M. Häger, M. Davies, Short fiber reinforced, high temperature resistant polymers for a wide field of tribological applications, in: K. Friedrich (Ed.), Advances in Composite Tribology, Elsevier, Amsterdam, 1993, pp. 107–157.

[49] K. Friedrich, Z. Lu, A.M. Häger, Recent advances in polymer composites' tribology, Wear 190 (1995) 139–144.

[50] A. Gebhard, F. Haupert. Reibungs- und Verschleißprüfung an Poly-paraphenylene (PPP), NanoProfile Bericht Nr. 09-1215, 3.7.2009.

[51] M.J. Neale, Drives and Seals – A Tribology Handbook, first ed., Butterworth-Heinemann Ltd, Oxford, UK, 1994.

[52] N. Ya. Radchenko, D.M. Krymskii, I.V. Kalinnikov, Piston rings made of thermoplastic materials, Chemical and Petroleum Engineering 10 (8) (1974) 706–707.

[53] D.F. Li, S.M. Rhode, H.A. Ezzat, An automotive piston lubrication model, ASLE Transactions 26 (2) (1982) 151–160.

[54] G. Knoll, H. Peeken, R. Lechtape-Grüter, J. Lang. Computer aided simulation of piston and piston ring dynamics – heavy duty engines, a look at the future. in: Proc. 16th Annual Fall Technical Conference of the Internal Combustion Engine, Division of the ASME, Lafayette, Indiana, October 2–6 1994, New York: ASME, United Engineering Center, ICE-Vol. 22.

[55] H. K. Work. United States Patent 1975818-Coating for Pistons.

[56] J. D. Fleming, A. O. De Hart. United States Patent 2817562: Coated Piston.

[57] N.L. McCook, B. Boesl, D.L.M. Burris, W.G. Sawyer, Epoxy, ZnO, and PTFE nanocomposite: friction and wear optimization, Tribology Letters 22 (3) (2006) 253–257.

[58] T. Kovacs, L. Dévényi, Investigation of wear process by a gradient method, Materials Science Forum 473–474 (2005) 213–218.

[59] D. Aleksendric, Č Duboka, Prediction of automotive friction material characteristics using artificial neural networks—cold performance, Wear 261 (2006) 269–282.

[60] Z. Zhang, K. Friedrich, Artificial neural networks applied to polymer composites: a review, Composites Science and Technology 63 (2003) 2029–2044.

[61] Z. Jiang, Z. Zhang, K. Friedrich, Prediction on wear properties of polymer composites with artificial neural networks, Composites Science and Technology 67 (2007) 168–176.

[62] F. Oster, F. Haupert, K. Friedrich, W. Bickle, M. Müller, New polyetheretherketone-based coatings for severe tribological applications, Materialwissenschaft und Werkstofftechnik 35 (10–11) (2004) 690–695.

[63] F. Oster, F. Haupert, K. Friedrich, W. Bickle, M. Müller, Tribologische Hochleistungsbeschichtungen aus neuartigen Polyetheretherketon (PEEK)-compounds, Tribologie und Schmierungstechnik 51 (3) (2004) 17–24.

[64] D. Wei, H.A. Spikes, The lubricity of diesel fuels, Wear 111 (1986) 217–235.

[65] G. Knothe, K.R. Steidley, Lubricity of components of biodiesel and petrodiesel. The origin of biodiesel lubricity, Energy and Fuels 19 (3) (2005) 1192–2000.

[66] G. Anastopoulos, E. Lois, D. Karonis, S. Kalligeros, F. Zannikos, Impact of oxygen and nitrogen compounds on the lubrication properties of low sulfur diesel fuels, Energy 30 (2–4) (2005) 415–426.

[67] A. Gebhard, S. Emrich, F. Haupert, M. Kopnarski, A. K. Schlarb. Mehrschichtverbundsysteme bei Grenzreibung (Vorprojekt)—Grundlegende Untersuchungen zur Bestimmung der relevanten Grenzreibungsmechanismen an polymeren Hochleistungsverbundwerkstoffen, Abschlussbericht. FVV-Nr. 891. FVV-Heft 825 (2006).

[68] ISO 4113:1998: Road vehicles; Calibration fluid for diesel injection equipment.

[69] B.J. Briscoe, T.A. Stolarski, G.J. Davies, Boundary lubrication of thermoplastic polymers in model fluids, Tribology International 17 (3) (1984) 129–137.

[70] Z.-Z. Zhang, W.-M. Liu, Q.-J. Xue, Effects of various kinds of fillers on the tribological behavior of polytetrafluoroethylene composites under dry and oil-lubricated conditions, Journal of Applied Polymer Science 80 (11) (2001) 1891–1897.

[71] F. Haupert, N. Knör, R. Walter, A.K. Schlarb. Optimization of mechanical and tri-biological properties of polyetheretherketone (PEEK) and polyphenylene–sulphide (PPS) nanocomposites, 3rd Vienna International Conference, Micro- and Nano-Technology, Vienna, March, 2009.

[72] N. Knör, A. Gebhard, F. Haupert, A.K. Schlarb, Polyetheretherketone (PEEK) nanocomposites for extreme mechanical and tribological loads, Mechanics of Composite Materials 45 (2009) 289–298.

[73] P. Carballeira, F. Haupert. Wear behavior of carbon nanofiber-epoxy composites. Admitted for poster presentation at the 10th Trends in Nanotechnology International Conference (TNT 2009) 07–11 September 2009, Spain, Barcelona.

[74] K. Friedrich, L. Chang, F. Haupert, Current and future applications of polymer composites in the field of tribology, in: L. Nikolais, M. Meo, E. Miletta (Eds.), Composite Materials, Springer, New York, USA, 2011 http://www.springer.com/materials/special+types/book/978-0-85729-165-3.

[75] D.B.S. Berry, B.I. Buck, A. Cornwell, L.N. Phillips, Handbook of Resin Properties, Part A, Cast Resins, Yarsley Testing Laboratories, Ashstead, 1975.

[76] Solvay Advanced Polymers, L.L.C. Solvay Advanced Polymers Technical Data Sheets. Alpharetta, Georgia.

# CHAPTER

# Friction and wear behavior of PEEK and its composite coatings

# 19

Ga Zhang*, Hanlin Liao[†], Christian Coddet[†]

*Institute for Composite Materials (IVW GmbH), Technical University Kaiserslautern, Kaiserslautern, Germany, [†]University of Technology of Belfort-Montbeliard, Belfort, France

## CHAPTER OUTLINE HEAD

Tribology of Polymeric Nanocomposites. http://dx.doi.org/10.1016/B978-0-444-59455-6.00019-2

## 19.1 **INTRODUCTION**

In order to improve the surface performance of metallic parts, for example, erosion resistance, friction coefficient, and wear resistance, polymer coatings are nowadays attracting more and more interest [1–11]. Lots of smart coating designs have been tailored, and many coating techniques, such as thermal spraying [1–4], painting [5], spinning [6,11], suspension spraying [7], and electroplating [8], were used to deposit polymer-based coatings on metallic substrates. According to the apparent forms of feedstock materials, these coating techniques can be categorized into two types: solid feedstock, such as powders, and liquid feedstock, such as suspensions or solutions. More recently, tribologically optimized polyetheretherketone (PEEK) composite films were thermally joined with steel and aluminum substrates by a thermal impacting welding process [12]. With extruded films, the structure of the composite coatings can be well controlled. This thermal impacting welding process shows great potential for manufacturing friction and wear-reduction coatings and sliding bearing materials.

Owing to advancements in material formulation and coating techniques, lots of remarkable results were achieved, in which the coated substrate showed a significantly improved surface performance. As a result, polymer coatings attract more and more industrial interest, especially in the automotive fields [13,14]. High temperature-resistant polymer coatings, for example, polyamide imide-based coatings [8], deposited on engine pistons permit tight clearance, noise, and friction reduction and thereby fuel economy. It is expected that long-term efforts will be carried out in the future on polymer coatings, in which coating process and material formulation will be the most important issues.

PEEK is a thermoplastic with an excellent toughness–stiffness combination, and therefore, it becomes one of the most attractive thermoplastics [15]. It is being more and more used for tailoring high-performance sliding materials due to its excellent thermal stability and good tribological performance, especially high wear resistance [16–18].

The good tribological performance of PEEK promotes the development of PEEK coatings on metallic substrates, especially lightweight alloy, which generally presents poor tribological performances. In recent years, PEEK and its composite coatings were prepared on such substrates using thermal spraying and painting techniques etc. [5,19–23]. PEEK and PEEK/SiC coatings deposited on aluminum were shown to significantly reduce the friction coefficient and wear rate of the substrate [5,23]. Previous studies showed that thermally sprayed PEEK coatings deposited on engine piston skirts led to a reduction in fuel consumption of about 3% [24].

In this chapter, the fabrication processes, crystalline structures, and adhesions of PEEK coatings are presented. It is shown how the crystalline structure, addition of nanoparticles, and the sliding conditions, such as applied load, sliding velocity, and ambient temperature, influence the dry sliding behavior of PEEK coatings. There is also an attempt to discuss the friction and wear mechanisms of an amorphous PEEK coating based on morphological analyses of the worn tracks.

## 19.2 **COATING PROCEDURES**

Thermal spraying and painting are two effective techniques for depositing PEEK coatings. In a previous study [1], PEEK coatings were deposited on metallic substrates by atmospheric plasma spraying (APS), high-velocity oxygen fuel (HVOF), and flame spraying (FS). The results showed that the APS process led to a degradation of the polymer owing to the high temperature of the plasma jet. PEEK coatings deposited using HVOF had an enormous roughness due to the high blowing pressure of the high-velocity flame. FS was observed to be more suitable to the preparation of PEEK coatings as homogeneous and dense coatings could be obtained.

In order to obtain a dense coating, before spraying, the substrate was preheated up to a temperature higher than the melting point of PEEK, for example, 400 °C. During spraying, the PEEK powder was injected into the flame jet where it was melted (or partially melted) and propelled to the substrate surface to build the coating. After spraying, the coating was quenched into cold water. The coating obtained exhibited an amorphous structure and a good adherence to the substrate [22]. The infrared spectra indicated that very little degradation of PEEK occurred after FS [25]. Due to the poor fluidity and agglomeration tendency of some fine fillers required for preparing composite coatings, a homogeneous composite coating can be difficult to be achieved by directly flame spraying the mixture of fillers and matrix powers. The homogeneity of the composite coating could be improved by a mechanical milling of the mixed powders before spraying [26]. With optimized milling parameters, thin fillers, for example, nanosized ceramic particles, could be embedded into polymer powders [26–28].

Painting is also an efficient method for depositing PEEK-based coatings on a substrate with regular configuration, for example, plate or cylinder shapes [29]. Employing this procedure, PEEK, or PEEK and fillers in the case of composite coatings, were firstly dispersed into an aqueous medium. The mixture was then continuously stirred and then subjected to an ultrasonic dispersion stage for reaching a uniform dispersion. Then, the slurry was applied evenly on the substrate. After being dried, the substrate-coating system was heated up to 400 °C and held at this temperature for 5 min, and then quenched in cold water. Figure 19.1 shows examples of cross-sectional structures of the obtained PEEK, PEEK/micron-SiC, and PEEK/micron-graphite coatings. Dense coatings were obtained using this process.

## 19.3 **CHARACTERIZATION OF THE COATING CRYSTALLINE STRUCTURE**

The crystalline structure of PEEK coatings was shown to be of great significance for their mechanical properties, for example, ductility and stiffness [22]. An amorphous PEEK exhibits a higher ductility but lower stiffness than does semicrystalline PEEK. Therefore, an investigation on the effect of the crystalline structure on the tribological behavior is important. However, it seems impossible to prepare

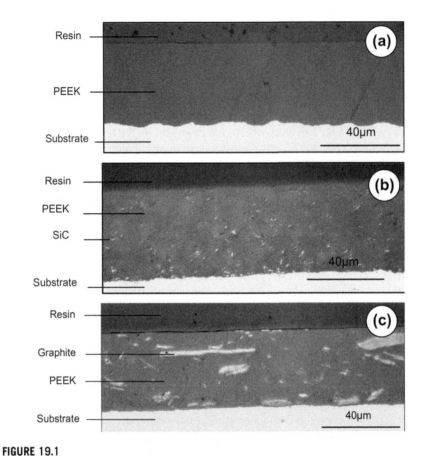

**FIGURE 19.1**

Cross-sectional structures of (a) PEEK, (b) PEEK/SiC, and (c) PEEK/graphite coatings.

amorphous bulk PEEK material because of its rapid crystallization speed and its low thermal conductibility [22,30]. Meanwhile, amorphous PEEK samples, with thicknesses <1000 μm, could be obtained using coating techniques, in which a quenching from the molten state of PEEK is performed [22]. Moreover, the crystalline structure of these samples could be controlled by general annealing treatments of these amorphous coatings. Thus, PEEK coatings with different crystalline structures were obtained by a combination of the FS process and thermal treatment procedures.

PEEK coatings, with final thicknesses of 500 μm, were prepared by an FS process. The PEEK feedstock powders had a mean diameter of 25 μm. The final coating roughness, $R_a$ was about 0.8 μm. In order to obtain different crystalline structures, the as-sprayed coating, referenced as $T1$, was subjected to heat treatments. The annealing temperatures and the holding times are listed in Table 19.1. The treated coatings are referenced from $T2$ to $T7$ correspondingly. The coating structures were

**Table 19.1** Annealing Temperatures and Holding Times of PEEK Coatings

| Sample Code | Annealing Temperature (°C) | Holding Time (min) |
|---|---|---|
| T2 | 180 | 30 |
| T3 | 220 | 30 |
| T4 | 260 | 30 |
| T5 | 300 | 30 |
| T6 | 260 | 1 |
| T7 | 260 | 10 |

characterized by differential scanning calorimetry (DSC) measurements and X-ray diffraction (XRD) analyses.

It is well known that in an amorphous polymer, the molecules range themselves randomly. When the polymer is heated at a temperature higher than its glass transition temperature ($T_g$), some of the molecules organize themselves to form crystalline grains, for example, spherulites in the case of PEEK [31–33]. Figure 19.2(a) shows the DSC scan of the as-sprayed coating (T1). A marked glass transition is apparent near the $T_g$ (143 °C), and it is followed by a marked exothermic crystallization peak. This confirms that the as-sprayed coating exhibits an amorphous structure [31]. Figure 19.2(b) and (c) show the DSC scans of coatings heat treated at different temperatures (T2–T5) and for different holding times (T4, T6–T7). Their glass transitions, compared to that of the as-sprayed coating, become less pronounced. These coatings do not exhibit any exothermic peak above $T_g$. Accordingly, it can be deduced that coatings T2–T7 exhibit semicrystalline structures. On the DSC scans of the heat-treated coatings, in addition to the main melting peak, a minor peak appears a few degrees below the main peak temperature. The location and the shape of the main melting peak remain unchanged, while those of the smaller peak change. With an increase in the annealing temperature, this small endothermic peak shifts to a higher temperature and exhibits a larger enthalpy. With an annealing temperature of 260 °C, an increase in the holding time leads to a similar behavior.

According to the stepwise crystallization model [32–35], two crystalline entities are produced in the spherulites of semicrystalline thermoplastics during an isothermal annealing. In the early stage of crystallization, lamellar stacks consisting of thick lamellae appear in the amorphous region, and they constitute the frame of the spherulites. However, the regions between these thick lamellae stay amorphous. Later, lamellar stacks with thinner lamellae form in these amorphous regions. The coexistence of these dual lamellae could explain the double melting behavior of coatings T2–T7. The low-temperature peak on the DSC diagram would result from the melting of the thinner lamellae in the semicrystalline coatings. Thus, the increase of the temperature and the enthalpy of this peak with the annealing temperature and the holding time would be related to the thickening of the lamellar stacks consisting of thin lamellae.

A structural evidence of this is provided by the XRD patterns of the coatings T1–T7 (Fig. 19.3). The X-ray patterns demonstrate that the as-sprayed and the

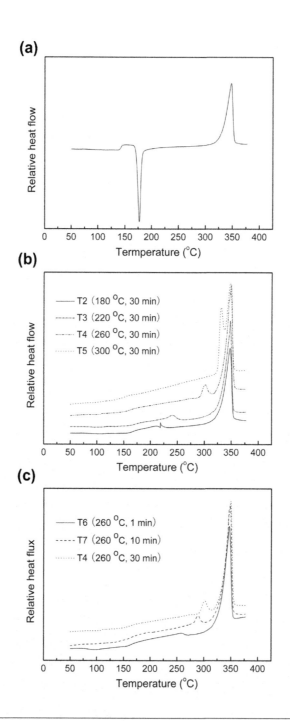

**FIGURE 19.2**

DSC scans of (a) as-sprayed coating, (b) coatings annealed at different temperatures
*T2–T5* and (c) coatings annealed with different holding times *T4, T6–T7*.

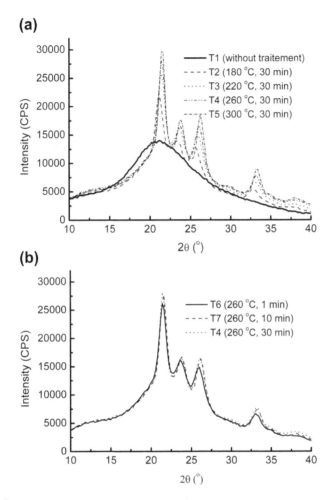

**FIGURE 19.3**

XRD patterns of the coatings (a) *T1–T5* and (b) *T4, T6–T7*.

annealed coatings possess amorphous and semicrystalline structures, respectively. Increasing the annealing temperature and holding time slightly increases the diffraction intensity. Thus, it can be concluded that the increase of the two factors, in the studied range, slightly increases the coating crystallinity.

## 19.4 ADHERENCE OF COATINGS

In real applications, the coating adhesion strength is one of the most important parameters. In order to avoid debonding failures of the coatings, they must adhere strongly to the substrates. Amorphous PEEK coatings exhibit a good adherence to metallic substrate [36]. However, annealed coatings with a semicrystalline structure adhere poorly to the substrate. This feature is more pronounced with a large coating thickness owing to significant residual stress in the coating.

Figure 19.4(a) shows the cross-section of the interface between an amorphous PEEK coating and an aluminum alloy substrate. It can be noticed that the coating has a close contact with the substrate. Figure 19.4(b) shows the coating–substrate interface obtained after a thermal treatment at 220 °C. After the heat treatment, the coating separates from the Al substrate in certain zones and cracks are observed at the interface. Moreover, some porosity near to the coating/substrate is observed. In order to confirm the interface morphologies, the coating was peeled off from the substrate, and the contacting surface was observed with a stereoscopic microscope (Fig. 19.4(c)). Numerous spherical pores are clearly observed, which is consistent with the observation of the cross-section structure (Fig. 19.4(b)). These pores are open to the interface, and therefore, the liquid resin filled them when mounting the

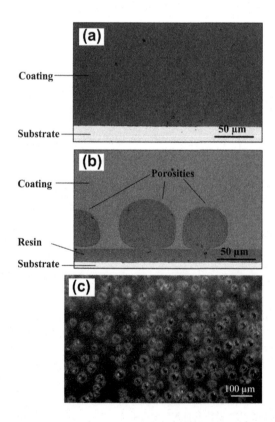

**FIGURE 19.4**

Coating–substrate interfaces of (a) amorphous coating (*T*1) and (b) semicrystalline coating (*T*4); contacting surface of semicrystalline coating (*T*4) (c). For color version of this figure, the reader is referred to the online version of this book.

cross-sectional samples with epoxy, which show a different contrast with the PEEK coating in Fig. 19.4(b).

The reduction in the coating adherence, after annealing, seems to be related to the residual stresses produced during the heat treatment. The densities of amorphous and semicrystalline PEEK are 1.26 and $1.32\,g\,cm^{-3}$, respectively [30]. Therefore, crystallization will provoke a volume contraction of the coating. With this contraction, tensile residual stresses will appear in the coating near the coating–substrate interface. Meanwhile, the mechanical resistance of PEEK, for example, the shear strength, decreases enormously above the glass transition temperature [37]. Accordingly, these residual stresses can be large enough to break the bonding between the coating and the substrate. Thus, a relative movement occurs at the interface. This might explain the cracks observed at the coating–substrate interface. When increasing the annealing temperature above 220 °C, the mechanical resistance of the coating is significantly reduced. At the early stage of annealing, the residual stresses might break the bonding between the coating and the substrate at some locations (Fig. 19.5(a)). Then, these zones might become free surfaces for the following volume contraction. As illustrated in the schematic (Fig. 19.5(b)), under an isotropic residual stress, the small debonded zones might develop into spherical pores at the end of the crystallization process. It is clear that when a high coating adhesive strength is required, an amorphous coating structure is important.

## 19.5 CORRELATION BETWEEN THE COATING CRYSTALLINE STRUCTURE AND ITS TRIBOLOGICAL BEHAVIOR

The friction tests were performed on a ball-on-disc (BOD) tribometer under a laboratory environment (temperature ~20 °C, humidity ~70%). The counterpart consisted of a 100Cr6 ball with a 15-mm diameter and a mirror-finished surface ($R_a$: 0.02 μm). The applied load and sliding velocity were, respectively, 15 N and $0.2\,m\,s^{-1}$. A 500-m relative sliding distance was performed, during which the friction force was measured with a sensor and was dynamically recorded into a computer. The friction coefficient was computed as the friction force divided by the applied load. The wear rate is defined as the worn volume per unit of applied load and sliding distance [15]. The inverse of

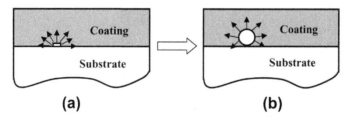

**(a)**    **(b)**

**FIGURE 19.5**

Schematic of the interfacial pore growth.

wear rate is wear resistance. In this chapter, the cross-section area of the worn tracks was measured with a Taylor-Hobson Surtronic 3P profilometer (Rank Tayloe Hobson Ltd, UK) after the completion of frictional sliding. The cross-section area of the wear tracks timing the perimeter permitted to obtain the total worn volume. At least 10 measurements were performed for computing the mean value of the wear rate.

Figure 19.6(a) and (b) show the friction coefficient and the wear rate of the coatings studied. The error bars represent the standard deviation of the obtained data. Compared with the amorphous coating, the semicrystalline coatings exhibit a slightly lower friction coefficient (0.325 on average vs. 0.358) and an obviously reduced wear rate (11.53 on average vs. $19.5 \times 10^{-6} \, \text{mm}^3 \, \text{N}^{-1} \, \text{m}^{-1}$). However, in the range studied, for the heat-treated coatings, the friction coefficient and wear rate show little dependence on the annealing conditions.

The increased coating hardness, as a result of the cold crystallization, could be responsible to the improved tribological performances. Figure 19.7(a) and (b) show the hardness of the as-obtained and the annealed coatings. The heat treatment significantly increases the coating hardness. However, in the studied range, the increase in the annealing temperature and holding time only leads to a slight increase in the coating hardness. It is clear that semicrystalline coatings exhibit a significantly higher hardness than the amorphous one does. However, in the range studied, the different crystallinities of the semicrystalline coatings only lead to a slight difference in the coating hardness. The dispersed spherulites in the amorphous matrix could restrict the motions and slippages of the polymer chains and thereby, the stiffness of the coating is increased [15,38].

Figure 19.8(a) and (b), respectively, show the worn surfaces of $T1$ and $T4$ coatings. Obvious plastic deformation and severe plows are observed on the worn surface of the amorphous coating (Fig. 19.8(a)). For the semicrystalline coatings, the plastic deformation is reduced, and a relatively smooth worn surface is observed (Fig. 19.8(b)). For a given load, compared with the amorphous coating, the semicrystalline coatings exhibit lower perpendicular deformations. The shear force between the two contact bodies and the plows on coating surface are decreased [39–41]. The enhanced coating stiffness after crystallization can reduce the mechanical loss module and the loss factor of polymers [38,42,43]. Under a shear force, during the relative sliding, the spherulites embedded in the amorphous matrix can restrict the motions and slippages of the polymer chains and thus decrease the internal energy dissipation in the semicrystalline surface layer involved in the frictional process.

## 19.6 EFFECTS OF SLIDING CONDITION ON THE TRIBOLOGICAL BEHAVIOR AND MECHANISMS OF AMORPHOUS PEEK COATINGS

The tribological behavior of a friction couple cannot be characterized just by intrinsic materials properties. The properties as well as the interactions of the system as a whole have to be taken into account [44]. Besides the material properties of the friction pairs, the frictional conditions can play an important role on the tribological behavior. Friedrich's [16,45] group found that the dynamics associated with the tribological testing was very important for the tribological behavior of polymers. The product of applied pressure and sliding velocity, $pv$, was thus

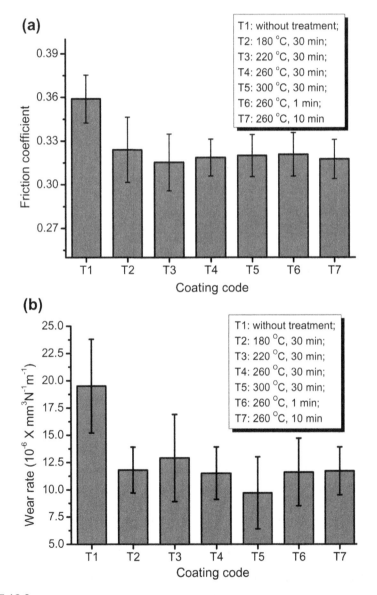

**FIGURE 19.6**

Friction coefficient (a) and wear rate (b) of the coatings studied.

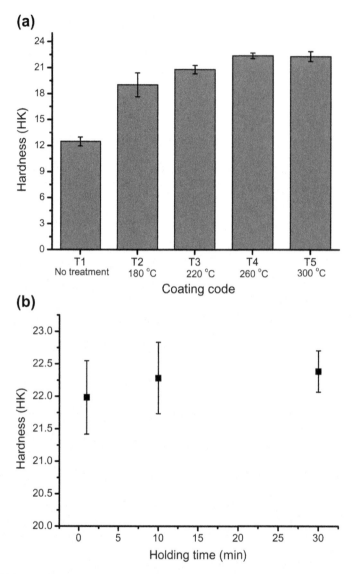

**FIGURE 19.7**

Evolutions of coating hardness versus (a) the annealing temperature and (b) the holding time.

regarded as an important factor for evaluating the tribological performance of polymers.

In addition, a number of factors, such as adhesion between the sliding pairs, the interfacial temperature, the strain rate at which the polymer surface layer is deformed etc., may affect the frictional work and the wear mechanisms [39]. A great diversity,

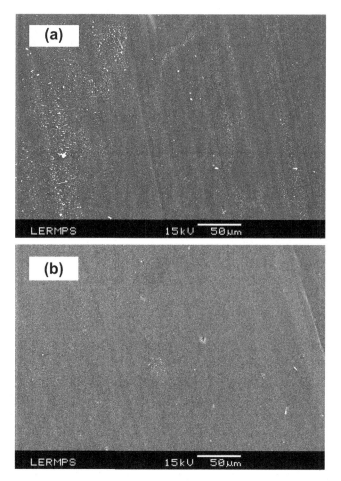

**FIGURE 19.8**

SEM micrographs of worn surfaces of (a) $T1$ and (b) $T4$.

with respect to the results obtained by several researchers, in investigating the effect of velocity on the tribological behavior of polymers, is observed [5,46–48]. Moreover, in polymer tribology, the relationship between the friction force and the applied load often deviates from a linear relationship [5,16]. Temperature is also an important factor influencing polymer tribological behavior. The results of Lu et al. [16] indicate that PEEK (semicrystalline) exhibits the largest friction coefficient and wear rate near its $T_g$. Hanchi et al. [49] also reported that PEEK and short carbon fiber-reinforced PEEK presented the highest friction coefficients near the $T_g$ of PEEK.

Therefore, the effects of the sliding conditions, that is, sliding velocity, applied load, and ambient temperature, on the tribological characteristics of PEEK coating were studied in this section.

### 19.6.1 Experimental

The amorphous PEEK coatings considered in this section were obtained using the painting technique. The feedstock powders (supplied by Victrex Scales Ltd) had a mean diameter of 10 μm. The coatings had a 40-μm thickness. The mean surface roughness, $R_a$, of the coating was 0.3–0.4 μm.

Friction tests were performed on a BOD CSEM tribometer (CSEM, Switzerland). The counterpart consisted of a 6-mm diameter 100Cr6 steel ball with a mirror-finished surface ($R_a$: 0.02 μm).

For investigating the influences of the sliding velocity and applied load on the tribological behavior of the amorphous PEEK coatings, the frictional tests were performed under laboratory environment (temperature ~20 °C, humidity ~70%). The applied load and sliding velocity varied from 1 to 9 N and from 0.2 to 1.4 m s$^{-1}$, respectively. The sliding distance in all the frictional tests was 2000 m.

When investigating the effects of temperature on the tribological behavior of the PEEK coatings, the applied load and sliding velocity were 9 N and 0.2 m s$^{-1}$, respectively. The frictional tests were performed at temperatures ranging from 20 to 200 °C. The sliding distance in the frictional tests was 1000 m. Before the tests, the samples were heated to the desired temperature in a furnace attached to the tribometer.

### 19.6.2 Dependence of the tribological behavior on the sliding velocity and applied load

Figure 19.9(a) and (b) illustrate the dependences of the friction coefficient and wear rate on the applied load and velocity, respectively. The tribological characteristics of amorphous PEEK are sensitive to the variations of the sliding velocity and applied load. Under 1 N, the increase of the sliding velocity, from 0.2 to 1.4 m s$^{-1}$, results in a practically monotonic decrease of the friction coefficient. Under 9 N, the increase of the velocity from 0.2 to 1.1 m s$^{-1}$ induces an increase of the friction coefficient. On the other hand, a sharp drop of the friction coefficient occurs when the sliding velocity is further increased from 1.1 to 1.4 m s$^{-1}$. However, under 5 N, the friction coefficient exhibits a maximum value at 0.8 m s$^{-1}$. From Fig. 19.9(b), it is seen that the wear rate does not follow the same tendencies as those of the friction coefficient. In contrast to the friction coefficient, <1 N, the wear rate monotonically increases when the sliding velocity is increased from 0.2 to 1.4 m s$^{-1}$. Under 9 N, however, the wear rate presents a sharp drop when the sliding velocity is increased from 1.1 to 1.4 m s$^{-1}$, following a similar tendency as the friction coefficient.

As shown above, the sliding velocity and applied load seem to play combined roles on the tribological characteristics. In order to clarify the role of each factor, the other has to be considered simultaneously.

#### 19.6.2.1 Tribological mechanism under 9 N

At a high load, the tribological behavior of amorphous PEEK can be closely related to its viscoelastic behavior. The increase of sliding velocity leads to distinct modifications

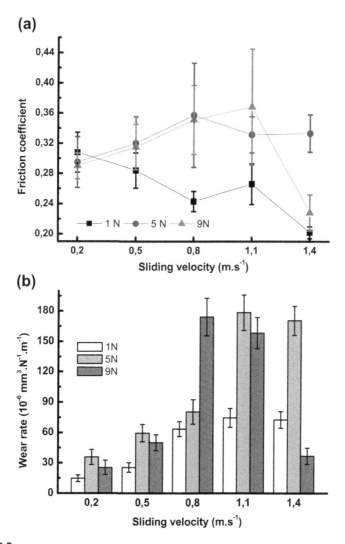

**FIGURE 19.9**

Dependences of the friction coefficient (a) and wear rate (b) on the applied load and sliding velocity. For color version of this figure, the reader is referred to the online version of this book.

of tribological mechanisms. Figure 19.10(a) shows the 3D morphology of the worn track (indicated by arrows) generated under 9 N at $0.2\,\text{m}\,\text{s}^{-1}$ (expressed thereafter as [9 N, $0.2\,\text{m}\,\text{s}^{-1}$]) after a completion of 2000 m sliding. The sliding direction of the counterpart ball is indicated on the image. It can be clearly seen that the worn track is not continuous and that some periodic material stacks with parabolic shapes appear along the sliding direction. Figure 19.10(b) shows the scanning electron microscope

(SEM) observation of the ridge of a stack. Figure 19.10(c) shows the top view of the worn track and the geometric profile of the track center along the sliding direction (along the dashed line). It can be noticed that the distance between two stack ridges (expressed thereafter as "period") is about 1.4 mm (7/5 mm). These morphological features suggest that viscoelastic deformation of PEEK took place during the friction process.

According to the two-term noninteracting friction model proposed by Briscoe [50], the surface layer, involved in the frictional process, was classified into two terms: the interface zone and the cohesive zone (Fig. 19.11). The former, with a depth range of about 100 nm, could mainly determine the friction work dissipation caused by the adhesion force between the two friction pairs. The dissipation caused by the plastic plow and hysteresis loss could be rather related to the cohesive zone, with a depth

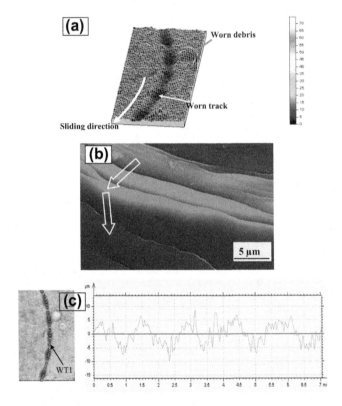

**FIGURE 19.10**

Characterization of a worn track generated at 9 N, 0.2 m s$^{-1}$: (a) 3D morphology of the work track; (b) SEM observation of a stack ridge, in which the arrows indicate the climbing and descending directions of the counterpart ball; (c) top view and geometric profile of the center of the track along the sliding direction. For color version of this figure, the reader is referred to the online version of this book.

comparable with the contact length. The intermittent periodic nature of the worn track, as revealed in Fig. 19.10, can be related to the result of a stick-slip motion of the counterbody ball, and a strain hardening effect of the polymer. For most of the polymers, the Van der Waals and hydrogen bonds are typical factors for the junctions occurring between the two counterparts [51]. Formation and rupture of these junctions control the adhesion component of friction. In the stick-slip process, due to the adhesion force in the interface zone, the PEEK surface presents a larger motion than the subsurface material, and thereby, material accumulation ahead of the counterpart occurs (Fig. 19.11). Meanwhile, the stress applied by the ball on the PEEK surface is such that the subsurface material is also deformed. During the stick stage, the tangential stress is smaller than the critical stress, which is related to the adhesion between the sliding pairs and the contact configuration, but it increases with time. Once the stress applied on the polymer surface exceeds the critical stress [52,53], the slip stage initiates and runs until the stress decreases to below the critical stress, when the ball and the PEEK surface stick again. The piling-up of the material ahead of the ball is completed during the slip stage. The material piling-up leads to a strain hardening effect on the polymer by possible entanglements and orientations of the molecular segments, which in turn increase the stiffness and tensile strength of the polymer in the stacking zone [52].

In molecular scope, at a low strain rate, the long molecules in the subsurface zone could, to some degree, orient themselves to the horizontal deformation of the surface. The low hysteresis loss could account for the low friction coefficient at this condition. With respect to the wear mechanism, it might be described as a "transfer" mode in which case the interfacial bonding between the two counterparts is stronger than the cohesive strength of the PEEK [39]. The polymer material might be transferred to the counterpart surface progressively in the form of very small fractions. Finally, the accumulated material on the counterpart surface could fall off in large pieces (wear debris marked in Fig. 19.10(a)).

When the sliding velocity is increased, the periodical feature of the worn track becomes less obvious. For 9 N, $0.8\,\mathrm{m\,s^{-1}}$, the worn track becomes practically

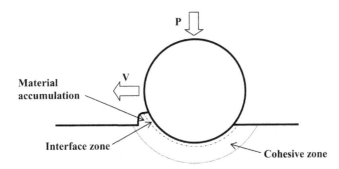

**FIGURE 19.11**

Schematic of material accumulation ahead of the sliding ball. The classification of surface layer is done according to Briscoe's model [41].

continuous. Figure 19.12(a) and (b) illustrate the 3D morphology of the worn track generated under these conditions and the distinct difference of the top views of this worn track (WT2) with that obtained for 9 N, 0.2 m s$^{-1}$ (WT1).

The worn surface of PEEK generated at 9 N, 1.1 m s$^{-1}$ is presented in Fig. 19.13(a). Figure 19.13(b) presents the framed zone with a higher resolution. Besides plows, ripple-like folds are observed on the worn surface (Fig. 19.13(b)). These folds are assumed to be formed by the significantly superior deformation of the surface parallel to the sliding, compared with that of the subsurface material.

When considering the aforementioned analyses, it can be deduced that, under 9 N, in the velocity range of 0.2–1.1 m s$^{-1}$, the speed dependence of the friction coefficient can be closely related to the viscoelastic property of the polymer. At an increased strain rate, the enhanced hysteresis loss can give rise to the friction coefficient. In the molecular scope, the high friction coefficients at intermediate velocities could be related to a less accordant motion of the molecules, in which case the entanglements and cuttings between the molecule segments could be significant. In such a case, the hysteresis loss could be important [54–56].

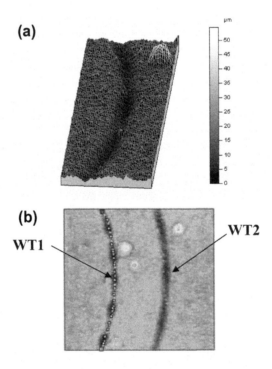

**FIGURE 19.12**

(a) The 3D morphology of the worn track generated under the conditions 9 N, 0.8 m s$^{-1}$; (b) comparison of top views of the worn track generated for 9 N, 0.8 m s$^{-1}$ (WT2) with that generated for 9 N, 0.2 m s$^{-1}$ (WT1). For color version of this figure, the reader is referred to the online version of this book.

Figure 19.14(a) shows the worn surface generated at 9N, 1.4 m s$^{-1}$, and Fig. 19.14(b) illustrates the framed zone with a higher resolution. The well-defined microplows on the worn surface imply that microcutting seems to be the dominant friction and wear mechanism. In the range of high velocities, elastic behavior is prevalent in the contact zone [39,57,58]. In this case, the shear strength contributes mainly to the friction force. Moreover, it is assumed that the increase in the sliding velocity might diminish the thickness of the cohesive zone by reducing the depth of stress distribution.

For a given pressure, there could be two critical sliding velocities leading to different tribological mechanisms of amorphous PEEK coating. The first critical

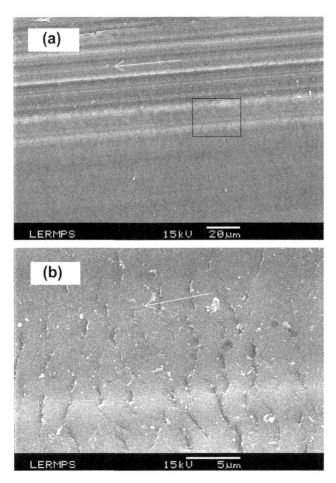

**FIGURE 19.13**

(a) Morphology of a worn surface generated at 9 N, 1.1 m s$^{-1}$; (b) observation of the framed zone in (a) with a higher resolution.

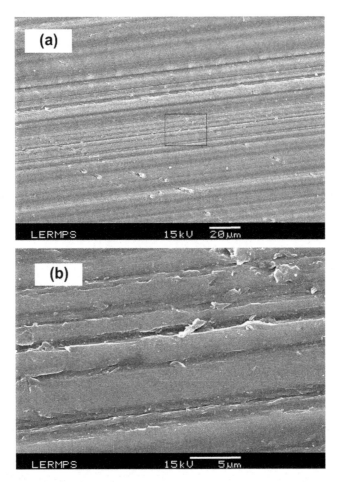

**FIGURE 19.14**

(a) Morphology of the worn surface generated at 9 N, 1.4 m s$^{-1}$; (b) observation of the framed zone in (a) with a higher resolution.

velocity corresponds to a velocity lower than that for which obvious plastic flow can take place; the second corresponds to a velocity higher than that for which the hysteresis loss is no longer a significant component contributing to the friction force. The critical speed could be related to the material nature, for example, molecule feature and material stiffness. Between these two critical values, there could be a velocity at which the polymer exhibits a high friction coefficient and wear rate.

### 19.6.2.2 Tribological mechanism under 1 and 5 N

Under a 1-N load, no obvious material stacking was observed on the worn track. Figure 19.15 illustrates worn surface morphology of the amorphous coating, after

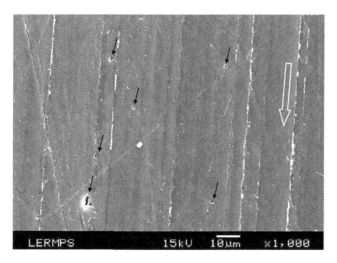

**FIGURE 19.15**

Morphology of a worn surface generated at 1 N, 0.2 m s$^{-1}$.

2000-m sliding at 1 N, 0.2 m s$^{-1}$. Numerous small holes are noticed on the worn surface, some of which are indicated by black arrows. At a small load, the "ironing" effect could be an important tribology mechanism [29,59]. During the sliding process, the asperities on the coating surface are sheared by the counterpart. With repeated effects, fatigue could occur on the surface around these asperities and finally, the corresponding microzones are peeled off from the matrix. Squeezed by the sliding ball under a low applied load, the debris can roll or be entrapped to the center of the contact zone, where the deformation of PEEK perpendicular to the surface is the largest [29]. Finally, some of them could be crushed in the plows (white phase in Fig. 19.15).

When increasing the sliding velocity, more severe plows are observed on the worn surface (Fig. 19.16). The interfacial temperature is a crucial factor determining the tribological characteristics [29]. As a result of the low thermal conductivity of PEEK, friction-induced heat surely provokes an increase of the interface temperature; thus, an increase of the sliding velocity can result in a higher contact temperature [40]. Due to the decreased mechanical strength of PEEK as a result of enhanced interface temperature, the friction coefficient decreases when the velocity increased, while the wear rate is increased.

Under an intermediate load, both the viscoelastic behavior and the contact temperature can play important roles simultaneously on the tribological behavior [29]. Figure 19.17 presents the SEM observation of a worn surface generated at 5 N, 0.2 m s$^{-1}$. Above a critical pressure, the worn debris can be pushed by the sliding counterpart (Fig. 19.18). A prow is pushed ahead of the debris, and the material on the polymer surface is continually displaced sideways to form ridges adjacent to the developing groove [39]. The increase of the sliding velocity leads to a higher

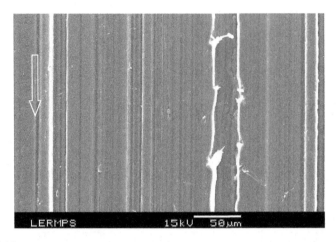

**FIGURE 19.16**

Morphology of a worn surface generated at 1 N, 1.1 m s$^{-1}$.

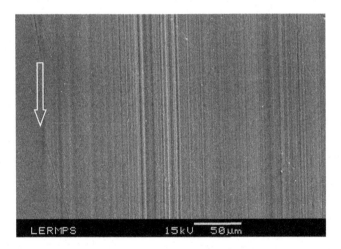

**FIGURE 19.17**

Worn surface morphology of PEEK coating generated at 5 N, 0.2 m/s.

**FIGURE 19.18**

Schematic of the formations of prows and ploughs.

interfacial temperature and in turn aggravates the surface plows. Moreover, the hysteresis loss could also contribute to the variation of the friction force, depending on the sliding velocity.

From the above discussion, it is assumed that there is a critical applied load, lower than that for which the "ironing" effect dominates the sliding process. In this case, the relative motion between the surface and subsurface materials is not obvious. The interfacial temperature plays an important role on the tribological performance. Above a critical value, the viscoelastic behavior of the polymer surface layer plays an important role on the tribological behavior. This critical value could be closely related to the contact geometry and the in depth deformation of the polymer. Such factors as polymer surface morphology, hardness, elastic modulus, and Poisson's ratio could affect this critical value.

### 19.6.3 Dependence of the tribological mechanism on the ambient temperature

Figure 19.19(a) and (b) show the dependences of the friction coefficient and wear rate on the temperature. It can be seen that the temperature has a significant effect on the coating's tribological characteristics. The friction coefficient increases with increasing the ambient temperature and reaches a maximum value at 160 °C (near $T_g$). A further increase in the temperature, from 160 to 200 °C, leads to a slight decrease of the friction coefficient. Similar to the friction coefficient, the preliminary increase in ambient temperature increases the wear rate. However, when the temperature is increased from 120 °C to 160 °C, a sharp drop of the wear rate was noticed. When the ambient temperature is further increased to 200 °C, the coating exhibits a maximum wear rate. It should be noted that the temperature on the interface between the sliding pairs is higher than the ambient temperature [60].

The 3D morphologies of the worn tracks obtained at 20 °C (WT3) and 50 °C (WT4) are compared in Fig. 19.20. It can be noticed that the increase of temperature enlarges the distance between the periodic material stacks (as an average from ~1.4 to ~2.2 mm). According to the stick-slip motion feature, the increase in the distance between two periodic material stacks is indicative of the greater time required for the critical stress to be reached before slip occurs [52]. This could be attributed to an increased adhesion, between the two sliding pairs, and to the decreased stiffness of the coating.

Figure 19.21(a) shows the SEM micrograph of the worn track WT4, generated at 50 °C. It can be observed that the ridge of the material stack exhibits a narrower width. Figure 19.21(b) shows the zone framed in Fig. 19.21(a) with a higher magnification. In this zone, scale-like parabolic folds, perpendicular to the sliding direction, are observed (indicated by dashed arrows). These folds might be formed by a significantly superior parallel deformation of the coating surface than that of coating subsurface. Strain hardening occurring in this zone due to molecular orientation and entanglement, which increases the stiffness of PEEK, is an indispensable factor for the formation of such folds. It should be noted that, outside the ridge zone, no

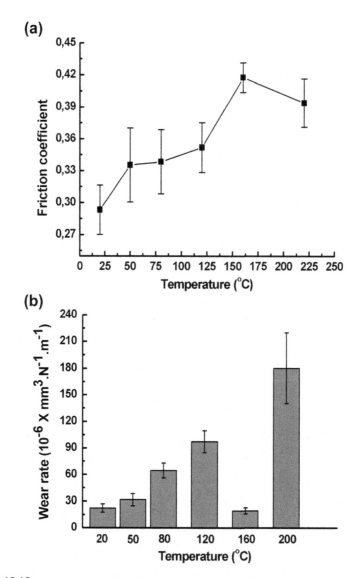

**FIGURE 19.19**

Dependences of the mean friction coefficient (a) and wear rate (b) on ambient temperature.

such fold is noticed on the wear track. The molecular entanglement occurring in this zone might limit the motions of the molecules. As a result, when the PEEK surface is deformed along the direction of the sliding ball, the subsurface presents a much smaller deformation.

When the temperature is further increased to 120 °C, the distance between two material stacks is increased to 3.8 mm, averagely. The increase of temperature also

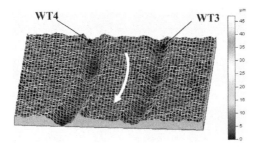

**FIGURE 19.20**

The 3D morphologies of the worn tracks obtained under 20°C (WT3) and 50°C (WT4). For color version of this figure, the reader is referred to the online version of this book.

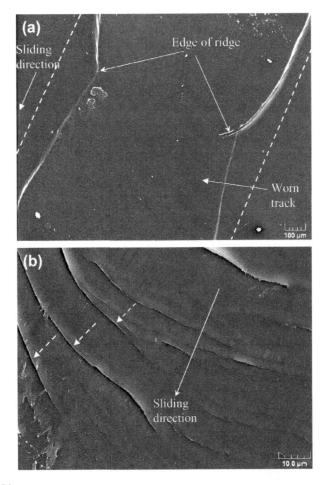

**FIGURE 19.21**

(a) SEM observation of the worn track T4; (b) ridge zone framed in (a) with a higher resolution.

results in more severe plows on the worn surface. The thermal softening effect of PEEK caused by the high temperature is responsible for this behavior.

At 160 °C, the coating exhibits the largest friction coefficient, but the smallest wear rate. The XRD analysis after the frictional test indicates that the coating possesses a semicrystalline structure [29]. At a temperature above $T_g$, the coating crystallizes rapidly following a stepwise mode [22]. As mentioned above, crystallization is of great significance for the frictional process. The stick-slip effect is alleviated, due to the increased coating stiffness.

Figure 19.22(a) and (b) show the SEM micrographs of the worn surface produced at 160 °C, and Fig. 19.22(c) illustrates the framed zone in Fig. 19.22(b) with a higher resolution. White lines, perpendicular to the sliding direction and corresponding to fatigue damages, are observed on the worn surface. In addition, parallel compact folds are noticed along the worn surface. From Fig. 19.22(c), it can be seen that friction fatigue initiates from the brim of the compact fold (indicated by the dashed arrow).

Fatigue is known to be a change in the material state due to repeated stressing, which results in progressive fracture. Its characteristic feature is an accumulation of irreversible changes, which give rise to generation and development of cracks [39]. For semicrystalline PEEK, the increased stiffness could reduce the material deformation in the subsurface zone. The spherulites could restrain the movements of polymeric chain segments; thereby stress concentration might develop around the spherulite. The shearing stress in the surface layer and the possible stress concentration around these spherulites could be responsible for the formation of the fatigue cracks. Friction fatigue occurs from the brim of the folds, where the maximum tangential stress or the tensile strain can take place (Fig. 19.22(c)).

Near $T_g$, the high friction coefficient could be related to the high hysteresis loss of PEEK. This is consistent with the fact that polymers exhibit the highest loss tangent in the vicinity of $T_g$ [59]. The increased mechanical properties caused by crystallization can be responsible for the reduced wear rate. Fatigue damage and mild plow constitute in this case the main wear mechanisms.

The increase in ambient temperature from 160 to 200 °C slightly decreases the friction coefficient, while it enormously increases the wear rate. Above $T_g$, an increase in the temperature sharply degrades the mechanical strength of PEEK [30]. The SEM observation of the worn surface gives evidence that enormous deformation occurs during the sliding process (Fig. 19.23). It is clear that, under a load of 9 N, the temperature of 200 °C is higher than the maximum service temperature range of the PEEK coating.

## 19.7 TRIBOLOGICAL BEHAVIOR OF NANO-SiC (7 WT.%)-FILLED PEEK COATINGS

In order to improve the friction and wear behavior of polymeric materials, typical ideas are to reduce their adhesion to the counterpart material and to enhance their hardness, stiffness, and compressive strength [15]. Numerous investigations have already been conducted for tailoring high-performance PEEK tribomaterials and understanding the

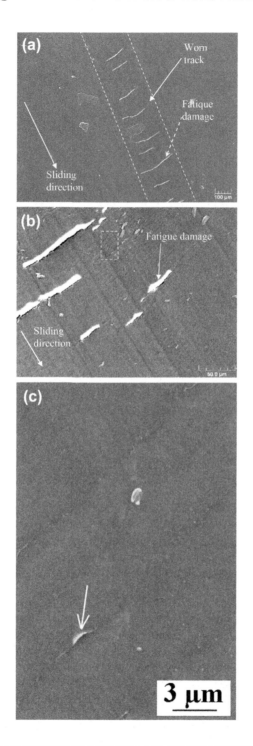

**FIGURE 19.22**

(a) SEM observation of a worn surface obtained at 160 °C; (b) SEM observation of the worn surface with higher resolution; (c) SEM observation of the framed zone in (b).

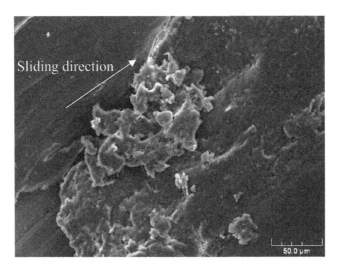

Sliding direction

50.0 μm

**FIGURE 19.23**

SEM observation of worn surface obtained at 200°C.

tribological mechanisms of the composites as well [15–17,42,48,61,62]. As a conventional route, the tribological performance of the polymer matrix can be significantly improved by adding combined short carbon fibers and solid lubricants, for example, graphite and polytetrafluorethylene [15,16,44,48]. In this hybrid composite system, short fibers improve the loading capability and the solid lubricants promote the formation of a transfer film. It was revealed in the past years that the incorporation of nanoparticles into the PEEK matrix can improve its tribological properties, especially its wear resistance [15,18,45,61,62]. Xue et al. [61,62] studied the tribological behavior of SiC-filled PEEK composites. Their results indicate that the addition of nano-SiC particles into the PEEK matrix reduces the friction coefficient and increases its wear resistance. With a concentration of 7–10% (wt.) nano-SiC, the composite exhibits the best tribological performance. Based on the conventional PEEK composites filled with carbon fibers and solid lubricants, the addition of nanoparticles into the conventional composite leads to significantly improved tribological performances [63–65].

Driven by industry requirements, polymer-based composite coatings were studied in the past years. It was revealed that the addition of different fillers can significantly improve the tribological performance, in comparison to the pure polymer coatings. Such fillers as $MoS_2$, graphite, SiC, $SiO_2$, and polyfluo wax improve the tribological behavior of polyurethane (PU) coatings [7,66]. The addition of nanoalumina particles into nylon-11 coating improves the scratch resistance of the polymer coating [26].

In order to improve the tribological performance of the amorphous PEEK coating, SiC nanoparticles were added, and the tribological behavior of the nanocomposite coating was studied in the authors' previous work.

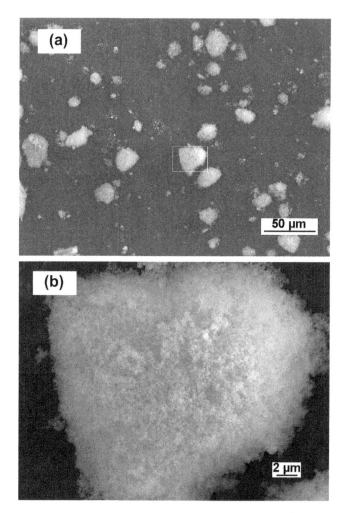

**FIGURE 19.24**

Morphology of the nano-SiC particles with low (a) and high (b) magnifications.

## 19.7.1 **Experimental**

The coatings were prepared by the painting technique using PEEK powders with a mean particle diameter of 10 μm. The SiC particles had a mean diameter of 50 nm. Figure 19.24(a) and (b) show the morphologies of these particles with different magnifications. It is clear that the nanoparticles have a strong agglomerating tendency. PEEK powders and the nanoparticles were mixed in an aqueous medium and then dispersed by mechanical stirring and ultrasonication before the slurry was applied on the aluminum substrate, as described above. In order to get good coating/substrate adhesion, the composite coating after being remelted was also quenched in cold water to get amorphous PEEK matrix. The concentration of the nano-SiC particles was 7 wt.%. The frictional test conditions were identical to those described in Section 19.6.2.

## 19.7.2 Results and discussion

### 19.7.2.1 Coating structure

Figure 19.25 shows the cross-section of a nano-SiC-filled PEEK coating. The white phase corresponds to SiC-rich zones. Even after mechanical and ultrasonic dispersing steps, obvious nanoparticle agglomerates still exist. During the coating preparation process, the nano-SiC particles are distributed on the surface of the PEEK particles and also in their interspaces. It is assumed that the dispersion state is of great significance for mechanical properties of coatings and their tribological performance as well. Further efforts to improve the dispersion of nanofillers in thermoplastic coatings are still a subject of current and future studies.

### 19.7.2.2 Tribological behavior

The dependence of the friction coefficient and wear rate of PEEK/nano-SiC coating on the applied load and sliding velocity is shown in Fig. 19.26. Compared with a pure PEEK coating, the composite coatings present lower friction coefficients under combined conditions of a high applied load and intermediate sliding velocity. Outside these ranges, the friction coefficient is slightly higher than that of the pure coating. Meanwhile, the dependence of the friction coefficient on the sliding velocity is less pronounced than that of the pure PEEK coating. The increase in the sliding velocity leads to a slight and linear decrease of the friction coefficient. The increase of the applied load also reduces the friction coefficient. On the contrary, an increase in the sliding velocity and the applied load monotonously enhances the wear rate. However, in comparison to the pure PEEK coating, the incorporation of SiC particles always significantly improves the wear resistance.

As reinforcing fillers in a soft matrix, the SiC particles improve the coating stiffness and creep resistance. Figure 19.27 shows the difference in hardness for a pure

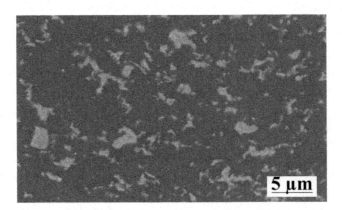

**FIGURE 19.25**

Cross-sectional structure of a nano-SiC filled PEEK coating.

PEEK and a PEEK/SiC coatings prepared by the painting technique. The 7 wt.% of SiC particles enhance by over 30% the coating hardness. Figure 19.28 shows a worn surface obtained under the conditions 1 N, 1.1 m s$^{-1}$. When compared with the pure PEEK coating (Fig. 19.16), it clearly appears that the plows created on the surface of the composite coating are significantly reduced. Due to the agglomeration of the SiC particles in the coating, some of them are probably torn out by the ball. Subsequently, they are spread on the surface of the worn track by the sliding ball. Thus, the direct contact and the adhesion between the ball and the coating can be reduced.

Under high loads, the material accumulation, occurring ahead of the sliding ball, is alleviated by incorporating nanofillers. Under the sliding conditions 9 N, 0.2 m s$^{-1}$, the period of the stacks is about 1.1 mm for the PEEK/SiC coating versus 1.4 mm for the pure

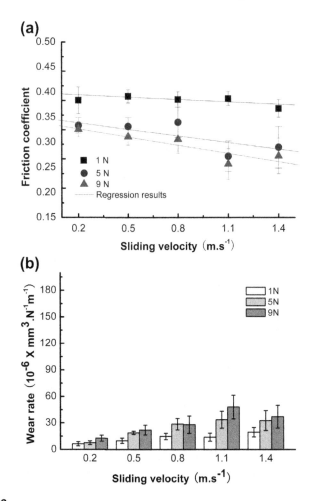

**FIGURE 19.26**

Dependence of the friction coefficient (a) and wear rate (b) of PEEK/nano-SiC coating on the applied load and sliding velocity.

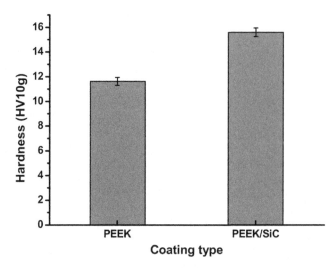

**FIGURE 19.27**

Hardnesses of PEEK and PEEK/SiC coatings.

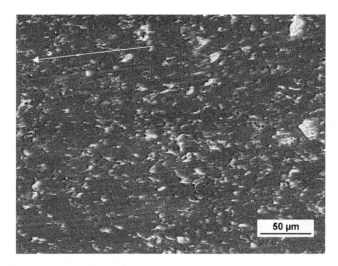

**FIGURE 19.28**

Worn surface of a PEEK/SiC coating produced under the conditions 1 N, 1.1 m s$^{-1}$.

coating. Based on the previous analysis, the decrease in the stack period corresponding to the incorporation of nano-SiC particles can be attributed to the decrease in the adhesion force between the sliding pairs and to the improvement of the coating stiffness.

Figure 19.29 shows the worn surface of a PEEK/SiC coating produced under the conditions 9 N, 1.1 m s$^{-1}$. The plows constitute the main surface characteristic. When

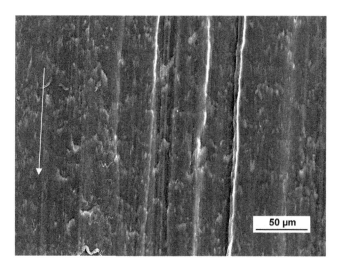

**FIGURE 19.29**

Worn surface of a PEEK/SiC coating produced under the conditions 9 N, $1.1 \, \mathrm{m \, s^{-1}}$.

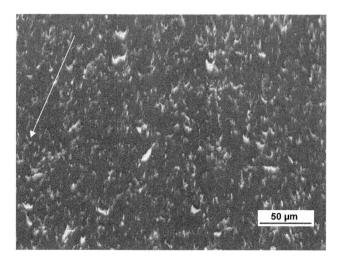

**FIGURE 19.30**

Worn surface of a PEEK/SiC coating produced under the conditions 9 N, $1.4 \, \mathrm{m \, s^{-1}}$.

comparing the surface morphology with that of a pure PEEK coating (tested in the same conditions, Fig. 19.13), ripple-like folds are not noticed on the worn surface. The stick-slip behavior is alleviated by the addition of nanoparticles. At higher sliding velocities, however, the plows on the worn surface are significantly reduced. Figure 19.30 shows a worn surface produced under the conditions 9 N, $1.4 \, \mathrm{m \, s^{-1}}$. The worn surface exhibits a typical morphology of mild abrasion effect. The grinding effect produced by the SiC particles entrapped on the contact interface, and the increased coating stiffness by the addition of the nanoparticles, can be important factors leading to different coating tribological behavior from that of the pure PEEK coating.

## 19.8 CONCLUSIONS

Being an effective lubricating layer, PEEK coatings are of great significance for improving the tribological performance of metallic counterbodies. According to the works conducted in the authors' laboratory, two efficient coating techniques, that is, FS and painting, were described. Using these two techniques, dense PEEK coatings with amorphous structures were obtained. Semicrystalline coatings were obtained when the amorphous coatings were subjected to an annealing treatment with temperatures higher than the glass transition temperature of PEEK. The amorphous coatings present a high adherence to aluminum substrate. Semicrystalline coatings adhere badly to the substrate.

The tribological behaviors of PEEK coatings under dry sliding conditions were investigated. First, the correlation between the coating crystalline structure and its tribological behavior was investigated. Compared with the amorphous coatings, the semicrystalline coatings exhibit a lower friction coefficient and wear rate. After crystallization, the increased coating stiffness contributes to the improvement of the tribological performance. Therefore, the coating adherence and its tribological performance must be considered simultaneously and be carefully controlled by controlling the coating crystallinity. In addition, the effects of the sliding conditions, such as sliding velocity, applied load, and ambient temperature, on the tribological behavior of an amorphous PEEK coating were investigated. The tribological behavior of the amorphous PEEK coating shows close dependence on the sliding conditions. It was assumed that the tribological behavior of the amorphous coating is closely related to its viscoelastic behavior and the temperature on the sliding interface.

Finally, the tribological behavior of an amorphous nano-SiC (7 wt.%)-filled PEEK coating was described. The incorporation of the nano-SiC particles significantly decreases the wear rate of the coating. The SiC particles increases the coating stiffness and could reduce the direct contact and thereby adhesion between the polymer and the counterpart. Moreover, at high sliding velocities, a grinding effect produced by the SiC particles could also be important for the tribological behavior of the coating.

## LIST OF SYMBOLS AND ABBREVIATIONS

$T_g$  Glass transition temperature
$R_a$  Mean roughness
$pv$  Product of applied pressure and sliding velocity

**SiC** Silicon carbide
**MoS$_2$** Molybdenum sulfide
**SiO$_2$** Silicon dioxide
**PEEK** Polyetheretherketone
**PFW** Polyfluo wax
**PU** Polyurethane
**APS** Atmospheric plasma spraying
**HVOF** High-velocity oxygen-fuel
**FS** Flame spraying
**IR** Infrared spectra
**DSC** Differential scanning calorimetry
**XRD** X-ray diffraction
**SEM** Scanning electron microscope

## Acknowledgments

The authors would like to acknowledge the company PSA Peugeot Citroën for the financial support. Thanks are due to Dr H. Yu for his assistance in the DSC and XRD analyses and to Ms P. Hoog for carefully checking the manuscript. The authors would like to thank their colleagues in LERMPS, University of Technology of Belfort-Montbéliard, for their help in the experimental work.

## References

[1] H. Liao, E. Beche, C. Coddet, On the microstructure of thermally sprayed "PEEK" polymer, in: C. Coddet (Ed.), Proceeding of ITSC 1998, ASM International, Materials Park, OH, 1998.

[2] L. Duarte, E. Silva, J. Branco, V. Lins, Production and characterization of thermally sprayed polyethylene terephthalate coatings, Surface and Coatings Technology 182 (2004) 261–267.

[3] S. Lathabai, M. Ottmüller, I. Fernandez, Solid particle erosion behaviour of thermal sprayed ceramic, metallic and polymer coatings, Wear 221 (1998) 93–108.

[4] T. Zhang, D.T. Gawne, Y. Bao, The influence of process parameters on the degradation of thermally sprayed polymer coatings, Surface and Coatings Technology 96 (1997) 337–344.

[5] G. Zhang, S. Guessasma, H. Liao, H. Coddet, J.-M. Bordes, Investigation of friction and wear behaviour of SiC-filled PEEK coating using artificial neural network, Surface and Coatings Technology 200 (2006) 2610–2617.

[6] Y. Wang, K. Brogan, S. Tung, Wear and scuffing characteristics of composite polymer and nickel/ceramic composite coated piston skirts against aluminum and cast iron cylinder bores, Wear 250 (2001) 706–717.

[7] H.J. Song, Z.Z. Zhang, Investigation of the tribological properties of polyfluo wax/polyurethane composite coatings filled with several micro-particulates, Materials Science and Engineering A 424 (2006) 340–346.

[8] M.Y. Chen, Z. Bai, S.C. Tan, M.R. Unroe, Friction and wear scar analysis of carbon nanofiber-reinforced polymeric composite coatings on alumina/aluminum composite, Wear 252 (2002) 624–634.

[9] V. Le Houérou, R. Robert, C. Cauthier, R. Schirrer, Mechanisms of blistering and chipping of a scratch-resistant coating, Wear 265 (2008) 507–515.

[10] D. Dascalescu, K. Polychronopoulou, A.A. Polycarpou, The significance of tribochemistry on the performance of PTFE-based coatings in $CO_2$ refrigerant environment, Surface and Coatings Technology 204 (2009) 319–329.

[11] L.Y. Lin, D.E. Kim, Tribological properties of polymer/silica composite coatings for microsystems applications, Tribology International 44 (2011) 1926–1931.

[12] T. Bayerl, A.K. Schlarb, Welding of tribologically optimized polyetheretherketone films with metallic substrates, TribologyInternational 43 (2010) 1175–1179.

[13] Z. Rasheva, T. Burkhart, A. Rehl, M. Janke, Reibungs- und Verschleißuntersuchungen von verstärkten Polyamidimid Harzen für Kolbenschaftbeschichtungen, Tribologie-Fachtagung, Gesellschaft für Tribologie e.V, Göttingen, 2010 (in Geman).

[14] A. Gebhard, F. Haupert, A.K. Schlarb, in: K. Friedrich, A.K. Schlarb (Eds.), Tribology of Polymeric Nanocomposites, first ed., Elsevier, Amsterdam, 2008, pp. 439–457.

[15] K. Friedrich, Z. Zhang, A.K. Schlarb, Effects of various fillers on the sliding wear of polymer composites, Composites Science and Technology 65 (2005) 2329–2343.

[16] Z.P. Lu, K. Friedrich, On sliding friction and wear of PEEK and its composites, Wear 181-183 (1995) 624–631.

[17] J. Voort, S. Bahadur, The growth and bonding of transfer film and the role of CuS and PTFE in the tribological behavior of PEEK, Wear 181-183 (1995) 212–221.

[18] Q. Wang, Q. Xue, H. Liu, W. Shen, J. Xu, The effect of particle size of nanometer $ZrO_2$ on the tribological behaviour of PEEK, Wear 198 (1996) 216–219.

[19] G. Zhang, H. Liao, H. Yu, S. Costil, S.G. Mhaisalkar, J.M. Bordes, C. Coddet, Deposition of PEEK coatings using a combined flame spraying–laser remelting process, Surface and Coatings Technology 201 (2006) 243–249.

[20] G. Zhang, S. Leparoux, H. Liao, C. Coddet, Microwave sintering of poly-ether-ether-ketone (PEEK) based coatings deposited on metallic substrate, Scripta Materialia 55 (2006) 621–624.

[21] B. Normand, H. Takenouti, M. Keddam, H. Liao, G. Monteil, C. Coddet, Electrochemical impedance spectroscopy and dielectric properties of polymer: application to PEEK thermally sprayed coating, Electrochimica Acta 49 (2004) 2981–2986.

[22] G. Zhang, H. Liao, H. Yu, V. Ji, W. Huang, S.G. Mhaisalkar, C. Coddet, Correlation of crystallization behavior and mechanical properties of thermal sprayed PEEK coating, Surface and Coatings Technology 200 (2006) 6690–6695.

[23] G. Zhang, H. Liao, H. Li, C. Mateus, J.M. Bordes, C. Coddet, On dry sliding friction and wear behaviour of PEEK and PEEK/SiC-composite coatings, Wear 260 (2006) 594–600.

[24] H. Liao, Application of PEEK Coating for Friction Reduction Purpose, Industrial Report of LERMPS, University of Technology of Belfort-Montbéliard, Belfort, 2001.

[25] L. Simonin, H. Liao, Characterization of flame-sprayed PEEK coatings by FTIR-ATR, DSC and acoustic microscopy, Macromolecular Materials and Engineering 283 (2000) 153–162.

[26] V. Gupta, T.E. Twardowski, R.A. Cairncross, R. Knight, HVOF sprayed multi-scale polymer/ceramic composite coatings, in: B.R. Marple, C. Moreau (Eds.), Proceeding of ITSC 2006, ASM International, Materials Park, OH, 2006.

[27] G. Zhang, A.K. Schlarb, S. Tria, O. Elkedim, Tensile and tribological behaviors of PEEK/nano-SiO$_2$ composites compounded using a ball milling technique, Composites Science and Technology 68 (2008) 3073–3080.

[28] M. Hedayati, M. Salehi, R. Bagheri, M. Panjepour, F. Naeimi, Tribological and mechanical properties of amorphous and semi-crystalline PEEK/SiO$_2$ nanocomposite coatings deposited on the plain carbon steel by electrostatic powder spray technique, Progress in Organic Coatings 74 (2012) 50–58.

[29] G. Zhang, Study on the Preceding and Properties of PEEK Based Coatings, PhD thesis of University of Technology of Belfort-Montbéliard, Belfort, 2006.

[30] PEEK, Properties Guide, Victrex, 2002.

[31] P. Cebe, S. Hong, Crystallization behaviour of poly(ether-ether-ketone), Polymer 27 (1986) 1183–1192.

[32] D.J. Blundell, B.N. Osborn, The morphology of poly(aryl-ether-ether-ketone), Polymer 24 (1983) 953–958.

[33] D.J. Blundell, On the interpretation of multiple melting peaks in poly(ether ether ketone), Polymer 28 (1987) 2248–2251.

[34] S. Tan, A. Su, J. Luo, E. Zhou, Crystallization kinetics of poly(ether ether ketone) (PEEK) from its metastable melt, Polymer 40 (1999) 1223–1231.

[35] A.J. Lovinger, S.D. Hudson, D.D. Davies, High temperature crystallization and morphology of poly (aryl-ether-ether-ketone), Macromolecules 25 (1992) 1752–1758.

[36] G. Zhang, H. Liao, M. Cherigui, J.P. Davim, C. Coddet, Effect of crystalline structure on the hardness and interfacial adherence of flame sprayed poly(ether-ether-ketone) coatings, European Polymer Journal, 2006. (Accepted for publication).

[37] J. Rault, Les Polymères Solides, Cépaduès-éditions, Toulouse, 1992 (In French).

[38] I.W. Ward, D.W. Hadley, An Introduction to the Mechanical Properties of the Solid Polymers, Chichester, Wiley, 1993.

[39] N.K. Myshkin, M.I. Petrokovets, A.V. Kovalev, Tribology of polymers: adhesion, friction, wear, and mass-transfer, Tribology International 38 (2005) 910–921.

[40] J. Li, H. Liao, C. Coddet, Friction and wear behavior of flame-sprayed PEEK coatings, Wear 252 (2002) 824–831.

[41] S. Wen, Tribological Pinciple, Tsinghua University Press, Beijing, 1990 (In Chinese).

[42] N.S. Eiss Jr., J. Hanchi, Tribological behavior of polyetheretherketone, a thermotropic liquid crystalline polymer and in situ composites based on their blends under dry sliding conditions at elevated temperatures, Wear 200 (1996) 105–121.

[43] J. Zhou, F. Yan, N. Tian, J. Zhou, Effect of temperature on the tribological and dynamic mechanical properties of liquid crystalline polymer, Polymer Testing 24 (2005) 270–274.

[44] K. Friedrich, Z. Lu, A.M. Hager, Recent advances in polymer composites' tribology, Wear 190 (1995) 139–144.

[45] S. Bahadur, C. Sunkara, Effect of transfer film structure, composition and bonding on the tribological behavior of polyphenylene sulfide filled with nano particles of TiO$_2$, ZnO, CuO and SiC, Wear 258 (2005) 1411–1421.

[46] Y. Yamamoto, T. Takashima, Friction and wear of water lubricated PEEK and PPS sliding contacts, Wear 253 (2002) 820–826.

[47] A. Schelling, H.H. Kausch, in: K. Friedrich (Ed.), Advances in Composites Tribology, Elsevier, Amsterdam, 1993, pp. 65–105.

[48] J. Hanchi, N.S. Eiss Jr., Dry sliding friction and wear of short carbon-fiber-reinforced polyetheretherketone (PEEK) at elevated temperatures, Wear 203-204 (1997) 380–386.

[49] B.J. Briscoe, in: K. Friedrich (Ed.), Friction and Wear of Polymer Composites, Elsevier, Amsterdam, 1986, pp. 25–59.

[50] D.H. Buckley, Surface Effects in Adhesion, Friction, Wear, and Lubrication, Elsevier, Amsterdam, 1981.

[51] R.D.K. Misra, R.S. Hadal, S.J. Duncan, Surface damage behavior during scratch deformation of mineral reinforced polymer composites, Acta Materialia 52 (2004) 4363–4376.

[52] R.S. Hadal, R.D.K. Misra, Scratch deformation behavior of thermoplastic materials with significant differences in ductility, Materials Science and Engineering A 398 (2005) 252–261.

[53] Z. Zhang, P. Klein, K. Friedrich, Dynamic mechanical properties of PTFE based short carbon fibre reinforced composites: experiment and artificial neural network prediction, Composites Science and Technology 62 (2002) 1001–1009.

[54] A.A. Goodwin, G.P. Simon, Dynamic mechanical relaxation behaviour of poly(ether ether ketone)/poly(etherimide) blends, Polymer 38 (1997) 2363–2370.

[55] B. Nandan, L.D. Kandpal, G.N. Mathur, Glass transition behaviour of poly(ether ether ketone)/poly(aryl ether sulphone) blends: dynamic mechanical and dielectric relaxation studies, Polymer 44 (2003) 1267–1279.

[56] W.C. Milz, L.E. Sargent, Frictional characteristic of plastics, Lubrication Engineering 11 (1955) 313–317.

[57] K. Tanaka, Kinetic friction and dynamic elastic contact behaviour of polymers, Wear 100 (1984) 243–262.

[58] B.J. Briscoe, Isolated contact stress deformations of polymers: the basis for interpreting polymer tribology, Tribology International 31 (1998) 121–126.

[59] C. Oudet, Polymères: Structure et Propriétés Introduction, Masson, Paris, 1993 (In French).

[60] G. Zhang, H. Yu, C. Zhang, H. Liao, C. Coddet, Temperature dependence of the tribological mechanisms of amorphous PEEK (polyetheretherketone) under dry sliding conditions, Acta Materialia 56 (2008) 2182–2190.

[61] Q. Xue, Q. Wang, Wear mechanisms of polyetheretherketone composites filled with various kinds of SiC, Wear 213 (1997) 54–58.

[62] Q. Wang, W. Shen, J. Xu, Q. Xue, The effect of nanometer SiC filler on the tribological behavior of PEEK, Wear 209 (1997) 316–321.

[63] L. Chang, Z. Zhang, L. Ye, K. Friedrich, Tribological properties of high temperature resistant polymer composites with fine particles, Tribology International 40 (2007) 1170–1178.

[64] G. Zhang, Structure–tribological property relationship of nanoparticles and short carbon fibers reinforced PEEK hybrid composites, Journal of Polymer Science Part B: Polymer Physics 48 (2010) 801–811.

[65] Y.J. Zhong, G.Y. Xie, G.X. Sui, R. Yang, Poly(ether ether ketone) composites reinforced by short carbon fibers and zirconium dioxide nanoparticles: mechanical properties and sliding wear behavior with water lubrication, Journal of Applied Polymer Science 119 (2011) 1711–1720.

[66] H.J. Song, Z.Z. Zhang, Investigation of the tribological properties of polyfluo wax/polyurethane composite coating filled with nano-SiC or nano-ZrO$_2$, Materials Science and Engineering A 426 (2006) 59–65.

# Surface engineering with micro- and nanosized solid lubricants for enhanced performance of polymer composites and bearings

# 20

Jayashree Bijwe*, Mohit Sharma†, Werner Hufenbach‡, Klaus Kunze‡,
Albert Langkamp‡

*Industrial Tribology Machine Dynamics & Maintenance Engineering Centre, Indian Institute of Technology, Delhi, Hauz Khas New Delhi, India, †Industrial Tribology Center, Indian Institute of Technology, New Delhi, India, ‡Technische Universität Dresden, Institute für Leichtbau und Kunstofftechnik, Dresden, Germany

## CHAPTER OUTLINE HEAD

Tribology of Polymeric Nanocomposites. http://dx.doi.org/10.1016/B978-0-444-59455-6.00020-9

## 20.1 INTRODUCTION

Reliable and satisfactory performance of materials in adverse operating conditions such as extreme temperatures, pressures, corrosive and biologically aggressive environments, nuclear radiation, etc. is a key criterion for material selection in engineering applications. New composites hence are continuously being developed and explored for desired applications. Among various engineering tribomaterials fiber-reinforced polymers (FRPs) form a very important class of tribomaterials because of their unique properties such as self-lubricity; high specific strength; resistance to wear, impact, corrosion, chemicals, solvents, nuclear radiation, contamination with oils, etc. apart from easy processibility in complex shapes and the capacity to absorb vibrations leading to quiet operation. The FRP-based tribocomponents can be used in extreme conditions of temperatures (cryogenic to moderate up to 300 °C), pressures (vacuum to high), etc. where liquid lubricants cannot be considered. Such tribocomponents fabricated from the polymer composites are used in typical situations where either hydrodynamic lubrication is not possible because of frequent starts and stops or under low pressure–velocity (PV) conditions. Polymeric bearings are the unique solution in the following situations where

- lubrication is a problem (tribological components in inaccessible equipment; for example, in nuclear reactors and in hazardous conditions in chemical plants or in vacuum or space);
- conventional lubricants cannot be employed (in space or cryogenic temperatures where liquid lubricants will either freeze or evaporate);
- lubrication is unacceptable because of the possibility of contamination of lubricant with product (plain bearings or gears in industries such as food, paper, pharmaceutical and textile, etc.);
- maintenance is spasmodic or impossible (bushes and seals in domestic appliances, toys and instruments);
- lubrication is sparse (aircraft linkage bearings) or as a safeguard in the event of failure of the lubrication systems (e.g. gears in trains) [1].

Literature review on the triboproperty enhancement by various techniques indicates that the efforts for tailoring the composite's surface and bulk simultaneously are not adequately reported. Lancaster [1] has described various general approaches for developing composite liners. In general, dry bearing materials are expected to have a low and steady coefficient of friction ($\mu$) and low specific wear rate with a good combination of other mechanical properties such as compressive strength, modulus, resistance to creep, fatigue, etc. The exact ranges of combination of properties depend on the typical tribosituation where the components are used. In case of bearings, $\mu$ in the range of 0.1–0.2 and specific wear rate ($K_o$) in the range of $10^{-15}$ to $10^{-17}\,m^3\,Nm^{-1}$ are supposed to be acceptable. However, these values especially in the case of polymers and FRPs highly depend on the operating conditions and hence cannot be strictly used as guidelines. Generally, in practice dry bearing materials are tailored with fibers (short or long), matrix and solid lubricants (generally, polytetrafluoroethylene (PTFE), graphite, etc. in particulate form) dispersed uniformly in the body of a composite. Although the enhancement in tribological performance of these composites is achieved to the desired level, this leads to two potential problems. First, since the solid lubricants are low surface energy materials, they lead to drastic reduction in almost all mechanical properties barring impact strength in general. Second, the cost of a material increases alarmingly and unnecessarily. In triboapplications surface properties are more important and decisive for friction and wear behavior, while bulk properties such as load carrying capacity, resistance to creep, compression, cracking, fatigue, etc. are more important for the bearing body. Since the requirements on the surface and bulk are different, both the parts should be designed separately and judiciously. Such composites containing solid lubricants on a working surface of the bearing along with the desired reinforcement on the surface and in the bulk need to be tailored. The literature highlights few such efforts on composite bearings [2,3]. A composite pipe included a PTFE inner layer and a covering layer of fiber-reinforced plastic material, which were joined by force-locking and form-locking through an intermediate layer. The intermediate layer was of a perfluoroalkoxy copolymer resin, and fibers of glass or carbon, in order to improve the adhesive and shear strengths between the PTFE inner layer and the covering layer [2]. In other cases of sliding bearing (composite material) a metallic porous carrier layer was sintered on an outermost metallic supporting layer. The sliding layer comprises a combination of polymer and fillers, which filled the pores. Polymers such as polyether sulfone (PES), polyvinyledene difluoride, polyphenylene sulfide or polyamides could be selected based on the severity of the operating conditions [3].

## 20.2 SURFACE-ENGINEERED COMPOSITES AND BEARINGS FROM POLYETHER-ETHER-KETONE AND CARBON FABRIC

### 20.2.1 Commingling of fibers for composites

Different processing techniques for composites with bidirectional reinforcement viz. impregnation, hand layup, pultrusion, sandwiching the fabric in thin films of matrix, etc. lead to difference in properties. In the literature, for example, impregnation (prepreg method) and film techniques were used for tailoring polyetherimide (PEI) composites with carbon fabrics (CFs) of different weaves in similar amount and significant differences in mechanical and tribological properties were reported. The impregnation technique proved superior to the film technique [4].

In the case of the film technique the reinforcing fiber strands below the crossover points are not wetted properly and remain points of weakness in the composite structure, which is a major concern to the researchers and practitioners in the industries.

In some cases such as for polyether-ether-ketone (PEEK) where the impregnation method is not applicable because it has no proper solvent, textile preforms made from the commingled hybrid yarns offer a simple yet elegant solution to this problem. Commingling of fibers of matrix and reinforcement followed by compression molding is a recent technique and enjoys special advantages such as controlling the contents accurately as per requirements, no need of solvent, no possibility of removal of sizing agent, no problems associated with nonwetting of reinforcing fibers with matrix in crossover points, etc.

Among the specialty polymers, PEEK is one of the most investigated tribopolymers during the past 20 years and numerous composites have been reported having a wide spectrum of triboproperties. Various fibrous reinforcements such as glass, carbon, aramid, etc. in various forms have been explored along with various combinations of solid lubricants and metallic fillers. However, this thermoplastic polymer is comparatively difficult to process because of its high melt viscosity. Moreover, the composites based on PEEK cannot be tailored with conventional techniques such as impregnation because of the hazardous nature of its solvents such as $H_2SO_4$.

Research papers reporting on in-depth investigations on the development of composites with commingled fibers of PEEK and carbon, followed by characterization and subsequent development of bearing and its performance evaluation are not available in the literature. Moreover, efforts on engineering the surface with the right type of solid lubricant in an optimized amount, which would lower the risk of losing the mechanical strength of the composites to a great extent, are not reported. The efforts in a similar direction are reported in the subsequent section. Thus, it mainly addresses two issues viz. the development of commingled yarn composites with engineered surface with an optimized amount of the right kind of solid lubricant and the development of bearing with and without tailored surfaces and to compare its performance with unmodified bearing to quantify the benefits of the surface tailoring.

### 20.2.2 Materials and methodology

#### 20.2.2.1 Commingled hybrid yarn fabrics

Individual fibers or fiber bundles can only be used on their own in a few processes such as filament winding. For most other applications, the fibers need to be arranged into some form of sheet, known as a textile fabric, to make handling possible. A hybrid yarn consisting of reinforcing fibers and matrix fibers integrated into the yarn structure in the form of staple fibers, filaments or split films enables manufacturing of the textile preform with the matrix component already distributed in the textile structure. The manufacturing of the composite part from such a textile preform is possible without any separate impregnation process. Figure 20.1 shows some of the hybrid yarn structures. Figure 20.2 shows the schematic of the commingling process for two types of yarns along with a photograph of the preform. In other hybrid yarns, inhomogeneous distribution of the matrix component hinders fiber impregnation, as under pressure the melted matrix component is unable to penetrate the reinforcement fiber bundles resulting in matrix-rich areas as is shown in the micrograph of the cross-section in Fig. 20.3. Commingled yarn, in contrast, has a potential for the homogenous distribution of reinforcement and matrix filaments across the yarn cross-section. The so-achieved very short flow paths during manufacture of the composite part allow fast and complete impregnation of the reinforcement filaments (Fig. 20.4). Thus besides these two principal advantages, commingled yarn is characterized by

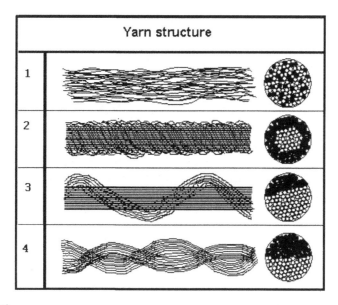

**FIGURE 20.1**

Different hybrid yarn structures; filled dots, matrix filament, and hollow dots, reinforcing filament: (1) commingling yarn, (2) cospinning yarn, (3) cowrapping yarn and (4) cotwisting yarn.

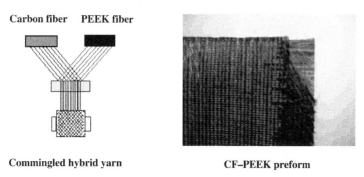

**FIGURE 20.2**

Process of commingling of fibers along with photograph of CF–PEEK preform.

- very less tendency to bleed the polymer during compression molding at temperatures little higher than melting.
- more precise control over the reinforcement/matrix ratio during yarn production and thereby in the subsequent composite.
- better processibility than the existing textile manufacturing technologies.
- improved mechanical properties of the composite part.
- rationalization of the composite manufacturing process and possibility to produce complex shaped parts due to better drape of the hybrid yarn.

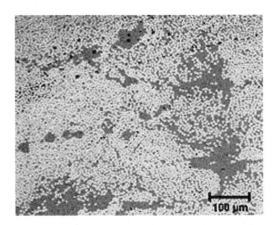

**FIGURE 20.3**

Cross-sectional view. The darker portion indicates matrix-rich areas.

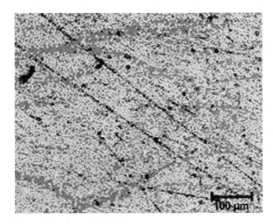

**FIGURE 20.4**

Composite with commingling hybrid yarn showing comparatively more homogenous distribution.

Institut für Textil und Bekleidungstechnik at TU Dresden reported development of such a yarn [5]. The details of individual fibers used for commingled CF–PEEK yarn and that of the commingling process are shown in Tables 20.1 and 20.2, respectively.

### 20.2.2.2 Fabrication of preliminary composites with tailored surfaces

Few composites as shown in Table 20.3 were developed to understand the efficiency of various fillers such as PTFE, bronze powder and their combinations [6].

#### 20.2.2.2.1 Process description

The samples were prepared using the autoclave curing facility. An autoclave system allows performing the process under defined temperature and pressure conditions

**Table 20.1** Details of Commingled CF–PEEK Hybrid Yarn

| Fiber | Supplier | Trade Name | Count (Tex) | Number of Filaments | Filament Diameter (µm) | Density (g cm$^{-3}$) |
|---|---|---|---|---|---|---|
| CF | Toray | T300JB-3K-40B | 198 | 3000 | 7 | 1.78 |
| PEEK | Zyex | Z4120 | 46 | 30 | 39 | 1.3 |

**Table 20.2** Processing Details of Commingled Yarn

| Fiber Count | 2× CF, 4× PEEK |
|---|---|
| Air nozzle | Taslan XV, Venturi 100, Nadel 86D |
| Air pressure | 5 bar |
| Overfeeding rate | 1.5% for CF, 3% for PEEK |
| Take-up speed | 100 m min$^{-1}$ |
| Preheating | No |

inside a pressure vessel according to a specified schedule (Fig. 20.5) in order to process a variety of materials. Internal autoclave pressure and vacuum assist the consolidation and bonding of the individual layers. The layup as shown in Table 20.4 was assembled on the autoclave tool. Figure 20.6 shows a schematic sketch of the laminate layup ready to be conveyed into the autoclave.

A 130×250-mm sheet of CF/PEEK commingled yarn preform was cut and seven such sheets were stacked on an iron base plate, which contained two vacuum outlet tubes. High-temperature sealant materials were applied along the edges of the iron plate. The stacked layers were covered with the PTFE sheet followed by placing of a thick glass wool breather layer on the PTFE sheet to eliminate the possible volatile materials. This breather layer was covered with the high temperature-resistant polymer (PEI) film. Then this assembly was fixed with the help of four rectangular iron rods and placed into the autoclave unit. Vacuum was applied through the vacuum outlet tubes.

A temperature of 385 °C and a pressure of 22 bar were used for making the composites. Prior to the autoclave run, the part was sealed with a vacuum bag, which also included layers of spongy material (breather, bleeder) to soak up excessive resin, if any, and carry away the volatile emissions and moisture. The duration of the cure cycle depends on the matrix system and generally lasts between 3 and 8 h. The sample was slowly heated to 385 °C during the warm-up cycle lasting 25 min; thereafter, the temperature was sustained for a duration of 1 h followed by a cooling cycle as is shown in Fig. 20.5. Table 20.5 summarizes the processing parameters.

The average thickness of the fabricated composite was 4 mm. A total of 11 composites with seven layers of CF–PEEK [6] were developed with and without modifiers on the surfaces. Few of them are described in Table 20.3. One composite was fabricated with PTFE (5 g cm$^{-2}$) on the top layer, while in others two and three top

**Table 20.3** Details of Composites with Selected Surface Layers

| Constituents on Surface | Layup Details | Modifier Weight (%) | Fiber Volume Fraction |
|---|---|---|---|
| CF/PEEK commingled yarn+PTFE powder | Seven layers CF/PEEK fabric 0° orientation with top layer impregnated with PTFE powder | 35 | 38 |
| CF/PEEK commingled yarn+PTFE powder | Seven layers CF/PEEK fabric 0° orientation with top layer impregnated with PTFE powder | 40 | 37 |
| CF/PEEK commingled yarn PTFE powder | Seven layers CF/PEEK fabric 0° orientation with top layer impregnated with PTFE powder | 50 | 34 |
| CF/PEEK commingled yarn PTFE powder | Seven layers CF/PEEK fabric 0° orientation with top three layers impregnated with PTFE powder | 55 | 31 |
| CF/PEEK commingled yarn | Seven layers CF/PEEK fabric 0° orientation and unmodified surface | Nil | 36 |
| CF/PEEK commingled yarn PTFE powder bronze powder | Seven layers CF/PEEK fabric 0° orientation with top three layers impregnated with 50% PTFE powder and 50% bronze powder | 30 | 48 |
| CF/PEEK commingled yarn bronze powder | Seven layers CF/PEEK fabric 0° orientation with top two layers impregnated with bronze powder | 10 | 33 |

layers were modified with PTFE. The other two composites were surface modified with bronze powder and a combination of bronze and PTFE powder in equal amounts ($2.5\,g\,cm^{-2}$). Tribological tests (Standard-DIN Deutsche Industrienorm ISO7148) were performed on the test setup as shown in Fig. 20.7 with the following parameters.

- Configuratio006e: shaft on plate
- Shaft: steel100Cr6 hardened and annealed (HRC Rockwell Hardness $59 \pm 1$; $R_a$, 0.2–0.3 μm)
- Specimen: CF–PEEK composite plate with 2 cm length and 1 cm width
- Operating parameters:
  - sliding speed: $0.5\,m\,s^{-1}$
  - normal load: incrementally 50–250N
  - temperature: 23 °C

### 20.2.2.3 Results on preliminary composites

Representative graphs for two composites (with and without PTFE powder on the surfaces) with three orientations of CF are shown in Fig. 20.7. It proved that the CF orientation at 45° to the sliding direction renders minimum μ. Fibers parallel to the sliding direction showed moderate performance while those in the transverse direction proved poorest. It also indicates that the unmodified surface leads to continuous increase in μ with sliding distance with quite high magnitude.

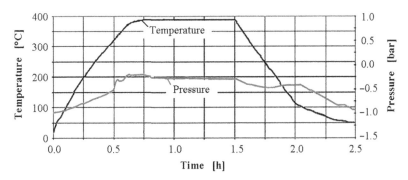

**FIGURE 20.5**

Specified schedule of processing of composites.

---

**Table 20.4** Arrangement of Various Layers of Fabric and Films before Conveying to the Autoclave (top to bottom)

Vacuum bag
Fiber glass fabric bleeder layer
Two layers of PEEK foil
Seven layers of commingled CF–PEEK fabric to achieve desired specimen thickness
Commingled fabric layer impregnated with modifying compounds
Two layers of PEEK foil
Bottom peel-off PTFE layer
Autoclave tool

---

The following conclusions were drawn about the surface modifiers based on the tribological studies [6]:

- unmodified CF–PEEK had the highest tensile modulus among all the composites and the addition of PTFE on the top layer resulted in a reduction in the off-axis tensile modulus (15–30%). The reduction was more severe with increasing PTFE content and inclusion in more layers.
- the coefficient of friction ($\mu$) for unmodified CF–PEEK varied between 0.4 and 0.6 with a lot of fluctuations with sliding time (unstable $\mu$), while the specific wear rate was in the range of $1$–$1.5 \times 10^{-15}\,\mathrm{m^3\,Nm^{-1}}$.
- addition of PTFE resulted in the disappearance of fluctuations along with a marked decrease in $\mu$ in a range of 0.15–0.25. The wear resistance also improved to nearly twice the initial value in the range of $10^{-16}\,\mathrm{m^3\,Nm^{-1}}$.
- bronze when used alone resulted in marked deterioration in both friction and wear performance.
- addition of PTFE in combination with bronze also did not lead to any noticeable improvement in friction or wear properties.

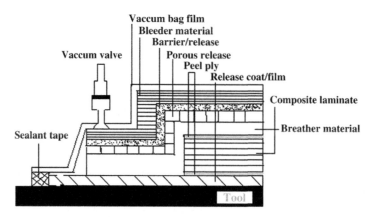

Vaccum bag layup

**FIGURE 20.6**

Schematic sketch of laminate layup ready to be conveyed into the autoclave.

**Table 20.5** Summary of the Consolidation Process

| Equipment | Autoclave |
|---|---|
| Process temperature | 385 °C (as per graph Fig. 20 5) |
| Heating rate | 12 K min$^{-1}$ |
| Cooling rate | 6.5 K min$^{-1}$ |
| Process pressure | 22 bar (as per graph) |
| Curing time | 385 °C for 50 min |

### 20.2.2.4 Fabrication of additional composites

Based on these preliminary results a new batch of composites with the same commingled yarns of CF and PEEK under the same processing conditions was developed. It was based on the placement of various types of solid lubricants viz. PTFE, MoS$_2$ and graphite but only in the top layer [6–8]. PTFE lubricant was used in various forms viz. long fibers, short fibers of various lengths, wool, etc. and was supplied by an Austrian company (Lenzing Plastics GmbH & Co KG, Austria); details are given in Table 20.6.

The details of composites are given in Table 20.7 and the stacks with the surface layer of lubricant are shown in Fig. 20.8.

### 20.2.3 Characterization of the composites

The composites were characterized for their mechanical properties (four-point bending test) according to the relevant DIN standards and results on various tests are shown in Fig. 20.9(a) and (b). Figure 20.10 shows the results of Young's modulus for these composites.

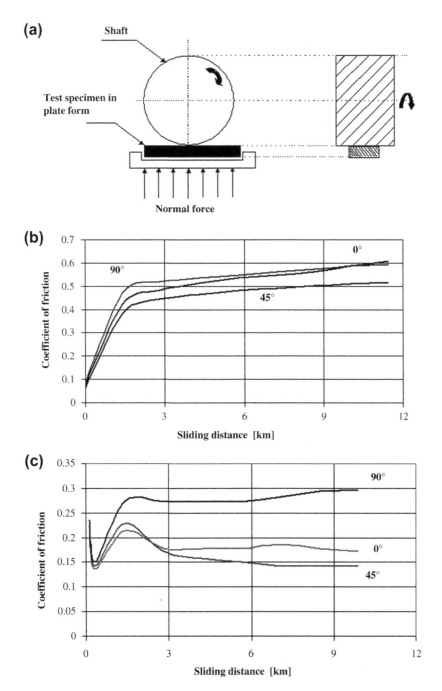

**FIGURE 20.7**

Friction coefficient as a function of sliding distance for CF–PEEK composite; (a) shaft sample interface in shaft-on-plate configuration; (b) without PTFE on surface and (c) with top layer with PTFE powder.

**Table 20.6** Details of PTFE Used as Surface Modifier (Lenzing Plastics GmbH & Co KG, Austria)

| S. No. | Nature of Modifier | Specification |
|---|---|---|
| 1 | 815/5 mm opened | PTFE 15 dtex with cut length 5 mm |
| 2 | 803/10 mm opened | PTFE short cut staple fibers 2.7 dtex, |
| 3 | Type FG 12 natural | PTFE long fibers, ribbon fibrillated, 1330 dtex, 1200 den, tenacity 15 |
| 4 | 803/60 | PTFE fiber fibrillated, 3 dtex, tenacity 13 |

**Table 20.7** Details of Surface Modifier (Solid Lubricants) and Surface-Tailored PEEK/CF composites

| S. No. | Code of Surface Modifier | Designation of Composites | Amount of Modifier (g) on Top Layer | Amount of Modifier (g cm$^{-2}$) |
|---|---|---|---|---|
| 1 | PTFE 803/5+10 mm opened | TA5 | 5 | 0.0154 |
| 2 | PTFE 803/15+20 mm opened | TB2 | 2 | 0.00615 |
| 3 | PTFE 815/5 mm opened | TC5 | 5 | 0.0154 |
| 4 | PTFE 803/10 mm opened | TD5 | 5 | 0.0154 |
| 5 | PTFE 815/5 natural bunched | TE5 | 5 | 0.0154 |
| 6 | PTFE natural long fibers (FG12) | TF5 | 5 | 0.0154 |
| 7 | PTFE wool 803/60 | TW2 | 2 | 0.00615 |
| 8 | PTFE wool 803/60 | TW3 | 3 | 0.00923 |
| 9 | PTFE wool 803/60 | TW4 | 4 | 0.01230 |
| 10 | PTFE wool 803/60 | TW5 | 5 | 0.0154 |
| 11 | PTFE wool 803/60 | TW5/5 | 5 each on top two layers | 0.0154 |
| 12 | Graphite powder | G5 | 5 | 0.0154 |
| 13 | MoS$_2$-OKS 100* | M5 | 5 | 0.0154 |

*Supplied by Thauer & Co. KG Siliconerzeugnisse-Mineralölprodukte, Dresden.*

It was also observed that inclusion of PTFE in the top two layers as in the case of composite $T_{W5/5}$ led to further reduction in strength. The results confirmed that the composite with long fibers of PTFE on the surface showed the highest values in almost all cases. PTFE in other forms such as wool was less effective. Powdery lubricants such as graphite, MoS$_2$ and PTFE proved poor in this case.

Hence, the tribological testing was done only on the composite with long PTFE fibers in adhesive wear mode. The bearing was also fabricated from this combination as shown in the schematic in Fig. 20.11.

**FIGURE 20.8**

Details of process of laying up of modifiers on the surface of CF–PEEK preforms. For color version of this figure, the reader is referred to the online version of this book.

## 20.2.4 Tribological testing of composites and bearings

Among 13 composites, tribotesting was carried out only on the composite with a modified surface with long fibers of PTFE (0.616 g cm$^{-2}$). Initial studies on this composite along with unmodified composite were conducted on tribotesting setup as shown in Fig. 20.7 in shaft-on-plate configuration. The studies showed that the unmodified composite showed scuffing behavior with a lot of fluctuations in μ. PTFE fiber inclusion on the top layer led to significant improvement in friction and wear performance in the adhesive wear mode. The coefficient of friction dropped to 0.12 from 0.5 to 0.6 and the wear rate reduced almost by 70 times.

The tribostudies on modified (PTFE $0.616 \, g \, cm^{-2}$) and unmodified bearings were done on a test setup as shown in Fig. 20.12. The results are shown in Fig. 20.13. It was observed that the wear rate depended both on inclusions of PTFE fibers and on the orientation of CF with respect to the sliding direction. CF in the bearing at 45° with respect to the sliding direction proved beneficial as compared to 90° in case of unmodified composite also. Shaft wear was 3.5 times lower, while bearing wear was approximately 25% lower due to the right selection of orientation of CF with respect to the sliding direction. In case of modified bearing, there was no observable wear of the shaft for both the orientations of CF. Bearing wear was reduced by almost 43 times in case of 45° and almost 10 times in case of 90° orientation.

Figures 20.14 and 20.15 show friction behavior with respect to the sliding distance for modified surface bearings with two fiber (CF) orientations viz. 45° and 90°, respectively. Figures 20.14(a) and 20.15(a) are for the coefficient of friction as a function of the sliding distance, while Figs 20.14(b) and 20.15(b) are for the increase in temperature of the bearing surface. It was observed that the bearing temperature was the highest (80 °C) in case of CF with 90° orientation as compared to that with 45° (55 °C). The μ in case of CF at 90° showed a stick-slip behavior initially indicating PTFE was not released on the shaft surface instantaneously. After 10 km of sliding, μ reduced from 0.7 to 0.3 (Fig. 20.15). In case of 45° on the other hand, μ was very low initially (around 0.15) and rose steadily to 0.2 as the sliding distance increased to 10 km indicating the extents of benefits of PTFE slowly decreased with increase in sliding distance. These facts showed that in this case PTFE was released very effectively as the sliding commenced.

Figure 20.16 shows micrographs of the original shaft and that slid against the bearing of the modified surface. It shows the traces of PTFE on the surface. Micrographs of the surface of the worn bearing and shaft are also shown in Fig. 20.16 indicating areas rich in PTFE and PEEK.

## 20.3 SURFACE ENGINEERING WITH MICRO- AND NANOSIZED SOLID LUBRICANTS

### 20.3.1 Nanofillers for enhancement of triboperformance

Nanocomposites (NCs) are the most researched materials today because of multifold potential of nanofillers. As the dimensions of a particle approach the atomic scale, it exhibits completely different properties. The prominent features of nanoparticles (NPs) when used as fillers are the following:

- a very low amount (2–3% by weight) provides an exceptional increase in very important performance properties such as mechanical strength, thermal properties, conductivity (thermal or electrical), flame retardancy, permeability, etc.
- a very high surface area to volume ratio provides a very large interfacial surface area as a driving force for enhanced interaction with other surfaces.

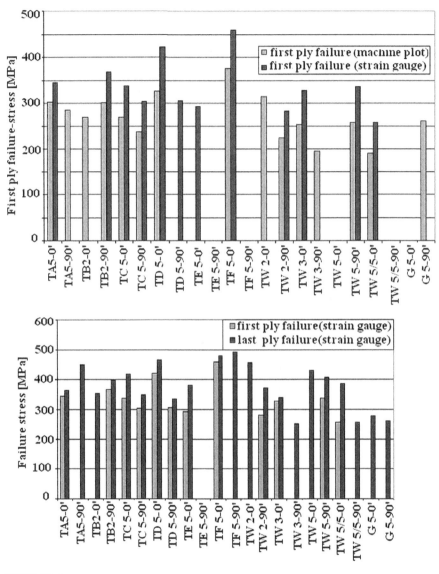

**FIGURE 20.9**

Comparative ply failure strength of the composites.

Macro/micro reinforcements improve the tensile strength and tensile modulus of polymers at the cost of their toughness. On the contrary, NPs generally improve both [7–13]. In general, NCs exhibit superior properties such as higher tensile strength, resistance against degrading temperature, flame retardancy, lesser permeability to solvents, etc. as compared to conventional fiber-reinforced composites. Various theories have been proposed for the improvement in the property of such composites [14].

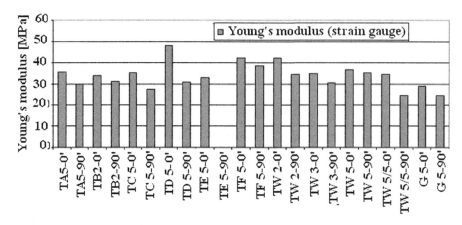

**FIGURE 20.10**

Young's modulus for the composites.

Successful exploration [15–47] of many nanofillers/NPs for the enhancement of triboperformance in the adhesive wear mode has been reported. The review by Briscoe and Sinha [15] has presented state-of-the-art position in a tabular form. The literature brings out the following facts about the role of nanofillers in the tribology.

- During the wear process, NPs are removed from the surface of a matrix and can act as third-body element in the contacting regions.
- The rolling effects of the NPs, especially at the edge of exposed fibers, reduce the shear stress in the contact region. This causes a reduction in μ and smoothening of the topography of a surface of a composite. It also protects the fibers adhering to the matrix and results in increased fiber thinning rather than their early brittle fracture, leading to reduced wear [16–20].

The rolling effect of NPs attributes the following:

- Reduction in μ and hence in the frictional heating at the interface of a tribopair.
- Spontaneous reduction of grooving/cutting wear by hard counterpart asperities.
- Protection of fibers from easy removal.

The rolling effect is also observed in the case of microparticulate inclusions, for which the small particles tend to tumble through the contact region and larger particles plow through it [22–24]. A critical size of a particle governs the transition from rolling to plowing. To achieve rolling, the ratio of the maximum particle dimension to the minimum gap of contacting bodies must be exceeding the critical value [18].

The hard microsized particles and fillers abrade the counterface that leads to the events of cascading problems. This also prevents the formation of a protective transfer film, which increases the counterface roughness and hence μ [23]. The NPs have the potential to reduce the abrasion. NPs (<100 nm) are of the same size as the counterface asperities and polish the highest asperities and promote the development of the transfer films [24].

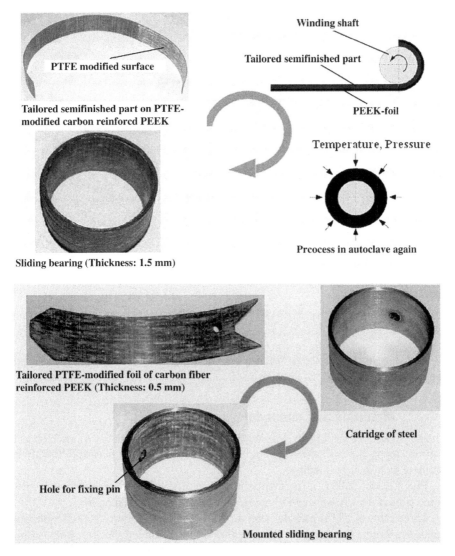

**FIGURE 20.11**

Schematic of the bearing fabrication process.

## 20.3.2 Nano- and micro-PTFE for surface modification of a tribocomposite

Since in our earlier work (Section 20.2) PTFE had shown the highest potential to improve surface properties of CF–PEEK composites, further work was focused on the exploration of PTFE particles of various sizes including that with nanometer size. Instead of the CF–PEEK composite system, the CF–PES system was selected

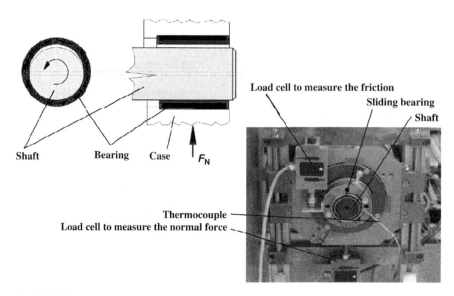

**FIGURE 20.12**

Tribotesting assembly for bearing.

for exploring the potential of PTFE as a surface modifier. The details of this work are reported in our recent work [48,49].

## 20.3.3 Materials and methodology

The CF (3 K, 2×2 twill weave) procured from Fiber Glast, USA, was selected as a reinforcement. It was treated with cold remote nitrogen oxygen plasma (CRNOP) to promote fiber–matrix bonding.

PES Veradel 3600 P, a thermoplastic high-performance specialty polymer procured from Solvay Advanced Polymers, India, was selected as a matrix.

Three sizes of PTFE particles (micro, 400–450 nm; submicro, 200–250 nm; and nano, 50–80 nm) as confirmed from field emission scanning electron microscopic (FESEM) studies (Table 20.8, Fig. 20.17) were selected for surface modification of composites.

### 20.3.3.1 Development of composites

Composites were developed by the solution impregnation technique. Twenty pieces of CF plies (CRNOP treated) (28 cm × 28 cm) were immersed in a solution of PES in dichloromethane (20 wt%) for 12 h. These prepregs were dried in an oven for an hour at 100 °C in a stretched condition and were stacked in a steel mold. The mold was then heated in a compression molding machine to a temperature of 380 °C–390 °C for 20 min under a pressure of 7.3 MPa. The composites were then cooled in a compressed condition.

| Fiber Orientation | PTFE Specification | Specific Wear Rate $W_{L/S}$ [µm km$^{-1}$] | |
|---|---|---|---|
| 45° | Without PTFE | Shaft | 1.10 |
| | | Bearing | 7.61 |
| | PTFE long fibers (Type FG12 natural, amount 0.615 g cm$^{-2}$) | Shaft | 0.00 |
| | | Bearing | 0.18 |
| 90° | Without PTFE | Shaft | 5.71 |
| | | Bearing | 9.10 |
| | PTFE long fibers (Type FG12 natural, amount 0.615 g cm$^{-2}$) | Shaft | 0.00 |
| | | Bearing | 0.82 |

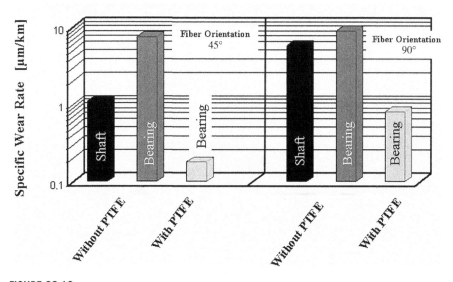

**FIGURE 20.13**

Comparative tribobehavior of bearings (modified and unmodified).

### 20.3.3.2 Technique for surface modification of composites

The surface designing was done for only the top two layers. A modified impregnation method was used to develop surface-tailored composites with PTFE of different sizes. PES and PTFE powders (2wt% of PTFE in PES) were mixed in a high shear ball mill using zirconia balls in an alcohol media for 16h followed by drying. The dried mix was then probe sonicated (probe diameter, 5mm) in an ethyl alcohol medium for 20min to achieve deagglomeration of NPs. The solution impregnation technique was then used to prepare two prepregs for the surface. The optimized (temperature from 280 °C to 380 °C in steps of 20 °C and pressure in the steps of 1MPa up to 6MPa) sequence of temperature and pressure was applied. The two tailored prepregs with 18

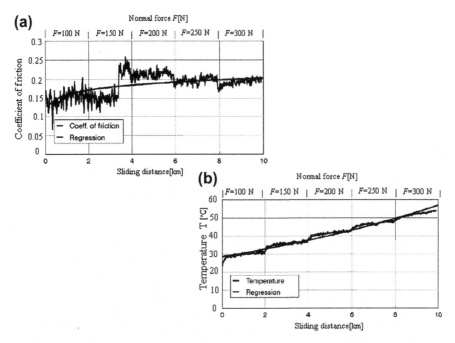

**FIGURE 20.14**

PTFE fiber-modified composite bearing with CF at 45° with respect to sliding direction; (a) μ as a function of sliding distance and (b) increase in bearing temperature as a function of sliding distance.

untailored prepregs were compression molded under optimized conditions followed by natural cooling under compressed state to the ambient temperature.

### 20.3.4 Characterization of composites

The composite $PES_{CFT}$ was characterized for physical and mechanical properties and the results are shown in Table 20.9.

### 20.3.5 Tribocharacterization of composites

The performance of composites was evaluated in the adhesive wear mode using UMT-3MT Tribometer supplied by CETR, USA, in a pin-on-disc configuration discussed elsewhere [48,49]. The $10\,mm \times 10\,mm$ composite pin was slid against a mild steel counterface disc at a constant speed of $1\,m\,s^{-1}$ and $1\,cm^2$ nominal contact area. The μ as a function of normal load and time during sliding was recorded with the help of viewer software. The specific wear rate $(K_o)$ was calculated using the equation [48]:

$$K_o = \frac{W}{\rho L d} \quad m^3 N^{-1} m^{-1}$$

(20.1)

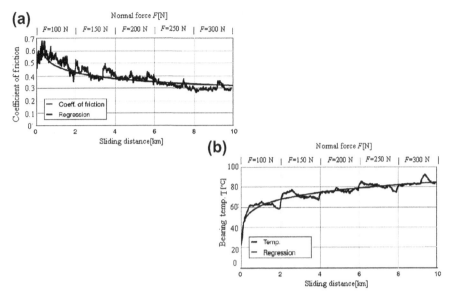

**FIGURE 20.15**

PTFE fiber-modified composite bearing with CF at 90° with respect to sliding direction; (a) μ as a function of sliding distance and (b) increase in bearing temperature as a function of sliding distance.

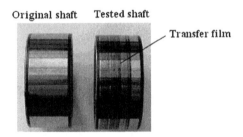

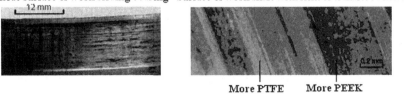

**FIGURE 20.16**

Micrographs of worn surfaces of shaft and bearing.

**Table 20.8** Details of Unmodified and PTFE-Modified PES–CF Composites [48,49]

| Designations of Composites* | Average PTFE Particle Size and Shape (FESEM Studies) |
|---|---|
| PES_CFT | – |
| PES_CFTN | 50–80 nm highly spherical |
| PES_CFTSM | 200–250 nm subrounded |
| PES_CFTM | 400–450 nm subangular |

*PES_CFT, composite with treated CF (unmodified); PES_CFTSM, composite with submicron-sized PTFE on the surface; PES_CFTM, composite with micron-sized PTFE on the surface; PES_CFTN, composite with nanosized PTFE on the surface.

**FIGURE 20.17**

FESEM micrographs of selected PTFE particles [48,49].

**Table 20.9** Physical and Mechanical Properties of CF–PES_T Composite [48]

| Properties/Materials | PES_CFT |
|---|---|
| Fiber weight % ASTM 3171-06 | 68.24 |
| Void fraction wt. % ASTM 2734 | 0.37 |
| Density ASTM D792 (g cm$^{-3}$) | 1.54 |
| HDT ASTM D648 (°C) | 233 |
| Tensile strength ASTM D638 (MPa) | 778 |
| Tensile modulus ASTM D638 (GPa) | 76 |
| Toughness ASTM D638 (MPa) | 4.3 |
| Flexural strength ASTM D 790 (MPa) | 835 |
| Flexural modulus ASTM D790 (GPa) | 68 |
| ILSS ASTM D2344 (MPa) | 46 |

Where $W$ is the weight loss in kg, $\varrho$ the density of the pin in kg m$^{-3}$, $L$ the load in N and $d$ the sliding distance in meters.

### 20.3.5.1 Influence of PTFE modification on the surface

The $\mu$ and $W_R$ were highly influenced due to PTFE modification as seen in the histogram in Fig. 20.18. As seen from Fig. 20.18 the major benefit endowed to the friction

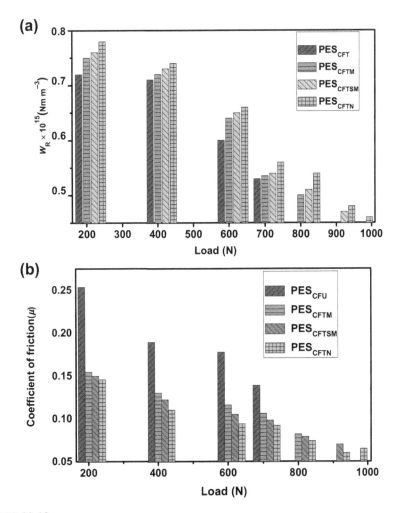

**FIGURE 20.18**

(a) Coefficient of friction; (b) specific wear rates as a function of increasing load for all surface-designed series of composites [48,49]. For color version of this figure, the reader is referred to the online version of this book.

and wear of a composite was due to the addition of PTFE. Almost 60% reduction in friction and ≈5% improvement in wear resistance ($W_R$) of the composites was observed due to inclusion of micron-sized PTFE on the surface. Thus the benefits to friction were higher than that for $W_R$.

Performance still improved with the decrease in the size of PTFE particles on the surface. Interestingly, further benefits to μ were marginal and to $W_R$ were noticeable. The benefits due to nano-PTFE increased as the load increased. Nano-PTFE particles at PES$_{CFTN}$ composites surface offers a large interfacial area, which leads to enhanced

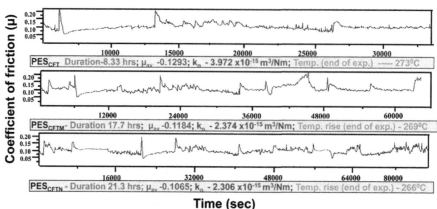

**FIGURE 20.19**

$\mu$ as a function of sliding time for long experiments for PES$_{CFT}$, PES$_{CFTM}$ and PES$_{CFTN}$ composites (recorded from viewer software) [48,49]. For color version of this figure, the reader is referred to the online version of this book.

adhesion with matrix and counterface and hence improved the quality of film transfer, which was the reason for improvement in triboperformance. The solid lubricants on the surface of the composites also influenced their long-term utility. Long wear experiments performed for CF-PES composites with and without solid lubricants on the surface against mild steel disc suggested the enhanced PV limit of the composites with tailored surfaces. PV$_{limt}$ can also be judged by measuring the limiting time under selected PV values. In such cases the experiments are continuously conducted till it records some of the unwanted features such as a sudden increase in $\mu$, vibrations, temperature, etc. In these experiments, a sudden increase in $\mu$ was observed. The higher the limiting time, the more is the triboutility of the surface. Surface designing enhanced the limiting load values of composites significantly from 700N to 950N, limiting running time from 8 to 21h, reduction in $\mu$ (from 0.12 at 700N load to 0.065 at 900N load) and wear resistance (from 5.5 to $5.8 \times 10^{14}$ Nm m$^{-3}$ at 700N load) especially in the case of PES$_{CFTN}$ composite (Fig. 20.19). Possibly the transfer of thin tenacious nano solid lubricants layer on the counterface steel surface was effective in improving the $W_R$ and $\mu$ of the modified composites especially at a higher load and experimental duration.

Figure 20.20 shows SEM micrographs of worn surfaces of surface-designed composites, (a) PES$_{CFT}$, (b) PES$_{CFTM}$ and (c) PES$_{CFTN}$. Micrographs (d), (e) and (f) are their respective high-resolution FESEM images ($\times$75 K). The fibers appear to be fully covered with the nano-PTFE particles. When a thin, coherent and uniform PTFE layer is transferred on the counterface, sliding is between two PTFE layers. First, on the composite surface and second, thin tenacious layer transferred on the countersurface. The existence on nanofillers (Fig. 20.20(f)) and microscale (Fig. 20.20(e)) is clearly visible with high-resolution FESEM images of worn surfaces of composites, which were absent for composites without a tailored surface (Fig. 20.20(d)).

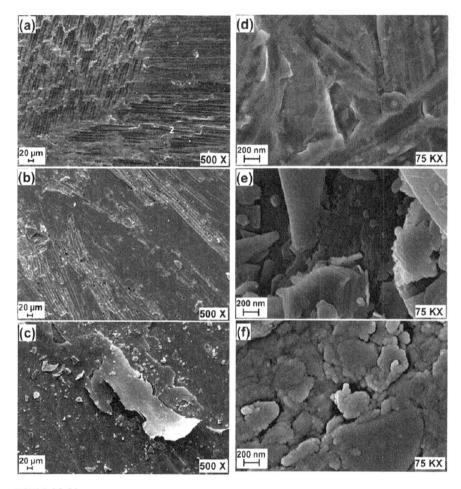

**FIGURE 20.20**

SEM micrographs (×500) of surface-designed composites after wearing; (a) PES$_{CFT}$, (b) PES$_{CFTM}$ and (c) PES$_{CFTN}$; (d), (e) and (f) are their respective high-resolution FESEM images (×75K) at 700 N load at 1 m s$^{-1}$ speed [48,49].

## 20.4 CONCLUDING REMARKS

### 20.4.1 Concluding remarks from studies in Section 20.2

Based on the studies carried out in Section 20.2 of this work, the following conclusions were drawn.

- A novel technique of using commingled fibers of carbon and PEEK to design a composite and bearing proved beneficial from several aspects.

- Instead of including solid lubricants in the body of a composite and deteriorating the strength significantly, surface modification proved the best choice that would retain the strength to a significant extent.
- Surface modification with solid lubricants led to some deterioration in the mechanical properties of the composites. The extent depended on the nature and type of modifiers. Among selected powdery materials (PTFE, MoS$_2$, Bronze, graphite, etc.), bronze proved poorest and PTFE was the best in this respect. Among various forms of PTFE such as fibers (short, stapled and long), powder and wool, long fibers proved most promising.
- The more the amount of PTFE on the surface, the better was the triboperformance and more was the deterioration in mechanical properties. Placement of PTFE in more than one layer also led to more deterioration in strength. It is the user who has to define the required set of strength properties before trying for surface modification.
- Modification of the surface with PTFE fibers led to a significant improvement in triboperformance. Both μ and wear performance improved significantly. The extent of improvement, however, depended on the orientation of carbon fibers with respect to the sliding direction.
- Among the three fiber orientations (0°, 45° and 90°), the surface with 45° orientation of CF showed the best results in terms of μ and wear resistance. For 0°, 90° and 45° orientations, it was in the range 0.2, 0.3 and 0.15, respectively, confirming 45° orientation of carbon fibers was the best suited for fabrication of bearing. The wear rate was approximately 75 times less than that of an unmodified surface confirming that long PTFE fibers on the surface with 45° orientation of long CF are the most suitable for significant improvement in the triboperformance at the cost of slight deterioration in strength properties (approximately 15%).
- Bearing surface when modified with PTFE fibers, the rate at which PTFE could be released on the shaft and would reduce friction depended on CF orientation. Composite with CF at 45° angle released PTFE very spontaneously reducing μ significantly and immediately. With increase in sliding duration, μ, however, slightly increased. For other orientation CF perpendicular to the sliding direction exactly opposite behavior was observed, although the extent of reduction in μ was not as high as in the earlier case.
- Based on the new technique developed for fabrication of composite bearings, with unmodified surface and modified surface with PTFE, it was concluded that the scuffing problem of the unmodified surface was completely removed due to inclusion of PTFE fibers. The coefficient of friction was very low (μ reduced from 0.5 to 0.15) without any fluctuations. For a good bearing, smooth operation and steady and low μ are very much required.
- Thus, long PTFE fiber-modified carbon fiber-reinforced PEEK (PEEK and carbon commingled) architecture proved to be a very efficient material for tribological applications. Such dry bearings have special advantages such as
  - maintenance-free (self-lubricity)
  - continuous service temperature up to 250 °C

- very good resistance to harsh chemicals and environment
- good mechanical properties.

### 20.4.2 Concluding remarks from the studies in Section 20.3

Based on the studies carried out in Section 20.3 of this work, the following conclusions were drawn.

Surface designing of $PES_{CFT}$ composites with micro-, submicro- and nanoscale PTFE fillers improved the overall triboperformance of composites, although the reduction in $\mu$ was significant rather than the wear resistance. The idea of surface treatment of a composite with solid lubricants to safeguard the composites from an unintentional reduction in strength properties and increase in the cost (if solid lubricant is employed in the whole composite rather than the surface both factors viz. strength and economics are affected significantly) proved successful. Surface designing enhanced the limiting load values of composites significantly from 700 N to 950 N, limiting running time from 8 to 21 h, reduction in $\mu$ (from 0.12 at 700 N load to 0.065 at 900 N load) and wear resistance (from 5.5 to $5.8 \times 10^{14}$ Nm m$^{-3}$ at 700 N load) especially in the case of $PES_{CFTN}$ composite. The increased surface area of contact due to the inclusion of nano-PTFE at the composite surface was responsible for enhanced triboperformance of the $PES_{CFTN}$ composite.

## Acknowledgments

The authors gratefully acknowledge the funding received from Deutscher Akademischer Austauschdienst (Germany) and the Department of Science & Technology, New Delhi, India, for the work in the first part and the Council of Scientific and Industrial Research (CSIR), New Delhi, India, for the second part of the work reported in this chapter.

## References

[1] J.K. Lancaster, in: K. Friedrich (Ed.), Composites for Aerospace Dry Bearing Applications, Friction and Wear of Polymer Composites Composites Materials Science Series, vol. 1, Elsevier Amsterdam, 1986, pp. 363–396.

[2] Composite Pipe Having a PTFE Inner Layer and A Covering Layer of a Fiber-Reinforced Plastics Material, United States Patent 6758245. <www.freepatentsonline.com>.

[3] Sliding Bearing Composite Material, International Patent F16C33/12, F16C33/20; F16C33/04. <www.freepatentsonline.com>.

[4] R. Rattan, J. Bijwe, Carbon fabric reinforced polyetherimide composites: influence of wear of fabric and processing parameters on performance properties and erosive wear, Materials Science and Engineering Part. A 420 (2006) 342–350.

[5] B.D. Choi, O. Diestel, P. Offermann, Commingled CF–PEEK hybrid yarn for use in textile reinforced high performance rotors, in: Proc. of the 12 Intl. Conf. on Composite Materials, Paris, 1999.

[6] R. P. S. Dhami, Development of Surface Tailored Carbon Fabric Reinforced Composites and Bearings M. Tech. Thesis Indian Institute of Technology, 2002 (under DST-DAAD sandwich program).

[7] Y. Wang, L. Zhang, C. Tang, D. Yu, Preparation and characterization of rubber-clay nano-composites, Journal of Applied Polymer Science 78 (2000) 1879–1883.

[8] T. McNally, W.R. Murphy, C.Y. Lew, R.J. Turner, G.P. Brennan, Polyamide 12 layered silicate nano-composites by melt blending, Polymer 44 (2003) 2761–2772.

[9] D. Gersappe, Molecular mechanisms of failure in polymer nano-composites, Physics Review Letter 89 (2002) 058301-1–058301-4.

[10] F. Yang, Y. Ou, Z. Yu, Polyamide 6/silica nano-composite prepared by in-situ polymerization, Journal of Applied Polymer Science 69 (1998) 355–361.

[11] Y. Ou, F. Yang, Z. Yu, A new conception on the toughness of nylon 6/silica nano-composite prepared via in-situ polymerisation, Journal of Applied Polymer Science 36 (1998) 789–795.

[12] J.M. Gloaguen, J.M. Lefebvre, Plastic deformation behaviour of thermoplastic/clay nano-composites, Polymer 42 (2001) 5841–5847.

[13] G.M. Kim, D.H. Lee, B. Hoffmann, J. Kressler, G. Stoppelmann, Influence of nano-fillers on the deformation process in layered silicate/polyamide 12 nano-composites, Polymer 42 (2001) 1095–1100.

[14] A. Usuki, A. Koiwai, Y. Kojima, M. Kawasumi, A. Okada, T. Kurauchi, O. Kamigaito, Interaction of nylon 6-clay surface and mechanical properties of nylon 6-clay hybrid, Journal of Applied Polymer Science 55 (1995) 119–123.

[15] S.K. Sinha, B.J. Briscoe, Polymer Tribology, Imperial College Press, 2009.

[16] K. Friedrich, Z. Zhang, A.K. Schlarb, Effects of various fillers on the sliding wear of polymer composites, Computer Science and Technology 65 (2005) 2329–2343.

[17] Z. Zhang, K. Friedrich, Tribological Characteristics of Micro and Nano-Particle Filled Polymer Composites – from Nano to Macro Scale, Polymer, Springer, US, 2005.

[18] L. Chang, Z. Zhang, L. Ye, K. Friedrich, Synergistic effect of nanoparticles and traditional tribo-fillers on sliding wear of polymeric hybrid composites, in: K. Friedrich, A.K. Schlarb (Eds.), Tribology of Polymeric Nano-Composites, Elsevier B.V, 2008.

[19] M.Q. Zhang, M.Z. Rong, K. Friedrich, Wear resisting polymer nano-composites: reparation and properties Ch-20, in: Y. W Mai, Z.Z. Yu (Eds.), Polymer Nano-Composites, CRC Press, Washington, DC, 2006.

[20] K. Friedrich, Z. Lu, A.M. Hager, Overview on polymer composites for friction and wear application, Theoretical and Applied Fracture Mechanics 19 (1993) l–11.

[21] X. Peng, H. Ma, Q. Zeng, Y. Lei, Tribological behavior of inorganic nano-particulates and polytetrafluoroethylene filled polyetheretherketone composites, Tribology International 24 (2004) 240–243.

[22] R.S. Dwyer-Joyce, R.S. Sayles, E. Ioannides, An investigation into the mechanisms of closed three-body abrasive, Wear 175 (1994) 133–142.

[23] S. Bahadur, The development of transfer layers and their role in polymer tribology, Wear 245 (2000) 92–99.

[24] D.L. Burris, B. Boesl, G.R. Bourne, W.G. Sawyer, Polymeric nano-composites for tribological applications, Macromolecular Materials and Engineering 292 (2007) 387–402.

[25] K. Tanaka, Effects of various fillers on the friction and wear of PTFE-based composites, in: K. Friedrich (Ed.), Friction and Wear of Polymer Composites, Elsevier B.V, 1986.

[26] Z.Z. Zhang, F.H. Su, K. Wang, W. Jiang, X.H. Men, W.M. Liu, Study on the friction and wear properties of carbon fabric composites reinforced with micro- and nano-particles, Material Science and Engineering: A 404 (2005) 251–258.

[27] F.H. Su, Z.Z. Zhang, K. Wang, W. Jiang, X.H. Men, W.M. Liu, Friction and wear properties of carbon fabric composites filled with nano-$Al_2O_3$ and nano $Si_3N_4$, Composites Part A 37 (2006) 1351–1357.

[28] F.H. Su, Z.Z. Zhang, K. Wang, W. Jiang, X.H. Men, W.M. Liu, Friction and wear properties of carbon fabric composites filled with several nano-particulates, Wear 260 (2006) 861–868.

[29] Q. Wang, X. Zhang, X. Pei, T. Wang, Friction and wear properties of solid lubricants filled/carbon fabric reinforced phenolic composites, Journal of Applied Polymer Science 117 (2010) 2480–2485.

[30] F.H. Su, Z.Z. Zhang, W.M. Liu, Mechanical and tribological properties of carbon fabric composites filled with several nano-particulates, Wear 260 (2006) 861–868.

[31] F.H. Su, Z.Z. Zhang, F. Guo, K. Wang, W.M. Liu, Effects of solid lubricants on friction and wear properties of nomex fabric composites, Material Science and Engieering: A 424 (2006) 333–339.

[32] F.H. Su, Z.Z. Zhang, W.M. Liu, Tribological and mechanical properties of nomex fabric composites filled with polyfluo 150 wax and nano-$SiO_2$, Computer Science and Technology 67 (2007) 102–110.

[33] Q. Xue, Q. Wang, Wear mechanisms of polyetheretherketone composites filled with various kinds of SiC, Wear 213 (1997) 54–58.

[34] Q. Wang, Q. Xue, W. Shen, The friction and wear properties of nanometre $SiO_2$ filled polyetheretherketone, Tribology International 30 (1997) 193–197.

[35] Q. Wang, Q. Xue, H. Liu, W. Shen, J. Xu, The effect of particle size of nanometer $ZrO_2$ on the tribological behavior of PEEK, Wear 198 (1996) 216–219.

[36] Q. Wang, Q. Xue, W. Shen, J. Zhang, The friction and wear properties of nanometer $ZrO_2$-filled polyetheretherketone, Journal of Applied Polymer Science 69 (1998) 135–141.

[37] M.C. Kuo, C.M. Tsai, J.C. Huang, M. Chen, PEEK composites reinforced by nano-sized $SiO_2$ and $Al_2O_3$ particulates, Materials Chemistry and Physics 90 (2005) 185–195.

[38] M.Z. Rong, M.Q. Zang, H. Liu, H.M. Zeng, B. Wetzel, K. Freidrich, Microstructure and tribological behavior of polymeric nano-composites, Industrial Lubrication and Tribology 53 (2001) 72–77.

[39] C.B. Ng, R.W. Siegel, Synthesis and mechanical properties of $TiO_2$-epoxy nanocomposites, Nanostructured Materials 12 (1999) 507–510.

[40] X.S. Xing, R.K.Y. Li, Wear behavior of epoxy matrix composites filled with uniform sized sub-micron spherical silica particles, Wear 256 (2004) 21–26.

[41] G. Tang, Z. Wang, Q. Ma, D. Zhao, Study on improving wear resistance of epoxy resin filled with nanometer alumina, Thermoset Resin 17 (2002) 4–8.

[42] L. Yu, S. Yang, H. Wang, Q. Xue, An investigation of the friction and wear behavior of micrometer copper particle and nanometer copper particle filled polyoxymethylene composites, Journal of Applied Polymer Science 19 (2000) 2404–2410.

[43] Q.L. Ji, M.Q. Zhang, M.Z. Rong, B. Wetzel, K. Friedrich, Tribological properties of nano-sized silicon carbide filled epoxy composites, Acta Materiae Compositae Sinica 21 (2004) 14–20.

[44] Y.H. Hu, H. Gao, F.Y. Yan, W.M. Liu, C.Z. Qi, Tribological and mechanical properties of nano ZnO-filled epoxy resin composites, Tribology International 23 (2003) 216–220.

[45] M. Shimbo, M. Ochi, N. Ohoyama, Frictional behavior of cured epoxide resins, Wear 91 (1983) 89–101.

[46] B. Wetzel, F. Haupert, M.Q. Zhang, Epoxy nano-composites with high mechanical and tribological performance, Computer Science and Technology 63 (2003) 2055–2067.

[47] B. Wetzel, F. Haupert, K. Friedrich, M.Q. Zhang, M.Z. Rong, Impact and wear resistance of polymer nano-composites at low filler content, Polymer Engineering and Science 42 (2002) 1919–1927.

[48] M. Sharma, Carbon Fabric Reinforced Polymer Composites: Development, Surface Designing by Micro and Nano PTFE and Performance Evaluation, PhD Thesis, Indian Institute of Technology Delhi, Aug 2011.

[49] Mohit Sharma, Jayashree Bijwe, Surface engineering of polymer composites with nano and micron sized PTFE fillers, Journal of Materials Science 47 (2012) 4928–4935.

# Novel nanocomposites and hybrids for high-temperature lubricating coating applications

# 21

**Namita Roy Choudhury, Aravindaraj G. Kannan, Naba Dutta**

*Ian Wark Research Institute, ARC Special Research Centre for Particle and Material Interfaces,*
*University of South Australia, Mawson Lakes, South Australia, Australia*

## CHAPTER OUTLINE HEAD

Tribology of Polymeric Nanocomposites. http://dx.doi.org/10.1016/B978-0-444-59455-6.00021-0

## 21.1 INTRODUCTION

Lubrication is a process that reduces the friction and wear between two moving solid surfaces with a film of solid, liquid, or gaseous material without causing any damage to the surfaces. Both friction and abrasion are very complex processes, which depend on bulk and surface properties of the material. Many dynamic systems are lubricated by fluids of various kinds. Lubrication of dynamic surfaces by fluids adds complexity, weight, and cost to the system, and it also breaks down at high temperature, which imposes various constraints and limits the performance of these systems. Also, in recent years, there has been increasing environmental regulations over the use of lubricating oils due to the problems associated with the waste lubrication oil disposal and the cleaning of the lubricated products surface [1]. Many applications such as aerospace, medical devices require the application of dry coatings that exhibit reduced friction and wear [2,3]. This has prompted research in lubricating thin films and coatings. The elimination of lubricating fluids by solid coatings poses additional demands such as increased wear and adhesion resistance, improved toughness, and lower coefficient of friction of the material. This can be achieved using various solid lubricant interactions, such as physisorption, chemisorption, chemical reaction, and also using higher quality matrix materials or self-lubricating coatings. Depending on the nature of interaction between a substrate and a lubricant, various modes of lubrication can be possible. Figure 21.1 shows the various aspects involved in such lubrication.

Among many different types of coatings, lubricating coatings exhibit reduced wear and friction without the use of tribological fluids, and they fall under the category of solid contact lubrication where either boundary lubrication or solid lubricant lubrication can occur. In boundary lubrication, a thin film of lubricant is adsorbed to the surface, which modifies the solid–solid interaction and prevents asperity welding, while solid film lubrication occurs through the action of a soft or hard film between the sliding surfaces. Such coatings with enhanced tribological performance can be formed from either metal or polymer nanocomposites when different nanoparticles are impregnated into either a metal or polymer or ceramic matrix. The matrix material serves as a reservoir of nanoparticles that are slowly released from the surface and provide easy shear and reduced oxidation of the coating or native substrate. In such cases, often, the effect of nanoparticles is attributed to the facile rolling of the nanoparticles like nanoball bearings.

Polymers and their composites represent an excellent class of tribological materials [4] due to their outstanding chemical stability, high-strength, lightweight, low-cost, excellent abrasion and corrosion resistance, shock absorbance, etc. However, their thermal conductivity is much lower than that of metals, and also, they are less tolerant to high temperatures. Such properties can be tailored using inorganic fillers of various architectures, for example, spherical, tube, cube, platelike, fiber, wire, and rods. Polymer composites formed by conventional methods do not have good dispersion and adhesion between polymers and fillers.

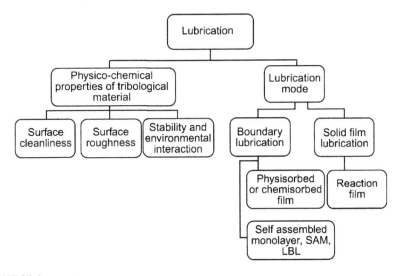

**FIGURE 21.1**

Various modes of lubrication.

This limits their applications. The adhesion between polymer matrix and inorganic filler can be improved largely by functionalizing the nanofillers. This leads to increased nanoscopic level dispersion thereby reducing the amount of fillers required to achieve the desired properties. The resulting hybrid material will have a higher strength-to-weight ratio than conventional composites.

Polymer nanocomposite coatings have the flexibility to impart lubrication in diverse conditions, yet maintaining the required hardness and toughness. These materials offer unique mechanical, thermal, and optical properties, which are useful in a wide variety of tribological purposes, due to their synergistic and hybrid properties derived from their individual components [5]. Unlike the conventional wet lubricating films that are environmentally benign and inefficient, the nanocomposite coatings can be solid, self-lubricating with good mechanical properties. Also, these coatings are superior to hard ceramic coating for tribological applications, since the ceramic coating is often brittle, exhibits high friction, and if the hard ceramic coating is rough, it can lead to high abrasion and rapid wear of the opposing surface.

The unique properties of the nanocomposite coatings not only arise from their individual components but also from the interaction between the individual components and the state of dispersion [6]. In a hybrid or nanocomposite material, as the length scale of the individual phases approaches the nanometer level, the properties of the hybrid are most strongly influenced by the interfacial properties rather than the bulk structures. Therefore, it is essential to understand the interfacial properties of these components to tailor the global properties of the hybrids [7,8]. Hence, a detailed nanolevel insight into synthesis/characterization along with the performance evaluation would give an insight into the structure–property–performance relationship of such nanocomposite coatings.

Therefore, this chapter will focus on various types of lubricating polymer nanocomposite coatings and their syntheses, characterization, and performance evaluation. It also establishes the critical parameters in nanocomposite lubrication, lubrication mechanism, and application with future outlook. For this chapter, high-temperature polymeric nanocomposites and hybrids are defined as those that will not oxidize or thermally degrade in air at temperatures <250 °C, above which lubricating oils, greases, and most of the polymeric materials degrade. Also, here high temperature does not encompass the temperature range in which the ceramic materials can withstand.

## 21.2 WHY POLYMER NANOCOMPOSITES?

Polymer nanocomposite is a newly developed concept introduced in the recent past in composite materials and polymer science. The concept consists of deriving a hybrid from two generically different constituent materials such as organics and inorganics at a molecular level dispersion. Hybridization of such organics and inorganics is an important and evolutionary route for the growth of a strong polymer–nanofiller interface. Practically, it offers an economic advantage and is versatile in nature. This can be explained by the low volume (1–5%) of nanofillers required to achieve properties that are comparable or even better than those achieved by conventional loadings (15–40%) of traditional fillers [9]. The lower loadings facilitate processing and reduce component weight. Also, this means a higher strength-to-weight ratio, which is very useful in the case of automobile and aerospace industries to increase fuel economy. A hybrid is defined as a system containing both inorganic and organic moieties in the backbone, and a chemical bond between the two moieties is formed. The resulting hybrid would possess an ideal combination of properties of both materials and a good surface finish. Polymeric nanocomposites are functional materials that can thus be used as coatings to reduce friction and wear.

They also confer abrasion resistance, corrosion resistance, thermal and electrical insulation characteristics. However, research on the tribological performance of such materials is much less advanced than it is for metals. Polymer nanocomposite tribology needs further significant development to replace metallic components by polymeric ones. Wear and poor scratch resistance are key problems in polymers, which for metals can be overcome by external lubrication. In polymers, strong interactions with the lubricant are often possible, leading to swelling. Thus, the application of lubricants is a much more complex problem.

Extensive efforts are now being made toward the development of materials possessing desirable properties of both organic polymers and glasses/ceramics by making organic–inorganic hybrid or hybrid nanocomposites. A revolutionary approach to enhancing inorganic particle dispersion is the in situ polymerization of nanostructured oxometallate cluster, metal alkoxide, or self-assembled platelet inorganics in organic polymer matrices via the sol–gel or intercalation process, producing molecular scale composite or nanocomposite or hybrid. Polymer hybrids represent a new class of alternative materials to conventionally filled polymers, which can overcome the problems associated with traditional composites [10,11]. The major advantages of producing a nanocomposite by hybridization are homogeneity at near molecular level, materials with a novel microstructure with a very low filler content, and tailoring the nanostructure by controlled hybridization. Such hybrid coatings can lead to novel functional materials with outstanding properties by careful choice of and combination of organic and inorganic components. The major advantage of nanoparticles/nanofillers is that they can be less abrasive due to their unique physicochemical properties. Increased dispersion and interfacial bonding of the nanofiller in a matrix result from the functionalization of the nanofillers, which often leads to enhanced performance. They exhibit a unique or novel functionality such as outstanding barrier and strength properties, enhanced solvent, scratch, abrasion, heat and chemical resistance, and decreased flammability due to highly dispersed nanoscopic inorganic particles, they are targeted for demanding parts such as gear systems in drilling systems, wear resistance materials, in tires [12,13]. Polymers used for such purposes are primarily various types of fluoropolymers, elastomers, and other engineering polymers such as polyamide (Nylon), polyimide, or amide–imide; polyphenylene sulfide; and polyurethane (PU). Various inorganics have been employed such as silica, alumina, titania, zirconia, barium titanate, and organometallates such as polyhedral oligo silsesquioxane (POSS), etc. [14–17].

However, for a variety of applications where mechanical devices are used to operate reliably in an extreme range of environments, it is crucial to develop materials that can operate in environments with extremes of temperature, pressure, radiation, and stress, and also the physical and chemical integrity of the materials be retained under such extreme conditions.

Fluoropolymer such as polytetrafluorethylene (PTFE) is known for its low frictional coefficient and good thermal and chemical stability. Polymer nanocomposites based on fluoropolymers are often considered for use as solid lubricants in many dynamic applications. Although submicron- or nanometer-sized metallic or ceramic particles are principal precursor materials for a range of advanced applications, including superior abrasion-resistant coating, electronics, and catalysis, etc., they often pose dispersion problems in the polymer matrix. The properties of the particles, such as shape, size, distribution, and flowability, also play an important role. The properties of such nanocomposites strongly depend on various parameters such as the degree of phase dispersion, the relative amount of each component, as well as the molecular weight of the polymer. Depending on the miscibility or level of interaction between the inorganic and polymer phases, various types of nanocomposites or hybrids result such as Class I nanocomposite material involving physical or weak phase interaction, for example, hydrogen bonding, van der Waals, or simple mechanical mixing and Class II hybrid material possessing strong chemical or covalent or ionocovalent bond between the phases. However, in any case, control of uniform particle size is extremely important, particularly at high loading of the nanoparticles. Several recently developed fabrication techniques, such as spin-assisted coating, dipping, plasma coating, and micropatterning, are employed to make the coating process more efficient and to impart unique tribological properties. Figure 21.2 shows various types of lubricating coating that will be discussed throughout this chapter. They are fabricated by either wet or dry technique and able to confer unique tribological property to a substrate.

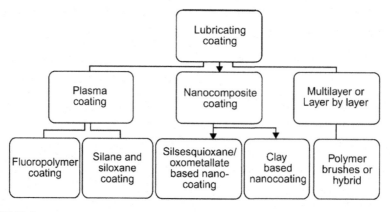

**FIGURE 21.2**

Thermostable lubricating coating.

## 21.3 THERMOSTABLE LUBRICATING COATINGS

### 21.3.1 Fluoropolymer coating

Fluorine-containing materials find considerable application in the field of coatings because of their interesting properties like water and oil repellency, low coefficient of friction, and chemical resistance. Fluorinated surfaces offer outstanding characteristics due to their unique molecular properties associated with the C–F bond that imparts a specific, physicochemical properties at interfaces. Their low surface tensions and low friction coefficients can play a key role in microelectronics, antifogging, antifouling applications and are also promising in biomedical applications. Materials, that have a lower surface tension than PTFE have surfaces predominantly covered by $CF_3$ groups. The $CF_3$ groups are more effective compared to the $CF_2$ groups in lowering the surface tension due to the bulky F atom [18,19]. Zisman established that the surface tension depends on the constituent groups and decreases in the order of $CH_2$ ($36 \, mN \, m^{-1}$) > $CH_3$ ($30 \, mN \, m^{-1}$) > $CF_2$($23 \, mN \, m^{-1}$) > $CF_3$ ($15 \, mN \, m^{-1}$). The properties of fluorinated surfaces thus depend not only on the coverage of the surface by the fluorocarbons but also on the degree of order on the surface [20]. A uniformly ordered array of $CF_3$ groups on a surface can result in a surface tension of $6 \, mN \, m^{-1}$ [19] and hence can be self-lubricating.

### 21.3.2 Hydrophobic plasma coatings

Fluorine-containing and silicon-containing monomers are two classes of monomers, which make good candidates for the plasma deposition of low-friction, hydrophobic nanocoatings onto polymeric, or metallic surface. For example, highly hydrophobic and frictionless plasma polymer films can be deposited from $CF_4$/acetylene and $C_2F_6$/acetylene mixtures or from hexamethyldisiloxane.

#### 21.3.2.1 Perfluorocarbon and hydrocarbon

Plasma polymerization of perfluorocarbons is generally slower than that of hydrocarbons, which is due to the difficulty of fluorine detachment to the plasma polymerization. The major difference observed between plasma polymerization of hydrocarbons and of perfluorocarbons may be reflecting the difference of bond energies of C–H and C–F and dissociation energy of F–F and H–H. The bond energy of C–C is $83 \, kcal \, mol^{-1}$. The relatively high dissociation energy of H–H ($104.2 \, kcal \, mol^{-1}$) compared to that of F–F ($37.7 \, kcal \, mol^{-1}$) may favor the detachment of hydrogen from hydrocarbons, but the detachment of fluorine from perfluorocarbon is not favored. This situation is clearly seen in plasma polymerization of $CH_4$ and $C_2H_6$ versus $CF_4$ and $C_2F_6$. Although their deposition rates are small, polymers are formed from methane and ethane, whereas no polymer is formed from $CF_4$ and $C_2F_6$. The breaking of C–C bonds leads to a various cycle rearrangement of molecules but effect polymer formation. If C–F does not break preferentially, then perhaps the most likely process would be

$$CF_3 - CF_3 \rightarrow 2CF_3 \qquad (21.1)$$

Since the recombination of $CF_3$ radicals leads to the original $CF_3$–$CF_3$ molecules, no polymer was formed from this reaction.

In the plasma polymerization of tetrafluoroethylene, glow characteristics play an important role. The glow characteristic is dependent on the combination of the flow rate of the monomer and discharge power. At a higher flow rate, a higher power is required. However, in addition, the decomposition of the monomer is dependent on a factor "$W$/FM". At a given flow rate ($F$), an excessive discharge power ($W$) enhances the decomposition of monomer and reduces the polymer deposition rate drastically. In fluorocarbon discharges, simultaneous etching and polymerization occur. These processes are highly interdependent, and which of these will dominate depends on a factor such as the F/C (fluorine/carbon) ratio of the monomer gas, reactor pressure, discharge power, flow rate of the gases, location of the substrate with respect to the close region, reactor geometry, and the mode of coupling the energy with the low pressure gas to form the plasma. Typically, gases with a high F/C ratio, for example,. $CF_4$ (F/C = 4) result in efficient etching, whereas a lower F/C ratio gas, for example, $C_2F_4$ (F/C = 2) polymerize easily. Plasma polymerization characteristics of saturated perfluorocarbons are viewed as a function of the F/C ratio of monomers, which clearly indicates the importance of the balance between ablation and polymer deposition. In our recent work, we have explored the plasma fluoropolymer coating of perfluoro-methylcyclohexane on automotive window-seal component. In such dynamic applications, elastomeric nanocomposite materials are often used. The elastomer surfaces, in those cases, must have a low friction coefficient, very high abrasion resistance and toughness. Ethylene–propylene–diene–monomer (EPDM) is the dominant individual elastomer of choice in many automotive applications, particularly in applications, which demand excellent chemical and thermal stability and ozone resistance. The automotive sealing components made from EPDM need specific coatings to obtain a low friction, high abrasion resistance, and release properties. For example, if the surface of the EPDM-based sealing section used in car windows are not modified by the application of a low friction coating, the resistance to movement of the glass window should be significant causing stress to both the window-winding mechanism and the passenger attempting to move the window [21–23].

As fluorine-containing organic compounds are very unique in nature to form hydrophobic surfaces, in our work, plasma polymerization of perfluoromethylcy-clohexane (F/C = 2) is used to form a nanocoating on the carbon black nano filler-reinforced EPDM substrate. Plasma fluoropolymeric coating offers low-friction, high abrasion resistance characteristics. The hydrophobicity may not depend on the plasma conditions. Such plasma polymeric coating has a Teflon-like chemical structure at short treatment time from 20 to 100 s. Crosslinking and branching are formed progressively for further increase in treatment time. Long chain and branches are contained in the bulk coating. The coating [24] adheres well to the EPDM substrate by covalent bonding. The coating is hard and inflexible at high plasma conditions (e.g. 10-min of treatment time). These properties depend on the plasma condition. The longer the plasma treatment times, the tougher and less flexible the plasma polymeric coating obtained. We have also fabricated fluorocarbon plasma polymer thin films on silicon substrates [25]. The thermal stability of the thin films formed on silicon substrates is very similar to that of Teflon. The combination of both visualization and characterization capability of scanning thermal microscopy has been employed to investigate the microstructure, surface morphology, thickness, physical properties, and homogeneity of these lubricating films [25]. The wear resistance test of these fluoropolymeric coatings deposited on the EPDM nanocomposite substrate were also examined extensively using a small load wear resistance tester (Fig. 21.3), which mimics the condition of a glass window-winding mechanism of a car.

In this work, the coated sample was cut into a rectangular shape and glued tightly on a steady metal plate. Standard glass was used to abrade on the coating at a specified speed. A small load was applied on the sample. The standard glass was abraded on the coating for 20 revolutions. The abrasion resistance of this fluoropolymer coating (Fig. 21.4(a)) is found to be excellent due to the chemical nature of the coating. The morphology of such an abraded sample (20-s coating) is shown in Fig. 21.4(b). The scanning electron microscopic (SEM) image shows the overall area where it is abraded. It appears that the film is very thin and squeezed hardly onto the substrate. There is some debris present on the surface (Fig. 21.4(c)). It is observed that the abrasion resistance of the 20-s coating is better than that of the EPDM due to the hydrophobic nature and the low surface energy of the coating, leading to a low friction in contact.

### 21.3.2.2 Siloxane plasma coating

Plasma polymerization of silicone-containing monomers has also received a great deal of attention to create low-friction, high abrasion-resistant surfaces. The first target of those studies was the production of $SiO_2$-like films devoted to

**FIGURE 21.3**

Small load wear resistance tester. For color version of this figure, the reader is referred to the online version of this book.

microelectronics applications [26–31]. In general, all organosilicon monomers have been successfully used as precursors for $SiO_2$-like films, and among them, hexa-methyldisiloxane (HMDSO) has gained increasing interest in the last decade owing to its lower cost and high deposition rate compared to those of similar silicone monomers [32,33]. Tetramethyldisiloxane (TMDSO) monomer has been reported to give a higher deposition rate than that of the HMDSO due to the hydrogen atom directly being bonded to the silicon atom. Despite all the advantages, there are only a few reports on this type of monomer compared to that of the HMDSO. Sharma and Yasuda studied the plasma polymerization of TMDSO on an aluminum sub-strate using magnetron glow discharge [34]. Recently, we have plasma-polymerized TMDSO on an ethylene–propylene–terpolymer elastomer (EPDM) nanocomposite substrate using an RF plasma discharge in order to produce coatings with pro-gressively lower surface energies. The density of such deposited plasma polymeric coatings at a power of 40 W could be calculated from the ratio of the deposition rate $(0.151 \, (\mathrm{mg} \, (\mathrm{mm}^2 {*} 1000)^{-1}) \, \mathrm{min}^{-1})$ to the film thickness rate $(0.002 \, \mu\mathrm{m} \, \mathrm{s}^{-1})$ [22,35]. The density of the coating is found to be approximately $1.25 \, \mathrm{g} \, \mathrm{cm}^{-3}$, which is simi-lar to that of the plasma polymerization of HMDSO $(1.16 \, \mathrm{g} \, \mathrm{cm}^{-3})$ [35] in the same formulation and to the conventional polymethyldisiloxane $(0.98 \, \mathrm{g} \, \mathrm{cm}^{-3})$ [36]. The film is hydrophobic and of a low surface energy. The plasma siloxane polymeric coatings possess the following properties: high water repellency [27,37] and ther-mal stability [35], low coefficient of friction [38] and surface energy [39], homoge-neous [40], tough [41], and relatively high permeability for most of the gases [42].

In this study, the abrasion resistance of the plasma polymeric coatings was also investigated, and the morphology of the abraded plasma siloxane polymeric coating

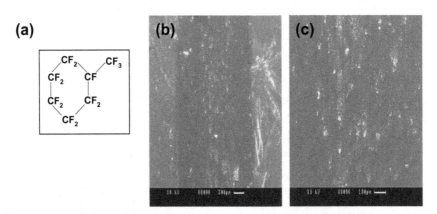

**FIGURE 21.4**

(a) Structure of the perfluoromethylcyclohexane; (b), (c) morphology of the abraded 20-s plasma fluoropolymeric coating by SEM: overall and abraded area, respectively.

*This figure was published by N.K. Dutta, N.D. Tran and N.R. Choudhury, Perfluoro(methylcyclohexane) plasma polymer thin film: Growth, surface morphology, and properties investigated by scanning thermal microscopy, Journal of Polymer Science: Polymer Physics 43 (2005) 1392–1400, Copyright Elsevier.*

was examined by SEM. The plasma polymeric coatings prepared from TMDSO plasma polymerization on the EPDM substrate at a power of 40 W for 20 s and 10 min of treatment times and 1.64 μl min⁻¹ flow rate were chosen. The structure of TMDSO and the morphology of an abraded 20 s coating are shown in Fig. 21.5(a), (b) and (c). The images are shown at two magnifications of 300 μm (b) and 100 μm (c). Fig. 21.5(b) does not show much change on the abraded area compared to that on the nonabraded area. There is some debris left on the surface indicating that the plasma polymer was fragmented. The coating appears to be torn rather than fragmented. The morphology of the abraded 10-min coating is shown in Fig. 21.5(d), (e). The images are shown at 300-μm (d) and 100-μm (e) magnifications. Fig. 21.5(d) shows the overall abraded area. The coating is fragmented evenly into pieces as shown in Fig. 21.5(e) [43].

A comparison of the observations made from the abraded 20-s coating and the abraded 10- coating shows the latter to be fragmented rather than torn. It is observed that the coating at low plasma conditions is more flexible than the coating at high plasma conditions. The flexibility of the coating is due to the extent of crosslinking in the plasma polymer. The 10-min coating is fragmented under an applied load indicating that it is hard and could be crystalline. Under this high plasma condition, the chemical structure of the coating could contain a high level of O–Si–O functional groups, which could indicate the formation of a silicalike structure. Therefore, the brittleness of the coating could be due to the silica structure. The coatings were not delaminated from the substrate surface, indicating strong adhesion of the coating to the substrate. It is noted that the bonding between the coating and substrate is dependent on the plasma energy. More importantly, the nanocoatings prepared are able to provide a low-friction and high-abrasion resistance to elastomer nanocomposite window-seal surfaces.

### 21.3.3 Oxometallate cluster-based nanocomposite coatings

In recent years, there has been considerable effort focused on the use of inorganic–organic structures for various applications from coatings to catalyst supports to

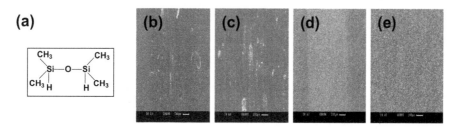

**FIGURE 21.5**

(a) Structure of TMDSO; (b), (c) morphology of the tested 20-s plasma organosiloxane polymeric coating; (d), (e) morphology of the tested 10-min plasma siloxane polymeric coating.

dendrimer cores. In the area of coatings in particular, these hybrid chemicals offer good application as thermally stable and abrasion-resistant coatings that give better performance organic polymers. One of the most important areas, worldwide, of the coatings industry is their tribological and self-cleaning applications. The incorporation of the inorganic component gives rise to coating properties (e.g. better adhesion to substrate) that can allow the coated substrate to perform its role for longer periods without reapplication of the coating. A great body of work has been performed widely on the use of sol–gel coatings using silicon-based hybrid molecules [13,44–47]. Much success has been accomplished using them over the years. Figure 21.6 shows the evolution of nanocomposite materials over several decades [48]. The nanocomposites in Generation I are often prepared by dispersing nanosized fillers into a polymer matrix (Fig. 21.6(a)). The nanofillers can be prepared via a sol–gel reaction. Generation II nanoreinforced systems are prepared by dispersing self-assembled nanoscale clay fillers (Fig. 21.6(b)) into a polymer matrix through intercalation or delamination. In these two systems, weak noncovalent interactions (e.g., hydrogen bonding or van der Waals attractions) are generally involved between the polymer matrix and the nanofiller. They fall under the category of class I hybrid. Generation III systems are called oxometallate clusters such as polyhedral oligomeric silsesquioxanes (POSS), which are prepared by a bottom-up approach with the use of nanostructured chemical building blocks to copolymerize, crosslink, or graft onto a polymer backbone (Fig. 21.6(c)). These blocks attach to the polymer matrix through covalent bonds (class II hybrid).

Of the great many types of hybrids available for coating application, perhaps the most exciting in terms of the offered physical and chemical properties is by the use of versatile well-defined POSS nanostructures. Recent improvements in their synthesis have resulted in their commercial viability as research candidate materials. The versatility of these molecules results from their high thermal stability, oxidation resistance, discrete molecular weight, and tailoring of the number of reactive sites that allow controlled functionalization.

POSS is a nanostructured material with a unique cagelike structure and nanoscale dimensions, represented by the general formula $(RSiO_{1.5})_n$ where R is an organo-functional group. They are molecularly precise and nearly isotropic molecules with a diameter of 1–3 nm, depending on the number of functional units attached to the

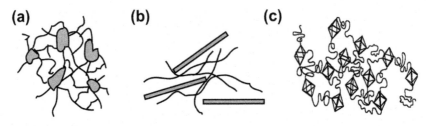

**(a)**     **(b)**     **(c)**

**FIGURE 21.6**

Various types of nanocomposites and hybrids.

cage structure. POSS can be prepared as either a completely condensed closed cage structure or a partially condensed open-cage structure with silanol groups, and it exists as two main derivatives, the fully functional octasubstituted POSS and the heptasubstituted trisilanol POSS. POSS has a rigid inorganic core surrounded by organic or reactive functionalities. POSS bearing covalently bonded reactive functionalities can react with polymer chains to form highly ordered unique hybrid materials that are suitable for various high-performance applications. Alternatively, POSS molecules can also be incorporated into a polymer matrix through blending or grafting to the polymer chain. The inorganic nature and multiple reactive functionalities of POSS material make it ideal for the preparation of organic–inorganic hybrids, with enhanced properties as a result of its characteristic architectural features. Molecular reinforcement in these hybrids primarily takes place by the addition of POSS in a blended, grafted, crosslinked, or copolymerized form (Fig. 21.7). Additionally, this material exhibits significant atomic oxygen and vacuum ultraviolet (UV) radiation resistance, degradation resistance through the formation of passive silica layer.

By changing the functional groups (R) on a POSS molecule, the characteristics and surface activity can be changed. Possible functional groups include almost any organic group such as methyl, isobutyl, cyclopentyl, or cyclohexyl. Several attempts have been made to produce linear polymers with POSS incorporated into the polymer backbone. This requires POSS molecule to have two identical functional groups at the opposite corner. Vinyl functional groups are also available that allow for the copolymerization of POSS with other monomers. Such a variety of readily available functionalization is the primary difference between POSS molecules and customary fillers. The presence of organic components improves the nanodispersion of inorganic components in the polymer matrix, which leads to the formation of multifunctional nanocomposites. Covalent POSS attachment results in polymers with higher glass transition temperatures ($T_g$) as well as increased hardness, toughness, abrasion resistance, thermal stability, fire retardance, and corrosion inhibition. POSS molecules

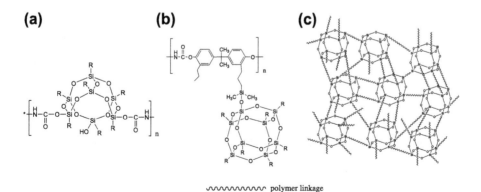

**FIGURE 21.7**

POSS-based nanocomposites with (a) linear, (b) grafted, and (c) crosslinked structure.

have been incorporated into conventional polymers such as epoxies [49,50], imides [51], methacrylates [52,53], carbonates [54], and urethanes [16,55,56]. Recently, Oaten et al. [16] prepared a silsesquioxane–urethane hybrid for thin film applications. They showed that a POSS-based PU hybrid forms a lamellar structure with a high degree of crosslinking on the metallic substrates (Fig. 21.8). The incorporation of POSS in the hybrid increased the thermal stability and the glass transition temperature ($T_g$) of the coating.

POSS can also be used as a nanocrosslinker in polymer systems. This can be accomplished through the functionalization of the POSS macromer with multiple functional groups and subsequently using them as nanocrosslinkers. Highly crosslinked methyl methacrylate (MMA) hybrids have been produced incorporating POSS crosslinkers. The required octamethacryl functionalized POSS was synthesized by hydrosilylating allyl alcohol to POSS. Subsequently, methacryl groups were introduced by reacting hydroxyl functionalized POSS with methacrylic acid anhydride (Fig. 21.9) [57] or methacryloyl chloride. Further polymerization of this functional POSS with MMA monomer enhances the nanofiller dispersion (as seen from AFM, Fig. 21.9), thermal stability of the hybrid sample by 43 °C compared to the pure PMMA.

Hybrid star polymer with octafunctional POSS as core has been reported to be used as the polymerization initiator. The arm of the star is grown from the core using atom transfer radical polymerization, ATRP of MMA [58], or by reaction with polyethylene glycol (PEG). Octaaminophenyl octasilsesquioxane has been used to produce thermostable organic–inorganic hybrid polymers with a high crosslink density by its reaction with various dianhydrides. The resultant polymers have a very

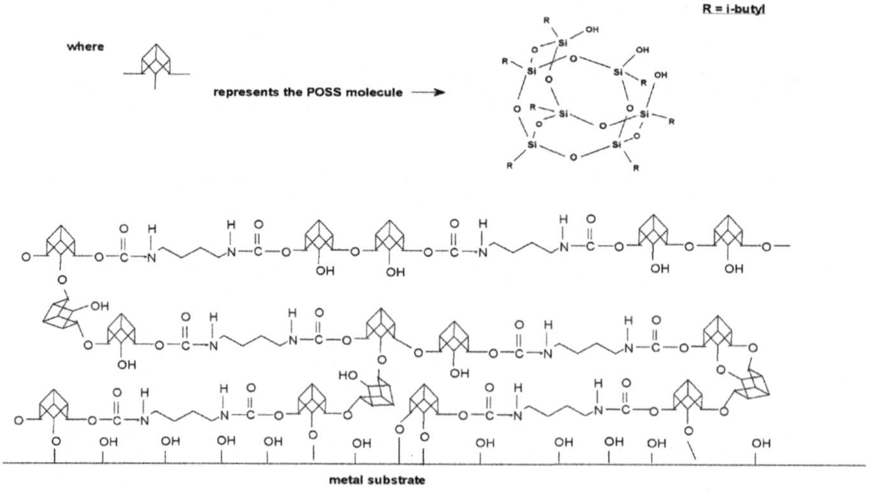

**FIGURE 21.8**

Structure of the POSS–urethane hybrid thin film on a metallic substrate.

*Reprinted with permission from Macromolecular 2005, 38, 6392–6401. Copyright 2005, American Chemical Society.*

high thermal stability; for example, polyimide hybrid crosslinked at 400°C displays only a 5% weight loss at 538 °C in nitrogen [59]. Novel polyimide–epoxy nano-composites [60] are also prepared by the in situ curing of polyamic acid with cubic silsesquioxane epoxides, where increase in the crosslink density shows little effect on compressive modulii, hardness, and coefficient of thermal expansion. More recently, the tribological performance of POSS-based nanocomposites is being explored. Rahul and Morgan [17] reported that the polypropylene nanocomposites containing 10% of octaisobutyl POSS showed an almost 60% reduction (Fig. 21.10) in the relative surface friction coefficient. The reduction in the coefficient of friction was attributed to the four-fold increase in the surface roughness (due to the addition

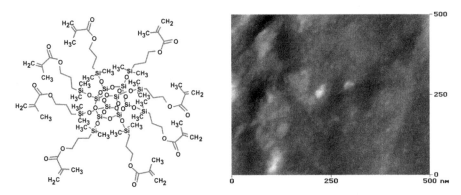

**FIGURE 21.9**

Inorganic POSS cluster, AFM of cluster crosslinked PMMA. For color version of this figure, the reader is referred to the online version of this book.

*Reprinted with permission from Chemistry of Materials, 2005, 17, 1027–1035. Copyright 2005, American Chemical Society.*

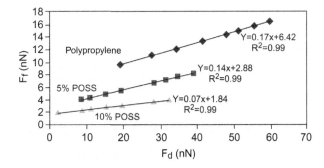

**FIGURE 21.10**

Comparison of the coefficient of friction of polypropylene and its nanocomposites with octaisobutyl POSS. For color version of this figure, the reader is referred to the online version of this book.

of POSS), which effectively reduces the surface contact area with the counterface. Figure 21.10 shows the lateral friction force ($F_f$) plotted against the loading force ($F_d$) for the polypropylene and its nanocomposites containing 5% and 10% POSS, where the slope of the straight line gives the coefficient of friction. In another study, Lichtenhan et al. [61] used various POSS molecules containing different organic functional groups at the periphery to improve the compatibility, increase surface area, and promote the lubricity between surfaces at a length scale not previously attainable. They showed that surface friction of the nanocomposite could be controlled by controlling the surface topology. Misra et al. used melt mixing to prepare nanocomposites based on nylon-6 and two structurally and chemically different POSS molecules, a close-cage nonpolar octaisobutyl POSS and an open-cage trisilanolphenyl POSS. Relative surface friction studies of these nanocomposites using AFM/LFM on nanoscale and pin-on-disc tribometry on a macroscale revealed significant friction reduction on incorporation of POSS at both scales. Reduction in the relative surface friction was attributed to the preferential migration and aggregation of POSS molecules on the surface resulting in increased surface roughness, increased surface hardness, and improved nanomechanical properties. Also, there has been an increasing interest in the development of a new class of fluoro-POSS fillers and its application in the polymer matrix due to its remarkable surface properties [62–64]. These fluoro-POSS-based nanocomposites may be potentially useful in tribological applications due to its low friction and ultrahydrophobic surface. Aravindaraj et al. [62] reported the synthesis of fluoro-PU based on fluoro-POSS. The prepared material exhibits ultrahydrophobic characteristics [65,66] and shows potential for lubricating coating application. In a recent study, Wang et al. [67] used a mixture of fluorinated-decyl POSS and fluorinated alkyl silane to prepare a self-cleaning superhydrophobic and superoleophobic polyester fabric with chemical, UV and abrasion resistance. The coated fabric could withstand 6000 cycles of abrasion damages without significantly altering its superhydrophobic and superoleophobic properties. Still, this type of nanocomposites needs a significant understanding on the structure–property–performance relationship and is in the early stages of research.

The molecular building-block approach to the synthesis of nanocoating materials involves the incorporation of other transition metal nanosized cluster (apart from POSS), for example, oxotitanate, oxozirconate, or mixed titanate/zirconate clusters into various polymeric matrices. Such a cluster is a typical discrete metal and oxygen framework functionalized with reactive organic group, which can subsequently be incorporated into the polymer matrix to significantly enhance the physical properties such as chemical and thermal stability. Hence, they are suitable for use in high-temperature, abrasion-resistant coating applications. Recently, Basch et al. and Schubert et al. [68–74] have prepared a series of acrylate- and methacrylate-substituted oxotitanate, oxozirconate, and oxo (titanate-zirconate) mixed clusters in different sizes and shapes by reacting the metal alkoxide with a calculated stoichiometric excess of the methacrylic acid or anhydride. Finally, the prepared clusters can be copolymerized with various monomers at various cluster/polymer ratios. The cluster dopants can also act as reinforcing agents and as nanocrosslinkers in creating nanoconfined hybrids. The organically modified transition metal oxide clusters (OMTOC) are the transition metal equivalents to the POSS materials. Figure 21.11(a) shows the structure of OMTOC with a $Ti_6O_4(OEt)_8(OMc)_8$ (OMc = methacrylate) structure. These materials hold the potential for abrasion-resistant coating on ophthalmic lenses. Thermogravimetric analysis (TGA) of the hybrid (Fig. 21.11(b)) shows that the thermal stability of the hybrid is significantly higher than that of the undoped polymethyl methacrylate (PMMA) even at a very low level of clusters. Other cluster structures have been reported using transition metal and/ or mixed cluster structures with different metals, such as $Ti_4O_2(Opr)_6(OOCR)_6$, $Zr_6(OH)_4O_4(OMc)_{12}$, $Zr_4O_2(OMc)_{12}$, and $Ta_4O_4(OEt)_8(OMc)_4$ [58].

### 21.3.4 Self-assembled, nanophase particle coating

Surface coating that enables strong covalent bonding at the substrate–coating interface can offer an enhanced performance and long-term stability to a substrate. Nanoparticles and nanostructured coatings, recently, have gained increasing interest for industrial application such as wear and abrasion resistance. This novel approach to robust interface bonding is based on a preformed, self-assembled, nanophase particle (SNAP)- [75] based sol–gel system that has the potential to produce a fully dense protective surface coating. The fundamental principle is to generate nanoscale particles from an aqueous solution with an inorganic core by arresting the particle growth using a surfactant and a subsequent creation of polymerizable or reactive organic exterior with an inorganic core. This is further crosslinked with an organic crosslinking agent to crosslink the nanoparticles to form a dense, protective film on the substrate (Fig. 21.12). Nanostructured surface coating based on SNAPs in a polymeric matrix can be a potential candidate for wear-resistant, lubricating coating.

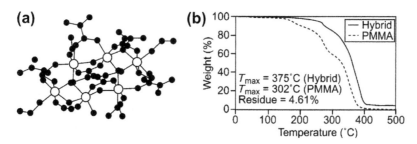

**(a)**    **(b)**

$T_{max}$ = 375°C (Hybrid)
$T_{max}$ = 302°C (PMMA)
Residue = 4.61%

**FIGURE 21.11**

(a) Structure of the oxotitanate cluster. Reproduced with permission from *G. Trimmer, B. Moraru, S. Gross, V.D. Noto and U. Schubert, CrossLinking of Poly(methyl methacrylate) by oxozirconate and oxotitanate clusters, Macromolecular Symposia 175 (2001) 357–366. Copyright Wiley-VCH Verlag GmbH & Co. KGaA* and (b) MTGA curves for the oxotitanate hybrid and pure PMMA.

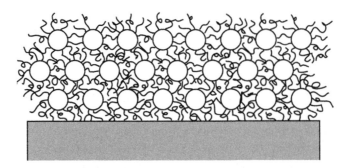

**FIGURE 21.12**

Organic–inorganic hybrid with SNAP coating.

Significant property enhancements have been reported for such hybrid systems including increased toughness, wear resistance, barrier characteristics, and increased oxidation resistance [75]. Therefore, a polymeric coating incorporating such a multi-functional nanostructured component is able to offer many demanding properties of a self-lubricating surface.

## 21.3.5 Clay-based nanocomposites

Apart from clusters or nanostructured cages, a wide variety of other nanoparticles of different geometries can be used to fabricate polymer nanocomposites with excellent tribological properties. Clay is a layered silicate, which belongs to the general family of 2:1 layered, or phyllosilicates with dimensions of a 1-nm layer thickness and lateral dimension from 30 nm to several microns, depending on the type of layered silicate. Montmorillonite, hectorite, and saponite are the most commonly used layered silicates. These nanofillers can be incorporated into a polymer matrix such as PMMA [76], polyamide [77], epoxy [78], and nylon [79] leading to either exfoliated or intercalated or phase-separated morphology (Fig. 21.13) depending on the nature of the components used and the method of preparation [79–83]. Among these three morphologies, exfoliated structures are of particular interest because of the level of polymer–clay interactions. The exfoliated structure leads to the increased interfacial interaction between the polymer matrix and clay layers, due to the availability of the entire surface of the layered silicates for interaction with the polymer matrix [76]. The clay-based nanocomposites can be prepared via the in situ polymerization or solution dispersion or melt mixing. In the solution dispersion method, clay is first exfoliated into single layers using a solvent in which the polymer is also soluble. The polymer then absorbs onto the delaminated clay surface, which forms a nanocomposite upon drying the solvent. In the in situ polymerization, the layered silicate is swollen with the monomer solution where polymerization occurs in between the intercalated sheets. In the melt intercalation process, the layered silicate is melt mixed with the molten polymer matrix. The extent of intercalation or exfoliation depends on the compatibility of the polymer matrix with the clay.

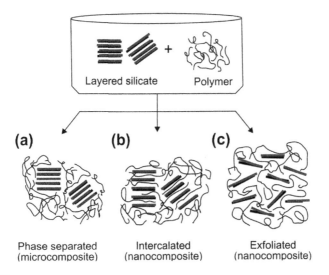

**FIGURE 21.13**

Scheme of interaction of layered silicates with polymer matrix and the resulting composites. For color version of this figure, the reader is referred to the online version of this book.

*This figure was published by M. Alexandre and P. Dubois, Polymer-layered silicate nanocomposites: preparation, properties and uses of a new class of materials, Materials Science and Engineering: R 28 (2000) 1–63, Copyright Elsevier.*

Polymer–clay nanocomposites with a fine dispersion of clay platelets combined with the strong interfacial interaction with the polymer matrix leads to superior wear performance. Dasari et al. [84] showed that mere exfoliation of clay platelets is not enough to prevent wear damage but that the strong interaction between the polymer and clay plays a significant role. Weak interaction between the polymer matrix and the clay platelets leads to the embrittlement of the nanocomposite resulting in poor tribological behavior. With 3 wt% organically modified clay loading in a thermoset polyester matrix, a maximum of 85% increase in the wear resistance and a 35% reduction in the friction coefficient can be achieved [85]. The reduction in the coefficient of friction is explained by the roller-bearing effect of nanoclay and nanoclay-wear debris and the formation of transfer film on the counterface surface. Also, clay has been incorporated into other polymeric systems such as epoxy [86], polycarbonate [87], PVDF [88], polyamide [89], and epoxy–PU interpenetrating networks [90], to enhance their tribological performances. Recently, the research in polymer–clay nanocomposite is also focused on the preparation of polymer–polymer–clay tertiary nanocomposites to enhance the tribological properties [89,91].

While most of those clay-based nanocomposites rely on effective intercalation through either solution or melt-based processes, this can pose some limitations. This can be alleviated by other approaches such as thermal spraying. In thermal spraying, the powders of polymer and nanoparticles or clays are heated in a thermal jet created either by plasma or by high velocity oxy-fuel combustion and sprayed onto a substrate. The benefits of thermal spraying over other techniques, such as fluidized bed, sol–gel, vapor deposition, etc., are that it can be readily applied to any intricate part of any dimension.

In addition to layered clay structures, clays can also form tubular structures such as halloysite and imogolite, which are naturally occurring and composed of aluminosilicates [92,93]. The dimensions of imogolite are approximately a length of 100 nm with an internal diameter of about 1 nm, which are considerably smaller than those of the halloysite tubes with dimensions of a length of 500–1000 nm and a diameter of 50 nm. Thus, imogolite nanotubes provide a much higher surface area than do halloysite nanotubes and consist of a silicon hydroxide interior and an aluminum hydroxide exterior. This nanotube-structured clay is gathering research interest due to its unique one-dimensional structure, high porosity, large surface area, and available surface functional groups. Furthermore, recent progress in the synthetic procedures [94] of these clay nanotubes has resulted in an increased interest in these materials. These nanotubes can be used as a filler in polymer matrices such as epoxy [95], polyamide [96], polypropylene [97], and PMMA [98] to form nanocomposite systems. The field of polymer nanocomposites with these nanotube structures for tribological applications is still not highly explored.

### 21.3.6 CNT-based coating

The other one-dimensional structure that has attracted significant interest is that of CNTs, which are cylinders of graphitic carbon with a diameter of only a few nanometers and a length of up to centimeters. They can be single walled or multi walled. Single-walled nanotubes (SWNTs) are sheet of graphene rolled into a seamless cylinder with a diameter of the order of 1 nm, whereas multiwalled nanotubes (MWNTs) consist of an array of concentrically formed cylinders with a diameter of 2–100 nm [99,100]. Due to their unprecedented mechanical, thermal, and chemical properties, CNTs are suitable for a wide range of applications, including as an advanced filler material in nanocomposites for tribological applications. For example, SWNTs reduce the coefficient of friction by an order of 10 in comparison with the conventional graphite lubricants in air and the coefficient of friction for graphitized MWNTs are one-fifth that of molybdenum sulfide lubricants [101]. Such an improvement in the tribological performance is also observed with various polymer–CNT nanocomposites, for example, thermoset and thermoplastic-based nanocomposites [102]. Figure 21.14 shows the plot of friction coefficient with different CNT contents for three different nanocomposites of epoxy, PMMA, and polystyrene matrices. Irrespective of the polymer matrix, the friction coefficient decreases with an increase in CNT loading and reaches a stable friction coefficient at 1% loading. These nanocomposites show a similar trend in the decrease in the wear rate. Similarly, polystyrene–acrylonitrile–CNT nanocomposite prepared by in situ polymerization showed a 30% decrease in the friction coefficient and a 28%

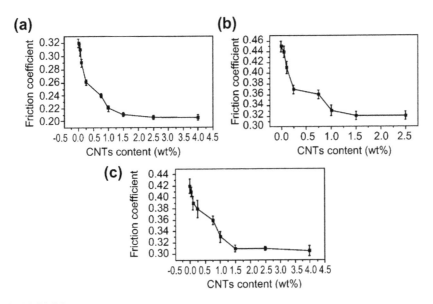

**FIGURE 21.14**

The friction coefficient of polymer–CNT nanocomposites with CNT content: (a) epoxy–CNT, (b) PMMA–CNT and (c) polystyrene–CNT.

*Reprinted with permission from Key Engineering Materials, 2007, 334–335, 661–664, Fig. 21.1. Copyright 2007 Trans Tech Publications.*

decrease in the wear rate at a 1% CNT loading in comparison with the copolymers [103]. When the amount of CNT loading was increased to >1%, the effect of CNTs on the tribological performance of the nanocomposites was not significant due to the agglomeration of CNTs. In another study, Giraldo et al. [104] showed the effect of the diameter of CNTs on the tribological properties of polyamide–CNT nanocomposites. With the decrease in the CNT diameter, the interfacial area between the CNT and the polymer matrix increased resulting in a better interaction and enhanced tribological performances. The favorable tribological performance of these nanocomposites can be attributed to either one or more of the following phenomena [102,103,105,106].

1. The incorporation of CNTs with a high aspect ratio enhances the mechanical stability. This increases the load-bearing capacity of the nanocomposites and resists the polymer structure plastic deformation and greatly reduces the adhesive and plow wear.
2. Because of its tubular structure, CNTs in the interface between two mating surfaces act as a spacer thereby effectively reducing the contact with the counterface.
3. In addition, CNTs possess excellent self-lubricating properties based on the sliding/rolling effect in the interface between mating surfaces.

Although CNT-based nanocomposites show potential performance, they are still in the early stages of tribological research, and only few studies have been carried out on the application of these nanocomposites as lubricating coatings. Samad and Sinha [107] studied the dry sliding and boundary lubrication performance of ultrahigh-molecular-weight polyethylene (UHMWPE)–CNT nanocomposite coating on steel substrates for a possible application in bearings and gears. Recently, Ribeiro et al. [108] examined the tribological and biomimetic performances of polyimide–CNT nanocomposites for cartilage replacement applications. The current research on CNT-based polymer nanocomposites focus on the following two areas [109]: (1) control of the alignment of the CNTs and its dispersion and (2) enhancement of interfacial interaction between CNTs and the surrounding polymer matrix.

Since the geometric structures of individual CNTs are highly anisotropic in nature, the tribological properties of CNT–polymer nanocomposites are orientation dependent. The frictional behavior of the nanocomposites can be engineered by either aligning the CNTs vertically to the substrate or dispersing flat. The vertically aligned nanocomposites show a high coefficient of friction, whereas the transversely aligned system shows the least coefficient of friction [110,111]. Thus, the tribological behavior of the CNT–polymer nanocomposite system can be manipulated according to the needs. Also, as in any nanocomposites, the dispersion of the nanotubes and their interaction with the polymer matrix play a critical role in their tribological performance. CNT–polymer nanocomposites can be formed via (1) solution processing [112,113], (2) melt processing [114,115], and (3) in situ polymerization [116–118]. This can lead to secondary interaction between polymer and nanotubes or nanotubes covalently bound to the polymer matrix. Strong interfacial interaction with the polymer matrix can be achieved by functionalizing the CNTs, which makes them more compatible with both the polymer host and solvent [119–122]. Lee and Rhee [123] showed that the treatment of CNTs with silanes enhanced the dispersion and interfacial strength in epoxy/CNT nanocomposites resulting in the improvement of tribological properties of the nanocomposite. Also, Zhang et al. [124] studied the tribological behaviors of PTFE/nomex fabric/phenolic composites reinforced with functionalized and unfunctionalized CNTs. They showed that the nanocomposite-containing PS-grafted CNTs outperformed the nanocomposite-containing unfunctionalized CNTs in pin-on-disc type wear tests. Similarly, Kim et al. [125] used PMMA-grafted CNTs as fillers in polyimide-based nanocomposite coatings on aluminum surfaces to enhance the interfacial interaction between the polymer matrix and CNT, thereby improving the tribological performance. Another possible way of increasing the interfacial interaction is by utilizing the nanomechanical interlocking of the nanotubes by changing their morphology and their configuration [126].

### 21.3.7 Graphene-based nanocomposites

Graphene is a flat monolayer of $sp^2$-hybridized carbon atoms packed into a two-dimensional honeycomb lattice. As shown in Fig. 21.15, it is the basic building block

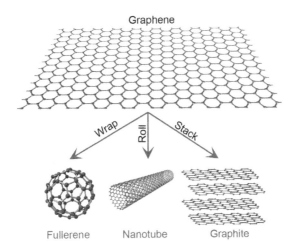

Graphene

Wrap    Roll    Stack

Fullerene    Nanotube    Graphite

**FIGURE 21.15**

Schematic showing that graphene is the basic building unit of fullerene, CNT, and graphite.

of all graphitized structures including zero-dimensional fullerenes, one-dimensional nanotubes, and three-dimensional graphite. Since the experimental discovery of single-layer graphene by Geim and Novoselov [127] in 2004, research on graphene has exploded exponentially and is currently one of the most studied materials among the materials science community. In addition, graphene exhibits advantages over other carbonaceous materials (including CNT) such as the ease and scalability of preparation methods at a relatively low cost and larger surface area, while maintaining similar or better properties. It possesses remarkable electronic, optical, mechanical, and thermal properties [128,129] and can be prepared via mechanical exfoliation, chemical vapor deposition, chemical exfoliation, and epitaxial growth. Although, mechanical exfoliation, chemical vapor deposition, and epitaxial growth can produce graphene with high quality for fundamental studies, large-scale production of graphene using these methods is not feasible. In addition, the hydrophobic nature of graphene along with its strong tendency to agglomerate in solvents presents challenges in the fabrication methods and severely restricts its applications. On the other hand, chemical synthesis of graphene oxide can be scalable at a relatively low cost and can also provide graphene with processability and new functions [130]. The presence of reactive oxygen functional groups in GO allows further chemical modifications, such as grafting of some functional molecules to enhance dispersibility and impart certain other properties.

Such a facile and scalable synthesis of GO has led to its application in various fields such as energy storage and conversion, optoelectronics, sensors, catalysis, electrochromic devices, and microelectronics. Also, graphene has been incorporated into polymer matrices such as polystyrene [131–133], epoxy [134,135], polyaniline [136,137], polyethylene [138], polyimide [139,140], and polycarbonate [141] matrices to improve the mechanical, tribological, thermal, and electronic conductivity of the polymer nanocomposites. Typically, such nanocomposites are prepared by solution blending, melt blending, and in situ polymerization and can be used in a variety of applications including tribological applications. As of with any other nanocomposites, dispersion and interaction of graphene nanosheets within polymer matrix plays an important role in determining its tribological behavior. Both covalent and noncovalent functionalization methods of GO have been established to improve the dispersibility of graphene in the polymer matrix. For example, GO/nitrile rubber composites were prepared by solution mixing and the prepared composite showed improved tribological properties in both dry and wet conditions [142]. It was reported that both frictional coefficient and wear initially reduced significantly with the increase in GO content, then increased with increasing GO content in dry sliding conditions. In the case of wet conditions, the frictional coefficient and wear rate decreased continuously with an increase in GO content. Also, Liu et al. [140] showed that the addition of GO in the polyimide matrix improves its friction and wear properties with the optimum GO addition of 3 wt%. It was proposed that the reduction in friction and wear could be attributed to the formation of transfer film, which reduce the contact between the counterface and the sample surface and the enhancement in load-bearing capacity. Recently, poly(sodium 4-styrene sulfonate)-mediated graphene and polyethyleneimine were electrostatically self-assembled into ultrathin layer-by-layer (LBL) structures [143], which exhibited a high load-bearing ability and long wear life. Even though such noncovalent interactions result in the homogeneous dispersion of graphene, it still exhibits poor stress transfer and slippage between two phases due to the weak interactions.

Ou et al. [144] studied the self-assembly of octadecyltrichlorosilane on graphene oxide using covalent interaction and the tribological behavior of the resulting film on a silicon substrate. They showed that the film exhibited reduced friction and enhanced antiwear properties, which were attributed to the strong covalent interaction as well as the self-tribological properties of reduced graphene oxide. In another study, Shin et al. [133] used octadecyltrichlorosilane as a coupling agent between graphene oxide and polypropylene leading to strong interfacial adhesion between graphene and polymer matrix. This resulted in a 189% increase in scratch deformation resistance at a low loading of 1% graphene oxide. Although these graphene-based nanocomposites show much improved tribological performance, they are still in the early stages of research and need further understanding on the efficient preparation of polymer–graphene nanocomposites. Furthermore, oriented distribution of graphene in nanocomposites and their effective fabrication as thin films and coatings is necessary to exploit this material to commercial success.

## 21.3.8 Thin films and multilayer coatings

The chemical, mechanical, and tribological properties of the surface can be improved in an even sophisticated way by adopting a multilayer or LBL coatings method. The purpose of using this method [145,146] is to either accommodate multilayers with specific functions (such as corrosion protection, wear protection, thermal barrier, diffusion barrier, and adhesion to the substrate) for each layer or to use a large number of alternating thin layers with differing mechanical properties (such as high shear modulus and low shear modulus materials) to alter the stress concentration in the surface region.

In a confined gap between mating surfaces such as boundary lubrication, various types of molecular lubricants are used to reduce friction. These are Langmuir–Blodgett films, self-assembled monolayers (SAMs), grafted polymers or brushes, and thin films. Table 21.1 gives a comparison of the nanotribological properties of various types of boundary and molecular lubricants [147]. Such molecular coatings offer a protective layer for many micro/nanoelectromechanical systems. However, the dynamics of the thin coating in the nanometer-confined space depends on the normal and shear stresses developed. Thus, a compliant film, which can undergo large deformation, can only sustain and exhibit nanotribological properties. Recently, Tsukruk described a novel approach toward the fabrication of a superior nanotribological coating based on elastomeric nanocomposite monolayers capable of very large elastic and reversible local deformation [148]. This results in entropy-driven variation in macromolecular conformation with very little energy dissipation. Such film (<10-nm thickness) can be immobilized to the substrate using three layers: an elastomeric layer with low shear modulus, a reinforcement layer with high compressive modulus, and a functionally reactive interfacial layer (such as SAM) between the rubber layer and the substrate, as an anchoring layer, which prevents dewetting or depletion from the substrate.

The unique characteristics of these nanocomposite thin films or multilayers that differentiate them from boundary lubricants are the random coil compliant configurations with short-range order that enable them to adopt any intricate surface topography and undergo large-scale reversible deformation. Also, the toughness and integrity of such nanocomposite coating result from the nanodomain structure, which can enhance thermal stability of such layers. Specifically, when the local pressure or temperature increases, the tribochemical stability of such a molecular coating becomes critical. Thus polymer nanocomposite is a prospective strategy to overcome such a concern. Moreover, self-healing characteristics can be imparted to such nanocomposites with an embedded oily or liquid nanophase, resulting in oily or gel nanocoating, where the liquid phase is strongly bound to the solid phase by secondary valence interactions. Recently, an elastomeric block copolymer-based nanocomposite monolayers have been reported for use in molecular tribology. In a typical process, an epoxy-terminated SAM was used as the reactive anchoring interface between the substrate and a nanocomposite layer. As an elastomeric material, a functionalized styrene–butadiene–styrene triblock copolymer with a maleic anhydride content of 2% was used for anchoring through epoxy–anhydride interaction from a dilute solution of the polymer. A schematic of such LBL film is shown in Fig. 21.16. The thus-formed tethered material is thermally stable upto 420 °C and can form a nanostructure through nanophase separation and does not dewet from the surface after extended high-temperature

**Table 21.1** Comparison of Nanotribological Properties of Various Types of Boundary and Molecular Lubricants [93]

| Coating | Thermal Stability | Surface Coverage | Adhesion | Friction Coefficient | Shear Strength | Wear Resistance |
|---|---|---|---|---|---|---|
| Langmuir films | Poor | Nonuniform/uniform | Low | Very low | Very low | Very low |
| SAMs | Good | Very uniform | Very low | Very low | Very low | Low |
| Composite SAMs | Good | Nonuniform | Low | Very low | Modest | High |
| Polymer films, physisorbed | Poor | Fairly nonuniform | Low | Modest | Low | Low |
| Polymer films, chemically tethered | Good | Uniform | Low | Modest | Low | Modest |
| Elastomeric films, physisorbed | Poor | Fairly nonuniform | Modest | Low | Very low | Low |
| Elastomeric composite films, chemically tethered | Good | Very uniform | Modest | Very low | Very low | Very high |

treatment (>150°C) in contrast to the physisorbed polymer film of similar composition. The films exhibit a very low coefficient of friction and shear strength under normal loads (Fig. 21.17) stress concentration in the surface region. The molecular structure and segmental mobility of such thin film coatings within a few nanometers of the surface has thus a significant influence on its application as lubricants, and protective coatings.

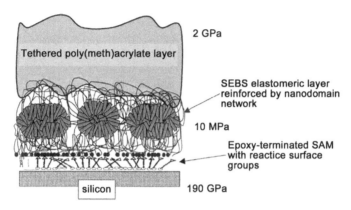

**FIGURE 21.16**

Trilayer architecture of multilayer coating with values of elastic modulus for different layers.

*This figure was published by A. Sidorenko, H.S. Ahn, D.I. Kim, H. Yang, and V.V. Tsukruk, Wear stability of polymer nanocomposite coatings with trilayer architecture, Wear 252 (2002) 946–955, Copyright Elsevier.*

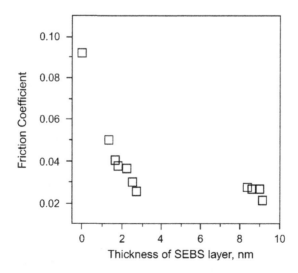

**FIGURE 21.17**

Friction coefficient vs layer of SEBS film.

### 21.3.9 **Polymer brushes**

Brushes based on block copolymers, covalently immobilized to a solid substrate, exhibit unique properties such as surface energy, friction, adhesion, and wettability due to wide-range morphology, that is, they can adopt under different conditions, and hence their properties can be tuned. Such a strategy can lead to the fabrication of wear-resistant robust compliant layers on any substrate or silicon surfaces. This can be performed via a simple double chemical self-assembling procedure from dilute solution followed by the spontaneous organization of the nanophase-separated systems. The tribological performance of such brushes can also be enhanced by exposing them to a solvent or environmental-induced assembly process.

The approach to the preparation of such an inorganic/polymeric composite brush from biodegradable and biocompatible constituents is a relatively new concept, and of interest particularly for tissue engineering and drug delivery applications. Recently, we described the synthesis of nanostructured porous silicon (pSi) and poly(L-lactide) (PLLA) nanocomposites. The composites were produced using tin(II) 2-ethylhexanoate-catalyzed surface-initiated ring-opening polymerization of L-lactide onto silanized pSi films. IR and XPS showed the successful functionalization of the pSi, by revealing C=O stretches and low atomic % of Si and Sn. The O/C ratio also confirmed the presence of a PLLA layer. Contact angle measurements

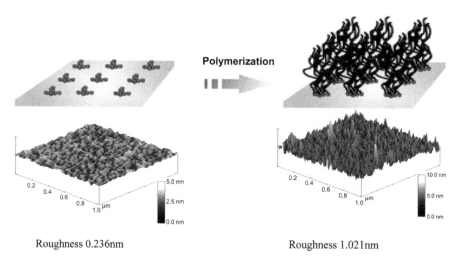

Roughness 0.236nm          Roughness 1.021nm

**FIGURE 21.18**

Hybrid from pSi. For color version of this figure, the reader is referred to the online version of this book.

*Reprinted with permission from S. McInnes, H. Thissen, N.R. Choudhury and N.H. Voelcker, Characterisation of pSi/poly(L-lactide) composites prepared using surface initiated ring-opening polymerisation in Proceedings of SPIE, BioMEMS and Nanotechnology II 6036, (2005).*

confirmed the presence of a PLLA layer, while AFM revealed the presence of PLLA nanobrushes (Fig. 21.18) [149].

## 21.3.10 **Hybrids**

Organic–inorganic hybrids prepared by LBL deposition of polyelectrolyte (PE) can be employed for superhydrophobic and lubricating coating of complex substrates and have shown promise to create hard coatings on surfaces. The properties of the film prepared by the LBL process, such as thickness, composition, ion exchange, surface morphology, can be tuned by varying the adsorption condition of the PE. The resultant film characteristics can be significantly influenced by factors like ionic strength, pH, and polymer molecular weight. This type of nanocomposite coating has the potential to create protective coatings with a greater strength and toughness, hardness than simple ceramic or polymer coatings. Recently, Cho et al. developed superhydrophobic surfaces by simple LBL deposition of 10 cycles of PAH [poly(allyl amine hydrochloride)] and PAA [poly(acrylic acid)]-coated zirconia nanoparticle and deposition of 1.5 bilayers of PAH and silica nanoparticles, followed by simple fluorination [150]. Many other nanocomposites have been reported in the literature based on poly(methylmethacrylate) [151,152], polyimide [153], poly(dimethylsiloxane), [154] etc. However, a very few nanocomposite materials are reported so far based on multiarm star polymers. Despite all their inherent advantages, the reports on such composite materials remain scanty. Recently, we reported star polymer–silicon carbide nanocomposites for abrasion-resistant application. The unique architecture of the star polymer with multiple compartments provides the ideal condition that drives these SiC particles to reside in the gaps between the arms and allows fabrication of thermodynamically stable polymer/SiC composites even at higher loadings (Fig. 21.19). This star polymer nanocomposite holds the potential for high abrasion-resistant coating due to the presence of SiC nanoparticles [155].

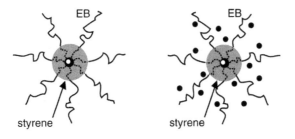

**FIGURE 21.19**

Schematic showing the SBC star polymer with SiC particles (dark dots within the EB arms).

*This figure was published by P. Majewski, N.R. Choudhury, D. Spori, E. Wohlfahrt, and M. Wohlschloegel, Synthesis and characterisation of star polymer/silicon carbide nanocomposites, Materials Science and Engineering: A 434 (2006) 360–364, Copyright Elsevier.*

## 21.3.11 Friction control in biomaterials using Polymer Nanocomposite

Tribology plays an important role in the human body in joints such as hip, knee, synovial joints, dental part, and jaw. Also, for artificial implants, vascular prosthesis, endoscopy, and catheter applications, where the device has to be placed or moved against the internal tissue, it is essential to consider the friction, lubrication, and wear properties. In all these cases, pain, injury, and after effects due to wear and friction pose a major clinical concern. Thus, tribology is the key parameter of consideration for the development of such advanced medical devices. A variety of prosthetic devices have been developed so far. However, simplified wear tests in which a polymeric specimen slides against a unidirectionally reciprocating ceramic or metallic counterface are not suitable to predict the wear of prosthetic joints. Although various types of low friction surfaces have been developed, most of them lose their activity in aqueous environments. UHMWPE has been widely used as a load-bearing material in orthopedic prostheses [156–158]. However, the functional lifetime of such devices is limited by adverse biological effects caused of their submicron-sized wear debris. In vitro simulation of its wear is thus critical in order to predict the role of wear debris particulates in osteolysis. Among many variables of wear studies, it is still not clear whether the selection of different lubricants, such as water or serum based lubricant, other parameter or biomaterial combination, could greatly influence the prediction. Thus, there is a distinct need for improved materials in artificial joints, which does not induce an osteolytic response in the body. The quest for more compliant materials is obviously a choice. In the human body, different types of cartilage found in different joints provides a stable movement with less friction than does any prosthetic replacement, and can alter its properties in response to differences in loading. The coefficient of friction values of normal joints are relatively low compared to those of any synthetic bearing due to a specific form of hydrodynamic lubrication, and also, boundary layer lubrication is another contributor of low-friction characteristics. In such a wet bioenvironment, unfilled polymer or conventional composite-based on carbon or glass fiber increases the volume of wear debris, enhancing third body wear in a joint, which often leads to a significant wear rate than that of two-body wear. Other simple mechanisms may be degradation of the polymer, creep, environmental stress cracking, etc. Thus, class II type organic–inorganic hybrid materials represent the ideal class of material that can be used to initiate bio-ingrowth of bone into porous surfaces without the use of any bone cement [159]. Recently, a new type of organic/inorganic nanocomposite has been developed through the use of a sol–gel method. This new material has the potential of an ideal bone substitute because it exhibits apatite-forming ability as well as mechanical properties comparable to those of human bone. However, it has the potential to leave behind the silica particles after the preferential degradation of biopolymer in vivo. In an attempt to reduce the silica phase, a novel strategy of material design has been proposed [159], which does not leave behind the silica particles after the preferential biodegradation of the polymer in vivo while not compromising other properties, a novel polycapralactone (PCL)–organosiloxane hybrid material was developed, and its apatite-forming ability in the SBF and tensile mechanical properties were evaluated. The schemes of PCL/silica nanocomposites and PCL–organosiloxane hybrids are given in Fig. 21.20. The advantage of the design of this novel hybrid material is that it does not contain silica particles at all, so it will not produce them even after the preferential degradation of PCL in vivo. Therefore, it is likely applicable to bioactive and degradable bone substitute material.

Friction control is also achieved on biomaterial's surfaces using a graft layer of functional polymer, whose surface properties could be changed by temperature, ionic density, or electrical potential. Lubrication mechanism and friction characteristics can also be induced using a thermosensing surface with a graft layer of a thermoresponsive polymer and its hydration force. For arterial bypass graft, microvessels, where high flow states primarily occur, the ability to withstand a high shear stress is the primary requirement of the selected material. Often, stiff and highly crystalline materials such as expanded PTFE are used; biocompatible nanocomposites may be alternatives in those applications provided the nanocomposites are biostable. However, for lower flow states, more compliant materials are needed for vascular prostheses and development of stents. PU, being a compliant material, is used for such applications. Recently, Kannan et al. [160] reported polyhedral POSS–PU nanocomposite, in which the silsesquioxane nanocores shield the soft segments of PU, strengthen it, and attribute compliance/elasticity and resist degradation. The authors highlighted the immense potential of the chemical integration of these nanocores

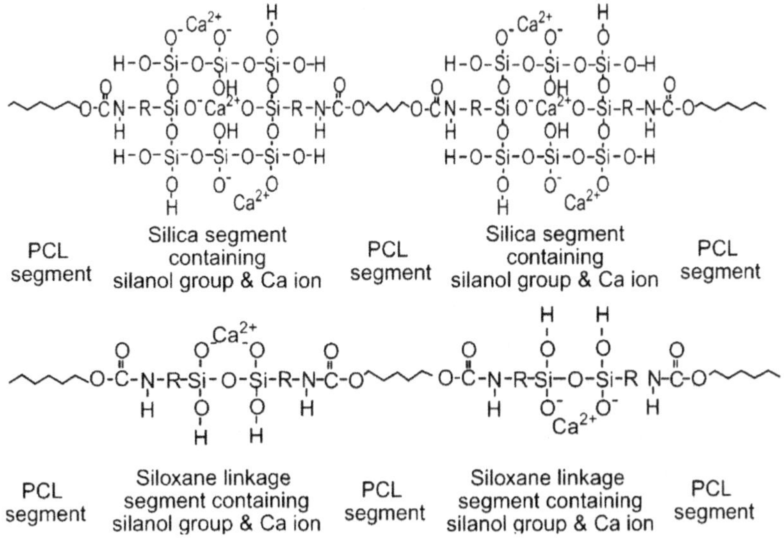


**FIGURE 21.20**

PCL/silica nanocomposite (top) and PCL–organosiloxane hybrid (bottom).

*Reprinted with permission from Biomacromolecules 2004, 5, 1575–1579. Copyright 2004 American Chemical Society.*

within the polymer matrix and optimization of the characteristics to make it suitable for biomedical devices. Additionally, incorporation of silicon repels platelet and fibrin adsorption and increased thromboresistance [161].

Polymeric materials are also used for optical purposes such as protective eyewear, ophthalmic, photographic, or other lenses. Unlike glass, they have a low abrasion resistance, and coatings are prospective candidates to address this issue. The coatings for such applications must possess optical functionality in addition to stringent physical properties. Thus, in ophthalmic lenses, scratch and abrasion-resistant functional coatings play a crucial role in the mechanical protection of the plastic lenses. Hybrid organic–inorganic sol–gel coatings offer outstanding mechanical properties to the lens material due to the dense inorganic–organic network in comparison to purely organic coatings. In order to obtain a high degree of condensation of the inorganic part, the hybrid network is designed to have the organic groups acting as short spacers between the inorganic domains. Such a design has developed commercial thiol-ene-based protective coatings (scheme below) on PMMA lenses [162].

$$-O-\overset{|}{\underset{|}{Si}}\diagup\!\!\diagdown\!\!\diagup\!\!\diagdown SH \; + \; H_2C\!\!=\!\!\overset{H}{\underset{|}{C}}\!-\!\overset{|}{\underset{|}{Si}}\!-\!O- \; \longrightarrow \; -O-\overset{|}{\underset{|}{Si}}\diagup\!\!\diagdown\!\!\diagup\!\!\diagdown S\diagdown\!\!\diagup\overset{|}{\underset{|}{Si}}\!-\!O-$$

Figure 21.21 shows the tumble test results of such uncoated and coated lenses (CR 39®) with a titanium boehmite nanoparticulate nanocomposite. Thus, the boehmite (AlOOH) nanoparticulate system (microhardness 306 MPa) is seen to protect the lenses better (Fig. 21.21) from mechanical wear and abrasion than does the molecularly dispersed oxotitanate clusters in other systems (Microhardness 260+/−10 Mpa). Also, another area of application is antireflective coatings on glass based on porous silica based sol–gel thin films. Achieving the abrasion resistance of such embossed coatings on curved ophthalmic lens surfaces still remains a challenge. However, for lenses with switchable photochromic activity, dyes (with chromophore) covalently

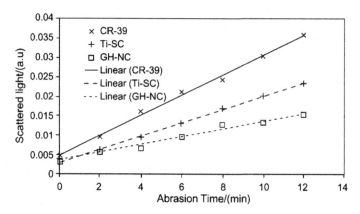

**FIGURE 21.21**

Haze of hybrid sol–gel coating on polymer lenses with abrasion time.

linked to inorganic networks represent a prospective approach to achieve high photo-chromic activity without compromising the coating's mechanical integrity.

### 21.3.12 Stimuli-responsive 'smart' polymer nanocomposites

Stimuli-responsive polymers and their composites are receiving increased atten-tion because of their ability to undergo conformational or chemical changes when exposed to external triggers such as pH, temperature, ionic strength, stress, or light. These changes are accompanied by variations in the physical properties of the sys-tems. They can be used in drug delivery applications, actuation, sensing, microfluidic devices, smart surface coatings, and biological substrates [163]. In the last decade, various sets of stimuli-responsive nanocomposites that comprise a combination of various polymeric systems with different nanofiller materials have been developed. These materials rely on various mechanisms such as modulation of filler–filler or filler–matrix interactions upon external stimuli and also could be matrix-dominated responsive systems. Various polymeric systems such as poly(methacrylic acid), poly(acrylic acid), poly($N$-isopropylacrylamide), and poly[2-(dimethylamino)ethyl methacrylate] have been used to prepare stimuli-responsive nanocomposites with fullerenes, CNTs, and cellulose nanofibers. Although, this type of material can be utilized in various applications, this section will focus on the application of stimuli-responsive materials for smart coatings.

## 21.4 FUNDAMENTALS OF POLYMER NANOCOMPOSITE TRIBOLOGY

### 21.4.1 Friction

Friction is a complex phenomenon measured by the coefficient of friction. In order to understand the significance of lubrication, it is necessary to gain an understand-ing of the phenomenon of friction and wear in relation to its origin. Like friction in polymers, friction in polymeric nanocomposites can be assigned to two sources; deformation term, relating to energy dissipation, and an adhesion term, which arises from the interface between the contacting surfaces. Several researches have demonstrated the influence of the chemical composition of the interfacial species and the morphology on the energy dissipation characteristics and hence frictional properties. Even under ambient conditions, surface oxidation can occur and change the tribological interfaces, which eventually lead to a significant change in fric-tional properties.

#### *21.4.1.1 Friction due to deformation*

When a hard sphere rolls over a lubricated polymer surface (where interfacial adhe-sion is minimum), the resistance to rolling originates from energy dissipation in the immediate layer of the polymer just underneath the sphere and is a direct result of its viscoelastic response. Relative motion continues on such lubricated surfaces due to minute displacement of the macromolecular chains. As the sphere rolls over, the macromolecules in its track become progressively deformed. Such a deformation depends on the interfacial interaction and can either be recoverable or elastic or permanent (plastic), in which case it incurs a loss of energy, and hence, the coefficient of friction changes due to deformation. The important parameter that characterizes the interfacial behavior of a material is the surface energy $\gamma_s$. It can be estimated from the contact angle measurement of various liquids on a given surface. The difference between the surface tension for wetting between the coating and the substrate, called surface energy mismatch, gives an estimation of the interfacial tension between them. Interfacial tension determines the degree of adhesion of the coating to the substrate; lower mismatch gives better adhesion. Thus, the viscoelastic behavior of a polymeric nanocomposite coating can significantly influence its lubrication/friction behavior in dynamic applications.

### 21.4.1.2 Friction due to adhesion

When two relatively smooth bodies slide against each other, with negligible contribution from friction due to deformation, the net frictional force originates from adhesion between the two surfaces. The adhesion component responsible for friction results from weak bonding forces such as van der Waals interaction, hydrophobic interaction, hydrogen bonding, etc. between two surfaces. Consequently, the composition of the outermost surface layer dictates the properties of the surface and magnitude of these forces defined by an important parameter surface energy.

## 21.4.2 Wear

Wear is the progressive loss of material from a solid surface due to mechanical action. In a polymer nanocomposite coating, although deformations occur in the polymer layer, the surface finish of the hard component will strongly influence the wear mechanism.

### 21.4.2.1 Interfacial wear

If the counterface is hard, wear will result from the abrasion between two surfaces, while with a smooth counterface, wear can result from various types of adhesion (electrostatic, van der Waals) with the deformation occurring primarily on the surface layers with the transfer of polymer to the harder counterface and generation of wear debris. In some cases, polymer chains undergo orientation in the direction of sliding. In polymer nanocomposites, the physicochemical nature of the nanoparticles gives rise to less abrasive characteristics and hence significantly reduces wear.

### 21.4.2.2 Cohesive wear

This type of wear originates from the deformation of the surface and subsurface layer of a coating caused by some counterface irregularity. It may eventually lead to abrasive type or fatigue type wear. When a rough hard surface rolls over a smooth surface, abrasive wear results leading to the plowing of the soft phase, which subsequently acts as loose wear particles and again contributes to abrasive wear. In applications involving repeated or cyclic stresses, the situation is rather different undergoing elastic deformation and causing fatigue wear. It starts with the initiation of a crack that propagates due to repeated deformation; finally, the material is fragmented from the surface.

### 21.4.3 Contact mechanisms

The tribological process of a contact involving surfaces in relative motion is very complex because it simultaneously involves friction, wear, and deformation mechanisms at different scale levels and of different types. In order to achieve the required tribological performance, it is essential to have a better and more systematic understanding of the mechanisms involved in a tribological contact. To establish the systematic understanding of the complete tribological process and to understand the interactions between mating surfaces, it is useful to analyze separately the tribological changes at five different scales/types: the macroscale and microscale mechanical effects, the chemical effects, the material transfer taking place during motion and the nanomechanical effects [146], as shown in Fig. 21.22.

The macromechanical and micromechanical effects describe the tribological phenomena by considering the whole contact area and asperity-to-asperity level contact, respectively. The interaction between the moving surfaces changes the surface condition such as the temperature and nature of surface, which can lead to tribochemical reactions. The chemical composition change that takes place as a result of tribochemical reactions influences the surface properties and its tribological properties. For example, the tribofilms formed during interaction on the counterface acts as a buffer layer between the surfaces in motion. This leads to the self-lubrication of the mating surfaces, and the self-lubrication process occurs in three stages as explained in Table 21.2. The first stage is the chemical modification of the surface and tribofilm embryo formation, followed by tribofilm formation in the second stage. The formed tribofilm may undergo deformation and may involve ejection of some materials as debris particles. In order to sustain this loss, the continuous formation of tribofilm is essential, which determines the life of the self-lubricating coating. The wear particles liberated from the interacting surface can have beneficial or detrimental effects on the subsequent tribological process depending on its hardness and shape in relation to the mating surfaces. For instance, if the wear particle liberated is a nanosized spherical particle, it can undergo rolling motion thereby acting as a ball bearing and improving the tribological properties, or if the wear debris is harder than the coating, it can increase the wear and friction through plowing. The development of understanding of the contact process at the atomic scale is essential to understand the tribological phenomena at the fundamental level and thereby controlling the macroscale lubrication.

### 21.4.4 Molecular dynamics simulations

On an atomic scale, tribological properties of polymer nanocomposites can be determined from molecular dynamics simulations. The approach is based on classical

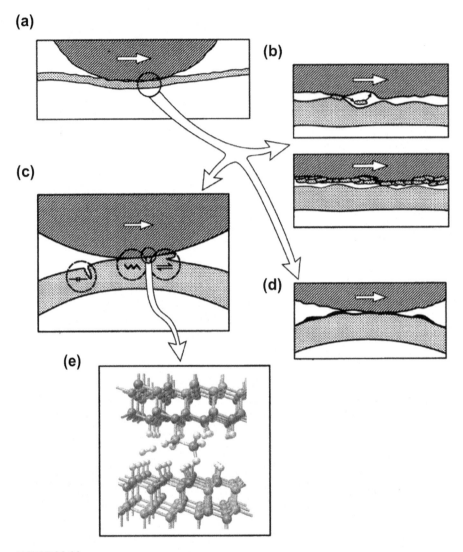

**FIGURE 21.22**

Tribological contact mechanisms. (a) Macromechanical, (b) material transfer, (c) micromechanical, (d) tribochemical, and (e) nanomechanical.

*This figure was published by K. Holmberg, A. Matthews and H. Ronkainen, Coatings tribology—contact mechanisms and surface design, Tribology International 31 (1998) 107–120, Copyright Elsevier.*

mechanics, thermodynamics monitoring motion of atoms with respect to time. While numerical solutions allow the motion of atoms to be calculated, it is important to first determine the interatomic forces (hence atomic motion) from the potential energy function. Thus, the derived results can then be compared with similar dimension

**Table 21.2** Stages of Self-Lubrication Mechanism for the Ideal Tribofilm

| Stage in the Self-Lubrication Process | Content |
|---|---|
| Chemical formation of tribofilm embryo | (a) Vicinity of the surface is chemically modified with little change in its physical properties |
| | (b) This modified zone is nonequilibrium to further change itself to a buffer layer in its usage |
| In situ formation of tribofilm | (a) Tribofilm is in situ formed from its embryo in the wear track or on the contact surface between surface and work materials |
| | (b) Due to its elastoplastic deformation, the frictional stress is relaxed |
| Sustainability of self-lubrication | (a) A new tribofilm is continuously formed on the contact surface to preserve low wearing and friction state |
| | (b) This sustainability is held on till the end of life limit |

*This table was published in Surface and Coatings Technology, 177–178, T. Aizawa, T. Akhadejdamrong, A. Mitsuo, Self-lubrication of nitride ceramic coating by the chlorine ion implantation, p. 573–581, Copyright Elsevier, 2004.*

experimental results such as those of AFM, which has different a time scale of measurement. Such a difference can lead to a substantial difference in results (a few orders of magnitude difference).

## 21.5 LUBRICATION MECHANISM IN NANOCOMPOSITE COATINGS AND THIN FILMS

Interfacial interactions between components in macroscopic composite create an interphase, whose properties are nonlinearly related to the properties of bulk components. The interphase is of the dimension of a few nanometers. However, in nanocomposites, the interaction is more pronounced due to the size resulting in higher interfacial surface areas, and the resultant material consists completely of an interphase. Thermodynamically, thus, there is only one phase in a nanocomposite [164]. It is widely known that nanoparticle can reduce friction and can have a ball bearing effect. While friction is related to adhesion, the exact relationship between friction and adhesion is yet to be understood. Although high adhesion energy between two surfaces can lead to high friction forces, often, the dissipative component of adhesion known as adhesion hysteresis can significantly influence friction. On the microscopic and molecular levels, friction is thus well represented by energy dissipation. Therefore, to reduce friction, the amount of energy dissipated should be diminished. Thus, the performance of a polymer nanocomposite as lubricant will depend on its dynamic properties and interaction with the shearing surfaces. The knowledge of fundamental dynamic properties of the coating or thin film and shear ordering of the nanoparticles, especially if anisotropic, is essential to understand the basic mechanism of friction and lubrication. Experimental data probing molecular level information and theoretical data can determine the mechanism of dissipation and establish correlation of energy loss with molecular events. For example, AFM has been very powerful to help measure frictional properties of SAMs of adsorbed alkyl trichlorosilanes on silicon surfaces, where alkyl chain length dependence has been attributed to the interplay of monolayer packing energy and local deformation of the film [165,166]. A similar effect could be envisioned in very thin film nanocomposite lubrication. The presence of interface introduces thermodynamic factors that can alter the morphology near the interphase. The understanding of the influence of the interface on the properties of the nanocomposite still needs an in-depth investigation, although it is well known that many stress-relieving processes, for example, adhesion, sharing of loads, or stress transfer occur at the interface. At the interface between a glassy and a rubbery substrate, some degrees of interpenetration of the constitutive chains can occur. Below the glass transition temperature or in crystalline state, the chain segments are in a solidlike frozen state; therefore, not much interdigitation and ordering can happen, and therefore, low friction can be observed at this stage. While in the amorphous state or above the glass transition temperature, a nanocomposite system can exhibit higher friction in comparison to that in its glassy state due to the amorphous chain interdigitation and entanglement during sliding, which is not easy to disentangle,

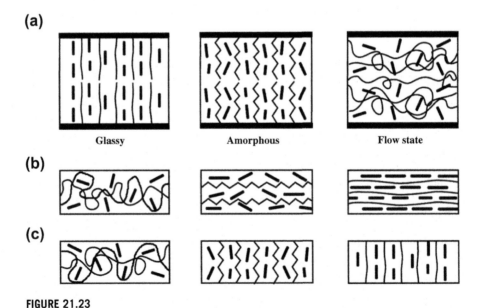

**FIGURE 21.23**

(a) Low friction nanocomposite surfaces: Dynamics of the system; (b), (c) switchable nanocomposites: ordering of anisotropic nanofillers.

hence resulting in higher energy dissipation. Figure 21.23(a) shows the behavior of such nanocomposite coatings under different states. At high temperatures, even if the degree of local chain interdigitation becomes high, the chains can readily disentangle. Thus, segmental mobility and interpenetration, local change in interfacial structure, can lead to changes in the ordering of the interfacial region on contact or during sliding. This becomes even pronounced with anisotropic nanoparticles such as clay or carbon nanotubes. The sliding/rolling of such nanoparticles act as a nanobearing between mating surfaces leading to the reduction in the friction and wear [167–170]. As polymer nanocomposites properties share a direct relationship with their chemical structure and morphology, the creation of frictional heat during sliding can lead to a change in the ordering of their morphology (Fig. 21.23(b), (c)). Such adaptable nanocomposite coating obtained via morphology tuning can reduce frictional forces acting within the system. The ability of such stimuli-sensitive nanofillers to undergo quick controllable change of position, wettability, or orientation offers tremendous potential to develop a switchable or targeted system. However, a major challenge with such colloidal anisotropic materials is orientation or organization in a highly viscoelastic medium. Preferential encapsulation of the anisotropic nanofiller in unique micelle-forming block copolymer and organizing them above the glass transition temperature of one of the phases may represent a powerful technique to overcome such problem.

## 21.6 CRITICAL PARAMETERS IN LUBRICATION
### 21.6.1 Surface preparation

Surface preparation is a crucial step to achieve strong adhesion of the bonded coating to the substrate. Often, the substrate is acid or alkali washed or grit blasted and coated. However, it is important to verify that the surface has the desired chemical composition. Also, rough surfaces will reduce the real contact area, although the asperities may be subject to abrasive or fatigue wear.

### 21.6.2 Nature of solid surface
#### 21.6.2.1 Molecular characteristics of the surface

The electronegativity of the bonds present in a system often determines resultant friction in that material, for example, aliphatic carbon–hydrogen or carbon–fluorine bonds in a self-assembled or ordered form offers lubricity, while aromatic structure produces a high coefficient of friction and adhesion. This factor gets compounded in a polymer as the molecular weight of the polymer increases.

### 21.6.3 Hardness of the coating

An important parameter in lubrication is the coating hardness and its relationship to the substrate hardness. Soft coatings reduce sliding-originated surface tensile stresses, which contribute to undesirable subsurface cracking and subsequently to severe wear. A hard coating on a softer substrate can decrease friction and wear by preventing asperity penetration through the coating both on a macroscale and a microscale.

### 21.6.4 Thickness of the coating

In the case of soft coatings, the thickness of the coating affects the degree of asperity penetration through the coating into the substrate, whereas a thick, hard coating can assist a softer substrate in carrying the load and can thus decrease the contact area and the friction. Thin hard coatings on soft substrates are susceptible to coating fracture or delamination because of stresses caused by substrate deformation.

### 21.6.5 Shear strength of the coating

The performance of a lubricating coating depends on its dynamic properties of the coating and its interaction forces with a shearing surface. The knowledge of shear properties of thin film coatings confined between two solid surfaces is crucial for understanding the basic mechanism of friction and lubrication. A surface force apparatus (SFA) is a powerful tool for measuring shear forces as both the area of contact and the distance of separation can be simultaneously determined through interferometric fringes. It can provide knowledge on the load-bearing capability and shear ordering of the confined film.

### 21.6.6 Coating/substrate interfacial adhesion

The interfacial adhesion between coating and substrate is one of the important factors that determine the tribological properties of the coating. Poor interfacial adhesion leads to delamination of the coating and is easily expelled from the working interface, thereby not serving the desired purpose of lubrication. Adhesion of the coating to the substrate is greatly influenced by surface preparation methods and the type of bond established. Formation of wear debris at the interface can often lead to the delamination of the coating.

### 21.6.7 Factors influencing lubricating coating performance

Several factors that determine coating performance are as follows:

#### 21.6.7.1 Porosity/defects

The major factor that undermines the performance of the coating is the presence of pores or defects. The presence of pores/defects increases the permeability of oxygen and water through coating, thereby reducing its strength, hence performance. Therefore, it is very crucial to form a coating free from pores and defects to achieve high barrier properties.

### 21.6.7.2 Multihydrogen bonding

Larger arrays of multiple hydrogen bonds that serve as physical crosslinks are useful building blocks for the reliable assembly of supramolecular structures, resulting in materials with improved properties [171]. This hydrogen-bonded supramolecules can be used as high-performance materials with their tunable and reversible characteristics. Hydrogen bonds increase the interchain cohesion and thereby reduce the free volume and flexibility of the polymer segments. This enhances the thermal and barrier characteristics of the coating. Oaten et al. [16] recently synthesized urethane–silsesquioxane-based hybrids for thin film application and showed that they possess better mechanical properties not only because of covalent link between the constituents but also due to intramolecular hydrogen bonding and strong van der Waals interactions. Such multiscale hydrogen bonding could be very powerful to enhance coating–substrate adhesion/orientation and hence performance.

### 21.6.7.3 Glass transition temperature ($T_g$)

The tribological property is highly influenced by the glass transition temperature of the coating. If the glass transition temperature is below the service temperature of the coating, molecular segment movement starts, leading to increased free volume, thereby increasing the dissipation. Therefore, it is important to use the right materials with a high glass transition temperature but either reinforced with nanoparticles or with crystalline regions to obtain high tribological performance.

### 21.6.7.4 Crosslinking and dispersion of nanofillers

Higher crosslinking and uniform dispersion of nanofillers lead to dense coatings, free of pores. They reduce the free volume and segmental mobility of the molecules thereby reducing the mobility of the coating. The effect of crosslinking on the tribological property depends on the type and extent of crosslinking. Therefore, a dense network system is desirable as tribological coating.

Apart from these factors, other factors, such as mechanical properties, viscoelasticity, hydrophobicity, and surface energy of the coating, can also influence the coating performance depending on the type of application.

## 21.7 FRICTION IN POLYMER NANOCOMPOSITE PROCESSING

The strong interest in nanoparticles and anisotropic nanofillers, such as clay or CNTs, stems from their ability to influence the overall properties of these nanocomposites at relatively low concentration. However, their use requires an understanding of how processing influences their organization and hence performance. The presence of such filler networks in a polymer matrix induces a large contribution to viscoelasticity. Their nanometer size dimension facilitates their deformation under shear despite their rigidity. Thus, in polymer nanocomposite processing, the friction characteristics of the mold and the plastic materials used in the production of parts is of great significance. Often, the tight tolerance of molded parts along with thin walled moldings leads to the production of deformed parts during ejection, increasing the number of reject and cost. One way to alleviate such a problem is to reduce frictional forces through easy part release. Friction has also other adverse effects. It creates heat and hence causes damage of the product and reduces performance and efficiency. The frictional heat also changes the orientation of the macromolecules present at the surface, leading to a high surface energy. The surface finish of the mold invariably determines the surface finish of the molded product. The wear resistance of a mold is a function of the mold material, its metal grade, heat treatment condition, microstructure, hardness, and the item being molded. The friction coefficient between the nanocomposite melt and the mold is one of the key controlling factors for surface defect formation. Friction creates excessive heat and produces surface damage such as wear, scuffing, pitting, and fretting.

While friction is independent of the apparent area of contact between the contacting bodies, the frictional force is proportional to the normal load between the bodies. The coefficient of friction ($\mu$) for a system for wear analysis is the ratio of frictional (tangential) force $F_t$ to the normal (vertical) force $F_n$ (Eqn (21.1)).

$$\mu = \frac{F_t}{F_n} \tag{21.2}$$

A high level of adhesion between the melt and mold can produce a problem in nanocomposite processing. This phenomenon is crucial in understanding flow involving high shear stress where factors like wall slip plays a dominant role. Even with a nanocomposite melt that completely wets the surface, the slip can occur due to the dynamics of long chains that glide past each other. Many problems in polymer processing and lubrication involve the use of additives to the polymer, which specifically modify the rheological and slip behavior at the wall or use a special category of tool materials, or both. In a recent study, we have investigated the wear and friction behavior of carbon nanoparticle-filled/rubber modified thermoplastics using a two-body abrasive wear tester, for example, a pin-on-disc apparatus where a pin is of the same material as the metal mold and the disc or the counterface is made of the polymer nanocomposite to be tested. The pin, in contact with the counterface, is held stationary with an appropriate load while the polymer sample is rotated. The weight loss experienced by the disc is used as a measure of the wear resistance to give the wear rate.

$$\text{Wear Rate} = \text{volume loss/sliding distance}(\text{mm}^3 \text{ mm}^{-1})$$

The wear rate of all samples show a strong dependence on both the applied load and temperature to the amount of wear taking place (Fig. 21.24). This is thought to be due to the lack of polycarbonate. However, this is more pronounced if the temperature of testing is close to the $T_g$ of one of the polymer phases. Of the two samples, the one without carbon black generally showed the greatest wear. It can be seen that the sliding speed also has a major effect on the coefficient of friction for all the test materials (Fig. 21.25). The greater the hardness of the system, less plowing occurs during sliding, hence a lower value of coefficient of friction.

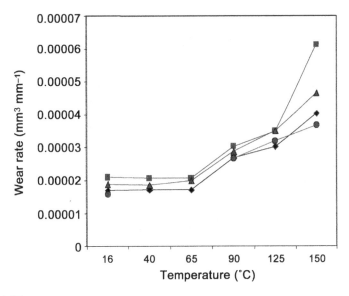

**FIGURE 21.24**

Wear rate vs temperature: ■ unfilled sample, ▲ filled sample, ◆ polycarbonate-blended rubber modified plastic, and ·polycarbonate blend with small rubber particle modified plastic.

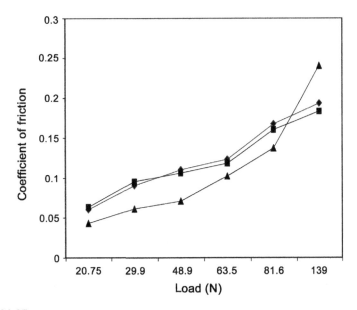

**FIGURE 21.25**

Coefficient of friction (16 °C, 370°) vs load ■ unfilled sample, ◆filled sample, and ▲-polycarbonate-blended filled sample. For color version of this figure, the reader is referred to the online version of this book.

It has been observed that the injection-molded samples show less wear than do the compression-molded samples due to densification and ordering. The unfilled sample owing to its smaller rubber particle size, and lack of carbon black as a reinforcing agent, showed the overall greatest wear rates. Filled rubber modified plastics when blended with polycarbonate showed the lowest wear rates over both load and temperature range. This is attributed to its high carbon black content, and most importantly the presence of polycarbonate in its structure. The polycarbonate provided the nanocomposite with a better strength and better temperature resistance over the temperature range of the experiments. In plastic packaging also, the friction coefficient plays the key role when the packaging film winds over the free running rolls, or product wrapping or bag formation.

## 21.8 TRIBOLOGICAL CHARACTERIZATION

The fundamental and applied studies of wear require quantification of wear volume, which is conventionally done by either measuring the weight loss after the tribology test or measuring the dimensional changes. However, these measurements are extremely difficult considering the small amount of worn-off volume in comparison to the worn component and the complexity of the tribological phenomena. Even small changes in the surface condition of the sample during tribology may significantly affect the performance of the surface before identifiable wear debris or wear scars form. Hence, it is important to monitor the surface condition prior to wear to establish a fundamental understanding of the tribological phenomena. It would be a difficult task to determine the onset of wear or study the surface condition prior to wear using conventional testing methods. Also, tribological studies at the nanometer level are essential for a fundamental understanding of the interfacial phenomena involving ultrathin films and for the further manipulations/modifications of such surfaces. This could be achieved with modern techniques, such as scanning kelvin probe (SKP), nanoindentation, and scanning probe microscopy, which includes atomic force microscopy (AFM), frictional force microscopy, and microthermal analysis (µTA). Coupling surface analysis to AFM is a powerful approach to investigate tribological interfaces as it can provide a direct relationship between the surface composition and interfacial forces. These techniques are reliable, relatively easy to use, allow one to study frictional and wear phenomena at the molecular level and can be used on any surface. Herein, the principle of operation of some specialized techniques and their application in nanotribology are presented.

### 21.8.1 Microthermal analysis

Microthermal analysis is a form of a localized characterization technique that is particularly useful for characterizing complex materials, such as nanocomposites, which are otherwise very difficult to characterize using already existing techniques. It has shown considerable promise for nanostructure/property characterization of complex materials and to detect its subsurface structure. It combines thermal analysis and imaging to reveal spatial variations in the thermal conductivity and diffusivity of the sample as well as surface topographic imaging by AFM [172–174]. It allows not only imaging the thermal conductivity and diffusivity of the surface but also any feature on the image at submicron scale can be characterized for its calorimetric or mechanical properties.

The source of thermal excitation is the near-field thermal probe itself, used both as a heater and as a temperature sensor and in the same time working as an atomic force microscope tip. For microthermal analysis, the resistive probe has proved to be the most useful device. The μTA also incorporates a cantilever whose deflection senses the force that acts between the probe and the sample, one or more feedback systems and actuators for spatial scanning (Fig. 21.26). The probe tip rasters over the sample surface and thermal conductivity images are obtained by measuring power required to maintain the tip at a constant temperature, whereas the thermal diffusivity images are obtained by applying modulation at isothermal temperature. The deflection of the cantilever with respect to the forces between the probe and the sample surface gives the topographic information.

Microthermal analysis allows the characterization of interaction of filler materials with the polymer matrix thereby providing an insight into the interfacial interaction and its relation to the resulting properties. Thus, the dispersibility of

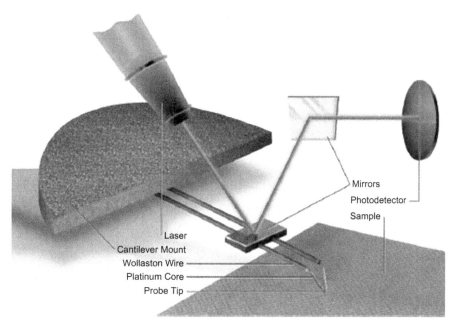

**FIGURE 21.26**

Schematic of the thermal probe. For color version of this figure, the reader is referred to the online version of this book.

*Reprinted with permission from TA Instruments, www.tainstruments.com.*

the filler materials in a polymer matrix, an important parameter in terms of nano-composites, can be studied using this technique. It also allows the determination of the thickness of thin films by creating a pyrolytic crater on the film to reach the substrate. The cross-sectional profile of the hole along the film gives the depth of the crater and thereby giving the film thickness [25]. The application of μTA to a tribological study was demonstrated on an injection-molded polycarbonate/acrylonitrile–styrene–acrylate blend system [175]. The topography and relative conductivity scans were performed on a specimen before and after the wear test using a pin-on-disc wear machine under a load of 50 N for 2 h. The results showed the interfacial wear mechanism deforming the rubber region leading to elongation in the direction of pin motion. In our recent work [176], we have mapped abraded and nonabraded areas by micro-TA. Figure 21.27 demonstrates the utility of such a tool to analyze the abraded area and shows the wear mapping of a polycarbonate–rubber-modified composite sample.

## 21.8.2 Atomic force microscopy

Among all the available microscopic techniques, AFM is the most suitable technique for the characterization of both the morphological and tribological properties of complex material surfaces with a nanometer-scale resolution. AFM can be used either in a contact or tapping mode depending on the sample properties and the experimental objectives. In the contact mode, a small probe attached to a cantilever is raster scanned across the surface and the cantilever deflection due to the force between the sample and the tip gives the morphological variations on the sample surface. In the case of the tapping mode, the probe tip is oscillated with an adjustable amplitude, which lightly taps over the surface. The change in the oscillation amplitude is used in the feedback signal for measuring the topographic variations across the surface. The AFM probe tip attached to the cantilever is a sensitive force sensor and can measure both the vertical and lateral force acting on the tip. This enables the simultaneous acquisition of both topographical and frictional mapping of the surface, which can

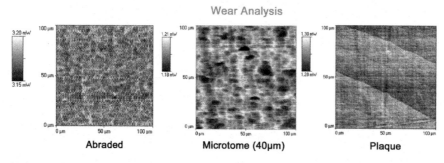

**FIGURE 21.27**

Wear mapping by micro-TA. For color version of this figure, the reader is referred to the online version of this book.

be used for the quantitative measurement of nanotribological and microtribological characteristics of the nanocomposite coatings. AFM has been widely used to study the surface features from microscale to nanoscale, and is increasingly used for the evaluation of tribological performance at atomic scale. This technique is useful to determine the surface roughness, mechanical properties, frictional force, and microscale wear mapping.

Apart from these capabilities, the influence of orientation of fillers, such as PTFE and CNT, on the tribological performance of the nanocomposites can be investigated using AFM. Liu et al. [177] used AFM to illustrate the dependence of friction coefficient of PTFE on the orientation direction and showed that the formation of a better friction-induced orientation results in a lower coefficient of friction. AFM is also capable of determining the height differences in the topographical images in subnanometer range, which helps one to quantify the wear volume [178]. Figure 21.28 shows the topographic and the corresponding bearing histograms of the steel surfaces before and after adding an extra Vickers diamond scratch. The difference between the bearing volume of steel before and after scratching gives the wear volume. This enables the quantification of minute wear confined within the

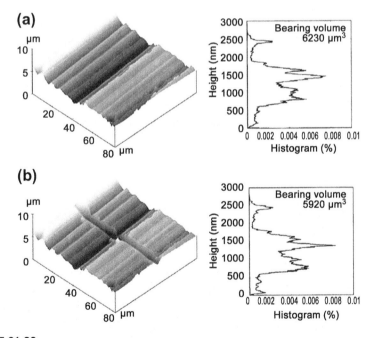

**FIGURE 21.28**

Topographic images and the corresponding bearing histograms of the steel surfaces before (a) and after adding an extra scratch (b).

*This figure was published by R. Gahlin and S. Jacobson, A novel method to map and quantify wear on a microscale, Wear 222 (1998) 93–102, Copyright Elsevier.*

original surface roughness, which is otherwise impossible using other techniques. Apart from the topographic and frictional force measurement, phase imaging using AFM can be used to detect and quantify changes in the composition across the polymer nanocomposites and molecularly thick lubricated surfaces [179]. In the case of fundamental tribological studies, it is essential to understand the contact mechanisms between the moving surfaces at the atomic scale to optimize the parameters to achieve desired friction and wear performance. This can be achieved using AFM tip, which allows the study of single asperity contact [180] and thereby characterizes the nanotribological properties. Hertel et al. [181] used AFM to manipulate the individual carbon nanotubes and their interaction with the surfaces, whereas Falvo et al. [169,170] used this technique to study the nanometer-scale motion of individual carbon nanotubes. This leads to the understanding of the relative motion of materials in contact for controlling macroscopic lubrication.

### 21.8.3 Kelvin probe technique

Even small changes in the surface condition of the sample during tribology may significantly affect the performance of the surface before identifiable wear debris or wear scars form. Hence, it is important to monitor the surface condition prior to wear to establish the fundamental understanding of the tribological phenomena. It would be a difficult task to determine the onset of wear or study the surface condition prior to wear using conventional testing methods. Due to its high sensitivity to the surface condition, the Kelvin probe (KP) can be used for this purpose.

The KP is a nondestructive, noncontact mode technique based on a vibrating capacitor to measure the surface work function distribution on the sample surface. The work function is defined as the amount of energy required to remove an electron from a material's surface. It is an extremely sensitive indicator of the surface condition and can be used to track changes in the surface such as surface contamination, corrosion, surface processing, chemical composition, and charge. This is based on the principle postulated by Lord Kelvin in 1898, which states that when two conductors are brought into electrical contact, a contact potential is generated across the two surfaces. Figure 21.29 explains the measurement of the work function difference between the scanning probe and the sample surface. Initially, the scanning probe and the sample exhibit different Fermi levels (Fig. 21.29(a)). When electrical contact is made between the scanning probe and the sample surface (Fig. 21.29(b)), charge flows from the surface with smaller work function to equalize the Fermi levels, which gives rise to surface charge and potential difference between the sample and the scanning probe. A backing potential $V_b$ (Fig. 21.29(c)) is applied to nullify the surface charge, which is equal to the work function difference between the sample surface and the scanning probe [182]. By using the scanning probes with an estimated work function, the absolute work function of the sample can be determined.

Li et al. [183] used the KP to study the kinetic response of the work function to wear phenomena and showed that the electronic work function of the surface decreased with time during sliding, whereas the surface roughness increased.

DeVecchio et al. [184] employed the KP to detect wear at extremely low loads where the wear debris or wear scars are absent. The authors demonstrated that this technique can be used in a range of materials including conductors, semiconductors, and insulators. Figure 21.30 shows the topographic and the surface potential maps of the wear region generated using a 1-$\mu$N load on the single crystal aluminum sample. The surface potential mapping using the KP shows a bright contrast on the worn region whereas the topographic image shows no detectable change in the worn region. Bhusan et al. [185] explained that the initial rise in surface potential is due to the removal of the surface contamination or oxide layer or lubricant during the initial one or two wear cycles. Further change in the surface potential with the increasing number of wear cycles is due to the subsurface fatigue wear resulting in the structural changes

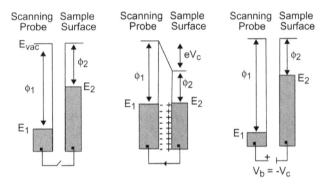

**FIGURE 21.29**

Schematic of the theory of KP measurements. For color version of this figure, the reader is referred to the online version of this book.

*Reprinted with permission from I.D. Baikie and G.H. Bruggink G.H., Characterization of oxides and thin film, Materials reliability in Microelectronics III, Ed. Kenneth P. Rodbell, William F. Filter, Harold J. Frost, Paul S. Ho, Materials Research Society Symposium Proceedings, vol. 309, Warrendale, PA, 1993, pp. 35–40.*

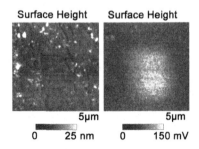

**FIGURE 21.30**

Surface height and surface potential maps of wear regions generated at 1 $\mu$N on a single crystal aluminum sample.

*Reprinted with permission from [184], Copyright (1998), American Institute of Physics.*

on the surface. Thus, this technique can be employed to study the wear precursor at extremely low loads which is almost impossible using other available techniques.

### 21.8.4 Nanoindentation

Indentation has been used for many years to evaluate the hardness of materials. This involves making an indent on the material surface and measuring the residual plastic impression in the specimen as a function of applied load, thereby measuring the area of contact of the indenter with the specimen. The hardness of the material is calculated from the Eqn (21.3), where $P_{max}$ is the load applied and $A$ is the area of contact.

$$H = \frac{P_{max}}{A}$$

(21.3)

With the recent advancements in nanoscale and microscale science, engineering and technology coupled with advances in instrumentation, indentation at the nanoscale combined with high-precision depth sensor, enables one to obtain other useful information such as elastic modulus, fracture toughness, strain hardening exponent, residual stresses, adhesion of thin films, and viscoelastic properties [186].

Nanoindentation is a powerful tool to measure the mechanical properties of the hybrid/nanocomposite coatings [187–189] and to determine the tribological behavior of these coatings [190]. Many nanoindentation instruments can be configured using a movable stage, to measure the scratch resistance and the coefficient of friction of thin films [191,192]. In the nanoscratch testing mode, a sharp tip is drawn across the sample under a specified normal force ($F_n$) to create a scratch of the desired length (lateral displacement). The corresponding lateral force ($F_t$) and the normal displacement are measured. The coefficient of friction ($\mu$) can be readily calculated from the normal and tangential force using Eqn (21.2). The measured normal displacement in response to the applied normal force is a measure of its scratch resistance. Nanoscratching has been extensively used to evaluate the friction, wear, and scratch behavior of the polymers and their nanocomposites. Dasari et al. [190] used a standard diamond indenter for scratch testing of a nylon-66-based ternary nanocomposite. They showed the effect of dispersion of fillers on the scratch behavior of the nanocomposite. Hur et al. [193] performed friction tests with the scratch mode of a nanoindenter and measured the friction properties as a function of friction speed. Nanoindentation tests can also be carried out at high temperatures to determine the mechanical and tribological properties of the nanocomposite coatings at which they are likely to operate [194]. This provides the opportunity to study the effect of temperature on the performance of these coatings at high temperatures.

## 21.9 APPLICATIONS

Polymer–inorganic hybrid and nanocomposite surfaces with improved tribological performances have significant potential applications ranging from microelectronic devices to machineries and moving parts such as pumps, engines, seals, bearings, power tools, automotive components, plastic and food processing machineries, aerospace, chemical process equipment, hardware, electronic components, screws, fasteners, plastic lenses, biomedical implants, low friction fibers for bandages, and sensors, membranes, biomedical devices.

## 21.10 FUTURE OUTLOOK

Polymer nanocomposites hold significant potential as green coatings for tribological applications in diverse fields. New functional possibilities can emerge from biomimetic self-organization of structures that result in completely new generation of lubricating materials. The key elements for controlled organization are the use of anisotropic building blocks such as nanotubes or clays and the introduction of additional functionality to build up the desired structures. The possibility of incorporating additional chemical functions on the surface of these particles will also provide a platform to control their organization during friction or sliding. This will give rise to a new generation of "smart" lubricating nanocomposites. It can be envisaged that such developments could lead to completely new avenues for fine tuning the nanocomposite and processing characteristics of the existing generation of polymer–inorganic nanocomposites. The success in fabricating such materials can be achieved only if the aggregation of the nanomaterials is avoided during application, or a high level of orientation or shear ordering can be achieved, which can impart many essential characteristics including low friction surface properties. However, understanding the lubrication mechanism and such shear ordering are also critical toward the development of such novel nanocomposite coatings. To understand the basic steps in lubrication, it is important to know the evolution of structure and morphology under lubricated conditions. This fundamental knowledge will significantly enhance our ability to develop new nanocomposites as well as their performance. Thus, the ability to develop switchable nanocomposites, which can undergo large-scale property or shape or morphology changes in response to an external stimulus such as stress, load, temperature, provides a unique platform to create materials with a unique combination of tribochemical and mechanical properties. Such nanocomposite coatings hold promise for successful lubrication in unusual situations, for example, high or low temperature, under vacuum, in nuclear applications, under extreme load or chemically reactive environments.

## References

[1] G. Byrne, E. Scholta, Environmentally clean machining processes—a strategic approach, Annals CIRP 42 (1993) 471–474.
[2] A.A. Voevodin, J.S. Zabinski, Nanocomposite and nanostructured tribological materials for space applications, Composites Science and Technology 65 (2005) 741–748.
[3] R.L. Fusaro, Self-lubricating polymer composites and polymer transfer film lubrication for space applications, Tribology International 23 (1990) 105–122.

[4] D.L. Burris, B. Boesl, G.R. Bourne, W.G. Sawyer, Polymeric nanocomposites for tribological applications, Macromolecular Materials and Engineering 292 (2007) 387–402.

[5] G. Schmidt, M.M. Malwitz, Properties of polymer–nanoparticle composites, Current Opinion in Colloid and Interface Science 8 (2003) 103.

[6] M.Q. Zhang, M.Z. Rong, K. Friedrich, Polymer nanocomposites as candidates for tribological applications, Materials Science Forum 539-543 (2007) 842–847.

[7] L.S. Schadler, S.K. Kumar, B.C. Benicewicz, S.L. Lewis, S.E. Harton, Designed interfaces in polymer nanocomposites: a fundamental viewpoint, MRS Bulletin 32 (2007) 335–340.

[8] L.S. Schadler, Nanocomposites: model interfaces, Nature Materials 6 (2007) 257–258.

[9] K.I. Winey, R.A. Vaia, Polymer nanocomposites, MRS Bulletin 32 (2007) 314–322.

[10] T. Saegusa, Organic–inorganic polymers hybrids, Pure and Applied Chemistry 67 (1995) 1965–1970.

[11] Y. Gao, N.R. Choudhury, Organic–inorganic hybrid from ionomer, in: H.S. Nalwa (Ed.), Handbook of Organic–inorganic Hybrid Materials and Nanocomposites, American Scientific Publisher, United States, 2003, pp. 271–295.

[12] C.M. Sanchez, B. Julian, P. Belleville, M. Popall, Applications of hybrid organic–inorganic nanocomposites, Journal of Materials Chemistry 15 (2005) 3559–3592.

[13] G. Schottner, Hybrid sol–gel-derived polymers: applications of multifunctional materials, Chemistry of Materials 13 (2001) 3422–3435.

[14] H.J. Glasel, E. Hartmann, R. Mehnert, D. Hirsch, R. Bottcher, J. Hormes, Physico-chemical modification effects of nanoparticles in radiation-cured polymeric composites, Nuclear Instruments and Methods B 151 (1999) 200–206.

[15] H.-J. Gläsel, E. Hartmann, D. Hirsch, R. Böttcher, C. Klimm, D. Michel, H.-C. Semmelhack, J. Hormes, H. Rumpf, Preparation of barium titanate ultrafine powders from a monomeric metallo-organic precursor by combined solid-state polymerisation and pyrolysis, Journal of Materials Science 34 (1999) 2319–2323.

[16] M. Oaten, N.R. Choudhury, Silsesquioxane–urethane hybrid for thin film applications, Macromolecules 38 (2004) 6392–6401.

[17] M. Rahul, S.E. Morgan, Nanoscale Surface Topography and Friction of Polypropylene/polyhedral Oligomeric Silsesquioxane (POSS) Hybrid Nanocomposites, Polymer Preprints (American Chemical Society, Division of Polymer Chemistry). vol. 47, American Chemical Society, 2006 410–411.

[18] R.E. Johnson Jr., R.H. Dettre, J.C. Berg (Eds.), Wettability, Marcel Dekker, New York, 1993.

[19] W.A. Zisman, Contact angle, wettability, and adhesion, Advances in Chemistry Series, vol. 43, American Chemical Society, Washington DC, 1964.

[20] J. Genzer, E. Sivaniah, E.J. Kramer, J. Wang, H. Korner, M. Xiang, K. Char, C.K. Ober, B.M. DeKoven, R.A. Bubeck, M.K. Chaudhury, S. Sambasivan, D.A. Fischer, The orientation of semifluorinated alkanes attached to polymers at the surface of polymer films, Macromolecules 33 (2000) 1882–1887.

[21] M. Dimopoulos, N.R. Choudhury, M. Ginic-Markovic, J. Matisons, D.R.G. Williams, Surface studies on the additive migration and diffusion in the windowseal rubber component influencing adhesion to coating, Journal of Adhesion Science and Technology 12 (1998) 1377–1390.

[22] T. Nguyen Duc, Molecular Tailoring of Elastomer Surface by Controlled Plasma Engineering, Ian Wark Research Institute, vol. Doctor of Philosophy, University of South Australia, Adelaide, 2003. pp. 203.

[23]  M. Ginic-Markovic, Bulk Modification of EPDM and its Adhesion to Polyurethane Coating, Ian Wark Research Institute, vol. Doctor of Philosophy, University of South Australia, Adelaide, 2000, pp. 289.

[24]  N.D. Tran, N.K. Dutta, N.R. Choudhury, Plasma-polymerized perfluoro(methyl cyclohexane) coating on ethylene propylene diene elastomer surface: effect of plasma processing condition on the deposition kinetics, morphology and surface energy of the film, Thin Solid Films 491 (2005) 123–132.

[25]  N.K. Dutta, N.D. Tran, N.R. Choudhury, Perfluoro(methylcyclohexane) plasma polymer thin film: growth, surface morphology, and properties investigated by scanning thermal microscopy, Journal of Polymer Science Part B: Polymer Physics 43 (2005) 1392–1400.

[26]  N. Bertrand, B. Drevillon, P. Bulkin, In situ infrared ellipsometry study of the growth of plasma deposited silica thin films, Journal of Vacuum Science Technology A 16 (1998) 63–71.

[27]  A. Hozumi, O. Takai, Preparation of silicon oxide films having a water-repellent layer by multiple-step microwave plasma-enhanced chemical vapor deposition, Thin Solid Films 334 (1998) 54–59.

[28]  R. Lamendola, R. d'Agostino, Process control of organosilicon plasmas for barrier film preparations, Pure and Applied Chemistry 70 (1998) 1203–1208.

[29]  S.Y. Park, N. Kim, Mechanism and kinetics of organosilicon plasma polymerization, Journal of Applied Polymer Science. Applied Polymer Symposium 46 (1990) 91–108.

[30]  R.K. Sadhir, H.E. Saunders, A.I. Bennett, Influence of substrate temperature on composition, structure and electrical properties of plasma-polymerized thin films of tetramethylsilane and ammonia mixture, Journal of Applied Polymer Science. Applied Polymer Symposium 46 (1990) 209–242.

[31]  A.M. Wrobel, M.R. Wertheimer, J. Dib, H.P. Schreiber, Polymerization of organosilicones in microwave discharges, Journal of Macromolecular Science and Chemistry A14 (1980) 321–337.

[32]  F. Benitez, E. Martinez, J. Esteve, Improvement of hardness in plasma polymerized hexamethyldisiloxane coatings by silica-like surface modification, Thin Solid Films 377-378 (2000) 109–114.

[33]  M. Morra, E. Occhiello, F. Garbassi, The effect of plasma-deposited siloxane coatings on the barrier properties of HDPE, Journal of Applied Polymer Science 48 (1993) 1331–1340.

[34]  A.K. Sharma, H. Yasuda, Plasma polymerization of tetramethyldisiloxane by a magnetron glow discharge, Thin Solid Films 110 (1983) 171–184.

[35]  V. Krishnamurthy, I.L. Kamel, Y. Wei, Fourier-transform infrared analysis of plasma-polymerized hexamethyldisiloxane, Journal of Applied Polymer Science 38 (1989) 605–618.

[36]  R.L. Miller, Solid state properties, in: J. Brandrup, E.H. Immergut, E.A. Grulke (Eds.), Polymer Handbook, John Wiley & Sons Inc, New York, 1999.

[37]  A. Hozumi, O. Takai, Preparation of ultra water-repellent films by microwave plasma-enhanced CVD, Thin Solid Films 303 (1997) 222–225.

[38]  H.J. Griesser, R.C. Chatelier, C. Martin, Z.R. Vasic, T.R. Gengenbach, G. Jessup, Elimination of stick-slip of elastomeric sutures by radiofrequency glow discharge deposited coatings, Journal of Biomedical Materials Research Part A 53 (2000) 235–243.

[39] W. Petasch, K. Baumgaertner, E. Raeuchle, M. Walker, Influence of plasma surface treatment on the adhesion of thin films on metals, Surface and Coatings Technology 59 (1993) 301–305.

[40] G. Grundmeier, M. Stratmann, Plasma polymerization. A new and promising way for the corrosion protection of steel, Materials and Corrosion 49 (1998) 150–160.

[41] A.M. Wrobel, Aging process in plasma-polymerized organosilicon thin films, Journal of Macromolecular Science, Part A 22 (1985) 1089–1100.

[42] S. Roualdes, N. Hovnanian, A. Van der Lee, J. Sanchez, J. Durand, Hybrid plasma polymerized membranes from organosilicon precursors for gas separation, Proceedings of the Twelfth International Conference on Chemical Vapor Deposition 9 (1999) 1147–1154. France.

[43] L.F. Thompson, K.G. Mayhan, Plasma polymerization of vinyl monomers. I. Design, construction, and operation of an inductively coupled plasma generator and preliminary studies with nine monomers, Journal of Applied Polymer Science 16 (1972) 2291–2315.

[44] C. Li, K. Jordens, G.L. Wilkes, Abrasion-resistant coatings for plastic and soft metallic substrates by sol–gel reactions of a triethoxysilylated diethylenetriamine and tetramethoxysilane, Wear 242 (2000) 152–159.

[45] U. Schubert, N. Husing, A. Lorenz, Hybrid inorganic-organic materials by sol–gel processing of organofunctional metal alkoxides, Chemistry of Materials 7 (1995) 2010–2027.

[46] C.J. Brinker, G.W. Scherer, Sol–Gel Science: The Physics and Chemistry of Sol–gel Processing, Academic Press, San Diego, CA, 1990.

[47] S. Hofacker, M. Mechtel, M. Mager, H. Kraus, Sol–gel: a new tool for coatings chemistry, Progress in Organic Coatings 45 (2002) 159–164.

[48] A. Lee, J.D. Lichtenhan, W.A.R. Sr, Epoxy–POSS and epoxy-clay nanocomposites: thermal and viscoelastic comparisons, ACS Proceedings of American Chemical Society – Division of Polymer Material Science and Engineering 82 (2000) 235–236.

[49] J. Choi, S.G. Kim, R.M. Laine, Organic/inorganic hybrid epoxy nanocomposites from aminophenylsilsesquioxanes, Macromolecules 37 (2004) 99–109.

[50] Y. Liu, S. Zheng, K. Nie, Epoxy nanocomposites with octa(propylglycidyl ether) polyhedral oligomeric silsesquioxane, Polymer 46 (2005) 12016–12025.

[51] M.E. Wright, B.J. Petteys., A.J. Guenthner, S. Fallis, G.R. Yandek, S.J. Tomczak, T.K. Minton, A. Brunsvold, Chemical modification of fluorinated polyimides: new thermally curing hybrid polymers with POSS, Macromolecules 39 (2006) 4710–4718.

[52] E.T. Kopesk, G.H. McKinley, R.E. Cohen, Toughened poly(methyl methacrylate) nanocomposites by incorporating polyhedral oligomeric silsesquioxanes, Polymer 47 (2006) 299–309.

[53] H. Weickmann, R. Delto, R. Thomann, R. Brenn, W.D. Doll, R.M. lhaupt, PMMA nanocomposites and gradient materials prepared by means of polysilsesquioxane (POSS) self-assembly, Journal of Materials Science 42 (2007) 87–92.

[54] Y. Zhao, D.A. Schiraldi, Thermal and mechanical properties of polyhedral oligomeric silsesquioxane (POSS)/polycarbonate composites, Polymer 46 (2005) 11640–11647.

[55] B.X. Fu, W. Zhang, B.S. Hsiao, M. Rafailovich, J. Sokolov, G. Johansson, B.B. Sauer, S. Phillips, R. Balnski, Synthesis and characterization of segmented polyurethanes containing polyhedral oligomeric silsesquioxanes nanostructured molecules, High Performance Polymers 12 (2000) 565–571.

[56] S. Zhang, Q. Zou, L. Wu, Preparation and characterization of polyurethane hybrids from reactive polyhedral oligomeric silsesquioxanes, Macromolecular Materials and Engineering 291 (2006) 895–901.

[57]  O. Toepfer, D. Neumann, N.R. Choudhury, A. Whittaker, J. Matisons, Organic–inorganic poly(methyl methacrylate) hybrids with confined polyhedral oligosilsesquioxane macromonomers, Chemistry of Materials 17 (2005) 1027–1035.

[58]  R.O.R. Costa, W.L. Vasconcelos, R. Tamaki, R.M. Laine, Organic/inorganic nano-composite star polymers via atom transfer radical polymerization of methyl methacrylate using octafunctional silsesquioxane cores, Macromolecules 34 (2001) 5398–5407.

[59]  R. Tamaki, J. Choi, R. Laine, Octa(aminophenyl)silsesquioxane as a building block for polyimide nanocomposites, Polymeric Materials Science and Engineering 84 (2001) 564–565.

[60]  J. Huang, Y. Xiao, K.Y. Mya, X. Liu, C. He, J. Dai, Y.P. Siow, Thermomechanical properties of polyimide–epoxy nanocomposites from cubic silsesquioxane epoxides, Journal of Materials Chemistry 14 (2004) 2858–2863.

[61]  J.D. Lichtenhan, X. Fu, M.R. Blue, P. Wheeler, R. Misra, S. Morgan, in: U. Patent (Ed.), POSS Nanostructured Chemicals as Dispersion Aids and Friction Reducing Agents, vol. 7723415, Hybrid Plastics Inc, US, 2010.

[62]  G.K. Aravindaraj, N.R. Choudhury, N.K. Dutta, Fluoropolyurethane hybrid for thin film applications, Polymer Preprints 48 (2007) 686–687.

[63]  J.M. Mabry, D. Marchant, B.D. Viers, P.N. Ruth, S. Barker, C.E. Schlaefer, Fluoropolymer property enhancement via incorporation of fluorinated polyhedral oligomeric silsesquioxanes (FluoroPOSS). Int. SAMPE Symp. Exhibition, Society for the Advancement of Material and Process Engineering 49 (2004) 1316–1328.

[64]  J.M. Mabry, A. Vij, B.D. Viers, R.L. Blanski, R.I. Gonzalez, C.E. Schlaefer, Fluorinated Polyhedral Oligomeric Silsesquioxanes, Polymer Preprints (American Chemical Society, Division of Polymer Chemistry). vol. 45, American Chemical Society, 2004 648–649.

[65]  A.G. Kannan, S.J.P. McInnes, N.R. Choudhury, N.K. Dutta, N.H. Voelcker, Designing Superhydrophobic Surfaces Using Fluorosilsesquioxane-urethane Hybrid and Porous Silicon Gradients, in: NH. Voelcker, HW. Thissen (Eds.), vol. 7267. Melbourne, Australia: SPIE, 2008. pp. 72670-72610

[66]  A.G. Kannan, N.R. Choudhury, N. Dutta, Fluoro-silsesquioxane-urethane hybrid for thin film applications, ACS Applied Materials Interfaces 1 (2009) 336–347.

[67]  H. Wang, Y. Xue, J. Ding, L. Feng, X. Wang, T. Lin, Durable, self-healing superhydrophobic and superoleophobic surfaces from fluorinated-decyl polyhedral oligomeric silsesquioxane and hydrolyzed fluorinated alkyl silane, Angewandte Chemie International Edition 50 (2011) 11433–11436.

[68]  A. Basch, S. Gross, N.R. Choudhury, J. Matisons, Inorganic–organic hybrid polymers from the polymerisation of methacrylate-substituted oxotantalum clusters with methylmethacrylate: a thermomechanical and spectroscopic study, Journal of Sol-Gel Science and Technology 33 (2005) 39–45.

[69]  Y. Gao, N.R. Choudhury, J. Matisons, U. Schubert, B. Moraru, Part 2: Inorganic–organic hybrid polymers by polymerization of methacrylate-substituted oxotitanium clusters with methyl methacrylate: thermomechanical and morphological properties, Chemistry of Materials 14 (2002) 4522–4529.

[70]  B. Moraru, G. Kickelbick, U. Schubert, Methacrylate-substituted mixed-metal clusters derived from zigzag chains of $[ZrO_8]/[ZrO_7]$ and $[TiO_6]$ polyhedra, European Journal of Inorganic Chemistry 2001 (2001) 1295–1301.

[71]  U. Schubert, New materials by sol–gel processing: design at the molecular level, Journal of Chemical Society Dalton Transactions 16 (1996) 3343–3348.

[72] U. Schubert, E. Arpac, W. Glaubitt, A. Helmerich, C. Chau, Primary hydrolysis products of methacrylate-modified titanium and zirconium alkoxides, Chemistry of Materials 4 (1992) 291–295.

[73] G. Trimmel, P. Fratzl, U. Schubert, Cross-linking of poly(methyl methacrylate) by the methacrylate-substituted oxozirconium cluster $Zr_6(OH)_4O_4$(methacrylate)12, Chemistry of Materials 12 (2000) 602–604.

[74] G. Trimmel, S. Gross, G. Kickelbick, U. Schubert, Swelling behavior and thermal stability of poly(methylmethacrylate) crosslinked by the oxozirconium cluster $Zr_4O_2$(methacrylate), Applied Organometallic Chemistry 15 (2001) 401–406.

[75] A.N. Khramov, V.N. Balbyshev, N.N. Voevodin, M.S. Donley, Nanostructured sol–gel derived conversion coatings based on epoxy- and amino-silanes, Progress in Organic Coating 47 (2003) 207–213.

[76] J.-M. Yeh, S.-J. Liou, M.-C. Lai, Y.-W. Chang, C.-Y. Huang, C.-P. Chen, J.-H. Jaw, T.-Y. Tsai, Y.-H. Yu, Comparative studies of the properties of poly(methyl methacrylate)–clay nanocomposite materials prepared by in situ emulsion polymerization and solution dispersion, Journal of Applied Polymer Science 94 (2004) 1936–1946.

[77] G. Srinath, R. Gnanamoorthy, Sliding wear performance of polyamide 6-clay nanocomposites in water, Composites Science and Technology 67 (2007) 399–405.

[78] J.-C. Lin, Compression and wear behavior of composites filled with various nanoparticles, Composites Part B: Engineering 38 (2007) 79–85.

[79] M. Alexandre, P. Dubois, Polymer-layered silicate nanocomposites: preparation, properties and uses of a new class of materials, Materials Science and Engineering: R 28 (2000) 1–63.

[80] A. Okada, A. Usuki, Twenty years of polymer–clay nanocomposites, Macromolecular Materials and Engineering 291 (2006) 1449–1476.

[81] S.S. Ray, M. Okamoto, Polymer/layered silicate nanocomposites: a review from preparation to processing, Progress in Polymer Science 28 (2003) 1539–1641.

[82] Y. Zhang, J.R.G. Evans, Approaches to the manufacture of layered nanocomposites, Applied Surface Science 258 (2012) 2098–2102.

[83] N. Bitinis, M. Hernandez, R. Verdejo, J.M. Kenny, M.A. Lopez-Manchado, Recent advances in clay/polymer nanocomposites, Advanced Materials 23 (2011) 5229–5236.

[84] A. Dasari, Z.-Z. Yu, Y.-W. Mai, G.-H. Hu, J. Varlet, Clay exfoliation and organic modification on wear of nylon 6 nanocomposites processed by different routes, Composites Science and Technology 65 (2005) 2314–2328.

[85] P. Jawahar, R. Gnanamoorthy, M. Balasubramanian, Tribological behaviour of clay—thermoset polyester nanocomposites, Wear 261 (2006) 835–840.

[86] S.-R. Ha, K.-Y. Rhee, H. Shin, Effect of MMT concentration on tribological behavior of MMT/epoxy nanocomposite, Journal of Nanoscience and Nanotechnology 8 (2008) 4869–4872.

[87] K. Nevalainen, J. Vuorinen, V. Villman, R. Suihkonen, P. Järvelä, J. Sundelin, T. Lepistö, Characterization of twin-screw-extruder-compounded polycarbonate nanoclay composites, Polymer Engineering and Science 49 (2009) 631–640.

[88] Q.-Y. Peng, P.-H. Cong, X.-J. Liu, T.-X. Liu, S. Huang, T.-S. Li, The preparation of PVDF/clay nanocomposites and the investigation of their tribological properties, Wear 266 (2009) 713–720.

[89] H. Hu, S. Yu, M. Wang, K. Liu, Tribological behavior of polyamide 66-based binary and ternary composites, Polymer Engineering and Science 49 (2009) 2454–2458.

[90] Q. Jia, S. Shan, Y. Wang, L. Gu, J. Li, Tribological performance and thermal behavior of epoxy resin nanocomposites containing polyurethane and organoclay, Polymers for Advanced Technologies 19 (2008) 859–864.

[91] M. Bo, W. Qihua, W. Tingmei, W. Honggang, J. Lingqi, P. Xianqiang, Preparation, characterization, and properties of polyamide 66/Maleic anhydride -grafted-polypropylene/clay ternary nanocomposites, Journal of Macromolecular Science: Physics 48 (2009) 55–67.

[92] M. Du, B. Guo, D. Jia, Newly emerging applications of halloysite nanotubes: a review, Polymer International 59 (2010) 574–582.

[93] W. Ma, W.O. Yah, H. Otsuka, A. Takahara, Application of imogolite clay nanotubes in organic–inorganic nanohybrid materials, Journal of Materials Chemistry, 2012.

[94] C.M. Levard, A. Masion, J.R.M. Rose, E. Doelsch, D. Borschneck, C. Dominici, F. Ziarelli, J.-Y. Bottero, Synthesis of imogolite fibers from decimolar concentration at low temperature and ambient pressure: a promising route for inexpensive nanotubes, Journal of the American Chemical Society 131 (2009) 17080–17081.

[95] L. Mingxian, G. Baochun, D. Mingliang, C. Xiaojia, J. Demin, Properties of halloysite nanotube–epoxy resin hybrids and the interfacial reactions in the systems, Nanotechnology 18 (2007) 455703.

[96] K. Hedicke-Höchstötter, G.T. Lim, V. Altstädt, Novel polyamide nanocomposites based on silicate nanotubes of the mineral halloysite, Composites Science and Technology. 69 (2009) 330–334.

[97] M. Du, B. Guo, D. Jia, Thermal stability and flame retardant effects of halloysite nanotubes on poly(propylene), European Polymer Journal 42 (2006) 1362–1369.

[98] W. Ma, H. Otsuka, A. Takahara, Poly(methyl methacrylate) grafted imogolite nanotubes prepared through surface-initiated ARGET ATRP, Chemical Communications 47 (2011) 5813–5815.

[99] S. Iijima, T. Ichihashi, Single-shell carbon nanotubes of 1-nm diameter, Nature 363 (1993) 603–605.

[100] S. Iijima, Helical microtubules of graphitic carbon, Nature 354 (1991) 56–58.

[101] K. Miyoshi Jr., K.W. Street, Novel carbons in tribology, Tribology International 37 (2004) 865–868.

[102] Z. Yang, S.-Q. Shi, Fabrication and tribological properties of polymer–carbon nanotubes nanocomposites, Key Engineering Materials 334–335 (2007) 661–664.

[103] C. Wang, T. Xue, B. Dong, Z. Wang, H.-L. Li, Polystyrene–acrylonitrile–CNTs nanocomposites preparations and tribological behavior research, Wear 265 (2008) 1923–1926.

[104] L.F. Giraldo, B.L. López, W. Brostow, Effect of the type of carbon nanotubes on tribological properties of polyamide 6, Polymer Engineering and Science. 49 (2009) 896–902.

[105] X. Li, W. Guan, H. Yan, L. Huang, Fabrication and atomic force microscopy/friction force microscopy (AFM/FFM) studies of polyacrylamide–carbon nanotubes (PAM-CNTs) copolymer thin films, Materials Chemistry and Physics 88 (2004) 53–58.

[106] B. Dong, Z. Yang, Y. Huang, H.-L. Li, Multi-walled carbon nanotubes/epoxy resin nanocomposites, Tribology Letters 20 (2005) 251–254.

[107] M.A. Samad, S.K. Sinha, Dry sliding and boundary lubrication performance of a UHMWPE/CNTs nanocomposite coating on steel substrates at elevated temperatures, Wear 270 (2011) 395–402.

[108] R. Ribeiro, S. Banda, Z. Ounaies, H. Ucisik, M. Usta, H. Liang, A tribological and bio-mimetic study of PI–CNT composites for cartilage replacement, Journal of Materials Science 47 (2012) 649–658.

[109] K.-T. Lau, C. Gu, D. Hui, A critical review on nanotube and nanotube/nanoclay related polymer composite materials, Composites Part B: Engineering 37 (2006) 425–436.

[110] P.L. Dickrella, S.B. Sinnottb, D.W. Hahna, N.R. Raravikarc, L.S. Schadlerc, P.M. Ajayanc, W.G. Sawyer, Frictional anisotropy of oriented carbon nanotube surfaces, Tribology Letters 18 (2005) 59–62.

[111] H. Wang, J. Feng, X. Hu, K.M. Ng, Tribological behaviors of aligned carbon nano-tube/fullerene–epoxy nanocomposites, Polymer Engineering and Science 48 (2008) 1467–1475.

[112] B. Safadi, R. Andrews, E.A. Grulke, Multiwalled carbon nanotube polymer compos-ites: synthesis and characterization of thin films, Journal of Applied Polymer Science 84 (2002) 2660–2669.

[113] M.S.P. Shaffer, A.H. Windle, Fabrication and characterization of carbon nanotube/poly(vinyl alcohol) composites, Advanced Materials 11 (1999) 937–941.

[114] T.X. Liu, I.Y. Phang, L. Shen, S.Y. Chow, W.D. Zhang, Morphology and mechanical properties of multiwalled carbon nanotubes reinforced Nylon-6 composites, Macro-molecules 37 (2004) 7214–7222.

[115] W.D. Zhang, L. Shen, I.Y. Phang, T. Liu, Carbon nanotubes reinforced Nylon-6 com-posite prepared by simple melt-compounding, Macromolecules 37 (2004) 256–259.

[116] S. Kumar, T.D. Dang, F.E. Arnold, A.R. Bhattacharyya, B.G. Min, X. Zhang, R.A. Vaia, C. Park, W.W. Adams, R.H. Hauge, R.E. Smalley, S. Ramesh, P.A. Willis, Syn-thesis, structure, and properties of PBO/SWNT composites, Macromolecules 35 (2002) 9039–9043.

[117] C. Zhao, G. Hu, R. Justice, D.W. Schaefer, S. Zhang, M. Yang, C.C. Han, Synthesis and characterization of multi-walled carbon nanotubes reinforced polyamide 6 via in situ polymerization, Polymer 46 (2005) 5125–5132.

[118] C. Park, Z. Ounaies, K.A. Watson, R.E. Crooks, J. Smith, S.E. Lowther, J.W. Connell, E.J. Siochi, J.S. Harrison, T.L.S. Clair, Dispersion of single wall carbon nanotubes by in situ polymerization under sonication, Chemical Physics Letters 364 (2002) 303–308.

[119] J.N. Coleman, U. Khan, W.J. Blau, Y.K. Gun'ko, Small but strong: a review of the mechanical properties of carbon nanotube–polymer composites, Carbon 44 (2006) 1624.

[120] S. Haojie, Q. Huan, L. Na, Z. Xueqiang, Tribological behaviour of carbon nanotubes/polyurethane nanocomposite coatings, Micro and Nano Letters 6 (2011) 48–51.

[121] I. Alig, P. Pötschke, D. Lellinger, T. Skipa, S. Pegel, G.R. Kasaliwal, T. Villmow, Estab-lishment, morphology and properties of carbon nanotube networks in polymer melts, Polymer 53 (2012) 4–28.

[122] M. Rahmat, P. Hubert, Carbon nanotube–polymer interactions in nanocomposites: a review, Composites Science and Technology 72 (2011) 72–84.

[123] J.H. Lee, K.Y. Rhee, Silane treatment of carbon nanotubes and its effect on the tribo-logical behavior of carbon Nanotube/Epoxy nanocomposites, Journal of Nanoscience and Nanotechnology 9 (2009) 6948–6952.

[124] H.-J. Zhang, Z.-Z. Zhang, F. Guo, Tribological behaviors of hybrid PTFE/nomex fabric/phenolic composite reinforced with multiwalled carbon nanotubes, Journal of Applied Polymer Science 124 (2012) 235–241.

[125] J. Kim, H. Im, M.H. Cho, Tribological performance of fluorinated polyimide-based nanocomposite coatings reinforced with PMMA-grafted-MWCNT, Wear 271 (2011) 1029–1038.

[126] M. Lu, H.L. Li, K.T. Lau, Formation and growth mechanism of dissimilar coiled carbon nanotubes by reduced-pressure catalytic chemical vapor deposition, Journal of Physical Chemistry B 108 (2004) 6186–6192.

[127] K.S. Novoselov, A.K. Geim, S.V. Morozov, D. Jiang, Y. Zhang, S.V. Dubonos, I.V. Crigorieva, A.A. Firsov, Electric field effect in atomically thin carbon films, Science 306 (2004) 666–669.

[128] A.K. Geim, K.S. Novoselov, The rise of graphene, Nature Materials 6 (2007) 183–191.

[129] A.K. Geim, Graphene: status and prospects, Science 324 (2009) 1530–1534.

[130] H. Bai, C. Li, G. Shi, Functional composite materials based on chemically converted graphene, Advanced Materials 23 (2011) 1089–1115.

[131] X.-Y. Qi, D. Yan, Z. Jiang, Y.-K. Cao, Z.-Z. Yu, F. Yavari, N. Koratkar, Enhanced electrical conductivity in polystyrene nanocomposites at ultra-low graphene content, ACS Applied Materials and Interfaces, 2011 (null-null).

[132] V.H. Pham, T.V. Cuong, T.T. Dang, S.H. Hur, B.-S. Kong, E.J. Kim, E.W. Shin, J.S. Chung, Superior conductive polystyrene—chemically converted graphene nanocomposite, Journal of Materials Chemistry 21 (2011) 11312–11316.

[133] W. Li, X.-Z. Tang, H.-B. Zhang, Z.-G. Jiang, Z.-Z. Yu, X.-S. Du, Y.-W. Mai, Simultaneous surface functionalization and reduction of graphene oxide with octadecylamine for electrically conductive polystyrene composites, Carbon 49 (2011) 4724–4730.

[134] M. Yoonessi, J.R. Gaier, Highly conductive multifunctional graphene polycarbonate nanocomposites, ACS Nano 4 (2010) 7211–7220.

[135] H. Yang, C. Shan, F. Li, D. Han, Q. Zhang, L. Niu, Covalent functionalization of polydisperse chemically-converted graphene sheets with amine-terminated ionic liquid, Chemical Communications (2009) 3880–3882.

[136] J. Xu, K. Wang, S.-Z. Zu, B.-H. Han, Z. Wei, Hierarchical nanocomposites of polyaniline nanowire arrays on graphene oxide sheets with synergistic effect for energy storage, ACS Nano 4 (2010) 5019–5026.

[137] S.H. Domingues, R.V. Salvatierra, M.M. Oliveira, A.J.G. Zarbin, Transparent and conductive thin films of graphene/polyaniline nanocomposites prepared through interfacial polymerization, Chemical Communications 47 (2011) 2592–2594.

[138] Z. Tai, Y. Chen, Y. An, X. Yan, Q. Xue, Tribological behavior of UHMWPE reinforced with graphene oxide nanosheets, Tribology Letters 46 (2012) 55–63.

[139] T. Huang, R. Lu, C. Su, H. Wang, Z. Guo, P. Liu, Z. Huang, H. Chen, T. Li, Chemically modified graphene/polyimide composite films based on utilization of covalent bonding and oriented distribution, ACS Applied Materials and Interfaces: Article ASAP, 2012.

[140] H. Liu, Y. Li, T. Wang, Q. Wang, In situ synthesis and thermal, tribological properties of thermosetting polyimide/graphene oxide nanocomposites, Journal of Materials Science 47 (2012) 1867–1874.

[141] M. Martin-Gallego, M. Hernández, V. Lorenzo, R. Verdejo, M.A. Lopez-Manchado, M. Sangermano, Cationic photocured epoxy nanocomposites filled with different carbon fillers, Polymer 53 (2012) 1831–1838.

[142] Y. Li, Q. Wang, T. Wang, G. Pan, Preparation and tribological properties of graphene oxide/nitrile rubber nanocomposites, Journal of Materials Science 47 (2012) 730–738.

[143] S. Liu, J. Ou, Z. Li, S. Yang, J. Wang, Layer-by-layer assembly and tribological property of multilayer ultrathin films constructed by modified graphene sheets and polyethyleneimine, Applied Surface Science 258 (2012) 2231–2236.

[144] J. Ou, Y. Wang, J. Wang, S. Liu, Z. Li, S. Yang, Self-assembly of octadecyltrichlorosilane on graphene oxide and the tribological performances of the resultant film, Journal of Physical Chemistry C 115 (2011) 10080–10086.

[145] K. Holmberg, A. Matthews, Coatings Tribology—Properties, Techniques and Applications in Surface Engineering, Elsevier, 1994.

[146] K. Holmberg, A. Matthews, H. Ronkainen, Coatings tribology—contact mechanisms and surface design, Tribology International 31 (1998) 107.

[147] A. Sidorenko, H.S. Ahn, D.I. Kim, H. Yang, V.V. Tsukruk, Wear stability of polymer nanocomposite coatings with trilayer architecture, Wear 252 (2002) 946–955.

[148] V.V. Tsukruk, Nanocomposite polymer layers for molecular tribology, Tribology Letters 10 (2001) 127–132.

[149] S. McInnes, H. Thissen, N.R. Choudhury, N.H. Voelcker, Characterisation of porous silicon/poly(L-lactide) composites prepared using surface initiated ring opening polymerisation, in: D.V. Nicolau (Ed.), Proceedings of SPIE; BioMEMS and Nanotechnology II, vol. 6036, SPIE, Brisbane, Australia, 2005. pp. 6036x-6031-6036x-6012.

[150] J.T. Han, Y. Zheng, J.H. Cho, X. Xu, K. Cho, Stable superhydrophobic organic–inorganic hybrid films by electrostatic self-assembly, Journal of Physical Chemistry B 109 (2005) 20773–20778.

[151] J.-M. Yang, C.-S. Lu, Y.-G. Hsuu, C.-H. Shih, Mechanical properties of acrylic bone cement containing PMMA–SiO$_2$ hybrid sol–gel material, Journal of Biomedical Materials Research 38 (1997) 143–154.

[152] J.M. Yang, H.S. Chen, Y.G. Hsu, F.H. Lin, Y.H. Chang, Organic–inorganic hybrid sol–gel materials. Part 2. Application for dental composites, Applied Macromolecular Chemistry 251 (1997) 61–72.

[153] J. Plojoux, Y. Leterrier, J.-A.E. Manson, F. Templier, Mechanical integrity analysis of multilayer insulator coatings on flexible steel substrates, Thin Solid Films 515 (2007) 6890–6898.

[154] M. Schonfelder, H. Steinberger, H. Schmid, in: Organisation USP (Ed.), Plastics Mouldings Finished with a Three-layer Coating and a Process for Producing the Finish, Bayer Aktiengesellschaft, United States, 1984.

[155] P. Majewski, N.R. Choudhury, D. Spori, E. Wohlfahrt, M. Wohlschloegel, Synthesis and characterisation of star polymer/silicon carbide nanocomposites, Materials Science and Engineering A 434 (2006) 360.

[156] S.M. Kurtz, O.K. Muratoglu, M. Evans, A.A. Edidin, Advances in the processing, sterilization, and crosslinking of ultra-high molecular weight polyethylene for total joint arthroplasty, Biomaterials 20 (1999) 1659–1688.

[157] D.W. Van Citters, F.E. Kennedy, J.P. Collier, Rolling sliding wear of UHMWPE for knee bearing applications, Wear in press, corrected proof 3130.

[158] P.V. Pavoor, B.P. Gearing, O. Muratoglu, R.E. Cohen, A. Bellare, Wear reduction of orthopaedic bearing surfaces using polyelectrolyte multilayer nanocoatings, Biomaterials 27 (2006) 1527–1533.

[159] S.H. Rhee, Y.K. Lee, B.S. Lim, J.J. Yoo, H.J. Kim, Evaluation of a novel poly(-caprolactone)-organosiloxane hybrid material for the potential application as a bioactive and degradable bone substitute, Biomacromolecules 5 (2004) 1575–1579.

[160] R.Y. Kannan, H.J. Salacinski, M. Odlyha, P.E. Butler, A.M. Seifalian, The degradative resistance of polyhedral oligomeric silsesquioxane nanocore integrated polyurethanes: an in vitro study, Biomaterials 27 (2006) 1971–1979.

[161] R.Y. Kannan, H.J. Salacinski, J. DeGroot, I. Clatworthy, L. Bozec, M. Horton, P.E. Butler, A.M. Seifalian, The antithrombogenic potential of a polyhedral oligomeric silsesquioxane (POSS) nanocomposite, Biomacromolecules 7 (2006) 215–223.

[162] G. Schottner, K. Rose, U. Posset, Scratch and abrasion resistant coatings on plastic lenses—State of the art, current developments and perspectives, Journal of Sol-Gel Science and Technology 27 (2003) 71–79.

[163] M.A.C. Stuart, W.T.S. Huck, J. Genzer, M. Muller, C. Ober, M. Stamm, G.B. Sukhorukov, I. Szleifer, V.V. Tsukruk, M. Urban, F. Winnik, S. Zauscher, I. Luzinov, S. Minko, Emerging applications of stimuli-responsive polymer materials, Nature Materials 9 (2010) 101–113.

[164] J. Choi, J. Harcup, A.F. Yee, Q. Zhu, R.M. Laine, Organic/inorganic hybrid composites from cubic silsesquioxanes, Journal of the American Chemical Society 123 (2001) 11420–11430.

[165] H. Yoshizawa, Y.L. Chen, J. Israelachvili, Recent advances in molecular level understanding of adhesion, friction and lubrication, Wear 168 (1993) 161–166.

[166] J.A. Harrison, S.S. Perry, Friction in the presence of molecular lubricants and solid/hard coatings, MRS Bulletin 23 (1998) 27–31.

[167] W.X. Chen, F. Li, G. Han, J.B. Xia, L.Y. Wang, J.P. Tu, Z.D. Xu, Tribological behavior of carbon nanotube-filled PTFE composites, Tribology Letters 15 (2003) 275–278.

[168] W.X. Chen, J.P. Tu, L.Y. Wang, H.Y. Gan, Z.D. Xu, X.B. Zhang, Tribological application of carbon nanotubes in a metal-based composite coating and composites, Carbon 41 (2003) 215–222.

[169] M.R. Falvo, R.M.T.A. Helser, V. Chi Jr., FPB, S. Washburn, R. Superfine, Nanometrescale rolling and sliding of carbon nanotubes, Nature 397 (1999) 236–238.

[170] M.R. Falvo, J. Steele, R.M. Taylor, R. Superfine, Evidence of commensurate contact and rolling motion: AFM manipulation studies of carbon nanotubes on HOPG, Tribology Letters 9 (2000) 73–76.

[171] F.H. Beijer, R.P. Sijbesma, H. Kooijman, A.L. Spek, E.W. Meijer, Strong dimerization of ureidopyrimidones via quadruple hydrogen bonding, Journal of the American Chemical Society 120 (1998) 6761–6769.

[172] R. Haler, Ez Muhlen, An introduction to $\mu TA^{(TM)}$ and its application to the study of interfaces, Thermochimica Acta 361 (2000) 113–120.

[173] M. Reading, D.M. Price, D.B. Grandy, R.M. Smith, L. Bozec, M. Conroy, A. Hammiche, H.M. Pollock, Micro-thermal analysis of polymers: current capabilities and future prospects, Macromolecular Symposia 167 (2001) 45–62.

[174] H.M. Pollock, A. Hammiche, Micro-thermal analysis: techniques and applications, Journal of Physics D: Applied Physics 34 (2001) R23–R53.

[175] S. Edwards, N.D. Tran, M. Provatas, N.R. Choudhury, N. Dutta, Visualization and characterization of bulk and surface morphology by microthermal analysis and atomic force microscopy, in: G.E. Totten, H. Liang (Eds.), Mechanical Tribology: Materials, Characterization and Applications, Marcel Dekker Inc, New York, 2004.

[176] M. Oaten, S. Edwards, J. Budin, M. Provatas, N.R. Choudhury, J. Matisons, Correlation of friction and wear phenomena to the structure and morphology of weatherable polymers, 24th Australasian Polymer Symposium, Royal Australian Chemical Institute, Beechworth, Australia, 2001. pp. PB. 20.

[177] X.X. Liu, T.S. Li, X.J. Liu, R.G. Lv, P.H. Cong, An investigation on the friction of oriented polytetrafluoroethylene (PTFE), Wear in press, corrected proof.

[178] R. Gahlin, S. Jacobson, A novel method to map and quantify wear on a micro-scale, Wear 222 (1998) 93.

[179] W.W. Scott, B. Bhushan, Use of phase imaging in atomic force microscopy for measurement of viscoelastic contrast in polymer nanocomposites and molecularly thick lubricant films, Ultramicroscopy 97 (2003) 151.

[180] C. Putman, R. Kaneko, Experimental observation of single-asperity friction at the atomic scale, Thin Solid Films 273 (1996) 317.

[181] T. Hertel, R. Martel, P. Avouris, Manipulation of individual carbon nanotubes and their interaction with surfaces, Journal of Physical Chemistry B 102 (1998) 910–915.

[182] I.D. Baikie, G.H. Bruggink, Characterization of oxides and thin films, Materials Research Society Symposium Proceedings 309, 1993.

[183] W. Li, D.Y. Li, A study on the kinetic response of the electron work function to wear, Wear 255 (2003) 333–340.

[184] D. DeVecchio, B. Bhushan, Use of a nanoscale Kelvin probe for detecting wear precursors, Review of Scientific Instruments 69 (1998) 3618–3624.

[185] B. Bhushan, A.V. Goldade, Measurements and analysis of surface potential change during wear of single-crystal silicon (100) at ultralow loads using Kelvin probe microscopy, Applied Surface Science 157 (2000) 373–381.

[186] A.C. Fischer-Cripps (Ed.), Nanoindentation, Second ed., Springer, New York, 2004.

[187] G.K. Aravindaraj, N.R. Choudhury, N. Dutta, Organic–inorganic Hybrid Coatings for Protective Applications, ACUN-5 International Composites Conference, Sydney, Australia, 2006, pp. 559–563.

[188] V.A. Soloukhin, W. Posthumus, J.C.M. Brokken-Zijp, J. Loos, G.D. With, Mechanical properties of silica–(meth)acrylate hybrid coatings on polycarbonate substrate, Polymer 43 (2002) 6169–6181.

[189] A.J. Atanacio, B.A. Latella., C.J. Barbe´, M.V. Swain, Mechanical properties and adhesion characteristics of hybrid sol–gel thin films, Surface and Coatings Technology 192 (2005) 354–364.

[190] A. Dasari, Z.-Z. Yu, Y.-W. Mai, Nanoscratching of nylon 66-based ternary nanocomposites, Acta Materialia 55 (2007) 635–646.

[191] K. Rau, R. Singh, E. Goldberg, Nanoindentation and nanoscratch measurements on silicone thin films synthesized by pulsed laser ablation deposition (PLAD), Materials Research Innovations 5 (2002) 151–161.

[192] L. Huang, J. Lub, M. Troyon, Nanomechanical properties of nanostructured titanium prepared by SMAT, Surface and Coatings Technology 201 (2006) 208–213.

[193] S. Hur, S.I. Hong, The characterization of adhesion and friction of PDMS polymer for nanoimprinting Lithography, Key Engineering Materials 297–300 (2005) 269–274.

[194] B.D. Beake, J.F. Smith, High-temperature nanoindentation testing of fused silica and other materials, Philosophical Magazine A 82 (2002) 2179–2186.

# A novel neural network approach for modeling tribological properties of polyphenylene sulfide reinforced on different scales

# 22

**Michael Busse\*, Alois K. Schlarb\*\***

*\*Leibniz-Institut fur Neue Materialien GmbH (INM), Saarland University, Saarbrucken, Germany, \*\*Chair of Composite Engineering, University of Kaiserslautern, Kaiserslautern, Germany*

## CHAPTER OUTLINE HEAD

## 22.1 INTRODUCTION

The application of neural network models are extremely useful with respect to valuable resources, time and money, not only in the study of a newly designed material's friction coefficient (*cof*) and specific wear rate ($w_s$), but also for identifying mechanical properties or even the study of material compositions that are not yet engineered. Similar to the comments already mentioned by Chang et al. in Chapter 3, artificial neural network (ANN) approaches combine the complexity of some techniques from statistics with the objective of machine learning that imitates human intelligence; however, this is done at a more "unconscious" level and hence there is no accompanying ability to make learned concepts transparent to the user. The ANN technology can be applied whenever the complexity of the problem is overwhelming from a fundamental perspective and where simplification is unacceptable [1,2].

Tribology of Polymeric Nanocomposites. http://dx.doi.org/10.1016/B978-0-444-59455-6.00022-2

In the past decades, quite a lot of studies were accomplished on the use of ANN in material science to investigate their capability to make predictions on mechanical and tribological properties; see [3–5] for more information. Regarding polymer composites, Velten et al. [6] first applied facile ANN technology to predict wear in short-fiber and particle-reinforced thermoplastics. This was followed by important studies such as those of Jiang et al. [7–10] and also Zhang et al. [4,11–14] that refined the ANN technology and explored the practicability of such networks in the field of engineering modern polymer composites. Gyurova et al. [15–19] employed ANN for the prediction of wear and friction properties in polyphenylene sulfide (PPS) composites filled with short carbon fibers and/or TiO$_2$ particles. Based on own methodology studies and results of [8,10] they found the gradient descent with momentum (GDX) method the best algorithm to train their ANN. Further they chose the mean relative error (MRE) as performance function. In respect of the best fitting network architecture, they applied a 7 – (9 - 3)$_2$ – 1 layer network for $w_s$ and a 7 – (3 - 1)$_2$ – 1 layer network for *cof* prediction issues [10,17–19]. In [17,19] the authors accomplished an importance analysis with the intention of gaining information about the relevance of adding characteristic mechanical—and thermomechanical—variables to the standard inputs to increase the prediction quality of their network. They found that adding a combination of tensile modulus and strain represented a strong correlation with $w_s$, followed by microhardness and damping properties. Regarding *cof*, microhardness, tensile properties, and compressive and storage moduli improved the simulation results.

In the present study we developed novel ANN architectures with the Levenberg–Marquardt (LM) training algorithm and mean squared error with regularization (MSEREG) as performance function and evaluated their performance in comparison of the achieved efficiency of Gyurova et al. [17,19]. Therefore, by applying the same database [10,15–19], we completed a comparative test of how the additional input variables may influence the prediction accuracy of our elaborated networks. Subsequently, we show an example of how our ANN approach can be employed to make highly accurate predictions on a test composite's friction and wear behavior. The developed ANN technology and the outcome of the additional input variable determination serves as basis for our future work, namely, predicting the tribological behavior of novel tribocompounds and also for the forecasting of potent material compositions.

## 22.2 METHODS

### 22.2.1 Experimental data

For this survey, data from pin-on-disc experiments published in [19] have been employed. The applied composites consist mainly of PPS as matrix material, where short carbon fibers, graphite, PTFE and/or submicro-TiO$_2$ was incorporated. Conventional pin-on-disc measurements were carried out for 20 h. The contact pressure and sliding speed were varied from 1 to 4 MPa and from 1 to 3 m s$^{-1}$. To define the *cof*, the normal and the frictional forces were measured. The mass loss of the specimen that could be determined after each experiment was applied to calculate the specific wear rate according to the equation

$$w_s = \frac{\Delta m}{\rho v t F_N} \left( \text{mm}^3 \text{ Nm}^{-1} \right)$$ (22.1)

where $\Delta m$ is the mass loss, $\rho$ denotes the material density, $v$ is the sliding speed, and $t$ connotes the duration of the experiment. $F_N$ labels the normal force that rests on the specimen during sliding.

Please refer to [17,19] for more details about the material composition and measurement setup.

The data of 115 *cof* and 110 $w_s$ measurements could be taken into account in the present study. Due to scaling issues and ANN restrictions, the $w_s$ data was normalized before feeding to the network as training-target and was transformed back later on.

### 22.2.2 **The artificial neural network approach**

The domains of ANN have arisen from multiple sources, ranging from the intention of mankind for understanding the human brain and adopting its functions into models, up to further issues where those smart models that have the ability to copy brain function, were applied to the practical commercial, scientific, and engineering disciplines of modeling, pattern recognition, and function predicting [1]. In brief, an ANN is a massively parallel distributed processor that has a natural ability for storing experimental evaluated knowledge and making this available for use. It resembles the brain in two aspects: first, a learning process lets the network acquire knowledge and second interconnection strengths known as synaptic weights are utilized to store the knowledge [20]. Thus, the before trained ANN is able to map from a set of inputs to a set of outputs. Generally, ANN consists of layers of interconnected nodes, each node producing a nonlinear function of its input. The feed-forward ANN nodes have one-way connections to other nodes, and the nodes can be labeled from inputs with low numbers to outputs with high numbers. Each node is only connected to following nodes with higher numbers. Figure 22.1 indicates an example architecture of a feed-forward ANN structure. Each node or unit within such a network sums up its inputs and adds a constant, the so-called bias to build a total input $x_j$. A function $f_j$ is then applied to the biased input $x_j$ to yield $y_j$. The links have weight factors $w_{ji}$, which are multiplied with the signals running through them. The input nodes have $f \equiv 1$; they are just there to distribute the inputs. The general definition allows more than one hidden layer, and further allows skip-layer connections from the input to the output layer. Thus, an ANN with the same transfer function $f_h$ or $f_o$ in a layer represents the general function

$$y_k = f_o \left( \alpha_k + \sum_{i \to k} w_{ik} x_i + \sum_{j \to k} w_{jk} f_h \left( \alpha_j + \sum_{i \to j} w_{ji} x_i \right) \right)$$ (22.2)

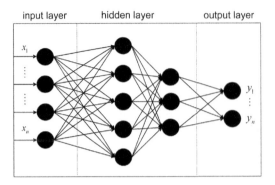

**FIGURE 22.1**

Example of a feed-forward ANN architecture with input layer, two hidden layers and output layer.

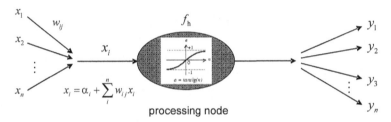

**FIGURE 22.2**

Schematic function of a single processing node within an ANN.

from inputs to outputs. The functions $f_h$ are almost invariably taken to be either linear, logistic threshold functions or hyperbolic tangent functions [1,21,22]. Figure 22.2 elucidates the general description of a single ANN node within a network.

Once an ANN architecture (number of input and output nodes, number of layers and layer nodes and type of transfer function applied to the nodes) is fixed, the network is ready for the training or learning process; in the context of employing ANN for function fitting issues, supervised learning is the learning strategy that performs best; hereby, learning is understood as the automatic extraction of usually implicit rules from example data. According to that, the learning procedure allows an ANN to discover patterns and their relationships, and to organize itself to perform associations. Two major implications can be derived from this capability for our purpose of predicting friction and wear: the ability to solve complex tasks whose rules are difficult to formulate and the ability to extract statistical models and knowledge-based rules from large data sets with numerous variables [23,24].

Generally, a finite set of to-be learned or to-be associated data (training dataset, divided into input values and the corresponding target values)

$$\left(\vec{x}^{(IN_1)}, \vec{y}^{(TARGET_1)}\right),...,\left(\vec{x}^{(IN_n)}, \vec{y}^{(TARGET_n)}\right) \tag{22.3}$$

is presented to the network together with the learning rule (performance function and training algorithm) that adjusts the nodes' weight factors and bias as a function of the training dataset (see next section). After learning, the ANN should be capable of associating every training input

$$\vec{x}^{(s)}, 1 \leq s \leq n, \tag{22.4}$$

with the exact or approximately exact target value

$$\vec{y}^{(s)}, 1 \leq s \leq n. \tag{22.5}$$

The network performance function defines now how to tune the values of the weights and biases of the network's nodes during the training process. Here, the network architect has yet again the freedom to choose the performance function and the training algorithm that will be applied to the ANN architecture. The application of the mean squared error (MSE) as a performance function was up to now an approved method for most feed-forward networks [25]. This one is defined as the mean sum of squares of the network errors between the network outputs $\vec{y}^{\text{OUT}}$ and the target outputs $\vec{y}^{\text{TARGET}}$:

$$\text{MSE} = \frac{1}{n} \sum_{i=1}^{n} \left( y_i^{\text{TARGET}} - y_i^{\text{OUT}} \right)^2 \tag{22.6}$$

For our application, we employed a modified performance function that was recently introduced to improve generalization by additionally adding a term that consists of the mean of the sum of squares of the network weights and biases, called MSEREG, denoted as

$$\text{MSEREG} = \lambda * \text{MSE} + (1 - \lambda) * \text{MSW}, \tag{22.7}$$

where $\lambda$ is the performance ratio, and

$$\text{MSW} = \frac{1}{n} \sum_{j=1}^{n} w_j^2, \tag{22.8}$$

where $n$ is the total number of weights and biases. Adopting now this advanced performance function causes the network to have smaller weights and biases, and this forces the network response to be smoother and less likely to overfit [26,27].

In forecasting, it is not crucial how well an ANN model fits the training data— only the quality of future predictions counts, i.e. the performance on novel data or the generalization ability: novel input values in the range of the training-$\vec{x}$-values that are presented to the proper trained ANN should lead to output values that are also in the range of the corresponding training-$\vec{y}$-values. This is what we utilized to evaluate our prediction performance. For noisy training data this recommends ANN that are not too large since networks with too many high-precision weights will lead to hypersensitivity and will not generalize well [22,23].

Hence, an adequate selection of a task-specific ANN including network architecture, performance function and training algorithm is essential to gain an inductive device whose prediction quality outside the training set, i.e. the performance on cases it has not yet seen, is satisfactory [25].

### 22.2.3 Training Algorithm

For multilayer feed-forward network training issues, any standard numerical optimization algorithm can be utilized to optimize the performance function, but there are a few key ones that have shown excellent performance for neural network training, which use gradient- or Jacobian-based methods, e.g. gradient descent, GDX, variable learning rate gradient descent, resilient backpropagation, LM, Bayesian regularization, etc. Please refer to [25] for a detailed description of several of these techniques.

For the application of forecasting the *cof* or the wear rate of tribological measurements in material science, existing studies applied several of the aforementioned algorithms on various architectures [4,7–13,15–19]. Even though the ANN simulation results for studying wear and friction properties of PPS composites in those articles sound proof and the prediction quality was quite good, the existing studies imply wide gaps due to the large variety of combinations in the design of an ANN for tribological prediction purposes. The authors in [9] tested different architectures and training algorithms with a fixed performance function by trail and error to identify an ANN that performs best on their data. Their most potent ANN was applied to detect additional (secondary) input parameters (mechanical and thermomechanical data) that are added to the standard inputs and that have the potential to increase the network's prediction quality. A more sophisticated methodology to identify an ANN that meets the problem-solving condition best is to theoretically elaborate this ANN on the basis of the underlying dataset: the inputs' and outputs' data values and margins, the data complexity and for this reason the ANN's requirements on architecture complexity. We could gain a better performance on prediction quality and on the identification of important input parameters by applying a different ANN architecture combined with a new performance function and a sophisticated training algorithm on the same data that was the basis for [10,15–19].

Gyurova et al. [17–19] employed the GDX methodology as the training algorithm because it offered the best results within their developed ANN architecture in the above-cited articles. Thanks to the fact that our to-be-solved problem results in a maximum of eight input nodes and just one output node, the network architecture that is worth to be considered does not exceed the amount of several hundred weights. This is popularly understood as a medium-sized problem. In general, gradient descent methods work fine for very simple models, but they are too unrefined for more complex implementations; to reach convergence can take an extremely long time. Therefore, it is more clever to apply the LM algorithm: LM is a nonlinear optimization method that combines the advantages of gradient descent (minimization along the direction of the gradient) and Newton's method (speed up the process of finding the minimum of a function by using a quadratic model). The benefit of LM optimization is that this significantly outperforms gradient descent and conjugate gradient methods for medium-sized problems [28–30]. It appears to be the fastest method for training moderate-sized feed-forward ANN and has also a very efficient MATLAB implementation, because the solution of the matrix equation is a built-in function, so its attributes become even more pronounced in a MATLAB setting [27]. More information about the application of LM to ANN training can be found in [25,29].

## 22.2.4  Procedure

Based on the dataset of [10,15–19], for this approach, we developed and tested different newly designed ANNs; according to the cited references we evaluated once more the input parameters that lead to the best network performance for the prediction of $cof$ and $w_s$. Besides the five different material ingredients (PPS, SCF (short carbon fibers), $TiO_2$, Gr, PTFE (PolyTetraFluorEthylen) and ZnS) and the testing conditions (sliding speed and contact pressure) that generate the essential dataset, referred to as BASE (= seven independent input parameters), additional input vectors, i.e. characteristic mechanical and thermomechanical PPS-composite properties that may have the potential to increase simulation accuracy were evaluated. Those have been added to BASE separately or also as smart aligned combinations (= 8 to maximum 10 independent input parameters). This procedure led to 23 different input parameter combinations that undergo this first performance testing. Table 22.1 lists those 23 combinations in more detail.

For our study we now applied our developed ANN architecture consisting of an input layer of 7 up to 10 input nodes (depending on the number of input parameters), one hidden layer incorporating 10 nodes for $cof$ and 12 nodes for $w_s$, and an output layer with one node for $cof$ or $w_s$. According to the previous section, our applied mathematical methodologies were the well-suited LM training algorithm with MSEREG as the performance function. Further, we utilized hyperbolic tangent sigmoid transfer functions in all hidden layer nodes and a linear transfer function in the output node. Simulations were accomplished following the input parameter order designated in Table 22.1. The employed dataset was randomly divided into 80% training and 20% testing data. Simulations were repeated 200 times and the input parameter-dependent ANN performances were recorded during repetition. The resulting best test performance for each of the 23 different inputs was expressed as mean squared and also as mean absolute error (MAE). To better compare our results with the simulation performances of [10,15–19] we additionally calculated the MRE that was chosen as a measurement tool by the referred authors.

Based on the results of the relevant input parameter identification, we extracted the most powerful and therefore proficient trained ANN to demonstrate its capability to make predictions on the tribology of composite materials. For this issue, we feed data to the chosen ANN that it has not yet seen (i.e. outside the training set) and modeled the speed and pressure-dependent course of $cof$ and $w_s$ for a fixed example composite, consisting of 85% PPS, 10% SCF, and 5% $TiO_2$.

**Table 22.1** Applied Input Parameter Compositions in Order of the Simulations Including the Corresponding Units and Later used Abbreviations

| No. | Input Parameters | Abbreviation |
|---|---|---|
| 1 | BASE: material composition, sliding speed $v$, contact pressure $p$ | BASE or B |
| 2 | B, microhardness | B & mh |
| 3 | B, tensile modulus $E_t$ | B & Et |
| 4 | B, tensile strength $\varepsilon_M$ | B & sigm |
| 5 | B, elongation at break $\varepsilon_M$ | B & epsm |
| 6 | B, tensile modulus $E_t$, tensile strength $\sigma_M$ | B & Et & sigm |
| 7 | B, tensile modulus $E_t$, elongation at break $\sigma_M$ | B & Et & epsm |
| 8 | B, tensile strength $\sigma_M$, elongation at break $\sigma_M$ | B & sigm & epsm |
| 9 | B, tensile modulus $E_t$, tensile strength $\sigma_M$, elongation at break $\sigma_M$ | B & Et & sigm & epsm |
| 10 | B, compression modulus $E_c$ | B & Ec |
| 11 | B, compression strength $\sigma_M^c$ | B & sigmc |
| 12 | B, elongation at break $\varepsilon_M^c$ | B & epsmc |
| 13 | B, compression modulus $E_c$, compression strength $\sigma_M^c$ | B & Ec & sigmc |
| 14 | B, compression modulus $E_c$, elongation at break $\varepsilon_M^c$ | B & Ec & epsmc |
| 15 | B, compression modulus $E_c$, elongation at break $\varepsilon_M^c$ | B & sigmc & epsmc |
| 16 | B, compression modulus $E_c$, compression strength, $\sigma_M^c$ elongation at break $\varepsilon_M^c$ | B & Ec & sigmc & epsmc |
| 17 | B, notched Charpy impact strength $a_n$ | B & imp |
| 18 | B, glass transition temperature $T_g$ | B & Tge |
| 19 | B, storage modulus E′ at $T_g$ | B & Etg |
| 20 | B, mechanical loss factor tan δ at $T_g$ | B & tan |
| 21 | B, mechanical loss factor tan $\delta_{max}$ | B & tanmax |
| 22 | B, glass transition temperature $T_g$, storage modulus at $T_g$, mechanical loss factor tan δ at $T_g$ | B & Tge & Etg & tan |
| 23 | B, glass transition temperature $T_g$, storage modulus at $T_g$, mechanical loss factor tan $\delta_{max}$ | B & Tge & Etg & tanmax |

## 22.3 RESULTS AND DISCUSSION

In all performed simulations to evaluate the combination of input parameters that lead to the best prediction quality for *cof* and $w_s$ the MSE served as a criterion for the network's performance. In addition, of the MSEs, we calculated the MAE as the most demonstrative error term since it has the same unit as the simulated value. Further, the determinations of the MRE described in [17,19] make our resulting ANN performance comparable to the MRE performance of [10,15–19]. Figure 22.3 shows the results of the *cof* simulations with the 23 input variations designated in Table 22.1. The black

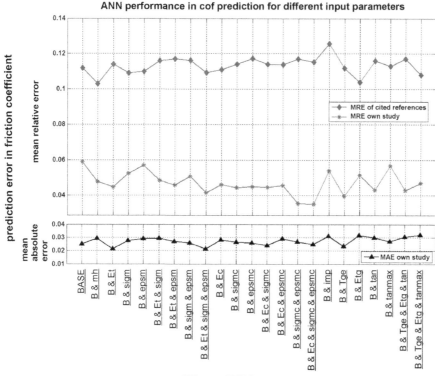

**FIGURE 22.3**

Network performance for the prediction of friction coefficient by adding extra mechanical- or thermomechanical input parameters, expressed in MRE and MAE. Input variable abbreviations can be found in Table 22.1. For color version of this figure, the reader is referred to the online version of this book.

(-▲-) dots depict the particular MAE in the *cof* prediction that can be achieved with the ANN we developed. The total range (in *cof*) of incorrect predictions within all tested input combinations reaches from 0.0213 to 0.0320. In comparison, the mean standard deviation of the tribological estimated *cof* values from the dataset is 0.6043. The purple (-♦-) dots in Fig. 22.3 illustrate the minimal MRE that were obtained in [17,19] with the same procedure but a different ANN approach. Here, Gyurova et al. identified microhardness, tensile strength or strain and the combination of all three tensile properties as well as the compressive modulus and the storage modulus as the most popular input parameters to improve the prediction quality [17,19].

However, with our ANN design it is possible to double the prediction quality on *cof* compared to the cited references: the green (-∗-) dots in Fig. 22.3 represent the best performance of our ANN on the different input variables as MRE. Irrespective of what is chosen as the input of the 23 possibilities, every simulation shows a much better prediction quality when compared to [17,19] (purple (-♦-) dots). Since our ANN's maximum absolute errors in predicting not-trained *cof* values are found to be 10 times smaller than the database's average standard deviation determined for *cof*, from the practical perspective, intrinsically it is not important to add parameters to the input BASE that exhibit a slightly better prediction performance. In total, the absolute error of *cof* for BASE could be improved about −0.0039 by adding chosen mechanical- or thermomechanical variables. Nevertheless, by investigating the added parameters that enhanced the simulation results for *cof* (we chose again MAE as the criterion), we also identified the tensile modulus $E_t$ and the combination of all tensile properties, namely, $E_t$, $\sigma_m$ and $\varepsilon_m$ as the most powerful additional input variables. These findings match the results of the comparative study of Guyrova et al. [17,19]. The microhardness that the authors also found to be important could not be verified by us. The ANN performance on predicting *cof* hinges heavily on the material composition, the sliding speed and the contact pressure since merely BASE also delivered very accurate results (MAE=0.0252).

Figure 22.4 illustrates the results of the $w_s$ simulations again with the 23 input variations designated in Table 22.1. Here, the prediction quality is again expressed as MRE for comparability reasons: the green (-*-) dots label our ANN performances on $w_s$ that arise on the basis of diversified input variables. In contrast, the purple (-♦-) dots represent the best performance (in MRE) of the ANN applied in the comparative study of Gyurova et al. [17,19]. Once more, the prediction quality that could be reached in our approach is much more potent; the average MRE of the cited reference was 0.72, whereas ours was 0.14. The absolute error values of the $w_s$ simulations are depicted separately in Fig. 22.4 since the dimension of $w_s$ does not fit the displayed MRE scale. The simulated 23 absolute error values for $w_s$ were all in the small range of $3.14 \times 10^{-7}$ and $5.04 \times 10^{-7}$ mm$^3$ Nm$^{-1}$. By way of comparison, the mean standard deviation in the database for $w_s$ was $2.48 \times 10^{-6}$ mm$^3$ Nm$^{-1}$, which is again 10 times more than our prediction quality. Similar to *cof*, the important input parameter identification for $w_s$ also demonstrated that the tensile properties appear to be beneficial for simulation accuracy. Here, tensile modulus $E_t$ and $\sigma_m$ seemed to be a good combination. In addition, the compression properties like the combinations $\sigma_M^c$ together with $\varepsilon_M^c$ or the combination $E_c$ with $\sigma_M^c$ and $\varepsilon_M^c$ also showed slightly better results than just BASE. However, by interpreting these results it is important to keep in mind that the performance benefit for applying additional input parameters in total is just $1.23 \times 10^{-7}$ mm$^3$ Nm$^{-1}$ where $w_s$ usually lies in the range around $1.0 \times 10^{-5}$ mm$^3$ Nm$^{-1}$. Interpreting now the outcome of the important input parameter estimation for both, *cof* and $w_s$, we verify that just the conventional assigned input variables (BASE) are deemed to satisfy the ANN's prediction quality requirements (achieved prediction quality <1/10 of measured $w_s$ and *cof* standard deviations).

Figure 22.5 shows an example simulation with our developed and trained ANN approach, employing *B & Et & sigm & epsm* as input variables: Here, a *cof* matrix for an example material composition was simulated for all sliding speeds between 1 and 3 m s$^{-1}$ and all pressures between 1 and 4 MPa; the step size was 0.1 in both directions. The chosen material combination was 85 vol% PPS, 10 vol% SCF and

5 vol% TiO$_2$. The red ($\blacklozenge$) dots are *cof* values from the measured data on this material. The mesh does now represent the different ANN-simulated *cof* values that theoretically would have occurred within a measurement setup where the sliding speed and contact pressure were modified. This would be the data of 651 single measurements of 20 h each. The mesh shows a very accurate fit to the five measured points. If the task would be now to find the combination of speed and pressure where the chosen material's *cof* is minimal, the simulation with a well-designed and trained ANN could

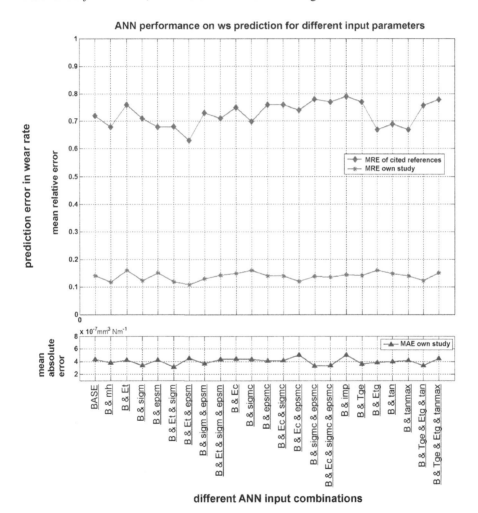

**FIGURE 22.4**

Network performance for the prediction of wear rate by adding extra mechanical- or thermomechanical input parameters, expressed in MRE and MAE. Input variable abbreviations can be found in Table 22.1. For color version of this figure, the reader is referred to the online version of this book.

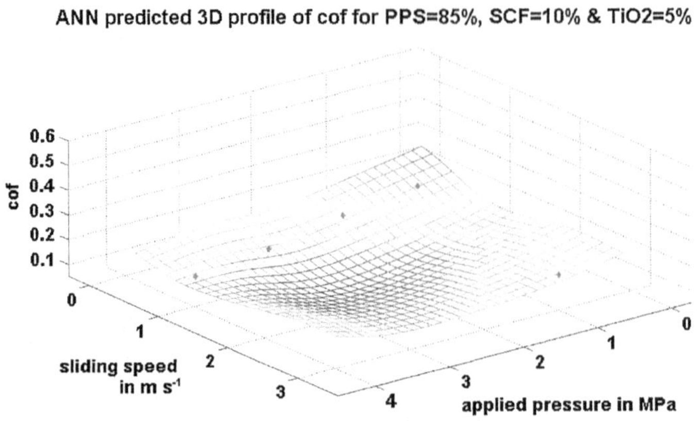

ANN predicted 3D profile of cof for PPS=85%, SCF=10% & TiO2=5%

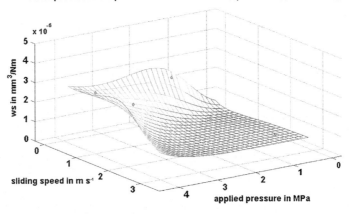

**FIGURE 22.5**

ANN-predicted 3D profile of the friction coefficient for the material combination PPS=85%, SCF = 10% and $TiO_2$=5%. The tribological measured values of *cof* are plotted as red (♦) dots. For interpretation of the references to color in this figure legend, the reader is referred to the online version of this book.

ANN predicted 3D profile of ws for PPS=85%, SCF=10% & TiO2=5%

**FIGURE 22.6**

ANN-predicted 3D profile of the wear rate for the material combination PPS=85%, SCF = 10% and $TiO_2$=5%. The tribological measured values of $w_s$ are plotted as green (♦) dots. For interpretation of the references to color in this figure legend, the reader is referred to the online version of this book.

be very beneficial; for the illustrated example, a pressure of 2.9 MPa with a speed of 1.7 m s$^{-1}$ is simulated to bear a very good *cof* for this material.

The simulation of $w_s$ in the same experimental setup is depicted in Fig. 22.6. Here, *B & Et & sigm* served as input parameters for the network. The green (♦) dots mark the material-specific $w_s$ from the measurements. Here, the simulation suggests a very small $w_s$ for low pressures with medium to high speed and large $w_s$ for high pressure with low to medium speed.

The mesh pattern of those simulations could now be employed to make real-world tribological tests in the areas of interest. Artificially changing or adding material ingredients makes it possible to predict how those, even not yet compounded materials, would behave in the real measurement setup. If the ANN was elaborated carefully, the training data was not noisy and the simulations were in the margin of the training set, the imaginable error should also be acceptably low.

## 22.4 CONCLUSIONS

In this study, we propose a novel ANN approach to make very accurate predictions on the *cof* and wear rate of PPS composites during tribological testing. Based on data from [19], we developed an ANN employing LM training algorithm as kernel with MSEREG as performance function. To yield an ANN with the highest performance quality, we tested how additional input variables, such as characteristic mechanical and thermomechanical properties, which are added to the conventional inputs (material composition, sliding speed, and contact pressure), may influence the prediction accuracy of our elaborated network. We compared those results with the outcome of a study that accomplished the same tests on that data but applied a different ANN. We found our networks' performance on the coefficient of friction prediction two times higher for every input option when compared to [17–19]. The quality of predicting the wear rate along with our ANN was about six times higher than the maximum quality that could be reached in the comparative study, no matter what the input combination was. Further, for both friction and wear, the resulting accuracy scale that could be achieved within our study was 10 times smaller than the tribological determined standard deviations in the database. Regarding the identification of additional input parameters that might improve the *cof* prediction, the tensile modulus $E_t$ and the combination of all tensile properties ($E_t$, $\sigma_m$, and $\varepsilon_m$) enabled increasing the network performance a little (in total −0.0039). With respect to the wear rate, the tensile modulus $E_t$ together with $\sigma_m$ followed by the compression properties $\sigma_M^c$ with $\varepsilon_M^c$ or also $E_c$ with $\sigma_M^c$ and $\varepsilon_M^c$ were detected to be a good input supplement. For this, the total enhancement was $-1.23 \times 10^{-7}\,\text{mm}^3\,\text{Nm}^{-1}$.

In both cases, for wear rate and for *cof*, the simulation improvement due to the incorporation of additional mechanical and thermomechanical input variables was relatively small. However, our approach was able to gain satisfactory high prediction accuracy by just applying the basic input variables, i.e. material formulation, sliding speed and contact pressure to our ANN.

Finally, we employed the best trained ANN with the most potent input variable combination to make predictions on the *cof* and wear rate of an example composite for illustration issues. Such simulations, if the ANN were elaborated carefully, are capable of predicting the outcome of tribological measurements for existing or just modeled composites (within the limits of the training margin). Based on this, the simulation results can then be applied to design a smarter tribological measuring plan that saves resources, time and money.

## Acknowledgments

The authors sincerely acknowledge the European Science Foundation (ESF) and German Research Foundation (DFG) for the financial support of this research work within the FANAS program of ESF.

## References

[1] D. Michie, D.J. Spiegelhalter, C.C. Taylor (eds): Machine Learning, Neural and Statistical Classification, vol. 37, Issue: 4, Publisher: Ellis Horwood.

[2] H.K.D.H. Bhadeshia, Encyclopedia of materials, Science and Technology (2008) 1–5.

[3] P. Zeng, Neural computing in mechanics, Applied Mechanics Reviews 51 (1998) 173–197.

[4] Z. Zhang, K. Friedrich, Artificial neural networks applied to polymer composites: a review, Composites Science and Technology 63 (2003) 2029–2044.

[5] H.E. Kadi, Modeling the mechanical behaviour of fiber-reinforced polymeric composite materials using artificial neural networks – a review, Composite Structures 73 (2006) 1–23.

[6] K. Velten, R. Reinicke, K. Friedrich, Wear volume prediction with artificial neural networks, Tribology International 33 (2000) 731–736.

[7] Z. Jiang, L. Gyurova, Z. Zhang, K. Friedrich, A. K. Schlarb, Artificial neural network based prediction of wear and mechanical properties of polyamide composites reinforced by short fibers, in: Proceedings of 47. Tribologie-Fachtagung, September 25-27, 2006, Goettingen, Germany, Reibung, Schmierung und Verschleiss: Forschung und praktische Anwendung, Gesellschaft fuer Tribologie e.V., Band 1, pp. 4/1-4/14, 2006.

[8] Z. Jiang, Z. Zhang, K. Friedrich, Prediction on wear properties of polymer composites with artificial neural network, Composites Science and Technology 67 (2007) 168–176.

[9]    Z. Jiang, L. Gyurova, Z. Zhang, K. Friedrich, A.K. Schlarb, Neural network based prediction on mechanical and wear properties of short fibers reinforced polyamide composites, Materials and Design 29 (2008) 628–637.

[10] Z. Jiang, L. Gyurova, A.K. Schlarb, K. Friedrich, Z. Zhang, Study on friction and wear behaviour of polyphenylene sulfide composites reinforced by short carbon fibers and sub-micro $TiO_2$ particles, Composites Science and Technology 68 (2008) 734–742.

[11] Z. Zhang, K. Friedrich, K. Velten, Prediction on tribological properties of short fibre composites using artificial neural networks, Wear 252 (2002) 668–675.

[12] Z. Zhang, K. Friedrich, Artificial neural network in polymer composites, in: D. Bhattacharyya (Ed.), Proceedings of the Third Asian-Australasian Conference on Composite Materials (ACCM-3), July 15–17, 2002, Department of Mechanical Engineering, University of Auckland, Auckland, New Zealand, 2002, pp. 105–118.

[13] Z. Zhang, P. Klein, K. Friedrich, Dynamic mechanical properties of PTFE based short carbon fibre reinforced composites: experimental and artificial neural network prediction, Composites Science and Technology 62 (2002) 1001–1009.

[14] Z. Zhang, N.M. Barkoula, J. Karger-Kocsis, K. Friedrich, Artificial neural network predictions on erosive wear of polymers, Wear 255 (2003) 708–713.

[15] L. Gyurova, Z. Jiang, K. Friedrich, Z. Zhang, A.K. Schlarb, Tribological and mechanical properties of selected PPS nanocomposites designed for tribological service, in: Proceedings of 2nd Vienna International Conference Micro-and Nano- Technology Viennano '07, March 14–16, 2007, Vienna, Austria. 2007, pp. 349–356.

[16] L. Gyurova, Z. Jiang, A.K. Schlarb, K. Friedrich, Z. Zhang, Study on the wear and friction of short carbon fiber and/or nano-$TiO_2$-reinforced polyphenylene sulphide composites using artificial neural networks, in: A. Fisher, K. Bobzin (Eds.), Friction, Wear and Wear Protection, International Symposium on Friction, Wear and Wear Protection 2008 Aachen, Germany, Wiley-VCH, Weinheim, 2009, pp. 417–422.

[17] L. Gyurova, P. Miniño-Justel, A.K. Schlarb, Modeling the sliding wear and friction properties of polyphenylene sulfide composites using artificial neural networks, Wear 268 (5–6) (February 2010) 708–714.

[18] L. Gyurova, K. Friedrich, Artificial neural networks for predicting sliding friction and wear properties of polyphenylene sulfide composites, Tribology International 44 (2011) 603–609.

[19] L. Gyurova, Sliding Friction and Wear of Polyphenylene Sulfide Matrix Composites: Experimental and Artificial Neural Network Approach, Dissertation, University of Kaiserslautern, Kaiserslautern, 2010.

[20] I.W. Sandberg, J.T. Lo, C.L. Fancourt, J.C. Principe, S. Katagiri, S. Haykin, Nonlinear Dynamical Systems: Feed-forward Neural Network Perspectives, 1 edition, Wiley-Interscience, February 2001.

[21] B.D. Ripley, Pattern Recognition and Neural Networks, Cambridge University Press, January 1996.

[22] B. Lenze, Einführung in die Mathematik neuronaler Netze, third ed., Logos Verlag, Berlin, 2009.

[23] J.A. Hertz, A. Krogh, R.G. Palmer, Introduction to the Theory of Neural Computation, Addison-Wesley, 1991.

[24] P. Treleaven, M. Vellasco, Neural computing overview, Computer Physics Communications 57 (1–3) (1989) 543–559.

[25] M.T. Hagan, H.B. Demuth, M.H. Beale, Neural Network Design, PWS Publishing, Boston, MA, 1996.

[26] L. Tian, A. Noore, Evolutionary neural network modeling for software cumulative failure time prediction, Reliability Engineering and System Safety 87 (2005) 45–51.

[27] Neural Network Toolbox for Use with MATLAB, User's Guide Releases. The Math Works, Inc., 2011.

[28] S. Roweis, Levenberg-Marquardt Optimization, at http://www.cs.toronto.edu/roweis/notes/lm.pdf.

[29] M.T. Hagan, M.B. Menhaj, Training feed-forward networks with the Marquardt algorithm, IEEE Transaction Neural Networks 5 (Nov. 1994) 989–993.

[30] N.N.R. Ranga Suri, D. Deodhare, P. Nagabhushan, Parallel Levenberg-Marquardt-based neural network training on linux clusters – a case study, Proc. 3rd Indian Conf. on Computer Vision, Graphics & Image Processing, Ahmadabad, India, 2002.

# Index

Note: Page numbers with "f" denote figures; "t" tables.

Printed in the United States
By Bookmasters